ELECTRONIC FAILURE ANALYSIS HANDBOOK

ELECTRONIC FAILURE ANALYSIS HANDBOOK

Electronic Packaging and Interconnection Series

Charles M. Harper, Series Advisor

Related Books of Interest

*To order or receive additional information on these or any other
McGraw-Hill titles, please call 1-800-822-8158 in the United States.
In other countries, contact your local McGraw-Hill representative.*

ELECTRONIC FAILURE ANALYSIS HANDBOOK

Techniques and Applications for
Electronic and Electrical Packages,
Components, and Assemblies

Perry L. Martin
National Technical Systems

McGRAW-HILL

New York   San Francisco   Washington, D.C.   Auckland   Bogotá
Caracas   Lisbon   London   Madrid   Mexico City   Milan
Montreal   New Delhi   San Juan   Singapore
Sydney   Tokyo   Toronto

McGraw-Hill

A Division of The McGraw·Hill Companies

Copyright © 1999 by The McGraw-Hill Companies, Inc. All rights reserved. Printed in the United States of America. Except as permitted under the United States Copyright Act of 1976, no part of this publication may be reproduced or distributed in any form or by any means, or stored in a data base or retrieval system, without the prior written permission of the publisher.

1 2 3 4 5 6 7 8 9 0 AGM/AGM 9 0 4 3 2 1 0 9

ISBN 978-0-07-162634-7

The sponsoring editor for this book was Stephen S. Chapman and the production supervisor was Pamela A. Pelton. It was set in Times by J. K. Eckert & Company, Inc.

Printed and bound by Quebecor/Martinsburg.

This book is printed on recycled, acid-free paper containing a minimum of 50% recycled de-inked fiber.

To my wife, Denise

Your support and understanding made this possible.

CONTENTS

Part 2 Electronic Failure Analysis Techniques

Part 3 Electronic Failure Analysis for Specific Technologies

Chapter 16. Switches and Relays

Chapter 17. Connection Technology

Chapter 18. Failure Analysis of Components

Chapter 19. Semiconductors

PREFACE

The rapidly evolving fields of semiconductor and microelectronic failure analysis have dominated international reliability and failure analysis conferences. This is quite understandable, given that the microelectronics industry was the first to integrate failure analysis into its corporate culture. This synthesis has resulted in huge increases in microelectronic reliability over the last several decades, and this success has spawned similar activities in other electronics related industries. Electronic failure analysis (EFA) is now embodied as part of the development and manufacturing process for parts suppliers, systems/subsystems designers and fabricators, the power industry, and others. Although the application of electronic and electrical failure analysis has increased, the documentation of these nonmicroelectronic efforts has for the most part been neglected. *The Electronic Failure Analysis Handbook: Techniques and Applications for Electronic and Electrical Packages, Components, and Assemblies* is an attempt to address this broadly based and neglected audience.

Electrical and electronic failures and failure analyses occur throughout a product's life-cycle. These failures and their associated analyses become more costly as the product progresses into each successive phase of technology and market maturation. The electronics industry is learning that aggressive use of failure analysis in the early stages of a product's life-cycle can increase reliability and will increase overall net profits. Yet, some manufacturers are not quick to learn, and this creates markets for the failure analyst to provide services in the insurance and legal areas. Just as with microelectronics, these areas are addressed in the handbook but not emphasized.

I have always felt that the subject area of electronic failure analysis could easily fill out a three-volume set. Market forces constrain us to a single volume, so as editor, some tough decisions had to be made about handbook emphasis and content. For reader convenience, this handbook is divided into three parts. Part 1 is the smallest section of the handbook and deals with the utilization of electronic failure analysis. Included is an introduction to the causes of electronic failures, their distribution throughout product life, failure avoidance, quality programs, product liability, and other related issues. Part 2 deals with analysis techniques and consists of two parts. Chapters 4 through 7 describe *nondestructive* techniques, whereas Chapters 8 through 12 pertain to *destructive* techniques. The application and optimization of these techniques to electronic materials is emphasized. Part 3 is about specific electronic packaging and component technologies. To the extent possible, the reader is introduced to the manufacturing materials and processes used to produce these technologies, typical failure modes and mechanisms, applicable failure analysis techniques, and application guidelines. Many of these chapters feature failure analysis case studies.

The *Electronic Failure Analysis Handbook* is a comprehensive reference and guide to electronic failure analysis, and I am confident that it will be a valuable addition and companion volume to the Electronic Packaging and Interconnection Series. This handbook could not have been produced without the hard work of the many outstanding chapter authors. Each of these scientists, engineers, and technologists is a recognized expert in a particular field. Together, their offerings form an inclusive and informative study of the field of electronic failure analysis.

Perry L. Martin

CONTRIBUTORS

Richard A. Blanchard
E^{x}ponent

William Denson
IIT Research Institute

John R. Devaney
HI-REL Laboratories

William G. Dunbar
Consultant

D. David Dylis
IIT Research Institute

J.J. Erickson
Hughes Space and Communications Company

Donald Galler
Department of Materials Science and Engineering
Massachusetts Institute of Technology

Duncan Glover
E^{x}ponent

Herbert Kaplan
Honeyhill Technical Company
L.W. Kessler
Sonoscan, Inc.

Alexander Kusko
E^{x}ponent

John D. Loud
E^{x}ponent

Ian Le May
Metallurgical Consulting Services Ltd.

Lawrence L. Ludwig, Jr.
Electronic Section, Failure Analysis and Physical Testing Branch, Materials Science Division, Logistics Operations Directorate, National Aeronautics and Space Administration

Perry L. Martin
National Technical Systems

Noshirwan K. Medora
E^{x}ponent

Gregory J. Mimmack
E^{x}ponent

Günter Müller
E^{x}ponent

David J. Roche
HI-REL Laboratories

J.E. Semmens
Sonoscan, Inc.

George Slenski
Air Force Wright Laboratory
Wright-Patterson Air Force Base

L.G. Vettraino
Electronic Failure Analysis Laboratory
National Technical Systems

Dennis H. Van Westerhuyzen
Raytheon System Company
Component and Materials Laboratory

ABOUT THE EDITOR

Mr. Perry L. Martin currently works for National Technical Systems (NTS), Sacramento Division, as the Group Manager for Reliability and Process Engineering. This position oversees all failure analysis activities for the division. Mr. Martin developed his extensive experience in electronic failure analysis and the electronics reliability field from over a decade of performing electronic failure analysis and environmental qualification testing for the U.S. Air Force at McClellan AFB in California. Mr. Martin provided direct support to the Air Force industrial repair shops (including avionics, radar, generators, and electrical accessories) at McClellan for process development and quality control, similar support (at Air Force request) to Air Force contractors, aircraft mishap (crash) investigative assistance, and support to the Air Force Office of Special Investigation (AFOSI) in contractor fraud investigations. Mr. Martin has given testimony as an expert witness in contractor fraud cases, has briefed congressmen and the Air Force Chief of Staff on his findings concerning problems with Air Force weapon systems, and was a consultant, contributor, and reviewer for the Air Force Wright Laboratory's *WL-TR-95-4004, Aircraft Mishap Investigation Handbook for Electronic Hardware.*

Introduction to Electronic Failure Analysis

CHAPTER 1
OVERVIEW OF ELECTRONIC COMPONENT RELIABILITY

D. David Dylis

IIT Research Institute (IITRI)

1.1 INTRODUCTION

Failure analysis is the process of determining the cause of failure, collecting and analyzing data, and developing conclusions to eliminate the failure mechanism causing specific device or system failures. Failures can occur throughout a product's life cycle. These failures can result from design, workmanship, components and manufacturing processes, overstress, and maintenance. Table 1.1 generically outlines factors affecting reliability.

Table 1.1 Distribution of Factors Affecting Reliability[1]

Factor	Failure percentage
Maintenance shop	17
Hardware reliability	16
Retest OK	28
Design	21
Quality	18

It is all too often that a failure analyst has limited information concerning the cause of a failure. Details surrounding the cause will not only aid the analyst in the failure analysis process, but the results of the analysis also will be more meaningful when completed. A failure reporting, analysis, and corrective action system (FRACAS), as described in Chapter 2, is used by many organizations to track failures associated with their products. A FRACAS can provide the valuable information necessary to perform a complete and accurate failure analysis. When data is not available, the analyst may take the role of a "private investigator" to determine details surrounding the cause of failure. A complete and accurate failure analysis will allow a designer or user to take the action necessary to remedy the problem. Cost savings associated with the corrective action are typically enough to pay for the failure analysis, and this savings can be multiplied many times over by eliminating future occurrences of the problem.

1.2 FAILURE TYPES

Electronic failures can be broken into basically three distinct and separate classes of failures; early-life, event-related, and wearout. Early-life failures, also known as *infant mortalities,* is the result of defects introduced during the manufacturing or assembly process.

The useful life period is characterized by a relatively constant failure rate caused by randomly occurring events. Wearout failure mechanisms occur as a result of prolonged exposure to operational or environmental stresses and will occur in the entire population of items if they are in service long enough.

Initially, an item typically exhibits a high failure rate. This failure rate decreases during the infant mortality period and stabilizes when weak and defective units have been eliminated. Failures occurring in the initial phase of deployment usually result from built-in flaws due to faulty workmanship, transportation damage or installation errors. Causes of infant-mortality failures include:.

- Inadequate quality control
- Uncontrolled manufacturing processes
- Inadequate component and system test specifications
- Component and system design deficiencies
- Material deficiencies
- Improper handling and packaging
- Incorrect setup, installation, or procedures
- Incomplete testing

Infant-mortality failures can be reduced by monitoring and controlling the manufacturing process. Failure analysis is vital during this phase to identify and correct the causes of problems so they can be eliminated prior to deployment.

The useful life period is normally characterized by relatively infrequent and unpredictable failures. The specifics vary among hardware types and components and will vary based on margins used in the design of an equipment. Failures occurring during the useful life period are referred to as *random failures* because of this infrequent and unpredictable nature. That is not to say, however, that the failures do not have a specific root cause, or that the occurrence is unavoidable. These failures may be the result of component or system design deficiencies or may be induced by users or maintenance personnel. Causes of useful life failures include:

- Inadequate component or system design margins
- Misapplication
- Latent component and system defects
- Excessive electrical, thermal or physical stress (beyond design limits)

The failure rate of a component will increase rapidly in the wearout phase. Wearout failures are due primarily to the deterioration of the design strength of a device resulting from the operation and exposure to environmental fluctuations. This deterioration may result from various physical and chemical phenomena that include:

- Corrosion/oxidation
- Insulation breakdown
- Frictional wear or fatigue
- Shrinking/cracking of plastic materials
- Metal migration

Wearout failures may be reduced or eliminated through preventive maintenance and by providing adequate design margins to extend the life of components.

1.3 ENVIRONMENTAL STRESSES CAUSING ELECTRONIC COMPONENT AND SYSTEM FAILURES

The environmental stresses to which an equipment is exposed during manufacturing, operation, shipping, and storage all have an effect on reliability. During manufacture, equipment can be subjected to a variety of stresses, including temperature extremes and cycling, humidity, dust, and so on. If the manufacturing processes are not under control or were not properly designed, the reliability of the equipment will be affected. Some of the effects take the form of latent defects: defects that are not readily noticeable but are virtual "time bombs" waiting to cause a failure. The types and levels of stresses to which an equipment will be subjected in operation depend on the environment in which it is used. Stresses arise from internal and external conditions, such as vibration, shock, radiation, and weather. These conditions seldom occur sequentially or alone; combinations of these conditions present the greatest stress.

The conditions and physics relating to the failure process provide an understanding of the mechanisms that cause failure. Electronic components are susceptible to various forms of damage that can be attributed to a limited number of physical or chemical mechanisms. Primary failure mechanisms include:

- Corrosion
- Distortion
- Fatigue and fracture
- Wear

The majority of electronic component failures analyzed are often corrosion related, but other mechanisms considered mechanical in nature also influence the reliability of electronic components. It is not unusual for connections and electronic elements that receive vibration and movement to fail from fatigue or fracture. Table 1.2 provides a description and types of mechanisms affecting the reliability of electronic and electromechanical components.

Failures relating to distortion, fatigue and fracture usually occur as a result of an equipment being operated outside of its design environment or beyond its intended life. Environmental factors that accelerate failures in electronic components and assemblies include vibration, thermal cycling, and thermal shock. Table 1.3 provides the distribution of failures with respect to environmental factors found in an airborne environment.

Temperature extremes at or beyond design limits of a component or system, and at high rates of temperature cycling, can occur in the operational environment. In some cases, the equipment might be installed in a temperature-controlled compartment. In others, the equipment will be directly exposed to temperature extremes and cycling. Cold soaking for hours prior to start-up will make equipment subject to thermal shock when it is powered up. The effects of temperature extremes, cycling, and thermal shock include parameter drift, mechanical deformation, chemical reactions, increased contact resistance, dielectric breakdown, and electromigration. To address the potential problems associated with temperature extremes, cycling and thermal shock, several approaches are commonly used. The

Table 1.2   Mechanisms Affecting the Reliability of Electronic and Mechanical Components

Distortion	Fatigue and fracture	Wear	Corrosion
Buckling	Ductile fracture	Abrasive wear	Corrosion fatigue
Yielding	Brittle fracture	Adhesive wear	Stress corrosion
Creep	Fatigue fracture	Subsurface—origin fatigue	Galvanic corrosion
Creep buckling	High-cycle fatigue	Surface—origin fatigue	Crevice corrosion
Warping	Low-cycle fatigue	(pitting)	Pitting corrosion
Plastic deformation	Residual stress fracture	Subcase—origin fatigue	Biological corrosion
(permanent)	Embrittlement fracture	(spalling)	Chemical attach
Plastic deformation	Thermal fatigue fracture	Cavitation	Fretting corrosion
(temporary)	Torsional fatigue	Fretting wear	
Thermal relaxation	Fretting fatigue	Scoring	
Brinnelling			

Table 1.3   Distribution of Environmentally Related Failures in an Airborne System[1]

Environmental factor	Failure percentage
Sand and dust	6
Moisture	19
Shock	2
Temperature	40
Salt	4
Altitude	2
Vibration	27

basic design can incorporate techniques to promote the even distribution of heat, thereby preventing hot spots and providing adequate cooling. Parts and components capable of operating in high temperatures can be selected. External cooling, insulation, and other design techniques reduce the periodicity and magnitude of temperature cycling and protect against thermal shock. Table 1.4 summarizes the effects, sources, and corrective actions required to eliminate the effects of temperature on components and systems.

Humidity and moisture can produce undesirable effects in electronics. It is not just those systems at sea or exposed to the weather for which moisture is a problem. Perfectly dry air is an exception. Humidity is nearly an ever-present condition, and it takes little to convert that humidity to moisture. For example, if a vehicle that has been operated in the cold outdoors is parked in a warm garage, condensation quickly occurs. Humidity can cause corrosion, electrical shorts, breakdown in insulation, and changes in resistance. Humidity can be controlled through design (hermetically sealed components, desiccants, and so on) and through environmental control systems that remove moisture. Controlling temperature changes can prevent condensation, and drains can be provided to eliminate any moisture that does condense inside equipment. Table 1.5 summarizes the effects, sources, and corrective actions required to eliminate the effects of humidity and moisture on components and systems.

Equipment located at sea or near coastal areas is subject to salt fog. Moisture and salt can combine to form a serious environmental condition. The effects on equipment reliability caused by moisture are exacerbated by the salt, which can accelerate contamination and

Table 1.4 Effects of Temperature on Components and Systems

Environment	Principal effects	Corrective action	Sources
High Temperature	■ Acceleration of chemical reactions ■ Deterioration of insulations ■ Outgassing of plastic materials ■ Reduced viscosity of lubricants ■ Dielectric breakdown ■ Electromigration ■ Electrical parameter drift	■ Minimize temperature through design and cooling ■ Eliminate heat source or isolate it from other potentially affected components or assemblies ■ Provide heat conduction paths	Sources of high temperature include ambient temperature, component self- heating, and friction of mechanical and electromechanical assemblies
Low Temperature	■ Materials become brittle ■ Viscosity of lubricants increases ■ Water condensation ■ Ice formation ■ Reduced chemical reactions ■ Shock/isolation mounts stiffen ■ Electrical parameter changes	■ Provide heating ■ Insulate ■ Incorporate materials designed for low temperatures	Typically experienced in uncontrolled environments
Thermal Cycling	■ Stress relaxation (electrical contacts) ■ Ductile and brittle fractures ■ Material deformation ■ Dielectric breakdown ■ Electromigration ■ Electrical parameter drift	■ Minimize coefficient of thermal expansion (CTE) differences between materials ■ Control or eliminate thermal cycling by introducing heating/cooling as required	Experienced in uncontrolled environments, as a result of power on/off cycling and when an equipment is cycled between two different environments during operation
Thermal Shock	■ Material cracking ■ Seal failures ■ Ruptures ■ Electrical parameter drift	■ Minimize CTE mismatches ■ Use appropriate mechanical design tolerances	Transition of an equipment between two temperature extremes

Table 1.5   Effects of Humidity and Moisture on Components and Systems

Environment	Principal effects	Corrective action	Sources
Humidity and moisture	▪ Corrosion of materials ▪ Breakdown in insulations ▪ Changes in resistance ▪ Fungus growth	▪ Control environment ▪ Select moisture-resistant or hermetic parts ▪ Use conformal coatings ▪ Provide drains to eliminate condensed moisture	Humidity is present in the natural environment. Rapid cooling of equipment can cause condensation.

corrosion of contacts, relays, and so forth. Equipment can be protected against salt fog by hermetically sealing components or by installing the equipment in environmentally controlled areas or shelters. Table 1.6 summarizes the effects, sources and corrective actions required to eliminate the effects of salt fog on components and systems.

Table 1.6   Effects of Salt Fog on Components and Systems

Environment	Principal effects	Corrective action	Sources
Salt fog	▪ Combines with water to form acid/alkaline solutions ▪ Corrosion of metals ▪ Insulation degradation	▪ Protective coatings ▪ Eliminate dissimilar metals ▪ Use hermetic seals	Salt is found in the atmosphere, the oceans, lakes, and rivers, and on ground surfaces.

Sand and dust can interfere with the mechanical components of electronic systems. Switches, relays, and connectors are vulnerable to the effects of sand and dust intrusion. Hermetically sealing components or installing equipment in environmentally controlled areas can eliminate or mitigate the effects of sand and dust. Table 1.7 summarizes the effects, sources, and corrective actions required to eliminate the effects of sand and dust.

Table 1.7   Effects of Sand and Dust on Components and Systems

Environment	Principal effects	Corrective action	Sources
Sand and Dust	▪ Long-term degradation of insulation ▪ Increased contact resistance ▪ Clogging of electromechanical components and connectors ▪ Acids formation when moisture is present ▪ Will collect at high static potential points and can form ionization paths	▪ Control environment ▪ Incorporate protective coatings and enclosures ▪ Filter air	Sand and dust are prevalent in all uncontrolled environments. Air turbulence and wind can force dust and sand into cracks and crevices in equipment.

Decreases, particularly rapid ones, in atmospheric pressure can cause failure in components if the design does not allow internal pressure to rapidly re-equalize. Decreases in pressure can also lead to electrical arcing, particularly in the presence of high voltage. Proper design and installation in environmentally controlled areas can avoid problems associated with pressure. Table 1.8 summarizes the effects, sources, and corrective actions required to eliminate the effects of atmospheric pressure on components and systems.

Table 1.8 Effects of High Altitude on Components and Systems

Environment	Principal effects	Corrective action	Sources
High altitude	• Induced damage due to pressure differential in a product • Evaporating/drying due to outgassing • Forms corona (arcing) • Ozone generation • Reduction of electrical breakdown voltages • Chemical changes within organic materials (e.g., rubber)	• Pressurize • Increase mechanical strength • Properly insulate high-voltage components • Minimize use of organic materials	Aircraft or other applications susceptible to rapid fluctuations in altitude or atmospheric pressure

As equipment is subjected to vibration, connections can fail, the substrates of microelectronic parts can fracture, and work hardening can occur in mechanical components. Vibration can be caused mechanically and acoustically and is either random or sinusoidal. Through proper design, vibration effects can be avoided (avoiding resonant frequencies through stiffening, orientation, and so on) or reduced (use of vibration isolators, and so forth).

The sudden or violent application of a force to an equipment causes mechanical shock. The equipment is subjected to rapid acceleration or deceleration, and the resultant forces can destroy connections, fracture structures, and break loose particles that can contaminate electronic components that incorporate a cavity in their design. Shock can occur when a mechanic drops the equipment as it is removed for maintenance or when an aircraft makes a hard landing. External or internal shock mounting, ruggedized packaging, and other techniques can reduce the level of shock actually experienced. Table 1.9 summarizes the effects, sources, and corrective actions required to eliminate the effects of vibration and mechanical shock on components and systems.

Table 1.9 Effects of Vibration on Components and Systems

Environment	Principal effects	Corrective action	Sources
Vibration	• Intermittent electrical contacts • Touching/shorting of electrical parts • Wire chafing • Loosening of hardware • Component/material fatigue	• Stiffen mechanical structure • Reduce moment of inertia • Control resonant frequencies	Vibration isolation is the controlled mismatch of a product's resonant and natural frequencies. It does not usually provide shock isolation. Shock mounts can increase vibration damage.
Mechanical shock	• Interference between parts • Permanent deformation due to overstress	• Use stronger material (as stiff and as light as possible) • Use shock mounts • Superstructure should be stiffer than supporting structure • Use stiff supporting structure if system natural frequency is >35 Hz • Transmit rather than absorb energy	The sudden application of force, measured in Gs of acceleration and milliseconds duration. Can be caused by handling, transportation, gunfire, explosion and/or propulsion.

The effects of solar radiation can be a significant thermal load beyond any load from the local ambient environment. This is especially true for automobiles and aircraft on runways. Sheltering and the use of different types of paint can help. Wind loading can be a significant stress, especially on large structures such as antennas.

Associated with electrical and electronic equipment are electrical and magnetic fields and emanations. These fields and emissions may be intentional (radio transmissions, for example) or unintentional (due either to poor design or to unavoidable laws of physics). The fields and emanations of two pieces of equipment can interfere (hence, the term *electromagnetic interference* or EMI) with the proper operation of each. Proper design is, of course, the best and first step in avoiding EMI problems. Shielding, grounding, and other techniques can also reduce EMI.

1.4 OTHER FACTORS INFLUENCING ELECTRONIC FAILURES

The manufacturing and maintenance environment can also affect the reliability of electronic components. Many of the failures reviewed by a failure analyst are a result of manufacturing errors. Table 1.10 outlines manufacturing defects typically uncovered during the manufacturing process.

Table 1.10 Manufacturing Defects Uncovered during the Manufacturing Process[12]

Manufacturing defect	Percentage
Open	34
Solder bridges	15
Missing parts	15
Misoriented parts	9
Marginal joints	9
Balls, voids	7
Bad parts	7
Wrong parts	7

Static electricity is also a serious threat to modern electronic devices and is prevalent in areas of low humidity. Electrostatic discharge (ESD) can degrade the operation of a component, cause component failures, or go undetected and result in latent defects. Certain devices are inherently more vulnerable to ESD damage. Parts should be selected with an eye toward ESD susceptibility. Electrostatic discharge susceptibility test and classification data on electronic components can be found in Ref. 15.

Precautions should be taken to reduce sources of electrostatic charges. Electrostatic charges can be brought into the assembly, repair, and part storage areas by people. They can also be generated in these areas by normal work movements. Clothing articles of common plastic such as cigarette and candy wrappers, styrofoam cups, part trays and bins, tool handles, packing containers, highly finished surfaces, waxed floors, work surfaces, chairs, processing machinery, and many other prime sources of static charges. These electronic charges can be great enough to damage or cause the malfunction of modern electronic parts, assemblies, and equipment.

Electrostatic discharge can cause intermittent upset failures, as well as hard catastrophic failures, of electronic equipment. Intermittent or upset failures of digital equipment are usually characterized by a loss of information or temporary disruption of a function or functions. In this case, no apparent hardware damage occurs, and proper operation may be resumed after the ESD exposure. Upset transients are frequently the result of an ESD spark in the vicinity of the equipment.

While upset failures occur when the equipment is in operation, catastrophic or hard failures can occur at any time. Most ESD damage to semiconductor devices, assemblies, and electronic components occurs below the human sensitivity level of approximately 4000 V. Other failures may not be catastrophic but may result in a slight degradation of key electrical parameters such as (a) increased leakage current, (b) lower reverse breakdown voltages of P-N junctions or (c) softening of the knee of the V-I curve of P-N junctions in the forward direction. Some ESD part damage is more subtle. It can remain latent until additional operating stresses cause further degradation and, ultimately, catastrophic failure. For example, an ESD overstress can produce a dielectric breakdown of a self-healing nature. When this occurs, the part can retest good, but it may contain a weakened area or a hole in the gate oxide. During operation, metal may eventually migrate through the puncture, resulting in a direct short through the oxide layer.

The natural environment in which maintenance occurs is often harsh and can expose equipment to conditions not experienced during normal operation. Maintenance itself can involve the disconnection and removal of equipment and subsequent reinstallation and reconnection of equipment. This "remove and reinstall" process is not always done because the equipment has failed. Often, equipment may be removed for test and calibration or to allow access to another failed item. This secondary handling (the primary handling required when the equipment has failed) and the natural environment make up the maintenance environment. The maintenance environment can degrade the reliability of equipment not properly designed for it. Maintenance-induced failures can be a significant factor in the achieved field reliability of an item. Proper design can mitigate both the probability and consequences of maintenance-induced failure.

Rain, ice, and snow can affect the temperature of equipment and can introduce moisture, with all its associated problems. These weather elements are a particular problem when maintenance is performed. During maintenance, access doors and panels are opened, exposing even those equipments located in environmentally controlled areas. Proper packaging, the use of covers or shrouds, and other techniques can be used to protect equipment from the elements. Internal drains, sealants and gaskets, and other design measures can also be used to protect equipment.

Equipment must be capable of surviving the rigors of being shipped via all modes of transport over intercontinental distances. The process of preparing for shipment, the shipment itself, the process of readying the equipment for use after delivery, and the handling associated with all these steps impose stresses sometimes more severe than the operating environment itself.

1.5 ELECTRONIC COMPONENT FAILURE MODES AND MECHANISMS

When designing systems containing electronic components, it is valuable to an engineer to have the knowledge of component failure modes and mechanisms and their probability of occurrence. These factors are also invaluable to an analyst when performing failure analyses and developing recommendations to eliminate the future occurrence of failures. Table

1.11 presents failure modes and mechanisms for a representative group of electronic components. The data used to prepare this table was collected by the Reliability Analysis Center in Rome, NY. Detailed failure distributions and specific source details are described in Ref. 11.

1.6 DORMANCY-RELATED FAILURE MECHANISMS

An area often overlooked is the affects of dormancy on electronic equipment. Dormancy can be important in the design of a product, since a predominant portion of its life cycle may be in a nonoperational mode. Moisture is probably the single most important factor affecting long-term nonoperating reliability. All possible steps should be taken to eliminate moisture from electronic devices and products. Hygroscopic materials should be avoided or protected against accumulation of excess moisture. Most failures that occur during nonoperating periods are of the same basic kind as those found in the operating mode, although they are precipitated at a slower rate. Dormancy failures may be related to manufacturing defects, corrosion, and mechanical fracture. When designing for dormancy, materials sensitive to cold flow and creep, as well as metallized and nonmetallic finishes that have flaking characteristics, should be avoided. Also, the use of lubricants should be avoided. If required, use dry lubricants such as graphite. Teflon® gaskets should not be used. Conventional rubber gaskets or silicone-based rubber gaskets are typically preferable.

When selecting components, it is recommended that only components with histories of demonstrated successful aging be used. Semiconductors and microcircuits that contain deposited resistors may exhibit aging effects. Parts that use monometallization are recommended to avoid galvanic corrosion. Chlorine or other halogen-containing materials should not be sealed within any circuitry components. Variable resistors, capacitors, inductors, potentiometers, and electromechanical relays are susceptible to failure in dormant environments and are generally not recommended.

It is recommended that components and systems in storage should not be periodically tested. Historical data shows that failures are introduced as a result of the testing process. Causes of many of the failures were test procedures, test equipment, and operator errors. Storage guidelines are:

- Disconnect all power.
- Ground all units and components.
- Pressurize all coax waveguides (use nitrogen to prevent moisture and corrosion).
- Maintain temperature at 50° F ± 5° F (10° C ± 2.7° C) (drain all equipment of water to prevent freezing or broken pipes).
- Control relative humidity to 50° F (10° C) ± 5 percent (reduces corrosion and prevents electrostatic discharge failure).
- Periodically recharge batteries.
- Protect against rodents: squirrels have chewed cables, mice have nested in electronic cabinets and porcupines have destroyed support structures (wood). Door/window seals, traps/poison, and frequent inspection protect against these rodents.

Failure mechanisms, failure modes and accelerating factors of electronic components subjected to periods of dormancy are described in Table 1.12.

Table 1.11   Failure Mode Distributions[1, 11]

Device type	Failure mode	a^*	Device type	Failure mode	a^*
Alarm, annunciator	False indication	.48	Battery, lead acid	Degraded output	.70
	Failure to operate on demand	.29		Short	.20
	Spurious operation	.18		Intermittent output	.10
	Degraded alarm	.05			
Battery, lithium	Degraded output	.78	Battery, rechargeable,	Degraded output	.72
	Start-up delay	.14	Ni-Cd	No output	.28
	Short	.06			
	Open	.02			
Capacitor, aluminum,	Short	.53	Capacitor, paper	Short	.63
electrolytic, foil	Open	.35		Open	.37
	Electrolyte leak	.10			
	Decrease in capacitance	.02			
Capacitor, ceramic	Short	.49	Capacitor, mica/glass	Short	.72
	Change in value	.29		Change in value	.15
	Open	.22		Open	.13
Capacitor, plastic	Open	.42	Capacitor, tantalum	Short	.57
	Short	.40		Open	.32
	Change in value	.18		Change in value	.11
Capacitor, tantalum,	Short	.69	Capacitor, variable,	Change in value	.60
electrolytic	Open	.17	piston	Short	.30
	Change in value	.14		Open	.10
Coil	Short	.42	Connector/connection	Open	.61
	Open	.42		Poor contact/intermittent	.23
	Change in value	.16		Short	.16
Circuit breaker	Opens without stimuli	.51	Crystal, quartz	Open	.89
	Does not open	.49		No oscillation	.11
Diode, general	Short	.49	Diode, rectifier	Short	.51
	Open	.36		Open	.29
	Parameter change	.15		Parameter change	.20
Diode, silicon control	Short	.98	Diode, small signal	Parameter change	.58
rectifier (SCR)	Open	.02		Open	.24
				Short	.18
Diode, Triac	Failed off	.90	Diode, thyristor	Failed off	.45
	Failed on	.10		Short	.40
				Open	.10
				Failed on	.05
Diode, zener, voltage	Parameter change	.69	Diode, zener, voltage	Open	.45
reference	Open	.18	regulator	Parameter change	.35
	Short	.13		Short	.20
Electric motor, ac	Winding failure	.31	Fuse	Fails to open	.49
	Bearing failure	.28		Slow to open	.43
	Fails to run, after start	.23		Premature open	.08
	Fails to start	.18			

Table 1.11 Failure Mode Distributions[1, 11] *(continued)*

Device type	Failure mode	a*	Device type	Failure mode	a*
Hybrid Device	Open circuit	.51	Keyboard Assembly	Spring failure	.32
	Degraded output	.26		Contact failure	.30
	Short circuit	.17		Connection failure	.30
	No output	.06		Lock-up	.08
Liquid crystal display	Dim rows	.39	Meter	Faulty indication	.51
	Blank display	.22		Unable to adjust	.23
	Flickering rows	.20		Open	.14
	Missing elements	.19		No Indication	.12
Microcircuit, digital, bipolar	Output stuck high	.28	Microcircuit, digital, MOS	Input open	.36
	Output stuck low	.28		Output open	.36
	Input open	.22		Supply open	.12
	Output open	.22		Output stuck low	.09
				Output stuck high	.08
Microcircuit, interface	Output stuck low	.58	Microcircuit, linear	Improper output	.77
	Output open	.16		No output	.23
	Input open	.16			
	Supply open	.10			
Microcircuit, memory, bipolar	Slow data transfer	.79	Microcircuit, memory, MOS	Data bit loss	.34
	Data bit loss	.21		Short	.26
				Open	.23
				Slow data transfer	.17
Microwave amplifier	No output	.90	Microwave attenuator	Attenuation increase	.90
	Limited voltage gain	.10		Insertion loss	.10
Microwave connector	High insertion loss	.80	Microwave detector	Power loss	.90
	Open	.20		No output	.10
Microwave, diode	Open	.60	Microwave filter	Center frequency drift	.80
	Parameter change	.28		No output	.20
	Short	.12			
Microwave mixer	Power decrease	.90	Microwave modulator	Power loss	.90
	Loss of intermediate frequency	.10		No output	.10
Microwave oscillator	No output	.80	Microwave voltage-controlled oscillator (VCO)	No output	.80
	Untuned frequency	.10		Untuned frequency	.15
	Reduced power	.10		Reduced power	.05
Microwave phase shifter	Incorrect output	.90	Microwave polarizer	Change in polarization	1.00
	No output	.10			
Microwave yttrium from garnet (YIG) resonator	No output	.80	Optoelectronic LED	Open	.70
	Untuned frequency	.15		Short	.30
	Reduced power	.05			
Optoelectronic sensor	Short	.50	Power Supply	No Output	.52
	Open	.50		Incorrect Output	.48

Table 1.11 Failure Mode Distributions[1, 11] *(continued)*

Device type	Failure mode	a^*	Device type	Failure mode	a^*
Printed wiring assembly	Open	.76	Relay	Fails to trip	.55
	Short	.24		Spurious trip	.26
				Short	.19
Resistor, composition	Parameter change	.66	Resistor, fixed	Open	.84
	Open	.31		Parameter change	.11
	Short	.03		Short	.05
Resistor, fixed, film	Open	.59	Resistor, fixed, Wirewound	Open	.65
	Parameter change	.36		Parameter change	.26
	Short			Short	.09
		.05			
Resistor, network	Open	.92	Resistor, thermistor	Open	.63
	Short	.08		Parameter change	.22
				Short	.15
Resistor, variable	Open	.53	Rotary switch	Improper output	.53
	Erratic output	.40		Contact failure	.47
	Short	.07			
Sensor	Erratic output	.59	Solenoid	Short	.52
	Short	.20		Slow movement	.43
	Open	.12		Open	.05
	No output	.10			
Switch, push-button	Open	.60	Switch, thermal	Parameter change	.63
	Sticking	.33		Open	.27
	Short	.07		No control	.08
				Short	.02
Switch, toggle	Open	.65	Synchro	Winding failure	.45
	Sticking	.19		Bearing failure	.33
	Short	.16		Brush failure	.22
Transducer, sensor	Out of tolerance	.68	Transformer	Open	.42
	False response	.15		Short	.42
	Open	.12		Parameter change	.16
	Short	.05			
Transistor, bipolar	Short	.73	Transistor, FET	Short	.51
	Open	.27		Output low	.22
				Parameter change	.17
				Open	.05
				Output high	.05
Transistor, GaAs FET	Open	.61	Transistor, RF	Parameter change	.50
	Short	.26		Short	.40
	Parameter change	.13		Open	.10
Tube, electron	Change in parameter	.53	Tube, traveling wave	Reduced output power	.71
	Open	.25		High helix current	.11
	Unstable output	.15		Gun failure	.09
	Short	.07		Open helix	.09

*a = failure mode probability

Table 1.12 Dormant Part Failure Mechanisms[1]

Type	Mechanism	Percent failure mode	Accelerating factor
Microcircuit	Surface anomalies	35–70 Degradation	Moisture, temperature
	Wire bond	10–20 Open	Vibration
	Seal defects	10–30 Degradation	Shock, vibration
Transistor	Header defects	10–30 Drift	Shock, vibration
	Contamination	10–50 Degradation	Moisture, temperature
	Corrosion	15–25 Drift	Moisture, temperature
Diode	Corrosion	20–40 Intermittent	Moisture, temperature
	Lead/die contact	15–35 Open	Shock, vibration
	Header bond	15–35 Drift	Shock, vibration
Resistor	Corrosion	30–50 Drift	Moisture, temperature
	Film defects	15–25 Drift	Moisture, temperature
	Lead defects	10–20 Open	Shock, vibration
Capacitor	Connection	10–30 Open	Temperature, vibration
	Corrosion	25–45 Drift	Moisture, temperature
	Mechanical	20–40 Short	Shock, vibration
RF coil	Lead stress	20–40 Open	Shock, vibration
	Insulation	40–65 Drift	Moisture, temperature
Transformer	Insulation	40–80 Short	Moisture, temperature
Relay	Contact resistance	30–40 Open	Moisture, temperature
	Contact corrosion	40–65 Drift	Moisture

1.7 DETERMINING THE CAUSE OF ELECTRONIC FAILURES

The course of action taken to determine the cause(s) of an electronic failure depends on the item to be evaluated (e.g., power supply, populated printed wiring board/module or discrete piece part). The process is outlined in Figure 1.1. This process can be applied to analyze either a component or printed wiring assembly. The problem analysis path for a discrete component incorporates an external visual examination and electrical test verification. The evaluation of a printed wiring assembly is a more complex procedure, requiring fault analysis to isolate a problem. This isolation often includes the review and evaluation of manufacturing and assembly processes, board materials and construction, component mounting techniques, and connectors. Board-level electrical testing is typically the first procedure used to localize the problem. Also, it is important that the failure analyst interface with the original equipment manufacturer (OEM), since it is important that diagnostic and corrective action interchange, as well as to provide an unbiased review of results.

Higher-level assemblies may require the use of circuit analysis techniques and methods (e.g., SPICE, Monte Carlo simulation) to isolate design problems. These procedures can either provide basic information concerning the sensitivity of a circuit to variability (deg-

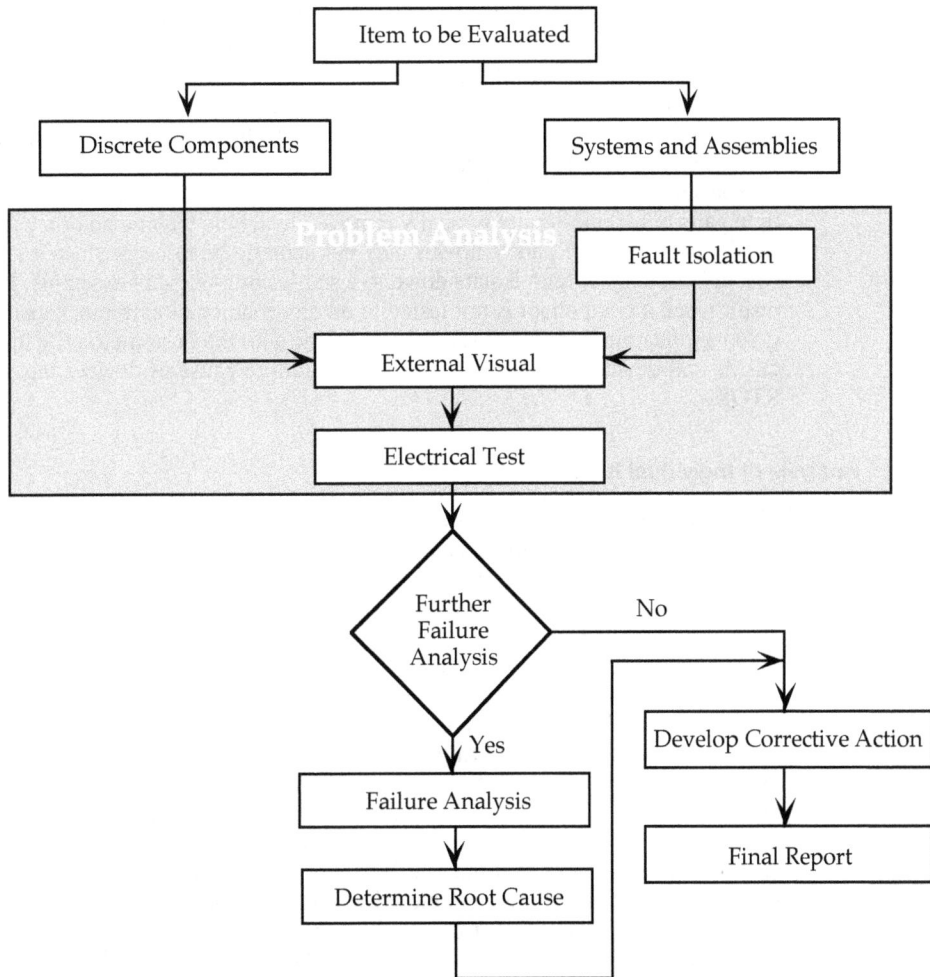

FIGURE 1.1   Electronic Problem Analysis Process Flow

radation) in the parameters of its component parts or use actual parameter variability data to simulate real-life situations and determine the probability of whether circuit perfor-mance is within tolerance specifications. In addition to providing information on the sensi-tivity of a circuit to degradation failures, these techniques are capable of giving stress-level information of the type needed to identify overstressed parts. Catastrophic failures can often be related to excessive environmental conditions, lack of design margin, or over-stress, which can provide valuable insight for corrective actions. These techniques will, therefore, be used to identify problems resulting from circuit designs that are not robust enough to tolerate the specific conditions being encountered.

Finite element analysis (FEA), a graphically oriented, automated technique for analyz-ing the effects of mechanical loads and thermal stress on a component or system, may also be utilized. It is a powerful method for modeling structures and assemblies with various levels of mechanical, vibration, and thermal stress, and producing visual evidence of any resulting deformation, harmonic resonance, or thermal flow and induced stress patterns.

In contrast to design-related problems, failures can also be due to defective parts. A background investigation is typically conducted to determine if similar problems have been encountered, and whether they are vendor, technology, or batch related. Such investigations on previous findings can often lead to quick solutions.

Special emphasis should be given to parts and assemblies that *retest OK* (RTOK). Multiple parts or assemblies are typically removed by a field technician or repair shop during a repair. Some of these part removals may not actually be failures, since it is typically more cost effective not to fault isolate down to a single component or assembly. RTOKs can also result when a component is not tested to an environmental extreme, causing it to fail in a given application. It is often necessary to work with the customer and with the equipment OEMs and device vendor(s) to ascertain the failure cause of devices that are found to be RTOK.

1.7.1   Analysis of Individual Items

The failure analysis process incorporates reliability to identify failure mechanisms for parts and materials. A failure analyst must go beyond the symptomatic defect identified during the failure analysis process and identify and correct the basic underlying part or process problem causing the problem (e.g., the root cause).

Once a discrete part or process step (product or fabrication) causing a problem has been fault isolated, a number of sophisticated failure analysis techniques may be required to correctly determine solutions to a problem. The specific techniques required to analyze a part or process depend on the problem under study.

When performing or selecting a source for failure analysis, it may be difficult for one laboratory to have the expertise in all failure analysis techniques required to perform an analysis. A given laboratory will often have their own area of expertise. The selection of a failure analysis laboratory needs to be based on the component type and the failure analysis techniques/methods that may be required to perform the analysis. In some instances, the original vendor will perform electrical testing of devices manufactured by them. Vendors and manufacturers are usually willing to analyze their own items at no cost just to determine how well these items function in the field. Additionally, participation by company engineers in the failure analysis process can be beneficial. Experience has shown that knowledge of the circumstances of the problem under study is necessary during the performance of a failure analysis. The elimination of this step has typically led to extraneous testing or the failure to perform key testing, resulting in a requirement for additional follow-up analyses. Also, a review of this step will remove any suspicion that results have been biased by a testing facility.

1.7.2   Failure Prevention and Elimination

With the increasing emphasis on identifying the source of a problem to ensure that the failure mode is eliminated, the failure analysis definition can be expanded to include failure prevention. Additionally, reliability physics practices can be used to verify the adequacy of a process or component in satisfying its desired application early in the design process. Root cause analysis, properly applied, can identify corrective action for many problems (i.e., a failed semiconductor device, a poor yield machining operation, and a nonrobust process).

Determining corrective actions for identified problems should address both catastrophic and chronic failures and defects, as illustrated in Table 1.13. A catastrophic example would be a change of flux in a standard soldering process. The result could be poor adhesion of component leads to printed wiring board contact pads. Usually, in cases like this, the cause and effect are easily identified, and corrective action determination is straightforward. In the chronic case, yield is usually considered acceptable.

Table 1.13   Catastrophic and Chronic Failures and Defects

Failure	Occurrence	Cause	Yield	Solution
Catastrophic	Sudden	Single	Unacceptable	Straightforward
Chronic	Over time	Compound	Acceptable	Difficult

The actual cause in this case is the result of various factors. If each is not addressed, the resulting corrective action will not suffice. Figure 1.2 illustrates the difference between catastrophic and chronic failures and the result of root cause analysis continuous improvement.

The traditional "quality control" scenario of an approach to improvement is illustrated in Table 1.14. In most cases, chronic defects are either not identified or are considered to be in the "noise level" and not addressed. However, to achieve high yields, low costs, and customer satisfaction, continuous improvement, through minimization of chronic defects, must occur, as illustrated in Fig. 1.3. Improvement starts with Pareto analysis, but chronic defects may not be eliminated, because the defect factor selection is arbitrary and not thorough.

An approach to the reduction of chronic defects is illustrated in Table 1.15. The process used is PM Analysis. This approach is described in the Japan Institute of Plant Maintenance PM Analysis Cause training course. "P" stands for phenomenon and physical, and "M" stands for mechanism, machine, man, material, and method. Basically, this is a much

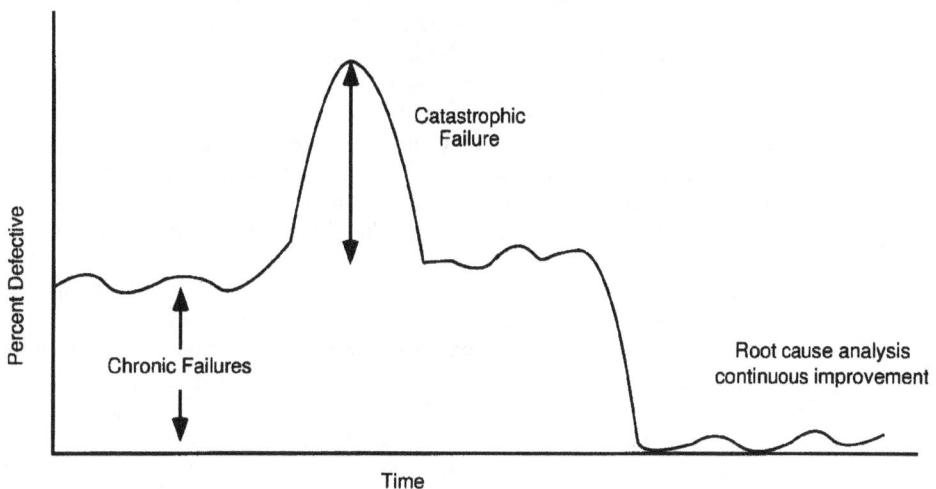

FIGURE 1.2   Difference between Catastrophic and Chronic Failures[1]

Table 1.14   Traditional Approach to Improvement

Steps	Description
1. Goal	• Defect reduction
2. Philosophy	• Prioritize • Analyze by priority • Identify problem factors and take corrective action
3. Approach	• Fishbone diagram (brainstorming)
4. Success	• Good for catastrophic problems

Table 1.15   Improvement Approach for Achieving "Zero" Defects

Steps	Description
1. Goal	• Zero defects
2. Philosophy	• Minimize prioritization • Maximize determination of defect factors • Investigate each factor • Correct all identified problems • Attempt to correct all defect factors • If not successful, change factor selection
3. Approach	• PM* analysis
4. Success	• Chronic defects

* P = phenomenon and physical; M = mechanism, machine, man, material, and method

more detailed analysis that attempts to detail all defect factors and tries to correct them to minimize chronic defects. It is important that you do not focus on only important items but also include those whose contribution is minor.

A successful continuous improvement plan would be first to apply the conventional quality control approach. If satisfactory results do not occur, then implement the more rigorous PM approach. However, it must be noted that cost and schedule factors can minimize the success of either approach. As improvements are made and yields rise, management support to continue analysis may decrease.

When the root cause of a failure has been determined, a suitable corrective action should be developed which will prevent further recurrence of this failure. Examples of corrective actions include, but are not limited to, changes in design, part derating, test procedures, manufacturing techniques, materials, and packaging.

After a corrective action has been proposed, its effectiveness should be established prior to its universal application. This may be performed on a prototype equipment, on the failed unit, in the engineering laboratory, or by various other appropriate means.

Failure analysis of items is necessary when prior problem analysis/fault isolation does not define the cause of a specific problem. It will also be used to verify problems uncovered by the fault isolation process. The following describes the capability, techniques, and expertise available to perform these failure analyses.

1.7.3 Analysis Techniques

Once the discrete part causing the problem has been fault isolated, a number of sophisticated failure analysis techniques may be required in order to correctly determine solutions
for the problem. The specific technique required to analyze a part depends on the problem
under study. The first step is to inspect the part using optical microscopy. This is a simple
but effective method of isolating certain problems. Many of the problems whose root
causes are vibration can be detected in this manner. Broken leads on an electronics board
subjected to high vibration can be isolated at this level of analysis. Massive electrical overstress, packaging integrity, and external corrosion defects can also be detected using visual
inspection.

The next step is to run electrical tests. This is to categorize the problem so that further
analysis can be channeled most efficiently. Test sets, curve tracers, and point-to-point
resistance measurements can be used at this step to determine into which of the following
categories the part would fall.

1. A marginal or degradation problem, where parts coming from the tails of a normal
   product distribution combine to cause malfunctions.

2. A "sport," where a part falls well outside of normal product distribution, or a particular
   critical parameter was never specified.

3. An environmental problem, where the equipment is stressed electrically, mechanically,
   thermally or chemically (intentionally or unintentionally) beyond the design limit.

4. RTOK, which could result from misapplication of a part, an intermittent problem such
   as cracked solder or loose particles, improper removal (troubleshooting printed wiring
   assemblies usually cannot isolate to a single part), poor electrical connections external
   to the part, or software problems in the printed wiring assembly tests.

Based on experience, most of the parts fall into the last two categories listed above. Typically, over one-third of the problems can be isolated during the first two steps. The test
sequence followed for analyzing the remaining two-thirds of the parts can take many different paths based on the findings during the first two steps. In any event, benign tests that
do not cause irreversible changes in the part should always be used first. Depending on the
findings at this point, they could include such tests as X-ray, fine and gross leak, loose particle detection (PIND), shock, vibration, centrifuge, thermal cycling, and dew point.

For hermetic cavity devices, residual gas analysis (RGA) should be run on some samples prior to opening. The analysis of the gaseous materials in the part cavity can often
lead to the root causes of a problem or the forming of hypotheses that lead to solutions.
This is particularly true if polymeric materials are used. RGA is also useful for package
integrity, surface contamination, and metal plating problems. Immediately after the part is
punctured for the RGA, a piece of tape should be placed over the hole for loose particle
recovery.

If no cause is found after all nondestructive evaluation has been completed, then the
destructive analysis begins. Again, the order of analysis must be carefully planned so that
the maximum information can be obtained before each irreversible step is completed.
Careful, detailed internal visual examination and photographic documentation is the most
effective method for either solving the problem or deciding which analysis technique to
use next. Often, the path of the overstress in a printed wiring assembly can be determined,
as well as approximate current/voltage levels and whether the source was high or low

impedance. Good parts can then be stressed to see if the field failures can be duplicated. Visual inspection can also provide information on contamination, poor electrical interconnections, conductive and nonconductive films, adhesive failures, and manufacturing and handling defects. At this point in the analysis, experience shows that about two-thirds of the part failure causes have been isolated to the point where corrective action can be initiated.

For the remaining one-third of the parts, many techniques may be used. These are selected based on information gathered by the previous analyses. Section II, Electronic Failure Analysis Techniques, details techniques used in the failure analysis process.

1.7.4 Development of Recommended Corrective Actions

Failure and engineering analyses will identify root causes for critical part or system deficiencies. Evidence of electrostatic discharge (ESD), software, maintenance, or handling induced failures may also be uncovered. To eliminate these problems, a corrective action recommendation is required. Based on results from prior analyses, and with assistance as required from the customer, the contractor and analyst should develop corrective action recommendations. These recommendations, when implemented on the equipment early in its life cycle, will reduce maintenance cost and improve the operational readiness of the equipment.

A report describing the corrective action recommendation as it pertains to a particular component/printed wiring assembly, system application, manufacturing process, device technology, and so on will be delivered to the customer for each problem studied. This report will include a description of the problem, a summary of failure circumstance information, results of failure testing, references to other test reports and source documents, corrective action recommendations, and a cost/benefit analysis to support the action recommendations, where applicable. Corrective actions will be implemented, as appropriate, by the customer with assistance from the contractor.

1.7.5 Failure Analysis Report

A written report of the failure analysis results documents the conclusions for all concerned parties and provides a permanent record to support historical failure tracking and corrective action verification. The report should include short statements of results supported by electrical test data (where appropriate), pictures and analysis actions taken to clearly identify the failure mechanisms. Clearly stated conclusions are preferred in a bullet format. The report should include the following, as a minimum:

- Background discussion
- Item/process description
- Analyses performance
- Graphic data
- Conclusions
- Recommendations
- References

1.8 STATISTICAL DISTRIBUTIONS

Statistics are typically used in the failure analysis process to analyze data pertaining to conditions surrounding a specific failure or group of failures. Statistics are of particular importance when modeling the time-to-failure characteristics of a component or system. Several distributions typically used for this purpose include the exponential, Weibull, normal, lognormal, and gamma. The exponential distribution is probably most often used because of its simplicity, and it has been shown in many cases to fit the time-to-failure characteristics of electronic devices. This distribution, however, is not always representative of specific failure causes, and other distributions are required to model the data. A description of distributions typically used to model electronic and electromechanical component time-to-failure characteristics are provided in this section. Discrete distributions such as the binomial and poisson are not covered but are described in numerous statistics and reliability texts.

1.8.1 Exponential Distribution

The exponential distribution is widely used to model electronic reliability failures during the useful life period, which tends to exhibit random failures and a constant failure or hazard rate. The probability density function for an exponentially distributed time-to-failure random variable, t, is given by:

$$f(t) = \lambda\, e^{-\lambda t}$$

where λ is the failure rate and $t > 0$. The reciprocal of the failure rate is the mean expected life, and it also referred to as the *mean time between failure* (MTBF).

1.8.2 Weibull Distribution

The Weibull distribution is very versatile and can be used to represent many different time to failure characteristics. It continues to gain popularity, particularly in the area of mechanical reliability, due to this versatility. The underlying physical phenomena causing the failure, however, must be clearly understood prior to the selection of this distribution. The two-parameter Weibull probability density function in terms of time, t, is represented by the following equation:

$$f(t) = \frac{\beta}{\eta^{\beta}} t^{\beta-1} \exp\left[-\left(\frac{t}{\eta}\right)^{\beta}\right] \qquad (\text{for } t \geq 0)$$

$$f(t) = 0 \qquad (\text{for } t < 0)$$

where β is the shape parameter and is the scale parameter, or characteristic life. When $\beta = 1$, the distribution is equivalent to the exponential distribution, and when $\beta = 3.44$, the distribution approximates a normal distribution.

1.8.3 Normal Distribution

The normal distribution is often used to model the distribution of chemical, physical, mechanical, or electrical properties of components or systems. Data conforming to a normal distribution is symmetrical about the mean of this distribution. When a value is subject to many sources of variation, irrespective of how the variations are distributed, the resultant composite of these variations has been shown to approach a normal distribution. The normal distribution is rarely used to model the time to failure of wearout mechanisms, but it can be used to model component design variations—for example, the tensile strength of materials. The probability density function is given by:

$$f(x) = \frac{1}{\sigma(2\pi)^{1/2}} \exp\left[-\frac{1}{2}\left(\frac{x-\mu}{\sigma}\right)^2\right]$$

where μ is the location parameter (which is equivalent to the mean), and σ is the scale parameter (which is equal to the standard deviation).

1.8.4 Lognormal Distribution

The lognormal distribution is a skewed distribution that can be used to model data where a large number of occurrences being measured are concentrated in the tail end of the data range. It is not as versatile as the Weibull distribution but has been shown to be applicable to the modeling of many wearout phenomena. The probability density function for the lognormal distribution is represented by:

$$f(x) = \frac{1}{\sigma x(2\pi)^{1/2}} \exp\left[-\frac{1}{2}\left(\frac{\ln x-\mu}{\sigma}\right)^2\right] \qquad (\text{for } x \geq 0)$$

$$f(x) = 0 \qquad (\text{for } x < 0)$$

Where μ and σ are, respectively, the mean and standard deviation of the data. The lognormal distribution is often used to describe the time-to-failure characteristics of electronic component failure mechanisms such as electromigration and time-dependent dielectric breakdown.

1.8.5 Gamma Distribution

The gamma distribution is used to model situations where a number of failure events must occur before the failure of the item of concern. The probability density function is:

$$f(x) = \frac{\lambda}{\Gamma(a)}(\lambda x)^{a-1} \exp(-\lambda x) \qquad (\text{for } x \geq 0)$$

$$f(x) = 0 \qquad (\text{for } x < 0)$$

where λ is the failure rate of a complete failure and a is the number of events required to cause a complete failure. $\Gamma(a)$ is the gamma function, which is represented by:

$$\Gamma[a] = \int_0^\infty x^{a-1} \exp(-x)dx$$

When $a = 1$, the gamma distribution is equivalent to the exponential distribution.

1.8.6 Choosing a Distribution

The specific statistical distribution best suited for a particular analysis can be determined empirically, theoretically, or by a combination of both approaches. A distribution chosen empirically is done by fitting data to a distribution using probability paper or computer software programs. Goodness-of-fit tests, such as the chi-square and Kolmogorov-Smirnov, can also be used to determine when a particular data set fits a distribution. A description of typical goodness-of-fit tests has been provided in this section.

Chi-Square Test. The chi-square test can be used to determine how well theoretical or sample data fits a particular distribution. A measure of the discrepancy between observed and expected values of a distribution is supplied by the chi-square (χ^2) statistic. The test is derived from the number of observations expected to fall into each interval when the correct distribution is chosen. The statistic is evaluated by subtracting an expected frequency from an observed frequency and is represented by:

$$\chi^2 = \sum_{j=1}^{k} \frac{(o_j - e_j)^2}{e_j}$$

where χ = number of events
  o = observed frequency
  e = expected frequency

Chi-square distribution tables are required to compare the value of to a critical value with the appropriate degrees of freedom. When data is limited, e.g., less than five observations from three or fewer groups or classes of data, it is recommended that a different test, such as the Kolmogorov-Smirnov (K-S) goodness-of-fit test be used.

Kolmogorov-Smirnov (K-S) Test. The Kolmogorov-Smirnov test is a nonparametric test that directly compares an observed cumulative distribution function (c.d.f.) to the c.d.f. of a theoretical or expected distribution. This test is easier to use than the chi-square test previously discussed and can provide better results with a small number of data points.

The K-S test, instead of quantizing the data first, is a continuous method, since it looks directly at the observed cumulative distribution function and compares it to the theoretical or expected c.d.f. The cumulative distribution function $F_O(t)$ may be evaluated from data at some point, t^*, by:

$$F_O(t^*) = \frac{\text{Number of components failing by } t^*}{\text{Total number of components on test} + 1}$$

One is added to the denominator to reduce bias. The theoretical cumulative density function $F_E(t)$ may be evaluated by integrating the probability density function $f(t)$. If the observed expected cumulative distribution functions are $F_O(x)$ and $F_E(x)$, then

$$D = \max |F_O(x) - F_E(x)|$$

is the K-S statistic for a one-sample statistic. A two-sample procedure also exists for comparing two samples. The K-S statistic represents the largest deviation of observed values from expected values on the cumulative curve. D is then compared to tables of critical values and, if it exceeds the critical value at some predetermined level of significance, it is concluded that the observations do not fit the theoretical distribution chosen.

1.9 ACKNOWLEDGMENT

The author wishes to acknowledge that much of the work forming the basis for this chapter was summarized from the *Reliability Toolkit: Commercial Practices Edition* developed by the Reliability Analysis Center and the U.S. Air Force Rome Laboratory in Rome, NY.

1.10 REFERENCES

1. *Reliability Toolkit: Commercial Practices Edition*. Rome, NY: Reliability Analysis Center, 1996.
2. Shooman, M. *Probabilistic Reliability, An Engineering Approach*. New York: McGraw-Hill, 1968.
3. Abernethy, Dr. R. B. *The New Weibull Handbook*. Houston: Gulf Publishing Co., 1994.
4. O'Connor, P. D. T. P*ractical Reliability Engineering*. New York: John Wiley & Sons, 1985.
5. Ireson, W. G. and C. F. Coombs. *Handbook of Reliability Engineering and Management*. New York: McGraw-Hill, 1966.
6. Amerasekera, E. A. and D. S. Campbell. *Failure Mechanisms in Semiconductor Devices*. New York: John Wiley & Sons, 1987.
7. Smith, D. J. *Reliability and Maintainability in Perspective*. New York: Halsted Div. John Wiley & Sons, 1985.
8. Villemeur, A. Reliability, *Availability, Maintainability and Safety Assessment, Volume 1, Methods and Techniques*. New York: John Wiley & Sons, 1992.
9. R*eliability Design Handbook*. Rome, NY: Reliability Analysis Center, 1976.
10. Dey, K. A. *Practical Statistical Analysis for the Reliability Engineer, SOAR-2*. Rome, NY: Reliability Analysis Center, 1983.
11. Chandler, G., W. K. Denson, M. J. Rossi, and R. Wanner. *Failure Mode/Mechanism Distributions, FMD-91*. Rome, NY: Reliability Analysis Center, 1991
12. Romanchick, D. Engineers put the quality test to electronics, *Quality*, Sept. 1996, 40–44.
13. Bennet, A. Electrolytic capacitors—tantalum vs. aluminum, *Electronic Products* 23(3), August 1980.
14. Farrell, J., D. Nicholls, W. K. Denson, and P. MacDiarmid. *Processes for Using Commercial Parts in Military Applications*. Rome, NY: Reliability Analysis Center, November 1996.
15. Crowell, W. H. *Electrostatic Discharge Susceptibility Data, VZAP-1995*. Rome, NY: Reliability Analysis Center, June 1995.

CHAPTER 2
OVERVIEW OF ELECTRONIC SYSTEMS RELIABILITY

Perry L. Martin
National Technical Systems (NTS)

Noshirwan K. Medora
E^xponent

2.1 INTRODUCTION

Any electronic product is an assemblage of various components and electronic packaging and interconnection technologies. The reliability of even a "well designed" product is usually reduced by some finite number of defects in both parts and workmanship imparted by "normal manufacturing processes." To achieve the level of reliability inherent in the design, these defects must be detected and fixed before the product leaves the factory. If not corrected, the defects will show up as fielded product failures that increase warranty costs and reduce the perceived quality and value to the customer. Additionally, manufacturing production costs will increase due to rework and repair if corrective actions are not taken. It is for these reasons that manufacturers have a requisite interest in the most effective means for the earliest elimination of defects.

Most manufacturers have traditionally depended on the final acceptance test to catch manufacturing defects. They have learned over time to add intermediate (board-level) electrical tests, screens, and inspections as part of the continuing course of manufacturing process improvement. Failure analysis can provide valuable feedback as to the type of test, inspection, or screen to use; duration or resolution required; and placement within the production process. The optimization of the overall reliability program will effectively cover the component and assembly fault spectrum while preventing redundant coverage and maximizing the resulting cost savings. It is hoped that the introductory material provided herein will prove beneficial to the failure analyst in the evolution of this essential feedback.

2.2 JUST WHAT IS RELIABILITY?

The most common measurement of reliability is *mean time between failure* (MTBF). MTBF is a statistical value that indicates how long, on average, an electrical unit can operate before failing. It should be realized that a specified MTBF depends on the method with which it is determined and the method's inherent assumptions. The assumptions vary widely according to the method and can cause very large differences in the final determination. The methods and assumptions used must be fully understood for the specified MTBF to be meaningful. This section will deal only in a general discussion of MTBF.

MTBF can be determined either by pure calculation or by a more direct demonstration. Demonstrated MTBF gives the most realistic results, but it is usually impractical due to the length of the testing and the high cost involved. If the testing is done before the manufacturing processes are mature, the variations due to manufacturing anomalies can make consistent results difficult.

A calculated MTBF, at a specified temperature, is often used as a verification that the design is relatively sound. This can be done using mathematical procedures for summing up the published failure rates of all the individual components to give a failure rate of the composite. Numbers generated in this manner should be used for comparative purposes only. This type of analysis may be useful for meeting contractual goals or marketing purposes, but it cannot provide any insight to corrective actions and often does not reflect the actual reliability measure of the equipment. The actual failure mechanisms in electronic assemblies and their frequency of occurrence are best identified using failure analysis.

2.3 FAILURE REPORTING, ANALYSIS, AND CORRECTIVE ACTION SYSTEMS

The military builds very expensive, one-of-a-kind, weapon systems that must perform reliably to ensure mission success. They realized that corrective action options and versatility are greatest during design evolution. At this point in the life cycle of the product, even major changes can be investigated to eliminate or significantly reduce susceptibility to known failure causes. The options become more restricted and expensive to implement as a design becomes firm. The earlier a failure cause is identified and positive corrective action implemented, the sooner both the producer and user realize the benefits of reduced failure occurrences in the factory and in the field.

It was for these reasons that a disciplined and aggressive closed-loop *failure reporting, analysis, and corrective action system* (FRACAS) was considered to be an essential element in the early and sustained achievement of the reliability and maintainability potential inherent in military systems, equipment, and associated software. The essence of the closed-loop FRACAS is that failures and faults of both hardware and software are formally reported, analysis is performed to the extent that the failure cause is understood, and positive corrective actions are identified, implemented, and verified to prevent further recurrence of the failure.[1]

Although FRACAS was a wonderful idea in theory, in practice the military received only limited benefits from these programs once the weapon system hit full-scale production. This was mostly due to personnel within the required bureaucracy overemphasizing design contributions and the perceived high cost associated with contractor compliance during manufacturing and life sustainment. The implementation of the military's "commercial off-the-shelf" policy should, in time, shift this balance.

In the commercial world, great emphasis is put on manufacturing and product sustainment costs. By design, the FRACAS must control these costs. The author is familiar with a system that utilizes a small IPC coded strip on each circuit board. The part number, serial number, lot number, date of manufacture, place of manufacture, machine operators, electrical test results, all manufacturing rework and repair required, digital copies of all visual and X-ray inspections, and warranty return and repair information are stored and continuously updated for each board and are accessible at the stroke of a reader pen. The company performs failure analysis on 100 percent of all returned boards for a period of two years following the introduction of a new product. It was demonstrated that, after a period of two years, most inadequate processes or suppliers have been identified for the associated technology, and additional feedback has limited value. Since the company utilizes a results-driven process, it would be difficult to question this policy.

Lessons learned should be used by engineers to write specifications to eliminate certain materials, components, or processes for safety or reliability reasons. Restrictions may be placed on using certain hazardous materials such as PCBs or beryllium insulators for safety reasons, and certain processes such as the use of halogenated solvents on unsealed electrolytic capacitors may be forbidden for reliability reasons. The specification should be discussed with any suppliers or subcontractors to ensure a proper understanding of the purpose of the restrictions.

When large quantities of units are purchased from an OEM, it may be possible for the user to exercise configuration control. If a manufacturer decides to "improve" a certain unit used in your product, you would need to be informed in advance of any form, fit, or functional changes. A viable supplier agreement will protect the user from major changes to a product or from a unit being "dropped" with little or no notice. Remember that, when buying an "off-the-shelf" unit, the manufacturer may be reluctant to grant configuration control to a customer unless the volume is substantial.

Formal product audits can be used as insurance against unforeseen process or product changes. These audits can be with the agreement of the supplier at the manufacturing plant, or they could consist of a combination of nondestructive and destructive physical analysis of a sample of product. Like most insurance, this can sometimes seem expensive when everything is going fine but is much appreciated when things do go awry.

2.4 QUALITY TESTS AND INSPECTIONS

When trying to determine which test or inspection technique is best, and most cost effective, at identifying a particular type of defect the following points must be considered:

- The complexity of the technology being tested/inspected
- The volume to be tested/inspected
- The per piece cost
- The required accuracy for defect location

In most cases, the test-and-inspection technology is rapidly evolving, and an accurate characterization of these criteria for each methodology is well beyond the scope of this writing. What will be presented is a general characterization from a slice in time. It is left to the reader to fill in the specifics based on the requirements of the situation.

2.4.1 Electrical Tests

There are a number of different intermediate, or board-level, electrical test techniques, and we will be able to present little more than general guidance for a few of the more universal types. For instance, boundary scan is a built-in component test accessible via a multipin test bus and specific to the device function. Like all functional test, it is an electrical test of a device or assembly that simulates the intended operating inputs and verifies the proper outputs. Even more basic are bare-board testers. These systems check for opens and shorts, test the current-carrying capacity of the traces, and verify the integrity of plated-through holes. We will now look at some of the techniques for testing populated boards, starting with those that require little or no fixturing, such as flying-probe testers.

Flying-probe testers have good accuracy but, because they have only a few simultaneous probe contact points, have limited measuring capacity and slow throughput because of repeated probe repositioning. Systems of this type can probe a variety of technologies down to fine-pitch surface mount technology but have limited fault coverage and, because of their slow pace, are unsuitable for higher volumes.

Manufacturing defect analyzers (MDAs) are a low-cost alternative for providing finite fault coverage. Faults are limited to analog defects and shorts. Moderately priced fixturing is required, and programming time is relatively short. Throughput is good, although the false reject rate is higher than with in-circuit testers. MDAs represent a good testing solution for simple consumer electronics.

In-circuit testers (ICTs) provide good fault coverage, finding opens, shorts, and defective or wrong components. Physical node access is required, as well as moderate to high cost fixturing. The high volumes that are possible can make the fixturing and programming investment more affordable. Throughput is high, and false reject rate is low.[2]

2.4.2   Automated Optical Inspection

Automatic optical inspection (AOI) improves data collection for in-process control, and the speed of the technique allows for total, rather than sample, inspections for solder joint defects. AOI techniques are also capable of finding misaligned, missing, or improperly oriented components. AOI requires line-of-sight access and can be subject to operator judgment in such areas as lighting and threshold adjustments to capture a useful image in situations with a wide variety of different solder joint types.

Automated machines are very useful for accept/reject decisions and as indicators for process variations in the PWB industry. Some systems have design rule verification software that can find design violations for minimum spacing of lines, lines that are too narrow, unintended shorts and opens, process induced pinholes and nicks, and the use of wrong-sized pads or holes. AOI systems have evolved to a level of sophistication suitable to be utilized the hybrid microcircuit, integrated circuit packaging, and component packaging industries.

2.4.3   X-Ray

Because of the high apparent density of solder to X-rays, X-ray systems perform very well on tasks involving solder joints or component placement. A simple two-dimensional X-ray of a circuit board can identify solder bridging, voids, excess solder, poor component alignment, bent pins, and marginal joints that may fail in the field. X-ray systems have an advantage over AOI systems in that they do not require special fixturing and do not require line-of-sight access.

Real-time (non-film) systems provide digital images that can be signal processed, providing automated and dependable identification of defects. X-ray laminography systems can produce cross-sectional X-ray images from several different angles, allowing assessment of critical solder-joint features. Since most X-ray exposure levels are very low, they do not harm most devices.

2.4.4   Acoustic Micro Imaging (AMI)

Acoustic micro imaging techniques are able to inspect solder joints that are out of the line-of-sight, examine plastic encapsulated devices for die attach and popcorn cracks, and perform any other application where discontinuities need to be identified. This is possible because acoustic techniques are sensitive to any sound impedance mismatches (voids and cracks). Until recently, AMI was mostly used in failure analysis applications because of the necessity for a coupling fluid and the low throughput that this constraint imposed. The use of a laser beam to couple the acoustic signal to the item being inspected has allowed increased throughput via part batches and automation.

2.4.5 Laser/IR Systems

Laser/infrared (IR) systems are based on the principle that "good" solder joints have a specified amount of solder and heat sinking and therefore will demonstrate a predictable characteristic heating and cooling profile. Usually a yttrium-aluminum-garnet (YAG) puls-ing laser is used to heat a solder joint to an expected temperature. After heating, the peak temperature and cool-down rate is monitored and compared to an equivalent joint on a known-good board. Defects such as cracks, non-wetting, excess solder or solder voids, and bridging completely modify either the temperature peak or the cooling profile. Variations from the known-good thermal profile can suggest various causes but cannot provide pre-cise defect identification. This method is incompatible with hidden solder joints, and the heating of visible joints must be controlled to prevent degradation (see Table 2.1).[3]

Table 2.1   Inspection Technique vs. Defect[*]

Defect	Visual	Machine optical	Laser IR	X-ray
Open solder joints	G	G	X	X
Shorted solder joints	G	G	X	X
Solder balls	M	U	U	X
Misaligned components	X	X	U	X
Incorrect components	X	M	U	U
Missing components	X	X	X	X
Damaged components	X	M	U	U
Incorrect orientation	X	U	U	U
Board damage	X	M	U	U
Incorrect board revision	X	X	U	U
Bent leads	G	M	M	M

* X = excellent, G = good, M = marginal, U = unacceptable

2.4.6 Other Techniques

Solder-paste inspection systems only measure paste deposits and are suitable for finding defects caused by the paste-printing process and providing measurement data for tight process control of paste-printing equipment. No fixturing is required. Throughput is medium, while fault coverage is limited to solder placement and amount.

The plating thickness on PWBs is a very important parameter that should be controlled both on the planar interconnections and in the vias and feed-throughs. The thickness of conductive traces is critical to guarantee sufficiently low impedance and to avoid excessive heat dissipation. Many measurement methods have been developed to monitor plating thickness; however, we will only discuss a few.[4]

Copper-adequacy inspection on PWBs can be performed using a contact probe and a combination of eddy current and microresistance techniques. When multiple layers of different plated materials are involved, a technique such as X-ray fluorescence can be used. It is applicable to coatings down to the microinch level on surfaces down to dimensions of several mils. The radius of curvature is limited to the order of the linear dimension of the measurement area.

X-ray fluorescence uses a beam of X-rays directed at the plated surface. The material absorbs X-rays and then re-emits X-rays of a frequency that is characteristic of the material. The thickness of the coating(s) can then be determined by comparison of the relative strength of signals to those of known thickness standards.

Cross sectioning is the most common procedure for determining PWB plating thickness. This technique is destructive and time consuming, requiring mounting, grinding, polishing, etching, and inspection. Suitable sample preparation requires substantial operator expertise, even with automation. Improper preparation can produce disastrous results such as the masking of obvious defects or the appearance of polishing artifacts.

2.5   ENVIRONMENTAL QUALIFICATION TESTING

2.5.1   Vibration

Almost every piece of electronic equipment will need to have some sort of vibration and/ or shock specified. If the equipment will be desk bound, such as a personal computer, the vibration specification is, as a minimum, tailored around shipping vibration. Most manufacturers will perform this type of testing, since equipment that can not survive shipment to the customer will surely cause a loss of customer goodwill as well as warranty problems. The testing may be performed on the equipment alone or may incorporate the shipping container.

Shock testing is sometimes specified if the unit is expected to meet with *bench handling,* a term used to describe an infrequent fall to the floor. Shock testing may use a drop test, which requires a minimal investment in equipment, or may require a long-stroke electromagnetic shaker.

If the equipment is in a high-vibration environment, such as a portable application, specific vibration specifications should be imposed. MIL-STD-810 offers many versions of vibration or shock tests for the many different types of environments to which military weapons are exposed.

2.5.2   Temperature

Temperature testing is accomplished to validate the overall thermal design, which includes the flow of cooling air and the thermal characteristics of the equipment. It is important to note that temperature tests can be misleading, since environmental chambers maintain a uniform thermal environment by utilizing high rates of airflow. This airflow may not exist in the natural environment of the equipment and has the effect of increasing the thermal efficiency of the equipment heatsinks. For example, semiconductor heatsinks can run 20° C cooler in a high rate of airflow at a given temperature.

2.5.3   Others

Many electronic consumer products, and most military and industrial ones, specify a humidity range. If condensation on or inside the equipment is possible, condensation must be considered for the environmental test specification. This may force the application of conformal coating to printed circuit boards, or the encapsulation of power supplies.

Although salt spray is generally considered to be a military specification, there are many commercial applications, such as marine uses, where the end product will be exposed to this sort of environment. MIL-STD-810 gives detailed procedures for salt spray testing; however, commercial products may not require such rigorous testing.

Altitude has the most dramatic effect on high-voltage insulation (see Chap. 20). Another consideration is the effectiveness of convection cooling at decreased air pressure. This effect can be mitigated in aircraft applications, which typically have plenty of air volume available. The decrease in air temperature at high altitudes will also help to increase the effectiveness of the cooling air. Ten thousand feet is a common specification called out if the unit is not expected to fly or is used in a pressurized environment. Space applications can be demanding, since air cooling is not available, and temperatures can vary widely due to solar heating.

2.6   ENVIRONMENTAL STRESS SCREENING

Environmental stress screening (ESS) of a product is a process that involves the application of one or more specific types of environmental stresses for the purpose of precipitating to hard failure, latent, intermittent, or incipient defects or flaws that would cause product failure in the use environment. The stress may be applied in combination or in sequence on an accelerated basis but within product design capabilities (see Tables 2.2 and 2.3).[5]

2.6.1   Temperature Cycling

An approximation of the types of failures detected in mature hardware by temperature cycling is:[6]

Design marginalities	5%
Workmanship errors	33%
Faulty parts	62%

Temperature ranges of –65 to +131° F (–54 to +55° C) are the temperatures most commonly used. Most parts will withstand temperature cycling with power off through a temperature range of –65 to +230° F (–54 to +110° C). The temperature rate of change of the equipment during the heating cycle will be facilitated with the power on and the chamber

Table 2.2 Screening Environments vs. Typical Failure Mechanics

	Screening environment		
	Thermal cycling	Vibration	Thermal or vibration
Type of failure	Component parameter drift	Particle contamination	Defective solder joints
	PWA opens/shorts	Chafed, pinched wires	Loose hardware
	Components incorrectly installed	Defective crystals	Defective components
	Wrong component	Mixed assemblies	Fasteners
	Hermetic seal failure	Adjacent boards rubbing	Broken component
	Chemical contamination	Two components shorting	Surface mount technology flaws
	Defect harness termination	Loose wire	Improperly etched PWAs
	Improper crimp	Poorly bonded component	
	Poor bonding	Inadequately secured parts	
	Hairline cracks in parts	Mechanical flaws	
	Out-of-tolerance parts	Improperly seated connectors	

temperature should be controlled to an appropriate value. Power to the unit under test should be turned off during the cooling cycle to facilitate the attainment of specified cooling rates. The maximum safe range of component temperature and the fastest time rate of change of hardware temperatures will provide the best screening (see Fig. 2.1).[7]

The rate of temperature change of the individual electronic parts depends on the chambers used, the size and mass of the hardware, and whether the equipment covers are taken off. In general, the rate of change of internal parts should fall between 1° F (0.55° C) per minute and 40° F (22° C) per minute, with the higher rates providing the best screening. A

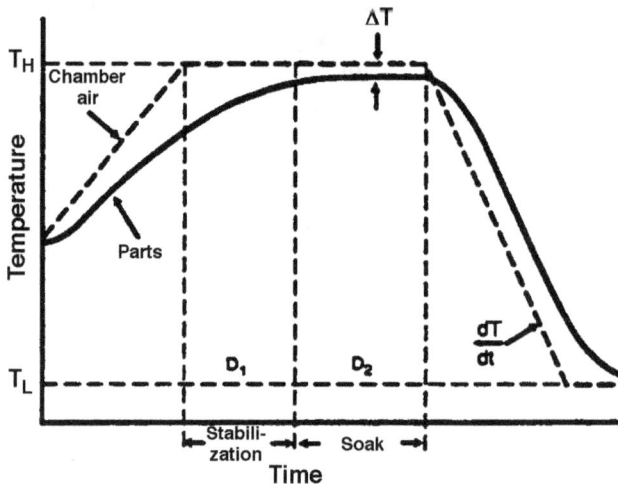

FIGURE 2.1 Typical temperature cycling profiles.

Table 2.3 Risks and Results of ESS at Various Levels

ESS conditions/trade-offs								Risks/effects		
Level of assembly	Power applied*		I/O†		Monitored‡		ESS cost	Technical		Comments
	Yes	No	Yes	No	Yes	No		Risk	Results	
Temperature cycling										
PWA		X		X		X	low	low	poor	Conduct pre- and post-ESS functional test screen prior to conformal coating.
	X			X		X	high	lower	better	
	X		X		X		highest	lowest	best	
Unit/box	X		X		X		highest	lowest	best	If circumstances permit ESS at only one level of assembly, implement at unit level.
	X			X	X		lower	higher	good	
		X		X		X	lowest	highest	poor	
System	X		X		X		highest	see comment		Most effective ESS at system level is short-duration random vibration to locate interconnect defects resulting from system integration.
Random vibration										
PWA	X		X		X		highest	low	good	Random vibration is most effective at PWA level if: Surface mount technology is utilized; PWA has large components; PWA is multilayer; PWA cannot be effectively screened at higher assemblies
	X			X	X		high	high	fair	
		X		X		X	low	highest	poor	
Unit/box	X		X		X		highest	low	best	Random vibration is most effective at this level of assembly. Intermittent flaws most susceptible to power-on with I/O ESS. Power-on without I/O reasonably effective. Decision requires cost/benefit trade-off.
	X			X	X		low	higher	good	
		X		X		X	lowest	higher	good	
System	X		X		X		low	low	good	Cost is relatively low because power and I/O normally present due to need for acceptance testing.

* Power applied—at PWA level of assembly, power on during ESS is not always cost-effective—see text.
† I/O—equipment fully functional with normal inputs and outputs
‡ Monitored—monitoring key points during screen to ensure proper equipment operation

thermal survey should be performed to evaluate the thermal response of the various elements in the hardware to changes in the temperature of the heat transfer medium (air for most, although electrically insulating fluids are used for higher [thermal shock] cycling rates).

Temperature cycling with good parts and packaging techniques is not degrading, even with several hundred cycles. However, the packaging design must be compatible with the temperature cycling program, or the acceptance test yield will be reduced. The key to selecting the appropriate temperature extremes is to stress the hardware adequately to precipitate flaws without damaging good hardware. The storage temperature (high and low) limits of hardware such as the materials in printed wiring assemblies and the maximum turn-on and operating temperatures of the electronic parts are important factors in determining the extreme values of temperature.

Some typical problems encountered are:

- Filters, motors, transformers and devices containing fine wire (#40 or #50) may develop problems.
- Because the coefficient of thermal expansion for plastics is about 8 to 30 times greater than Kovar transistor leads or Dumet diode leads, transistors mounted on plastic spacers and coated with conformal coating will produce cracked solder joints in a few temperature cycles if the leads are not stress relieved.
- Large multipin modules soldered into the printed circuit board may result in solder joint cracking, especially if the conformal coating bridges between the module and the board.
- Breakage of glass diodes can be expected if great attention is not given to the encapsulating material and the process.

Temperature cycling is most compatible with printed circuit board construction and least compatible with large, complex, potted modules where failure means scrapping the entire module. The equipment should be closely monitored during the operating portions of the cycle.

2.6.2   Vibration

The effectiveness of any vibration screen should be evaluated by engineering analysis of the equipment and the expected flaws, using factory and field failure data, and the failure history of the equipment during and subsequent to the screen, adjusting the screen parameters as the screen matures. Four methods are described in order of descending analytical complexity. Selection depends on such factors as: (1) hardware availability, (2) number and production rate of items to be screened, (3) availability of vibration equipment (shakers, data acquisition analysis, etc.), and (4) availability of experienced dynamic test or screening personnel.

A step-stress approach should be used to determine the tolerance limit or design capability of the hardware for the screen. By knowing this limit, a safe screening level can be determined and changed as required to obtain satisfactory screening results. The overall input level is tailored to the produce. Ideally, the final vibration input screening level should be greater than or equal to one-half the design capability of the hardware and greater than the operating level or possibly even the design requirement[8] (see Fig. 2.2[9] and Table 2.4).[10]

2.7   BURN-IN

Burn-in is the process of applying an electrical load to components or assemblies at elevated temperature to weed out defective parts such as bad solder connections and other assembly related defects. Combined with failure analysis, burn-in can be a cost-effective way to find weak components and monitor suppliers. *Cycled burn-in* is the continuous on-

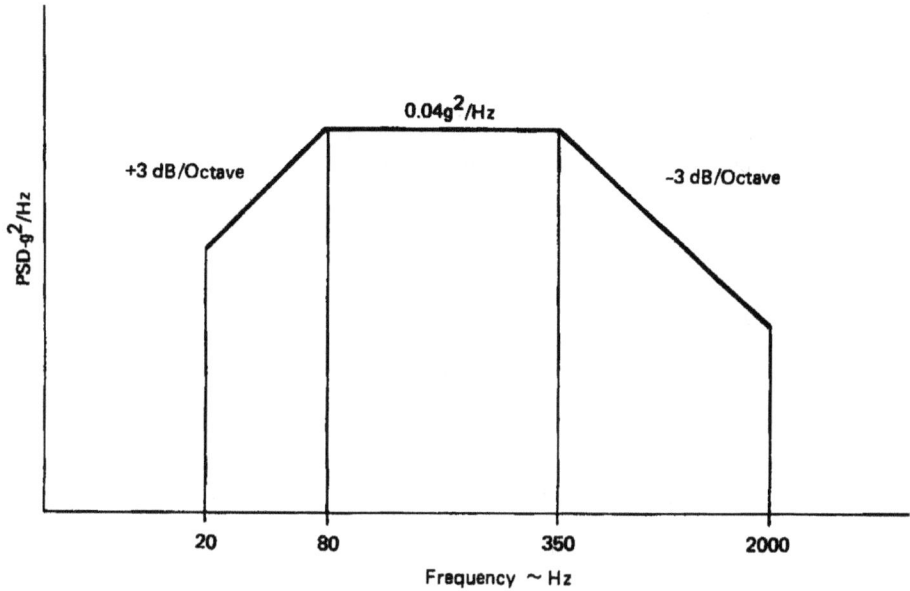

FIGURE 2.2 Random vibration spectrum.

Table 2.4 Baseline Vibration Profiles

	Level of assembly		
Characteristic	PWA*	Unit	System
Overall response level†	6 g$_{rms}$	6 g$_{rms}$	6g$_{rms}$
Frequency‡	20–2000 Hz	20–2000 Hz	20–2000 Hz
Axes** (sequentially or simultaneous)	3	3	3
Duration axes sequentially axes simultaneously	10 min/axis 10 min	10 min/axis 10 min	10 min/axis 10 min
Product condition	unpowered	powered, monitored	powered, monitored

* When random vibration is applied at the unit level, it may not be cost-effective at the PWA level. However, PWAs manufactured as end item deliverables or spares may require screening using random vibration as a stimulus. However, at the system level, when a response survey indicates that the most sensitive PWA is driving the profile in a manner that causes some PWAs to experience a relatively benign screen, that PWA should be screened individually. Each PWA screened separately should have its own profile determined from a vibration response survey.

† The preferred power spectral density for 6 g$_{rms}$ consists of 0.4 g^2/Hz from 80 to 350 Hz with a 3 db/octave rolloff.

‡ Vibration input profile for each specific application should be determined by vibration response surveys that identify the correlation between input and structural responses. Higher frequencies are usually significantly attenuated at higher levels of assembly.

**Single-axis or two-axis vibration may be acceptable if data shows minimal flaw detection in other axes.

Note: Pure random vibration or quasirandom vibration are considered acceptable forms of vibration for the purpose of stress screening, the objective is to achieve a broadband excitation.

and-off switching of the current that forces the unit to cycle thermally. Cycled burn-in is usually more effective than a constant-load burn-in. Cycling may also surge circuitry that contains inductive components, providing an even more efficient screen. Most failures will occur relatively early in the cycle; however, burn-in duration should be determined by tracking the failures over time and finding the most cost efficient cutoff point. The duration of burn-in may need to be adjusted periodically.

2.8 *EQUIPMENT SAFETY*

A variety of electrical products require accredited testing to ensure that a manufacturer can make products meeting specified safety requirements according to nationally recognized test standards. In the U.S., the Occupational Safety and Health Administration (OSHA) now runs a Nationally Recognized Testing Laboratories (NRTL) program for accreditation of independent test laboratories. The most widely known is Underwriters Laboratories (UL). Original equipment manufacturers should note that having agency approvals on a product's components or subassemblies can make the process of obtaining approval for the entire system much easier. This is usually done under the UL-recognized component program for use in a particular type of equipment or application. The main point to keep in mind is that the cost of obtaining all these certifications is relatively high, and it may be prohibitive to production if the product volume is relatively low. An off-the-shelf product that already carries the approvals may be the only viable alternative.

There may also be product test requirements derived from where the product is to be marketed (an example is the IEC standards) or the of equipment (FCC standards). An approval that must be sought if the product is to be marketed in Canada is that of the Canadian Standards Association (CSA), which is the Canadian equivalent of the U.S. National Bureau of Standards. CSA product requirements are usually quite similar to UL and CSA may request a copy of applicable UL reports. CSA has at least one facility that is a member of the NRTL. It is important for managers to realize that these tests are performed to meet government regulations concerning product safety and only indicate a valid design. These tests usually provide no insight into production anomalies, offer no protection from product liability claims, and have no influence over warranty costs.[11]

2.9 *EXAMPLES OF SAFETY- AND RELIABILITY-RELATED INVESTIGATIONS*

2.9.1 Cruise Control Module—ESD/Surge Tests

Sudden acceleration is alleged to have been caused by a cruise control module. It was suggested that either an electrostatic discharge (ESD) or a high-energy voltage surge on the power cables caused a malfunction of the cruise control module. Tasks involved the design and construction of a bench setup to test the cruise control module and actuator. A Schaffner NSG 432 static discharge simulator was used to generate an electrostatic discharge. A surge test was conducted for positive and negative surges using a Schaffner NSG 500 B/11 automotive interference generator. Testing was conducted per Society of Automobile Engineers (SAE) standards.

Figure 2.3 shows the test setup and instrumentation used to perform surge tests and ESD tests on a cruise control unit. From left to right: A Tektronix 2236, 100 MHz oscillo-

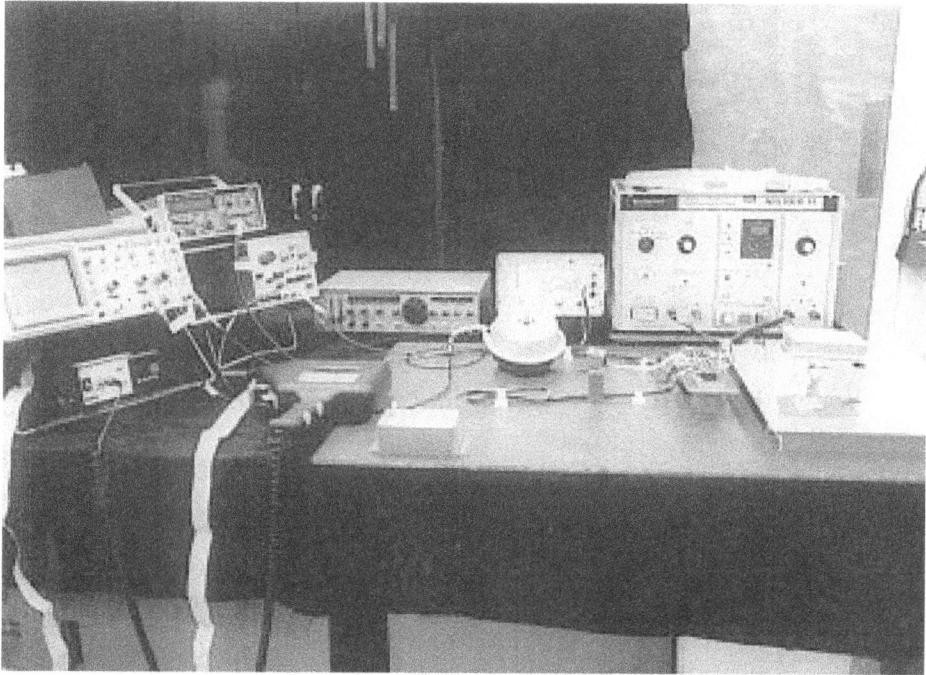

FIGURE 2.3   Test setup for performing ESD an d surge tests on a cruise control module.

scope was used to display the test waveforms. Under the oscilloscope is the power supply for the Schaffner NSG 432 static discharge simulator. To the right of the Tektronix oscilloscope is the Tektronix FG503 function generator, used to generate the RPM pulses for the cruise control module under test. A Philips PM 6671 120 MHz high resolution timer/counter was used to display the frequency of the speed and RPM pulses transmitted to the cruise control module under test. A Krohn-Hite model 1600 lin/log sweep generator was used to generate the speed pulses. A precision dc source was used to supply the dc voltage. To the right of the precision dc source was the Schaffner NSG 500 B/11 interference generator for automotive electronics. This interference generator was used to generate the surge pulses for the cruise control module (see Fig. 2.4).

2.9.2   Heat Tape Investigation

Extensive electrical tests were conducted on sample heat tapes to determine if simulated fault conditions would create an arcing fault in the heat tape. Grounding systems for marine applications, which generally have an ungrounded system, were investigated. Various simulated fault conditions were generated to determine the short circuit current and its corresponding effect on the heat tape. The investigation involved the review of several designs and circuits for ground fault interrupters for the system.

Fig. 2.5 is a photograph of the arcing observed during heat tape testing. The arcing was very intense and, under certain test conditions, the peak current exceeded several hundreds of amperes.

FIGURE 2.4   Cruise control module connected to the test fixture.

2.9.3   Electrical Testing of Aluminum Ladders

The electric utility primary distribution conductor is typically 13.8-kVac line to line and 8.0-kVac line to ground. If an aluminum ladder makes contact with an energized distribution conductor, the high voltage can result in electric shock and possible electrocution of the individual. Severe burns, resulting in death, have also been reported to be caused by high fault currents.

Leakage current tests were conducted to determine the feasibility of modifying aluminum extension ladders to reduce electrocutions on contact with a primary distribution conductor. Electrical tests were conducted using more than 150 typical ladder contaminants such as latex paints, oil-base paints, caulks, paint thinners, cleaners, and so on to determine the magnitude of the leakage current as a function of time.

High-voltage contamination tests were conducted to determine the arcing fault current using a 20-kVA high-voltage power supply. Figures 2.6 and 2.7 present the arcing and flames observed during the high voltage contamination tests for two different contaminant lengths.

Real-world tests were conducted on aluminum ladders using an instrumented mannequin, in contact with a ladder, which was allowed to make contact with a distribution conductor with high-current, short-circuit current capability. The ladder was modified to include an insulating link in the fly section. The insulated link was purposely contaminated with a typical contaminant. The mannequin was instrumented with a monitoring circuit, designed to detect hazardous levels of electric current and to provide visual indication by

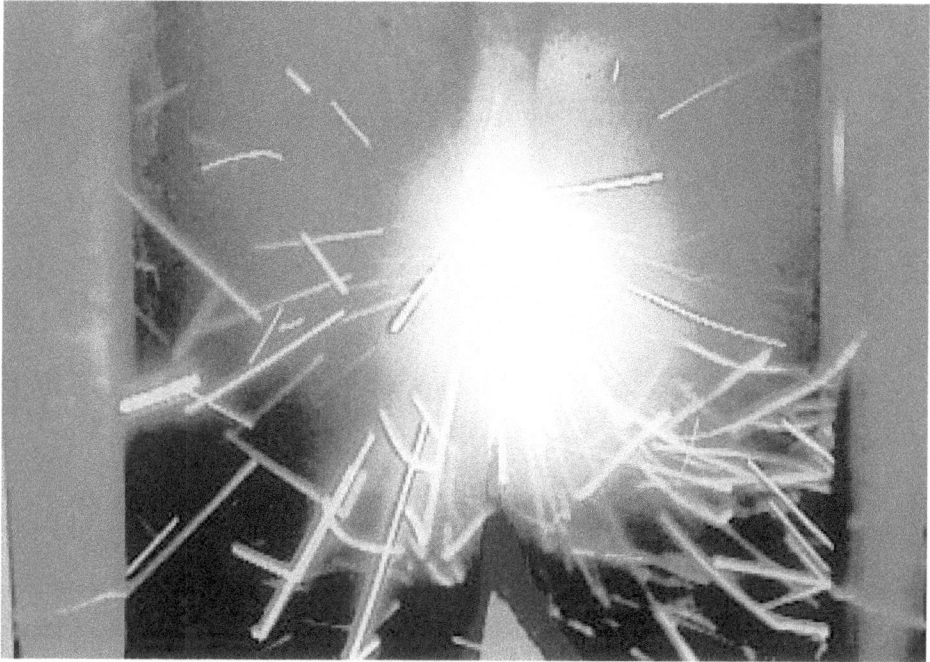

FIGURE 2.5   Arcing observed during the electrical tests conducted on the heat tape.

FIGURE 2.6   Arcing observed during high-voltage contaminant tests. Base material is Delrin®, contaminant is aluminized paint (10 in. long × 1 in. wide), and drying time was seven days.

FIGURE 2.7 Flames observed during high-voltage contaminant tests. Base material is fiberglass, contaminant is aluminized paint (10 in. long × 1 in. wide), and drying time was seven days.

energizing a photo flash bulb. The results of the tests were presented to the Consumer Product Safety Commission (CPSC).

Figure 2.8 presents the test setup for the Real World Tests. Figure 2.9 shows a close-up of the instrumented mannequin at the base of the ladder. Figures 2.10 and 2.11 present the close-up views of the arcing observed on the contaminated insulating link for two different contaminants.

2.9.4 Failure Analysis of Rectifier Transformers at a Steel Mill

A detailed failure analysis was conducted on three 9,029-kVA, 13,800/650-Vac rectifier transformers, two of which failed in service at a large steel mill. The first photograph (Fig. 2.12) shows the detanking operation of one transformer. It was determined that, during a retrofit operation, the new transformer insulating fluid contained a small quantity of water, which subsequently caused an arcing fault and a failure. The work involved performing a risk assessment of 163 other transformers in the facility as well as developing a computer database for future maintenance, tests, and inspections of the transformers. Figures 2.12 and 2.13 present two views of the transformer.

FIGURE 2.8   Test setup for real-world tests. The mannequin is visible at the base of the ladder. The distribution conductor is supported by wooden poles and is located at a height of 30 ft. above ground.

FIGURE 2.9 Close-up of the mannequin at the base of the ladder during real-world tests. The power components for the mannequin are located on its back. A photoflash bulb located on the top of the mannequin's head visually indicates when the mannequin current exceeds hazardous levels.

FIGURE 2.10 Arcing observed on contaminant during real-world tests. Base material is Delrin®. Contaminant is latex caulk.

FIGURE 2.11 Arcing observed on contaminant during real-world tests. Base material is Delrin®. Contaminant is latex paint.

FIGURE 2.12 View of a 9029 kVA, 13.8 kV/650 V transformer detanked, showing the windings.

FIGURE 2.13 Head-on view of the transformer in Fig. 2.12.

2.10 REFERENCES

1. MIL-STD-2155, *Failure Reporting, Analysis and Corrective Action System,* 24 July 1985.
2. Oresjo, Stig. Part 9—Step by step SMT; Test and inspection, *SMT Magazine,* October 1996.
3. Harper, Woychik, & Martin. Seminar, "Electronics Manufacturing Fundamentals." Technology Seminars Incorporated, 14–15 August, 1996.
4. Palmer, David. Component and board-level physical test methods, *Electronic Materials Handbook, Vol. I—Packaging.* ASM International, 1989.
5. Tri-Service Technical Brief 002-93-08, "Environmental Stress Screening Guidelines," July 1993.
6. NAVMAT P-9492, "Navy Manufacturing Screening Program," May 1979.
7. Tri-Service Technical Brief 002-93-08.
8. *Ibid.*
9. *Ibid.*
10. NAVMAT P-9492.
11. See 29 CFR 1910.7 for OSHA regulations.

CHAPTER 3
PRODUCT LIABILITY

Ian Le May

Metallurgical Consulting Services Ltd.

3.1 INTRODUCTION

The subject of product liability must be of concern to any individual or company involved in the design and manufacture of products. If a failure occurs as a result of a defect in design or in the manufacturing process of the product itself, or from a defective component supplied by a subcontractor, and if an injury or loss occurs as a consequence, the costs to the manufacturer, designer, or both can be very high.

The identification of the causes of failures and of the defects responsible is achieved by means of failure analysis, and it is important that those involved in conducting such analyses have an understanding of the law as it relates to product liability as well as of their responsibility to ensure that the investigations made will stand up to scrutiny in court in the event of a product liability action.

Another separate but related aspect of failure analysis is that design and production procedures are improved as a result of a knowledge of prior failures. Unfortunately, there is no substitute for experience, and it is disappointing that there is almost always inadequate feedback to designers—at least concerning failures that have taken place. The concept of failure is very important when design is undertaken in that the designer must consider the various ways in which failure can be avoided. While design codes may be followed, they cannot take account of all the possible outside influences on the product in service: these include environmental factors to which the product may be exposed and misuse that may occur either inadvertently or foreseeably. For anyone who doubts that design is refined and improved as a result of failure investigations, the classic case of the de Haviland Comet aircraft may be cited. In this case, the very complete investigation and subsequent testing[1,2] generated valuable knowledge for design against fatigue. The book by Petroski[3] also provides valuable illustrations regarding the improvement of design from consideration of prior failures. Finally, in this connection, it is appropriate to quote the words of the great Victorian engineer, Robert Stephenson, who said, in a discussion at the Institution of Mechanical Engineers in London in 1856:[4]

...he hoped that all the casualties and accidents, which had occurred during their progress, would be noticed in revising the Paper; for nothing was so instructive to the younger Members of the Profession, as records of accidents in large works, and the means employed in repairing the damage. A faithful account of these accidents, and of the means by which the consequences were met, was really more valuable than a description of the most successful works. The older Engineers derived their most useful store of experience from the observations of those casualties which had occurred to their own and to other works, and it was most important that they should be faithfully recorded in the archives of the Institution.

If a product is identified as unreasonably dangerous after being put on the market, the manufacturer has a duty to recall it for either modification or replacement. However, if a product liability claim has been filed against a product, a manufacturer should not make improvements or modifications to it without first discussing the matter fully with counsel and the insurance company. It would be prudent to withdraw the product from the market, at least temporarily, if there is a prima faci case indicating that the product liability suit might put the manufacturer at risk: again the matter is one for advice from legal counsel and the insurer.

The foregoing discussion has been a general one and not specific to electronic failure analysis, and the same is largely true of the remainder of the chapter. The legal aspects and principles apply equally, whether the product is a pressure vessel, power station, bridge, automobile, control circuit, or microprocessor. However, it is believed that, with the sophistication of modern methods, potential failures in electronic components can be analyzed and tested for in advance to a much greater extent than has been possible with the more traditional products.

In this chapter, product liability law is reviewed, with the basic actions involving negligence, strict liability, and express warranty being examined. The historical development of product liability (particularly in the United States) is described at the same time as these topics are discussed. The establishment of a product liability case is outlined, along with the definition of a defect. The cost of product liability is then reviewed, and the final section deals with electronic failure analysis and court proceedings, covering aspects of the failure analyst's work in conducting an analysis and preparing for court and presenting some relevant court cases.

As some of the terminology may be unfamiliar to readers, definitions of a number of the legal terms are included in the glossary on p. 3.16.

3.2   PRODUCT LIABILITY LAW

Product liability suits have mushroomed over the past 20 years, and the law relating to product liability has changed substantially. This is not so much because of new laws being established, but rather because of precedents being established from other cases: most interpretations come from case law developed from prior judgments. Negligence may be defined as "conduct which involves an unreasonably great risk of causing damage"[5] and, in the past, evidence of negligence on the part of the designer or manufacturer needed to be developed by the plaintiff in order to succeed in a product liability suit—but this is no longer true today, at least in the United States It is appropriate to review the history of product liability actions before examining the current situation.

Product liability is not absolute liability: the occurrence of an injury in which a product played a part does not necessarily mean that the manufacturer or supplier of the product is responsible and can be held liable. For an action to prevail, it must be established that a manufacturer or supplier violated a legal responsibility to the injured party. The responsibilities are based on the principles of law against which the conduct of the defendant or the quality of the product can be tested. These principles are:

- Negligence, which tests the conduct of the defendant
- Strict liability and implied warranty, which test the product's quality
- Express warranty and misrepresentation, which test the performance of the product against the explicit representations made by the manufacturer or supplier on its behalf

3.2.1 Negligence

Since the early 19th century at the latest, a plaintiff has been able to sue his immediate seller for negligent preparation, inspection, or misrepresentation of a product. In general, the negligence was more probably the fault of the manufacturer, who was for many years protected by the rule of privity, which required privity of contract between the buyer who had suffered injury and the seller. Thus, the injured party could not take action for recovery for a manufacturing defect in a product that had been purchased through an intermediary. The privity rule was breached in 1916 when, in the precedent-setting case of MacPherson v. Buick Motor Company,[6] the court held that an injured party could recover from the manufacturer who was remote from him if negligence on the part of the manufacturer could be shown. The court stated that if "the nature of a thing is such that it is reasonably certain to place life and limb in peril when negligently made, it is then a thing of danger." In this particular case, the injury had been caused to the plaintiff when a wooden wheel collapsed, which was constructed of defective wood by its supplier. The defendant was found negligent because of failure to inspect the wheel. Subsequently, this rule has been adopted generally, so that the ultimate user can sue the remote manufacturer for injury caused by a defective product where negligence on the part of the manufacturer can be proved.

Although the MacPherson ruling was originally adopted for automobiles, it has been extended subsequently to cover other cases of a defective product. In the early 20th century, there was a large increase in the number of consumers injured by contaminated food and drink and, as a result, the courts eliminated the privity rule for food and drink. It was then abolished for personal products and later, in 1960, following the Henningsen decision,[7] the rule of privity was eliminated for general products. In the Henningsen v. Bloomfield Motors case, the plaintiff, Mrs. Henningsen, was driving a new 1955 Plymouth automobile when a loud noise occurred. The car swerved and struck a highway sign and brick wall, causing injury to the driver. Action was taken against the manufacturer, Chrysler Corporation. The court ruled as follows:

Under modern conditions the ordinary layman, on responding to the importuning of colorful advertising, has neither the opportunity nor the capacity to inspect or to determine the fitness of an automobile for use; he must rely on the manufacturer who has control of its construction, and to some degree on the dealer who, to the limited extent called for by the manufacturer's instructions, inspects and services it before delivery. In such a marketing milieu his remedies and those of persons who properly claim through him should not depend "upon the intricacies of the law of sales. The obligation of the manufacturer should not be based alone on privity of contract. It should rest, as was once said, upon the demands of social justice." Mazetti v. Armour & Co. (1913)[8] "If privity of contract is required," then, under the circumstances of modern merchandising, "privity of contract exists in the consciousness and understanding of all right-thinking persons."

Accordingly, we hold that under modern marketing conditions, when a manufacturer puts a new automobile in the stream of trade and promotes its purchase by the public, an implied warranty that it is reasonably suitable for use as such accompanies it into the hands of the ultimate purchaser. Absence of agency between the manufacturer and the dealer who makes the ultimate sale is immaterial.

The rule subsequently followed is that, if a product is inherently dangerous and is later found to be defective, the user (if injured) can sue the manufacturer without there having been any contractual agreement in place, provided the defect caused the injury, there was reasonable means existing to find the defect, and the user had used the product in a reasonable manner. The MacPherson rule was included in the Restatement (Second) of Torts § 402A (1966) and has been extended in recent years to include bystanders.[9]

In the above discussion, the adjective "reasonable" has been used several times. Because all human activity involves some element of risk, the conduct of a defendant is considered negligent only when it falls below the standard of what would have been done by a "reasonable" person under the same circumstances. Although this may appear to be an imprecise standard, it provides flexibility for the courts. If a manufacturer has followed sound procedures of quality control and inspection and has taken "reasonable care," the plaintiff would not be able to recover for his injuries on the basis of negligence on the part of the manufacturer. There are, however, three factors that the court must take into account in a negligence action. These are (1) the probability of harm resulting, (2) the seriousness of the harm that occurred, and (3) the difficulty and cost of taking precautions to minimize, or eliminate the risk of harm occurring. This does not eliminate the responsibility of the manufacturer if he follows established codes. The reasonable manufacturer is expected to be cognizant of current industry standards and technology. In other words, if an improved inspection technique has been developed that could have been used to detect the latent defect, and the manufacturer could reasonably have been expected to know of it and to have employed it, then negligent behavior may be adjudged.

The *res ipsa loquitur* doctrine (the thing speaks for itself) has been used to prove negligence in many cases. This doctrine may be used when there is no evidence of negligence except for the inference that the occurrence itself could not have occurred without negligence. Generally, the doctrine will apply only if the plaintiff can show that the injury was caused by something under the sole control of the defendant and that ordinarily it would not have occurred in the absence of negligence. If the plaintiff can establish these two factors, then the burden will shift to the defendant to rebut the inference. The majority opinion in the case of Escola v. Coca Cola Bottling Co.[10] in which Justice Traynor gave a concurring opinion advocating strict liability is of particular interest. The court upheld, by a majority opinion, a jury verdict in favor of a waitress who was injured by an exploding bottle. The court stated:

Although it is not clear in this case whether the explosion was caused by an excessive charge or defect in the glass there is a sufficient showing that neither cause would ordinarily have been present if due care had been used. Further, defendant had exclusive control over both the charging and inspection of the bottles. Accordingly, all the requirements necessary to enable plaintiff to rely on the doctrine of *res ipsa loquitur* to supply an inference of negligence are present.

It is true that defendant presented evidence tending to show that it exercised considerable precaution by carefully regulating and checking the pressure in the bottles and by making visual inspections for defects in the glass at several stages during the bottling process. It is well settled, however, that when a defendant produces evidence to rebut the inference of negligence which arises upon application of the doctrine of *res ipsa loquitur,* it is ordinarily a question of fact for the jury to determine whether the inference has been dispelled.

3.2.2 Strict Liability and Implied Warranty

Product liability actions can be based on a second theory, that of strict liability. As already noted, Justice Traynor, in his concurring opinion in Escola v. Coca-Cola Bottling Co.[10] advocated strict liability. He noted that violation of pure food statutes imposed criminal liability, a liability "without proof of fault, so that the manufacturer is under the duty of ascertaining whether an article manufactured by him is safe. Statutes of this kind result in a strict liability in tort to a member of the public injured." Additionally, the application of the doctrine of *res ipsa loquitur* in products liability cases approaches the strict liability concept. Even though the inference of negligence may be dispelled by evidence of due care, Justice Traynor stated that an injured party "is not ordinarily in a position to refute such evidence or identify the cause of the defect, for he can hardly be familiar with the manufacturing process as the manufacturer himself is. In leaving it to the jury to decide whether the inference has been dispelled, regardless of the evidence against it, the negligence rule approaches the rule of strict liability."[11]

The landmark case of Greenman v. Yuba Power Products, Inc.[12] in 1963 established strict liability in tort for product defects as the law in California. The majority judgement was written by Justice Traynor, whose earlier opinions have been referred to above. The action arose because of an accident to the plaintiff while using a Shopsmith power tool as a lathe in 1957. The power tool had been purchased at Christmas 1955, and the attachments so that it could be used as a lathe were bought in 1957. After several times using the unit as a lathe, a piece of wood being worked on flew out and seriously injured the plaintiff. The retailer and manufacturer were given written notice by the plaintiff of breach of warranty some 10 1/2 months later, and action was commenced alleging this as well as negligence.

If the strict liability theory under which the action was to be heard was implied warranty of merchantability, then the notice had not been filed in a timely manner as required by the California Commercial Code, and the court decided that a theory of strict liability should be recognized based on the law of torts rather than one based on contract.

Consequently, the California court abolished any privity requirements, allowing action to be taken against a remote manufacturer. A plaintiff was no longer required to give notice of an alleged breach of warranty to a manufacturer, and the statute of limitations for contracts, which commenced at the time of purchase, no longer applied. Instead, the statute of limitations was that for the law of torts, beginning at the time of the injury. Subsequently, strict liability in tort has been adopted by the courts of most states in the United States.

The concept of strict liability is defined clearly in the Restatement (Second) of Torts, § 402A (1966), which states:

(1)   One who sells any product in a defective condition unreasonably dangerous to the user or consumer or to his property is subject to liability for physical harm thereby caused to the ultimate user or consumer, or to his property if

   (b) the seller is engaged in the business of selling such a product, and

   (c) it is expected to and does reach the user or consumer without substantial change in the condition in which it is sold.

(1)   The rule stated in Subsection (1) applies although

   (a) the seller has exercised all possible care in the preparation and sale of his product, and

   (b) the user or consumer has not bought the product from or entered into any contractual relation with the seller.

Although the definition of strict liability in tort appears clear, there are still a number of questions that need to be resolved by the court in particular cases. These include: What is the definition of seller? What is a product? What is a defect? What is the meaning of *unreasonably dangerous*? What is meant by *all possible care*? There are no fixed rules by which these questions can be answered, but the question of reasonableness in strict liability can best be answered in terms of balancing risk against utility.

To determine if a product is unreasonably dangerous, Wade[13] suggests seven questions need to be asked, as follows:

1. How useful and desirable is the product?

2. Are other, safer products available that will meet the same needs?

3. Is the product likely to cause injury, and is the injury likely to be severe?

4. How obvious is the danger?

5. How much knowledge about the danger does the public have, and what could it normally expect regarding the danger (particularly with respect to established products)?

6. Would care in using the product (i.e., following instructions or heeding warnings) avoid the injury?

7. Could the danger be eliminated without seriously impairing the usefulness of the product or making it unduly expensive?

The answers to these questions will vary depending on the environment in which the product is used. They are likely to be different for the same product being used in an industrial setting and in the home. The risk that is considered reasonable is likely to be much greater when the product is being handled by a skilled craftsman, for example, than if the same product is being used by amateurs or by children in the home.

In a product liability case, defects can be of two types: production defects or design defects. Manufacturing or production defects occur during the manufacture or fabrication of the product. On the assumption that the manufacturer has adequate standards for product quality, this implies that the product did not meet the manufacturer's standards, and the defect is a production one. On the other hand, if the product met the manufacturer's standards for production and contained a defect that led to injury, then either the design was defective or the manufacturer's standards were insufficiently high. In either case, the defect would be classed as a design one.

The concept of a design defect was set out by the California Supreme Court in the case of Barker v. Lull Engineering Company, Inc.[14] It was held that a product would be defective:

(a) if the product failed to perform as safely as an ordinary consumer would expect when used in an intended or reasonably foreseeable manner, or

(b) if the benefits of the challenged design are outweighed by the risk of danger inherent in the design.

Thus, a consumer expectation test provides a base minimum below which no product must fall. However there are products that are unreasonably dangerous even though they may meet consumer expectations.

3.2.3 Express Warranty and Misrepresentation

Another theory for the recovery of damages for an injury does not require the plaintiff to establish the presence of a product defect. If a seller expressly warrants a product as having certain characteristics or as performing in a particular way, and the product fails to perform in a manner that fulfills these criteria, then the defendant will be held liable if the plaintiff can prove that an injury occurred as a result of the product's failure to meet the warranty.

An express warranty is defined in Section 2-213 of the Uniform Commercial Code (UCC) as including any affirmation of fact or promise made by the seller to the buyer, or "any description or promise" or "sample or model" that relates to the goods and which becomes "part of the basis of the bargain." It is not necessary that the seller has used the formal words "warrant" or "guarantee."

3.2.4 Punitive Damages

In the United States, punitive damages have been awarded in some product liability cases in recent years. Punitive damages are distinguished from compensatory damages, which are based on strict liability. The award of punitive damages requires that the conduct of the defendant be proved to have been more than negligent, terms such as gross negligence, and willful, reckless, and wanton disregard for the safety of others having been used.

Kirkwood[15] notes that punitive damages are a penalty for past wrongs and a deterrent to others who may be tempted to commit a similar wrong in future. He notes further that the U.S. Supreme Court has held that in a recent case punitive damages of 526 times the amount of compensatory damages was not per se unreasonable.

3.2.5 Product Liability Law in Canada and Other Countries

In Canada, the situation is somewhat different from that in the United States in that strict liability in tort has not been adopted as the law. This is largely because Canadian laws are based on English common law, and it was only with the coming of the European Product Liability Directive[16] that strict liability has become a part of the English law of tort. There are two bases for action in Canada: the first is under contract principles, and the second depends on demonstration of negligence. They will be reviewed briefly here.

Under contract theory, a seller is strictly liable for defective products. The implied warranties of reasonable fitness for purpose and of merchantable quality, which had been established in the 19th century, were embodied in the English Sale of Goods Act of 1893, and this was adopted throughout the British Commonwealth, including Canada.[17] Historically, if there was no direct connection between the seller and the injured party (the consumer), contract law would not apply. Now, some provinces have enacted consumer products warranties legislation to protect individual consumers from unfairness arising from a lack of privity.[18] In other circumstances, the Canadian courts have used some ingenuity in extending contract law to third parties, but contract law, and its normal requirement for privity, is not a satisfactory means for the protection of an injured party and for recovery for loss and injury. Thus, action may need to be commenced under tort law.

As already noted, the case of MacPherson v. Buick Motor Company[6] caused a breach in the requirement for privity in the United States in 1916. In Canada, the breach was made in 1920 in the case of Buckley v. Mott[19] in which the plaintiff was injured by powdered glass in a chocolate bar. Mr. Justice Drysdale treated the chocolate bar as a dangerous article, stating:

In the American Courts it is held that where defendants manufacture and put a dangerously faulty article in its stock for sale, they are therein negligent and liable to an action for such negligence, it being the proximate cause of injury to plaintiff without any reference to contract relation existing between him and the plaintiff. . . . [T]here was a duty to the public not to put on sale such a dangerous article as the chocolate bar in question; that defendants were guilty of negligence in this respect, which was the proximate cause of plaintiff's injuries.[20]

The English courts did not apply negligence principles in product liability cases until more than a decade later, in 1932, in the case of Donoghue v. Stevenson.[21] In this action, it was alleged that a decomposed snail was found in a bottle of ginger beer bought by someone for a friend. The friend, having drunk the ginger beer and only then having discovered the remains of the snail, sued the manufacturer for damages. Although there was no privity between the plaintiff and the manufacturer, the House of Lords set out the principle which still applies in the Canadian law of product liability:

> a manufacturer of products, which he sells in such a form as to show that he intends them to reach the ultimate consumer in the form in which they left him with no reasonable possibility of immediate examination, and with the knowledge that the absence of reasonable care in the preparation or putting up of the products will result in an injury to the consumer's life or property, owes a duty to the consumer to take reasonable care....

Whereas, in the United States, the doctrine of strict liability in tort has been adopted, this still is not the case in Canada, although it is surely only a matter of time before it takes place. Thus, in Canada, a plaintiff must establish that the other party was negligent and that this negligence caused the injury. This applies for design defects, production defects, or a failure to warn of danger. An action may be based on the doctrine of *res ipsa loquitur*, which provides an "inference of negligence," but only when the actual cause is unknown.

In the European Union, product liability laws have been rationalized between the various member countries, based on the EC Product Liability Directive of July 25, 1985[16] This Directive is based on "liability without fault of the part of the producer," and the producer cannot escape responsibility by proving "no fault."[22] The European legislation may be likened to U.S. product liability law, although its operation is very different in a number of ways, as, for example, the mechanism for awarding punitive damages does not exist, trials do not involve a jury, and the level of compensation is far below that of the U.S.[22] Other countries are gradually adopting similar strict liability in tort laws. For example, on January 1, 1994, Switzerland adopted the federal law on product liability of June 18, 1993, which conforms to the EU guideline except in one very minor point relating to compensation.[23] In Japan, there has been, until recently, no product liability law, as such:[24] actions concerning product liability caused by a defective product were based on the Civil Code. They were generally based on "negligence in tort" (Civil Code Article 709) and, if there was a contractual link, on the basis of "imperfect performance of obligations" (Article 415). In either case, the onus was on the plaintiff to prove the existence of a defect in the product and that there was a causal relationship between the defect and the injury.[22] A product liability law was enacted in late 1994[25] but does not embrace strict liability.

3.3   *ESTABLISHING A PRODUCT LIABILITY CASE*

To establish a case in products liability, the plaintiff must establish three facts, as follows:

1. The product was defective, i.e., it contained a defect.
2. The defect was present when the product left the defendant's hands.
3. The defect was the proximate cause of the injury suffered by the plaintiff.

These three factors must be established independent of the theory of law on which the case is to be based. If negligence is alleged, it must be shown additionally that the defendant's conduct was unreasonable.

Conversely, if all three points cannot be proved, the plaintiff's case will normally fail, and judgement will be in favor of the defendant. Some specific exceptions constituting special cases are discussed in Section 3.6.

3.3.1 The Defect

In any investigation to determine the cause of a failure from which it is expected to initiate a product liability action, a defect must be detected. However, finding a defect in a product is not sufficient: the defect may or may not have played a part in the accident that caused injury. In a substantial number of cases, there are many failed parts or components present after an accident, and it is necessary to reconstruct the sequence of events that took place prior to and during the accident.

Assuming that a defect is found and that it can be determined that it led directly to the occurrence of the accident, i.e., it is the proximate cause of the accident, it must still be determined if the defect was one that was originally present or if it had been caused during operation and service of the failed equipment. For example, we may consider a failure resulting from the separation of a wire at a soldered joint in a printed circuit board. Having found this proximate cause of failure, the question then becomes one of why the soldered connection separated. Was it because of an initially poor joint, one in which there was insufficient wetting of the surfaces by the solder, for example? Or was it because of misuse during service in a manner that would not have been reasonably foreseeable by the designer and manufacturer? Alternatively, was it a failure that developed during service, such as by a fatigue mechanism brought about by vibration? If the answer to this last question is "yes," then it is necessary to consider whether the designer or manufacturer, or both, should have considered such conditions as likely to arise, in which case the design may have had an initial defect.

Alternatively, it may be that the design itself was sound, and the fault was one of production, with a production defect having occurred. In designing a system, the designer must consider the consequences of failure of the various components and must build in the appropriate safeguards, indicators, back-up systems, and so on, as appropriate. Clearly, the extent of the backup systems depends on the extent of risk involved and the cost of the extra safety measures, and we return to the concepts of reasonableness discussed previously. The safety measures and systems required in the event of a failure in a engine component in an aircraft are much greater than those required for an automobile.

3.3.2 The Conduct of the Plaintiff

The defendant in a product liability case who is defending himself against the allegations will attempt to prove that the conduct of the plaintiff has been at fault in that he misused the product. At the very least, the defendant will attempt to show that there was contributory fault on the part of the plaintiff. In this case, the basis of the action becomes important, i.e., whether it is based on negligence, strict liability, or express warranty.

If the action is based on negligence, in some states, the action for recovery will fail if the defendant can show "contributory negligence" on the part of the plaintiff. This would be the case if the plaintiff had behaved in an unreasonable manner and had abused the product or failed to discover an obvious defect in it. In other jurisdictions, the award of damages will be reduced on the basis of a comparative negligence rule, i.e., the court may hold the defendant 40 percent responsible and the plaintiff 60 percent responsible, for example. If the plaintiff assumes an obvious risk, then this can be considered as the fault or responsibility of the plaintiff.

Where an action is based on strict liability, the situation is different. A defendant cannot defeat the claim of a plaintiff by showing that the latter unknowingly misused the product or failed to discover an obvious defect. The defendant must show that the plaintiff "voluntarily and unreasonably encountered a known risk." Thus, the defendant must show that the plaintiff knew of the risk before the event causing injury, and chose to take that risk, which could be classed as being unreasonable.

In the case of an action based on express warranty, the plaintiff's conduct is judged on the same basis as for strict liability. In other words, the defendant must prove that the plaintiff knew that the product did not meet the warranty claimed and then voluntarily and unreasonably chose to take the risk. Serious abuse of the product by the plaintiff (in an unreasonable manner) is a defense, as in strict liability cases.

3.4   COST OF PRODUCT LIABILITY

Product liability is an expensive business. The costs involved in defending against a product liability suit are considerable, but they are small compared with the amount that may be awarded to a plaintiff or plaintiffs in the case of serious injury if the case is decided in their favor. Although there has been a mushrooming of the number of product liability cases over the past 20 years, there are indications that the number of product liability suits is declining at present, in the U.S. at least.[15,26] This is probably, in part, because of the very large losses that have been suffered by the insurance industry in recent years, making them more cost conscious of and more resistant to making settlements, and also because of the tendency to mediate settlements rather than go to trial.

Nevertheless, when an action is commenced, investigation and preparation must be done carefully and thoroughly, because it is never certain that a settlement will be reached prior to a trial, although a considerable number of settlements are made on the courthouse steps. If it is perceived by either side that the other has not made sound and adequate preparations, settlement will either be very one sided, or a court action will take place with predictable results that may not be a true reflection of the justice of the respective cases of plaintiff and defendant. All relevant evidence must be introduced through discovery and trial: it is not normally possible to introduce additional evidence in an appeal to a higher court, except in very special circumstances.

Settlements in product liability cases may exceed \$20 million and, in some cases, may greatly exceed this. For example, the settlement for a vehicle fire after a crash, which involved the placement of the gas tank on a GMC pickup truck, was \$105 million.[15] Awards are not only for personal injuries sustained, although the medical costs and the costs of support and rehabilitation may be very high, and may leave the victim in a precarious financial situation (setting aside the physical and mental problems and discomfort). The largest cost factor is often the loss of production or loss of market share suffered by a company having experienced a failure in some part of its manufacturing plant.

From the point of view of manufacturers and designers of electronic equipment, the consequences of the failure of a single component or connection could be very serious, as discussed in Section 3.3.1, and the reliability of components must be taken into account at the design stage so that allegations of defective design in the event of a failure can be mitigated. Adding to the up-front production costs may be very prudent later.

3.5   ELECTRONIC FAILURE ANALYSIS AND COURT PROCEEDINGS

In conducting a failure analysis, it should be realized that court proceedings may result at some later stage. In some cases, particularly where an accident has been very serious, it may be clear from the outset that an action is likely to follow. However, it does at times appear that the most simple and apparently trivial cases are those that end up in court. Thus, there should always be sufficient recording of information during any investigation so that proper preparations for a court action can be made. This applies to all failure analyses and not only to those involving electrical or electronic components or systems.

3.5.1   Continuity of Evidence and Documentation

In conducting an examination of a failed component, a record should be kept that will show from whom the component was obtained and the various stages of the investigation, particularly where these involve destructive testing. Depending on the particular circumstances, it may be necessary for all parties to be present while tests are made. This is likely if the investigation is being made at the instigation of counsel; where the examination is being made as a result of an insurance claim or directly on behalf of the manufacturer in the absence of any legal action, it is probable that other parties will not be present. In either case, photographs and sketches showing the component and any alterations that have been required during the investigation are a "must."

It is also very important that the investigator obtain as much background information as possible at an early stage, as this may not be easily available later. How the component or system was being operated may be crucial in identifying the causes for failure. Factors such as environment are also very important, and it may be that the failed component was being used under conditions that differed substantially from those for which it was designed. For example, short circuiting may have been produced by exposure to a very humid or wet environment. Whether the component should have been designed to operate in such conditions is a separate issue to be resolved later. The first essential in the investigation is to determine the cause of failure.

If tests are to be made, for example to determine a response time or simply to measure a physical property such as resistance, then the test procedures used should be recorded, along with details of instruments employed. The test methods should preferably be standardized ones, for example as set out by the American Society for Testing and Materials (ASTM) or other recognized standards-writing bodies. In the event of court action, and where the results of tests are important, the investigator, who will be called in the role of an expert witness, can expect to be cross-examined either during deposition or at trial, or both, and must be able to defend the test procedures used and results obtained.

Parts should be tagged or marked for identification. This is important for their display and identification at deposition or in court. It is useful to prepare a collage showing details of the dismantling of a larger component or circuit and of the cutting up of any parts for detailed examination. This can be assembled from photographs, sketches, or both. Using such procedures avoids the possibility of long arguments as to exactly where parts came from and allows the attorneys, judge, and jury to be more readily persuaded of the investigator's competence, care, and concern.

Sometimes components, circuits, or parts of them, must be shipped from one laboratory to another, either for more detailed examination or to enable other parties to examine them. These must be labeled before being shipped; they should be photographed, and identification numbers should be recorded before shipping, in case of loss somewhere en route. It is best if critical parts are shipped by hand, but they should otherwise be sent by a secure system, such as courier or registered mail. A dated transmittal note should always be prepared and records made of the sending and receiving of parts. A complete record of the movement of parts should be available.

As it is often a matter of years between an investigation and court proceedings, critical parts must be secured safely during that time. Although this may be inconvenient and may take significant space, it is something that must be done by those who are seriously involved in failure analysis work. The writer has had personal experience of several cases in which components were not secured by one of the parties and were discarded, leading to considerable embarrassment on the part of those responsible, to say the least. An action to recover may fail if the plaintiff or his representative has lost evidence; similarly, the defendant does not have a very credible case if he or his agent has mislaid critical evidence. If the investigator is responsible, he may be subject to a civil action to recover losses from the defendant or plaintiff, as the case may be.

Sometimes a report on the cause of the failure must be prepared; on other occasions the attorney may ask that no report be prepared until it is specifically requested. In some cases a statement of opinion may be all that is required, this setting out the conclusions reached and the general basis on which they have been reached. It is important to remember that all material prepared is potentially "discoverable," i.e., its existence will be disclosed to the opposing attorneys, and the documents themselves may need to be provided prior to trial, unless privilege can be established. In general, reports of failure investigations will need to be provided, and this will apply also to draft reports. In consequence, if a report is to be prepared, it should be in an agreed format and with the opinions well established prior to its being prepared for submission to the attorney responsible for handling the case. To prepare a draft that is submitted, reviewed by the attorney, and rewritten with changes in the opinions, suggests that the final report has been prepared under the instructions of the attorney and does not accurately reflect the opinions of the investigator. The credibility of the expert can be lost under cross-examination when more than one report on the failure, with different conclusions or different emphases, can be shown in court.

3.5.2 Preparation for Litigation and Court Appearance

When a product liability case is being prepared for a possible trial, and indeed throughout the investigation if this is being conducted in conjunction with an attorney with the expectation of litigation, the failure analyst should keep in close touch with counsel so that he or she knows the opinions of the analyst, who can be expected to be called as an expert witness at trial. It is very important that the expert goes over the case thoroughly with counsel, and the relevant questions that counsel should ask are known and understood—and that questions that are likely to be asked by the opposing attorney are considered and can be answered. Frequently, it is the experienced expert who can foresee the possible lines of attack that the other side may pursue, rather than the attorney, who is not so well versed technically and inclined to accept the view put forward by the analyst working with him or her as the only probable failure scenario. All alternative scenarios must be considered and rejected if appropriate. If one of them appears possible or likely, then this must be stated to the attorney. The role of the expert is to provide honest and correct advice as to failure cause, not to take a biased view. At the same time, it is appropriate to point out weaknesses in the case that the other side has made or is likely to present. The attorney's task is to do the best possible job for the client so that the best possible financial result is obtained. In many cases, this will involve settling the case out of court, and any weaknesses in the case must be identified at as early a stage as possible.

If, because of financial or other constraints, the investigation has not been as thorough as it might have been, counsel should be informed. It may be better to accept a reduced out-of-court settlement than to run the risk of losing the case because of inadequate investigation. Alternatively, counsel may be able to present the case in such a manner that the critical points only are concentrated on and that the expert investigation constitutes a reduced part of the investigation and case, rather than having a key role.

In some jurisdictions in the United States and Canada, the expert, who has conducted a failure analysis and will be called upon to testify, will be required to prepare a report on the investigation; in others, a written statement of the opinions to be presented is all that is required. To reduce the time taken in a trial, reduce the costs, and encourage a settlement out of court, a discovery process is used in which as much as possible of the background information can be provided by each side prior to trial. Depositions may be taken from potential witnesses and from the experts. These are provided under examination by the opposing side's attorney in the absence of a judge, but with the testimony recorded by a court reporter. The evidence is given under oath and can be read into the proceedings at a subsequent trial as part of the evidence. Good preparation by the expert for this is just as important as preparation for the trial itself.

It is important to emphasize that the expert witness has a very privileged role in court. A witness can normally testify only with respect to facts to which he or she was a direct witness. The expert witness can provide opinion evidence. Thus, the expert is almost taking the place of judge and jury, but clearly they need to be convinced of the expert's credibility before they can be expected to accept the expert's views. The first hurdle for the potential expert is to be accepted by the court as having expertise in a specified area or areas. It is important that the areas of expertise required for the testimony are established between attorney and expert prior to the trial, and that there is adequate backup in terms of qualifications, publications, or experience on the part of the expert to justify the claim of expertise in the areas of concern. It would be inappropriate to claim expertise in "engineering" or probably even "electrical engineering." The categories are too broad. Expertise might be claimed in, for example, "electric motor design and manufacture" or in "integrated circuits and instrumentation." If there is a need for specific expertise dealing with one aspect of the failure and the investigator does not have this, then another expert witness who can testify with authority in this particular area may need to be briefed and called.

3.6   *SOME LEGAL CASES*

When the literature dealing with product liability cases is reviewed, there are many examples involving the failure of mechanical components and relatively few examples dealing with electrical and electronic failures. Similarly, when the failure analysis literature is examined, there are relatively few cases dealing with electrical and electronic components: for example in the *Handbook of Case Histories in Failure Analysis, Volume 1,* published in 1992, out of 115 cases, only 2 deal with electrical equipment, and both are concerned with failures in electrical transmission line cables.[27,28] The situation is largely a reflection of the types of products in general use. Today, with miniaturization and the greatly increasing dependence on electronic components and computers, the picture is starting to change. Hence, the importance of strict quality control, fail-safe design, and the need to consider electronic failure analysis at the outset by manufacturers. In this connection, studies have been reported of the effects of thermal and stress cycling on integrated circuit packages and on soldered joints used in electronic circuits to predict crack propagation rates and reliability.[29] Similarly, because of the failure of 21 hermetically sealed metal-can transistors out of 100 million manufactured since 1981, an extensive investigation including accelerated corrosion tests has been reported.[30]

However, it is advisable to discuss a few of the cases reported in the open literature and to mention some other investigations made in connection with cases that have been settled out of court.

An interesting case is that of Chestnut v. Ford Motor Company.[31] The background is that a Vietnam veteran in his late twenties claimed to have been driving a new car on a winding road when the car's headlights stopped providing illumination, and the car plunged into a ravine, seriously injuring the driver. The claim was that the retractable covers over the lights had closed as a result of design and production defects.

The headlight covers were actuated by a linkage that was operated by vacuum. This was, in turn, actuated by a solenoid valve that was energized when the light switch was operated. When current was shut off from the solenoid, either by switching off the lights or by an electrical failure, the headlight covers would close.

The design defect theory related to a design such that, if failure of the electrical circuit occurred, the road illumination was eliminated, and a more logical approach would have been to make the headlight covers open in the event of a circuit failure. It was also alleged that the poor design of the electrical connectors could lead to premature failure if bent or subjected to vibration.

The production defect theory related to the connectors, one of which had separated and evidently been partially cracked at the time of manufacture. However, the question of when the connector had cracked through completely was not addressed, so there was no conclusive evidence that the connector had fractured prior to the accident and not as a result of the crash. Without such evidence, the case was lost, regardless of whether the theory of a design defect or a production defect was considered. In fact, on appeal, the decision in favor of the defense was overturned, albeit on a technicality, and the case was subsequently settled out of court.

Powell[32] describes the investigation of a case involving an accident between a car and a motorcycle, in which the motorcyclist received serious injuries. The driver of the car was sued and, in his defense, stated that the headlight of the motorcycle was not on. A report prepared for the defense by an engineering consulting firm indicated that the headlight was off because "neither filament was burned and there was no oxide coating on the filaments." No supporting photographs or background analysis was given. Subsequent examination conducted by Powell[32] of the headlight, whose glass had been shattered and filaments bent, showed that the lamp had been on, one of the filaments having small particles of glass fused to it, the other being unoxidized at its ends but having its center portion, which had made contact with the posts of the other filament, oxidized and having produced melting of the glass insulator with which it had made contact. The analysis showed conclusively that one filament had been on at the time of the crash when the glass shattered, while the other had been off but became energized in its center region only, as a result of the deformation that took place in the accident. The case was settled in favor of the plaintiff.

Powell comments on his knowing of other similar lawsuits in which the defense was based on the same claim, namely a nonfunctioning headlight, and he also makes the point that there are many consulting companies, such as the one in the case described, that depend on relationships that have been established with insurance companies to survive financially. He comments further that it may be difficult for them to provide unbiased opinions. This is something that all failure analysts and potential witnesses must guard against. Not only is it unethical and unprofessional to act in such a way, but the result of its being shown that a consultant is not providing an unbiased opinion but is acting as a "hired gun" can be a total loss of reputation and credibility.

The writer has been involved in several actions concerning explosions in lead-acid automotive batteries in which defects have been alleged to be the cause. Although such batteries are widely used and largely taken for granted, they contain sulfuric acid in the electrolyte, and hydrogen is produced during charging or as a result of internal corrosion. Hydrogen can build up to the extent that ignition will produce an explosion, and ignition sources exist in terms of possible arcing at electrical connections. On several occasions, when injuries have resulted from such an explosion, the writer has been able to identify specific sources of excessive hydrogen production as a result of corrosion and of specific sources of ignition. Several of these are described elsewhere,[33] all having led to out-of-court settlements.

It has previously been indicated that the identification of the defect that led to the injury or loss on the part of the plaintiff is a critical factor. If a defect cannot be identified as being the proximate cause of the loss, and if it cannot be shown that the defect was present when the product left the defendant's hands, then an action to recover will normally fail. However, there are circumstances where a product liability case may be won by the plaintiff without a specific defect having been identified. These include cases in which the product has apparently been handled carefully and was then destroyed in the accident, as in the explosion of a bottle. The *res ipsa loquitur* doctrine is a negligence doctrine by definition, but in the case of Lee v. Crookston Coca-Cola Bottling Co.,[34] in which the plaintiff was injured when a Coca-Cola bottle exploded in her hand, the court ruled that the case was one of strict liability and considered that *res ipsa loquitur* could prove the existence of a defect as well as negligence. In applying the ruling the court noted that the basic reason was:

> Plaintiff's inability to determine the specific defect or cause, coupled with the fact that the defendant is in a better position to present evidence as to the cause of the accident.... The same reasoning applies with equal force to excuse plaintiff from proving a specific defect when pursuing recovery on the theory of strict liability. This is especially true in exploding-bottle cases and similar situations where the product is destroyed by reason of the defect, which is also obliterated, or where the defective part alone is destroyed by the incident. In short, under the theory of strict liability plaintiff should not be required to prove specifically what defect caused the incident, but may rely upon circumstantial evidence from which it can reasonably be inferred that it is more probable than not that the product was defective when it left defendant's control.[35]

A similar argument to that outlined above has been applied in cases termed fact-of-the-accident cases. The term *res ipsa loquitur* has been avoided so as not to provide a suggestion of negligence. Such cases are generally where a new product, such as an automobile that has not been handled improperly, suffers a failure such as a fire or loss of control. One case in particular may be mentioned, in which circumstantial evidence alone was sufficient to establish the presence of a defect at the time of sale, although the precise defect could not be identified. In this case, there had been numerous repairs for electrical problems to a relatively new vehicle. A fire occurred under the dashboard on the day following one of these repairs, and it was adjudged that the fire would not have occurred if there had not been a manufacturing defect present.[36]

A final case concerns failure of a relay to operate in a mine hoisting system, resulting in the mine cage striking the bottom of the shaft. After the accident, the relay was found to operate intermittently, and a fracture was detected in a soldered joint connecting a wire from the relay to another wire from the printed circuit board on which it was mounted. The intermittent operation related to periodic surface-to-surface contact across the break. The question was whether the soldered joint had been defective from the time of manufacture or whether it had failed as a result of service.

Detailed examination of the soldered connection indicated that, although it was a poor joint, there had been a continuous path of solder across the connection at the time of manufacture, and that failure had occurred through this during operation, most probably as a result of vibration and consequent fatigue. The connection was not one that would resist such fatigue loading. Settlement was reached out of court.

3.7   GLOSSARY

Defect   The nonconformity that should not be present in the product if it had been designed, produced, and handled in a correct manner.

Defendant   The supplier, designer, manufacturer, distributor, or seller of the product that is alleged to be defective and against whom action is being taken.

Express warranty and misrepresentation   Where a seller expressly warrants or claims that a product will perform in a specific manner or has certain characteristics and the product fails to fulfill these claims, the buyer can take action to recover as the seller is then liable, based on his misrepresentation of the product.

Implied warranty   The theory of implied warranty of merchantability implies a contractual right. However, this has been broadened so that it is not simply the immediate purchaser who can take action, but action for recovery in the event of a failure leading to loss can be taken by the ultimate user against the original supplier or manufacturer.

Negligence   This has been defined as "conduct which involved an unreasonably great risk of causing damage."

Plaintiff   The individual or party who is making the complaint that a deficiency in a product has caused injury, damage, or both.

Privity   "Privity of contract" relates to there being a direct relationship between buyer and seller. In the past, this provided an almost absolute immunity to manufacturers because users normally dealt with or purchased products through an intermediary.

Product liability   This refers to the liability of a manufacturer, supplier, or designer for the consequences of malfunction of the product.

Res ipsa loquitur   This Latin phrase, meaning "the matter speaks for itself," refers to the doctrine that, if a failure has occurred, then this is a sufficient demonstration of the product's having a fault and of the liability of the manufacturer or supplier.

Strict liability   This doctrine has been developed and applied on the basis that a manufacturer, lessor, or other supplier of a defective product (including a service) is held to be responsible for the damage resulting, independent of the care taken and the inspection procedures used during its production.

Tort   This can be defined as a civil wrong. An action under the law of tort involves an action for recovery for damages inflicted on a plaintiff arising from a failure of a defective product.

3.8 REFERENCES

1. Walker, P.B. The scientific investigation of aircraft accidents. *Proc. I. Mech. E.* 179(1), 997–1014, 1964–1965.

2. de Haviland, Sir G., and Walker, P.B. The Comet failure. *Engineering Progress through Trouble,* 51–53, R.R. Wayte, ed., London: I. Mech. E., 1975.

3. Petroski, H. *Design Paradigms: Case Studies in Error and Judgement in Engineering,* Cambridge: Cambridge University Press, 1994.

4. Wayte, *R.R. Engineering Progress Through Trouble (V),* London: I. Mech. E., 1975.

5. Terry. Negligence. 29 *Harvard Law Review* 40 (1915).

6. MacPherson v. Buick Motor Co. 217 N.Y. 382, 111N.E. 1050 (1916).

7. Hennigson v. Bloomfield Motors, Inc. 32 N.J. 358, 161 A.2d 69 (1960).

8. Mazetti v. Armour & Co. 75 Wash. 622, 135 P.2d 633 (Sup. Ct. 1913).

9. Codling v. Paglia, 32 N.Y. 2d 330, 298 N.W. 2d 622, 345 New York Supplement 2d 461 (1973).

10. Escola v. Coca-Cola Bottling Co. of Fresno, 24 Cal. 2d 453, 150 P.2d 436 (1944).

11. Escola v. Coca-Cola Bottling Co. of Fresno, 24 Cal. 2d 453, 463, 150 P.2d 436, 441 (1944).

12. Greenman v. Yuba Power Products Inc., 59 Cal. 2d 57, 377 P.2d 897, 27 Cal.Rptr. 697 (1963).

13. Wade, D. Strict liability of manufacturers, 19 *Southwestern Law Journal* 5 (1965).

14. Barker v. Lull Engineering Co, Inc. 20 Cal. 3d 413, 573 P. 2d 443, 143 Cal. Rptr. 225 (1978).

15. Kirkwood, J.S. The status of product liability in the USA, *Structural Failure: Technical, Legal and Insurance Aspects,* ISTLI Special Publication 1, 33–42, H.P. Rossmanith, ed., E & FN Spon., London, 1996.

16. EC Product Liability Directive, Brussels, July 25, 1985.

17. Linden, A.M. *Canadian Tort Law,* 5th ed., 536. Toronto: Butterworth, 1993.

18. Province of Saskatchewan, Consumer Products Warranties Act R.S.S. 1978, C. C-30 and also Business Practice Acts of Newfoundland, Ontario and Prince Edward Island, Unfair Trade Practices Acts of Alberta, and The Trade Practice Act of British Columbia.

19. Buckley v. Mott, 50 D.L.R. 408, N.S. (1920).

20. Buckley v. Mott, 50 D.L.R. 421, N.S. (1920).

21. Donoghue v. Stevenson, A.C. 562, 101 L.J.P.C. 119, 147L.T.281 (1932).

22. Bisanz, H. Comparison of product liability law cases in the EC, U.S. and Japan: Status quo and beyond. *Structural Failure, Product Liability and Technical Insurance IV,* 28–33, H.P. Rossmanith, ed., Amsterdam: Elsevier, 1993.

23. Packman, T. Present-day Swiss legislation about product liability. *Structural Failure: Technical, Legal and Insurance Aspects,* ISTLI Special Publication 1, 59–66. H.P. Rossmanith, ed., E & FN Spon., London, 1996.

24. Yamada, T. The relationship of technology and law in civil suits in Japan. *Structural Failure: Technical, Legal and Insurance Aspects,* ISTLI Special Publication 1, 43–47, H.P. Rossmanith, ed., E & FN Spon., London, 1996.

25. Japanese Law No.85, 1994.

26. Ross, B. Ever diminishing concentric circles: The business outlook for engineering consultants in the litigation area. *Structural Failure: Technical, Legal and Insurance Aspects,* ISTLI, Special Publication 1, 25–28, H.P. Rossmanith, ed., E & FN Spon., London, 1996.

27. Chakrapani, D.G. Fatigue fracture of aluminum wires in high-voltage electrical cables in Alaska. *Handbook of Case Studies in Failure Analysis,* Vol. 1, 424–427, K.A. Esaklul, ed. Materials Park, OH: ASM International, 1992.

28. Furtado, H.C., and Mannheimer, W.A. Failure of an aluminum connector in an electrical transmission cable. *Handbook of Case Studies in Failure Analysis,* Vol. 1, 428–430, K.A. Esaklul, ed. Materials Park, OH: ASM International, 1992.

29. Shimizu, T., Nishimura, A., and Kawai, S. Structural reliability of electronic devices and equipment. Fatigue 90, *Proc. Fourth Int. Conf. on Fatigue and Fatigue Thresholds,* 993–1003, H. Kitagaura and T. Tanaka, eds., Materials and Component Engineering Problems, Birmingham, 1990.

30. Drake, J., Beck, C., and Libby, R. Transistor bonding wire corrosion in type TO cans. *J. Materials Engineering and Performance,* Vol. S, 39–45, 1996.

31. Chestnut v. Ford Motor Company, 445 Federal 2d 967 (4th Circuit, Court of Appeals 1971).

32. Powell, G.F. Forensic metallurgy. *Microstructural Science,* Vol. 5, 11–26, J.D. Braun, H.W. Arrowsmith, and J.L. McCall, eds. New York: Elsevier, 1977.

33. Le May, I., and Brown, S.J. Explosive failure of rectangular vessels: the case of automotive batteries. *Integrity of Structures, Hazardous Release Protection, Piping and Pipe Supports, and Pumps and Valves,* PVP-Vol. 333, 183–191, H.H. Chung et al., eds. New York: ASME, 1996.

34. Lee v. Crookston Coca-Cola Bottling Co., 290 Minn. 321, 188 N.W. 2d 424 (1971).

35. Lee v. Crookston Coca-Cola Bottling Co., 290 Minn. 333, 188 N.W. 2d 424 (1971).

36. Anderson v. Chrysler Corp., 403 S.E. 2d 189 (W. Va. 1991).

3.9 ADDITIONAL READING

Brown, S.J., ed. *Forensic Engineering, Part I.* Humble, TX: ISI Publications, 1995.

Phillips J.J., and Prior, R.E. *Products Liability,* 2d ed. Charlottesville, VA: Mitchie. Vol. 1, 1993; Vol. 2, 1995.

Weinstein, A.S., Twerski, A.D., Piehler, H.R., and Donaher, W.A., *Products Liability and the Reasonably Safe Product.* New York: John Wiley and Sons, 1978.

Colangelo, V.J., and P.A. Thornhill. *Engineering Aspects of Product Liability.* Metals Park, OH: American Society for Metals, 1981.

Electronic Failure Analysis Techniques

CHAPTER 4
PHOTOGRAPHY AND OPTICAL MICROSCOPY

Lawrence L. Ludwig, Jr.
Electronic Section, Failure Analysis and Physical Testing Branch,
Materials Science Division, Logistics Operations Directorate,
National Aeronautics and Space Administration

4.1 INTRODUCTION

Electronic failure analysis (FA) is complex process with many intermediate steps to the final objective. Photography and optical microscopy are the techniques used to document (photodocument) that process, providing the analyst with a permanent visual record.

4.1.1 The Purpose of Photography and Optical Microscopy in Failure Analysis

The purpose of photography and optical microscopy in FA is to provide a clear, concise set of views that document an FA investigation from start to finish. This photographic documentation leaves no portion of the process without visual support or validation. The statement "the subject part was severely weathered with a high degree of exterior corrosion" can mean faded paint and light surface rust (if the part is in Arizona) or green and white chloride/salt deposits along with a rust scale 0.25 in. in depth (if the part is at Kennedy Space Center). A thorough visual trail documenting the whole FA process validates the conditions found and the testing performed. This validation process leaves no room for misinterpretation of the conditions found, and it aids in explaining intricate anomalies difficult to describe in written text. Photographic documentation is therefore the cornerstone of good FA work.

4.2 OVERVIEW

4.2.1 Overview of Photography and Optical Microscopy in Failure Analysis

This chapter is intended as a practical guide into the general use of *photodocumentation* within electronic component failure analysis. The term *photodocumentation* accurately describes the purpose of photography and optical microscopy in electronic component failure analysis. The easiest and fastest way of documenting any physical characteristic is via pictures (optical photography). The largest and most time consuming portion of any FA investigation is in documenting physical conditions related to the component.

4.2.2 Photodocumentation Nomenclature

Within this realm of photodocumentation, views of objects/components at magnifications of one (1×) and below are considered to be *photographs*. Views at magnifications between

1 and 50 (1 to 50×) are considered to be *photomacrographs*, and magnifications greater than 50 (50×) are termed *photomicrographs* (This standard may differ slightly at different institutions). From these definitions, it can be seen that photography and optical microscopy are in fact both elements of photodocumentation, running the range from photograph to photomicrograph. The practical realm of optical microscopy typically ends in the magnification range of 50 to 100 times (50 to 100×) due to the loss of *depth of field* (DOF) within the picture. Depth of field is defined loosely as the photographic characteristic range of objects at the front (closest to the camera) and at the back (furthest from the camera) of the photograph that are in focus. At higher magnifications, wider lens apertures are needed to let in additional light, but this wider aperture narrows the DOF so that, in many cases, the front and back of an object are not both in focus. The alternative to wider lens apertures is to allow longer exposure times. With high magnification photography, the random vibration of the camera stand from normal activities (such as lab personnel walking through the room) caused a defocusing effect or blurring of the area of concern. This makes long exposure times as impractical as wider apertures, but vibration can be attenuated through the use of special dampening equipment such as pneumatically isolated photography stages. Typically, metallograph work or work conducted on an electronic integrated circuit (IC) probing station makes use of such pneumatically isolated stages, and these devices can photodocument objects optically at magnifications up to and beyond 700 (700×) times.

When the lack of DOF limits the ability to photodocument the subject component characteristic, *scanning electron microscope* (SEM) documentation of the object typically becomes necessary.

4.3   THE PHOTODOCUMENTATION PROCESS

4.3.1   Incoming Inspection

The first portion of the photodocumentation process is the incoming inspection. This is where the FA analyst documents the "as received" condition of the component. The analyst should never take the reported field conditions as gospel. In a vast number of cases, the part is routed by personnel who are intermediate in the management chain. These people cannot accurately report the operating environment or field conditions of the component (despite the fact that they may erroneously document these conditions by recording them on the FA request form). Obtain an accurate description of the environmental conditions from a good technical contact, close to the components actual use, *then* perform the incoming inspection. A good source, typically, are the technicians involved in removal or installation. Often, problem sources or data held as common knowledge among the technicians never makes it into the engineering staff. Data obtained about the environmental conditions may influence the incoming inspection and the type of photodocumentation desired on the component. Figure 4.1 shows the "as received" condition of an Atlas Centaur (AC67) main engine after it was recovered from the ocean floor. This shot graphically illustrates the engine post recovery condition and greatly influences the types of shots used for any subcomponents removed from the engine.

A thorough visual inspection of the component should be made prior to any other work, and every possible anomalous condition should be noted then photodocumented. This visual inspection is typically unaided optically; however, if areas of concern are noted that

FIGURE 4.1   Atlas Centaur main engine recovered from the ocean floor.

require magnification, these areas should be viewed with a stereomicroscope (or whatever optical tool is deemed necessary) and then photodocumented as required.

Every component should have an overview shot that documents the general condition, and a series of follow up shots documenting particular details that were noted in the incoming inspection. Some of these overview shots can be omitted from the final report if they are not pertinent to the conclusions or the final outcome of the analysis. However, it is always better to have and not need than to need and not have.

During the course of an analysis, the number of photographs taken of a component has a tendency to grow at an astonishing rate. Even the most organized of investigators will have shots pulled out of sequence for show-and-tell presentations to investigation board members, lead engineers, and so on, with the photodocumentation becoming a shuffled mess. I have used several methodologies to alleviate the confusion factors with photodocumentation. For those using "instant" film or video hard-copy prints, the simplest and most useful method is to label each photograph as it is printed with an indelible marker. The print can be labeled with a short caption, the date, and time. The date and time may not seem significant; however, when an investigator has to backtrack through several hundred prints during the final report writing process, date and time, combined with a well kept lab notebook, are a tremendous help. For those using film that requires development, the only solution is to maintain a photo log. This log tracks the roll number, a shot caption, and the date and time taken. The odds are that all those shots you have taken over the last week will be sent out for development and come back at once, usually with only the lab processing information on the package. Engineering time is too valuable to waste resorting photos back into chronological order. The optimum method for maintaining photodocumentation is to have each photograph placed in a three-ring binder as it is taken and numbered sequentially with a short caption below each shot. This method is a typically warranted only for investigation boards or situations where several hundred shots are to be tracked and referenced. The point here is this: we are investing large amounts of time and financial resources to document the failed component condition, and accurate chronological sequencing of these photographs is as much a part of the documentation process as are the photographs themselves.

Some of the main errors to avoid with overview shots are oblique views, lack of magnification references, and lack of follow-up shots on "high" detail areas. Oblique views are shots of the component based around a corner (showing two sides of the object) or an elevated corner shot (showing top and two sides). Figure 4.2 shows an oblique view of an ac bias board removed from a hardware interface module (HIM) power supply, part of the launch processing *command and control* system (LPS), at KSC. The reason to avoid oblique views is that true magnifications of elements within the picture cannot be determined with accuracy. Figure 4.3 shows a planer shot of the same component. Flat planer shots are preferred, especially if a scale or ruler is placed within the photograph, in the plane of interest, so the magnification can be determined. Oblique views can be used effectively to circumvent having several "as received" photographs when the magnification of items within the photograph are of no concern. Each photograph within a report should have a title or caption and a posted magnification. Oblique views should be labeled as such in the magnification block (i.e., *Magnification: Oblique View* as opposed to *Magnification: 1.0×*).

If the component has areas of high detail, each of these areas should be shot at a higher magnification for future reference. Figures 4.4 and 4.5 illustrate this technique. Figure 4.4 is an overview of an area under a removed capacitor that had been epoxied to the printed wire board (PWB), and Fig. 4.5 in a magnified view of the highly detailed area (arc path between LAN and solder pad) first shown in the overview.

4.3.2   Test Set and Process Documentation

During the course of an FA investigation, testing usually proceeds in a controlled manner, progressing from nondestructive evaluation to destructive testing and, finally, to dissection and disassembly, sometimes with simulation testing thrown in for good measure. Test set and process documentation can require that test fixtures, test sets, and unusual processes

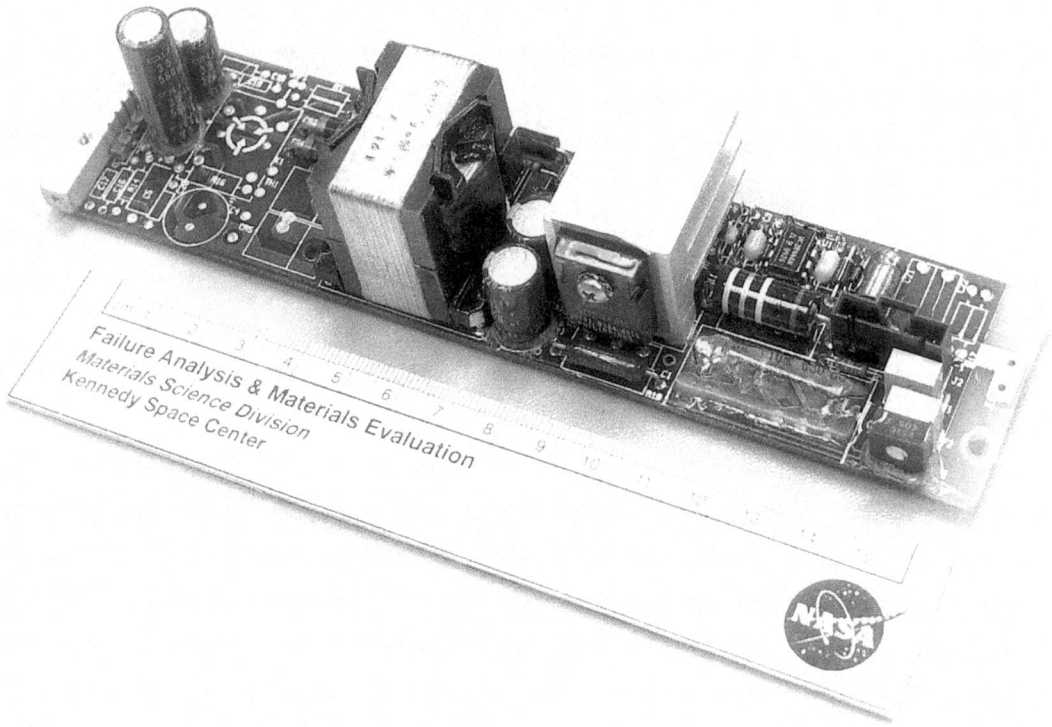

FIGURE 4.2   View of ac bias board from a hardware interface module power supply. Note the lack of consistent magnification from right to left.

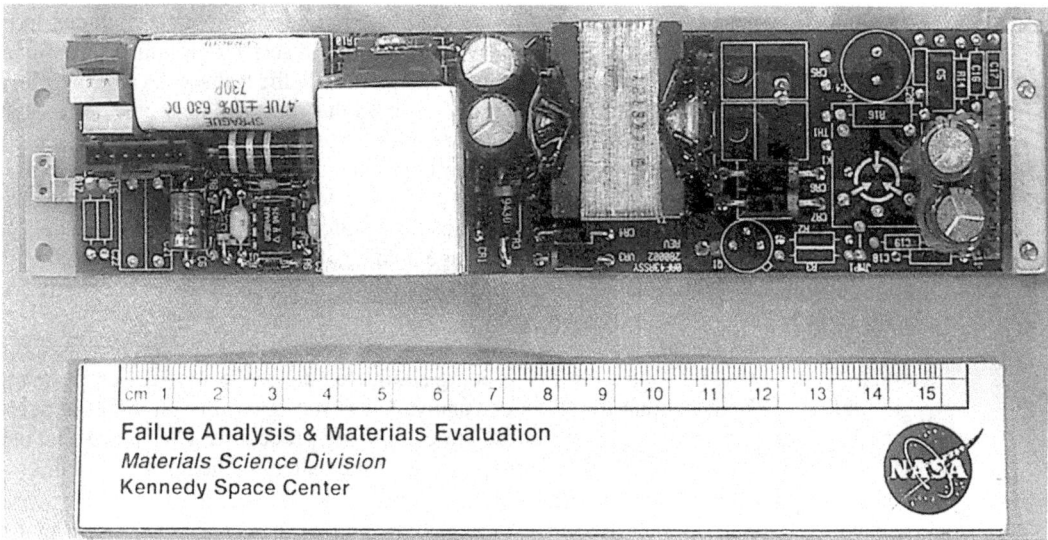

FIGURE 4.3   Planer view of the ac bias board of Fig. 4.2. Note the uniformity in magnification in this view.

FIGURE 4.4   Overview of the removed capacitor area on an ac bias board. The arrow points to an arc track path between runs.

be photodocumented. As far as test sets are concerned, a good schematic included in the FA report usually is quite sufficient. However, elaborate test sets and/or unusual configurations should be photodocumented to show the uniqueness of the test set. As an example, an incident involving the Magellan flight battery required that each cell be monitored individually. A discharge test was performed, paying close attention to the corner voltage (the point at which the cell goes into rapid discharge) of each cell, and ensuring that no individual cell went into reversal. The actual test set had 11 voltmeters stacked in 2 columns with 2 video cameras used to document the cell discharge rate. Each camera ran alternately for an hour with a five-minute overlap. The voltage readings from each test were pulled off the videotape and translated into a spreadsheet of discharge times and voltages. The test set was crude but very effective. Figure 4. 6 shows the test set configuration.

Using this same incident as an example, special procedures can also require photodocumentation, especially if the FA work is being conducted under the auspices of an investigation board. The Magellan flight battery analysis was run under the direction of such a board and required documentation for the battery safeing procedure. The battery had been involved in a fire that burned away large portions of the battery cable harness and the conductor insulation (see Fig. 4.7).

Before the battery could be moved, the battery harness had to be "safed" so that no additional loss of energy from the battery could occur. Figure 4.8 shows an overview of the analysis team members in a clean-room environment performing the safeing operation,

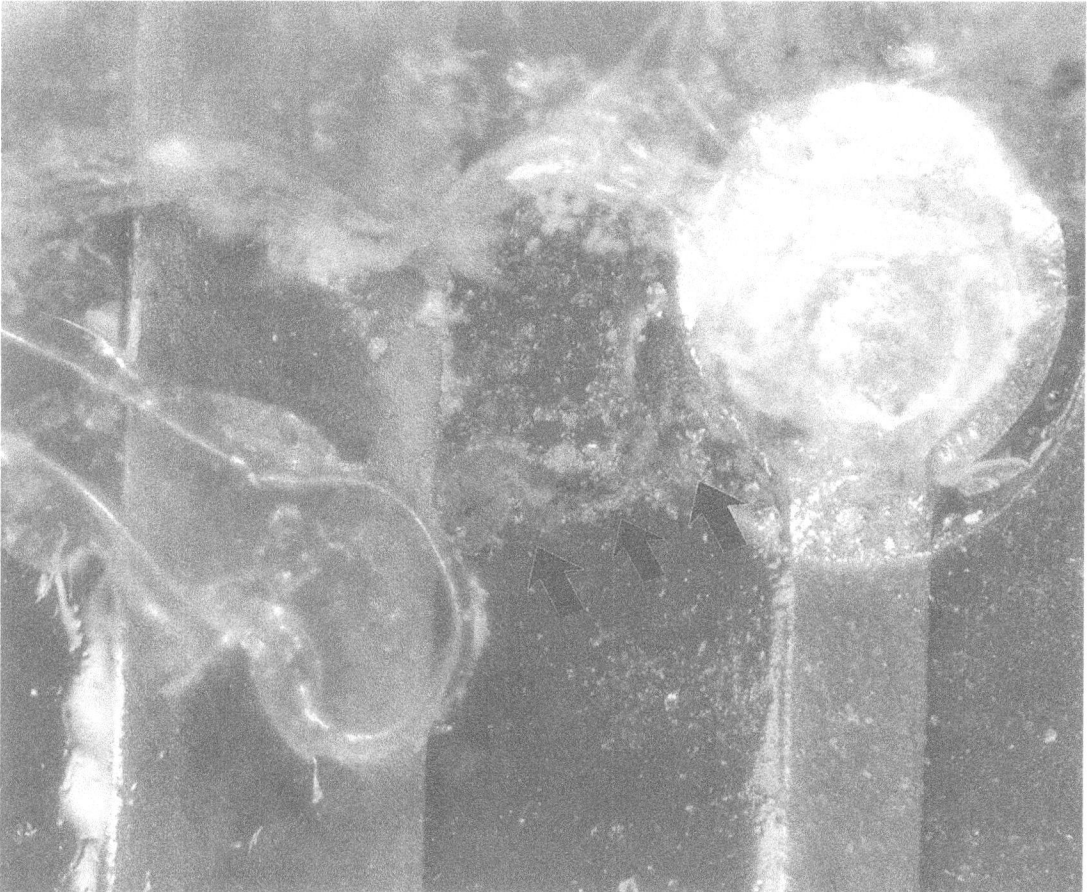

FIGURE 4.5   Highly magnified view of the arc damage on shown in Fig. 4.4. The arrows point to the arc track path.

and the safed battery is shown in Fig. 4.9. Note the presence of the video cameras in these shots. Videotape is best used for documenting procedural events and processes that the investigator finds difficult to explain. Although video prints typically lack the resolution of report-quality photographs, they can be used to show the action resulting from a testing process or from simulation testing. Many video printers can store 2, 4, 8, or up to 12 video frames on a print. This allows the investigator to easily show sequential processes. Figure 4. 10 shows such a sequential shot of a liquid oxygen and kerosene rocket engine test firing, the igniter flame is shown first followed by the motor ignition sequence. Video documentation will be covered in detail later in the chapter.

During the course of the FA process, any changes to the component brought about by the analysis process should be photodocumented. If a wire segment is to be stripped to facilitate electrical measurements, shots of that area in an undamaged condition *must exist before* the wire is stripped. Whether a shot of the stripped wire is taken is the investigator's decision. If the stripped wire can be described easily in the written text and has no future impact on testing, then shots are not needed. However, it is always better to have and not

FIGURE 4.6 Magellan flight battery discharge test configuration. The voltmeters are arranged sequentially, measuring battery cells 1 through 11.

need, than to need and not have. Likewise, if needle probes are used to check continuity on insulated wire strands, shots of that area in an undamaged condition *must exist before* the measurements are made. You can rest assured that any photograph taken after the wires are probed (and have needle puncture marks) will raise questions concerning whether this was an existing condition. To clarify these issues, photographs of the "before" condition can be cropped and placed in the same view as the "after" photos taken later in the analysis. Figure 4.11 shows a solenoid valve electronic position indicator package with three wires removed from the solder attached points. The view directly below it is an "as received" shot confirming that these wires were attached when delivered to the lab.

This brings another point to light. A lot of physical handling occurs during an analysis, so care to preserve the original condition must be a priority. However, it is unrealistic to expect no changes in the component to occur during the testing portion of the analysis. Therefore, careful measures to document the more fragile areas of the component should be made first. Photodocumentation of these fragile areas should be the investigator's first priority. Later testing and changes in the component condition are then quantified and can be successfully related to their impact on the failure mode of the component.

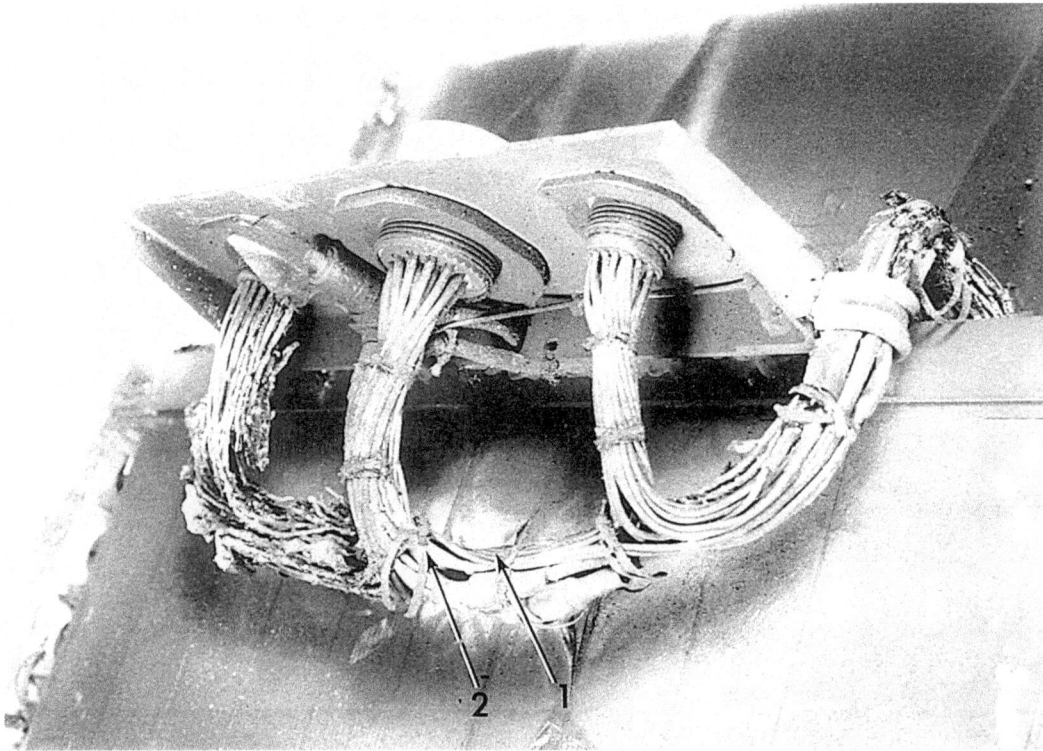

FIGURE 4.7   Magellan flight battery condition after fire. Arrow 1 points to undamaged section of wire; arrow 2 points to undamaged bundle tie.

4.3.3   Dissection and Disassembly

Dissection and disassembly is typically defined as the process by which the internal parts of the component are exposed for observation and individual analysis. Dissection usually relates to packages being machined open, whereas disassembly usually denotes component housings and so on, which are taken apart by conventional methods. Dissection requires one style of photodocumentation, and disassembly requires another.

Dissection is usually performed to provide access to internal subcomponents. Radiographs of the component housing, especially providing information about internal clearances, are required. The dissection process is typically a machining or abrasion type processor in which the outer housing is cut open. If the component has cleanliness issues attached to the housing internal cavity, then photodocumentation of the machining process becomes quite necessary. NASA metrologists have developed several dissection techniques that are used to prevent particulate from machining operations from entering the machined open housing. One such technique was to machine the part until just a foil-thin thickness of the housing was left and to perform the final opening operation under a clean bench (laminar airflow bench) separating the housing using an Exacto™ knife or scalpel. In this case, the machined part should be photographed prior to the final opening operation

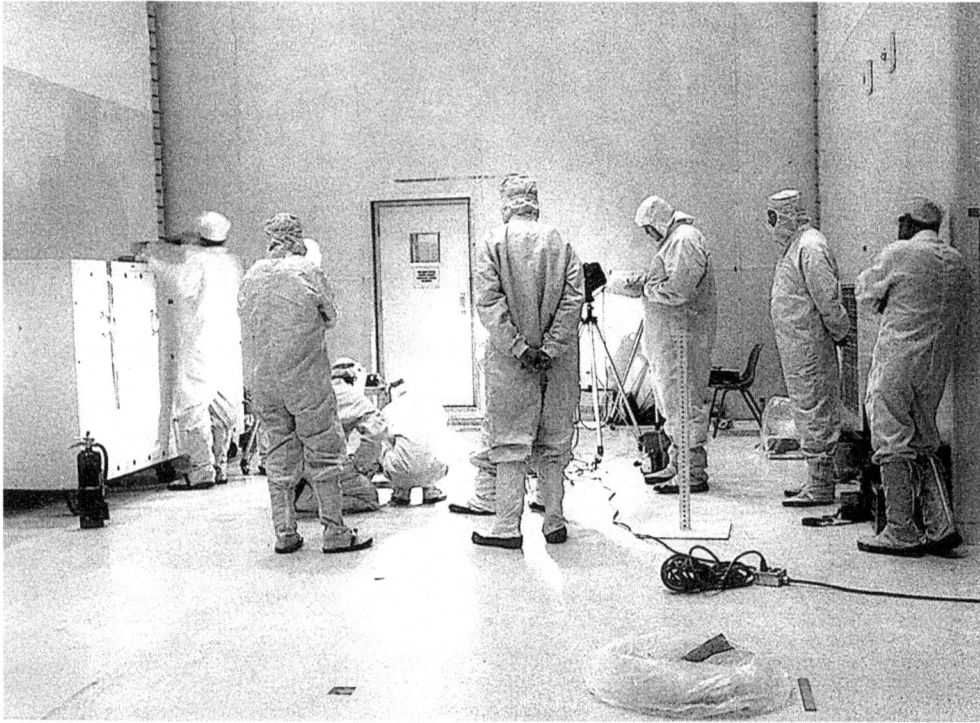

FIGURE 4.8 Magellan flight battery safeing procedure in the spacecraft encapsulation and assembly facility (SEAF 2), clean room facility.

and photographed again on the clean bench just after the initial series of cuts. A final shot showing the separated housing sections can be made, with all three photographs documenting the clean, nonintrusive methodology of outer housing removal. This type of photodocumentation validates that the source of particulate contamination is not from the opening operation and is related to some other process. Chemical analysis of the internal particulate can also help substantiate the source of internal contamination. Figures 4.12 through 4.14 illustrate a typical dissection process on a hybrid integrated circuit package. An overview of the part to be dissected is shown, a close-up view is shown depicting the result of the method of dissection, and, finally, the part is displayed with the housing removed. These three shots identify the part in question, the method of dissection (corners filed at a 45° angle), and the condition of the part when the housing was removed.

In laboratories such as this one, the subject part would then be turned over to chemical analysis personnel to sample and analyze internal contamination. It is particularly helpful to obtain high-magnification photographs of the subject contamination. This has two significant results, the first being that it documents the type and appearance of the contamination prior to sampling, and the second that it provides the investigator and the chemical analyst with a visual reference of the contamination location. This may allow the chemist to identify the source of the contamination. This point can be expanded to encompass most other external operations: a photograph of the area of concern helps to identify what is really required.

FIGURE 4.9 Magellan flight battery assembly.

Disassembly usually denotes component housings and such that are taken apart by conventional methods. As with the dissection process, which places emphasis on careful dissection to prevent disturbing the interior of the component, the disassembly process requires care to annotate parts removed in a particular order and to document any abnormalities with the parts or housing. For example, if a top cover plate is held on with four machine screws, and three of the screws come out in pristine condition but the fourth is covered in rust, a shot relating the abnormal screw to its original position on the cover plate needs to be taken. Later, if the top cover is removed to expose an area within the component damaged by moisture intrusion and corresponding rust stains on the bottom of the cover plate, the investigator has the original orientation of the outer cover and screws. This may allow determination of the component mounting orientation. The investigator can use this information to check the field conditions and track down the source of the moisture intrusion.

As useful information about the component is revealed in the disassembly process, photographs documenting this information should be taken. If the disassembly of the component housing is extensive, logical stopping points should be used to photodocument the process. The parts removed up to that point should be laid out in a logical manner—one that helps describe disassembly (or reassembly, if later simulation testing is to occur) and then photographed. If high-magnification shots of individual subcomponents are required to understand their function or condition, then these should be taken as well. The final objective of photodocumenting the disassembly of a component should be to ensure an easy understanding of component structure, and that each part of that structure can be visually referenced as to condition and location. Figures 4.15 through 4.17 illustrate a disassembly process on a 250-ton crane electrical system potentiometer.

FIGURE 4.10 Video still ignition sequence of a liquid oxygen rocket engine test firing

These shots consist of an overview, followed by an initial disassembly, and finally by a breakdown shot of all the internal components. Figure 4.18 shows a magnified shot of a second crane potentiometer interior, which has an unusual V wire contact spring that engages the wiper shaft. This shot focuses the reader on the spring contact arrangement (upper arrow), providing a clear view of how the V spring is tied to the housing outer resistance path (lower arrow). Careful use of these photographs can circumvent several paragraphs of verbiage describing this physical configuration.

4.4 PHOTODOCUMENTATION EQUIPMENT

4.4.1 Film-Type Photodocumentation

Thirty-five millimeter film is a reasonable choice, due to the wide availability of cameras, microscope adapters, lenses, and film processing centers. The larger-format films such as

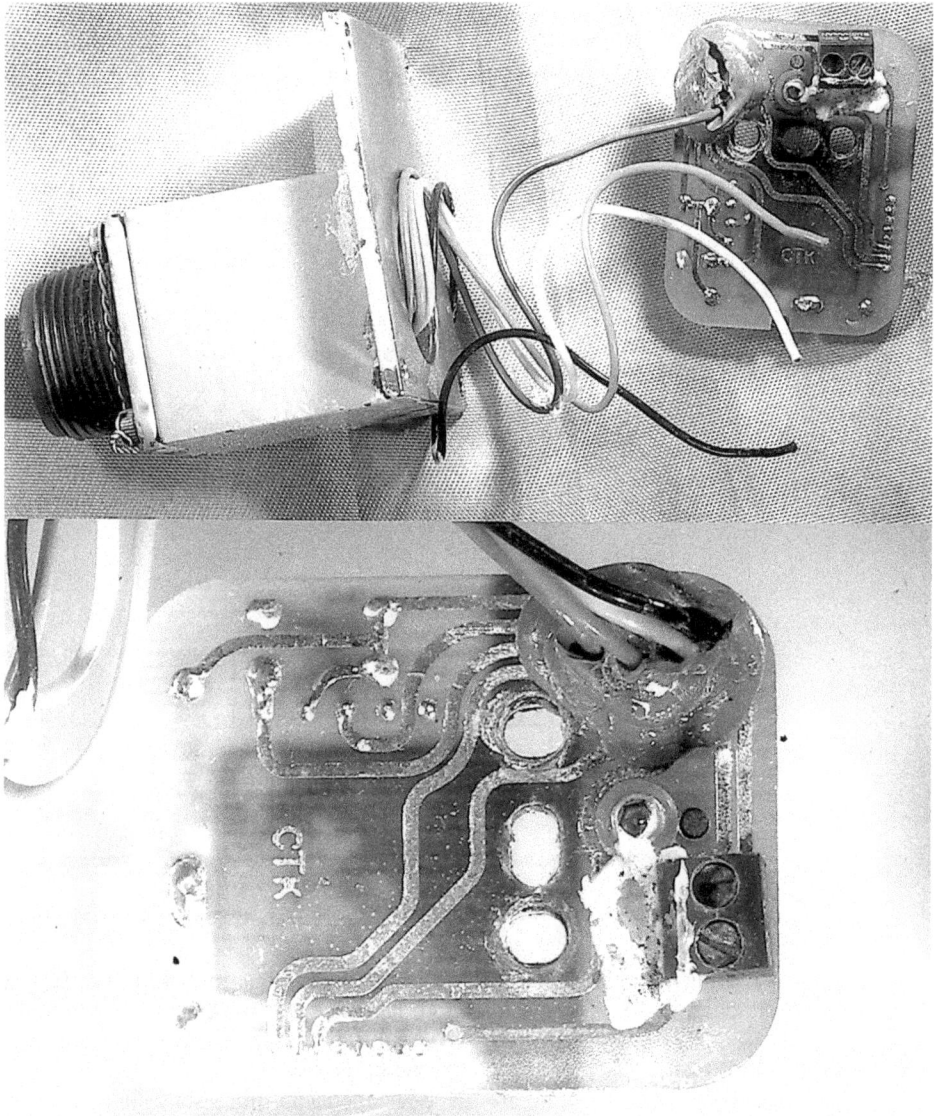

FIGURE 4.11   Solenoid valve electronics package, upper view depicting the separation of interface wiring, lower view showing the "as-received" condition.

the Hasselblad™ format provide better quality negatives, which greatly enhance image enlargement. However, the availability of microscope adapters and other such paraphernalia is not as good with the large-format cameras. In addition to this limitation, the large-format camera systems are much more expensive, which can add considerably to the total cost that a lab invests in its photodocumentation system. Both the 35-mm and the Hasselblad™ formats have a fundamental flaw when used for FA purposes. The investigator has to overcome a large learning curve, becoming a camera/photography expert. Then, after

FIGURE 4.12   Solenoid valve hybrid integrated circuit package, view 1 of 3 depicting the dissection process.

suffering through the learning curve, the investigator must wait for film development from either an in-house or outside photo lab. Electronic component FA photodocumentation is extremely difficult when you have real-time access to your shots, let alone when you have to crank in internal or external processing time.

Several years ago, Polaroid™ developed a 35-mm instant slide film that proved reasonably acceptable. The problem with this system was how to transfer a slide easily to a report picture. The instant slide system did have two nice advantages:

1. Only the report photographs would have to be sent out for prints (or made into prints in house).

2. The slide process has video camera adapters for video freeze-frame units, which allows videotape to be used for field documentation.

Still, the film has to be developed, which means removing the camera from the photography stage, removing the film, waiting for development, and possibly resetting for a second set of photos. That is not conducive to job completion.

FIGURE 4.13 Solenoid valve hybrid integrated circuit package, view 2 of 3 depicting the dissection process.

Going back to the learning curve issue, an investigator, having become comfortable with a camera and lens system, can use procedural techniques to overcome the lack of real-time film development. For example, using the camera's internal light meter, if a shot looks good at 1/125 s and a particular lens aperture, the investigator can then vary the exposure time to 1/250 and 1/60 s. This yields three shots, with the 1/125 exposure time being optimum, the 1/60 shot brighter, and the 1/250 shot darker. However, if the investigator is off on the meter reading, then one of the other shots will suffice. If none of the three is acceptable, a second set of photos will be required. Again, this just is not conducive to completing the task. The same procedure can be used with varying lens apertures; however, using the highest lens aperture number (i.e., the smallest opening) possible helps to maintain the depth of field (DOF) of the shot. One of the advantages of 35-mm and large-format film camera systems is that they produce a negative from which any number of prints can be made.

My final comment on film type photography is that both 35-mm and the Hasselblad™, large-format systems are quite workable. I have used 35-mm to a great extent, both inside and outside the lab, without an unacceptable amount of duplicated work.

4.4.2 Instant Film Photodocumentation

Many available "instant" cameras are acceptable for FA photodocumentation. One of the largest manufacturers of instant film is Polaroid™. A widely used instant film for FA purposes is the Polaroid™ 50 series films.

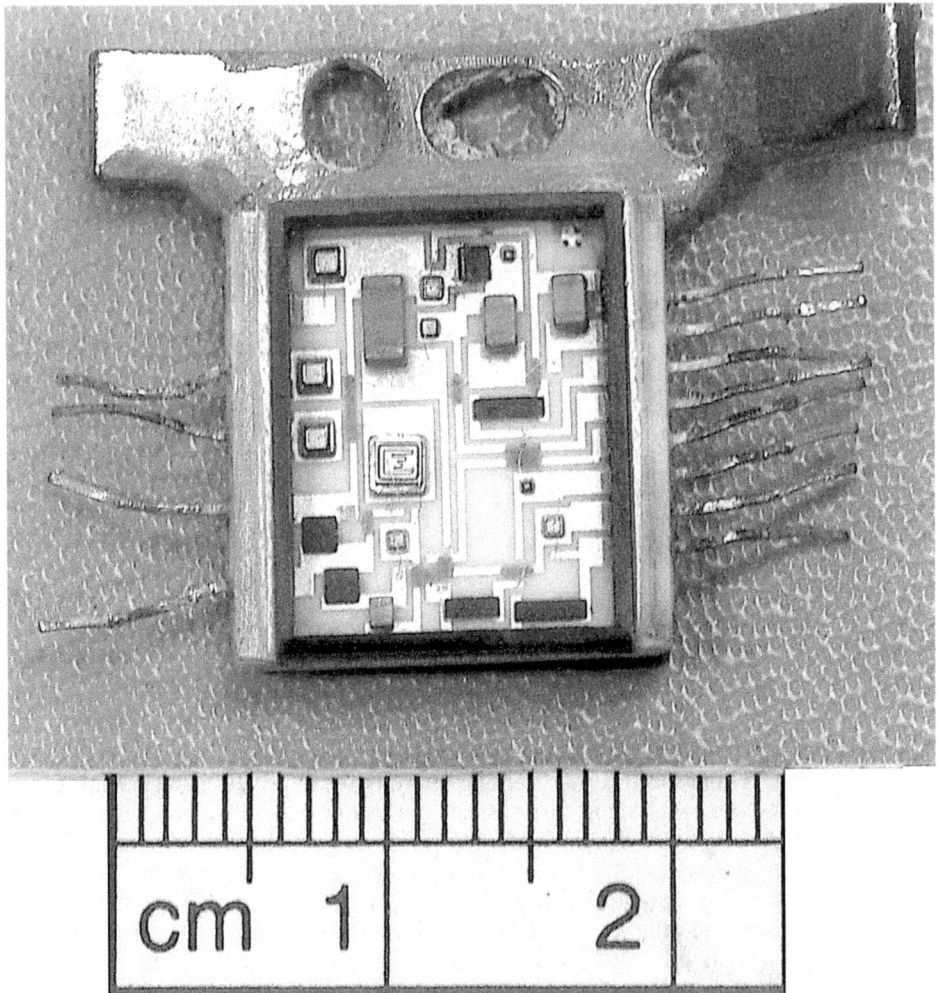

FIGURE 4.14 Solenoid valve hybrid integrated circuit package, view 3 of 3 depicting the dissection process.

The type 52 film is a medium–high-contrast black-and-white film (ASA/ISO 400/27)[*] that yields a report-quality print, 3.5 × 4.5 in. The type 59 film is a color film of ASA/ISO 80/20 that yields report-quality prints, also 3.5 × 4.5 in. Type 51 film is a high-contrast black-and-white film of ASA/ISO 320/26, same dimensions. The types 51 and 52 films described above need a coater or image fixer to permanently set the image on film. Chemical coaters are provided with the film, and the print is coated after development by applying a thin film using the supplied chemical coater. The coating dries in 2 to 3 min. Type 53 film is a medium-contrast film of ASA/ISO 800/30, and it does not need a coating.

---

[*]ASA and ISO are recognized standards for film speed, and the numbers indicate how quickly the film reacts to light. The number indicates a degree of sensitivity to light; therefore, for a high-number film (for example, 400 ASA), less light is required for a well exposed shot.

FIGURE 4.15   A 250-ton crane electrical system potentiometer overview, view 1 of 3.

FIGURE 4.16   A 250-ton crane electrical system potentiometer overview, view 2 of 3.

FIGURE 4.17 A 250-ton crane electrical system potentiometer overview, view 3 of 3.

The above-mentioned films have many advantages. The first is that the film is very inexpensive—approximately $0.50 per shot, and $1.00 per color print. The second is that the camera systems using these film types are typically very forgiving in operation. A couple of practice shots will allow most users to find the correct aperture and shutter speed. A vague familiarity with the subject of which lenses are used for which magnification shot, and the FA analyst is in business. With the use of a professional photography stage, medium-high magnification shots can be accomplished by using repetitive exposures. A six-second exposure time can be accomplished by setting the shutter speed to one second and clicking six times. As the magnification increases, the random table vibration affects the focus to a greater extent, so this multiple-exposure procedure has a limited range of use.

The most significant advantage for these film types is the normal film development time, which is 30 s. The film is normally processed by removing it from the camera, while the camera and focus remain fixed above the subject component. This allows the investigator to determine whether the documentation is sufficient half a minute after the shot is taken. It also allows the shot setup to remain intact. One major disadvantage to films and camera systems of this type is that you do not have a negative to use for printing additional shots. This means that, for each view you take, you need to make two prints—one for the final report and one to archive. Duplicate photographs can be made (by shooting a photograph of the photograph), so you can make a second print from the first; however, the second print will be of a lesser resolution. A second disadvantage is that some of the films require coating, and even with coating do not have an infinite storage time. The Type 51 and 52 films tend to fade with time. This disadvantage can be circumvented by digital archiving of the photographs.

NASA labs have used these film series for a long time, with a great deal of success. Even today, with the push toward digital photography, we maintain two full camera systems that use the 50 series films. The Polaroid™ camera stages and cameras (MP4™ sys-

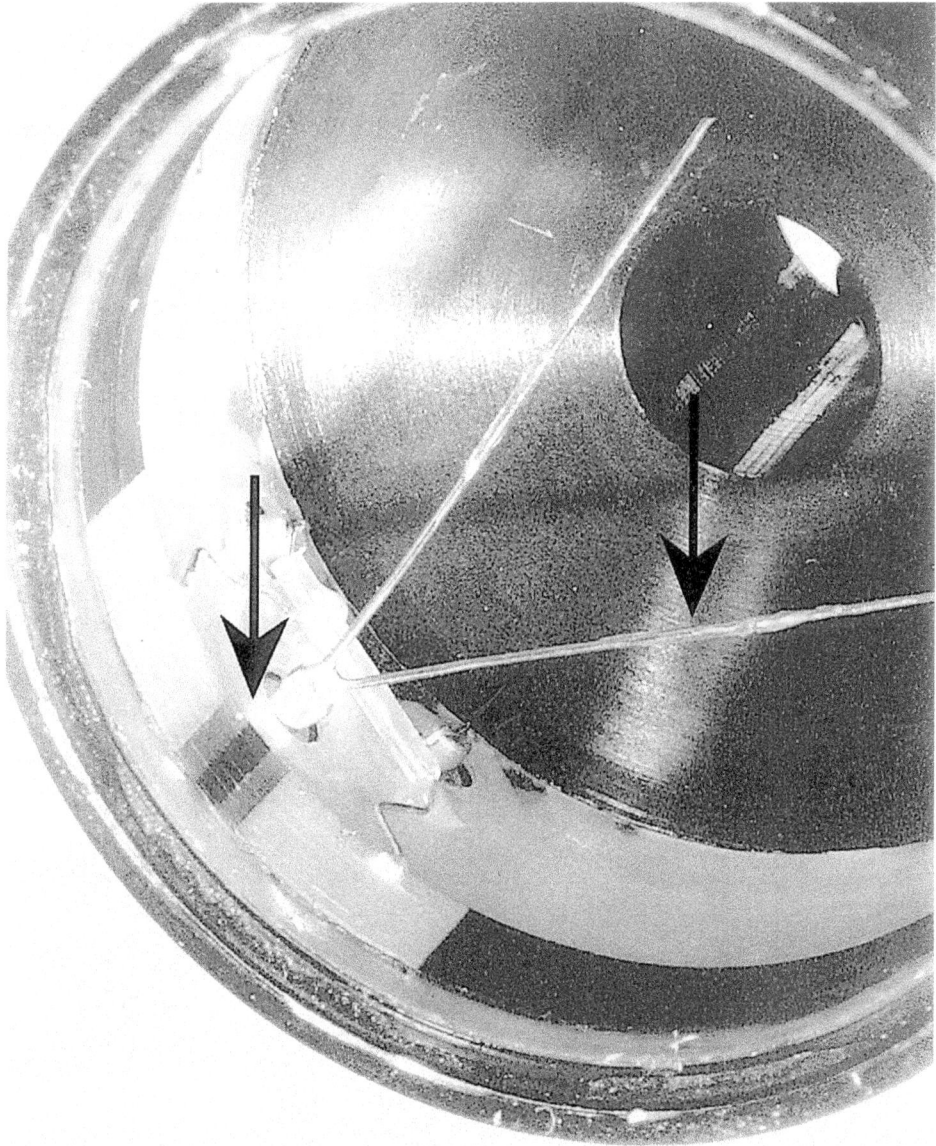

FIGURE 4.18 A 250-ton crane electrical system control potentiometer, magnified view of V-spring and contact assembly.

tem) can be used as the cornerstone of a low-cost documentation system that can handle the needs of a small lab very efficiently.

Another widely used film series is the Polaroid™ 600 series films.

The Polaroid Propack™ cameras use various formats of the Polaroid™ instant film, typically based around the particular type Polaroid™ back with the camera is equipped.

The 664, 667, and the 669 films are typical of the pack film, and they have an image area of 2 7/8 × 3 3/4 in. The 664 and the 667 are both black-and-white films, the 664 having an ASA/ISO of 100/21 and the 667 an ASA/ISO of 3000/36. One can see that the 667 is a very fast film, which enhances its usefulness in low-light conditions. The 669 is a medium-contrast color film with an ASA/ISO of 80/20. Many oscilloscope cameras use these same films. This film works very well for high-contrast situations, such as an oscilloscope or curve tracer output. The Polaroid Propack™ camera is a flexible camera which has a minimal focal distance of 3 1/2 ft. It produces a reasonable 3 1/4 × 4 1/4 in. print in black and white or color.

I believe that both the older SX70™ and the still older Propack™ cameras both were available with a microscope adapter that allowed use in the microscope trinocular position or through an eyepiece objective. Figure 4.19 shows the Polaroid Microcam™ SLR micro-

FIGURE 4.19 Polaroid Microcam™ microscope camera.

scope camera. The new Polaroid Microcam™ SLR is a much improved configuration over these old adapters. Its only drawback is that it is a dedicated camera that shoots only through a microscope. The SX70™ and the Propack™ cameras, and their new equivalent, are highly useful for field photodocumentation. The old Propack™ camera system is a nice field camera that can also double as a lab camera for a limited number of shots. Both cameras have limitations due to their use of a single fixed-lens. These cameras would not be sufficient as a laboratory's only means of photodocumentation.

The final instant film discussed is the SX70™ style film, mentioned previously. This film is a 640/29 ASA/ISO film that is relatively easy to use and works well with an automated camera and flash system. The SX70™ film is defined by Polaroid™ as an *integral-type* film. It has a very fine grain structure that lends itself well to being scanned and enlarged.

4.4.3   Video Documentation

I started using videotape for analysis documentation back in 1986 on the Delta 178 and Atlas Centaur 67 investigations. Both investigations were conducted under the auspices of an investigation board and required each and every operation/procedure to be videotaped. The first thing I noticed then (and it is still true today) is that they require a lot of light. Most commercially available "camcorder"-style recorders require bright light for indoor operation. There are some pretty amazing new video cameras that operate very well in low light conditions, but this is professional quality equipment with price tags to match. The labs here started with the standard VHS camcorders manufactured by Panasonic™ and Magnavox™. Using videotape for FA photodocumentation is slightly different from video documenting a test sequence and/or lab analysis procedure. I have already commented on VHS tape not being report-quality documentation. A standard frame of VHS tape contains two fields that are combined to make one frame. Since each frame has slightly differing video information (depicting movement and so on), when the frames are combined, you get a slight blurring effect. The human eye compensates for this (or ignores it) when videotape is run at full speed (30 frames per second). For report-type photodocumentation, this makes a bad situation worse.

Typical videotape has a horizontal resolution of roughly 220 to 250 lines per field, and a vertical resolution of approximately 480 lines vertical. You actually have more visual data in the vertical mode. Due to the lack of horizontal resolution, you get a raster line effect on VHS video stills. The combining of fields and the raster line effect lead me to consider VHS stills unacceptable for report-quality documentation. You want the image presented in the report to be free of visible artifacts that might cause confusion about the image. SVHS tape has vertical and horizontal resolutions that are almost the equal (approximately 400 to 480 lines of resolution), which greatly reduces the raster line effect. SVHS tape is much better suited for report photodocumentation, although it is not completely up to standards yet. Hi8 has roughly the same resolution as SVHS tape and can produce nearly report-quality documentation.

For photodocumentation purposes, you will need a method of reviewing the tape one frame at a time, and a method of pulling a single frame from the tape and reproducing it in a high-quality format. This means editing VCRs and a freeze-frame unit or video printer with these capabilities. For editing VCRs, we are currently using units with built-in time base correctors. This allows correction of color anomalies due to extended cable runs and so forth. We have a master editing controller system that controls up to four recorders. We have also used a video capture and editing program for a high-end personal computer (PC) that allows us to insert titles and arrows, or highlight areas of concern on a video print. You can see that, in terms of equipment, using videotape as a medium becomes intense. Figure 4.20 shows the video editing station described above.

We use our video system as a full video editing lab. We have found over the years that five minutes of well edited videotape is worth hours of briefings, especially when it comes to upper management. A scaled-down system could consist of one video printer, one editing VCR, and a monitor. One of the disadvantages to videotape is the learning curve. Professional video camera men are well paid for a reason. Two of the biggest tips for using video as a photodocumentation system is to *always use a tripod,* and take at least 10 full seconds of every view or scene. Most novice camera operators pan and tilt much too quickly. If you recall how a frame is built out of two fields, then you can see why 10 seconds of motion-free tape is necessary. One of the final disadvantages of videotape is the lack of selection in lenses. Most camera systems come with a professional-quality zoom lens arrangement, which does not easily allow for highly magnified shots. The two high-end video cameras we use both come with a fixed zoom lens arrangement.

So far, I have exclusively discussed the downside of video documentation. Videotape documentation has several advantages. One is that videotape is easily archived: pull the record tab, label it, and file it away. The exact location of each shot printed from videotape can be annotated on the label or case using the chronological order of the tape counter. All the procedures and various stages of FA on a component can be placed on the same tape, and all the photodocumentation for a job can be duplicated by coping the original tape. Videotape is a widely used medium, and tapes of an analysis can be sent to the requester as

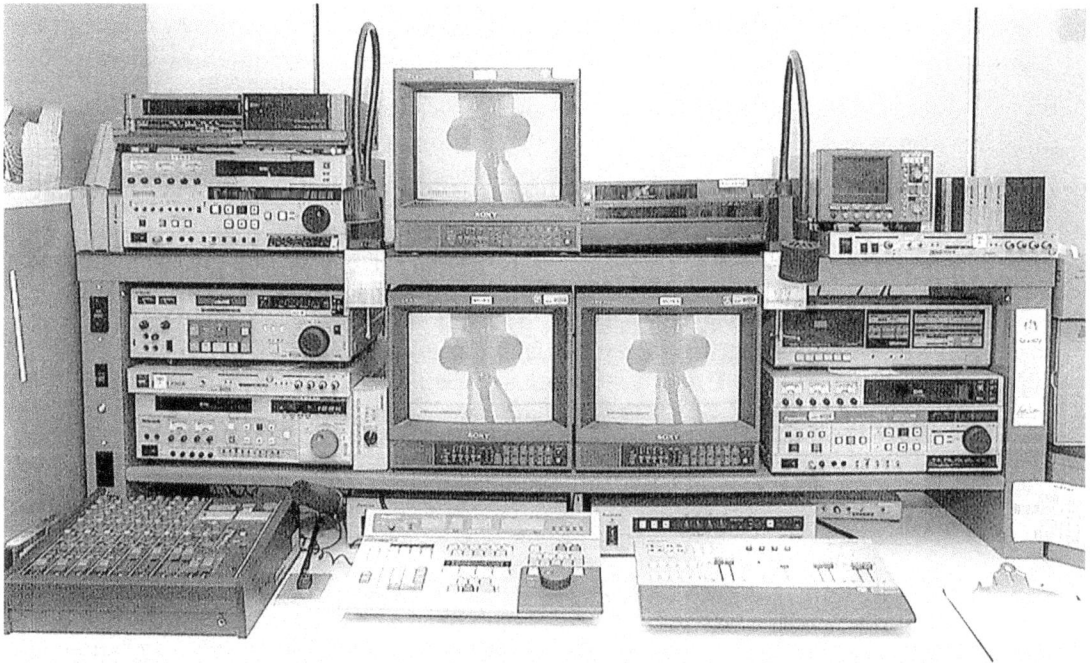

FIGURE 4.20 A laboratory video editing station.

part of the report. All of these advantages make the argument for videotape photodocumentation quite convincing.

4.4.4   Video Still Photodocumentation

Video still photodocumentation can be defined as the process whereby a video camera signal/image is sent directly to a print, without the use of recording tape. The lab has two video cameras for still photodocumentation. One is used for both video and still work, and the other is used only for still photodocumentation. The video camera is used in conjunction with a video printer for still photography. The camera is mounted on a T-stand (an elaborate horizontally mounted camera stand) and is placed over a professional photography light stage. This arrangement yields the same kind of effective photography stage as the Polaroid MP4™ camera stand. Figure 4.21 depicts the camera, camera stand, and light stage. The camera has RGB, SVHS, and composite video outputs. This particular camera uses a Sony™ bayonet proprietary lens mount, which we use with a Fujinon TV-Z™ zoom lens for shots up to 6× (the lens has a screw-on close-up lens adapter). For shots at higher than 6× magnification, we have two microscope tube lens systems. One is a standard tube microscope, and the other is an extended depth of field system. These are easily attached to

FIGURE 4.21   Effective photodocumentation workstation consisting of SVHS camera, camera T-stand, and light stage.

the Sony by using an adapter that converts the Sony™ bayonet mount to C mount. We can typically work at magnifications up to 100×, although work above 50× has an extremely small depth of field. Microscopes and tube microscopes will be covered in a later section. The video camera used in this system has proven to be quite functional due to the selectable light intensity gain control. The operator has the option of 0, 9, or 18 dB of gain for light intensity (not to be confused with automatic/manual iris modes). This helps significantly when trying to get magnified shots where direct lighting washes out or reduces details. The light intensity gain is not something you want to use instead of appropriate lighting. There is a trade-off between image contrast and use of the light intensity gain.

A second feature of this camera is the shutter speed control. The camera has the ability to act like a 35-mm camera with a fixed shutter speed. The camera has shutter speeds of 1/125, 1/250, 1/500, 1/1000, 1/2000, 1/4000, and 1/10,000 s. This means that a sample period of as short as 1/10,000 s is used for the next frame (out put at 30 frames/s). In situations where light is washing out detail, setting a higher shutter speed can eliminate the washed-out effect. The most useful application of the shutter speed function is to help stop motion on video shots. The faster shutter speed can allow videotaped operations to have some printable, in-focus images to be used for photodocumentation. Between the light intensity gain and the shutter speeds, this camera has a great deal of flexibility when used for photodocumentation purposes. The only down side to this camera (other than the price tag, approximately $25,000) is the printed image resolution. The image resolution is slightly better than SVHS tape when run straight to the video printer. Recorded shots subsequently pulled from videotape are the same image resolution as SVHS tape. Figures 4.22 and 4.23 are identical shots of an exposed IC die, one sent straight to the camera and the other recorded and printed later, You can judge the difference in image quality for yourself.

The other video camera we use for video stills is a specialized frame-capture camera. This is kind of a hybrid camera, more like a true digital camera than a video camera; however, the initial image for capture is viewed from an RGB monitor or through a recorder that converts RGB to composite. The camera is controlled from the PC over an RS232 bus, and all the image information is downloaded to PC through a image-capture board. When an image is selected for capture, it is then scanned by the camera (using a four-pass process) and downloaded to PC hard drive. The camera functions are controllable only via the provided software and video-capture card. The images are stored in a TGA format for later printing and/or processing. The image resolution for this system is as good as any I've seen on the market. The exact image sizes is 4416 × 3456 pixels at 24-bit per pixel color depth and can be processed in Photostyler™ or Photoshop™ (two of the more popular image processing programs). This camera is provided with a C mount, a mounting fixture that allows a wide variety of video and microscope lens arrangements. The camera is roughly the same size as the SVHS camera and can be used on the same camera stand system. Some of the advantages to this system are the image quality and the ease of use. The camera has a high adaptability from macrophotography to microphotography through the use of various lenses. Most importantly, the image capture time is only a few seconds. A full-size view of the captured image is available for inspection on the computer 17-in. monitor at the end of the image capture sequence.

The disadvantages to this system are few but significant. The first is that this system is usable only through the PC. In most cases, this eliminates it as a field system and makes it hard to move around the lab. There will always be cases in which you need to bring the camera to the part rather than vice versa. The second disadvantage is that focusing accurately on a part at high magnification is difficult. The video output used to focus has

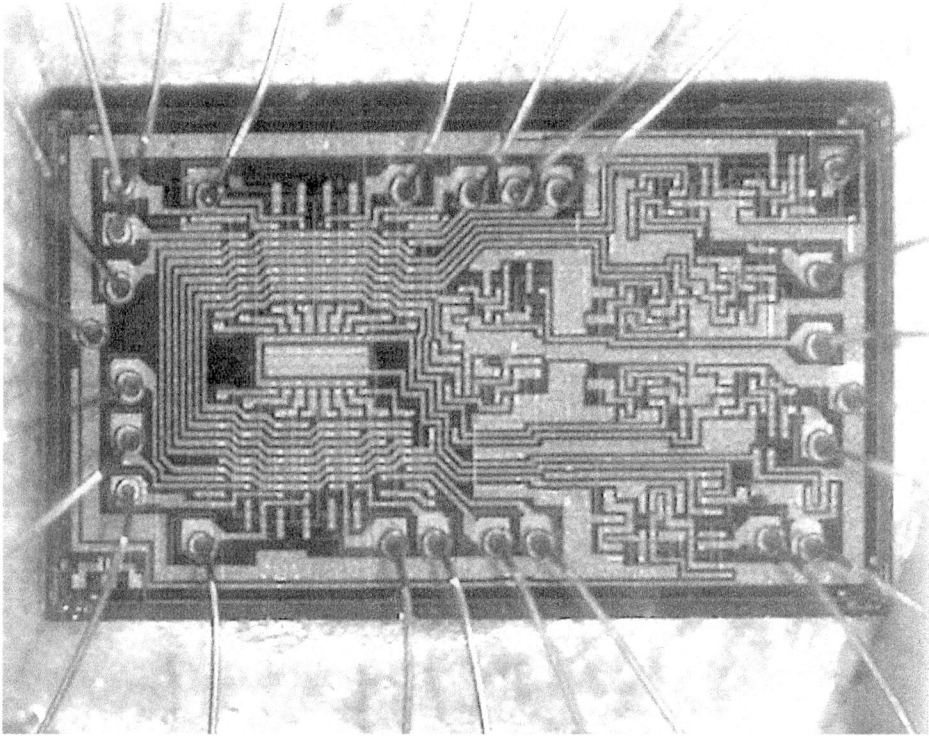

FIGURE 4.22 IC circuit die photographed with a high-resolution SVHS camera, print sent straight to the video printer.

roughly four times less resolution than the final image. This makes for a trial-and-error focusing routine. Luckily, image capture times are very, very, short. The last disadvantage for this system is that all images are processed through the PC and stored on the PC if saved. This is not as convenient as the true digital cameras discussed next.

4.4.5 Digital Photodocumentation

Quite a few digital cameras have sufficient resolution for electronic component failure analysis. The general configuration for these cameras is a high-end 35-mm camera system front end coupled with a high-resolution ccd sensor back and electronic recording media. One of the high-end cameras discussed in this section is based around a Nikon™ 35-mm camera front and a proprietary hard drive back and image storage system (Kodak DCS200™). The other is a high-resolution digital camera, also designed in partnership with Nikon™, that uses a digital memory card and a proprietary image format for saving images (Fujix DS-505™). The one camera design downloads images from the on-board hard drive (which can hold 50 images) via a SCSI interface connector on the camera back. This proved to be a significant advantage over the video still camera (preceding section), which uses a full-size image-capture board. The Kodak™ can be (and often has been) sent into the field with a portable laptop computer with full-color LCD display. This allows the user to view, in real time, the photodocumentation taken. This camera and laptop arrange-

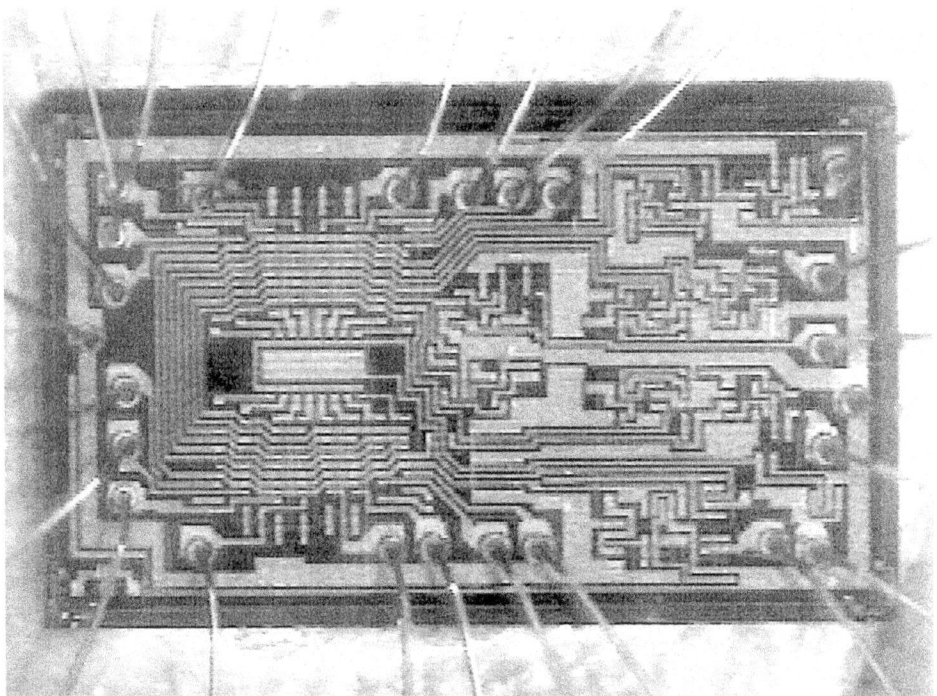

FIGURE 4.23 IC circuit die photographed with a high-resolution SVHS camera, recorded to SVH tape, then printed on the video printer.

ment was recently used on the Tethered Satellite Failure Investigation. The investigation was run by an investigation board headed up by Marshall Space Flight Center personnel. Thorough photodocumentation was required by the board members before the satellite was subjected to nondestructive and destructive analysis. Figures 4.24 through 4.27 show shots of the separated tether end were taken with the DCS200™. Both the Kodak™ and the Fujix™ use a standard F mount (Nikon mount), which allows the use of the normal 35-mm lenses. In addition, an F-to-C-mount adapter allows the use of microscope adapters and tube microscopes.

The Fujix™ camera uses a replaceable card memory storage system that provides a small advantage over the DCS200™. When the DCS200™ hard drive is full, the camera must download all the images to an image storage file and then delete all the images off the camera hard drive. With the Fujix™ camera, you can simply remove the memory card and put in an empty one. The Fujix™ cards (depending on memory size) hold 15 images, and the card reader is interfaced via a SCSI port and is just as portable with a laptop as the DCS200™. The only difference is that the Fujix™ camera would require you to remove the card from the camera to view the images. The Fujix™ does come with a composite video output and a digital image out port. The DCS200™ has an image size of 1.5 million pixels (1524 × 1012) and camera ISO ratings of "El(ectronic) equivalent to ISO" 50, 100, 200, and 400. The Fujix DS-505™ has an image size of 1.3 million pixels (1280 × 1024) and a camera ISO rating of 800. Both of these cameras are report-quality imaging devices that work extremely well in the lab environment. The DCS200™ has had a few problems

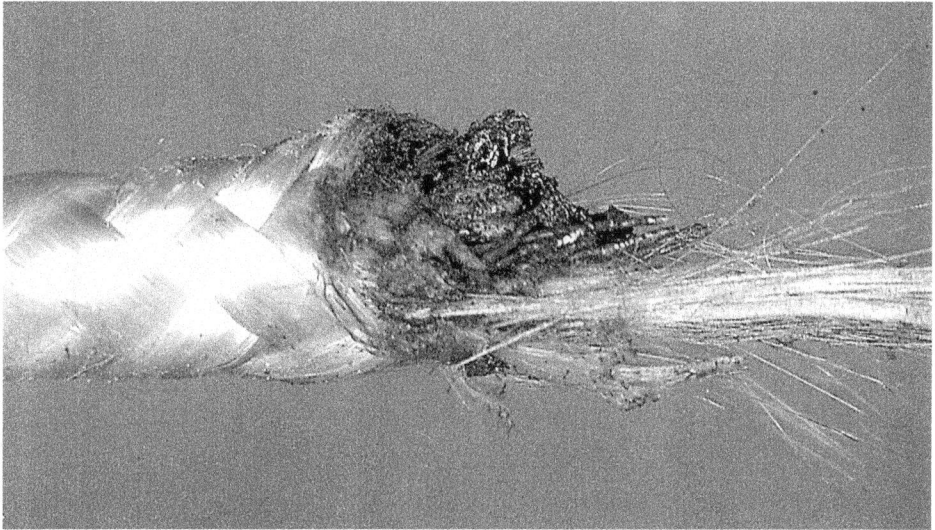

FIGURE 4.24   Digital photograph overview of the tethered satellite separated end.

FIGURE 4.25   Digital photograph overview of the tethered satellite separated end, melted area.

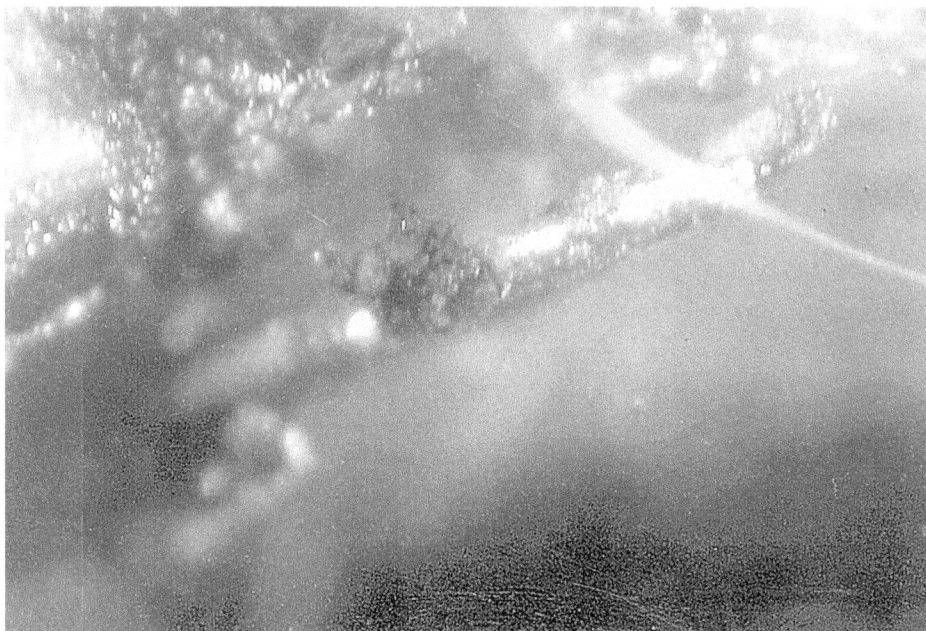

FIGURE 4.26 Digital photograph overview of the tethered satellite separated end depicting the tensiled condition of the wire ends.

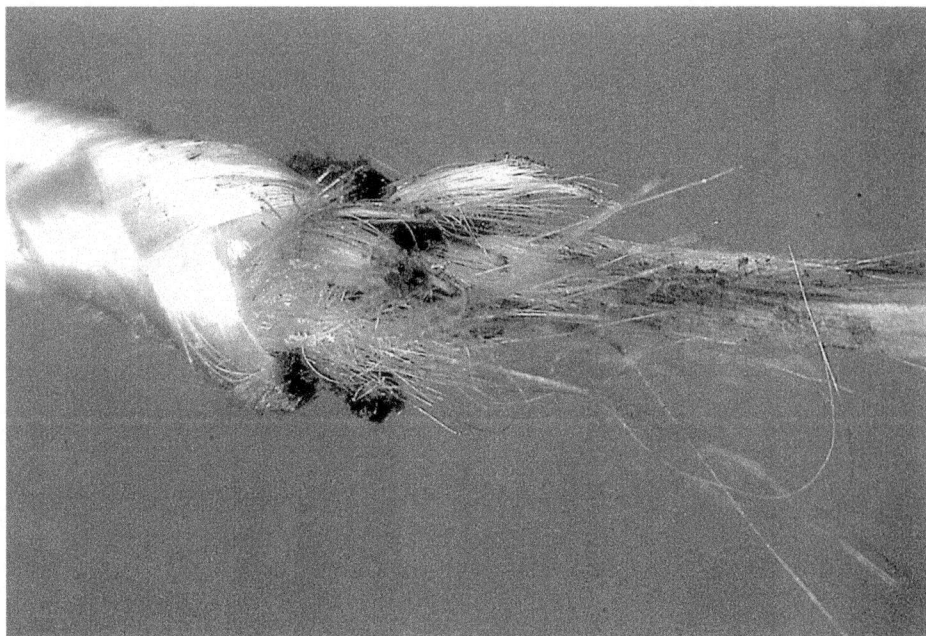

FIGURE 4.27 Digital photograph overview of the tethered satellite separated end, opposing view to Fig. 4.24.

with the camera back connections (quite possibly due to rough handling) that were fixed in the DCS460™ model. The support obtained from Kodak was exceptional. The slower ASA ratings of these cameras make the use of a tripod mandatory for most lower-light situations. Other than this, the only possible disadvantage related to these camera systems consists of the price tags.

4.4.6   Digital Image Processing

The following is a generalized description of the laboratory image handling system. This system is based on five working engineers performing failure analyses and a photodocumentation specialist handling photography for this section, with the rest of the 50-engineer division available on request. The system is the result of years of synthesis in a government working environment and as such may not be directly applicable to those setting up an FA lab; however, it should give you some ideas about image handling and storage. The heart of the labs system is a network which allows us to share digital files in 6 GB of hard drive space. Images taken on a job are placed in a directory by the photodocumentation specialist, and the directory label is investigator specific. Each photograph is stored under the official job number (i.e. directory Eng1/962E0034/). When the engineer has reviewed these image files, the needed ones are then transferred to his personal directory, again under the official job number. Once the engineer has finished with the images in that directory, they are placed the whole directory (in an archive directory), and a CD ROM of these image directories is made. The other aspect of this system, which has expedited the exchange of FA information between the labs and the line engineering staff, is Internet access. An open directory of image files accessible from the Internet to the line engineering staff was built, and the various outside agencies interested in our work were given directions about how to log on to this open directory. Personnel from the west coast have tracked FAs on flight hardware by accessing the photodocumentation nearly in real time. We have shared both visual information and written communication with all the other NASA centers. This system helps to provide data to those who need it in a user-friendly method. The investigator can focus on the work at hand while the requester keeps himself updated on the job progress. Recent *external tank FAs* involving the ET separation pyro harnesses were handled in this manner. The investigation board, as well as the line engineers, were able to access and discuss the lab findings nearly in real time. As a result, one launch was delayed until confidence in installed harness assemblies was assured. In days past a 4, 5, or even 10 GB hard drive space was a serious resource, but in today's computer environment, it is hardly anything to mention. The system we have in place can be duplicated by an inexpensive local area network, a dedicated server, Windows for Work Groups™, or Windows 95™, plus a single Internet connection. Anytime you place a computer on the Internet, there are some security issues, but these can be handled by physical isolation of the Internet machine within your system.

The other aspect of digital image processing is digital image manipulation. Several image processing programs are available that handle the basic requirements of digital image manipulation. Image Pals™, by Ulead Systems; Photostyler™, by Aldus; and Photoshop™, by Adobe are three of the most widely recognized. Image pals™ is a lower-end image album and editing system that the engineers like using. They keep track of their FA report images using the album function of this program. Photostyler™ and the new Photoshop™ are two of the top-end image processing software systems. All three are adequate for FA purposes, with advanced users preferring either of the high-end systems.

The primary need for digital image processing is color correction and brightness/contrast corrections. Anything other than this should be looked at with a seriously negative attitude. Color correction can and is addressed in several video/digital photography magazines. There are standards for color correction in digital images; however, in this lab, each image is used or not used at the investigator's discretion. Colors are corrected only to the degree that the printed hard copy looks as close as reasonably possible to the actual component. The purpose of FA photodocumentation is to validate existing conditions or conditions resulting from testing. As such, color and brightness/contrast are adjusted to provide the most accurate display of the component condition.

Figures 4.28 and 4.29 are shots of a rotational transducer used on a Transporter Crawler (the vehicle that picks up the Mobile Launch Platform and Shuttle and transports them to the pad). The first shot (Fig. 4.28) depicts an overview of the component, and you can see that the component is being supported by an alligator clip for this shot. Figure 4.29 is the

FIGURE 4.28 Untouched photograph of a crawler transporter steering rotational transducer.

FIGURE 4.29 Digitally manipulated photograph of a crawler transporter steering rotational transducer.

same shot with the alligator clip digitally removed. This illustrates the power, and the potential for abuse, of digital photography. A relatively inexperienced user can digitally manipulate images to show things that never exited.

Some of the image manipulation functions that can prove to be valuable are automated tonal corrections for bright and contrast, image rotate, and the image cropping functions. Often, an image can be intensely detailed, with the object of concern mounted in neither a horizontal nor vertical mode. Performing a partial rotate (by degree) and then cropping around the component allows you to produce an image that is easily understood. In addition to these functions, a composite photograph can be produced using the image copy and paste commands. For example, if before-and-after shots of a component are necessary, they can be placed in a single image to clarify the actual change in the component. The digital image processing capabilities of an application such as Photoshop™ are much too detailed to delve into in this chapter. The applications mentioned have sufficient power and provide the investigator with all the necessary tools for handling and processing digital images.

4.4.7 Photographic Paraphernalia

Some standard photographic peripherals should be mentioned. Optical light pipes, diffusers and component stages, and items of this nature are necessary tools for the FA investigator. Most of these items have to do with mounting a component for a shot or providing sufficient illumination. One of the most useful tools I have come across is a professional light stage for small components (Bogen Light Modulator™). This device comes with a full set of optical light pipes, a backlit mounting stage, three fully adjustable light sources, a set of color filters, a polarized light filter, a diffuser cone, a reflector, and a fourth arm to use for holding scales or additional fiber optic lighting sources. This lighting/mounting system goes a long way toward providing all the peripherals a FA investigator needs.

If addition to these items, you should have a pair of flexible fiber optic lights cables, a series of scales (depicting millimeters, centimeters, and meters) and some white translucent paper for making light diffusers. A set of flexible arm holders also helps in placing lights or scales in the field of view. This minimal set of paraphernalia should be enough to get the job done. Figure 4.30 shows a professional light stage for small components and most of the items mentioned above. Figure 4.31 shows a optical light source with flexible cable and end fixtures.

4.5 *LENS AND MICROSCOPE FUNDAMENTALS*

4.5.1 Lens Fundamentals

This discussion on lens fundamentals is by no means a thorough narrative on lens theory. Hopefully, it provides enough basic information on lenses and how they function to assist the failure analyst in the photodocumentation of his work.

There are two basic lens types: positive and negative. A *negative* lens is thinner in the center than at the edges and diffuses or scatters light. If parallel light rays strike a negative lens, the lens construction forces these parallel light rays to angle outward as they pass through the lens body. If the parallel rays are so far apart that they strike the lens on either side of the lens' center, then they would be forced to diverge from each other. A negative lens, therefore, has no focal point or point at which the light rays converge.

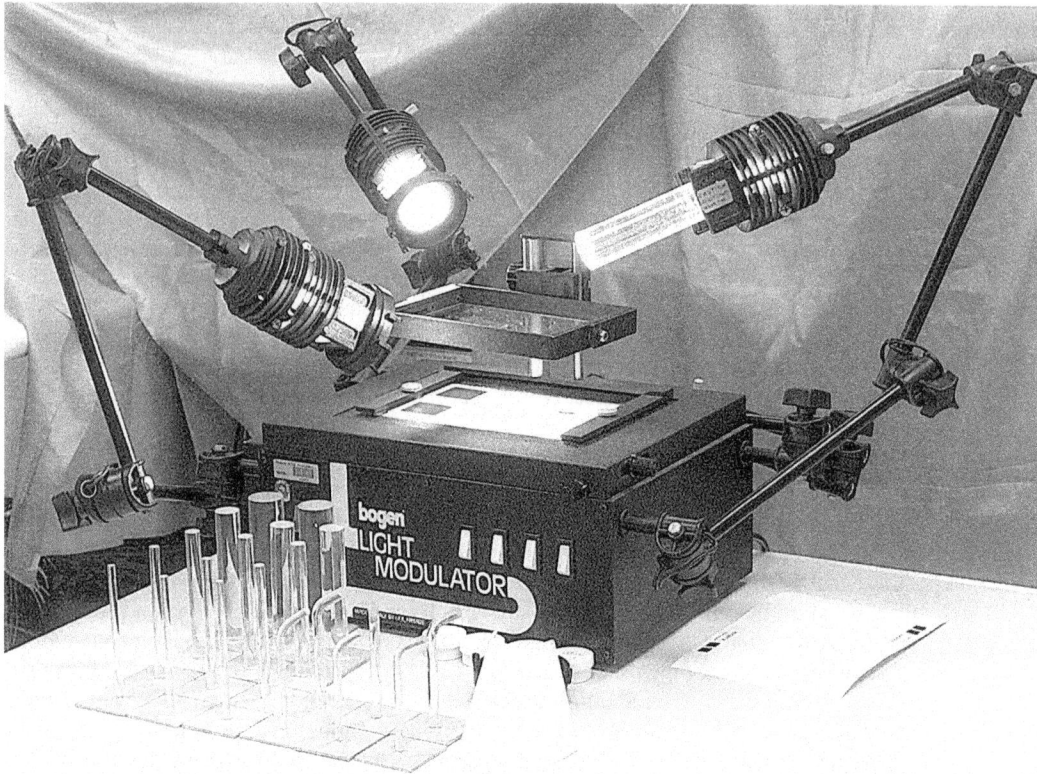

FIGURE 4.30 Professional component light stage with accessories.

A *positive* lens is thicker in the middle and converges light rays. The point at which parallel light rays (coming from infinity through the lens) would converge is called the *focal length* or *focal point* of the lens. A symmetrical positive lens (a lens ground to the same shape on both sides) has essentially two focal points, one at each of the front and back of the lens. The focal length of the lens (F) has a direct relationship with an object some distance in front of the lens (u) and the image focus point for that object at the rear of the lens (v). This relationship is defined as

$$\frac{1}{F} = \frac{1}{u} + \frac{1}{v} \tag{4.1}$$

An object placed 10 times the focal length in front of the lens would be focused in back of the lens at 10/9 F. The magnification of the object is given by $m = F/(u - F)$, where (m) stands for object magnification. Using the previous equation, the magnification can also be expressed in terms of (F) and (v). The magnification at 10 times the focal length in front of the lens would be 1/9×. An object placed in front of the lens at 2 times the focal length (F) would have an image focus (v) at the back of the lens at 2F. The magnification would be 1×. As the object is brought closer and closer to the lens, the focal point at the rear of the

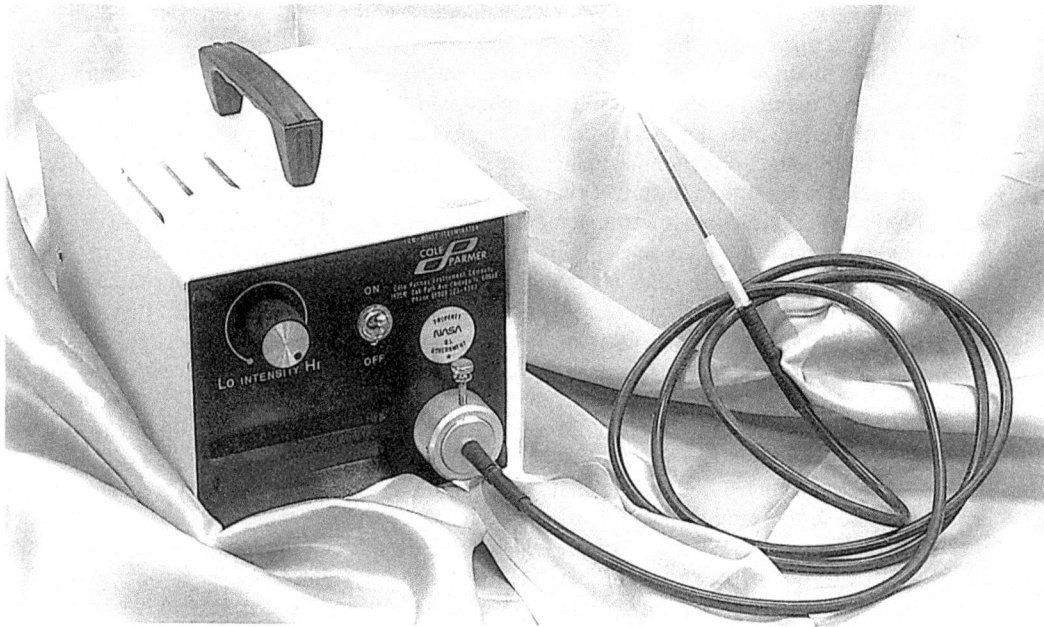

FIGURE 4.31   Optic light set.

lens gets farther and farther away, and the actual magnification becomes greater and greater. In theory, this would allow us to use a single lens for all our magnification and optical photography work. There are two problems with this theoretical situation. The first is that, to accommodate the magnifications, we would need a bellows several yards in length to house the film holder. The second is related to a term that has more to do with microscope lenses than camera lenses: *empty magnification.* Each lens has a set maximum resolution defined by its *circles of confusion.* This circle of confusion is directly related to the aperture (or diameter) of the lens. The circle of confusion is defined by the magnification where a point source in front of the lens becomes a blurry circle when focused at the rear of the lens. For a perfect lens (corrected for all aberrations), the circle of confusion is directly linked to the aperture of the lens. Two lenses with the same focal length and different apertures (diameters) will have different size circles of confusion, with the smaller aperture lens having the larger circle of confusion. The resolution of a lens is typically given in *line pairs per millimeters.* This is the maximum number of line pairs a specific distance (in millimeters) apart that can be accurately resolved or counted at a given magnification. The higher the number of line pairs per millimeter, the better the resolution. At some magnification, the circle of confusion principle sets in, and no higher number of line pairs can be counted because of the blur in the image. Physical magnification beyond this point is useless, because it only magnifies the blur. Magnification beyond this point is termed *empty magnification.*

The primary characteristic of a lens is its focal length. Typical compound optical photography lenses are constructed as nonsymmetrical (e.g., the Tessar and Sigmar lenses) with a focal plane at the rear (film side of the lens) that is much shorter than the front focal point. This provides a big advantage when focusing and shooting shots from infinity up to the front focal distance. Shots any closer than that will be out of focus due to the lack of rear focal plane adjustment.

The second lens characteristic is the aperture (diameter) of the lens. The aperture is typically represented as a ratio of the lens maximum diameter to the focal point. A lens aperture of 4.5 indicates the focal point (for infinity) is 4.5 times the diameter of the lens. This aperture number is sometimes called the *speed* of the lens. This is somewhat confusing, because the aperture adjustment ring allows you to adjust the aperture setting by opening (and closing) a diaphragm arrangement within the lens. The lens speed refers only to the maximum aperture of the lens.

These two characteristics are combined with the shutter speed, another factor that the investigator can control to make photographic images. The shutter speed is an adjustable parameter on most cameras, set up in partial increments of a second. A typical MP4™ arrangement would be B, 1, 2, 4, 8, 15, 30, 60, 125. The B value is a manual shutter release (press to open the shutter, release to close), the 1 would be a 1-s exposure time, the 2 would be a 1/2-s exposure time, and so on and so forth until reaching 125, which is a 1/125-s exposure.

The three controlling factors an investigator has are the lens selection (focal length), the aperture setting, and the shutter speed. The focus, aperture setting, and shutter speed work in combination to allow the correct amount of light through the lens, to the film or image capture media. In the case of film, a specific film speed requires a given amount of light for correct exposure. Films are numbered for exposure speed in accordance to two main standards, an ASA number or an ISO number. ISO is a newer numbering system (in the U.S.) and, for this chapter, film speed will be discussed using ASA. An ASA 200 film is a good all-purpose film. It is fast enough to shoot indoors with low light and, combined with a medium-fast shutter speed, works well outdoors. For very fast outdoor events (such as auto racing), move to a 400 ASA film. This film can handle a fast shutter speeds of 1/500 s or greater without opening the aperture up all the way. Remember, the larger the aperture, the smaller the depth of field. Film speed for 35-mm cameras ranges from 50 to 1500 ASA for color and 3000 ASA for black and white. There is a trade-off between image quality with faster speed films and slower speed films. Faster films may appear grainier when magnified.

Within this chapter, *depth of field* has been discussed several times without really explaining what affects it. DOF is a finite factor: only a portion of a photographic image will be in sharp focus. The investigator strives to capture as much useful data as possible about a component with the photodocumentation. The investigator would prefer that the front (portion of the component closest to the lens) and back (portion of a component farthest away from the lens) of the component both be in sharp focus. To do this, the correct lens and aperture must be used to maximize depth of field. A formula for DOF is

$$\frac{2cf(m+1)}{m^2} \tag{4.2}$$

where c is the maximum allowable circle of confusion in millimeters, $f(m+1)$ is the effective aperture (or effective f number), and m is the magnification. A reality check on this equation was performed to see if it would correlate to practical experience with camera lenses and depth of field. For this check, we used $f = 11$, $m = 0.1$, and a 50-mm lens with a circle of confusion of 0.07 mm. (The human eye can differentiate circles of confusion only if they are larger than 0.07 mm.) The 0.1 magnification worked out to an object distance of approximately 21.6 in., and the depth of field worked out to 6.6 in. These numbers appeared to fit well with practical experience. The DOF changes rapidly as the magnification increases. For $f = 11$, a magnification of 7, the DOF would be less than 0.3 mm.

As mentioned, standard photography lenses are usually nonsymmetrical in design, optimizing the longer focal distances (low magnifications). Macro lenses are usually symmetrical in design or are optimized for the focal distances below $2F$. These lenses are usually used with a bellows type camera (e.g., the MP4™) or with an adjustable focal tube to shoot photographs in the low magnification numbers, 1 to 18×. All of the factors mentioned above still apply: lens focal length, aperture setting, film speed, and shutter speed.

4.5.2 Microscope Fundamentals

As in the case of a modern camera lens, microscope lenses consist of a compound system. The microscope itself is a compound system consisting of an objective lens, the microscope tube (or body), and the eyepiece (or ocular). These components requires a minimal amount of discussion, primarily to address some characteristics of each, which the investigator needs to know about.

One of the first characteristics of objective lenses is *field curvature*. Field curvature can be defined as an optical effect in which the focal point for the outer portions of the lens is at a different point from the center of the lens. This visually results in an image where the center is in sharp focus, and the outer edges get progressively fuzzier (more and more out of focus). Field curvature is caused by the difference in length that light rays at the outer edge of a lens must travel compared to the path of a light ray passing through the center of the lens. Modern microscope lenses deal with this field curvature with a design called a *plan* or *plano* lens. These lenses are flat-field objectives and are designed to correct field curvature.

An optical aberration common in microscope lenses is *spherical aberration* of the lens. Spherical aberration is defined as a lack of consistency in the focal point of a lens from center to outer edge due to the spherical shape of the lens. Most modern microscope objectives are corrected significantly for spherical aberration, and any remaining aberration can be corrected in the ocular. Another type of optical aberration is *chromatic aberration*. This aberration is due to the lens being made of a medium (glass) through which light must pass. White light is made up of light of various colors (wavelengths) (e.g., red, green, and blue) that are affected differently as they pass through the lens. In a lens uncorrected for chromatic aberration, the various light colors would have differing focal points, and color *fringing* (an optical effect that displays the colors like a rainbow around the edges of the lens area) would result.

Three types of microscope lenses are in use: *achromat*, *semiapochromat* (fluorite), and the *apochromatic* lenses. The achromatic lens has been corrected for chromatic aberrations for two wavelengths of light, most often red and blue, and corrected for spherical aberration for one wavelength of light (most often, yellow-green). The achromatic lens is the most common and least expensive type of lens. The semiapochromatic lens is typically constructed using fluorite (and is sometimes called a fluorite lens) and has been corrected chromatically for two wavelengths of light and corrected for spherical aberrations for two wavelengths of light. This gives it an advantage over achromatic lenses, but the addition of the fluorite layer in the lens adds to the lens cost, making it a more expensive lens. The third type of lens is the apochromatic lens, which has been corrected chromatically for three wavelengths of light and for two wavelengths of light for spherical aberration. Apochromatic lenses are better optically than the other two type lenses and are also more expensive.

Each of the above-mentioned lenses can be purchased in a flat-field version that corrects for field curvature, and each would be designated with the *plan* or *plano* prefix. The lens names would then be planachromat, planfluorite, and planapochromat. For the FA investigator, the image captured through the microscope needs to be as correct in color and focus as possible. Therefore, it is most useful to work with the planapochromat-type lens objectives. If highly corrected lenses are not available, perfectly valid photodocumentation can be achieved using black-and-white film and by limiting the field of view to the center portion of the microscope image.

Optical photography lenses are designated by a focal length and a lens speed, which is the ratio of the aperture to the focal length. Microscope objectives are not designated in the same manner. Objective lenses are typically marked with a degree of correction (achromat and so on), a magnification, a numeric aperture number, and a focal length (in millimeters). We have discussed the types of correction, and the magnification is self-explanatory, as is the focal length. An explanation for numerical aperture follows.

The *numeric aperture* (NA) is a calculated number that expresses the aperture as an angular measurement ($NA = n \sin u$) where n is the refractive index number of the media through which light is being passed (1.0 for air), and where the angular aperture u is one-half the angle made by the light entering the objective lens when the lens is in focus (on a point source). If a lens has an numeric aperture greater than 1, then it is set up as an immersion lens to be used with oil or water between it and the object of interest.

The resolution of a lens was discussed in the previous section on lenses. The resolution of an objective lens is related to the numeric aperture in the same way that the resolution of a optical photography lens was related to its aperture. The principle of the circle of confusion defines the interrelationship with magnification and resolution, given in line pairs per millimeter.

The numeric aperture is also related to the *working distance*. The working distance is the distance from the end of the objective lens to the top of the object being viewed. The higher the magnification and numeric aperture number, the shorter the working distance.

The microscope body is a mechanical tube that allows the objective lens to be screwed in at the bottom and a viewing eye piece (ocular) to be attached at the top. Most modern microscopes have a pair of oculars and as such are referred to as *stereomicroscopes*. For any stereomicroscope, the objective images are typically split (and offset by some angle, usually through a prism) and then directed to the bottom side of the oculars.

The distance from the attached objective to the attached ocular is called the *mechanical tube length.* The microscope tube body allows the needed focal length for correctly projecting the magnified image from the objective onto the ocular. Modern microscopes are typically standardized at a mechanical tube length of 160 mm, which typically is not adjustable. Objective lenses are corrected for optical aberrations based on this standard mechanical tube length. Therefore, when used on microscopes with nonstandard tube lengths or an adjustable tube length, either overcorrection or undercorrection of the optical aberrations will result. Also, the marked magnification on an objective lens is not correct at a different tube length. For advanced microscopists, there are some advantages to changing the mechanical tube length. These advantages typically pertain to biological and chemical microscopy where the use of slides, slide covers, and slide illumination techniques are necessary.

Oculars or eyepieces are the final component in our discussion of microscope fundamentals. The objective lens, combined with the tube body, focuses an image on the ocular diaphragm. For different types of oculars, this diaphragm is either inside the ocular (e.g., Huygenian oculars, simple negative, two planoconvex lens ocular, with plane side up) or outside at the bottom of the ocular (e.g., positive Ramsden ocular, two planoconvex lenses with the plane side outward). Oculars can be much more complex than the two simple lens arrangements described above. Typically, a microscope matches the ocular's construction to the type and level of compensation of the objective lens.

The ocular diaphragm defines the field of view seen through the microscope. A graticule (scale) or grid can be fixed to the ocular diaphragm and becomes superimposed visually on the magnified image from the objective lens. This can aid in determining the magnification of a photomicrograph or in counting particles and so on. For those FA investigators who wear glasses, most standard microscopes provide oculars designed for people who wear glasses while viewing through the microscope. These oculars are sometimes called *high-point eyepieces.*

4.6   *OPTICAL MICROSCOPY*

4.6.1   Overview of Electronic FA Microscopes

The electronics FA investigator does not use the microscope in the same manner as the chemical or biological microscopist. The most common microscope type is the *bright-field* or transmitted light microscope. This type of microscope is so named because light is usually transmitted from a source through a condenser and a clear specimen (slide and stained sample) to the objective lens. The sample is then viewed as a lightly stained (mostly transparent) sample against a bright background. The magnifications on this type of microscope are large (60 to 200×), and therefore the working distances are small. This type of microscope is not practical for the electronics FA investigator. Most electronic samples are solid and are not transparent/visible with transmit light. Even in the case of viewing integrated circuit dies, a form of incident light is used through a microscope that is closer in design to a metallurgical microscope rather than a transmitted-light microscope.

The type of microscope used for the majority of electronics FA work is a *low-power stereomicroscope.* Typically, it has an objective capable of being adjusted through a range of magnifications using a *zoom ring.* The range of magnification is usually 0.6 to 4.0 times the power of the oculars used. For a 10× ocular set, the range of magnification would be 6 to 40×. For 20× oculars, the range would be 12 to 80×. This makes the stereomicroscope very simple to use. The stereomicroscope has three useful characteristics: large working distances, good depth of field, and eyepieces with a wide flat field of view.

Light to illuminate objects under the stereomicroscope is usually provided by a light ring attached at the same level as the bottom of the objective. Other forms of indirect lighting can be used, such a fiber optic light source. The stereomicroscope is not typically equipped with a transmitted light system (lighting from the bottom of the sample) or with an incident (reflected) light system (light reflected down through the objective to the sample and back up through the microscope).

The other type of microscope used for electronics is an incident-light microscope. These microscopes were called *metallurgical* microscopes for years, because they were used to examine metal or opaque samples that were not visible by standard bright-field transmitted-light microscopy. This type of microscope is used when the investigator needs the higher magnifications not available in the low-power stereomicroscope and is willing to accept the smaller depth of field and shorter working distances. Magnifications up to 700× can be achieved with these microscopes.

These microscopes are typically used for highly specialized purposes. In the case of electronic FA, that purpose is for use on an electronic probing station. Typically, for magnifications above 120×, an SEM is used because of its greater depth of field and clarity. Unfortunately, SEM examination produces images in black-and-white only. The component (for SEM) is usually in a vacuum chamber during examination and not accessible for electrical parameter measurements. To overcome these difficulties, we use the high-power electronics microscope and probing station. In performing an integrated circuit (IC) FA, the package (either ceramic or plastic) is opened to expose the die, and then the entire IC can be examined, photographed, and probed to find the output of each subsection of the IC. This is the primary reason for the high-power electronics microscope.

There are some subtle differences in this type microscope. The mechanical tube lengths are longer (185 to 250 mm) to house the reflected lighting apparatus. The objectives are also not corrected for spherical aberrations resulting from the use of glass coverslips (which are necessary on slides for standard transmitted-light microscopes). In most cases, the objectives are of a different thread size on the electronic/metallurgical microscope and will not thread into a standard Royal Microscopical Society (RMS) objective opening. For reflected-light microscopy work, a condenser-type arrangement (also used on transmitted-light systems) is used, which allows full manipulation of the light source before it is passed into the microscope tube body and is reflected downward to the objective lens onto the sample.

The standard techniques of sample imaging with a transmitted light microscope are also available with the electronics/metallurgical microscope and its reflected lighting system. The term *incident* is typically used preceding these lighting terms to indicate that light is generated from a reflected or incident lighting system. *Incident bright-field illumination, incident dark-field illumination,* and *interference contrast illuminations* are all techniques used with the metallurgical microscope. These imaging techniques will be discussed in a subsequent section.

4.6.2   Low-Power Stereo Microscope Equipment

The manufacturers of stereomicroscopes are too numerous to mention. For present purposes, were limit this section to a small subset of equipment and discuss the overall characteristics, advantages and disadvantages of each. The first unit is a Nikon SMZ-2T™. It is shown in Figs. 4.32 (front) and 4.33 (back). This unit is an excellent choice for discussing the various elements of a modern low-power stereomicroscope. Figure 4.32 shows an overview of the unit from the front, the top arrows are pointing at the 10× ocular (eyepiece) set. The next arrows down (from the ocular arrows) point to the adjustable ocular settings. Very few people have perfect vision, and most people's vision differs from eye to eye. This feature allows the user to center one eye on the object and focus, then adjust the other eyepiece setting so that each eye is focused on the image when the master focusing adjustment is used. This feature greatly enhances the comfort levels for using the microscope, allowing for longer viewing times. The third set of arrows (from the top) indicate the ocular positioning system. This allows the user to adjust the width the oculars are set apart. The variation in pupil-to-pupil width in individuals is rather large. The need for a sliding nosepiece (ocular width adjustment) is self-evident. The fourth set of arrows from the top point to the zoom (magnification adjustment) ring/knob. For this microscope, it varies the magnification from a low of 1 × the oculars (10×) to 6.3 × the ocular magnification (63×). The fifth set of arrows from the top point to the microscope body set screw. The whole microscope is seated on the ring housing, the set screw is threaded through. The

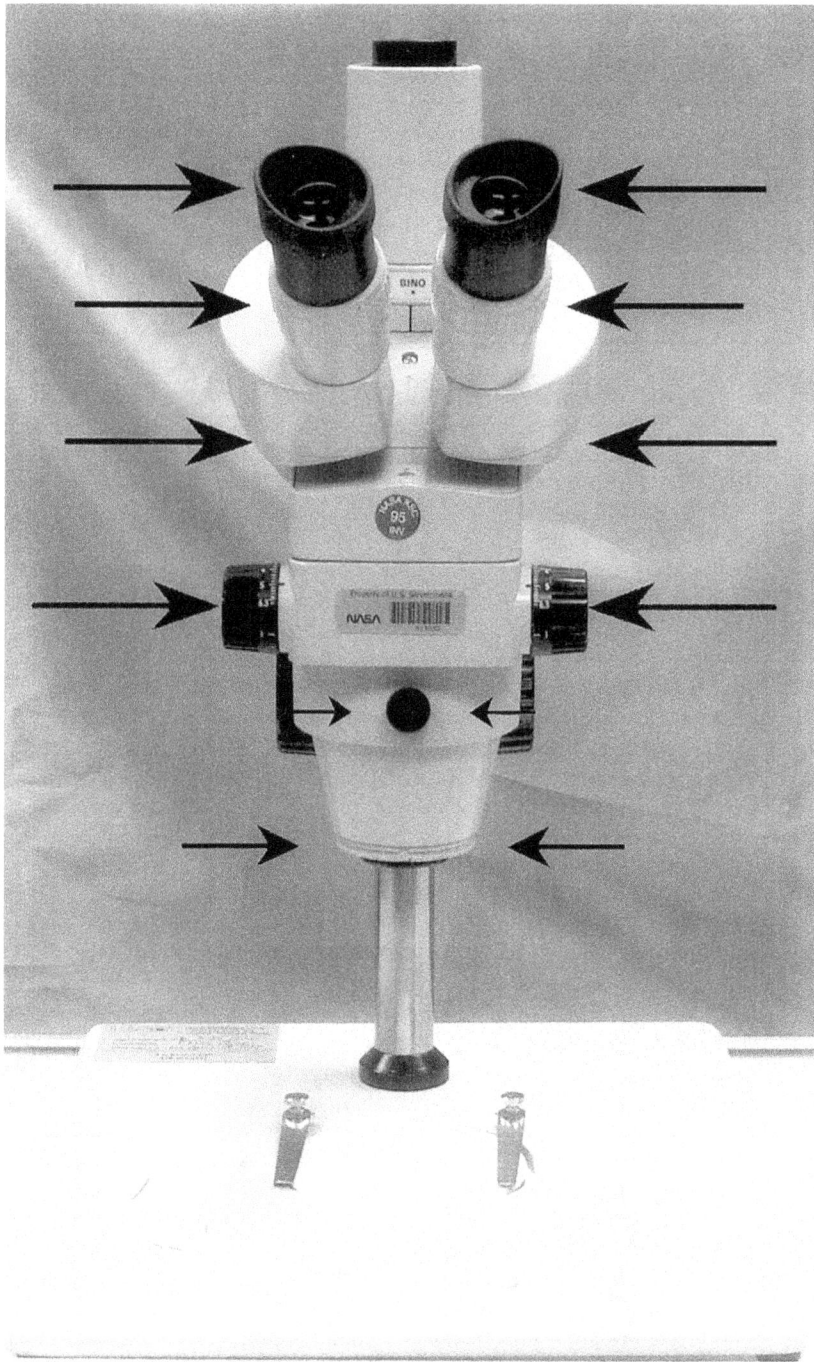

FIGURE 4.32 Front view of a typical stereomicroscope. The arrows indicate microscope features described in the narrative.

microscope field of view can be rotated left or right to aid in component examination simply by loosening the set screw. The sixth set of arrows point to a slotted ring down near the objective lenses. This slotted ring arrangement is used for attaching optical ring lights or other types of illumination.

Figure 4.33 depicts the back of the Nikon SMZ-2T™. The top set of arrows point at the trinocular opening that houses the relay lens for the camera system. The Nikon™ microscopes have been manufactured with a trinocular arrangement that allows the user to switch an objective image through the trinocular opening (and through the relay lens) to a camera. The camera is mounted to a "C" mount tube that slides down over the relay lens. The second set of arrows from the top point to the trinocular set screw, which is used to fix the camera mount tube in a set position. This trinocular set screw allows the user to make small vertical adjustments on the camera so that the objects viewed through the oculars are in focus at the same time the image is focused at the camera. This greatly expedites the microphotography process. The third set of arrows point to the slotted microscope assembly post. This post, along with the base unit, supports the microscope assembly, allowing a vast range of vertical adjustments to be made for the accommodation of electronic components. The fourth set of arrows point to the trinocular opening ring. This ring opens the trinocular optical path from one of the objective lenses to the camera, switching the optical path between an ocular and the trinocular opening. The fifth set of arrows point to the focus adjustment knobs. These knobs raise and lower the microscope head to facilitate focus. One of the knobs also has a tension adjustment, providing a frictional force that holds the microscope weight at the adjusted level. The sixth and seventh set of arrows point to the height adjustment set screw (upper) and the safety slip ring. The safety slip ring is a *very important tool.* This ring is to be set at some reasonable level above the component so that, when the microscope adjustment ring is loosened, the microscope head cannot be dropped onto the component, damaging it or destroying a FA investigation. Trust me—good operating procedures require the setting of this device for all microscope operations. The professional reputation of an investigator who drops the microscope head onto the component under examination will suffer. There is virtually no way to eloquently describe such an event in an FA report.

The two figures and discussion above provide the novice microscopist with a quick overview of the low-power stereomicroscope and its components. This microscope is a well built tool. Its advantages lie in flexibility. It can be used for most small components with the minimal use of external lighting sources. It has a relatively small footprint and can be stationed close to an investigators work bench/test bench for quick observations on components under test. The disadvantage of this system is the lack of adjustment for larger components. The vertical adjustment requires physical lifting of the microscope head and repositioning it vertically on the assembly post. This is a two-handed operation that cannot be done while looking at an object. Larger objects may not fit easily under the microscope head or there may be insufficient room for focus adjustment.

The other stereomicroscope discussed in this chapter is the boom-mounted stereomicroscope (Nikon SMZ-10™). Figures 4.34 and 4.35 show the microscope configuration, including the mechanical boom assembly, optical ring light system, adjustable sample stage, and video camera mount with video printer. This microscope shares some of the good characteristics of the SMZ-2T™ as well as one of it faults. It has a major advantages in two areas. The first is in the area of magnification. The zoom (magnification adjustment) ring/knob on this microscope is variable from a low of $0.4 \times$ the oculars ($4\times$) to $4 \times$ the ocular magnification ($40\times$). For the vast array of electronic components, starting at a lower magnification is a major advantage, especially if you are photodocumenting. The other

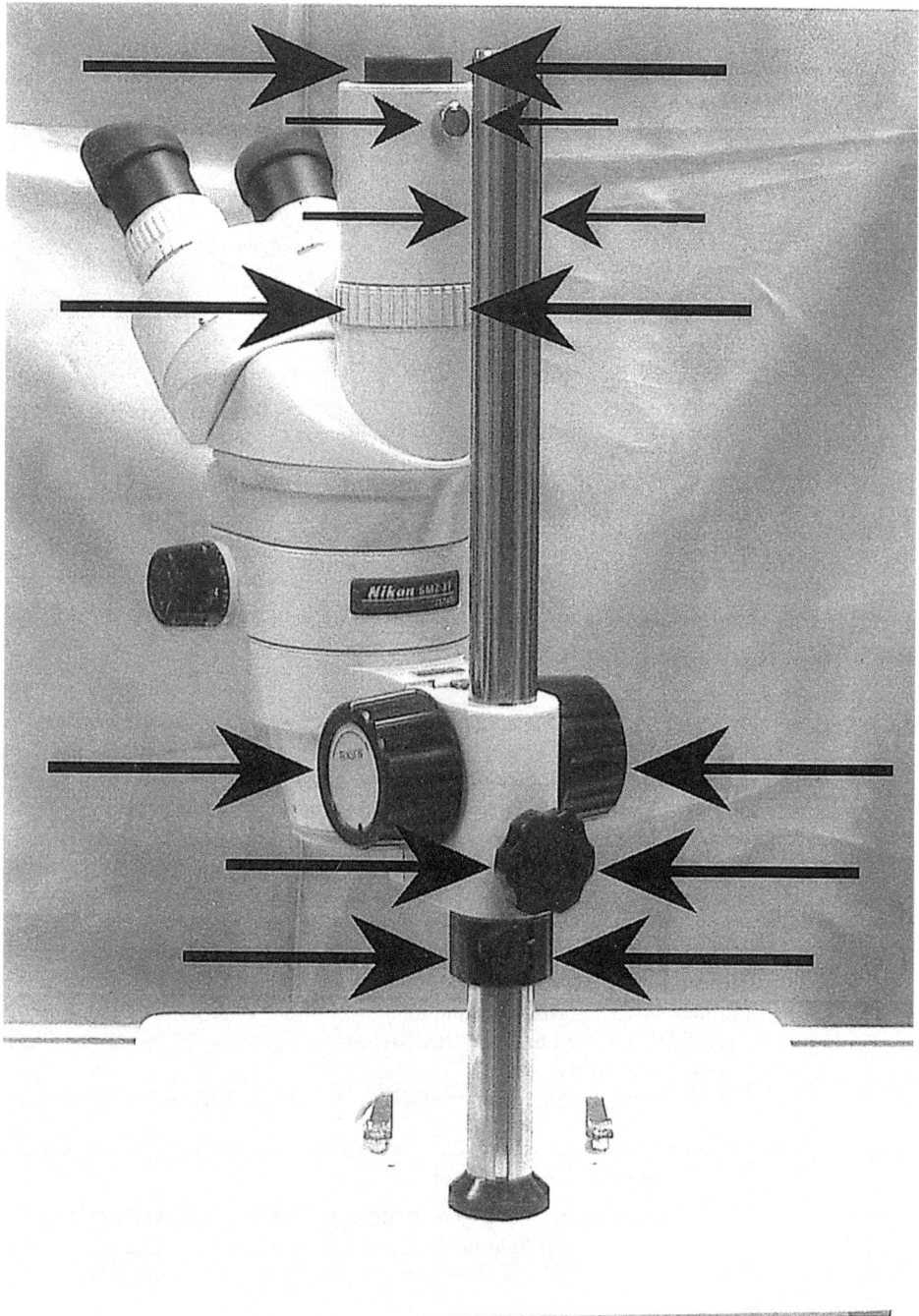

FIGURE 4.33 Back view of a typical stereomicroscope. The arrows indicate microscope features described in the narrative.

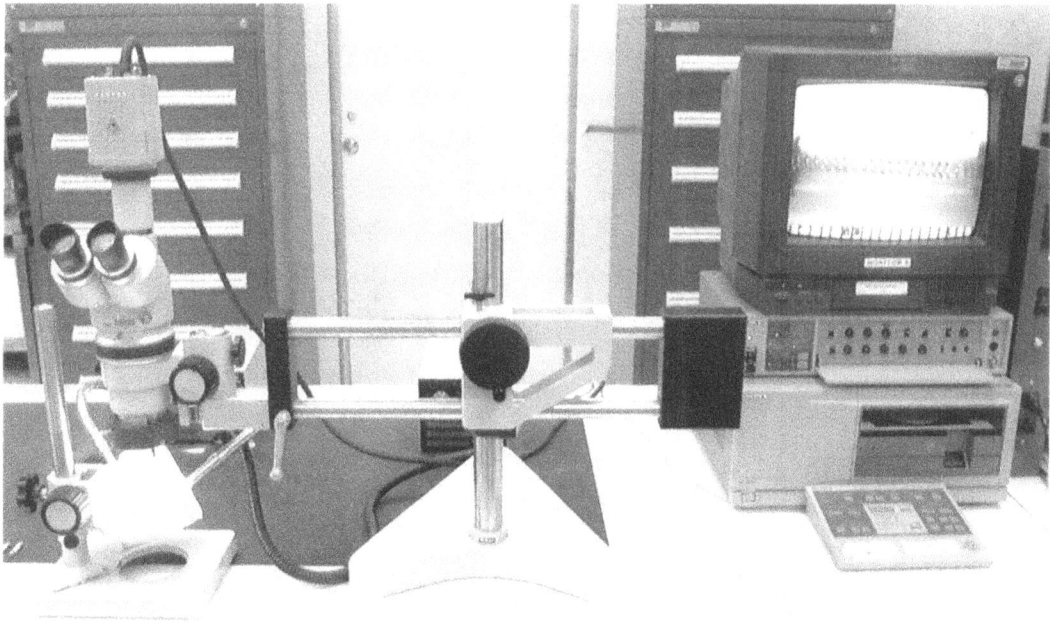

FIGURE 4.34   Front view of the typical boom-mounted stereomicroscope.

advantage is the boom assembly with which this microscope has been equipped. It has infinitely more adjustment for small to medium-large components. The boom allows movement in a circular motion (arc) and has a geared assembly providing motion forward and back. This allows the unit to be easily adjusted for objects large to small. The disadvantage this microscope assembly shares with the SMZ-2T™ is that no accommodation has been made for adjustments (outside of the normal focus adjustment) in the vertical direction. With the addition of a video camera, the weight of the microscope optical stage is quite great, and any adjustment is a considerable operation.

I have use these two stereomicroscopes as my examples in this text; however, this in no way implies that either is the microscope of choice. These two microscopes have the most complete set of microscope functions and options of any in our lab. Equivalent options are available on most all the other microscopes on the market, including Zeiss™, Wild™, and American Optics™.

Two final comments about low-power stereomicroscopes are as follows:

1. Never buy a stereomicroscope without a trinocular arrangement (even though the ocular photography is available).
2. Photomicrographs never match the optical quality of the stereoscopic view (due to the enhanced depth of field the dual optical views of the component yield).

4.6.3   Advanced Microscope Techniques

The preceding section on stereomicroscopes was blatantly devoid of microscopic technique. The reason for that is fairly obvious: the large working distances available using the

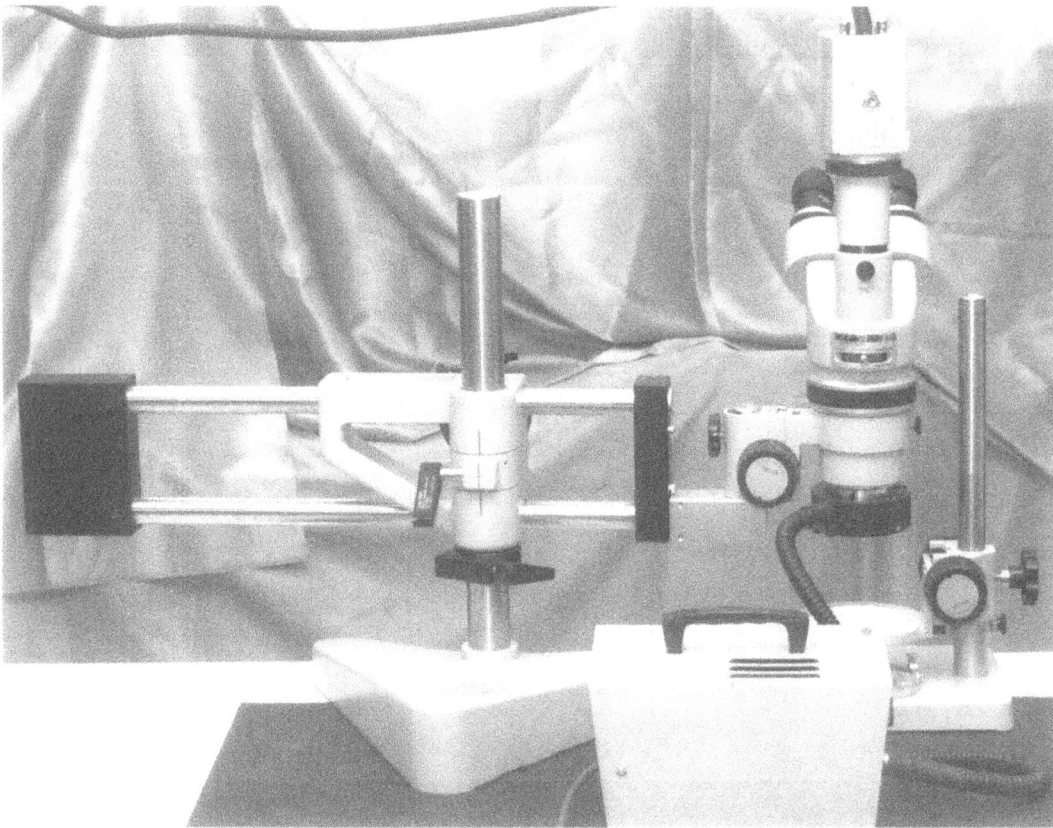

FIGURE 4.35   Back view of a typical boom-mounted stereomicroscope.

stereomicroscope allow one to use many of the standard lighting techniques from optical photography. Where additional lighting is needed, a fiber optic ring light can be fixtured to the objective end of the microscope. For getting light into nooks and crannies or for lighting topographical features, the small fiber optic light cables and tubes can be placed using various support devices. Working with the low-power stereomicroscope is one of the easier tasks the electronic failure analyst faces. The same cannot be said for the analyst who endeavors to use the high-power electronic probing station.

The first obstacle to overcome when working at magnifications from 50 to 700× is the effect of random vibration. A fellow investigator, walking through the next room, can destroy your ability to capture high-magnification images if you lack some form of isolation platform. Earlier in the chapter, isolation platforms were mentioned, typically used in conjunction with an electronic probing station. Almost every microscope company selling electronic probing stations either sells isolation platforms or has an arrangement with a company that does.

To remove some of the confusion about exactly what an electronic probing station is, a description of an electronic probing station follows, with Fig. 4.36 illustrating a typical system. The top fixture is a high-definition TV camera (SVHS) with matching relay lens

FIGURE 4.36   Isolation platform and an electronic probing station with incident light microscope. The arrows indicated the microscope features described in the narrative.

(top arrow), this fixture could also accommodate a still photography system. The second arrow points to the stereo ocular set, which has the standard adjustments for width and eye to eye focus adjustment. This feature proves to be critical when setting up optimal illumination (Kohler illumination). The next arrow points to the incident lighting fixture. It contains a section for inserting filters, adjustments for the aperture and field diaphragms, and a fixture for inserting polarized light filters. The next arrow points to the focus adjustment ring (this ring has a fine and gross focus ring). The next arrow points to the objective set. The objectives shown in this photograph are a Leitz Wetzlar™ L 50X/0.60NA DF lens, an LL 20X/0.40NA lens, an NPL 10X/0.20NA lens, an NPL 5X/0.09 DF, and a PL 3.2X/0.06NA lens. These objectives screw into the rotating ring assembly and can be replaced with other objectives for specialized viewing needs such as dark-field objectives. Further down (next arrow) is the IC zero insertion force (ZIF) socket and card assembly mounted to the IC stage. This assembly pulls fully out from under the objective set to allow insertion of a decapsulated IC, and it facilitates electrical connections to the lead frame pins via the edge connector at the back of the assembly. On both sides of the IC socket are smooth finished metal platforms. The end of a placed needle probe is evident in the lower right section of the monitor in Fig. 4.36. These metal platforms provide a bonding surface for the IC needle probes, which have magnetic bases. Figure 4.37 shows a view of the probe assembly highlighting the three-axis adjustment knobs. These needle probe devices allow the analyst to make contact on a decapsulated IC (which has had the glass passivation layer chemically or mechanically etched away). Electrical parameters can be measured via a curve tracer, oscilloscope, or other instrument to quantify the condition of a small portion of the IC. By selectively cutting through the LANs/runs interconnecting the IC devices with a needle probe, small portions of the IC die can be electrically isolated and analyzed for electrical response. In this manner, a highly populated IC can be troubleshot and the fault area isolated. Advanced metallurgical techniques can then be used to lap and polish through a fault site, looking for anomalous conditions. These techniques are to be discussed in detail in other sections of this book.

The next arrow down points to the IC stage. This stage has a height adjustment ring that allows the focus and illumination of the microscope to be optimized and to have the IC

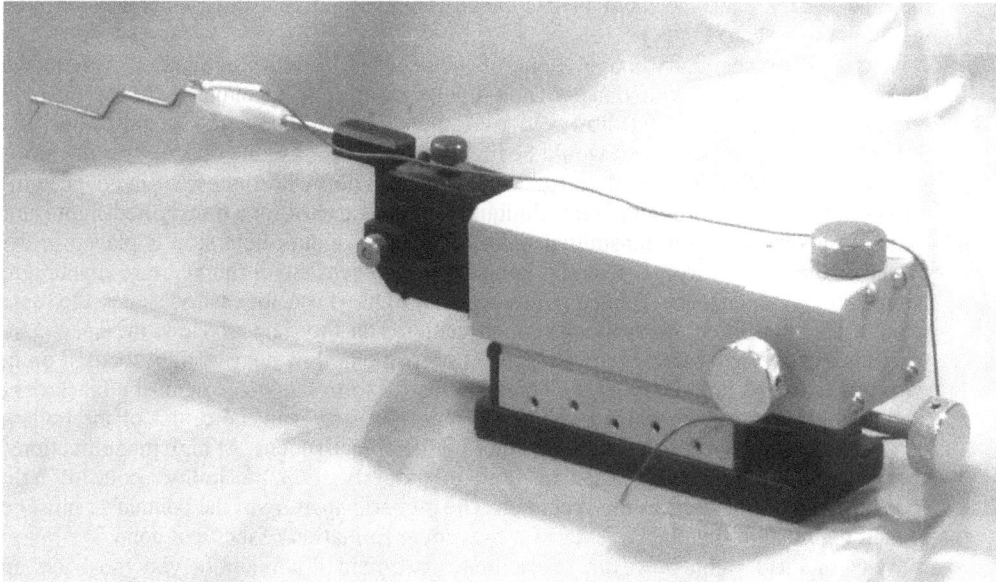

FIGURE 4.37   Electronic probing station needle probe. The view highlights the three-axis control knob configuration.

adjusted in the vertical plane to the correct focal point. This adjustment is critical for a probing station, and this will become evident in the next section, on lighting and advanced techniques. The IC stage also has adjustments for the x and y movements (in the horizontal plane).

The subject of advanced microscopic techniques is a difficult one to discuss from a microscope or tool standpoint, due to the vast differences in microscope design and configuration. Just as Ford and Chevrolet do not design and use compatible parts, microscope manufactures tend to have proprietary configurations for nearly all of the necessary attachments and options. To work effectively around this, it is necessary to describe in general terms some of the techniques with which the electronic failure analyst needs to be aware.

Lighting/illumination for the advanced electronic probing station is the first order of business. The standard illumination technique is Kohler illumination, which allows the microscopist to makes use of non-homogenous light sources like electric lamps. The setup is much like the set up for Kohler illumination in the transmitted light (biological) microscope. The goal is to provide focused, uniform lighting to the object being imaged. The manufacturer of your specific microscope will provide step-by-step instructions on how to accomplish the correct lighting setup. The light path from the microscope lamp fixture passes through a lamp condenser lens, an aperture diaphragm, one (or more) relay lens, through the field diaphragm, through an additional relay lens and, finally, into the microscope tube body. The light rays then strike a reflecting/transmitting surface within the microscope tube body and are reflected through the objective lens to the specimen. The illuminated image of the specimen is then passed back through the objective lens, along the microscope tube body (including the optics for the stereoscopic view) and, finally, through the ocular to the eye point. The basic principle of Kohler illumination is to center the light source in the aperture diaphragm and then to focus the image of the bulb filament in the plane of the aperture diaphragm. This provides a uniform illumination present at the plane of the field diaphragm. The resultant uniform illumination at the field diaphragm is then focused on to the specimen. (This typically happens automatically when the specimen is brought into sharp focus.)

The resultant specimen image is one of incident bright-field illumination. Due to the conditions of dealing with opaque, non–light-transmitting specimens, the bright-field effect is not readily apparent; however, were the specimen mounted to a glass slide, the field surrounding the specimen would be highly lit in contrast to the opaque specimen.

Incident dark-field illumination (dark-field illumination with the use of incident lighting) requires different illumination techniques from those used for a transmitted-light (biological) microscope. In a transmitted-light microscope, a dark-field stop is placed in the center of the light source, before the lamp condenser lens (with the aperture diaphragm opened to maximum), producing a hollow cone of light whose apex is focused in the same plane as the specimen. With the specimen centered in the field of view, the specimen, which is usually transparent or translucent, scatters the hollow cone of light focused on it, with this scattered light entering the objective lens to form the image against a black/dark background. With no specimen present, the objective lens is in the center of the hollow cone of light and therefore has no light entering the objective lens. At high magnifications, special substage condensers must be used to properly form the hollow cone of light focused in the same plane as the specimen. The numeric aperture of the condenser must be greater than that of the objective lens for the proper formation of the light cone.

With incident light, however, to facilitate dark-field illumination, you must use an objective lens that has been fitted with an annular condenser or with an objective lens containing a dark-field stop fitted to the center of the light path. Dark-field illumination, therefore, requires special objective lenses used specifically for dark-field microscopy. (The Leitz Wetzlar™ L 50X/0.60NA DF lens mentioned earlier is a dark-field objective, the DF indicating *dark field*.)

The final technique to be discussed is that of *interference contrast* illumination. A specialized form of interference contrast illumination is (Nomarski) *differential interference contrast illumination* (DIC). This technique is quite popular with integrated circuit analysts because of the amount of very fine detail that becomes evident using this technique. It also sometimes yields an appearance of surface relief. Like dark-field illumination, DIC requires special microscope equipment and adapters, and the specifics for setting up each microscope will be left to the microscope manufacturer. The appropriate use of DIC in integrated circuit analysis is described in the chapters on advanced IC failure analysis techniques and is not repeated here. The following is a discussion on the principles of interference contrast illumination. The basic principal behind interference contrast illumination is that a light beam from a homogeneous source can be split, one beam being used as a reference and the other beam being modified by a specimen, then recombined to form a complete image of the specimen. Due to the principal of constructive/destructive interference, extremely small changes in light wavelength are evident when combined with the reference beam. The detail within the image beam (one modified by the specimen) causes interference in the recombined beam, displaying the detail of the specimen. This principal is extremely sensitive to small changes in the image beam and depicts very fine detail at higher magnifications. This technique eliminates halos in edge structures, which are prevalent in phase-contrast microscopy. It produces very fine surface detail, introduces a color effect to the specimen, enhancing contrast, and gives the appearance of a three-dimensional surface to the specimen.

The interference contrast illumination effect is produced by an optical system that splits the light source into a reference beam and a object beam. In some systems, these beams are focused one (object beam) in the plane of the specimen the other focused above or below the specimen. This type system is typically referred to as a double-focus system. The (Nomarski) differential interference contrast illumination system uses a modified shear system to produce the reference and object beams. In a shear system, both beams are focused in the plane of the specimen; however, they are displaced horizontally, with the object beam illuminating the specimen, and are recombined prior to being presented to the ocular. The methodology for interference contrast illumination for the incident light (metallurgical) microscope was first worked out by M. Francon, making use of savart plates (double flat plates of quartz crystal) to generate split beams. This incident-light microscope interference contrast illumination process allows for the microscopic examination of three-dimensional objects such as metallurgical samples, plastics, and glass, as well as the composite construction of an integrated circuit die. This technique can be very effectively used by the electronics failure analyst to examine a diverse array of electronic samples.

4.7 REFERENCES

1. H.A. Petersen. "Failure analysis of astronaut EVA headset," FAME 91-3050, Failure Analysis and Materials Evaluation Laboratory, Kennedy Space Center, Florida, June, 1991.
2. John G. Delly. *Photography through the Microscope*. Eastman Kodak, 1988, 9th ed.
3. Brian Bracegridle. *Scientific Photomacrography*. BIOS Scientific Publishers, 1995 (part of the Royal Microscopical Society, *Microscopy Handbook 31*).
4. L.C. Martin. *The Theory Of The Microscope*. New York: American Elsevier Publishing Company, 1966.

5. Roy Morris Allen. *Photomicrography.* New York: Van Nostrand, 1958.

6. Sidney F. Ray. *Applied Photographic Optics.* Focal Press, 1988

7. Otto R. Croy. *Creative Photo Micrography.* Focal Press, 1968

8. R. C. Gifkine. *Optical Microscopy of Metals.* New York: American Elsevier Publishing Company, 1970

9. Photographic support for this chapter provided by Stephen W. Huff, photodocumentation specialist.

CHAPTER 5
X-RAY/RADIOGRAPHIC COMPONENT INSPECTION

Lawrence L. Ludwig, Jr.
Electronic Section, Failure Analysis and Physical Testing Branch,
Materials Science Division, Logistics Operations Directorate,
National Aeronautics and Space Administration

5.1 INTRODUCTION

5.1.1 Definition of X-Ray

X-rays can be defined as electromagnetic radiation with a wavelength from 0.1 to 10 angstroms (Å). The term X-ray is typically used to describe some form of *radiographic inspection* (RI). Radiographic inspection consists of an X-ray source, natural or man-made, which is directed at an object under inspection, with some means of viewing the resultant X-ray beam after its modified by the object under inspection. There are many methodologies for using X-rays for inspection as well as a vast many uses for X-rays. X-rays can be used for spectrometry, fluorescence analysis, X-ray crystallography, medical radiography and radiation therapy, absorptiometry, and industrial radiographic inspection, just to mention a few. The area that best describes what failure analysts use is that of *industrial radiographic inspection*. The analysts uses RI as a nondestructive aid in the failure analysis (FA) process.

5.1.2 The Purpose of X-Rays in Electronic Failure Analysis

X-ray or radiographic inspection, as used in the field of electronic failure analysis, is a multipurpose tool. It can be used as a purely nondestructive technique to determine the internal condition of a component, as a diagnostic tool to determine the degree of internal damage/degradation, or as an aid to dissection, depicting the relative proximity of each component to the other and the component housing.

In the guise of a nondestructive tool, radiographic inspection of a component prior to disassembly/dissection allows the investigator to evaluate the internal condition of a component and document that condition. By taking a pair of views 90° apart, the investigator has basic documentation of the components internal geometry. When the component is opened for visual inspection, an anomalous condition may then be recognized. This may be due to a fuller understanding of the components internal operation when viewed in three dimensions or because of the complexity of the construction obscuring the anomaly on its initial viewing, radiographically. The views taken prior to opening confirm that the anomalous condition was not due to the opening procedure but existed prior to the destructive phase of analysis. In this manner, the radiographic inspection becomes a nondestructive tool that validates the findings of the investigator.

As a diagnostic tool, radiographic inspection allows the direct examination of printed wire board assemblies (PWBs) for cracked solder joints, separated LANs/runs, and a vast array of other manufacturing anomalies. Components such as resistors/capacitors can be examined directly for correct internal construction, and stressed components sometimes are evident from direct examination such as bond wire separation on ICs. Ceramic-body fuses can be examined for electrical overstress, and indications of mechanical damage or the result of a fuse holder anomalies can be seen from radiographic inspection.

Radiographic inspection (RI) can be a tremendous aid to dissection/disassembly. If an object is to be windowed so a visual inspection can be performed, then RI can locate the best area for the window. If a component is to be opened for inspection yet kept in a functional condition, RI can identify the parts of the external housing that are used as support for the internal components mechanism. RI allows the investigator to check the clearance of internal components prior to dissection. Removal of the outer housing by cutting through an internal subcomponent is very difficult to describe eloquently in the final report (oops!) and can have a huge impact on the outcome of the investigation.

5.2 HISTORY OF X-RAYS

5.2.1 The Discovery of X-Rays

In the late 1890s, the scientific community felt that all the advancements in the realm of materials were known and that only improvements in accuracy of experimental data were to be gained. As is often the case with science, a plateau had been reached—a barrier blocked further scientific findings, and the door to the next level was not apparent. A paradigm shift would have to occur, and that door would have to be recognized before it could be opened.

Preceding the discovery of X-rays, Volta (1800) had developed the electric battery, Oersted (1820) had linked electricity and magnetism; Ampere (1820) had mathematically formulated the electrical/magnetic relationship; Ohm (1827) had developed the relationship between current, resistance, and voltage; and Faraday (1831) had discovered electromagnetic induction. By 1850, Plücker had observed fluorescence opposite the negative electrode in a charged vacuum tube, Gassiot (1859) and Crookes (1879) had experimented with "cathode rays" in vacuum tubes. Maxwell (1873) had published his "Treatise on Electricity and Magnetism," and Hertz (1885) had proven Maxwell' s equations experimentally and had demonstrated the passage of cathode rays through thin metal foils (1892). You can see from these scientific accomplishments that the scientific community could become complacent. They had a working model of the "atomic" structure of materials and had a real handle on most of the fundamental electrical principles. The experimentation with "cathode rays" proved to be the method of identifying the door to the next level. By the late 1890s, the door had finally been identified; now the paradigm shift was needed before Wilhelm Röentgen and J.J. Thompson could open the door. Jean Perrin (1895) and J.J. Thompson (1897) both showed that "cathode rays" carried an negative charge and could be deflected with a magnet. The scientific community was just not willing to accept a new form of matter, and the vast majority viewed "cathode rays" as a wave-like vibration. It was not until Thompson, with a much improved vacuum system, proved that "cathode rays" could be manipulated with a magnetic field, that support for "cathode rays" as energetic charged particles could be developed. In 1895, Wilhelm Röentgen, during experimentation with "Crookes tubes" and "cathode rays" discovered X-rays. He had carefully repeated experimentation with protected (covered) photographic plates and was baffled by their fogging (being exposed even though covered). One evening, he returned to a darkened room where he had forgotten to turn off his "Crookes tube." He found that a sheet of paper covered in barium platinocyanide was glowing. Papers covered in barium platinocyanide were known to fluoresce in strong direct light. The fact that the paper was glowing indicated that some form of ray was striking the paper and was easily traced to the "Crookes tube." Röentgen quickly put the data from the photographic plates and the glowing paper together and was soon well on his way, conducting experiments on the nature and properties of X-rays. He was awarded the first Nobel prize in physics for his work on X-rays in 1901.

The two properties that Röentgen discovered about X-rays were that (a) they can easily penetrate opaque bodies and (b) that they can ionize a gas into its two oppositely charged parts. Today, we know that X-rays are just the same as the electromagnetic radiation we call light, only with a much shorter wave length. J.J. Thompson and his young assistant Ernest Rutherford found that X-rays split gasses into an equal number of positively and negatively charged atomic particles. The negatively charged particles were eventually labeled electrons. Thompson and Rutherford eventually viewed the positively charged particles as atoms with their electrons removed and concluded that the X-ray "photon" had given up its energy in removing the electron from the atom. From this beginning, the whole of particle physics was born. X-rays and ionized gas opened the door, and Rutherford and his contemporaries walked through it into the scientific particle theories we know today.

5.2.2 X-Ray Fundamentals

The generation of X-rays is a very understandable process from a molecular viewpoint. An electron has a known charge, and it also has mass. If it is in motion, it then has an electrical energy potential and a kinetic energy potential. When such a particle strikes a solid (atom stationary in a homogeneous surrounding), it gives up its energy in the form of light and heat. If the speed of the electron is sufficiently large, then the light generated is of such a short wavelength that it is in fact in the form of X-rays. Therefore, the generation of X-rays requires a source of electrons that can be accelerated to specific level (speed) and a target material with which to collide. Electrons generated in a standard environment would collide with air molecules shortly after having been generated. As such, trying to accelerate the electrons to a specific speed in air would be useless. Therefore, an evacuated chamber or tube is necessary to accelerate the electrons without causing them to strike or recombine with the atoms of air. The method for accelerating an electron is to provide a large voltage potential (strong E field) aligned along the longitudinal axis of the tube. The electrons feel the E field due to their charge and accelerate in the appropriate direction.

Electron generation is a somewhat erratic process—electrons coming from the source may start with an initial velocity in any direction. It is desirable for the electrons to achieve peak acceleration and strike the target at the opposing end of the tube rather than striking the sides of the tube. To achieve this, the use of a device to focus the direction of the electrons on the target is required. A series of orthogonal magnetic fields would allow you to focus the electrons as they move down the tube to the target. At this point, we have an evacuated tube with an electron source, a target material, an accelerating voltage, and a focusing device to ensure that the electrons strike the target. If the tube is made of a large, thick material, the X-rays generated when the electrons strike the target will be absorbed to some degree by the tube itself. The target should be positioned or orientated in such a manner that X-rays being generated exit the tube through a window (thin foil of a specific metal) that will absorb a minimum on the X-rays generated. As mentioned earlier, when electrons strike a solid, they give off light and heat, with more energy producing heat than light. Therefore, the target material must be of a material with good thermal characteristics (high melting point and high thermal conductivity), such as tungsten, which melts above 2540 K. It is also desirable to have a relatively small focal point (for image resolution purposes) as the point of impact for the electrons. Even target materials such as tungsten may then need some method of cooling. Having a portion of the target extend outside the tube would allow the target to be hollow and for water to be jetted into the interior of the target, cooling the target material. At this point, it appears all the requirements for an device to generate X-rays have been identified, as follows:

- an evacuated tube with an electron source
- an accelerating voltage
- a focusing device to ensure a small focal area
- a target material, cooled and correctly oriented to a window for emitting the X-ray beam

5.2.3 X-Ray Spectra

So far, the discussion has been centered around manufacturing of X-rays for RI. There are natural sources of radiation: radioactive materials (isotopes) that emit gamma radiation or neutron radiation. Gamma rays are essentially the same as X-rays, the differences lying in that gamma radiation has an even shorter wavelength than X-rays. As such, gamma radiation can be used as an X-ray source for field and lab radiographic inspection under appropriate conditions. Gamma radiation is normally used in field equipment where X-ray sources and power supplies would be cumbersome. It has the additional advantages of being monoenergetic (having a well defined spectral distribution), it requires less shielding, and typically has very stable sources. The disadvantages lie in that gamma radiation sources are always "on," are usually of a lower intensity, have a large effective spot size (poor focus), and require the administrative hassles of additional licenses. The smaller wavelengths do not affect photographic media as well and generally provide a lower contrast image radiograph. Gamma radiation sources work well, because of their wavelength, in industrial radiographic inspection where large metal/dense parts need to be inspected and long exposure times can be arranged.

Neutron radiation (n-ray) is an entirely different method of radiographic inspection. The process of absorption for neutron radiation is much different from that of X-ray absorption. From practical experience, the value of n-ray is that less dense materials (rubbers/plastics) encased in dense metal housings can successfully be imaged.

So far, in this discussion on X-ray fundamentals, wavelength has been mentioned repeatedly but not specifically discussed. The wavelength of X-rays determine the degree to which they penetrate materials. What determines the wavelength of an X-ray beam?

Two considerations affect the spectrum of the radiation emitted from an X-ray generator. The first is the line spectra of the target material. Each pure target material will yield a characteristic radiation or X-ray wavelengths based on its molecular structure. In most cases, only two of these spectra will have sufficient intensity to be apparent. They typically appear as high-intensity pulses or emissions at two characteristic wavelengths, $K\alpha$ or $K\beta$, with the Kβ wavelength being 15 percent smaller than the Kα. There is a critical voltage level below which the characteristic radiation peaks do not emit.

The second consideration that affects the wavelength of the X-ray beam is the accelerating voltage of the tube. The waveform of the X-ray spectra (the continuous spectrum) is based purely on the potential applied at the tube and the numbers of electrons accelerated by this voltage. For a constant applied voltage and fixed tube current (electron stream down the tube), the spectrum is fixed at a specific short wavelength value, with the spectra tailing off in intensity toward the longer wavelengths. The maximum intensity of this waveform is approximately proportional to the atomic number of the target. For very light (atomic weight) targets, the intensity of the characteristic spectra may be much greater than that of the continuous spectrum. This information is more applicable to those doing X-ray diffraction than those of us performing RI. Most RI systems have a heavy (atomic weight) tungsten target due to the heat considerations. What this spectra information means to the FA investigator is that, in an adjustable voltage and tube current system, as kilovolt levels increase, X-rays of shorter wavelength are applied. These shorter wavelength rays have greater penetration power. As tube current increases, intensity also is increased via the higher number of X-ray "photons" that are radiated to the object under inspection. This is analogous to using the kilovolt level for penetration power and milliampere settings for image contrast.

5.2.4   X-Ray Tubes

Previously mentioned was the ability to adjust a tube for voltage and tube current. Depending on the age and design of the X-ray tube, the operational parameters will vary. The first types of tube designed were called *ion tubes.* These devices used a voltage potential applied across the tube to ionize a gas at low pressure. The positive ions would then collide with the cathode releasing electrons that would be accelerated to the target at the opposing end. From the preceding discussion of X-ray spectra, it is apparent that, in an ion tube, the level of vacuum (gas content) affects the current through the tube and is also interrelated to the applied voltage. None of these three parameters are adjustable independently. This vastly limited the usefulness of these tubes for RI. They did find a widespread application in precise spectroscopic and radiographic diffraction equipment where a controlled spectra was required.

The other type of tube is the *hot filament tube* or *electron tube.* This type requires an independent source of electrons and an extremely high level of vacuum, roughly below 10 millibars. In this type of evacuated tube, the electron source is based on thermionic emission. A hot filament positioned directly in front of a parabolic shaped cathode expels electrons, with the cathode focusing the electron beam onto the target/anode. The first such tubes were called *Coolidge tubes* (designed by Dr. W. D. Coolidge at General Electric), were of the permanent sealed variety, and had an operational life of approximately 2000 hr. The main advantage to these tubes was that the level of vacuum was fixed, and the tube current and applied voltage were independent of each other. During the late 1960s and early 1970s, RI equipment used the sealed type tube and the crystallographic X-ray analysis equipment used the demountable X-ray tube. The demountable tube allowed for changes in the target material as well as the foil X-ray window on the tube. This allowed those doing crystallographic X-ray analysis to use targets and matched filtered windows to obtain monochromatic radiation.

Later advancements in the sealed tube made use of flat, thin discs or plates as targets, which would aid in the dissipation of heat. As an advancement in decreasing the focal spot size, line filament emitters were used in conjunction with angled rectangular anodes. The angle (approximately 19°) on the rectangular node makes the incident X-ray beam look like it is emitting from a square anode. Another major breakthrough in tube design occurred with the advent of the rotating anode tube. The rotating anode is a large circular disk with an angled outer radius on the end of a sealed rotor winding. An induction stator winding placed around the lower section of the glass tube caused rotation in the rotor and anode. The focal spot would be aligned to the center of the angled radius, and the incident beam would be projected out the tube window at the object under inspection. The size of the anode, and the fact that the target edge was being uniformly heated around all 360° of rotation, allowed a much larger amount of power (higher tube current) to be dissipated in the anode. Subsequently, a great deal more X-ray radiation was generated. This type of tube is popular to day where high-wattage X-ray generation is needed.

The early demountable tube designs were mainly concerned with changing of the target and windows. With advancements in vacuum technology (the development of ion and turbo pumps), very high levels of vacuum could now be used for demountable tubes. Better anode designs for demountable tubes were derived that made use of recirculated liquid cooling. This meant that more heat could be dissipated by the anode, and therefore more X-rays per target area could be developed. With advancements in electronic control systems, the focusing field coil systems could develop uniform, small focal spots. Adding all these together led to the advent of the *microfocus tube.* Microelectronic components have very small geometries. The average bond wire is of 0.005 in. dia. Spot sizes of less than 10 μm (3 to 5μm typical) microns can be generated and therefore relatively large magnifications can be achieved. For RI work involving electronic components, the microfocus system is the tool of choice.

The next section looks at the relationship between spot size and resolution.

5.2.5 The Geometry of Radiography

The geometries involved in obtaining a radiographic image using film, real-time radiography, or fluoroscopic inspection all hinge around the lack of "optics" for X-rays. Once an X-ray has been generated, it progresses in a straight line until absorbed, i.e., the X-ray photon energy is transferred to some form of matter at an atomic level. Some minor reflection of X-rays can be accomplished but, due to the X-ray wavelength being of the same order of magnitude as the atomic spacing in matter, no real collimation of an X-ray beam can be obtained. All forms of electromagnetic radiation (light, radio waves, X-rays) are deflected by neither magnetic nor electrical field. This means that the only control available for X-ray beams is the size of the spot used to generate them. The following figures show the difference in resolution based on focal spot size. Figure 5.1 shows an object under inspection in an X-ray system with a large focal spot size. The focal spot is of a diameter D, having a uniform brightness (X-ray intensity). The rays generated by the focal spot are emitted and to some degree absorbed by the object being inspected, casting a "shadow" at the imaging device. The rays pass the corner of the object, given by distance A, and then proceed to the image screen at a distance B. It can be seen that combined rays, rays originating from opposing sides of the focal spot, cast an indistinct shadow or penumbra that obscures the detail of the corners of the object. The width of the penumbra is given by $P = D(B/A)$. For a spot size of 0.1 in., a distance A of 5 in., and a distance B of 10 in., the width of P would be 0.2 in. The magnification is given by the formula $M = 1 + (B/A)$, or in this case 3L, L being the original object size. If we consider the distinct, sharp edge of the object and its width, L, the corners on the object are only evident (resolved) if the distance between the

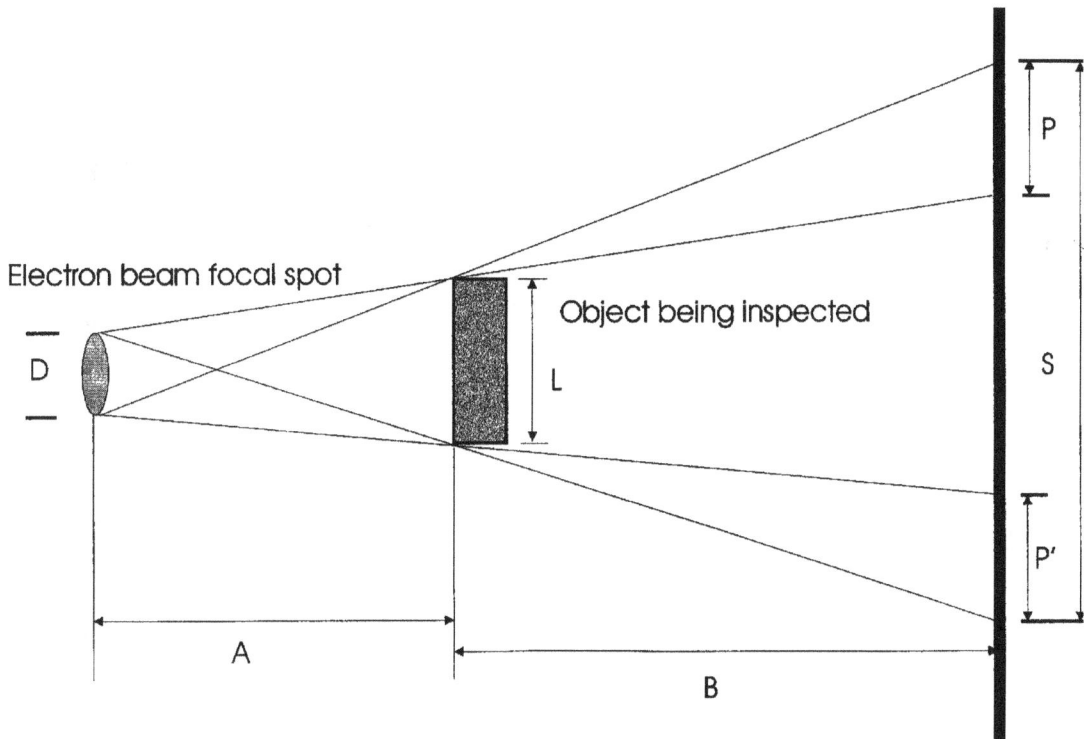

FIGURE 5.1 Radiograph resolution with a large focal spot size.

centers of the two prenumbras (P and P′), given by L(B + A)/A is larger than the width of the prenumbra D(B/A). From this relationship, the minimum resolution R can be calculated by the equation R = D[B/(A + B)]. For this particular example, the minimum resolution is R = 0.067 in.

If the spot size were to shrink by an order of magnitude (D = 0.01 in.), the minimum resolution would change to R = 0.0067. The minimum resolution is therefore directly proportional to the spot size. Contrasting Fig. 5.2 with Fig. 5.1 illustrates this relationship graphically. The most important information concerning the geometry of a radiograph is that the magnification is directly related to the objects distance from the generator (focal spot) and its distance to the imaging device, and that the resolution or sharpness of an image is directly proportional to the focal spot size and interrelated to the amount of magnification.

5.2.6 X-Ray Generators

In Sec. 5.2.2, the basic requirements for an X-ray generator were discussed, with these requirements identified as an evacuated tube with an electron source, an accelerating voltage, and a focusing device to ensure a small focal area, and a target material, cooled and correctly oriented to a window for emitting the X-ray beam.

Within the scope of the discussions on tubes (Sec. 5.2.3), the advancements in vacuum systems, anodes, anode cooling, and focusing systems were all presented. The image

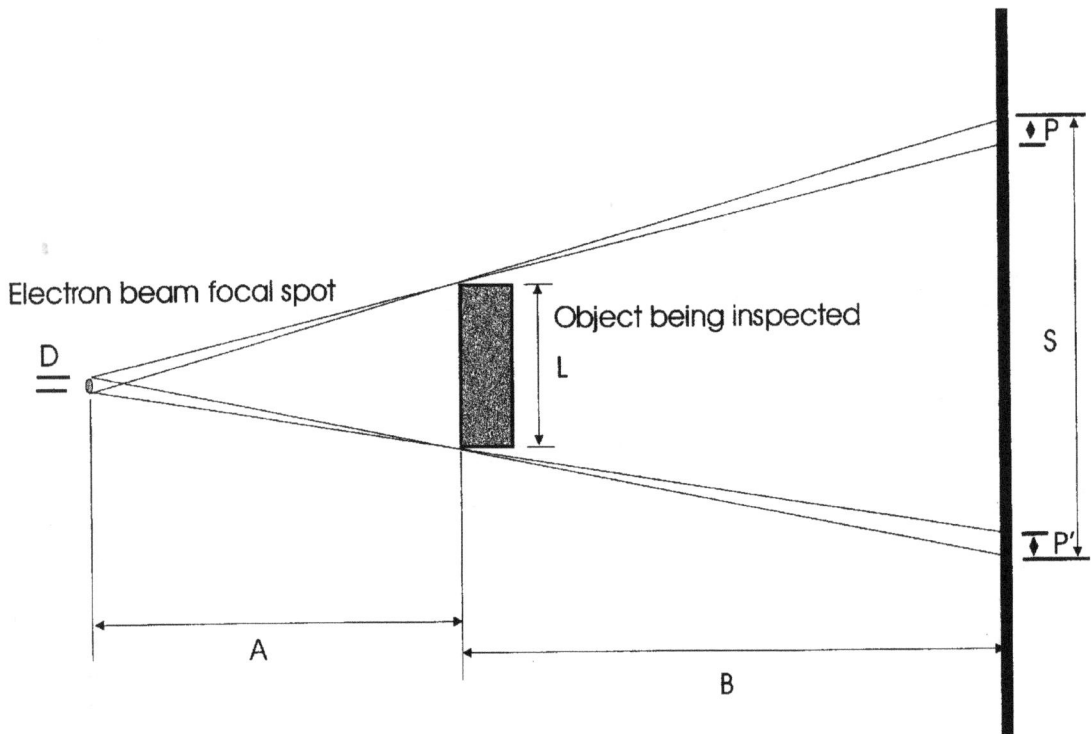

FIGURE 5.2 Radiograph resolution with a small focal spot size.

geometry related to the focal spot was discussed in the preceding Sec. 5.2.5. Information pertaining to the X-ray spectra and selection of target materials was presented in Sec. 5.2.4. This section will discuss the remaining aspect of X-ray systems: the electrical generator systems that apply the accelerating voltage to the X-ray tube assembly.

The first types of generators were tubes that were single phase, self-rectified, and as such operated on ac power, usually stepped up from a voltage step-up transformer. Under normal operation, on the positive half-cycle, the electrons from the filament/cathode assembly were accelerated toward the anode, and on the negative half-cycle the system did not conduct. The problem with these systems was that they emitted X-rays only during the positive half-cycle and that the they were relatively low-power devices. If they were operated at high voltages, there was the danger of the anode becoming too hot and emitting electrons. These electrons could end up bombarding the cathode during the negative half-cycle and generating X-rays at the cathode end (in a direction never intended and possibly not shielded). This could shorten the cathode life, the filament life, and the operator's life.

The next type generator developed was the single-phase, half-wave rectified unit. Early devices used rectifier tubes or "valve tubes." These devices were very similar in design to X-ray tubes and had to be operated below a saturation current level or they too would generate X-rays. Their function is the same as today's high-voltage diodes or SCRs. The problem, again, with half-wave rectified generators is that they generate X-rays only on the correct polarity half-cycle. This produced a nonuniform X-ray beam and, due to the pulse of the tube current to the target, caused anode heating problems. The higher pulsed currents heated the target more than lower steady-state currents. With the advent of the full-wave rectified circuit (first called the *Gratz circuit*), a more stable X-ray output was developed. Later full-wave circuits made use of "condensers" (capacitors) to filter the voltage waveform, applying a smoothed dc potential to the tube.

Many advanced rectification schemes (half-wave and full-wave) were used to apply a smoothed potential to the tube. Some used capacitors in combination with "valve tubes" to double or triple the transformer voltage. As high-voltage technology and solid state technology developed, the three-phase, full-wave rectified generator came into being. These generators can provide very large operating potentials. Systems have been developed with 1000 kV and higher potentials. The state-of-the-art generators of today make use of three-phase full-wave rectification. The more advanced generators often convert 60 Hz input power to 720 Hz potentials to take advantage of the increase in stability it provides in the output X-ray intensity.

5.3 RADIOGRAPHY

5.3.1 Common Nomenclature

The following is a list of terms and their definitions as pertaining to RI. These terms should help the electronic component failure analyst to be comfortable around the RI of a component, understanding the operations and adjustments made during the examination.

Penetrating radiation Radiation, either particles or electromagnetic waves, that penetrate through a measurable thickness of matter. Examples of electromagnetic waves include the high-energy photons like X-rays and gamma rays; lower-energy photons like infrared, ultraviolet and visible light, and radio waves (LF, HF, and μwave). Examples of particles include cathode rays (electrons), which only penetrate through thin materials, neutrons, and other energetic charged subatomic particles. Also included as penetrating radiation are sound waves (or phonons).

Intensity A measurement of the energy carried by the X-ray photons over a fixed time into an object of known surface area. Typically refereed to as *beam intensity,* it is a reference for the amount of X-rays generated. The unit of measure for intensity is the röentgen, although it can be measured in ergs per square centimeter per second or watts per square centimeter. Intensity is a measurement of energy flux in the beam and as such is also a measure of the X-ray generator output.

Dosage The amount of X-ray energy delivered to an object under exposure to the X-ray beam. The unit of measure can be the röentgen or the rad, which is defined as an energy absorption of 100 ergs per gram of irradiated material. The dosage typically refers to amount of X-rays absorbed by the object under inspection. When dosage concerns the levels of radiation to which a human being has been exposed, the unit of measure is the röentgen equivalent man (REM). Realizing that the differing types of radiation will have differing effects in human tissue, a normalizing factor must be used to standardize the dosage. The radiation biological effectiveness (RBE) is the method used to standardize radiation dosages. The RBE is the ratio of rads of 250 kV X-rays to rads of applied radiation, to give the same effect (rads 250 kV X-rays/rads of applied radiation). Thus, if a known radiation exposure is normalized by the RBE, the unit of measure is given in REM.

Absorption The reduction in beam intensity as an X-ray beam travels through a given media. Each material has an absorption coefficient μ and a mass absorption coefficient μ/ρ, where ρ is the density of the material.

Scattering   The mechanism for deflection of X-rays as they pass through a given media. The basic principle of scattering is that X-rays are electromagnetic waves that, when striking an molecule with an electron, accelerate that electron in an oscillatory manner that is directly proportional to the energy of the absorbed X-rays. Electromagnetic theory requires the electron to radiate electromagnetic energy at the same frequency as that of the incident beam—however, not necessarily in the same direction. Therefore, the X-rays strike an electron and are reemitted in a new direction, or scattered.

Image   A two-dimensional pattern of visual information, typically representing some characteristic of the object imaged.

Radiograph   Conversion of X-rays to a visible image formed by passing the radiation through a specimen. Radiographs make use of the differing absorption levels of materials, depending on their density, blocking the incident X-ray beam just as light is blocked by opaque bodies. As such, a radiograph can be viewed as a shadow image projecting the varying density of the specimen.

Real-time radiography (RTR)   Radiography in which the visual image of the penetrating radiation passing through the specimen is presented to the eye within a time interval that is smaller than the integration time of the eye, or at the speed of the imaging system (typically 33 to 200 ms). This means that, as the object is moved, translated, or rotated, the image is updated at a speed that permits visual examination of the component for anomalies.

Digital radiography   Radiograph in which the image area is divided into small elements (pixels) with an integer number used to represent the intensity (in gray levels) of the pixel. The whole image is saved in this manner to some form of computer media (disk, tape), thereby representing the whole image as a digital image.

The usefulness of digital radiography lies in the computer enhancements that are available to reveal detail about the object inspected.

Sensitivity   The ability to detect a small anomaly in the image or radiograph. The measure of sensitivity is based around the actual size of an anomaly detected.

Resolution   The ability to differentiate or identify between two or more small, closely spaced anomalies in the image or radiograph; a measured characteristic in which line pairs are differentiated visually up to a given number in line pair per mm.

Contrast   The difference in visual intensity (gray level) between two points on the image or radiograph and their average intensity. It can also be defined as the difference between the average intensity of the two points and the highest intensity of the points. *Contrast* implies that a sharp delineation between differing levels of density in the object under inspection is readily apparent in the radiograph or image.

5.3.2   Image Quality/Image Quality Indicators

The quality of a radiograph is a purely objective determination. The radiograph imaging system must have the sensitivity to detect the possible anomalies in the area of interest. The image system must also have the resolution to differentiate between anomalies located close together. It must have sufficient penetration to image the area of concern yet not wash out the details sought. The contrast between dense and less dense areas must be such that both areas are correctly displayed visually.

If any of these factors are poor, or lacking, then the quality of the radiograph is questionable. To ensure that a radiograph can image a defect of a given size, known standards are used that have artifacts of a known dimension. These devices are called *image quality indicators* or *penatrameters*. The different styles include line penatrameters, hole or thickness penatrameters, step-wedge penatrameters, and *electronic image quality indicators* (IQIs). Line penatrameters consist of wires of a known medium and diameter that are fixed at a set distance apart. These lines can be set on an object under inspection of the same material, and the resultant radiograph gives a clear indication of the object and, hopefully, the penatrameter. Hole penatrameters are similar to wire penatrameters but are read in a different manner. The hole penatrameter has a specific number that represents its thickness. It also has a series of holes that represent 1T, 2T, and 4T, these being the thickness of the penatrameter with the diameter of the 1T hole the same as the thickness of the penatrameter material.

Figure 5.3 shows a no. 5 stainless penatrameter on a 0.125-in. thickness of aluminum. The 4T and 2T holes are evident. The 1T hole is not visible, determining that resolution can only be related to the 2T hole size. The 2T diameter is twice the penatrameter thickness, and the 4T is four times the thickness. The wedge-step penatrameter has stepped wedges of the material under test with holes drilled into each wedge in the shape of a number that represents the step thickness. The electronic IQI is a blend of the line, hole, and step penatrameter. The electronic IQIs come in a set of eight, with an outer thickness of stainless steel that varies from 0 to 1/16 in. In this regard, the electronic IQI models the step penatrameters. The electronic IQI also has a series of wires crossed at 90° angles. There are three wires in each direction (six total), with their diameters decreasing in size so that the bottom pair of crossed wires are the smallest in diameter. The wires are also offset in the thickness with the largest diameter being roughly on one face of the IQI and the smaller two wires offset at the center and at the opposing surface.

The same is true for the crossing set of wires. Figure 5.4 shows an optical photograph of a no. 1 and no. 2 electronic IQI, a no. 10 stainless hole penatrameter, a lead line pair per millimeter phanom, and an eight-step aluminum step gage. In addition to these two func-

FIGURE 5.3   Number 5 hole penatrameter shown through 0.125 in. of aluminum. Note that only the 4T and 2T holes are evident.

FIGURE 5.4 Optical photograph of a no. 1 and no. 2 electronic IQI, a no. 10 stainless hole penatrameter, a lead line pair per millimeter standard, and an eight-step aluminum step gage.

tions, the electronic IQI has a series of spherical particles placed above and below the crossed wire set with the maximum diameter being the same as the largest wire and incrementing downward in size for five additional steps (for a total of six spherical particles).

A penatrameter serves two functions in radiography. The first of these is to determine if the selected wavelength is sufficiently penetrating the object under inspection. The penatrameter is placed in the vicinity of the thickest/most dense portion of the object under inspection. If this penatrameter is visible on the radiograph then X-ray penetration through the specimen did occur.

The second thing a penatrameter does is to determine the sensitivity of the radiograph. The smallest hole, wire, or other item imaged, expressed as a percentage of the object thickness, yields the penatrameter sensitivity of the radiograph. The smallest hole, wire, step, or particle evident in the penatrameter determines the size of the smallest anomaly that can be detected from the radiograph. Figure 5.5 shows a no. 3 electronic IQI behind a populated PWB. The three arrows point to two sample particles and a visible line, all indicating a specific image resolution level. Figure 5.6 shows the smallest particle sizes on the IQI. The smallest particle (left arrow) is just beyond the level of perception on this print.

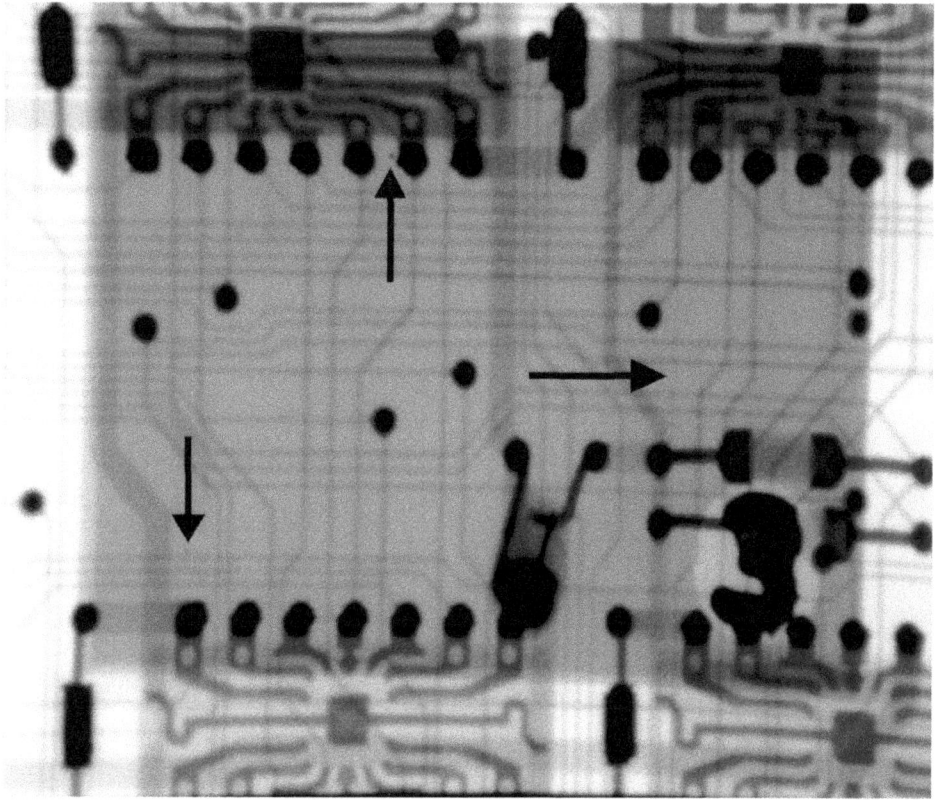

FIGURE 5.5 Magnified radiograph of a no. 3 electronic IQI in use. Arrows note the locations of two sample particles and a visible line.

5.3.3 An Overview of the Types of Radiographic Devices

A basic discussion touching on isotropic sources and types of X-ray tubes has been conducted, along with discussions and definitions on image quality, sensitivity, resolution, and contrast. In this section, a brief discussion of the various types of X-ray imaging devices will be conducted.

In all types of imaging systems, a standard process must be performed. Some device must intercept the X-ray beam and convert the X-ray photons to some form of electrical excitation. This electrical excitation will most likely be amplified, stored/accumulated, converted to visible light, and then, in the case of digital radiography, digitized and processed.

The types of imaging devices can be broken down into three categories; X-ray image intensifiers (XRII), video fluoroscopy (VF) systems, and X-ray videocon (XRV).

For the XRII system, in addition to the image intensifier, a collimator lens, mirror, camera lens, and video camera are needed. The XRII consists of an input window (thin metal window), an initial X-ray sensor combined with a photocathode, an anode, and an output phosphor device with output window. The initial X-ray sensor translates the X-ray energy

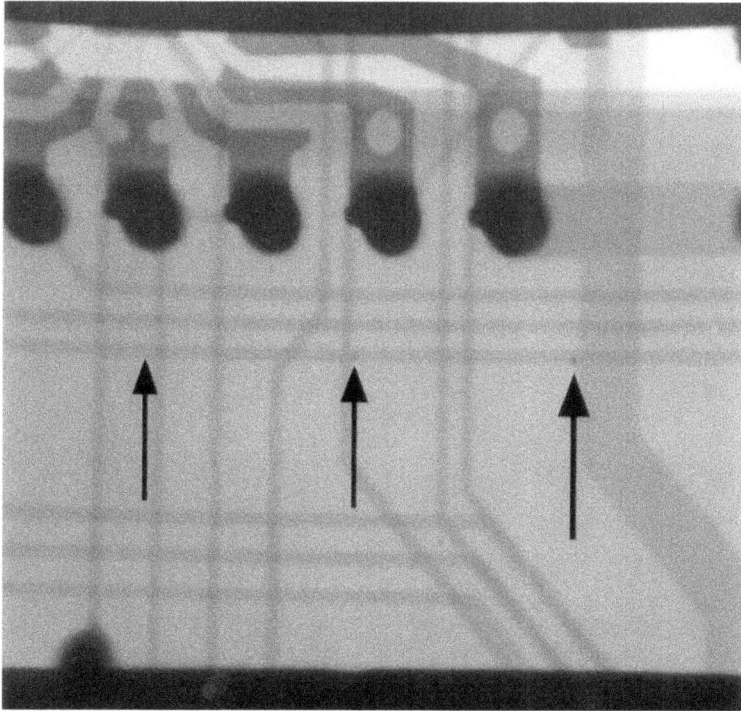

FIGURE 5.6 Magnified radiograph of a no. 3 electronic IQI depicting the lower particle set. Note that only two of the three are evident, determining the level of resolution.

to visible light photons that are used as stimulus to the photocathode, which the photocathode converts to a free electron. The free electrons are accelerated by a series of focusing electrodes that compress the field of view, also giving a gain factor in light intensity. The electrons strike the output phosphor, producing a larger visible light image. This light image is collimated and presented to an output video camera. The XRII has the ability, by changing the focusing electrode potentials, to change the field of view so that an effective magnification of the image is produced.

For the video fluoroscopy system, the main component is the screen (scintillator). In addition, a mirror, intensifier lens, optical image intensifier (optional), camera lens, and video camera are used. In the video fluoroscopic system, a direct conversion of X-ray energy to visible light is made by the fluorescent screen, and a mirror is sometimes used to relay the image to a intensifier lens (optional) and then to an optical image sensor (optional) and/or camera system. The advantages of video fluoroscopic systems is that they are typically highly sensitive and have excellent resolution and relatively low cost. There are also many different types of systems. The basic system makes use of the fluoroscopic screen and nothing else. Mirrors and relay lenses can be added. The light-intensified fluoroscope makes use of a light intensifier transducer before presenting the image to the observer. The light-intensified fluoroscope can be used with optics to magnify the field of view or shape the image prior to presentation to a camera system. A low-level light TV (LLLTV) fluoroscope uses the more sensitive lower-light-level fluoroscopic screens in conjunction with a low-light-level camera system. In combination with this system is the light-intensified LLLTV fluoroscopes, which add an optical light intensifier when the screen output is below a certain threshold value.

For the XRV system, all that is required is the video camera with XRV tube. An advantage to XRV is the resolution of the image. The disadvantage is the field of view (from the camera), 3/8 × 1/2 in.

In addition to these types of imaging systems, new technology is being developed and coming onto the market continuously. The microchannel plate system makes use of a fluorescent screen, a fiber optic system coupled to a microchannel plate. The microchannel plate output is applied to a cathodoluminescent screen with the result sent to a video camera system or LLLTV system. The advantages for the microchannel system lie in superior image resolution and image contrast. The main disadvantage lies in the 3-in. dia. screen size.

The scintillator array systems use a scintillator array along with photomultipliers to scan an object under radiographic inspection and use conversion circuits to build the final image. The advantage is the image resolution; the disadvantage is the scan time and processing time for the image. The scintillator also does not have as good an image contrast as other systems. The last imaging system discussed is the semiconductor (diode) array (with and without a fluorescent screen). The diode array is similar to the scintillator array in that it may be an area array or a line array. If it is a line array, then the object needs to be translated to build the scanned image. A fluorescent screen can be used with diode arrays, the diode array being sensitive to light photons rather than X-ray photons. The advantages to the diode array is the high-resolution image; the disadvantage is the field of view, 1 × 1 in.

5.3.4 Film Radiography

The first experimental use of X-rays, back in 1895, involved the use of photographic plates as sensing media. The use of film to capture the X-ray photon (to depict the relative change in intensity due to objects under inspection as a shadow graph of proportional darkness) has been a tool to both the medical and industrial professions for the last 100 years. It is highly doubtful that film radiography will go away.

For the electronic component failure analyst, the use of film radiography most likely will not be a regular necessity. Today, there is an abundance of real-time radiographic systems. The advent of the microfocus, real-time system has made nondestructive examination of electronic components a much easier and productive process. However, at some point, you may find film radiography a necessity. There are still some applications were long-term exposures in the field or radiography using isotopes may become necessary, mandating the use of film.

One of the primary differences between standard photography film and X-ray film is the thickness of the film. Radiographic film is thicker because X-rays produce a more equal distribution of the grain throughout the film thickness, thereby producing a darker exposure. In addition to being thicker, radiographic film is usually coated with the emulsion medium on both sides of the film.

Three factors of film radiography are critical to successful imaging. These are frequency, intensity, and exposure. Most emulsions of a silver composition have a distinct sensitivity to X-rays of a frequency higher than the absorption frequencies of silver, and they are less sensitive to X-ray frequencies lower than these absorption frequencies. The analyst must remember that elevating the kV potential shortens the wavelengths of the X-rays produced. At some very high energy level (kilovolts), the X-ray beam has a lesser effect on X-ray films because the high-frequency X-rays affect radiographic film to a lesser extent. Therefore, the choice of X-ray frequency (kilovolt level) used by the analyst is important to the production of useful radiographs and must be balanced against the need for object penetration.

For film radiography, as with regular photography, the intensity of the "light" source is a critical factor. The exposure time and the X-ray beam intensity are interrelated. If you define the exposure of a film to be the time exposed to the X-ray beam multiplied by the intensity of that beam, you have established a primary characteristic of a developed radiograph. The result of exposure (beam intensity × time) is the darkness of the radiograph or density, where density is defined as the $\log_{10}$ of the incident light applied to the front of a developed radiograph, divided by the light passed through the radiograph ($D = \log_{10} L_{incident}/L_{passed}$). By plotting a graph of the $\log_{10}$ of exposure (E) on the x-axis and the density (D) on the y-axis, the characteristic curve of the film is displayed. The lower portion of that curve starts out with an area at which the exposure is insufficient to cause any density change (no exposure). As the exposure increases, a slight asymptotic rise occurs. This area is the underexposed area of the film. The next portion of the curve has constant slope and is a straight line upward toward a maximum density, D_{max}. This is the correct exposure range of the film. The horizontal distance between the start and end of this section of the curve is the latitude in which correct exposures can be made. The top end of the curve appears like the top portion of a gradual sloped parabola. This is the region of over exposure, and the area with a negative slope is the area of reversal.

Figure 5.7 shows the characteristic curve of radiographic film for a typical film. D1 is the minimum density level in the correct exposure part of the curve, and D2 is the maximum density in the correct exposure part of the curve. The vertical difference between D1 and D2 is the maximum contrast of the radiograph for correct exposure. The slope of the curve in the correct exposure region is the contrast of the film. The speed of the film is

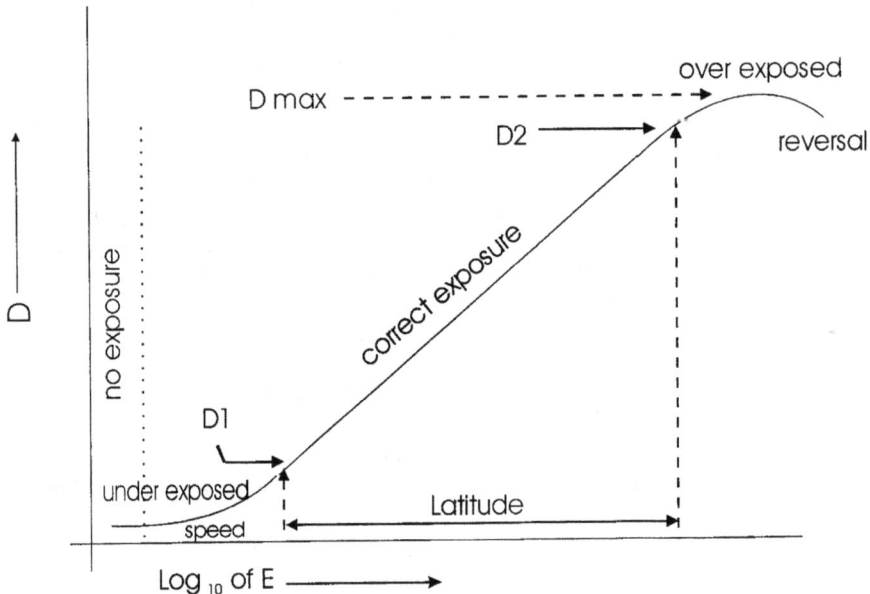

FIGURE 5.7   Characteristic curve of radiographic film.

defined by the left-hand portion of the curve between the no-exposure region and the point D1. At either the upper section of the curve or the lower section of the curve, changes in density are almost directly related to either time or X-ray intensity, and as such, small density variations cannot be detected. The slope of the curve is the measure of contrast for a film, because a film with a large slope the contrast may rise so sharply that denser materials may not be imaged with one single exposure. In this case, multiple radiographs must be made to correctly image the areas of concern. Some films do not have a linear slope. The gradient is the value of the slope at a single point. For some X-ray films, the gradient may increase over the entire curve. Films of this type are designed to give the best differentiation in high-density materials.

Some X-ray films are designed to use intensifying screens. Kodak™ has industrial inspection X-ray films that make use of lead screens. These screens are typically thin foils placed in direct contact with the film. Under exposure to X-rays, the foils emit electrons, lower-energy radiation, and characteristic radiation. Films have a tendency to be more reactant to these forms of radiation and, as such, enhance the radiographic effect. Some films may make use of fluorescent screens. Dental radiographs currently make use of them. These screens fluoresce when irradiated. The screens are placed in direct contact with film that is sensitive to the visible light that they emit. In this manner, the fluorescent screens augment the X-ray intensity.

Two techniques that help with film radiography are shielding and blocking. Shielding consists of shuttering down (closing the exposure windows) of the system so that only those X-rays that can directly strike the area of interest on the object under test are passed to the object. This technique helps minimize the amount of partial scatter and prevents "fogging" of the radiograph.

The other technique is blocking. Blocking is the filling of any voids in the object under test to eliminate scatter. If the object under test has a hole drilled through the object and is next to an area of concern, the scatter from the edge of the drilled hole can obscure or fog the image, destroying the detail sought in that area. If a material such as lead, or one similar in composition to the object of interest, is used to fill that hole, then the scatter is reduced or eliminated, and the detail in the adjacent area becomes evident.

There is considerably more information available about film radiography; however, with this basic overview and a small amount of trial and error, the analyst should have a good chance of getting the required shots.

5.3.5 Real-Time Radiography

Real-time radiography was defined in Sec. 5.3.1 as being related to the speed of the radiograph formation such that the image presented to the eye is within a time interval smaller than the integration time of the eye, or at the speed of the imaging system (typically 33 to 200 ms). That is a relatively strict, formal definition. Several of the imaging systems covered in Sec. 5.3.3 consist of array devices that either scan an object under inspection or have the object translated in front of the array so as to build a scanned radiographic image. These systems meet most of the requirements of real-time systems except for the image presentation time. Special-purpose X-ray systems such as computer tomography (two- and three-dimensional rendering of an object from X-ray scanning) make use of advanced computer processing and display the image a short time after it has been scanned. These type systems are nearly real time, and the discussion that follows has been expanded to include them.

Figure 5.8 is a block diagram of the typical components within a real-time radiographic system. The command and control for the X-ray tube and generator set regulate some things previously discussed, such as applied potential (kilovolts) and tube current (milliamperes). It also controls and monitors things like vacuum level, X-ray tube focus, cooling (water flow to target) and, in some cases, it monitors and blocks power dissipation into the target that exceeds some threshold value. It can also provide setup adjustments for the filament current. On some systems, the actual focal spot position can be readjusted on the target by (electron) beam steering. Beam steering can also be used to translate through very small image areas on objects at high magnifications. Additionally, the generator performance is sometimes monitored, and operation can be shut down if overvoltage or overcurrent conditions start to exist.

The manipulator command and control system typically controls the exact position within the X-ray beam that the object is located. Most X-ray systems are located inside protected (radiation shielded) housings or rooms. The inside dimensions of the housing or

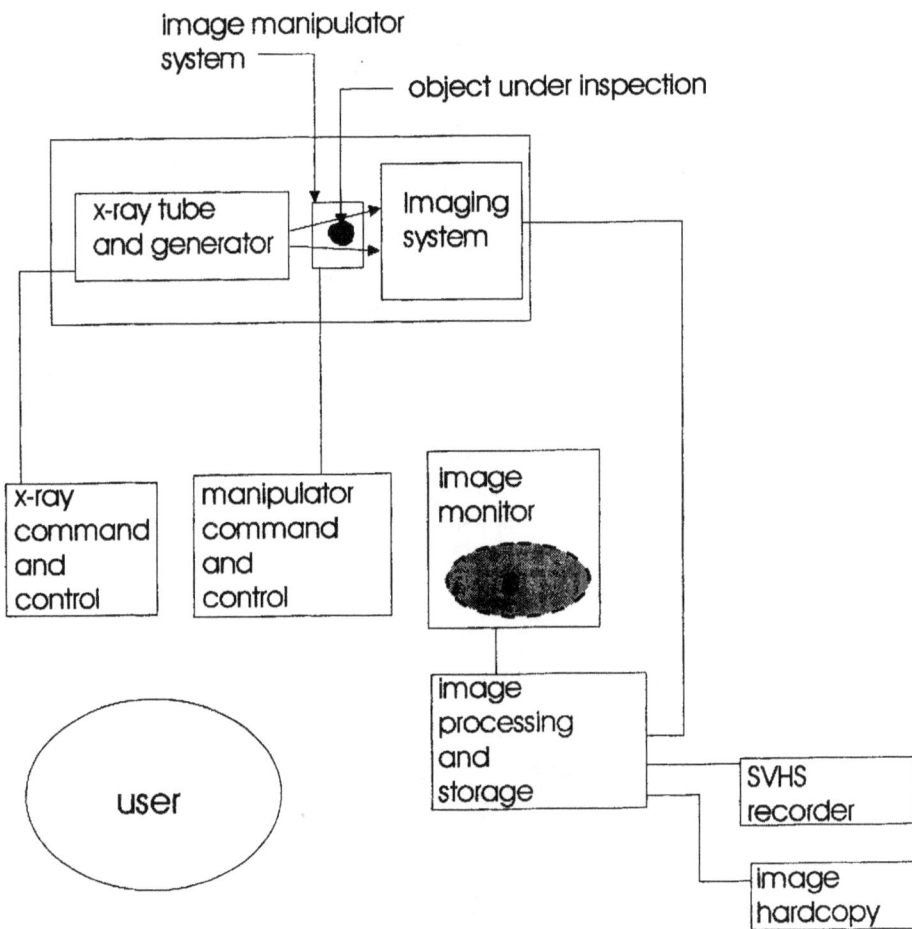

FIGURE 5.8 Block diagram of a generic RTR system.

room determines the x, y, and z translation of the part. The x direction is defined as perpendicular to the beam, the z direction is defined as parallel with the beam, and the y direction is the vertical direction. For a given system, a part can be translated some plus and minus distance in the x-axis (back and forth across the beam) and some plus and minus direction in the y-axis (up and down within the beam), and some plus and minus distance in the z-axis (closer to the tube and further away from the tube). The x- and y-axes position the part for best image viewing within the beam. The z-axis determines the relative distances from the imaging device and the tube and as such determines the magnification of the part (see Figure 5.1). If these are all the directions (x, y, and z) that the manipulator controls, the manipulator is said to be a three-axis manipulator system. Some manipulator systems have a rotation capability, typically around the y-axis in both a positive and negative direction. This would make such a system a four-axis manipulator system. In addition to rotation, some systems have a tilt function that tilts the part some maximum and minimum angle (typically $\pm15°$) toward the tube and away from the tube. A manipulator system with x, y, x, rotate, and tilt would be called a five-axis manipulator system. The reason why a five-axis manipulator is needed is not immediately apparent; however, due to safety and health reasons, radiographic systems shut themselves down when the access portal or door is open. To reposition a part due to an unleveled condition (not parallel with the x-axis) or off center condition (not centered on the y-axis) wastes time. Objects not centered rotate off center and cannot be viewed easily when rotated at magnification. Objects not parallel to the x-axis may obscure data due to the blending of a front and back structural edges. The usefulness of rotate and tilt becomes self-evident when used within a system.

In addition to the movement in five axes, the manipulator system most often has a variable motion control for each axis. Movement of the object at magnification must be much slower than that at a unity magnification. The manipulator system must control the rate of motion in each axis. The final control function the manipulator system determines is the maximum weight of the part under inspection.

Section 5.3.3 discussed the varying types of imaging devices. Common to most of them is the need for a computer imaging system, if only for capture and storage of the image information. The image processing and storage device shown in Figure 5.8 can be both a processing and storage device and a command and control device. For array type image systems, it controls the construction of the image. For systems using a video camera, it can control the gain and offset of the camera and all the other camera parameters. It projects the radiographic image to the system monitor for the operator to see and may provide disk (digital) storage of the image after it is saved. Any image manipulation and or enhancement would be performed by this system. Also controlled by this system are alternative image storage devices such as SVHS tape and image hard copy.

These three subsystems, X-ray command and control, manipulator command and control, and image processing and storage, provide an overview of the of the real-time radiographic system.

The difference between real-time radiography and other forms of radiographic inspection the ease of use. Once a part has been fixtured for inspection, the object is examined end to end under beam intensities and frequencies determined and varied by the operator as needed. Areas of concern are positioned for optimal viewing, the magnification of the area, and the X-ray parameters are adjusted, all in real time, then hard documentation (film, digital, or video) of the object generated. The other advantage to RTR over other types of systems is that, if you are unsuccessful at imaging a part, you know it immediately.

5.4 THE RADIOGRAPHIC PROCESS

5.4.1 General Inspection

The process of radiographic inspection starts with the component under inspection. The investigator should have some idea about the reported failure mode of the component. The investigator should also have some idea of the benefit from radiographic inspection of the component. If a semiconductor device has been damaged by a possible electrostatic discharge (ESD), then close attention to the bond wires must be paid. If the bond wires are of aluminum, and the devices internal cavity is gold plated for reliability reasons, the chances of directly viewing the bond wires is slim. However, the part might be oriented in such a manner that the bond wires are viewed from a horizontal plane, and the upper and lower gold platings are above and below the field of view. This type of thought process is the starting point for each radiographic analysis/inspection. A close visual examination of the component is the first step. Areas of highly contrasting densities should be noted and handled by blocking or shielding. The best orientation for inspection should be determined. Later, the part is fixtured in the manipulator system precisely so that translation through the areas of interest can be accomplished. Comparative samples of known-good and known-bad components can also be a large help in determining subtle differences radiographically. Obtaining comparative samples ahead of time and getting information from the component manufacture about areas of interest are good background steps in preparation for a radiographic analysis.

The second step is to consider the choice of radiation used. With experience, an investigator should get a feel for the size and density of objects imaged and know the approximate range of applied voltage and tube current levels to be used. Careful thought about imaging extremely dense components should be given. If a large wattage, dissipated in the tube, is needed to penetrate a component, adjustments to cooling and or increases to the spot size can be made to offset the large power dissipated. If the component is constructed from very light materials such as plastics, then much lower potentials must be used. For systems with video cameras, a manual adjustment to the camera lens iris may be needed to operate effectively in low kilovolt ranges. Camera gain and offsets may also have to be adjusted electronically from the image processing system.

The third step in the radiographic process is the fixturing of the component in the X-ray system for examination. At first glance, this appears redundant—we mentioned part orientation to obtain the best views in step 1; however, fixturing is a more than orienting the part. Larger parts require holders, clamps, or vises that are firmly mounted to the manipulator stage. Usually, these parts are of metal construction and can interfere with the radiographic examination. Care must be used in positioning the part. Foam pads can be used between the clamps and component to facilitate some degree of rotation when examining a part. Tape can be used to secure wire bundles and loose connectors out of the field of view. Copper foil tape can be used on ESD-sensitive parts to ground them to the chassis. Lead foil can be used to cover subcomponents that might be damaged by direct exposure to X-rays. All of these considerations should be kept in mind in fixturing the part for examination. The time to find that the clamping system is not fastened to the manipulator stage is not after you have tilted the stage 15°.

The final step in radiographic inspection is the actual examination. After proceeding through the component examination, considering the adjustments required for the radiation level and fixturing the part, the investigator is ready to optimize the radiographic image. At this point, good operating procedure should be discussed.

The component should be examined in a position closest to the imaging device (least magnification) first. Optimize the field of view so that a hard copy at this position gives an overview of the part. It is assumed that the generated radiographs will go into the failure analysis report and that, to correctly interpret the radiographic data, an overview is needed to orient the reader. Even if no useful data is evident at this magnification, a second view 90° (or some reasonable amount of rotation) from the first still should be taken. Often, internal structures are complex and will obscure a defect from immediate recognition. The defect may not be actually discovered until after disassembly/dissection (see Case Study no. 3). Having a second, unused view is no problem. Not having the second view when it could have identified a preexisting anomaly *is* a problem. From this point, the area of concern should be examined by gradually moving the part closer to the X-ray tube, magnifying the part. As with optical photography, a series of shots should be taken at reasonable increments, going from lowest magnification to highest magnification. This gives a documentation path to the area of concern, which can be used in the report to avoid lengthy discussions. At this point, we are at the area of concern and are making the normal adjustments to optimize the radiographic image. A good rule of thumb is to start with the best possible live image you can get before trying to apply image enhancements. The order of precedence is as follows:

1. Maximize the tube current to gives the maximum contrast.
2. Adjust the voltage to optimize the penetration of the part. If a great deal of scattering/blooming is evident, reduce the voltage, even though the image is on the dark side.
3. Shutter the image so that only the area of interest is in the field of view, and readjust the kilovolt level to just below the level at which the blooming effect occurs.
4. Use the camera gain and offset to adjust the image brightness for optimum viewing. If the gain and offset function are not available you can adjust the camera lens iris incrementally toward a lower number. (For many RTR systems, this may be as far as you can go. The remaining steps are most applicable to XRII and fluoroscopic imaging systems.)
5. The application of an average function may be helpful. A recursive average does a real-time average of several (user selectable) video frames as they are presented. This helps eliminate random noise due to scattering or other artifacts.
6. If still more optimization is needed, then a contrast stretch may be applicable. A contrast stretch translates the input gray levels to a sharper output slope, clipping the unused range at either end of the gray scale table.

At this point, we have worked from generator and tube, to imaging device and camera, and to image processing system, in sequence, to optimize the radiographic image. Further image enhancements are available and will be presented in the next section, as well as a discussion on averaging and contrast stretch.

5.4.2 Simple Image Enhancements

This section will cover three basic image enhancements that are useful to the operator in optimizing a radiographic image. These three functions are the *recursive average* (2 to 16 frames), the *average function* (up to 256 frames), and the *contrast stretch*. The starting point for these functions is an understanding of digitized images. As an image processing system captures the video image of a radiograph from the imaging system camera, it converts the image to a pixel array in computer memory with a image location and a gray-level intensity number. The location is most likely based on the order it was stored into memory. If a digitized image of 500 × 500 pixels is viewed as an array of 8-bit numbers, then you have 250,000 memory locations of 8-bit integers. This array can accurately represent a radiograph of 256 distinct gray levels. It turns out the human eye cannot differentiate any more than 256 gray levels. If several (4) memory locations of 250,000 bytes are reserved as image arrays, then some basic image operations can be performed. The first radiographic image can be digitized and stored at the same time it is being displayed. The next radiographic image can be digitized and stored, the first image gray levels can be added to the second image, and the resultant sum can be divided by two. This new image is then written back to the image one storage space and simultaneously displayed. In this manner, a two-frame recursive average is performed. For a four-frame recursive average, the first two frames are averaged as just described, and the next two frames are digitized and placed in memory locations three and four, then averaged, and the resultant average is placed back in memory location four. Memory locations one and four are added, averaged, written to image location one, and simultaneously displayed. In this manner, a four-frame recursive average is performed. The process has the same general procedure for frame averaging in increments of 2, 4, 8, or 10 frames. As the number of frames increases, the delay time to display the next image goes up.

The recursive frame average is a tool to help eliminate background noise from a radiograph. Image intensifiers and fluoroscopic systems have a tendency to react to scatter, random system noise, and other high-frequency artifacts, displaying this noise within a real-time radiograph. The recursive frame average quickly eliminates high-intensity pixel variation (white one frame and black the next) by forcing the pixel to an average value that is closer to the actual value, which should be displayed.

Where the recursive average eliminates random noise, the average function helps to reveal actual radiographic information that may be buried in the noise. The average function is very easy to understand: you digitize the first frame and store it, digitize the second frame and add it to the first frame, and so an and so forth until, at the very end, the total summation of the frames is divided by the number of frames. This method can be performed with an 8-bit system for up to 256 frames by using two memory location for each pixel. Figures 5.9, 5.10, and 5.11 show a radiograph of an electronic IQI. Figure 5.9 is live video, Figure 5.10 is a four-frame recursive average, and Figure 5.11 is a 256-frame average. The four-frame recursive average eliminates the random noise shown in Figure 5.9, bringing out more detail in the smaller particle set. The smaller crossed line pairs become visible in the 256-frame average in Figure 5.11.

In an a digitized system, the displayed radiographic information is typically presented as 256 distinct gray levels. This is because most people can differentiate only 256 gray levels. The radiographic information, however, can be derived as 16- or 32-bit gray level information. For a system that digitizes up to 16 bits, a display translation table or look up table (LUT) must be used to map the gray levels to the 8-bit display. Figure 5.12 shows two 8-bit LUTs. The first LUT is a normal input gray level to output gray level function. The second LUT is a contrast stretch, translating the normal input 8-bit information into a region with a much sharper contrast slope, and clipping (ignoring) the gray levels on each end. This procedure is analogous to using a faster, high-contrast film in film radiography. In this manner, discontinuities in the radiograph that were relatively close together in terms of gray level are now much further away (stretched) and have become more visible. Figure 5.13 is a gray level contrast stretch of the IQI from Figure 5.11. Note how much more visible the smaller crossed line pair have become.

FIGURE 5.9   Electronic IQI shown in live video mode.

These three simple image enhancements can provide the analyst with some serious tools to help analyze electronic component radiographs and maximize the amount of information derived.

5.4.3 Advanced Techniques

Digital image enhancements cover a broad spectrum of operations. Image manipulation software packages run from a few hundred dollars to tens of thousands. To cover all the available types of advanced image enhancements would in itself require enough material for a book. This section is limited to identifying some of the most applicable image enhancements for radiography.

For the electronic failure analyst, the types of defects usually examined are solder joints (for cracks), bond wires (for continuity), and internal component structure. In each of these examinations, the fault or area of interest appears as an edge of some finite thickness against a background. This background may be obscuring the information/geometry we are looking for. At an edge on a radiograph, a shift in gray levels occurs. If the shift in gray levels is dramatic, the edge is well imaged. If the shift is of only one to two gray levels, the change may not be visible. Add some random noise to this scenario, and the edge is totally

FIGURE 5.10 Electronic IQI shown after a four-frame recursive average.

obscured. From this, it would appear that an edge enhancement function would make the analyst's life easier.

Edge detection or gradient detection is a function that can easily be implemented with the use of matrix algebra and the availability of a stored image. For a stored image of an edge, over the full length of the edge, a shift in gray levels occurs. This shift in gray level may be partially obscured by recorded pixel information of random noise. A user who measures gray level values might be able to use the small shift in gray level to identify an edge. If a direction of travel is determined for these gray levels by looking at a pixel, the pixel behind it, and the pixel ahead of it, the user might confirm that a pattern of similar gray levels was forming. By looking at pixels on both sides of this pattern (edge), which would most likely be at different gray levels, one can determine which pixels are noise but really should be a specific gray level that matches the gray level of the edge. This would be a very tedious operation for an analyst, checking the image for possible edges obscured in the noise using only the gray level value. This is, however, the perfect type of operation for a computer.

An image enhancement operation that applies a predefined matrix array designed for detecting edges can be used on the image array as a whole. The form of such an array and its actual mathematical operation are shown in Table 5.1. Table 5.2 depicts some 3 × 3

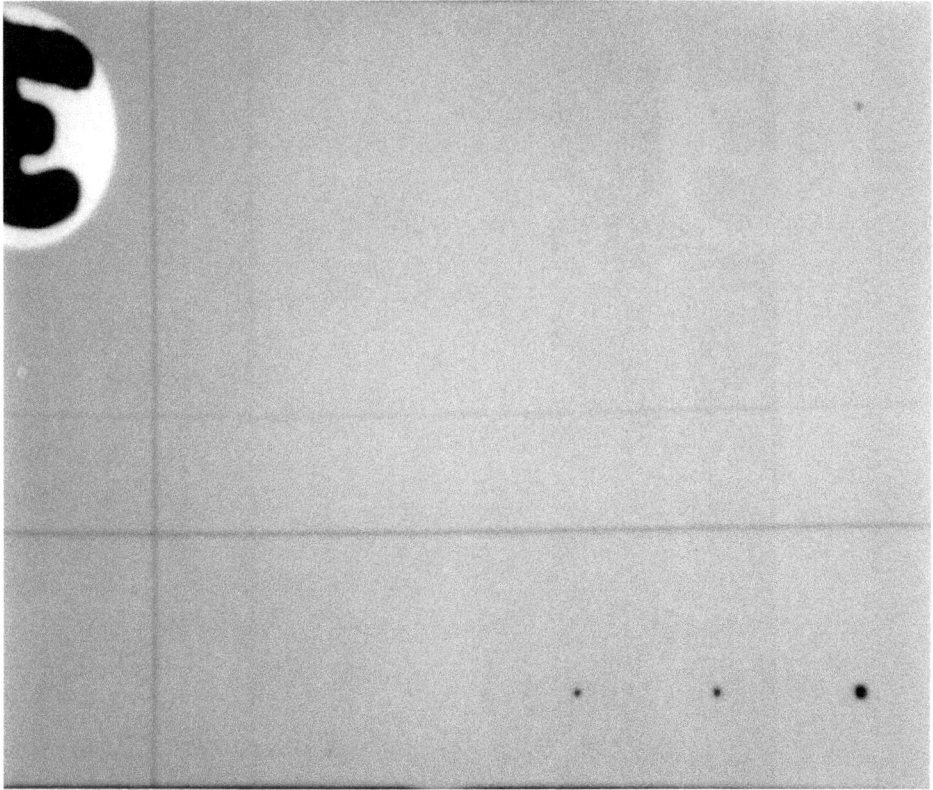

FIGURE 5.11   Electronic IQI shown after a 256-frame average.

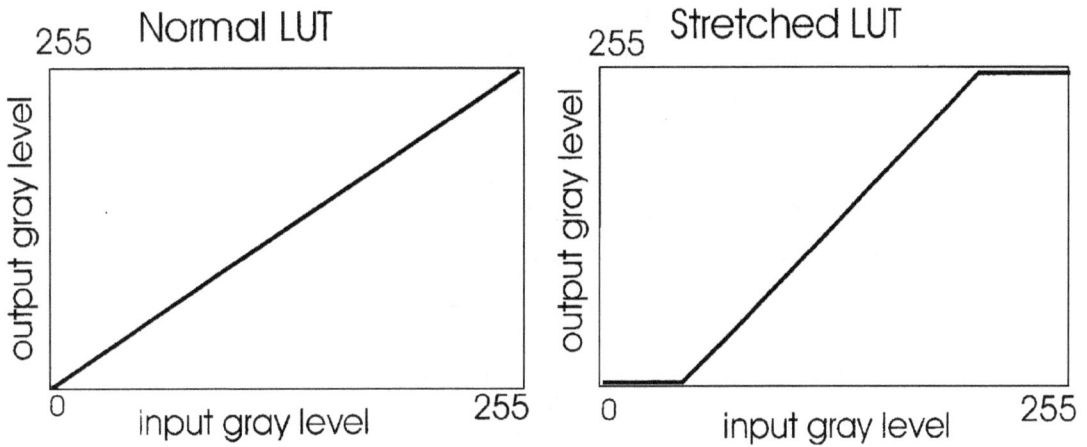

FIGURE 5.12   Input/output LUTs depicting a normal input/output relationship and a contrast-stretched relationship.

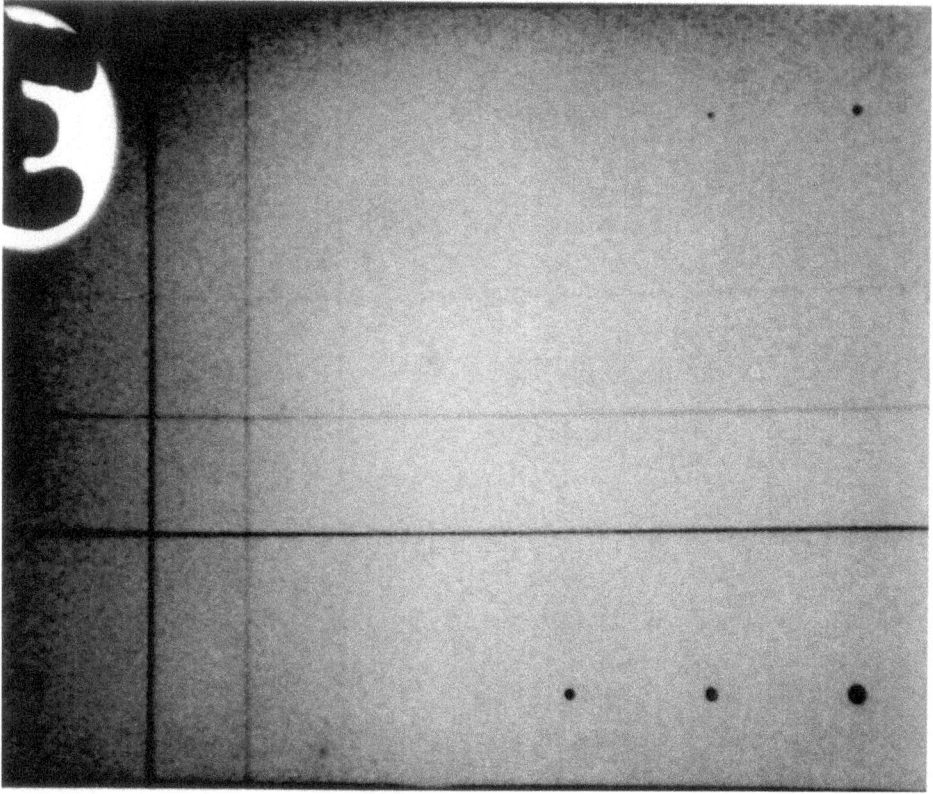

FIGURE 5.13 Gray-level contrast stretch on the IQI of Fig. 6.11. Note how the crossed line pairs have become much more visible.

Table 5.1 Image Enhancement Matrix Operation

$$P5 = \sum_{i=1}^{9} M_i * P_i$$

P5 becomes interdependent on all its surrounding pixels. Each pixel in the entire image pixel array is evaluated in this manner.

Enhancement Matrix	Image Pixel Array

$$\begin{bmatrix} M1 & M2 & M3 \\ M4 & M6 & M7 \\ M8 & M9 & M10 \end{bmatrix}$$

P	P	P	P	P
P	**P1**	**P2**	**P3**	P
P	**P4**	**P5**	**P6**	P
P	**P7**	**P8**	**P9**	P
P	P	P	P	P

Table 5.2 3×3 Edge Enhancement Matrices

Directional matrices			
North	South	East	West
$\begin{bmatrix} 1 & 1 & 1 \\ 0 & 0 & 0 \\ -1 & -1 & -1 \end{bmatrix}$	$\begin{bmatrix} -1 & -1 & -1 \\ 0 & 0 & 0 \\ 1 & 1 & 1 \end{bmatrix}$	$\begin{bmatrix} -1 & 0 & 1 \\ -1 & 0 & 1 \\ -1 & 0 & 1 \end{bmatrix}$	$\begin{bmatrix} 1 & 0 & -1 \\ 1 & 0 & -1 \\ 1 & 0 & -1 \end{bmatrix}$
NW	SW	NE	SE
$\begin{bmatrix} 1 & 1 & 0 \\ 1 & 0 & -1 \\ 0 & -1 & -1 \end{bmatrix}$	$\begin{bmatrix} 0 & -1 & -1 \\ 1 & 0 & -1 \\ 1 & 1 & 0 \end{bmatrix}$	$\begin{bmatrix} 0 & 1 & 1 \\ -1 & 0 & 1 \\ -1 & -1 & 0 \end{bmatrix}$	$\begin{bmatrix} -1 & -1 & 0 \\ -1 & 0 & 1 \\ 0 & 1 & 1 \end{bmatrix}$
Nondirectional matrices			
$\begin{bmatrix} -1 & -1 & -1 \\ -1 & 8 & -1 \\ -1 & -1 & -1 \end{bmatrix}$	$\begin{bmatrix} -2 & -3 & -2 \\ -3 & 20 & -3 \\ -2 & -3 & -2 \end{bmatrix}$	$\begin{bmatrix} 0 & -1 & 0 \\ -1 & 4 & -1 \\ 0 & -1 & 0 \end{bmatrix}$	$\begin{bmatrix} 1 & -3 & 1 \\ -3 & 8 & -3 \\ 1 & -3 & 1 \end{bmatrix}$

edge detection matrices. Figure 5.14 shows the before and after effects of an edge detection function on a radiograph. The edge enhancement was done in a northerly direction. The three horizontal lines from the penatrameter are quite distinct; however, the vertical traces are lost, except those portions positioned diagonally across the image.

The line detection function is very similar to the edge detection function; however, it does not produce such a high degree of contrast at the line/background interface. Table 5.3 shows some typical 3×3 matrices for line detection. Figure 5.15 depicts same object as in Figure 5.14 using the line detection function.

Table 5.3 3×3 Edge Enhancement Matrices

Vertical	Horizontal	Diagonal + 45	Diagonal − 45
$\begin{bmatrix} -1 & 2 & -1 \\ -1 & 2 & -1 \\ -1 & 2 & -1 \end{bmatrix}$	$\begin{bmatrix} -1 & -1 & -1 \\ 2 & 2 & 2 \\ -1 & -1 & -1 \end{bmatrix}$	$\begin{bmatrix} -1 & -1 & 2 \\ -1 & 2 & -1 \\ 2 & -1 & -1 \end{bmatrix}$	$\begin{bmatrix} 2 & -1 & -1 \\ -1 & 2 & -1 \\ -1 & -1 & 2 \end{bmatrix}$
Vertical and horizontal		Diagonal ± 45	
$\begin{bmatrix} 0 & 1 & 0 \\ -1 & 0 & -1 \\ 0 & 1 & 0 \end{bmatrix}$		$\begin{bmatrix} -1 & 0 & 1 \\ 0 & 0 & 0 \\ 1 & 0 & -1 \end{bmatrix}$	

There are three additional functions which may prove to be of value to the analyst in the realm of digital image enhancement. The first two of these are the *sharpen* (highpass) function and the *smooth* (lowpass) function. On radiographs where clear delineation between objects/edges of darker intensities is necessary, the *smooth* function may prove to

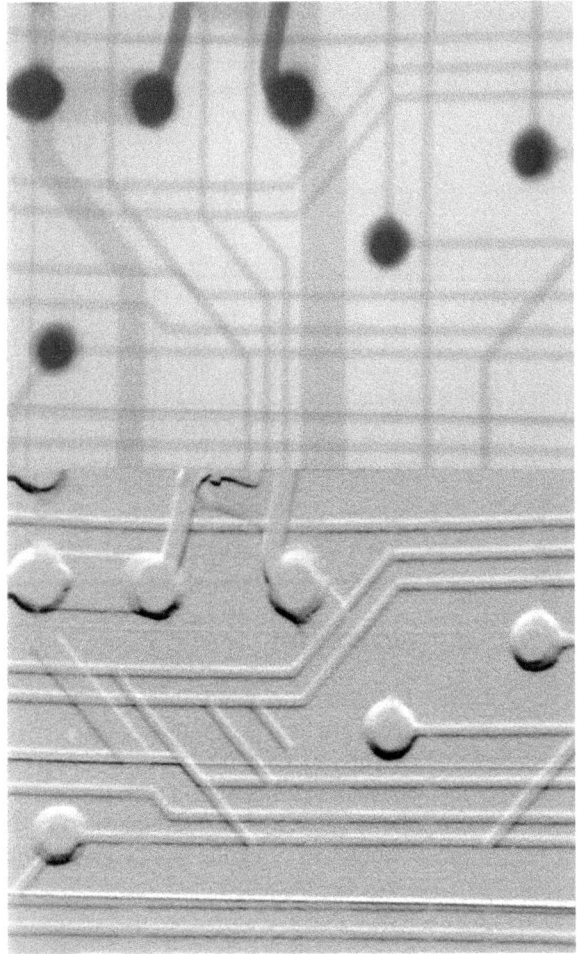

FIGURE 5.14 Radiograph before (top) and after (bottom) the north direction edge detection function. Note the increased visibility in the penatrameter lines and the loss of the vertical traces.

be of help. On areas where lighter contrast are needed, a sharpen operation may help provide the differentiation needed. Table 5.4 shows the sharpen and smooth 3×3 matrices.

The last function mentioned is the median function. This function helps with the random noise or scattering problems on radiographs. The median function replaces outliers (random extreme high or low value pixels) with the median value within a comparative sample. Table 5.5 shows the median functions effect (mathematically) to a segment of digitized radiograph. Its obvious that large gray level increases due to scattering may be significantly improved by use of the median function.

5.4.4 On-Screen Metrology/Dimensional Analysis

Referring to Figure 5.2, a top-down image of an object being inspected is shown. It is apparent that the magnifications at the front (closest to the beam source) of the object and the back of the object are different. The radiograph will show both corners, with the rear

FIGURE 5.15 Radiograph of the line detection function performed on the same radiograph shown in Fig. 5.14, upper view.

Table 5.4 3 × 3 Matrices for Sharpen and Smooth Enhancements

Sharpen

$$
\begin{bmatrix} 0 & -1 & 0 \\ -1 & 5 & -0 \\ 0 & -1 & 0 \end{bmatrix}
\quad
\begin{bmatrix} -2 & 1 & -2 \\ 1 & 5 & 1 \\ -2 & 1 & -2 \end{bmatrix}
\quad
\begin{bmatrix} -1 & -1 & -1 \\ -1 & 9 & -1 \\ -1 & -1 & -1 \end{bmatrix}
\quad
\begin{bmatrix} -2 & -3 & -2 \\ -3 & 21 & -3 \\ -2 & -3 & -2 \end{bmatrix}
$$

Smooth

$$
\begin{bmatrix} 0 & 1 & 0 \\ 1 & 0 & 1 \\ 0 & 1 & 0 \end{bmatrix}
\quad
\begin{bmatrix} 1 & 1 & 1 \\ 1 & 0 & 1 \\ 1 & 1 & 1 \end{bmatrix}
\quad
\begin{bmatrix} 1 & 4 & 1 \\ 4 & 12 & 4 \\ 1 & 4 & 1 \end{bmatrix}
\quad
\begin{bmatrix} 3 & 3 & 3 \\ 3 & 5 & 3 \\ 3 & 3 & 3 \end{bmatrix}
$$

corner inside the front corner toward the interior of the radiograph. A measurement made from this radiograph would indicate that the distance from corner to corner in the rear was shorter than the distance from corner to corner in the front. Both distances are the same. This error is due to the magnification of the object. A collimated beam of light of equal diameter as the width of the object would produce a shadow graph that was an exact dupli-

Table 5.5   Median Image Enhancement Function

	Stored image pixel array				
The highlighted values are ranked lowest to highest	27	15	15	7	5
	26	17	14	9	7
	28	18	121	10	9
	28	19	9	14	12

9   9   10   (14)   114   17   18   18   ~~121~~

	Stored image pixel array after median function				
The median value is selected, the outlying value is dumped and replaced with the median value.	27	15	15	7	5
	26	17	14	9	7
	28	18	14	10	9
	28	19	9	14	12

cation of the objects measurement. Such a device is called an *optical comparator* and is used specifically to make precision measurements. With real-time radiography, attempts at measuring objects on screen can generate erroneous data quickly if the operator is not careful. Using the previous example, if the front surface of the object shown in Figure 5.2 is a decapsulated IC, and the lead frame pins are extending from the front corner of the object, then we can use the front corner-to-corner measurement as a standard and then make measurements of the lead frame pins in the horizontal direction with a relative degree of accuracy. The lead frame pins are in the same z-plane of reference as the front corner. An investigator who measures the front corner-to-corner distance with a microme-ter and the height of the front corner would have a vertical and horizontal reference. Mea-suring the number of pixels on the radiographic image for these two lengths would allow conversion from pixels to inches or millimeters. At this point, any measurement made in the same z-plane would be accurate (almost as accurate as the micrometer readings used for calibration).

The degree of error for measuring a defect within the object but not in the same z–plane can be calculated based on the difference in magnification from the front plane of the object to the back plane of the object. In Figure 5.2, the values for A and B were 5 and 10 in., with the magnification being given by $M = 1 + (B/A)$. The front face of the object was therefore at a magnification of 3×. By changing the z–plane by the depth of the object (lets give the object a depth of 0.2 in.), the magnification changes to 2.88× and the percent change is 3.84 percent. If a defect presumed on the front surface measured 0.0330 in. and was really on the back plane, its actual size would be $1.0384 \times (0.0330)$ or 0.0342 in. In many cases, the difference (error) can be tolerated. The key issue here is that the range of error must be known.

This discussion describes the most rudimentary of RTR on-screen measurement sys-tems. A standard RTR on-screen measurement system would have inputs for vertical and horizontal calibration, would give back both pixels and units of measurements, and would give the x, y, and vector measurements between points. It would have dual cursors to act as points of reference and would give a gray level readout at each cursor.

By limiting the plane of measurement to one defined plane that was calibrated to known standards, very precise measurements can be taken. There are some tricks of the trade that will help with on-screen dimensional analysis. The edges of objects have a tendency to bloom or wash out when using sufficient energy to penetrate the whole object. The way a radiographer thinks is to maximize internal detail. The person making on-screen measurements does not care what the radiograph looks like as long as his reference points are distinct. Lowering the kilovolt levels will often help to delineate edges while leaving the internal reference points visible. The other thing the on-screen dimensional analyst should do is closely watch the gray levels. As a cursor is approaching an edge, there is typically a distinct transition or gradient of gray levels. The outside edge may be at a gray level of 40 and, as the cursor transition across the edge, the sequential pixel gray levels might read 40, 43, 50, 55, 55. The transition range 43, 50 can be used as the actual edge of the object. By being consistent with the use of the gray levels, better measurements can be made.

The final word with regard to on-screen analysis: measure only in the calibrated plane, and know your maximum error.

5.5   *CASE STUDIES*

So far, we have discussed history, theory, and technique. The discussion now shifts to real jobs and the radiographic inspection of them.

5.5.1   Electrical Overstress on JAN TXV2N222A NPN Transistors

Preanalysis conditions.   A high-current problem was experienced on payload bay floodlight ballast assemblies S/N 507 and S/N 660. During the troubleshooting process, it was discovered that the Q9 transistor, part no. JANTXV2N2222A, in each ballast was internally electrically shorted from collector to base. The transistor documented in FL4144 was reported to have a resistance of 10.0 Ω from collector to base. The transistor documented in FL4156 was reported to have a resistance of 6.0 Ω from collector to base.

Description of the analysis.   The two subject transistors were visually examined. The electrical characteristics of the transistors were measured with a Tektronix model 370 curve tracer. The collector-to-base characteristic of both transistors was a nearly linear resistive characteristic in both polarities. The emitter-to-base condition of both transistors was electrically open (less than 0.1 μA at 25 V, both polarities). The collector-to-emitter breakdown ($V_{(BR)CEO}$) characteristic of both transistors was less than 0.1 μA at 100 Vdc. The current gain (h_{fe}) of both transistors at I_B = 8 mA and V_{CE} = 8 V measured −1.0. Fine and gross leak checks were performed. The fine leak test results were acceptable No gross leaks were detected.

Radiographic examination.   Radiographic inspection revealed that the emitter bond wire in each transistor was no longer present between the emitter post and the transistor chip, as shown in Figure 5.16. In each transistor, a few small dense particles approximately the size of the bond wire diameter were observed in the vicinity of where the emitter bond wire should have been, as shown in Figure 5.17.

Decryption of the analysis (continued).   The transistors were delidded. Internal visual examination revealed that all that remained of the emitter bond wire in each transistor was a short portion of the wire at each bond. On each portion of wire, a shiny gold sphere was found at the opposite end from the bond. A few gold beads were found on the inside of the

FIGURE 5.16 Radiograph of missing emitter bond wire.

lid of each transistor. Visible damage to the transistor chips was also observed. No discoloration was observed at the base ball bond; however, a halo was seen in the aluminum film around more than half of the bond. The halo suggested that the area of aluminum film outlined by the halo had been melted by resistive heating. Similar halos were also observed at the emitter ball bonds on both chips. The emitter-to-base electrical characteristics of the transistors were measured with the Tektronix model 370 curve tracer by contacting the emitter ball bond on each chip with a microprobe. The emitter-to-base characteristic of both transistors was a resistive characteristic in both polarities which measured 1.6 Ω (8 mV at 5 mA) on LDC CRP8118 and 2.0 Ω (10 mV at 5 mA) on LDC CRP8131.

Analysis findings.   The electrical characteristics and the radiographic observations indicated that the emitter bond wire was no longer present between the emitter post and the chip in either transistor. Internal examination indicated that the emitter bond wire in each transistor had been destroyed by current overstress that had fused the bond wire. The excessive current had damaged the aluminum film on each chip at the emitter ball bond. Excessive current had also flowed through the base bond wire and damaged the aluminum film on each chip at the base ball bond. The fault current that flowed through the base bond wire was significantly less than what flowed through the emitter bond wire. However, it could not be determined whether the collector fault current was the sum or the difference of the emitter and base fault currents. The electrical characteristics indicated that both p-n junctions of each transistor were shorted and had no current rectifying characteristics. Both transistors were damaged by EOS (electrical overstress) that melted the emitter bond

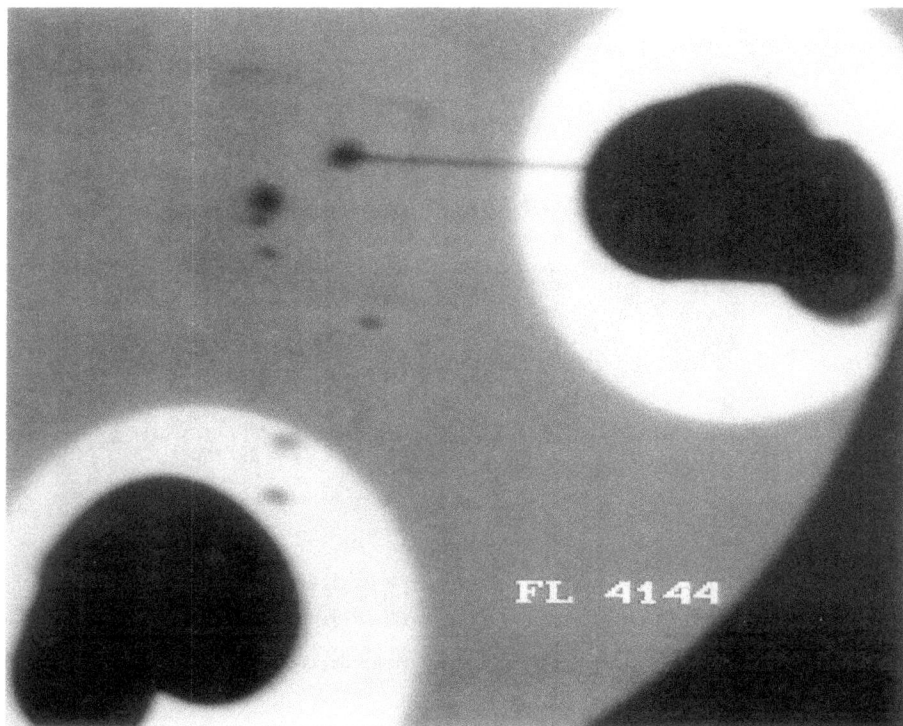

FIGURE 5.17 Emitter bond wire debris.

wire and caused both of the p-n junctions in each transistor to become shorted. The fault current that flowed through the base bond wire of each transistor was significantly smaller than that which flowed through the emitter bond wire.

5.5.2 Liquid Oxygen Temperature Transducer Hardware Anomaly

Preanalysis conditions.   It was reported that this probe produced a faulty liquid oxygen (LOX) temperature measurement during cryogenic load. The probe is designed to measure temperatures between –170° C and –185° C. The boiling point of LOX is approximately –183° C.

Radiographic examination.   Radiographic examination indicated that loose hardware was present inside the temperature probe housing. This hardware is shown in Figs. 5.18 and 5.19.

Description of analysis.   The transducer underwent a weld analysis. Evaluation of the welded sections on the subject components revealed no anomalies. The probe tip was submerged in liquid nitrogen (LN^2), which boils at approximately –196° C, to precool the element. After several seconds, the probe tip was precooled, and the output of the probe was recorded as –0.7 Vdc. The probe was removed from the LN^2 and then placed in LOX. The

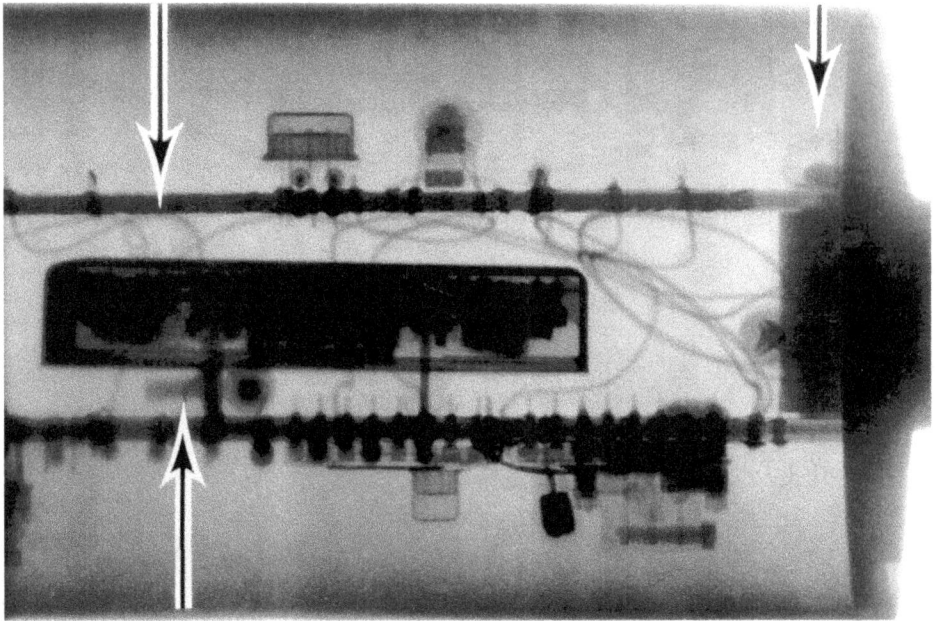

FIGURE 5.18 Radiograph of temperature probe housing depicting loose hardware.

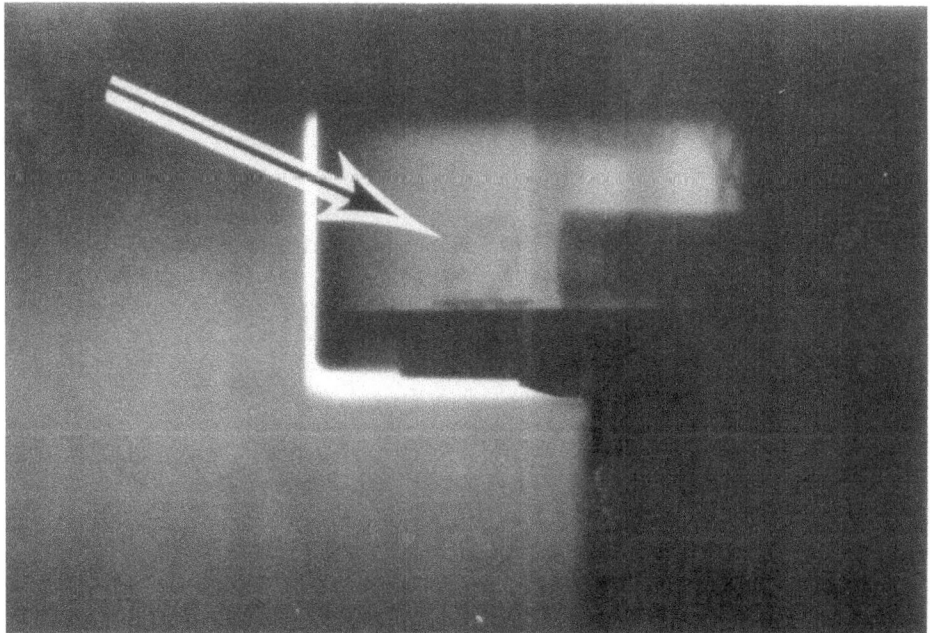

FIGURE 5.19 Radiograph depicting loose hardware at bottom of housing after rotation of the housing to the vertical orientation.

output remained at –0.7 Vdc (the probe did have the reported anomaly). The internal electronics were examined. The examination confirmed the loose hardware condition revealed during radiographic inspection. The three screws and washers used to mount the internal PC board to the end plate were backed out of their threaded studs and were floating loose inside of the electronic module enclosure. It was noted that there was no thread locker used on these screws. Electrical analysis revealed that an operational amplifier (opamp) IC used in the output stage of the power converter board was not functioning properly. The IC was replaced, and the LOX temperature transducer functioned as designed

Analysis findings.   The manufacturer uses red GLPT insulating varnish in conjunction with a lock washer to prevent the PC board mounting screws from becoming loose in a high-vibration environment and stated that the use of this varnish was implemented in 1987 after it was noted that some probes were being returned with loose hardware. However, the inspection procedure made no provision to ensure the presence of thread locker on internal threaded components. It was suggested that their inspection procedure incorporate instructions to visually examine internal threaded components for the presence of thread locker. This suggestion was implemented immediately by the manufacturer. The cause of the abnormal LOX temperature transducer output was a faulty opamp IC located on a PC board inside the electronic module enclosure. The loose hardware found inside the unit was most likely caused by vibration in conjunction with the absence of thread locker on the mounting screws used to secure the PC board inside the enclosure. The loose hardware was most likely the cause of the defective IC.

5.5.3 Circuit Breaker Spring Anomaly

Preanalysis conditions.   A new circuit breaker was installed in the HUMS rack as the main power breaker and was operated only a few times before it failed. The breaker failed in the open or tripped position and could not be reset. The lever was unusually loose, and moving it to the OFF position would not reset the breaker.

Radiographic examination.   Radiographic inspection revealed two armatures that are designed to lock together when the breaker is not tripped and slide past each other when the breaker is tripped. As received, these two armatures slid past each other, causing the breaker to be in the tripped condition regardless of the position in which the arming switch was placed. The radiograph shown in Figure 5.20 shows the armatures in both the straight position (switch off, upper view) and the bent position (switch on, lower view). Had these armatures been locked together when the switch was moved to on, the armatures would have pushed the contact lever down (large arrows) to make electrical contact with its mating contact as designed. On closer radiographic examination, a loose wire spring was discovered in the area between the two armatures. Had this wire spring been hooked around an adjacent surface, the locking mechanism would have worked properly. Figure 5.21 shows a radiograph, enhanced with an outline function, depicting the loose wire spring location.

Description of analysis.   On dissection of the breaker, the loose wire spring was verified. No other anomalies were discovered, and the area where the wire spring should have been attached was visually inspected under a 40× microscope. No scratches or wear marks were found that might indicate that the spring had been mounted either correctly or incorrectly.

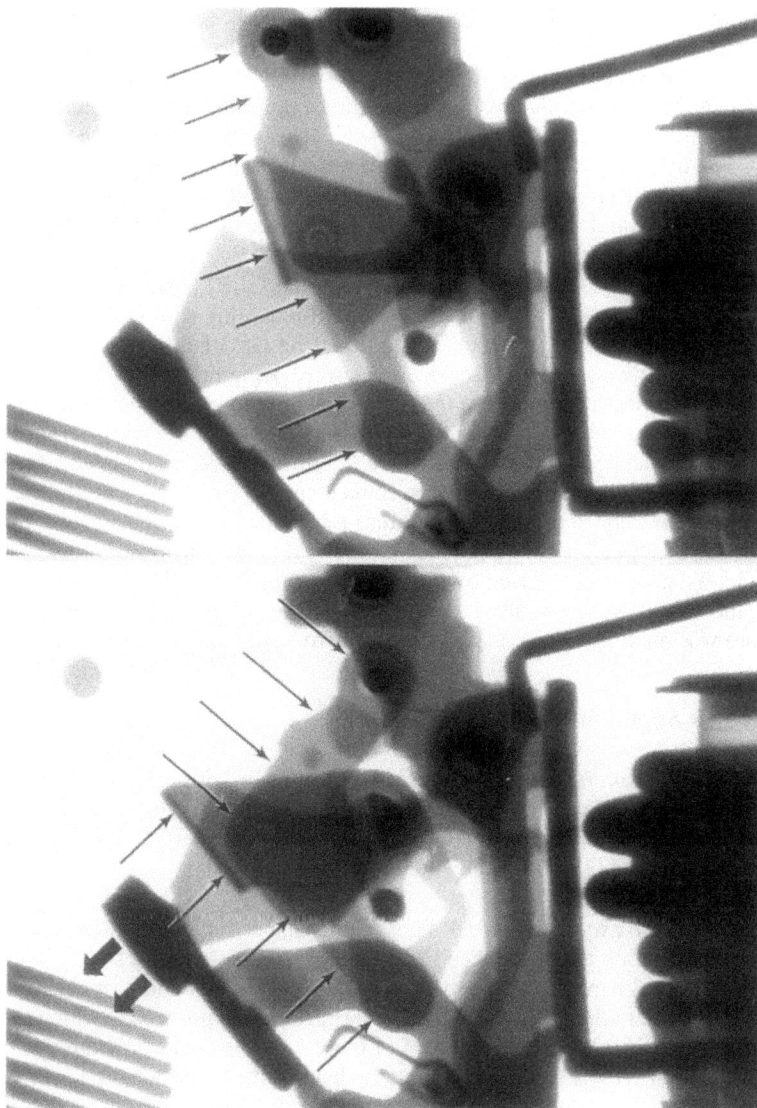

FIGURE 5.20 Radiographs of the circuit breaker off (top) and on (bottom) positions.

The wire spring was then placed around the edge on which it was designed to rest, and the breaker was given extensive functional tests. The breaker operated normally.

Analysis findings. A small wire spring was found to be loose in the tripping mechanism, therefore causing the breaker to be continuously in the tripped position. No evidence was found that would indicate whether the spring was ever mounted correctly or incorrectly. The breaker was working properly before the incident, so it is believed that the spring came loose during operation of the mechanism.

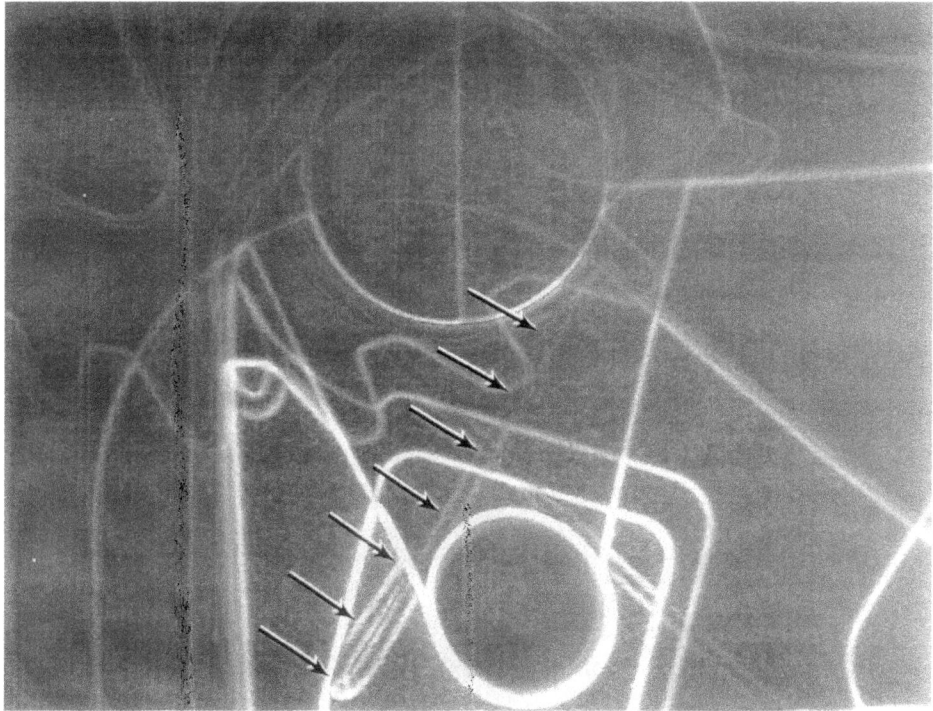

FIGURE 5.21   Radiograph, enhanced with an outline function, depicting the anomalous spring attachment.

5.5.4   Selected Radiographs

Figure 5.22 shows a composite radiograph of a hybrid integrated circuit. The IC base was a ceramic material that was laid out with ten IC packages to a wafer and then separated using a laser cutting operation. The laser cutting operation left a residual metal film between pads in an area where pads from one IC would line up with pads from the next IC over. After the soldering operation to attach the external package pins, solder adhered to the residual metal film, almost bridging the gap between pins (less than a 0.005-in. gap). After a short exposure to high humidity and operating voltages, the bridge became continuous via a carbon path shorting the pins. The manufacturer allowed an additional spacing on the ceramic wafer to eliminate the anomaly.

Figure 5.23 shows two views of the failed tethered satellite tether. The left-hand view depicts the normal configuration approximately 18 in. back from the end. The right-hand view shows the deformation to the failed end and the spherical globules resulting from melting. The longest strand has the appearance of being tensiled.

Figure 5.24 shows three high-density connector sockets. All three were from the same vendor. It is immediately apparent that the right-hand socket spring contact is severely enlarged. Closer examination indicates that the barrel has been assembled upside-down, pinching the socket spring open. Note the lack of rolled edge at the top of the right-hand socket. Reports from the manufacturer indicated the use of an automated process that sorts the barrels prior to assembly. It was conjectured that, most likely, an employee found an expelled/misplaced barrel and inserted it into the assembly machine upside-down.

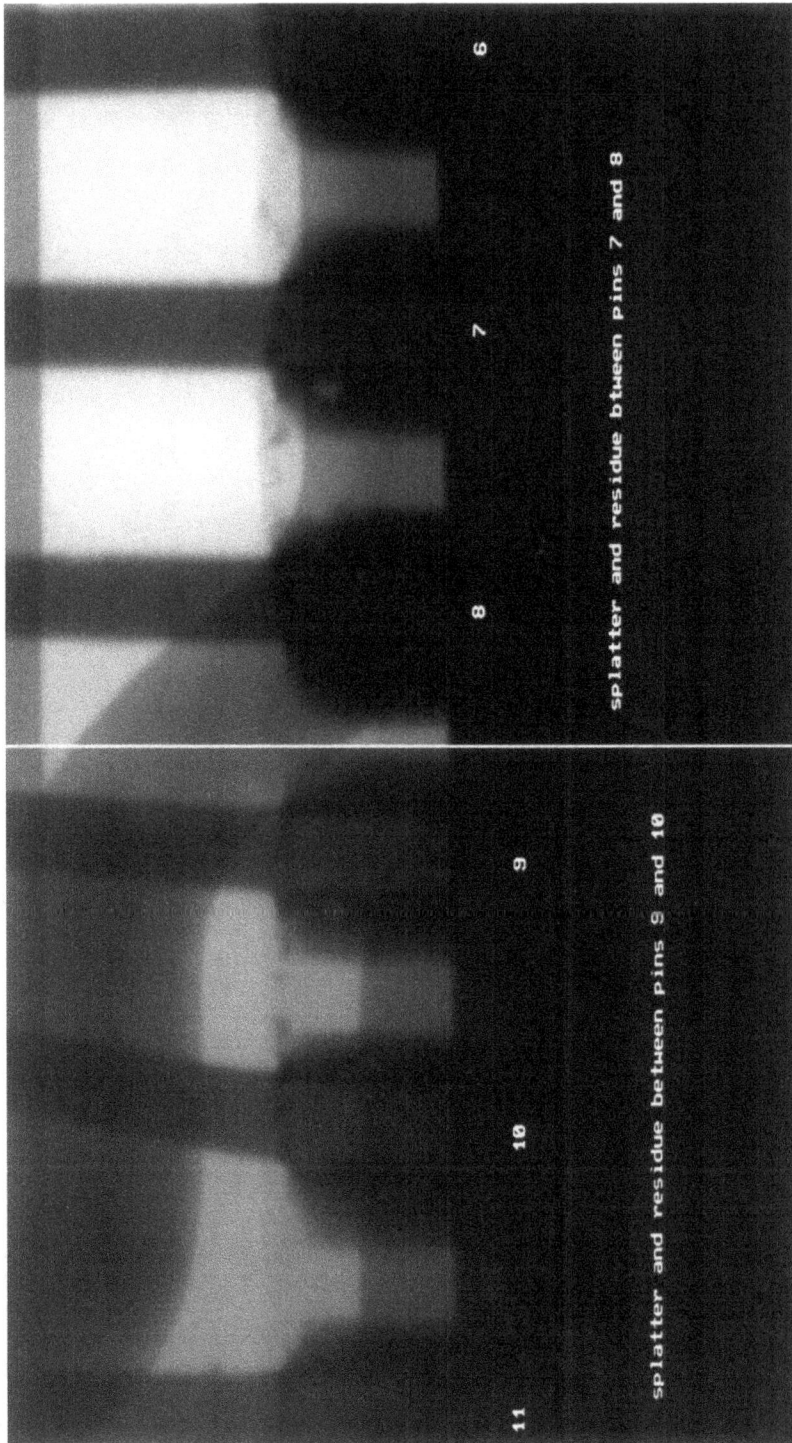

FIGURE 5.22 Radiograph of a hybrid IC, depicting splatter and metal residue between shorted pins.

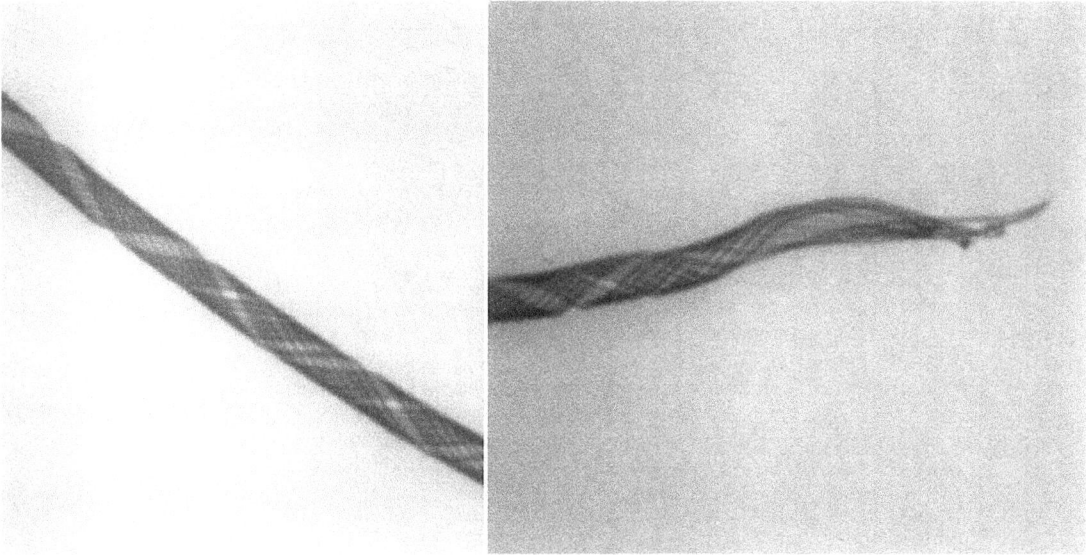

FIGURE 5.23 Tethered satellite failed tether. Left-hand view depicts the normal configuration, and the right-hand view shows the deformation and melting at the failed end.

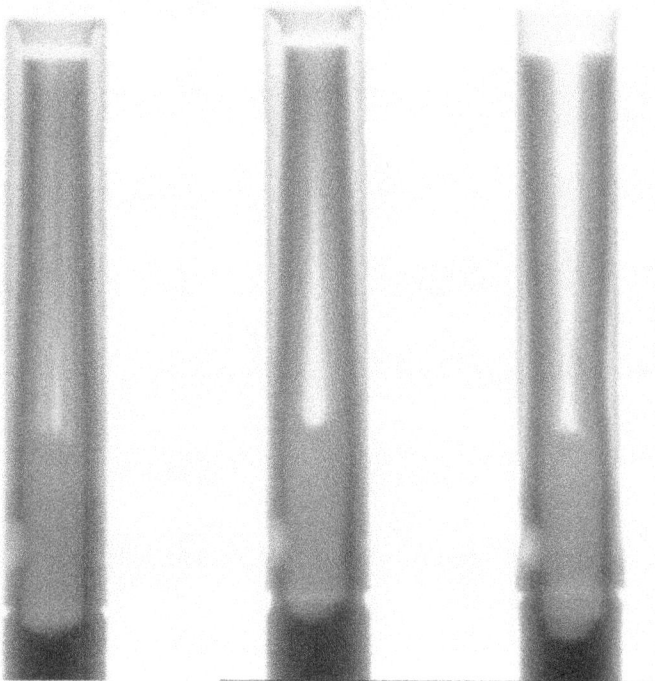

FIGURE 5.24 High-density connector sockets. The right-hand socket is the result of a manufacturing anomaly.

5.6   REFERENCES

1. Close, F., M. Marten, and C. Sutton. *The Particle Explosion.* Oxford University Press, 1987.

2. Guinier, A. and D.L. Dexter. *X-ray Studies of Materials.* Interscience Publishers, 1963.

3. Brown, J.G. *X-rays and Their Applications.* New York: Plenum Press, 1966.

4. Clark, G.L., *Applied X-rays.* New York: McGraw-Hill, 1955.

5. Greco, C, and S. Greco. *Piercing the Surface X-rays of Nature.* Harry N. Abrams, Inc., 1987.

6. Csorba, I.P. *Image Tubes.* Indianapolis: Howard W. Sams, 1985.

7. Anton, H. *Elementary Linear Algebra.* New York: John Wiley & Sons,1984.

8. Ray, S.F. *Applied Photographic Optics.* Focal Press, 1988.

9. Rauwerdink, J. Failure analysis of two low power, high frequency, silicon NPN transistors, *FAME 94-3009,* Kennedy Space Center, FL: Failure Analysis and Materials Evaluation Laboratory, June 1994.

10. Bayliss, J.A. Failure analysis of a liquid oxygen temperature transducer, *FAME 91-3056,* Kennedy Space Center, FL: Failure Analysis and Materials Evaluation Laboratory, June 1992.

11. Leucht, K.W. Failure analysis of a hydrogen umbilical mass spectrometer rack circuit breaker. *FAME 93-3045,* Kennedy Space Center, FL: Failure Analysis and Materials Evaluation Laboratory, May, 1995.

12. Photographic support for this chapter provided by Stephen W. Huff, photodocumentation specialist.

CHAPTER 6
INFRARED THERMOGRAPHY

Herbert Kaplan

Honeyhill Technical Company

6.1 INTRODUCTION: WHY THERMOGRAPHY?

Infrared thermography is the use of infrared (IR) thermal imaging instruments to produce thermal maps of the surfaces at which the imaging instruments are aimed. This surface is called the *target* or the *target surface*. The thermal maps are called *thermograms* and can be presented in monochrome (black and white) or color. The gray shades or color hues can be made to approximate closely the temperature distribution over the target surface.

A thermal imager measures the infrared radiant exitance from a target surface and displays a raw or qualitative thermogram. Menus and controls provided on the instrument allow the operator to introduce the corrections necessary to convert this image to a quantitative thermogram, which is a close approximation of the real temperature distribution over the surface of the target.

The physical mechanisms within electronic circuits usually involve the flow of electrical current which, in turn, produces heat. Since thermal behavior is closely related to the flow of current, the temperature distribution in an operating package, component, or assembly can be a good, repeatable indication of its operational status. Figure 6.1 is a quantitative thermogram of a section of a printed circuit board showing an overheated device.

Not all electrical and electronic packages, components, and assemblies lend themselves to failure analysis by means of infrared thermography. The extent to which thermal signa-

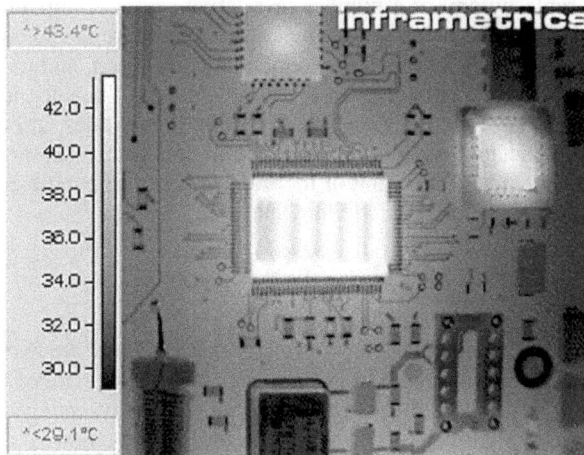

FIGURE 6.1 Thermogram of an overheated device on a hybrid electronic assembly. *Courtesy of Inframetrics, Inc.*

tures are useful for failure analysis depends on the physical configuration and the materials involved in the particular test article.

The advantages of IR thermography that apply to the inspection of electronics are that it is fast, noncontact, nonintrusive, and can produce thermograms in real time (TV rates or faster), which can be saved in a variety of media for further analysis. The major restrictions are that a direct line of sight to the target is required and that, in most cases, only the surface temperature can be mapped.

The relationship between surface temperature distribution on an electronic test article and its normal "acceptable" operating condition has been studied for many years. As a result, numerous programs of IR thermography have been established to develop standard thermal profiles (STPs) of components, circuit cards, and other electronic modules and to use these STPs as benchmarks against which similar test articles can be evaluated.

It has been learned through these programs that, in many cases, a detected deviation from standard thermal behavior is an early precursor of failure. It has also been noted that deviation from "normal" is often detectable in the first few seconds after power is applied to the test article. Thus, a powerful tool for detection of failure mechanisms is available that is noncontacting, nondestructive, global, and fast. In some cases, it is capable of detecting failure mechanisms long before they can cause serious damage to test articles and associated components.

Although IR thermography is based on radiative heat transfer, a working familiarity with all modes of heat flow is necessary for the development of a successful program of IR thermography.

6.2 HEAT, HEAT TRANSFER, AND TEMPERATURE

Heat is a form of energy that can be neither created nor destroyed. *Heat flow* is energy in transition, and heat always flows from warmer objects to cooler objects. *Temperature* is a measure of the thermal energy contained by an object—the degree of hotness or coldness of an object measurable by any of a number or relative scales. Temperature defines the direction of heat flow between two objects when one temperature is known.

There are three modes of heat transfer: conduction, convection, and radiation. All heat is transferred by means of one or another of these three modes—usually, in electronic elements, by two or three modes in combination. Of these three modes, infrared thermography is most closely associated with radiative heat transfer, but it is essential to study all three in order to understand the meaning of thermograms and to pursue a successful program of IR thermography for electronic failure analysis.

6.2.1 Temperature and Temperature Scales

Temperature is expressed in either absolute or relative terms. There are two absolute scales called *Rankine* (English system) and *Kelvin* (metric system). There are two corresponding relative scales called *Fahrenheit* (English system) and *Celsius* or *centigrade* (metric system). Absolute zero is the temperature at which no molecular action takes place. This is expressed as *zero Kelvins* or *zero Rankines* (0 K or 0 R). Relative temperature is expressed as *degrees Celsius* or *degrees Fahrenheit* (°C or °F). The numerical relationships among the four scales are as follows:

$$T_{Celsius} = 5/9(T_{Fahrenheit} - 32)$$

$$T_{Fahrenheit} = 9/5(T_{Celsius} + 32)$$

$$T_{Rankine} = T_{Fahrenheit} + 459.7$$

$$T_{Kelvin} = T_{Celsius} + 273.16$$

Absolute zero is equal to –273.1° C and also equal to –459.7° F. To convert a change in temperature or *delta T* (ΔT) between the English and metric systems, the simple 9/5 (1.8:1) relationship is used:

$$\Delta T_{Fahrenheit\ (or\ Rankine)} = 1.8\ \Delta T_{Celsius\ (or\ Kelvin)} \tag{6.1}$$

Table 6.1 allows the rapid conversion of temperature between Fahrenheit and Celsius values. Instructions for the use of the table are shown at the top.

6.2.2 Conductive Heat Transfer

Conductive heat transfer is the transfer of heat in stationary media. It is the only mode of heat flow in solids but can also take place in liquids and gases. It occurs as the result of atomic vibrations (in solids) and molecular collisions (in liquids) whereby energy is moved, one molecule at a time, from higher-temperature sites to lower-temperature sites. An illustration of conductive heat transfer is what happens when a transformer is mounted to a heat sink. The thermal energy resulting from the I^2R loss in the transformer windings is conducted into the heat sink, warming the heat sink and cooling the transformer.

The Fourier conduction law,

$$Q/A = \frac{k(\Delta T)}{L} \tag{6.2}$$

expresses the conductive heat flow Q per unit area A through a slab of solid material of thickness L and a thermal conductivity k with a temperature drop of $\Delta T = T_1 - T_2$ across it, as shown in Fig. 6.2a. In real terms, this expression means that the rate of heat flow increases with increasing temperature difference, increases with increasing thermal conductivity, and decreases with increasing slab thickness. Thermal conductivity is, generally, highest for metals and lower for nonmetals such as silicon and carbon composites. This makes the manner in which heat is conducted within semiconductor devices and circuit modules dependent on the structural materials.

6.2.3 Convective Heat Transfer

Convective heat flow takes place in a moving medium and is almost always associated with transfer between a solid and a moving fluid (such as air). Free convection takes place when the temperature differences necessary for heat transfer produce density changes in the fluid and the warmer fluid rises as a result of increased buoyancy. Forced convection takes place when an external driving force, such as a cooling fan, moves the fluid. An illustration of convective heat transfer is what happens at the interface between the cooling fins of a heat sink and air moved by a cooling fan. The thermal energy stored in the heat sink is convected into the surrounding air, warming the air and cooling the heat sink.

Table 6.1 Temperature Conversion Chart

Instructions for Use:
1. Start in the "Temp." column and find the temperature you wish to convert.
2. If the temperature to be converted is in °C, scan to the right column for the °F equivalent.
3. If the temperature to be converted is in °F, scan to the left column for the °C equivalent.

°C	Temp.	°F	°C	Temp.	°F	°C	Temp.	°F
−17.2	1	33.8	−2.8	27	80.6	11.7	53	127.4
−16.7	2	35.6	−2.2	28	82.4	12.2	54	129.2
−16.1	3	37.4	−1.7	29	84.2	12.8	55	131.0
−15.6	4	39.2	−1.1	30	86.0	13.3	56	132.8
−15.0	5	41	−0.6	31	87.8	13.9	57	134.6
−14.4	6	42.8	0.0	32	89.6	14.4	58	136.4
−13.9	7	44.6	0.6	33	91.4	15.0	59	138.2
−13.3	8	46.4	1.1	34	93.2	15.6	60	140.0
−12.8	9	48.2	1.7	35	95.0	16.1	61	141.8
−12.2	10	50	2.2	36	96.8	16.7	62	143.6
−11.1	12	53.6	2.8	37	98.6	17.2	63	145.4
−10.6	13	55.4	3.3	38	100.4	17.8	64	147.2
−10.0	14	57.2	3.9	39	102.2	18.3	65	149.0
−9.4	15	59	4.4	40	104.0	18.9	66	150.8
−8.9	16	60.8	5.0	41	105.8	19.4	67	152.6
−8.3	17	62.6	5.6	42	107.6	20.0	68	154.4
−7.8	18	64.4	6.1	43	109.4	20.6	69	156.2
−7.5	19	66.2	6.7	44	111.2	21.1	70	158.0
−6.7	20	68	7.2	45	113.0	21.7	71	159.8
−6.1	21	69.8	7.8	46	114.8	22.2	72	161.6
−5.6	22	71.6	8.3	47	116.6	22.8	73	163.4
−5.0	23	73.4	8.9	48	118.4	23.3	74	165.2
−4.4	24	75.2	10.0	50	122.0	23.9	75	167.0
−3.9	25	77	10.6	51	123.8	24.4	76	168.8
−3.3	26	78.8	11.1	52	125.6	25.0	77	170.6

Table 6.1   (continued)

°C	Temp.	°F	°C	Temp.	°F	°C	Temp.	°F
25.6	78	172.4	54.4	130	266	193	380	716
26.1	79	174.2	60.0	140	284	199	390	734
26.7	80	176.0	65.6	150	302	204	400	752
27.2	81	177.8	71.1	160	320	210	410	770
27.8	82	179.6	76.7	170	338	216	420	788
28.3	83	181.4	82.2	180	356	221	430	806
28.9	84	183.2	87.8	190	374	227	440	824
29.4	85	185.0	93.3	200	392	232	450	842
30.0	86	186.8	98.9	210	410	238	460	860
30.6	87	188.6	104.0	220	428	243	470	878
31.1	88	190.4	110.0	230	446	249	480	896
31.7	89	192.2	116.0	240	464	254	490	914
32.2	90	194.0	121.0	250	482	260	500	932
32.8	91	195.8	127.0	260	500	288	550	1022
33.3	92	197.6	132.0	270	518	316	600	1112
33.9	93	199.4	138.0	280	536	343	650	1202
34.4	94	201.2	143.0	290	554	370	700	1292
35.0	95	203.0	149.0	300	572	399	750	1382
35.6	96	204.8	154.0	310	590	427	800	1472
36.1	97	206.6	160.0	320	608	454	850	1562
36.7	98	208.4	166.0	330	626	482	900	1652
37.2	99	210.2	171.0	340	644	510	950	1742
37.8	100	212.0	177.0	350	662	538	1000	1832
43.3	110	230.0	182.0	360	680	566	1050	1922
48.9	120	248.0	188.0	370	698	593	1110	2012

Conversion Factors

$0°\text{ C} = (°\text{F} - 32) \times 5/9$ $0\text{ Kelvin } = -273.16°\text{ C}$

$0°\text{ F} = (°\text{C} \times 9/5) + 32$ $0\text{ Rankine} = -459.69°\text{ F}$

Courtesy of EPRI NDE Center.

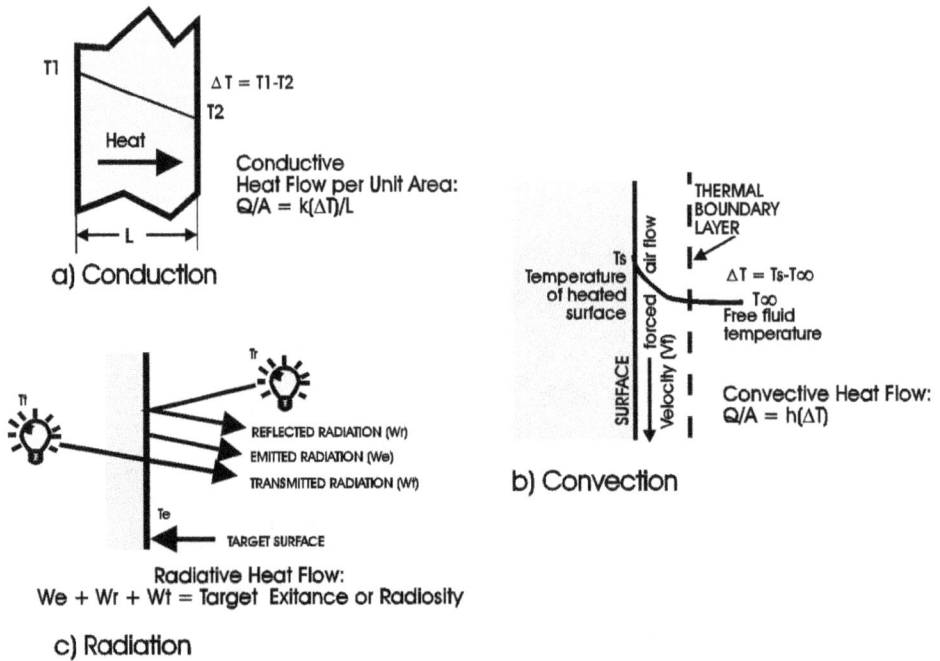

FIGURE 6.2 The three modes of heat transfer.

Figure 6.2*b* demonstrates the forced convective heat transfer between a flat, heated plate and a moving fluid. In convective heat flow, heat transfer takes effect by means of two mechanisms: the direct conduction through the fluid and the motion of the fluid itself. Newton's cooling law defines the convective heat transfer coefficient, *h*, which expresses the combined effect of these two mechanisms:

$$h = \frac{Q/A}{T_s - T_\infty} \qquad (6.3)$$

The presence of the plate in Fig. 6.2*b* causes the free fluid velocity, V_f, to decrease to zero at the surface and influences its velocity throughout a variable distance called the *boundary layer*. The thickness of the boundary layer depends inversely on the free fluid velocity and is greater for free convection and smaller for forced convection. The rate of heat flow, in turn, depends on the thickness of the convection boundary layer and the temperature difference, ΔT, between T_s and T_∞ (T_s is the surface temperature, and *T* is the free fluid temperature outside the boundary layer).

Convective heat transfer per unit area is expressed as a rearrangement of Newton's cooling law:

$$\frac{Q}{A} = h(\Delta T) \qquad (6.4)$$

In real terms, this expression means that the rate of heat flow increases with increasing temperature difference, with increasing fluid velocity, and with increasing fluid heat transfer coefficient.

6.2.4 Radiative Heat Transfer

Radiative heat transfer is unlike the other two modes in several respects.

- It can propagate through a vacuum.
- It occurs by electromagnetic emission and absorption.
- It occurs at the speed of light and behaves in a manner similar to light.
- While conductive and convective heat transferred between points is linearly proportional to the temperature difference between them, the energy radiated from a surface is proportional to the fourth power of its absolute temperature. The radiant thermal energy transferred between two surfaces is proportional to the third power of the temperature difference between the surfaces.

An illustration of radiative heat transfer is what happens when a power resistor is mounted adjacent to a diode on a circuit card. The thermal energy resulting from the I^2R loss in the resistor radiates from the surface of the resistor, and some of it is absorbed by the surface of the diode, warming the diode and cooling the resistor.

Thermal infrared radiation leaving a surface is called radiant *exitance* or *radiosity*. It can be emitted from the surface, reflected off the surface, or transmitted through the surface, as illustrated in Fig. 6.2c. The total exitance is equal to the sum of the emitted component, (We), the reflected component, (Wr), and the transmitted component, (Wt),. The surface temperature, however, is related only to We, the emitted component.

6.3 INFRARED BASICS

The measurement of thermal infrared radiation is the basis for noncontact temperature measurement and infrared (IR) thermography. The location of the infrared region in the electromagnetic spectrum is illustrated in Fig. 6.3.

Like light energy, thermal radiation is a photonic phenomenon that occurs in the electromagnetic spectrum. While light energy transfer takes place in the visible portion of the spectrum, from 0.4 to 0.75 μm, radiative heat transfer takes place in the infrared portion of the spectrum, between 0.75 and ≈1000 μm, although most practical measurements are made out to only about 20 μm.[*]

All target surfaces warmer than absolute zero radiate energy in the infrared spectrum. Figure 6.4 shows the spectral distribution of energy radiating from various idealized target surfaces as a function of surface temperature (T) and wavelength (λ). Very hot targets radiate in the visible spectrum as well, and our eyes can see this because they are sensitive to light. The sun, for example, has a temperature of about 6,000 K and appears to glow white hot. The heating element of an electric stove at 800 K glows a cherry red and, as it cools, loses its visible glow but continues to radiate. This radiant energy can be felt by a hand placed near the surface, but the glow is invisible because the energy has shifted from red to infrared. Also shown are two key physical laws regarding infrared energy emitted from surfaces.

The heat radiated from the surface is expressed by the Stephan-Boltzmann law:

---

[*]μm = micrometer or "micron" = one millionth of a meter. It is the standard measurement unit for radiant energy wavelength.

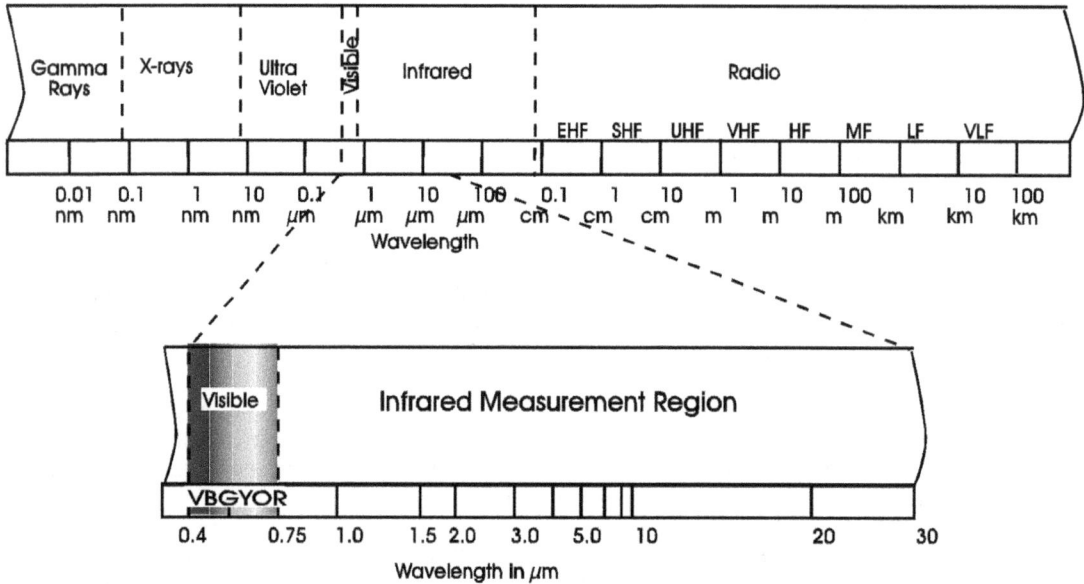

FIGURE 6.3 Infrared in the electromagnetic spectrum.

$$W = \sigma \varepsilon T^4$$

where W = radiant flux emitted per unit area (W/cm^2)
ε = emissivity (unity for a blackbody target)
σ = Stephan-Boltzmann constant = 5.673×10^{-12} W cm^{-2} K^{-4}
T = absolute temperature of target (K)

The peak wavelength, λ_m, (in μm) at which a surface radiates is determined by Wien's displacement law:

$$\lambda_m = b/T$$

where λm = Wavelength of maximum radiation (μm)
b = Wien's displacement constant = 2897 (μK)

From the point of view of IR radiation characteristics, there are three types of target surfaces; blackbodies, graybodies and non-graybodies (also called real bodies or spectral bodies). The target surfaces shown in Fig. 6.4 are all perfect radiators (or blackbodies). A blackbody radiator is defined as *a theoretical surface having unity emissivity at all wavelengths and absorbing all of the radiant energy impinging upon it.* Emissivity, in turn, is defined as *the ratio of the radiant energy emitted from a surface to the energy emitted from a blackbody surface at the same temperature.* Although blackbody radiators are theoretical and do not exist in practice, the surfaces of most solids are graybodies; that is, surfaces with high emissivities that are fairly constant with wavelength. Figure 6.5 shows the comparative spectral distribution of energy emitted by a blackbody, a graybody, and a non-graybody, all at the same temperature (300 K).

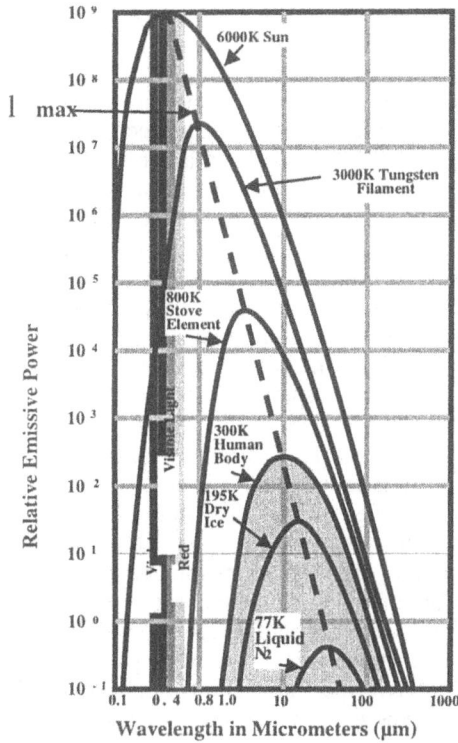

Stephan-Boltzmann Law:

Radiant flux per unit area in watts/cm,

$$W = \sigma \varepsilon \, T^4$$

Wein's Displacement Law:

$$\lambda \, max = \frac{b}{T} = \frac{2897}{T} \cong \frac{3000}{T}$$

FIGURE 6.4   Spectral distribution of radiant energy from blackbody targets.

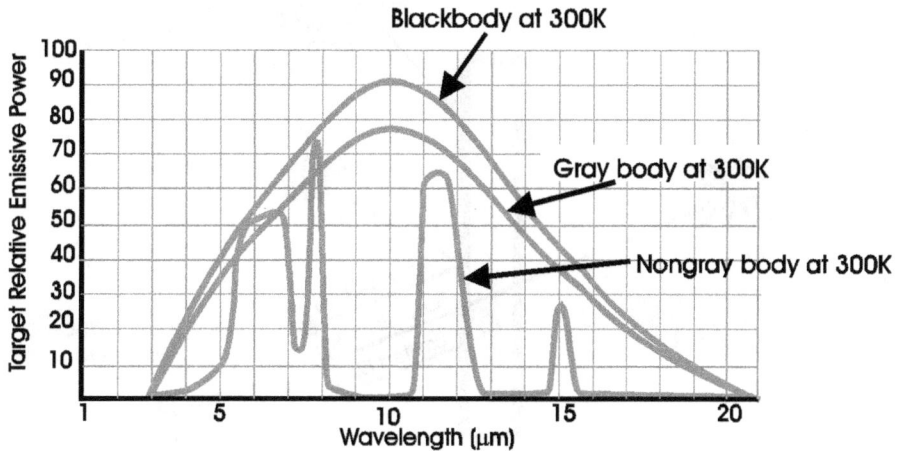

FIGURE 6.5   Spectral emission of a blackbody, graybody, and non-graybody.

Referring back to Fig. 6.4, the total exitance available to a measuring instrument has three components: emitted energy (We), reflected energy (Wr) from the environment and other reflecting sources, and energy transmitted through the target surface (Wt) from sources behind the target. Since a blackbody has an emissivity of 1.00, it will reflect and transmit no energy. A graybody resembles a blackbody in spectral distribution and in that it transmits no energy but, since its emissivity is less than unity, it may reflect energy. A non-graybody may emit, reflect, and transmit energy. To illustrate this, Fig. 6.6 shows the three components that constitute Wx, the total exitance that an instrument "sees" when aimed at a target surface. For a blackbody, Wx = We; for a graybody, Wx = We + Wr; and for a non-graybody, Wx = We + Wr + Wt. Since only the emitted component, We, is related to the temperature of the target surface, it becomes apparent that a significant problem in making IR temperature measurements is eliminating or compensating for the other two components. Since virtually all electronic packages, components, and assemblies are nontransparent graybodies, this problem is usually reduced to avoiding, or compensating for, reflected artifacts that have no relation to the surface temperature of the test article.

Infrared radiation from the target passes through some transmitting medium on its way to the infrared instrument. Through a vacuum, there is no loss of energy, but most infrared measurements are made through air. The effect of most atmospheric gases can be ignored for short distances, such as a few meters. As the path length increases, however, atmospheric absorption can be a source of error. Most instruments used in IR thermography provide the user with a means for correcting temperature measurements for atmospheric errors by entering the estimated measurement path length into the computation. Figure 6.7 depicts the spectral transmission characteristics of a 10 m path of 25° C ground-level atmosphere at 50 percent humidity. Two spectral intervals can be seen to be relatively free from absorption losses. These are known as the 3–5 μm and the 8–14 μm atmospheric "windows." Almost all infrared sensing and imaging instruments are designed to operate in one of these two windows.

Since most IR thermal measurements of electronics for failure analysis are made through a relatively short atmospheric path, errors due to atmospheric losses can generally be ignored. Solid media, however, such as a glass or quartz viewing port used in some

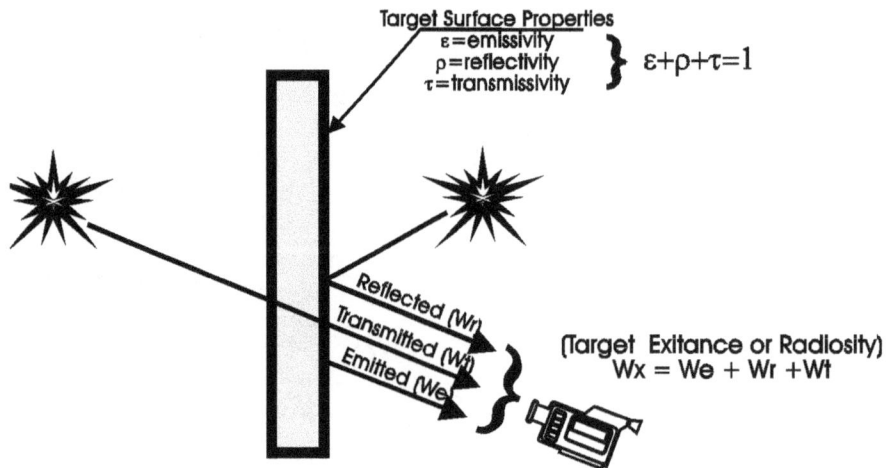

FIGURE 6.6   The three components that constitute total exitance from a surface.

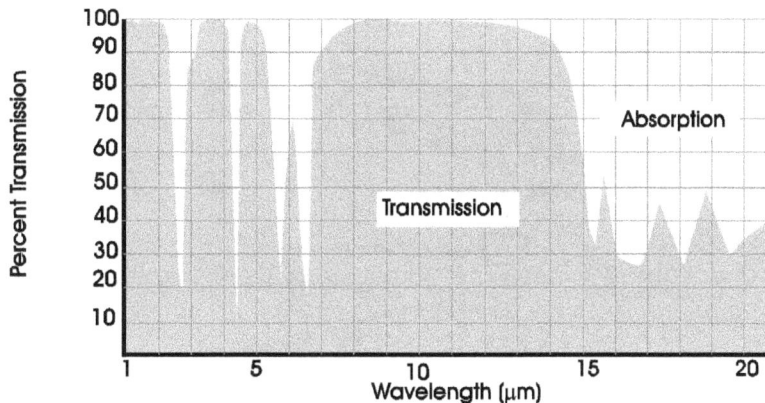

FIGURE 6.7   Spectral transmission characteristics of a 10 m path of 25° C ground-level atmosphere at 50% humidity.

manufacturing processes, can seriously attenuate infrared energy if situated between the target and the infrared instrument and can make temperature measurement difficult. In those cases, the spectral transmission and absorption characteristics of the medium must be taken into consideration in calibrating the measurement instrument.

6.4 HOW THERMAL IMAGING INSTRUMENTS OPERATE

Thermal imaging instruments map the infrared thermal exitance by using optics and an infrared detector. There are currently two types of thermal imaging instruments appropriate for use in electronics failure analysis: *scanning imaging radiometers* and *focal plane array imaging radiometers*. In the first type, moving optical elements are used to scan the energy from the target surface onto a single detector or a line of detectors; in the second type, the optics are fixed, and a staring multielement detector array is placed at the focal plane of the optics. The performance of the two imager types is comparable but, because of inherent advantages that will be discussed later in this chapter, the focal plane array (FPA) imagers are gradually replacing scanning imagers in most electronics applications.

6.4.1 Elements of Thermal Imagers

The schematics of Fig. 6.8 illustrate the operation of a scanning imager (top) and a staring FPA imager (bottom). In the scanning imager, the optics collect the energy from each spot (resolution element) on the target surface in sequence and focus it onto the detector. In the FPA imager, each detector in the array receives energy from one assigned resolution element on the target surface, and scanning is accomplished electronically by means of multiplexer circuitry. The infrared detector is a transducer that converts the radiant energy impinging on its sensitive surface to some form of electrical change. Depending on the type of detector used, this can be in the form of a change in EMF or impedance or capacitance, or in the release of photons. The electronic processor converts these changes to use-

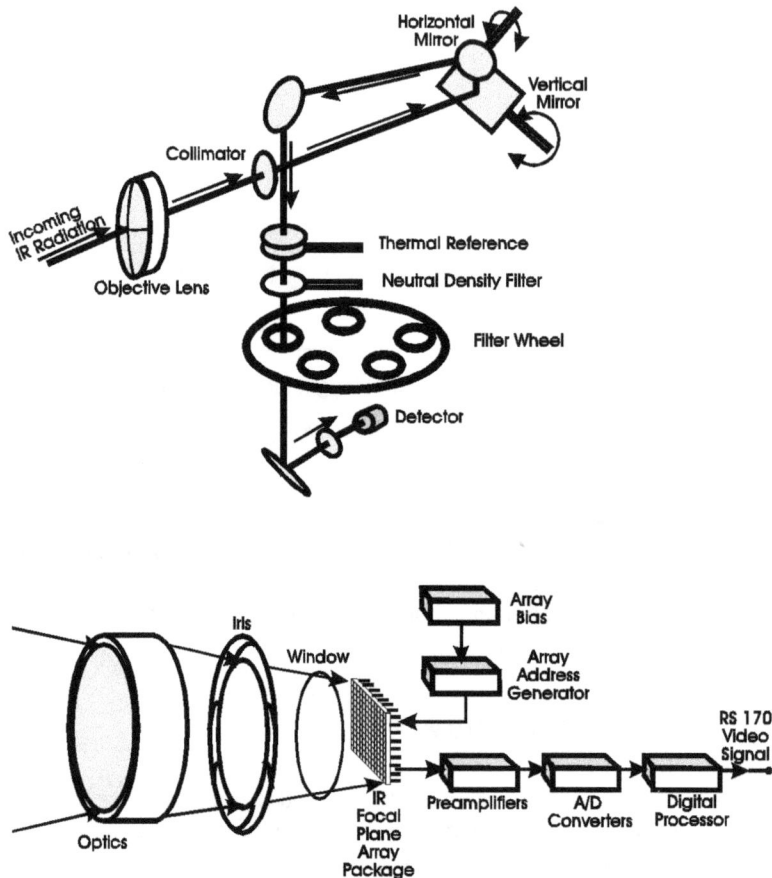

FIGURE 6.8 Typical infrared scanning radiometer (top, courtesy of Inframetrics, Inc.) and infrared staring focal plane array radiometer (bottom, courtesy of Honeywell).

ful signals, introduces the appropriate corrections, and produces the thermogram in analog or digital form, which is then available for display, processing and storage.

6.4.2 Performance Parameters of Thermal Imagers

The significant performance parameters for an infrared thermal imager to perform failure analysis on electronic packages, assemblies, and components are

- Temperature range
- Temperature absolute accuracy and repeatability
- Thermal resolution (minimum resolvable temperature difference)
- Total field of view (picture size)
- Imaging spatial resolution (instantaneous field of view)

- Measurement spatial resolution (instantaneous measurable field of view)
- Frame repetition rate
- Spectral region of operation

Temperature Range.   Temperature range is a statement of the high and low limits over which the target temperature can be measured by the instrument. Most imaging radiometers cover the temperature range of 0 to 300° C, which is adequate for most electronic analysis applications.

Temperature Absolute Accuracy and Repeatability.   Absolute accuracy, as defined by the National Institute of Standards and Technology (NIST) standard, describes the maximum error, over the full range, that the measurement will have when compared to this standard blackbody reference. Repeatability describes how faithfully a reading is repeated for the same target over the short and long term. Since absolute accuracy is based on traceability to the NIST standard, it is difficult for a manufacturer to comply with a tight specification for absolute accuracy. An absolute accuracy of $\pm 0.5°$ C $\pm 1\%$ of full scale is about as tight as can be reasonably specified. Repeatability, on the other hand, can be more easily assured by the manufacturer and is usually more important to the user. A typical short- and long-term repeatability specification, acceptable for most electronic analysis applications, would be $\pm 0.25°$ C.

Thermal Resolution.   Thermal resolution defines the smallest target temperature change the instrument will detect. Thermal resolution is also called "minimum resolvable temperature difference" (MRTD). It is the smallest temperature change at the target surface that can be clearly resolved on the displayed image. An acceptable specification for temperature sensitivity for most electronic analysis applications would be 0.1° C at a target temperature of 25° C. In this case, the thermal resolution of the instrument would improve for targets hotter than 25° C.

Total Field of View.   Total field of view (TFOV) defines the image size in terms of total vertical and horizontal scanning angles, y and x, respectively, for any given lens. An example of a typical TFOV specification would be 20° vertical × 30° horizontal (with standard 1× lens) and would define the total target size encompassed by the thermogram, V (height) and H (width) in terms of working distance (d) by means of simple trigonometric relationships as follows:

$$V = \text{total vertical image size} = d[2 \tan(y/2)]$$

$$H = \text{total horizontal image size} = d[2 \tan(x/2)]$$

Changing the lens will change the TFOV. A 2× lens in the above example, substituted for the standard 1× lens, would change the TFOV to 10° vertical × 15° horizontal.

Imaging Spatial Resolution.   Imaging spatial resolution represents the size of the smallest picture element that can be imaged with any given lens. It is also called *instantaneous field of view* (IFOV) and is the angular projection of the detector element at the target plane. An example of a typical IFOV specification would be 1.7 mrad @ 35% MTF with standard 1× lens. The "35% MTF" refers to the *modulation transfer function* test that is used to check imaging spatial resolution. The simple expression $D = \alpha d$ can be used to estimate the actual imaging spot size at the target plane from manufacturer's published data by substituting the published IFOV for α.

Changing the lens will change the IFOV. A 2× lens in the above example, substituted for the standard 1× lens, would change the IFOV to 0.85 mrad.

Measurement Spatial Resolution. Measurement spatial resolution (IFOVmeas) is the spatial resolution describing the minimum target spot size on which an accurate measurement can be made in terms of its distance from the instrument. An example of a typical IFOVmeas specification would be 3.5 mrad @ 95% SRF.

The "95% SRF" refers to the *slit response function* test used to check measurement spatial resolution. The simple expression: $D = \alpha d$ can again be used to estimate the actual measurement spot size at the target plane from manufacturer's published data by substituting published IFOVmeas for α.

Changing the lens will change the IFOVmeas. A 2× lens in the above example, substituted for the standard 1× lens, would change the IFOVmeas to 1.75 mrad.

Frame Repetition Rate. Frame repetition rate is defined as the number of times every point on the target is scanned in one second. An example of a typical frame repetition rate specification for an imager would be 30 frames/s.

Spectral Range. Spectral range defines the portion of the infrared spectrum over which the instrument will operate. The spectral range is determined by the detector and the instrument optics. The operating spectral range of the instrument is often critical to its performance and, in many applications, can be exploited to solve difficult measurement problems. In most electronic diagnostics applications, however, the spectral range is not critical. Most thermal imaging radiometers operate in either the 3 to 5 μm or the 8 to 12 μm spectral region, both of which are suitable to most electronic diagnostic applications.

Suitability of IRFPA Imagers to Electronic Failure Analysis Applications. The four parameters of greatest importance in most electronic failure analysis applications are thermal resolution, imaging spatial resolution, measurement spatial resolution, and frame repetition rate. While the thermal resolution capability of scanning imagers and staring FPA imagers is comparable, the spatial resolution of FPA imagers is somewhat better, and the frame repetition rate of FPA imagers can be faster. For this reason, the FPA imager is becoming the instrument of choice for most thermographic electronic failure analysis applications.

6.4.3 Diagnostic Software and Image Processing

Depending on manufacturer and model, some of the software required for electronic failure analysis applications is now incorporated into instruments, and some is still available only on computer-driven software packages. Diagnostic software can be classified into five groups as follows:

- Correction for errors and artifacts—true temperature measurement
- Image processing and diagnostics
- Image recording, storage, and recovery

- Image comparison (differential infrared thermography)
- Database and documentation

Correction for Errors and Artifacts—True Temperature Measurement. Correction for errors and artifacts provides the capability of converting qualitative thermograms to quantitative thermograms, providing the user with the true radiance or apparent temperature value of any or all points on the target surface. This is an important feature for most electronic reliability studies. Most imaging instruments include corrections for system throughput variations such as detector response characteristics and attenuation through the optical elements. As previously mentioned, atmospheric attenuation is negligible for most electronic measurement applications. The important corrections that must be made are for target emissivity and for reflection artifacts.

For most systems, the displayed temperature readings are based on the assumption that the entire target surface has the same effective emissivity. Effective emissivity is defined as "the measured emissivity value of a particular target surface under existing measurement conditions (rather than the generic tabulated value for the surface material) that can be used to correct a specific measuring instrument to provide a correct temperature measurement." For the most accurate temperature measurement of a specific area, then, it is necessary for the operator to insert the effective emissivity of that material. Some systems allow the assignment of several different emissivities to different areas of the target selected by the operator with the resulting temperature correction.

Although look-up tables of emissivities for various materials are available (see Table 6.2), they are generic and should only be used when a measured effective emissivity of the material cannot be obtained. Emissivity can vary with spectral range, temperature, view angle, and surface features. Appendix 6.A, at the end of this chapter, illustrates a procedure that may be used to measure the effective emissivity of a sample material using the instrument with which measurements will subsequently be made.

For limited applications, specifically microimaging for semiconductor devices, some manufacturers offer automatic emissivity correction routines that measure and correct for the emissivity variations over the entire target. These will be discussed later in this chapter. Normally, a color scale or gray scale is provided along one edge of the display with temperatures shown corresponding to each color or gray level in the selected range. The operator can place one or more spots or crosshairs on the image and the apparent temperature value at that location will appear in an appropriate location on the display. The "isotherm" feature allows the operator to select a temperature "band" or interval, and all areas on the target within that band then appear enhanced in a predetermined gray shade or color hue.

The two types of reflection artifacts are *point source reflections* and *ambient background reflections.* A point source reflection is usually apparent by inspection of the thermogram. This may be the reflection of a heater or an incandescent lamp, but most commonly it is the reflection of the operator's body off low-emissivity portions of the target surface. These can be eliminated by baffling or placing a screen behind the instrument when making critical measurements. For ambient background reflections, the operator must insert a value for ambient background temperature. This is the average temperature of the ambient background behind the instrument. The software will then provide the appropriate correction to the apparent temperature.

Image Processing and Diagnostics. Image processing and diagnostics relies on software that allows manipulation and analysis of each pixel in the thermogram presenting

Table 6.2 Normal Spectral Emissivities of Common Materials *(courtesy of AGEMA Infrared Systems, Inc.).*

Material	Temperature °C	Wavelength μm	Emissivity (ε)
Alumina brick	17	2–5	0.68
Aluminum, polished	0	8–14	0.05
Aluminum, rough surface	0	8–14	0.07
Aluminum, strongly oxidized	0	8–14	0.25
Aluminum, heavily weathered	17	2–5	0.83–0.94
Aluminum foil, bright	17	2–5	0.09
Aluminum disk, roughened	17	2–5	0.28
Asbestos board	0	8–14	0.96
Asbestos fabric	0	8–14	0.78
Asbestos paper	0	8–14	0.94
Asbestos slate	0	8–14	0.96
Asbestos board	17	2–5	0.96
Brass, dull, tarnished	0	8–14	0.22
Brass, polished	0	8–14	0.03
Brick, common	0	8–14	0.85
Brick, common	17	2–5	0.81–0.86
Brick, facing, red	17	2–5	0.92
Brick, facing, yellow	17	2–5	0.72
Brick, waterproof	17	2–5	0.87
Brick, glazed, rough	0	8–14	0.85
Brick, refractory, rough	0	8–14	0.94
Bronze, porous, rough	0	8–14	0.55
Bronze, polished	0	8–14	0.10
Carbon, purified	0	8–14	0.80
Cast iron, rough casting	0	8–14	0.81
Cast iron, polished	0	8–14	0.21
Cement	0	8–14	0.54
Charcoal, powdered	0	8–14	0.96
Chromium, polished	0	8–14	0.10
Chipboard, untreated	17	2–5	0.90
Clay, fired	0	8–14	0.91
Concrete, dry	36	2–5	0.95
Concrete, rough,	17	2–5	0.92–0.97
Copper, polished, annealed	0–17	8–14	0.01–0.02
Copper, commercial burnished	0	8–14	0.07
Copper, oxidized	0	8–14	0.65
Copper, oxidized to black	0	8–14	0.88
Enamel	0	8–14	0.90
Fiberboard, untreated	17	2–5	0.85
Filler, wood, white	17	2–5	0.68
Formica	27	8–14	0.93
Frozen soil	0	8–14	0.93
Glass	0	8–14	0.92
Glass, chemical ware	35	2–5	0.97
Glass, frosted	0	8–14	0.96

Table 6.2 continued

Material	Temperature °C	Wavelength μm	Emissivity (ε)
Gold, polished	0	8–14	0.02
Granite, natural surface	36	2–5	0.96
Gravel	0	2–5	0.28
Hardwood, end grain	17	2–5	0.82
Hardwood, face	17	2–5	0.68–0.73
Ice	0	8–14	0.97
Iron, heavily rusted	17	2–5	0.91–0.96
Iron, hot rolled	0	8–14	0.77
Iron, oxidized	0	8–14	0.74
Iron, sheet, galvanized, burnished	0	8–14	0.23
Iron, sheet, galvanized, oxidized	0	8–14	0.28
Iron, Shiny, etched	0	8–14	0.16
Iron, wrought, polished	0	8–14	0.28
Lacquer, bakelite	0	8–14	0.93
Lacquer, black, dull	0	8–14	0.97
Lacquer, black, shiny	0	8–14	0.87
Lacquer, white	0	8–14	0.87
Lampblack	0	8–14	0.96
Lead, gray	0	8–14	0.28
Lead, oxidized	0	8–14	0.63
Lead, red, powdered	0	8–14	0.93
Lead, shiny	0	8–14	0.08
Limestone, natural surface	36	2–5	0.96
Mercury, pure	0	8–14	0.10
Mortar	17	2–5	0.87
Mortar, dry	36	2–5	0.94
Nickel, on cast iron	0	8–14	0.05
Nickel, pure polished	0	8–14	0.05
Paint, alkyd enamel	40	2–5	0.95–0.98
Paint, silver finish	25	2–5	0.26
Paint, silver finish	25	8–14	0.31
Paint, Krylon, flat black	50	2–5	0.95
Paint, Krylon, flat white 1502	40	2–5	0.99
Paint, Krylon, ultra–flat black	36	2–5	0.97
Paint, 3M 9560 optical black velvet coating,	40	2–5	>0.99
Paint, oil, average	17	2–5	0.87
Paint, oil, average	0	8–14	0.94
Paint, oil, flat black	17	2–5	0.94
Paint, oil, black gloss	17	2–5	0.92
Paint, oil, gray flat	17	2–5	0.97
Paint, oil, gray gloss	17	2–5	0.96
Paper, cardboard box	17	2–5	0.81
Paper, white	17	2–5	0.68
Paper, black, shiny	0	8–14	0.90
Paper, black, dull	0	8–14	0.94
Paper, white	0	8–14	0.90

Table 6.2 continued

Material	Temperature °C	Wavelength μm	Emissivity (ε)
Perspex, plexiglass	17	2–5	0.86
Pipes, glazed	17	2–5	0.83
Plaster	17	2–5	0.86–0.90
Plasterboard, untreated	17	2–5	0.90
Plastic, acrylic, clear	36	2–5	0.94
Plastic, black	17	2–5	0.84
Plastic paper, red	36	2–5	0.94
Plastic, white	17	2–5	0.84
Platinum, pure, polished	0	8–14	0.08
Plywood	17	2–5	0.83–0.98
Plywood, commercial smooth finish, dry	36	2–5	0.82
Plywood, untreated	36	2–5	0.86
Polypropylene	17	2–5	0.97
P.V.C. (polyvinyl chloride)	17	2–5	0.91–0.93
Porcelain, glazed	0	8–14	0.92
Quartz	0	8–14	0.93
Redwood	17	2–5	0.83
Roofing metal, various colors and textures	0	2–5	0.51–0.70
Rubber	0	8–14	0.95
Shellac, black, dull	0	8–14	0.91
Shellac, black, shiny	0	8–14	0.82
Shingles, asphalt, various colors and textures	0	2–5	0.74–0.96, avg. 0.79
Shingles, fiberglass, various colors and textures	0	2–5	0.74–0.98, avg. 0.86
Shingles, solid vinyl, various colors and textures	0	2–5	0.75–0.93, avg. 0.84
Snow	0	8–14	0.80
Steel, galvanized	0	8–14	0.28
Steel, oxidized strongly	0	8–14	0.88
Steel, rolled freshly	0	8–14	0.24
Steel, rough surface	0	8–14	0.96
Steel, rusty red	0	8–14	0.69
Steel, sheet, nickel-plated	0	8–14	0.11
Steel, sheet, rolled	0	8–14	0.56
Styrofoam insulation	37	2–5	0.60
Tape, electrical insulating, black	35	2–5	0.97
Tape, masking	36	2–5	0.92
Tar paper	0	8–14	0.92
Tile, floor, asbestos	35	2–5	0.94
Tile, glazed	17	2–5	0.94
Tin, burnished	0	8–14	0.05
Tungsten	0	8–14	0.05
Varnish, flat	17	2–5	0.93
Wallpaper, average	17	2–5	0.85–0.90
Water	0	8–14	0.98
Wood paneling, finished	36	2–5	0.87
Wood, polished, spruce	36	2–5	0.86
Zinc, sheet	0	8–14	0.20

information in a wide variety of qualitative and quantitative forms for the convenience of the user. Some of these capabilities are described below.

In addition to the spot measurement capability discussed previously, line profiles may be selected. The analog trace (in x, y, or both) of the lines on the image intersecting at the selected spot will then appear at the edge of the display. Some systems allow the operator to display as many as seven sets of profiles simultaneously. Profiles of skew lines can also be displayed on some systems. Selected areas on the thermogram, in the form of circles, rectangles or point-to-point freeforms, can be shifted, expanded, shrunk or rotated or used to blank out or analyze portions of the image.

Detailed analysis of the entire image or the pixels within the area can include maximum, minimum, and average values, number of pixels, or even a frequency histogram of the values within the area. Color scales can be created in almost infinite variety from as many as 256 colors stored within the computer. Electronic zoom features allow the operator to expand a small area on the display for closer examination or to expand the colors for a small measurement range. Autoscale features provide the optimum display settings for any image if selected. 3-D features provide an isometric "thermal contour map" of the target for enhanced recognition of thermal anomalies.

Image Recording, Storage, and Recovery. The phrase *image recording, storage, and recovery* describes the capability of recording and retrieving images and data in memory, hard disk, diskette, videotape and, most recently, personal computer memory/computer industry association (PCMCIA) cards. Most newer commercial thermal imaging systems incorporate some means, such as a floppy disk drive or a PCMCIA card (also called a *PC card*) to store images in the field. Usually, about 40 images, with all accompanying data, can be stored on a 3.5-in diskette or a PC card. Some analysis usually can be done with on-board software, but more extensive diagnostics usually require a separate computer with both a hard disk drive and one or more floppy disk drives.

Options include IEEE or RS232 ports for access to additional storage plus a VCR option so that an entire measurement program can be recorded on videotape through an RS170 connector. Videotapes can be played back into the system, and images can be saved to disk. Images can be stored from a "frozen frame" thermogram of a live target on operator command, or the operator can set up an automatic sequence, and a preset number of images will be stored at preset time intervals. Stored images can be retrieved, displayed, and further analyzed.

Image Comparison with Differential Infrared Thermography (DIT). Image comparison with differential infrared thermography allows the automatic comparison of thermograms taken at different times. This includes time-based DIT (comparison of images taken of the same target) as well as comparative DIT (comparison of images taken of different but similar targets). A special software program allows the operator to display two images side by side or in sequence and to subtract one image from another, or one area from another, and to display a pixel-by-pixel "difference thermogram." Comparison (subtraction) of images can be accomplished between two images retrieved from disk, between a live image and an image retrieved from disk, and between a live image and an image stored in RAM. In this way, standard thermal images of acceptable electronic packages, components, and assemblies can be archived and used as models for comparison to subsequently inspected items. It is also possible to subtract a live image from a previous baseline image for subsequent time-based thermal transient measurements. This image comparison capability is critical to many past and current electronics applications as will be illustrated later in this chapter.

Database and Documentation. The term *database and documentation* describes the universal capability of saving records, files, data, and documents in an orderly manner. This capability provides the thermographer with a filing system so that records of all measurement programs can be maintained on magnetic media, including actual thermograms, time, date, location, equipment, equipment settings, measurement conditions, and other related observations.

Most manufacturers of thermal imaging equipment have developed comprehensive report preparation software to facilitate timely and comprehensive reporting of the findings of infrared surveys and other measurement missions. These packages provide templates that allow the thermographer to prepare reports in standard word processor formats, such as Word for Windows®, into which TIFF (tagged image file format) or bitmap images, imported from various imaging radiometers, can be directly incorporated. Additional diagnostic software is customarily provided in these packages so that postanalysis and trending can be added to reports.

6.5 IR THERMOGRAPHY TECHNIQUES FOR FAILURE ANALYSIS

Various techniques have been employed to optimize the effectiveness of infrared thermography in electronics applications including routine thermographic inspection of powered test articles, power stimulation of bare multilayer boards, differential infrared thermography, spatial emissivity correction, and the use of neural networks for fault detection and isolation. These techniques will be discussed and illustrated in the remainder of this chapter.

6.5.1 Thermal Behavior of Electronics Modules

It has been established that most failures in electronics packages, components, and assemblies are heat related; that is, they are caused by excessive heating, they generate excessive heating, or both. Even in cases where failures result in open circuits with no heating, the absence of heating itself may reveal the failure or malfunction. As previously mentioned, standard thermal profiles can be developed for generic samples of test articles under repeatable test conditions, and these STPs can be used as benchmarks for future comparison. In some cases, these STPs are distinctive, with clear, repeatable thermal gradients. In other cases they are not.

6.5.2 Applicability of Differential Infrared Thermography to Electronic Failure Analysis

Background. During the 1960s, infrared thermal images were used, to a limited extent, to verify the thermal design of some printed circuit boards (PCBs) slated for critical space hardware. It was observed that each PCB design appeared to produce a distinct, unique, and repeatable thermogram when power was applied. It also appeared that malfunctioning PCBs caused variations to appear on these thermograms. Because the IR imaging equipment available at that time did not have the image processing power necessary to pursue the implications of these findings, it wasn't until the early 1980s that PC-compatible IR imagers were used to verify the concept of the standard thermal profile for printed circuit boards and other electronics modules.

In 1983, a prototype infrared automatic mass screening system was completed under an Air Force contract and installed at McClellan Air Force Base in Sacramento, California. The system combined a high-resolution infrared imager with a mechanically calibrated test mounting bench, programmable power supplies, and a diagnostic computer and software. Its function was to compare, rapidly and automatically, the thermogram of a PCB unit under test (UUT) with the STP of that PCB, called up from mass storage, to detect gross faults. This was accomplished by means of differential infrared thermography (DIT), whereby the STP thermogram was subtracted, pixel by pixel, from the UUT thermogram, and the difference thermogram was compared to a preset acceptance tolerance matrix. This technique of comparing two images derived at different times is known as *comparative DIT*. Several military contracts soon were awarded, based on the promising results obtained from this first system. The original purpose of these programs was to develop screening techniques that would reduce the work load on automatic test equipment (ATE) and reduce testing time by culling out obvious "failed" modules.

Effectiveness Criteria for DIT.   For the most effective results, the two images to be compared should have been taken on the element under the same conditions of ambient and load, and with the same imager scale and sensitivity settings. Also, if the images are not properly registered with one another prior to subtraction, the result will not be meaningful. This means that the unit under test (UUT) must be viewed at the same distance and view angle as the element from which the stored standard thermal profile (STP) was derived. It also means that the two elements must be at the same location on the display prior to subtraction. If it is unlikely that these conditions can be met in the investigation of a particular element, then comparative DIT is probably not appropriate.

Until the development of dynamic subtraction software, the usefulness of DIT was limited by the necessity for precision mechanical registration between the two images prior to subtraction. This required a costly and cumbersome positioning fixture such as the mechanically calibrated test mounting bench used on the early PCB screening programs. The advent of the new software allowed the user to view both live and stored images simultaneously and superimposed. This allows the stored image to be used by the operator as a positioning template to align the two images prior to subtraction—the fixture is no longer required. Care is still necessary, however, to match carefully the target image sizes and viewing angles.

The types of elements on which comparative DIT can be most effective are those that exhibit complex thermal signatures on which visual comparison of thermograms can miss subtle details, and thermal artifacts can go unnoticed. Experience has shown that small relay assemblies and electronic elements such as power supplies, printed circuit cards, and hybrid modules fall into this category. By comparison, the thermograms of large, isothermal elements have a less complex structure, and thermal artifacts and anomalies indicating defects or operational problems are usually quite apparent to the thermographer without the use of DIT. The use of time-based DIT, on the other hand, is equally useful for trending studies on all elements, regardless of size or thermal complexity.

Despite the limitations discussed, there are numerous advantages to DIT when correctly and appropriately applied. The basic difference between DIT and conventional thermography is that, while conventional thermography presents a thermogram of a target with its full dynamic range of temperatures, DIT presents only differences, thereby highlighting changes or variations from a norm. Situations where DIT is particularly useful are those where a difference measurement is significant. They include

- Comparison of a complex STP with a UUT thermogram
- Trending over time
- Transient events or events of short duration
- Situations where surface artifacts such as emissivity and reflectivity differences tend to mask results
- Nondestructive testing of materials

In circuit cards (and in most other electronic devices), it was found that the most effective comparisons can be made within 30 seconds of power-up. At this time, all components will have reached more than 50 percent of their ultimate temperature rise. Also, co-heating effects between components (due mainly to radiative heat transfer) will not as yet have become significant enough to caused thermal "wash-out" of the image.

Establishing Baselines. The key to a meaningful comparison is the validity of the STP. If it is noted or anticipated that typical operating load conditions for the element vary widely, a set of STPs may be necessary, with images recorded for a series of operating loads. In each STP, the load condition should be part of the recorded image data. However, it should be noted that, among elements most suited to comparative DIT (small relay assemblies, power supplies, printed circuit cards, and hybrid modules), operating load conditions are usually consistent from element to element, and the need for multiple dynamic ranges is seldom encountered.

A stored difference thermogram file serves four distinct purposes. It serves as

- A baseline reference when no significant deviation from the STP is noted
- An event in historical trending when small deviations begin to appear
- A signature of a particular failure mechanism for future image diagnostics
- A file that can be inserted directly into a report using available report-generation software

6.5.3 Augmenting Thermography with Artificial Intelligence

In many cases, infrared thermography can be used in electronic fault diagnosis to isolate malfunctions on a circuit card to specific areas where ATE cannot. Given a series of thermograms of an "acceptable" UUT taken at specific intervals after power-up, the heating rate of components on a UUT can be computed and compared with a mathematical model based on design parameters. Permissible temperature deviations can be estimated for each component and confirmed through subsequent testing. Using neural networks, the test system can "learn" to recognize and classify variations and combinations of "out-of-tolerance" measurements and use the information to assign a probable root cause of failure and a suggested repair protocol.

For more than five years, studies of the effectiveness of neural networks combined with infrared thermography have been conducted, primarily at the Ogden Air Logistics Center, Hill Air Force Base, Utah. The result of these studies promises to bring high confidence failure prediction closer to reality.

6.5.4 Spatial Emissivity Correction for Semiconductor Devices and Other Microtargets

As previously discussed, the emissivity of a point on a surface depends, in general, on its material and texture and can vary between zero (perfect reflector) and 1.0 (perfect emitter or "blackbody"). The wide variation of emissivities on the surface of most semiconductor devices can result in misleading results when thermography is used to characterize temperature distribution on a powered device under test (DUT).

Fortunately, for some very small targets, it has become possible to produce "true temperature" thermograms of target surfaces by measuring the infrared radiant energy leaving a surface, isolating the component of this energy related to the surface temperature, and applying pixel-by-pixel corrections to this energy component for the emissivity variations on the target surface, thus computing surface temperature.

This can be accomplished by

1. Placing the unpowered DUT on a heated, temperature controlled substage
2. Measuring the radiant exitance from the target at two known temperatures obtained by heating uniformly
3. Assuming that the transmitted component is negligible
4. Isolating and nullifying the reflected component
5. Computing the emissivity of every pixel, and using this emissivity matrix to calculate the true temperature at every pixel
6. Using this matrix to correct the thermogram of the subsequently powered DUT, thus displaying a "true temperature" thermogram

This procedure has been reduced to practice and is currently used in quality control and failure analysis by many manufacturers of semiconductor devices. There are stringent conditions, however, for the successful implementation of spatial emissivity correction. They are as follows:

1. The device must be fully stabilized at each temperature before images are stored.
2. The ambient conditions must remain unchanged throughout the procedure.
3. The device must not be moved throughout the procedure.

6.5.5 Thermal Regression Analysis

Most operating printed circuit boards (PCBs) are in a rack of boards where they are parallel to one another and interact thermally. Since infrared thermography requires a clear line of sight between the target and the imager, viewing the components on any given operating board requires the use of an extender card. This makes the board accessible for thermography but changes the thermal environment and, as a result, may change the thermal profile.

The use of thermal regression analysis allows the in-situ thermal profile of a PCB to be synthesized. This requires that the PCB be de-energized and placed immediately in an environmentally stable container where a series of at least two thermograms are taken at precisely measured time intervals after power-off. The procedure is described in greater detail in a specific case study in Sec. 6.6.5.

6.6   CASE STUDIES

The case studies presented in this section are typical of previous and current applications of infrared thermography to electronic failure analysis.

6.6.1   Conventional thermography

Figure 6.9 shows the thermogram of a programmable logic device obtained using conventional thermal imaging and a close-up lens. The purpose of the study was to verify the presence of small epoxy droplets (dark circles) that appear as cool patterns around the logic in the center of the image.

In addition, the analysis software capability previously mentioned in Sec. 6.4.3, under *Image Processing and Diagnostics,* is illustrated. Several displays are provided including quantitative data at four cursor locations and frequency histograms for two area scans (circle and square) and a line scan.

6.6.2   Bare Multilayer Boards

For more than ten years, thermal imaging has been used for the location of shorts and plated through holes on bare multilayer circuit boards immediately after fabrication and

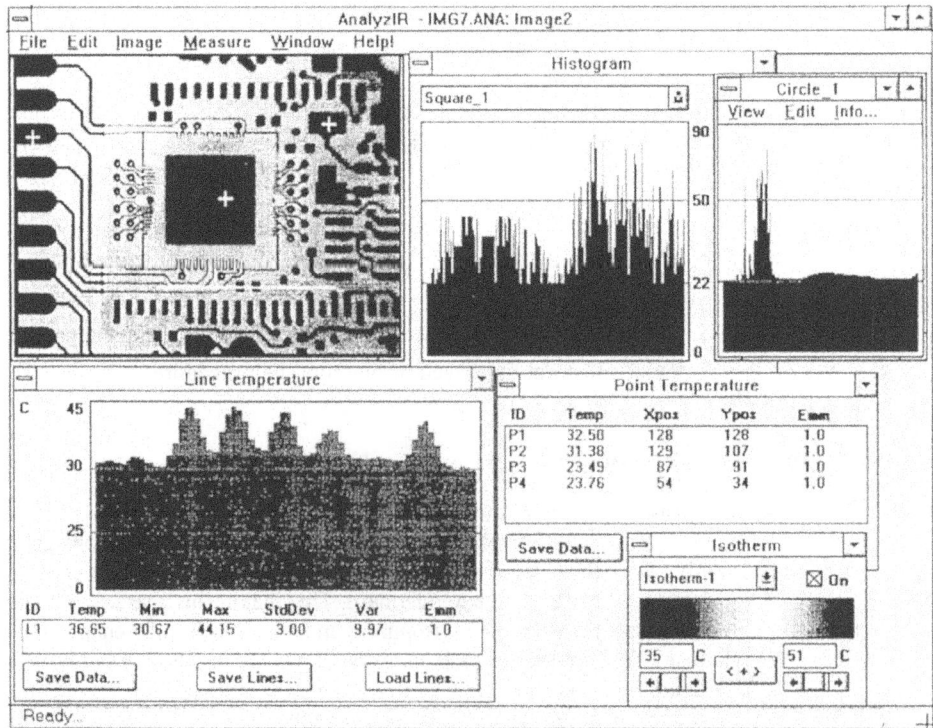

FIGURE 6.9   Thermogram and software analysis of a programmable logic device (courtesy of FLIR Systems, Inc.).

prior to assembly. While the existence of these interplane short circuits can be readily detected by electrical testing and the planes involved identified, the specific address on the circuit board cannot. Many of these boards are larger than 4 ft^2 in area and contain in excess of 300,00 clearance holes. The cost invested in each board is substantial, even before assembly, and the use of thermography in production reportedly has been 100 percent effective in salvaging boards with interplane shorts due to plated-through holes and other buried defects.

In preparation for test, a circuit board is placed in a test fixture such that the side viewed is the one closest to the nearest plane involved in the short, as indicated by electrical testing. The test is predicated on the fact that, if current is passed between the two planes, the shorted areas will be the first to heat up.

A control thermogram is taken to serve as a reference baseline image before any stimulus is introduced. This characterizes the emissivity and reflection patterns of an unstimulated board. A controlled current is then passed between the two planes involved, with the maximum wattage limited to a safe value. The current is increased until the operator observes the appearance of the defect as a hot spot, or until the safe wattage limit is reached.

The recovery success results reported after less than one year of testing at the IBM facility in Endicott, New York, and published in 1987 are shown in the following tabulation. The testing was performed by production personnel with the safe maximum power dissipation established at 2 W. Results are shown for all production and prototype boards tested to that time, in terms of the power required to detect the defect on each board.

- All boards (production and development)

  82% < 2W
  58% < 1W
  45% < 0.5W
  23% < 0.3W

- Production boards

  95% < 2W
  73% < 1W
  60% < 0.5W
  33% < 0.3W

The boards on which the defects could not be located with the 2-W limit were then tested by engineering personnel who were permitted to increase the power levels until the defects were located. No lost boards resulted, and the final recovery rate was reported at 100 percent.

It should be noted that this is a classic example of qualitative thermography. The basis for fault detection is a difference (rise) in temperature detectable by the imaging system as the result of a controlled stimulus. No temperature measurement is necessary.

6.6.3 Applying DIT to Circuit Cards

Under a program funded by (the Electric Power Research Institute (EPRI), a visit to the Wolf Creek Nuclear Facility in Kansas early in the program resulted in a finding involving circuit cards in an instrument power supply battery inverter system. One of the four *inverter gating and synch* cards had failed six months prior to the initial visit, resulting in a costly outage. The replaced card was compared with the other three identical cards using comparative DIT. It was found that the same failure mechanism deemed responsible for the outage in the subsequent root cause analysis appeared to be once again at work on the remaining three cards. This is illustrated in Fig. 6.10.

The photograph of the card (top) shows the location of the key components in the analysis. The thermogram of the recently replaced card, NN11 (left), was subtracted from the thermogram of one of the remaining cards, NN13 (right). The resultant difference thermogram is shown at the bottom.

The root cause was identified as the failure of a capacitor that was surge protected by a pair of zener diodes, D10 and D11, which had been thermally overstressed and exhibited heat damage. On the initial DIT comparisons, the zener diodes on the other cards were the only anomalies that appeared, exhibiting temperature differentials as high as 120° F when compared with the corresponding zener diodes on the recently replaced card (NN11). These were operating at or near room ambient (about 78° F), which is assumed to be their normal operating temperature.

The remaining three cards were replaced during the next scheduled outage, and the thermal behavior of the replaced cards was examined on a test bench under controlled, simulated power-up conditions using an imager that incorporated a real time dynamic subtraction capability. The performance of the replaced cards was confirmed, with the old cards exhibiting excess heating similar to that observed on site; the heating in the vicinity of zener diodes D10 and D11 was observable within a few seconds after power-up.

A follow-up visit was scheduled 15 months later to record thermograms of all the replaced inverter gating synch cards in operation to determine whether the same failure mechanism that caused the original outage was detectable 15 months after replacement.

Live image subtractions were performed using the NN11 thermograms as STPs, since NN11 appeared to exhibit the greatest heating at zener diodes D10 and D11. All four cards exhibited some heating in the vicinity of zener diodes D10 and D11, clearly indicating that the failure mechanisms that caused the original outage were, once more, at work. A temperature measurement reference point (1) was placed over the center of D11, which appeared to be at a higher temperature than D10 in every case, and a temperature measurement point (2) was placed over the center of the card where there were no components, to serve as a background ambient reference. Temperature rises above reference (ΔTs) were observed on all 4 cards (T1 – T2) ranging from 30 to 60° F with the emissivity set at 0.95. The measurements on the full images were as follows:

Element	Image #	T1 (at D11)	T2 (ambient reference)	ΔT (T1 – T2)
NN11	NN1194.00	140.9°F	87° F	53.9° F
NN12	NN1294.00	122.0°F	93° F	29.0° F
NN13	NN1394.00	130.1°F	98° F	32.1° F
NN14	NN1494.00	125.6°F	84° F	41.6° F

The recently replaced NN11 which, during the initial visit, showed no detectable ΔT, now exhibited a ΔT of 53.9° F (140.9 – 87) at that location.

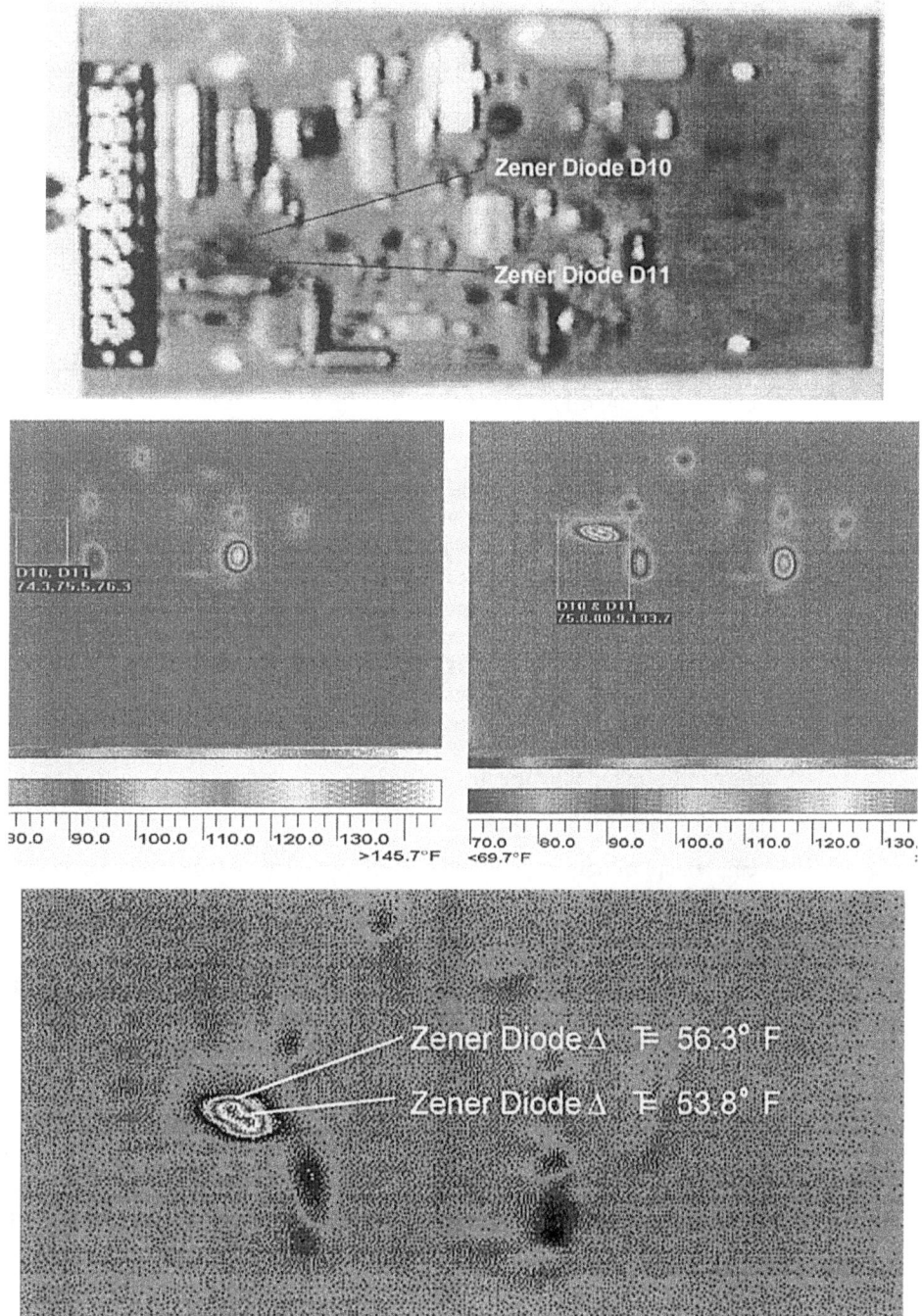

FIGURE 6.10 Application of differential thermography to an inverter gating and synch card. Top: photo of inverter gating and synch card. Center: Thermograms of NN11 and NN13. Bottom: difference thermogram, NN13 and NN11. *Courtesy of EPRI NDE Center.*

The reason for the repeated occurrence of failure mechanisms is now believed to lie in the design layout of the card. This places D11 and D10 in very close proximity (less than 1/4 inch for D11) to resistor R14, which appears to operate at temperatures in excess of 400° F. The continuous "baking" of the diodes in radiant heat from R14 is believed to be the cause of thermal stresses, which then cause them to break down and overheat on their own. It was recommended that, pending a design revision to the inverter gating and synch cards, all inverter gating and synch cards of the current design be monitored frequently and replaced periodically. This application also established that comparative DIT can be a useful tool for facilitated trending.

6.6.4 Semiconductor Devices, Chips, and IR Microimaging

The importance of accurate temperature measurements is illustrated by the image sequence of Fig. 6.11. This study involved the design evaluation of a new ignition system device developed by a major auto manufacturer. One of the design goals of this particular device was that the temperature would not exceed 150° C anywhere on the device. The top image is the initial powered test thermogram with the emissivity assumed to be unity over the entire surface. The cursor was placed at the location of the highest apparent temperature located at the center right portion of the thermogram (cursor 1). With the emissivity set at 1.0, this apparent temperature was 126.9° C, well within the thermal design limit.

The center image is the emissivity matrix for this device produced by following the procedure described in Sec. 6.5.4. It can be seen that the emissivity was low (about 0.23) at the location where the highest temperature appeared to be. The bottom image is the "true temperature" thermogram, corrected by the emissivity matrix, and indicating a maximum temperature of 183° C at the same location. This temperature was well in excess of the design limit and, as a direct result of this study, the device was redesigned to provide a more uniform temperature distribution.

The two composite displays in Fig. 6.12 illustrate an overheated print head (arrow) on an ink-jet printer, which is a likely precursor to failure. The typical maximum operating temperature of normal print heads of this type should be no more than 20° C above ambient. The "true temperature" image is shown in the lower right quadrant of each display.

The maximum temperature on the upper image appears to be 162.7° C at the location of the defective print head. The lower image is a 10× magnification of the upper image, and the high spatial resolution of the lower image shows that the maximum temperature at the defective print head is actually in excess of 185° C. The physical size of the hot area is less than 30 μm across. Microimaging thermography equipment is currently capable of spatial resolutions down to 3 μm.

This spatial emissivity compensation technique has been extended to printed circuit boards (PCBs) by means of a temperature-equalized box and an innovative software routine, as illustrated by the three-image sequence of Figs. 6.13 through 6.15. Figure 6.13 is a thermogram of a powered PCB with emissivity set at 1.00 throughout (no correction). Two devices on the PCB have shiny aluminum surfaces and appear colder than the other components covered with graphite or plastic.

Figure 6.14 is a thermogram of the same board, uniformly heated in a temperature equalized box covered with infrared-transparent film. This isothermal image is converted to an "emissivity" thermogram by means of an "equalization" software routine. It can be seen that the emissivity varies widely, from about 0.16 to 0.96. When all these emissivity values are applied, the software generates the "true temperature" thermogram of Fig. 6.15.

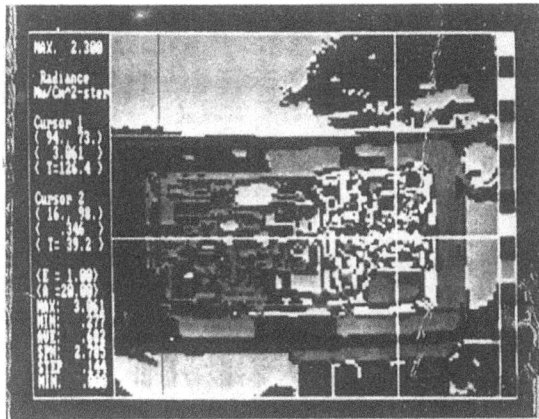

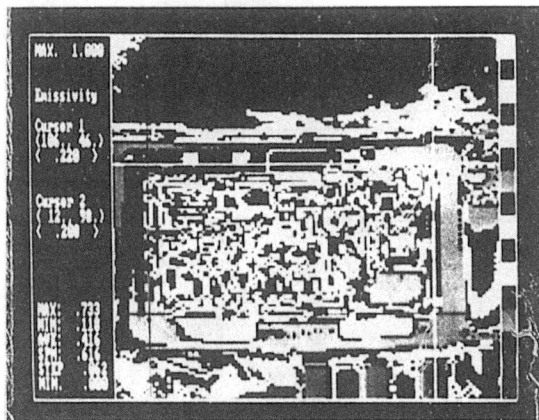

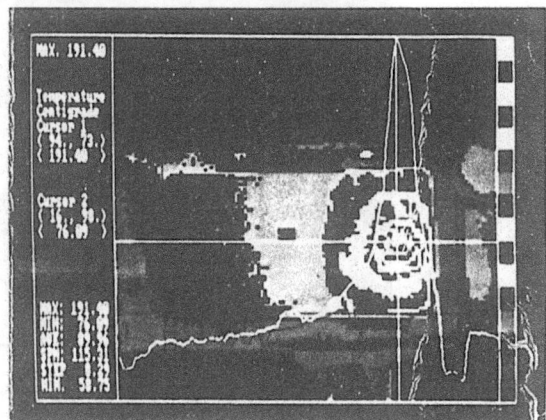

FIGURE 6.11   Top: thermogram of an ignition system device assuming unity emissivity (apparent max. temperature = 126.9° C. Center: computed spatial emissivity matrix of the device. Bottom: "true temperature" thermogram of the device corrected by the emissivity matrix (true max. temperature = 183° C). *Courtesy of Barnes Engineering Division, EDO Corp.*

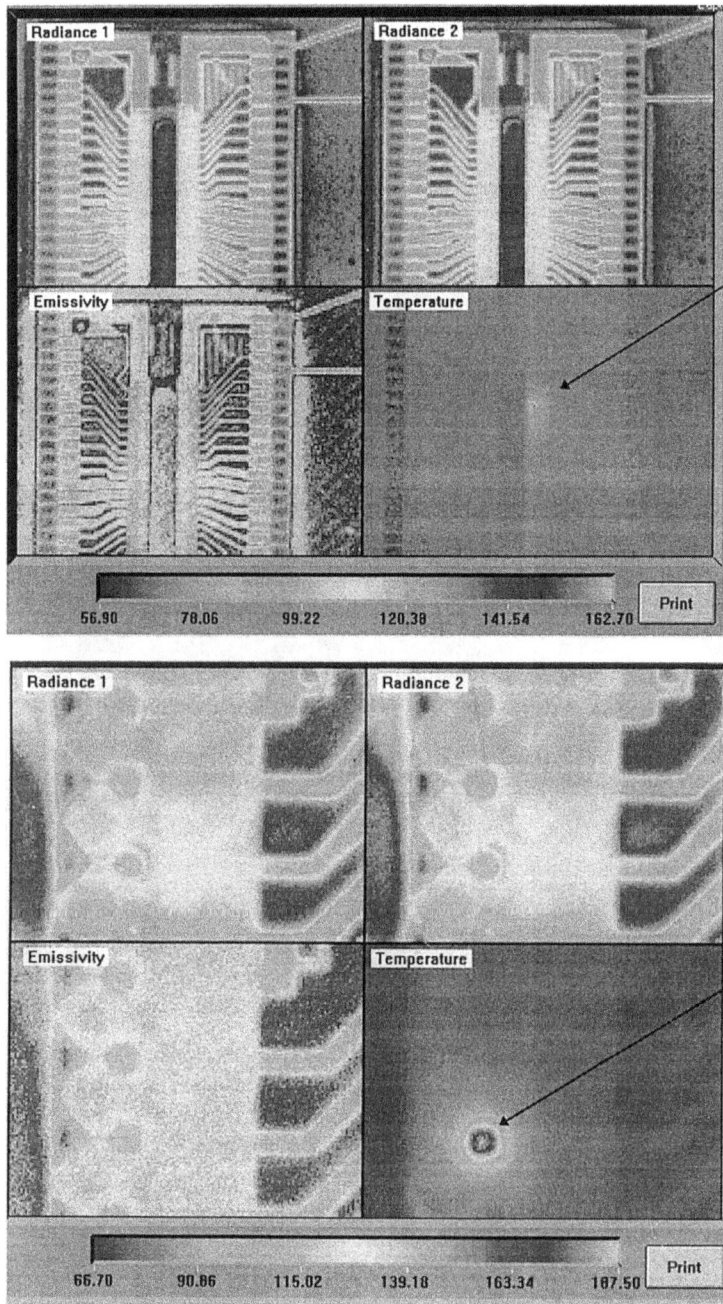

FIGURE 6.12 Two composite displays of thermal microimages of an ink-jet printer showing a defective (overheated) print head (arrow). The "true temperature" image is shown in the lower right quadrant of each display. The lower image is a 10× magnification of the upper image and shows a maximum temperature in excess of 180° C. *Courtesy of Barnes Engineering Division, EDO Corp.*

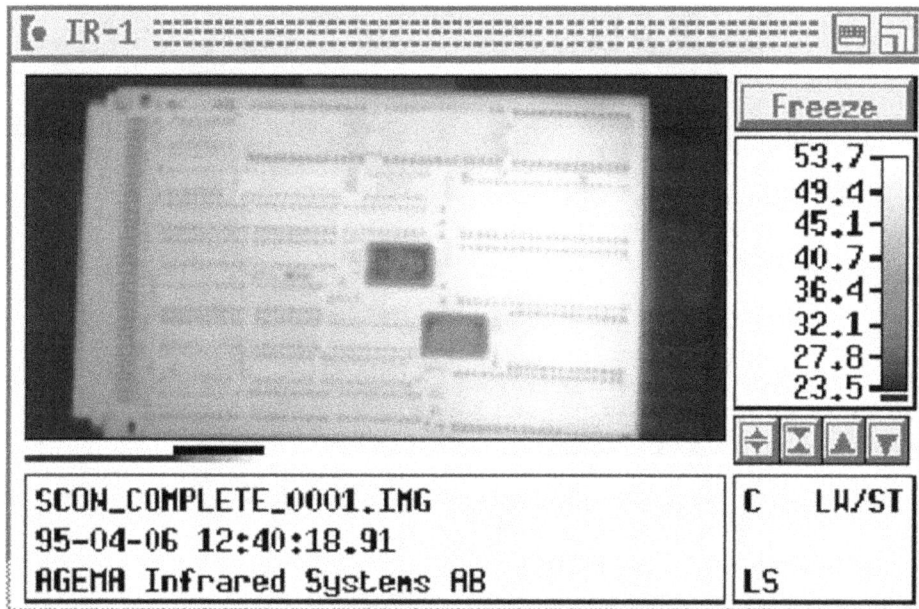

FIGURE 6.13   Thermogram of a powered PCB with emissivity set at 1.00 throughout (no correction). *Courtesy of AGEMA Infrared Systems.*

6.6.5   Thermal Regression Analysis for Determination of In-Situ Temperatures

Since IR thermography requires a clear line of sight from the imager to the target, it is not possible to directly view the components on a PCB mounted in a rack of cards. The use of an extender card allows this viewing but alters the convection and radiation of heat to the PCB so that the thermogram may be quite different from that in its assigned slot. Thermal regression analysis is used to overcome this, as illustrated in Fig. 6.16.

The left-hand thermogram is a reproduction of Fig. 6.13 and is actually a synthesized image that is the end result of a regression analysis sequence. This sequence is initiated by deenergizing the powered, in-situ PCB, removing it from its assigned location, placing it in the equalization box, and recording a thermogram (center image). A short time later, another thermogram is recorded (right-hand image). The software routine starts a clock at the instant of power-off so that the actual times at which the two subsequent images were acquired are known. Since the rate of cooling in air can be defined by two points on the cooling curve, this function is incorporated into the software routine and used to extrapolate backward to "zero" time. Thus, the "apparent" in-situ thermogram of the PCB (left-hand image and Fig. 6.13) is reconstructed.

6.6.6   Neural Networks on Circuit Cards

The automatic test equipment used for the seven circuit cards of the FLCP (flight control panel) for the F-16 aircraft at Hill Air Force base in Utah experienced operating problems over an extended period, and a program of infrared thermography augmented with an "on-

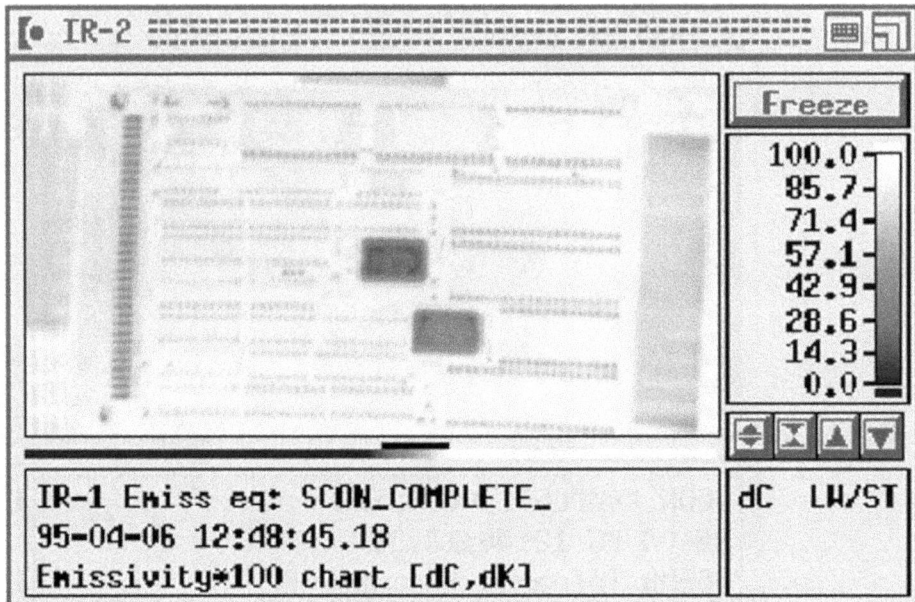

FIGURE 6.14 Thermogram of the board shown in Fig. 6.13, uniformly heated in a temperature-equalized box and converted to an "emissivity" thermogram by means of an "equalization" software routine. *Courtesy of AGEMA Infrared Systems.*

the-fly" neural network paradigm was introduced. Researchers at Ogden Air Logistics Center have developed a system called Neural Radiant Energy Detection System (NREDS), which was deployed to test the circuit cards.

The first step in the testing was the establishment of STPs for acceptable cards. This was in the form of a series of thermograms taken at intervals immediately after power-up to characterize the thermal rise patterns. Because of thermal saturation, as described in Sec. 6.5.3, this was accomplished during the first few minutes after power-up. Because heating patterns could vary to some extent for different test sequences, a series was run for each type of test run on each card.

Once acceptable limits were established for each component, on each card, thermograms were run on cards scheduled for repair. For each test, the neural network generated a flag which was superimposed over each component on the thermogram, to ascertain whether the component was operating within the acceptable temperature range, as illustrated in Fig. 6.17. It also assigned brightness shades to each flag, with the more severe discrepancies assigned the brightest shades.

In most instances the faulty component turned out to be the one with the brightest flag. This is shown at component U1 at the upper right hand corner of Fig. 6.17. The flag for this component shows a heating rate of 1.228° per second which is substantially greater than the 0.0289° value established as the allowable maximum. As it turned out in the case illustrated, U1 was shorted and replacement was mandatory. In some cases, however, the replacement did not return the card to functionality, but shifted the malfunction indication to another component which had been damaged by the incorrect installation of U1.

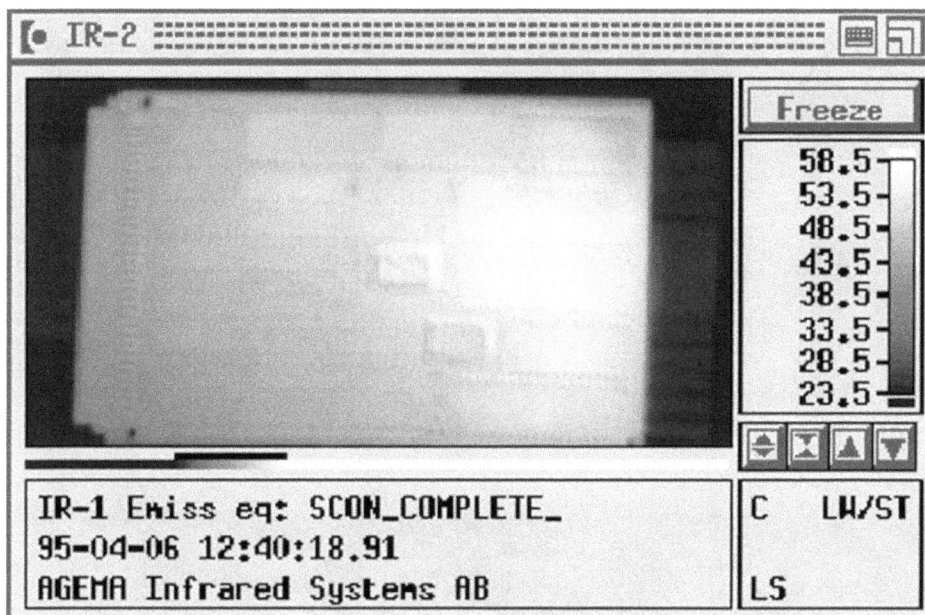

FIGURE 6.15 Software-generated "true temperature" thermogram of the board shown in Fig. 6.13. *Courtesy of AGEMA Infrared Systems.*

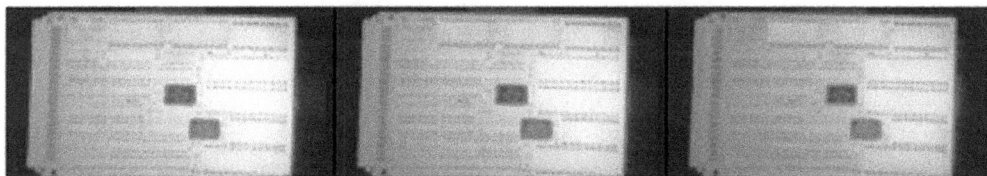

FIGURE 6.16 Synthesized powered in-situ thermogram (left) generated by applying thermal regression analysis to the center thermogram (taken 18 seconds after power-off) and the right-hand thermogram (taken 50 seconds after power-off). Courtesy of AGEMA Infrared Systems.

6.6.7 High-Resolution Images Made Possible by IRFPA Imagers

Figure 6.18 is a typical high-resolution thermogram of an operating device on a hybrid assembly. It illustrates the details that can be detected with currently available thermal imaging equipment using modern cooled IR focal plane array detectors. The warming in the center is due to normal operating current. The lettering on the device is visible due to the difference in emissivity between the ink and the surface material of the device.

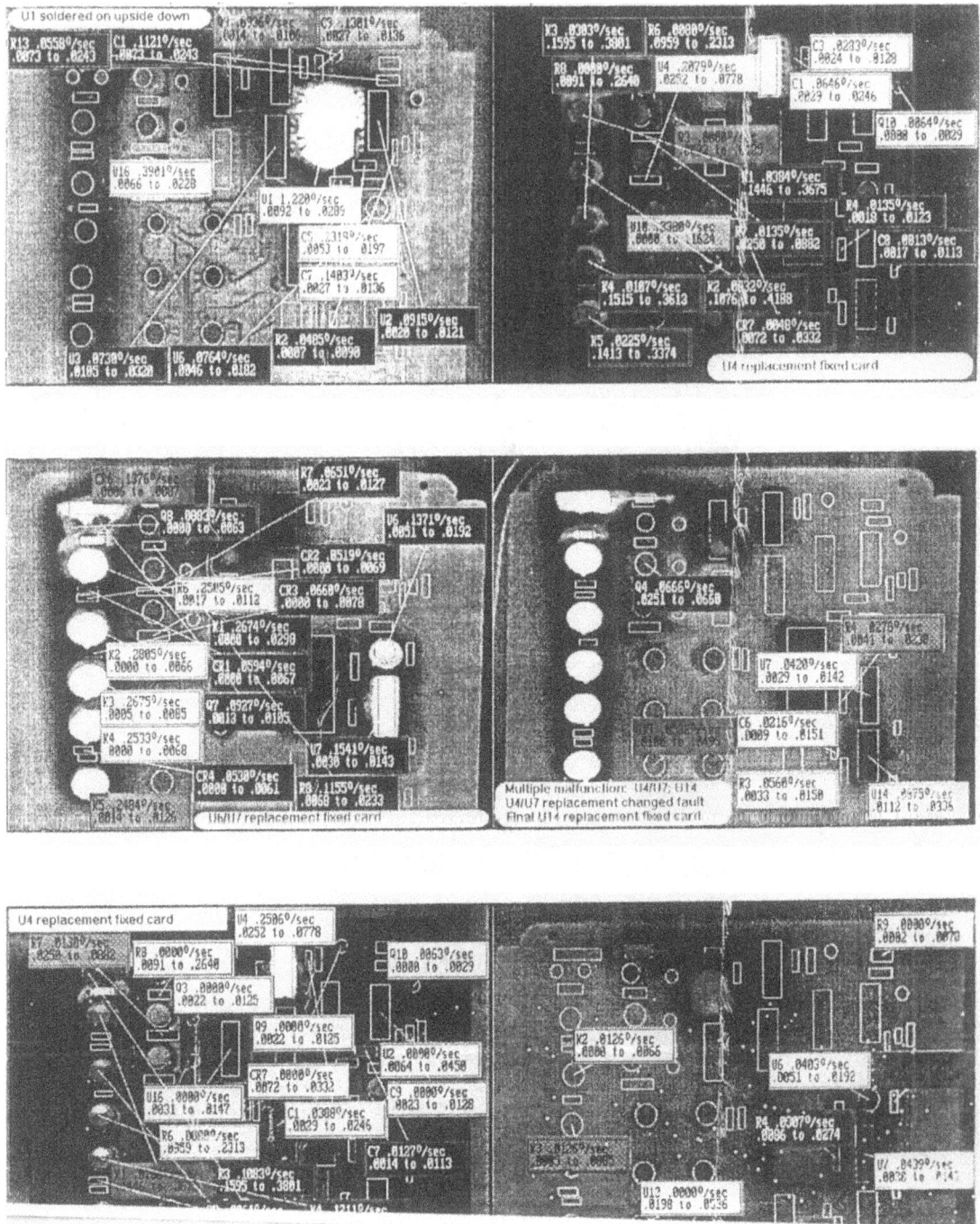

FIGURE 6.17 Circuit card thermogram with operating temperature deviation status flags assigned by a neural network. *Reprinted from SPIE Proceedings No. 2766, p. 287.*

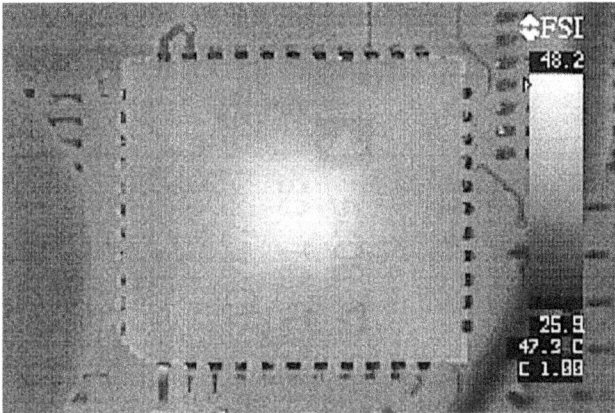

FIGURE 6.18 High-resolution thermogram of an operating device on a hybrid assembly. *Courtesy of FLIR Systems, Inc.*

CHAPTER 7
ACOUSTIC MICRO IMAGING FAILURE ANALYSIS OF ELECTRONIC DEVICES

L.W. Kessler
J.E. Semmens
Sonoscan, Inc.

7.1 INTRODUCTION

Acoustic micro imaging (AMI) is a relatively new analytical tool that has rapidly become an integral part of the modern FA laboratory. The technique is nondestructive in nature and gives important information about samples that is not easily obtained by any other techniques—for example, the integrity of the of the bonding between internal layers of an electronic component or the bonding of components to a substrate. In an electronic device, bond quality could have a significant impact on reliability, especially when a mismatch of thermal expansion coefficients between materials causes disbanded areas to grow larger over time.

In the FA laboratory, AMI is useful at various stages of analysis and especially prior to performing any type of destructive analysis such as cross-sectioning or peel testing. AMI helps to elucidate internal construction, precisely locate anomalous areas, and monitor physical changes within components during accelerated life test protocols. AMI is not always a substitute for destructive physical analysis (DPA); rather, it is a useful adjunct for making DPA more efficient by revealing and locating suspicious areas on which to perform a cross section.

In other usage, AMI technology is employed for screening many, even thousands, of samples. The basic principles of the analytical and screening modes of AMI are the same but, in the screening mode, computer analysis and automation of parts handling greatly speed up the inspection process. Now, parts belonging to a suspicious batch or a batch with an unacceptable reject rate can be segregated into accept/reject categories with little or no operator intervention.

The types of defects that AMI is particularly sensitive to are usually associated with unintentional air gaps between materials. Disbonds and delaminations are simple examples. Voids, cracks, porosity, and so forth exhibit similar high-contrast detail when imaged with AMI. Since high-frequency ultrasound does not propagate well in air or vacuum, maximum image contrast is obtained when such interfaces are encountered. Ultrasound does, however, propagate well through most solid materials used in electronic components. To produce acoustic images, the ultrasound must be effectively delivered (coupled) to the sample. Since air is not a good ultrasonic energy conductor, this is accomplished through a fluid (such as water) in which the sample is immersed. The lens portion of the ultrasonic transducer is also immersed.

The instruments that employ AMI technology are generally referred to as acoustic microscopes, and there are several different types. Although these have been fully described in the literature,[1] a brief review is included here for completeness.

7.2 METHODS

The most important types of acoustic microscopes which are in use today are the scanning laser acoustic microscope (SLAM) and the C-mode scanning acoustic microscope (C-SAM). Both instruments utilize high-frequency ultrasound to nondestructively detect internal discontinuities in materials and components. Typical frequencies employed range from 10 to 200 MHz. The higher frequencies are associated with higher resolution because of the shorter wavelength, and the lower frequencies are needed to penetrate thicker or more lossy materials. The frequency chosen for analyzing a sample is a compromise between resolution and penetration to the depth of interest. The penetration depends strongly on the specific material. For example, at any specific frequency, crystalline materials typically might be the most transparent to ultrasound, followed by ceramics, glasses, metals, and with polymers being the most absorbing. Table 7.1 is a listing of typical penetration depths of high-frequency ultrasound into various materials. The values are approximate and do not take into account specific instrument and transducer characteristics.

7.2.1 Scanning Laser Acoustic Microscope

The scanning laser acoustic microscope (SLAM) is a shadowgraph through-transmission instrument (Fig. 7.1). A planar continuous-wave beam is directed toward a sample through a fluid. The spatial pattern of transmitted ultrasound is then detected by a rapidly scanning, finely focused laser beam that acts like an ultrasensitive point-by-point "microphone." The transmission of ultrasound through the sample is affected by its elastic properties and construction. Defects, discontinuities, and interfaces influence the ultrasound transmission and are thus readily visible as shadow areas. A SLAM produces images in real time, 30 pictures per second, and simultaneously reveals bond integrity of all interfaces throughout the entire sample thickness. Because of its high speed and "entire sample" inspection mode, the SLAM is particularly useful for screening a large number of parts.

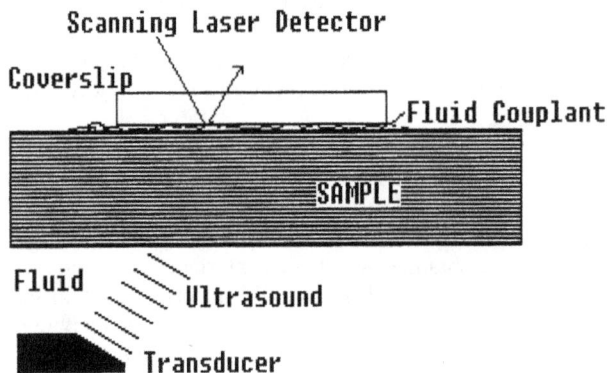

FIGURE 7.1 Typical ultrasound penetration into common materials.

Table 7.1 Typical Thickness Limits for Acoustic Micro Imaging Analysis

Material type	100 MHz	30 MHz	10 MHz
Ceramics:			
Silicon nitride, SiN			
Hot pressed	25 mm	75 mm	250 mm
Reaction sintered	5 mm	15 mm	50 mm
Silicon carbide, SiC			
Fully dense	25 mm	75 mm	250 mm
Molded	3 to 6 mm	9 to 18 mm	30 to 60 mm
Alumina, Al_2O_3			
Electronic grade	3 to 6 mm[*]	9 to 18 mm[*]	30 to 60 mm[*]
Fully dense	10 mm	30 mm	100 mm
Refractory bricks (high porosity)	25 to 500 μm	75 to 1500 μm	250 to 5000 μm
Metals:			
Steel, type 4140			
Annealed	2 mm	6 mm	20 mm
Quenched and temp'd	5 mm	15 mm	50 mm
High-speed steels	3 to 10 mm	9 to 30 mm	30 to 100 mm
Titanium	10 mm	30 mm	100 mm
Aluminum	10 mm	30 mm	100 mm
Powdered metals	1 to 10 mm[*]	3 to 30 mm[*]	10 to 100 mm[*]
Tungsten carbide	5 mm	15 mm	50 mm
Polymer Composites (directional-FRP):			
Parallel to fibers[†]	0.25 to 3 mm	0.75 to 8 mm	2.5 to 30 mm
Across fibers[†]	0.1 to 0.5 mm	0.3 to 1.5 mm	1 to 5 mm
Random fibers[†]	0.1 to 0.5 mm	0.3 to 1.5 mm	1 to 5 mm
Biological Tissues:			
Liver, kidney	2 mm	6 mm	20 mm
Liquid:			
Water (conductor)	Attenuation 2.5 dB/mm of travel	Attenuation increases with the square of the frequency.	Attenuation increases with the square of the frequency.

[*] Depends on porosity
[†] Depends on fiber type

In the SLAM acoustic image, features throughout the thickness of the sample are projected onto the detection plane—an optically mirrored plastic block called a *coverslip*. To obtain through-transmission images, it is necessary to have "acoustic access" to both sides of the sample; that is, there can be no intervening air gaps or undesirable layers of material between the ultrasound source (transducer) and receiver (coverslip–laser beam).

7.2.2 C-Mode Scanning Acoustic Microscope

The C-mode scanning acoustic microscope (C-SAM) is primarily used in the reflection (pulse-echo) mode to produce images of samples at specific depth levels within the sample. A focused ultrasonic transducer alternately sends pulses into and receives reflected pulses (echoes) from discontinuities within the sample. Echoes are produced at interfaces between two materials that have different elastic properties. The echoes are separated in time based on the depths of the reflecting features in the sample. An electronic gate is used to "window out" a portion of the echo signals, thereby selecting a specific depth range to view. A very high-speed mechanical system scans the transducer over the sample, and point-by-point data are collected. The output image is digitally constructed from the data point collection.

Although the mechanical scanning of the transducer is inherently slower than the electronic scanning of the laser beam in the SLAM, the C-SAM is preferred for analytical studies where layer-by-layer analysis is needed. C-SAM images are produced in tenths of seconds, typically.

A C-SAM is frequently operated in the through-transmission mode. The C-SAM transmission images can be enhanced somewhat over the SLAM images, since the ultrasound can be focused to a specific depth within the sample instead of relying on the shadowgraph projections alone. The same penetration and contrast limitations of through-transmission imaging apply to SLAM and C-SAM.

7.3   IMAGING MODES

7.3.1   A-Mode and C-Mode

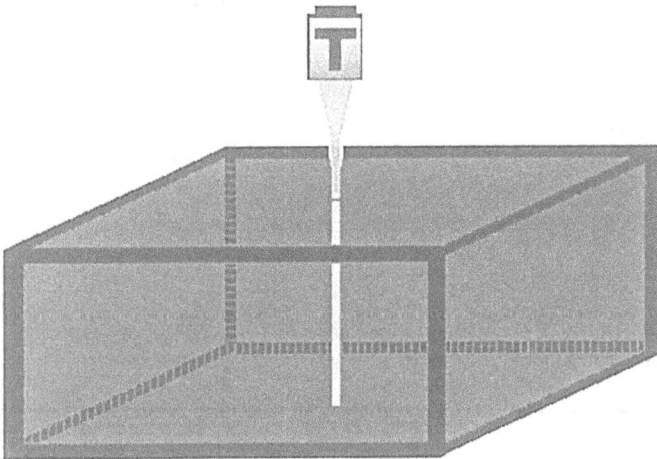

FIGURE 7.2   The A-mode is an oscilloscopic display of echoes returning from a sample when the transducer is in a single, stationary position.

In reflection-mode acoustic micro imaging, the fundamental information is contained in what is called the A-mode, an example of which is shown in Fig. 7.2. The A-mode is an oscilloscope display of the received echoes as a function of their arrival times (depths) at each x-y coordinate of the transducer position. Echoes displayed in the A-mode correspond to different interfaces in the device being examined. The times between echoes relate to corresponding feature depths in the sample by means of the velocity of sound in the material.

The amplitude and phase (polarity) information contained in the echoes are used to characterize the bond condition at the interface. The equation that describes the fraction of the wave amplitude (R) reflected from an interface is as follows:

$$R = \frac{Z_2 - Z_1}{Z_2 + Z_1} \tag{7.1}$$

where Z_1 and Z_2 are the characteristic acoustic impedances of the materials through which the ultrasound is traveling and the material encountered at the interface, respectively. The acoustic impedance value of a material is determined by the product of its elastic modulus and mass density.

Whereas the A-mode refers to echoes from a single transducer position, the C-mode refers to a compilation of A-mode data for all x-y transducer positions over the sample. Thus, the C-mode is an image derived from scanning the transducer over the area of interest and compiling the A-mode data.

7.3.2   Interface Scan Technique

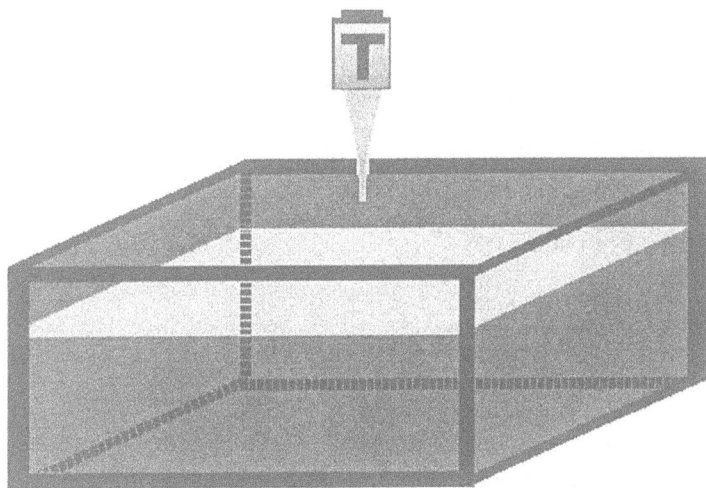

FIGURE 7.3   The interface scan is an x-y plot of echo data at a specified depth, Z, where a specific interface is located.

The most common imaging method used to evaluate samples for delaminations is the interface scan. As shown in Fig. 7.3, this method involves gating the A-mode signal for the appropriate interface echo of interest. The geometric focus of the acoustic beam is optimized for the interface as well. The acoustic image displays the amplitude and phase (polarity) of the gated echoes via the AIPD (acoustic impedance polarity detector).[2]

In this mode, images can be displayed of just the positive echoes, just the negative echoes, or a combination of both in a unified display. The usual image displays the echo amplitudes as different gray scales or colors. If the positive echoes and negative echoes are to be differentiated in the unified display, as is common when searching for delamination, the negative echoes will be displayed on a different color scale from the positive echoes. To avoid confusion, however, the instrument operator must carefully define the color assignments.

7.3.3   Bulk Scan Technique

The bulk scan, shown in Fig. 7.4, is used to examine a volume of material instead of a specific interface. The gating of the acoustic signal within the sample begins immediately after an interface echo and includes all of the interval up to (but not including) the next interface echo. The geometric focus of the transducer is usually placed in between the two interface echoes. If the material being inspected is homogeneous and without discontinuities there will be no echoes within the gate and, therefore, no features displayed in the image. Voids or other discontinuities will cause echoes to appear in the gated region and will appear as bright areas in the image. This imaging mode is typically used in material characterization applications, such as fingerprinting plastic encapsulation compounds or ceramic material. However, it also has other important screening applications such as searching a volume of material for voids whose locations are uncertain. The bulk scan technique is somewhat analogous to dark-field optical microscopy in which scattering features appear bright against an otherwise dark background.

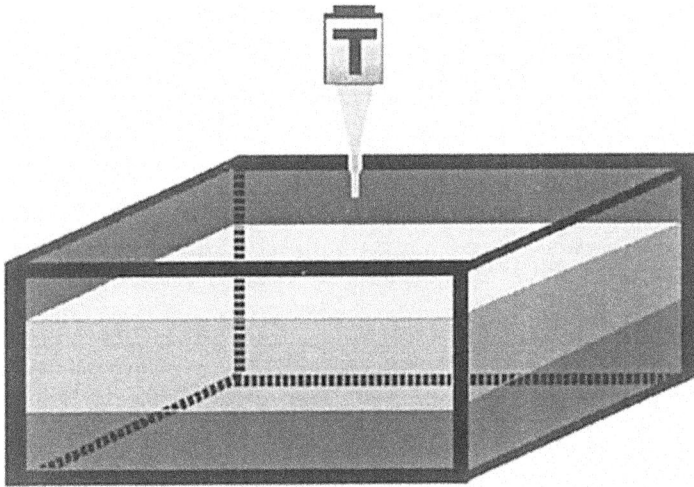

FIGURE 7.4   The bulk scan is an x-y plot of echo data over a range of Z depths that do not encompass specific interfaces.

7.3.4   Through-Transmission Mode

Through-transmission acoustic micro imaging (Fig. 7.5) relies on sending ultrasound into one surface of a sample, propagating it through its entire thickness, and detecting the transmitted signal on the other side using a separate receiving transducer. The SLAM uses an ultrasound plane-wave transducer to generate the signal and a scanning laser as the detector. The C-SAM THRU-Scan™ mode uses a second piezoelectric transducer as the

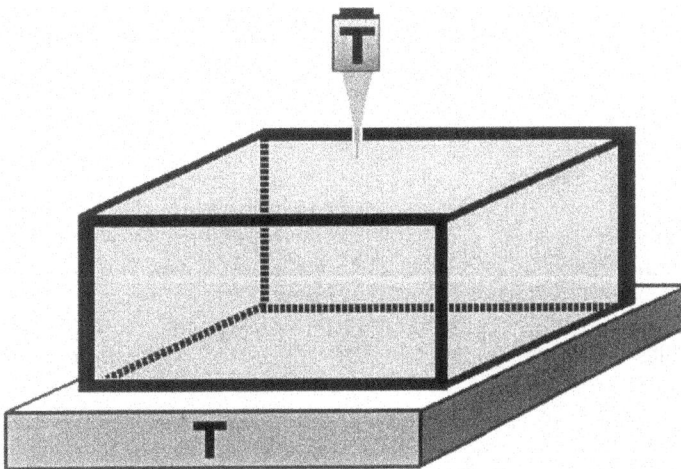

FIGURE 7.5   The through-transmission mode is an *x-y* plot of the acoustic pulse that exits the sample and therefore has traveled through its entirety.

detector. Defects, if present, typically block the ultrasound from reaching the detector and appear as dark areas in the acoustic image. Inherent to the through-transmission techniques is the requirement for acoustic access to both sides (surfaces) of the sample, as discussed earlier.

7.3.5 Time-of-Flight Profile Mode

The time-of-flight (TOF) profile mode (Fig. 7.6) provides a three-dimensional projection of depth information within a sample. Instead of detecting the amplitude and polarity of the echo signals received, the acoustic microscope tracks the time (distance) between the top surface of the sample and an internal interface. The variations in depth are displayed in gray or color scales in which the colors near the top of the scale correspond to features closer to the surface of the part. Colors lower on the color scale correspond to features deeper in the part. This mode is commonly used to follow the contour of an internal crack or to detect nonplanarity of a surface—for example, the geometric distortion and tilt of a die and lead frame as a result of improper injection molding.

7.3.6 Quantitative B-Scan Analysis Mode

Q-BAM™ is an imaging mode that provides a virtual cross-sectional (x-z) view of the sample along a designated plane through the device (Fig. 7.7). By sequentially indexing the depth position of the transducer focus during the data acquisition, each pixel of the cross section is in the correct focus for that depth in the sample. The distances between the internal depth levels is related to the ultrasound velocity in the material and the echo return times. The Q-BAM produces a view of the sample that is perpendicular to the top surface. This is in contrast to the interface and bulk scan modes, which produce views that are parallel to the sample surface (orthopedics views).

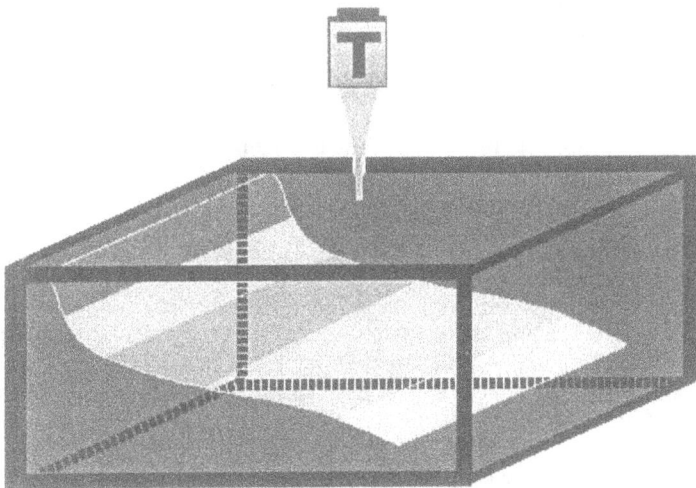

FIGURE 7.6   The time-of-flight mode is an x-y plot of received echo arrival times relative to the top surface of the sample. There is no echo amplitude information in this image.

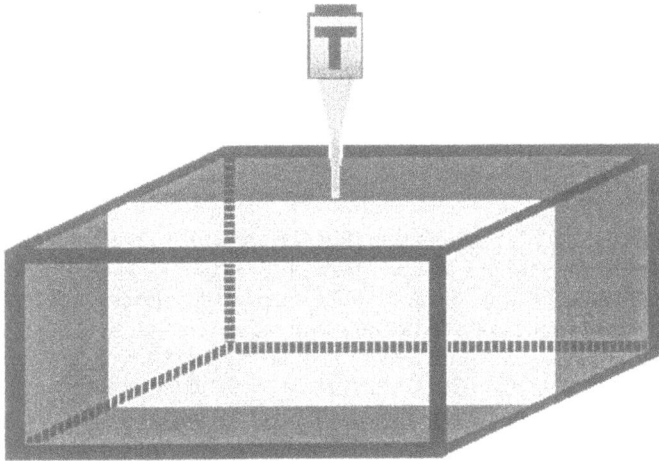

FIGURE 7.7 The quantitative B-scan mode is a cross-sectional view, an *x-z* plot of echo data at a specific *y* position.

7.4 EXAMPLE STUDIES

Early studies with acoustic micro imaging centered on inspecting relatively simple-geometry components such as ceramic capacitors, plastic encapsulated dual in-line package (DIP) ICs and semiconductor die attachment to ceramic packages. In general, the materials were easy to penetrate with ultrasound, the interfaces of interest were easy to access, and the defects were relatively large and easy to detect. Now component packaging has become more complex. Plastic packages have become much thinner, heat spreaders and heat slugs may be included within the encapsulation, multichip modules are increasingly used, and flip-chip interconnections are replacing conventional wire bonds for high-density interconnections. AMI techniques have evolved to accommodate the changes. The push to make packages thinner and interconnects smaller has required higher ultrasonic frequencies to obtain improved axial (in the direction of sound travel) and spatial (in the plane of the surface—*x-y*) resolution while maintaining sufficient penetration and working distance to access the various internal layers of the devices.

7.4.1 Ceramic Chip Capacitors and Multilayer Ceramic Substrates

Ceramic capacitors consist of multiple layers of ceramic and metallization and are cofired to their final state. Delamination between layers, as well as porosity and cracks, can cause electrical failure. Figure 7.8 illustrates the construction of a typical ceramic capacitor. Multilayer ceramic substrates used as multichip module and hybrid substrates have geometric construction similar to that of capacitors and, except for size, have similar structural problems.

Both C-SAM and SLAM acoustic micro imaging are routinely used to screen large numbers of unmounted ceramic chip capacitors for internal defects. Since through-transmission imaging is not specific to a particular depth, a view through the entire part is avail-

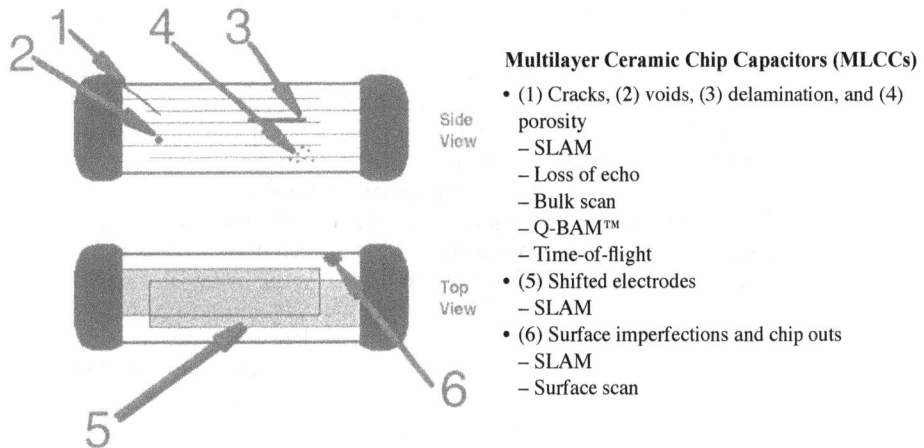

Multilayer Ceramic Chip Capacitors (MLCCs)

- (1) Cracks, (2) voids, (3) delamination, and (4) porosity
  - SLAM
  - Loss of echo
  - Bulk scan
  - Q-BAM™
  - Time-of-flight
- (5) Shifted electrodes
  - SLAM
- (6) Surface imperfections and chip outs
  - SLAM
  - Surface scan

FIGURE 7.8   Diagram of ceramic capacitors and typical construction defects.

able in one scan. An ultrasound frequency is chosen that will permit transmission of the signal through a nondefective part. Defect-free areas will appear bright in the gray scale acoustic image. By contrast, thin air gaps (smaller than 0.1 μm) created by delaminations or cracks will obstruct the ultrasound and appear dark in the through-scan images. Voids will be revealed as dark spots, and porosity will cause signal loss (attenuation) from scattering of the ultrasound. For this reason, through-transmission imaging is useful to obtain a rapid assessment of the components.

Referring to the images presented in Fig. 7.9, an end user was concerned about the quality of incoming ceramic chip capacitors from various manufacturers. Although all of the components passed stringent electrical tests, some would not survive the surface mount assembly process, and others would fail unexpectedly after the assembly was in use by a customer. An inspection protocol of 200 parts per incoming lot was suggested by the reliability engineers. If internal defects were found to be at an unacceptable level, the information was fed back to the capacitor manufacturers for corrective action. By instituting SLAM screening of incoming lot samples, a very significant failure rate reduction was achieved over a period of just a few months Previously, a comprehensive and time-consuming DPA protocol was used to determine the quality of the capacitors. And, since the

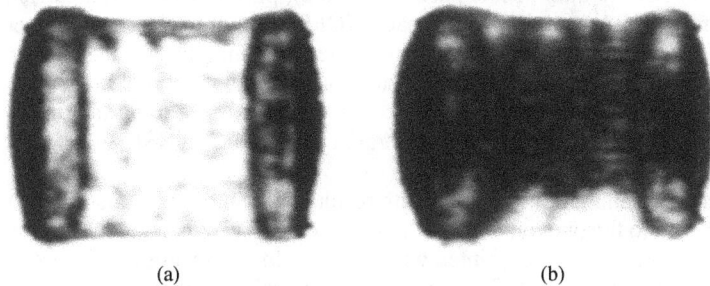

(a)   (b)

FIGURE 7.9   SLAM images of (a) a defect-free capacitor and (b) delamination in a ceramic capacitor.

DPA reveals defects only at the cross-section sites, a full area/volume inspection could not be performed on each sample.

Figures 7.9*a* and 7.9*b* show 100-MHz still-frame SLAM images of two ceramic chip capacitors, one of which (a) is defect free. The other (b) contains a large delamination. In the "defect-free" part, the entire active layer of the capacitor appears bright, indicating good transmission of the ultrasound across all the internal layers. The terminations appear darker due to the additional termination material in these areas. In contrast, the delaminated part (Fig. 7.9*b*) appears dark where the delamination blocks the ultrasound from the detector. The delamination can be seen in the center of the part and under the terminations. Small areas where there is no delamination, at the edges of the capacitor, show transmission of the ultrasound. The SLAM technique continues to be used to screen capacitors as well as ICs and other components whenever large volumes of parts and rapid turnaround are necessary. The SLAM also can be used to inspect components that are not flat, such as capacitors with terminations and leads. Because of the real-time nature of the technique, the operator can hold the sample in the viewing field and orient it geometrically to inspect under the terminations and to reduce or interpret artifacts arising from surface nonplanarity. From an analytical perspective, SLAM and C-SAM through-transmission techniques are valuable to identify the presence of delaminations. In cases where the interpretation of echo waveforms (reflection mode) is ambiguous or imprecise, through-transmission techniques are useful to confirm delamination.

Figure 7.10 illustrates the comparison of SLAM, C-SAM, and DPA on a typical capacitor. In Fig. 7.10*a*, a the SLAM through-transmission (same as C-SAM through-transmission) reveals a small circular defect in the upper left corner of the sample. The defect is dark against a bright (but textured) gray scale background. In Fig. 7.11*b,* the C-SAM bulk scan mode reveals the same defect as bright against a dark background, with obvious enhanced contrast. Note that this enhancement cannot be achieved with "image processing" of the through-transmission data. Rather, it is the gating differences of the echo waveforms that produce the enhancement. Figures 7.10*c* and 7.10*d* are cross-sectional views made by DPA at the site indicated by the acoustic microscope data.

7.4.2   Plastic Packaged ICs

A typical plastic-packaged IC (PIC) is constructed of a silicon semiconductor die adhesively bonded to a metal lead frame structure. The assembly is then encapsulated (on both sides) with an epoxy compound that is highly filled with silica particles to reduce shrinkage during cure and to add dimensional stability as a function of temperature. Different styles of PICs are dual in-line packages (DIPs), plastic leaded chip carriers (PLCCs), ball grid arrays (BGAs), and quad flat packs (QFPs). Figure 7.11 shows the construction of a typical PIC along with the typical package problems and AMI solutions. Many of the defects are associated with the encapsulation material itself and its bonding (or lack of) to the die, lead frame, and paddle. Defects of these types are important to identify, and there are several new standards being generated by the IPC, ANSI, and JEDEC organizations dealing with plastic IC reliability.[4]

Internal delaminations coupled with residual moisture in the plastic encapsulation material can lead to the phenomenon of "popcorn cracking" when the IC is temperature stressed during its soldering to the printed wiring board. In this example, a popcorn-cracked PLCC is examined with C-SAM. First, the parts were evaluated in the through-transmission mode to get an overview of the existing problems. Figure 7.12 shows a THRU-Scan image in which a large dark area is evident over the central area of the part, indicating defects that

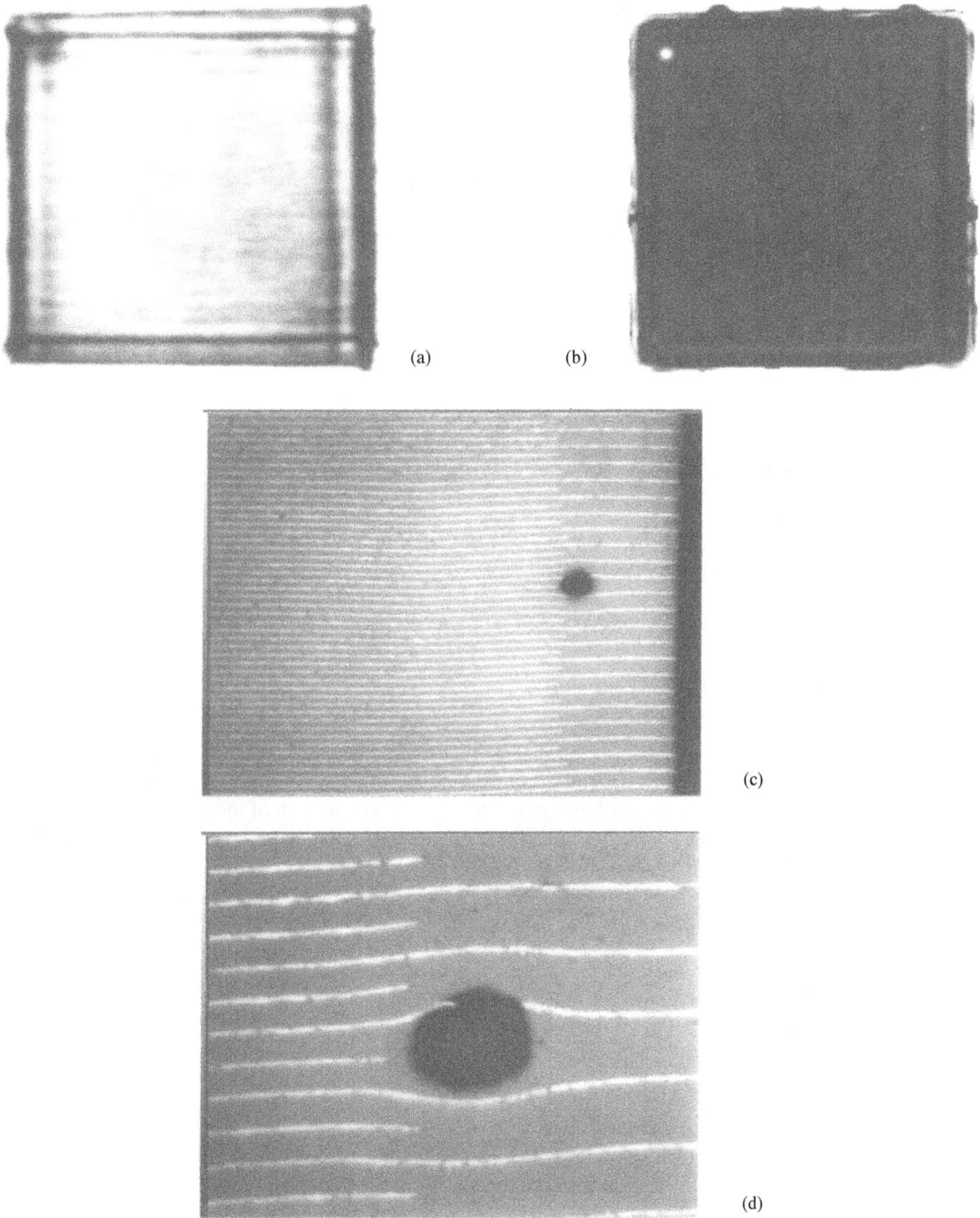

FIGURE 7.10 (a) SLAM image of ceramic capacitor with void (dark circle), (b) C-SAM image of same capacitor showing void (bright circle), (c) cross section obtained after AMI location of defect, and (d) cross section of higher magnification.

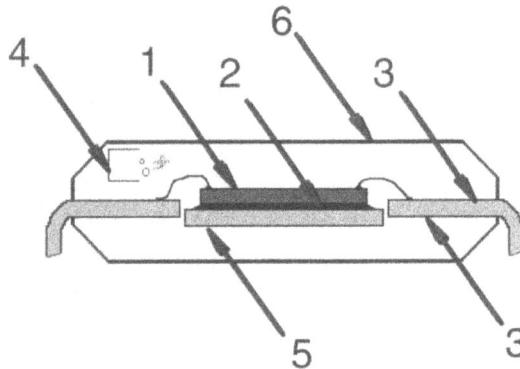

Plastic Packaged ICs (PICs)

- (1) Die surface/mold compound
  – Interface scan
- (2) Die attach
  – Interface scan
- (3) Lead frame/mold compound
  – Interface scan
- (4) Plastic encapsulant (mold compound)
  – Void and filler distribution
  – Characterization
- (5) Paddle/mold compound
  – Interface scan

- (6) Surface image
  – Surface scan
- (7) Package verification/screening
  – SLAM
  – THRU-Scan™
- (8) Nondestructive cross section
  – Q-BAM™
- (9) Crack/tilt profile
  – Time-of-flight (TOF)

FIGURE 7.11 Diagram of plastic integrated circuit and typical construction defects.

block the ultrasound from the detector. The bright regions correspond to areas where the sound travels through the entire thickness relatively unobstructed. In the center area of this sample, there is one small, bright (bonded) area. The pattern of the lead frame can be seen at the outer areas of the package in the image. Mold marks on the top and bottom surfaces of the part cause the five circular features in the image.

The through-transmission image is used as an overall indicator that a defect is present and as a map to guide further analysis. Switching now to the reflection mode, analysis can be performed at individual layers by geometrically moving the transducer focus to the correct depth level and electronically gating the echo that arises from that interface.

Figure 7.13a shows an interface scan at the molding compound to die interface. The image is a black-and-white print of a color-coded image whose brightness corresponds to the echo amplitude, and whose color corresponds to the polarity of the echoes. Positive echoes, which occur in bonded areas, are shown in the gray scale. Large amplitude, negative echoes characteristic of delaminations are color coded in red (these areas appear dark in the black-and-white print). The die surface is light gray, indicating that it is mostly bonded. A diagonal shadow is apparent across the die, which corresponds to a localized inhomogenity of the molding compound. This can be caused by "knit lines" or "jetting" of the material in the mold before curing. A small delamination (darker area) of the die surface is present at the lower right corner of the die. The bond wires can be seen surrounding the edges of the die. Portions of the popcorn crack occur at the same level as the die surface and can be seen next to the wires.

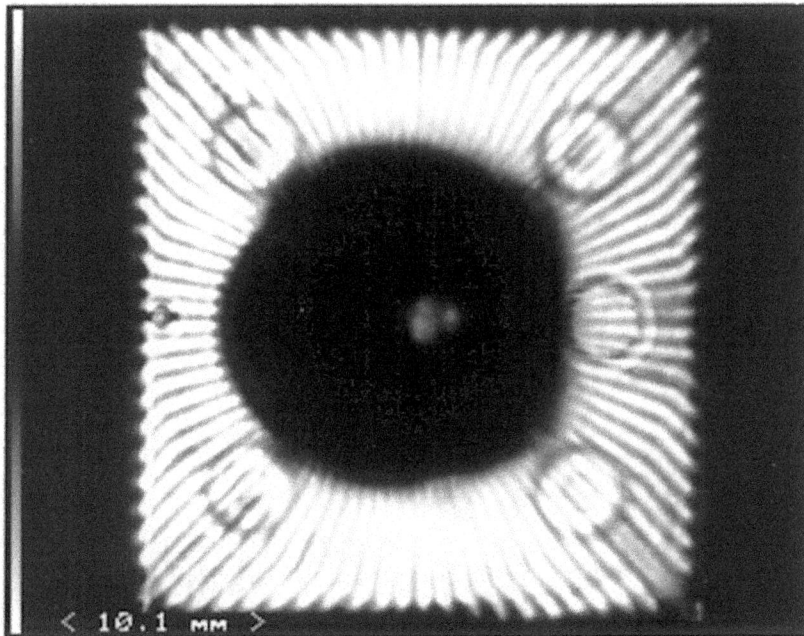

FIGURE 7.12   SLAM image of PLCC with large popcorn crack.

The die attach interface image is displayed in Fig. 7.13b. This is also a color-coded image printed in black and white. The majority of the die attach is disbanded. Only a small portion of the die attach (the light area below the center of the part) is bonded. Segments of the popcorn crack and some of the lead frame, which are at the same level as the die attach, are seen around the edges of the part.

Figure 7.13c shows an interface scan with a wide gate that includes several depth levels in the part (die attach, lead frame, and die pad) at a magnification that shows the entire device. The bonded areas of the lead frame are white in this image. Delaminated portions of the lead frame appear dark. The die pad surrounding the die is delaminated, as is most of the die attach. The only bonded area of the die attach appears bright in the image. Part of the popcorn crack can be seen between the edge of the die pad and the lead frame. A crescent-shaped shadow is seen at the right side of the die pad. This corresponds to the segment of the crack, which is at a level above the gated interfaces. The defect blocks the ultrasound from reaching this level and, therefore, is dark.

Figure 7.13d shows a bulk scan of the molding compound between the top surface of the part and the die top surface. A grayscale map that displays amplitude information (only) of the echoes was used for this image. In this mode, large-amplitude reflections corresponding to cracks and voids are white. Notice the correspondence of the crack features in Figs.7.8c and 7.8d. As a material, the molding compound itself appears fairly uniform except for a few white spots which indicate voids. Mold marks on the surface of the part appear as faint circular features.

The quantitative B-scan analysis mode (Q-BAM) produces a nondestructive cross section (x-z) of a sample. In Fig. 7.14, the depth relationship of the internal features in this sample can be seen. The interface scan of the part showing the die attach, lead frame, and

(a)

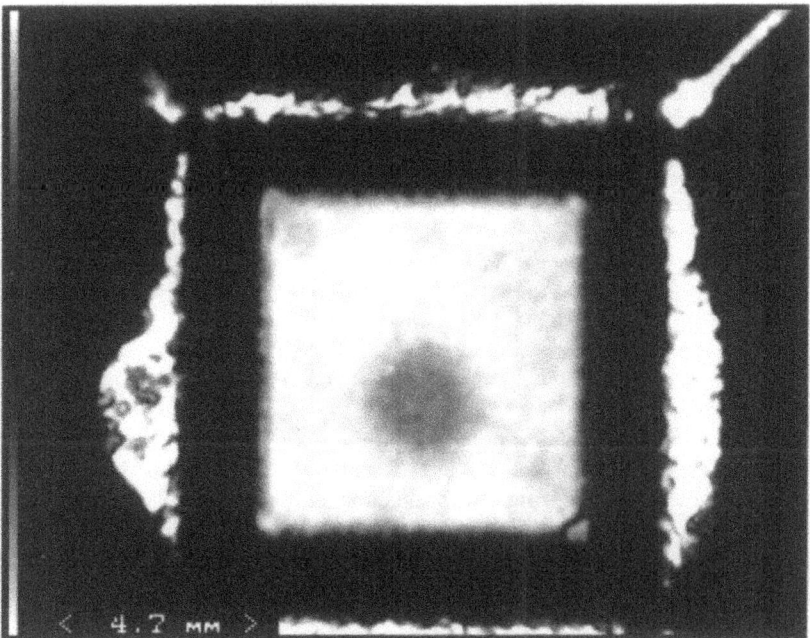

(b)

FIGURE 7.13 C-SAM images of PLCC (Fig. 7.12) showing various interfaces. (a) Die-to-mold compound interface and (b) die attach interface. *(continues)*

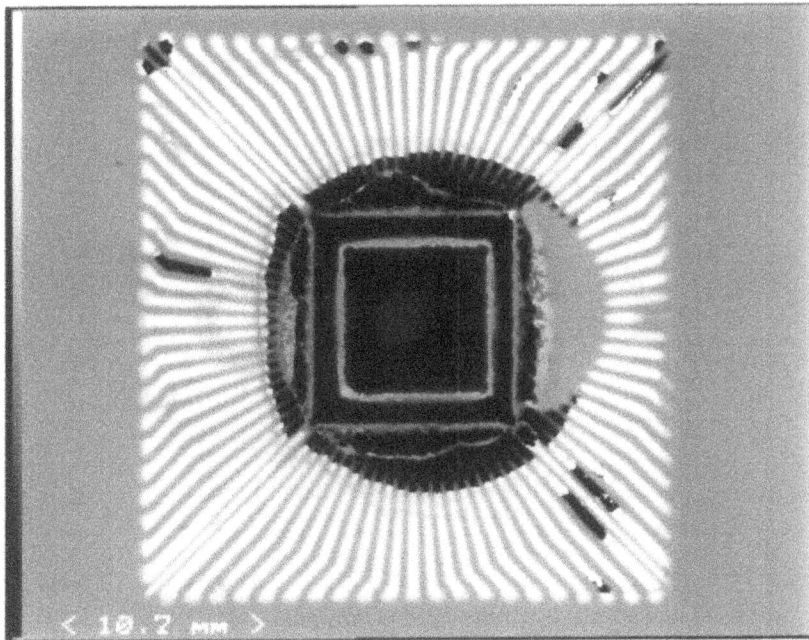

(c)

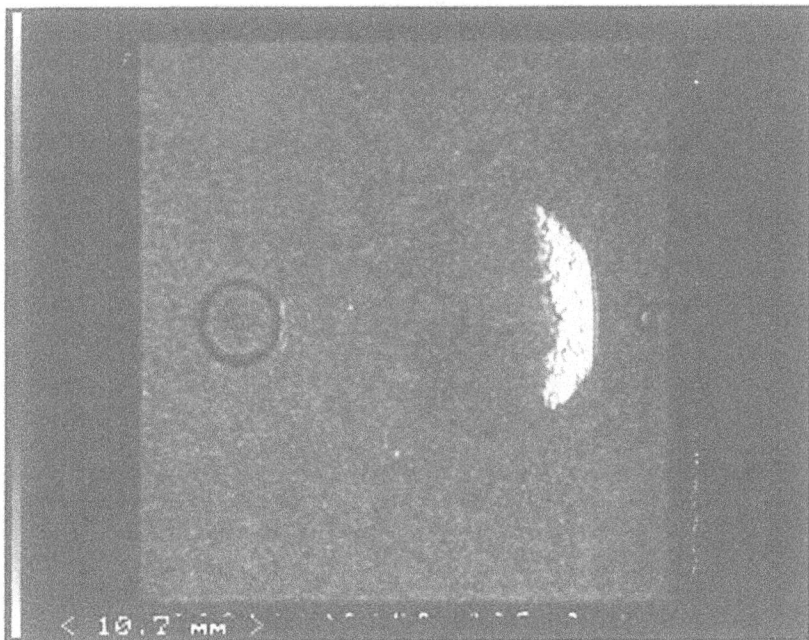

(d)

FIGURE 7.13 *(continued)* (c) Interface scan with a wide gate and (d) bulk scan of the molding compound.

FIGURE 7.14 Q-BAM image of PLCC (Fig. 7.12).

die pad levels is shown in the upper half of the image for reference. The image is shown in a grayscale amplitude map. The narrow horizontal white line across the image (just above center) indicates where the cross section Q-BAM is taken. The indicator line can be positioned anywhere over the full-screen interface image, and the Q-BAM is then performed at that location.

The lower half of the display shows the cross-sectional view defined by the indicator. Scale markers are used for feature depth measurements. The scale increments are defined at the bottom of the screen and are based on the acoustic velocity in the encapsulation material. The different levels of the die surface, die attach, lead frame, and die pad can be seen in the Q-BAM section. The popcorn crack can be seen originating near the die pad level and extending to the surface of the part.

Another method for viewing the internal depth relationship of features in a device is the time-of-flight profile mode. This gives a three-dimensional depth view of the internal features. The PLCC with the popcorn crack is shown using the TOF Profile mode in Fig. 7.15. The varying depths associated with the internal construction of the part are displayed in three dimensions. The crack can be seen to project above the plane of the lead frame at one side of the part.

7.4.3 Evaluation of Gallium Arsenide Die Attach

Gallium arsenide semiconductors are used for higher-frequency electronic applications than silicon. They are typically made much thinner than silicon die—100 vs. 500 μm. Therefore, higher acoustic frequencies are required to separate the echoes from the top and back surfaces of the die. This separation is required to image the die attach layer bonding.

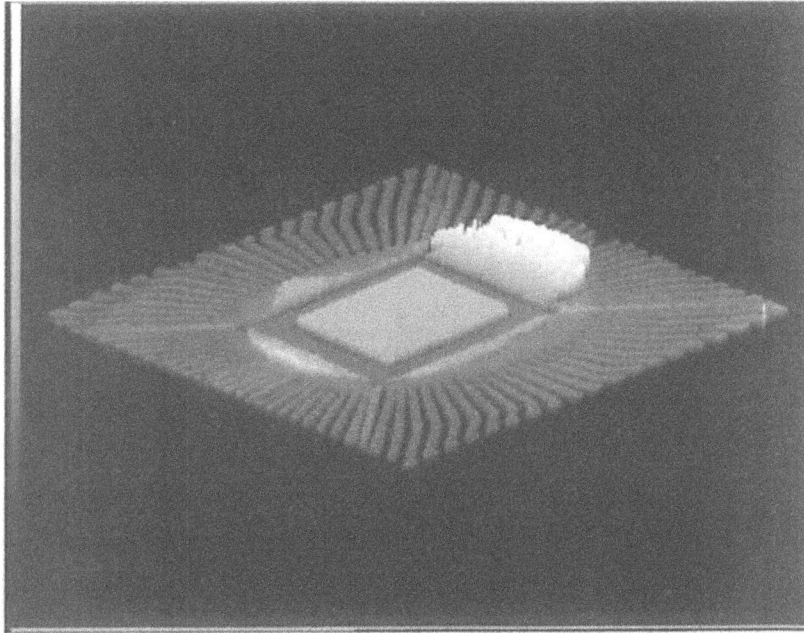

FIGURE 7.15   Time-of-flight image of PLCC (Fig. 7.12).

As in other cases, the importance of the die attach is thermal dissipation and, in the case of microwave devices, rf impedance. Figure 7.16a is an acoustic image of a large GaAs microwave device bonded onto a ceramic substrate and viewed from the top surface of the die. The ultrasound transducer was geometrically focused and the A-mode was gated at the device surface. The features seen on the image are due to the metallization layer(s) on the GaAs. Figure 7.16b shows the die attach interface. Here the transducer was focused at the die attach interface and that specific echo was selected for the display. The white patches are clear, well defined die attach disbonds. Note that the metallization pattern on the top of the die influences the ultrasound being transmitted to the other side. Therefore, this pattern can still be viewed at the lower depth level. The pattern is essentially shadowed onto the lower level.

7.4.4   Tape Automated Bonding

Tap automated bonding (TAB) is a method of interconnection in which all the interconnection leads are simultaneously bonded to the chip (inner lead bonds) or to the substrate (outer lead bonds). A disbanded or poorly bonded lead in a TAB device will cause an electrical fault. The interconnect sizes in these samples are typically between 50 to 100 μm^2. Figure 7.17 displays a 200 MHz through-scan SLAM image of four TAB inner lead bonds. One edge of the die can be seen toward the right of the screen. The bond sites are to the left of the edge. The bond sites are approximately 75 μm (3 mil). The leads extend off to the right toward the tape carrier (which is out of the field of view). The well bonded leads appear bright in the through-transmission mode. The poorly bonded/disbonded lead appears dark. In this application, the acoustic data were quantified using image analysis

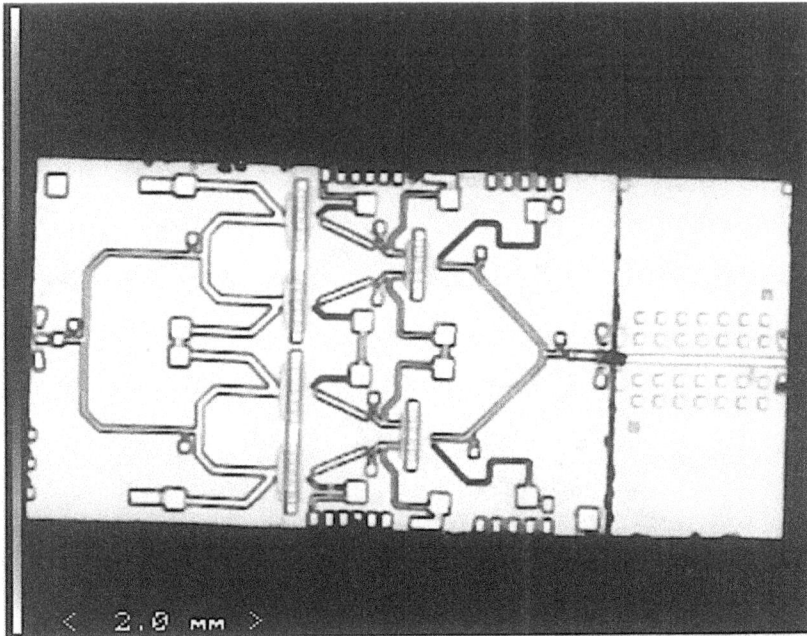

(a)

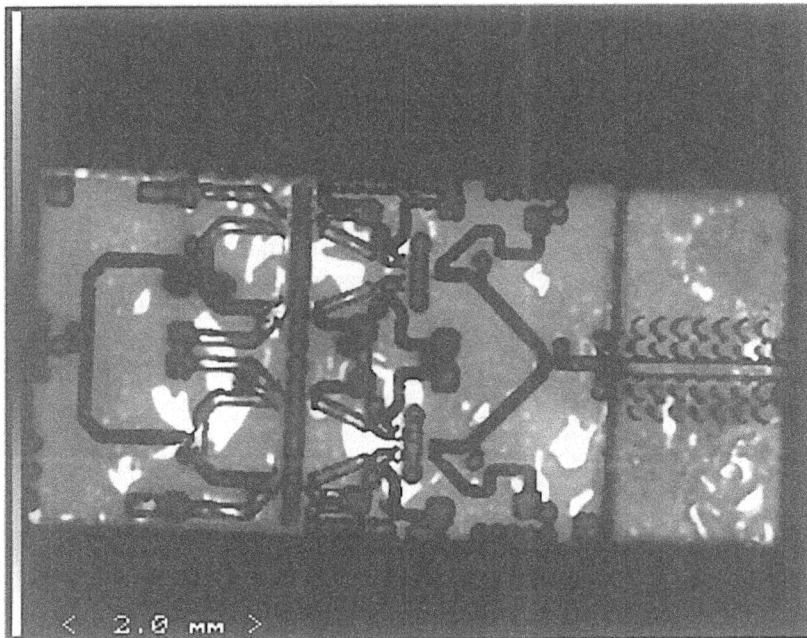

(b)

FIGURE 7.16   (a) Surface-only image of gallium arsenide die and (b) die attach image.

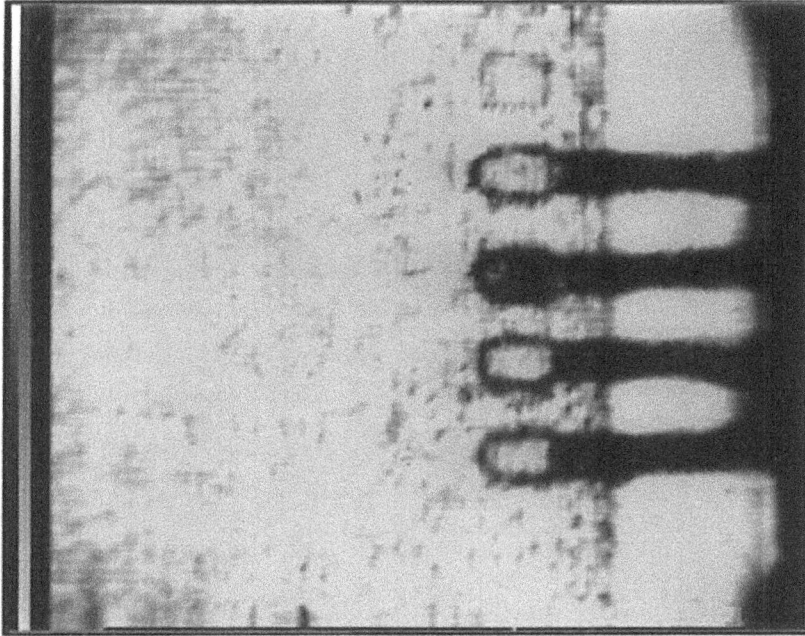

FIGURE 7.17   SLAM image of TAB bonds onto silicon die.

techniques to measure the area of bond per lead. This correlates with pull strength values and accept/reject criteria MIL-STD-883D, Method 2035,[5] which was developed under a USAF Rome Laboratory contract. It shows the correlation between bond area measured nondestructively and pull test measured destructively Now, with AMI, nondestructive inspection of TAB bonds can be performed instead of pull testing.

7.4.5   Flip-Chip Interconnect Evaluation

An interconnect technology that is becoming very important in microelectronics is flip-chip attach. As shown in Fig. 7.18, silicon die are mounted circuitry-side down (flipped) directly to the substrate connection sites by means of solder bumps. The space between the chip and substrate is typically "underfilled" with an epoxy encapsulant to seal the device and give the chip attach additional mechanical strength. A badly bonded bump or voids in the solder bumps can lead to faulty operation. Voids in the underfill material can also cause problems because of localized stresses that occur due to moisture absorption by the underfill material and by differential thermal expansion.

Flip-chips are typically evaluated using the reflection mode technique due to the fact that the devices are mounted to multilayer ceramic substrates or composite boards. Multi-layer substrates often limit acoustic access to the bond interfaces of the interconnects in the through-transmission mode. Therefore, this mode is not usually employed for flip-chip inspection. The reflection mode allows for single-sided access to the bumps and underfill interfaces through the back of the silicon die. The small size of the bonds requires evaluation at frequencies beyond 200 MHz. Figure 7.19 shows an example of a C-SAM image of the chip-to-solder-bump interface on a flip-chip sample. Small voids at the interface

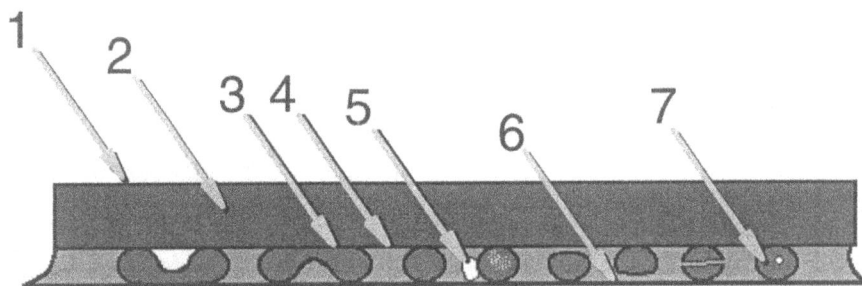

Flip-Chip/C4 (Controlled Collapse Chip Connection)

- (1) Surface image
  – Surface scan
- (2) Die or chip
  – Cracking
- (3) Chip/bump
  – Interface scan
- (4) Chip/underfill
  – Interface scan

- (5) Underfill uniformity
  – Interface scan
- (6) Bump/substrate
  – Interface scan
- (7) Solder bump integrity
  – Bulk scan

FIGURE 7.18   Diagram of flip-chip and typical construction defects.

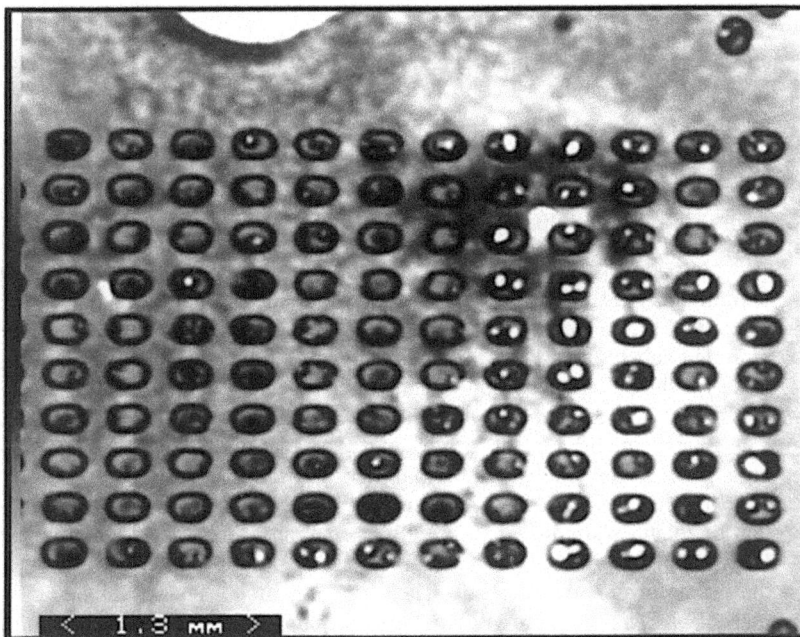

FIGURE 7.19   180-MHz C-SAM image of flip-chip at chip-to-bump interface, showing voids in bums (white).

appear as white features. Bonded bumps appear dark. A void (white area) is also present in the epoxy underfill region between the bumps.

7.4.6 Multichip Modules

Multichip modules (MCMs) represent a broad range of electronic packaging and include ceramic, polymer composite, and silicon substrates onto which may be bonded several semiconductor die, as well as passive components. The miniaturized assembly may be further packaged into a hermetic package or encapsulated with an organic compound. Figure 7.20 illustrates one MCM configuration and typical investigations that are performed with acoustic micro imaging. Many of the sites of interest are similar to examples shown previously in this chapter, and no further images are shown.

7.5 CONCLUSIONS

Acoustic micro imaging has proven itself to be a valuable nondestructive testing technique for the inspection and evaluation of electronic components and assemblies. The most frequent utilization is for bond evaluation, but quantitative AMI methods are also being used for materials characterization. The same acoustic microscope instruments are used to collect these data. AMI inspection is now required for many devices according to standards

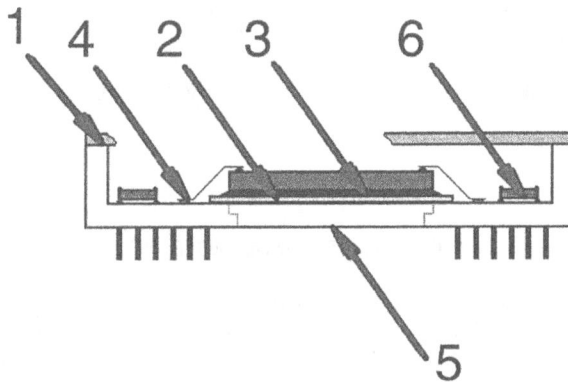

Multichip Modules

- (1) Lid seal   – Interface scan
  – Interface scan   – Bulk scan
- (2) Substrate attach – Loss of echo
  – Interface scan   – Time-of-flight
- (3) Die attach   • (6) Passive components
  – Interface scan   – Integrity and attachment
- (4) Lead attachment   • (7) Nondestructive cross section
  – Interface scan   – Q-BAM™
- (5) Multilayered package • (8) Surface image
  – SLAM   – Surface scan

FIGURE 7.20   Diagram of multichip module and typical defects.

and specifications issued by the JEDEC, DOD Mil-Spec-883, and IPC, as well as specific procurement needs of industry and government.

Continued developments in microelectronic devices create an almost continuous flow of new materials and assembly problems to be solved by nondestructive AMI. Further developments in failure analysis AMI involve higher resolution and penetration to accommodate the shrinking of device geometries. Additional AMI developments are underway for a new class of production-mode inspection instruments that offer higher speed, on-line access, automatic operation, and no required operator interpretation.

The successful application of AMI to an inspection problem depends on the knowledge and skill of the failure analyst, the knowledge of the sample construction, and the proper application of the inspection techniques. AMI is a valuable part of the FA process and is usually performed prior to decapsulation, cross sectioning, or other destructive methodology.

7.6 REFERENCES

1. L.W. Kessler. *Acoustic Microscopy, Metals Handbook,* Vol. 17, 9th Edition: Nondestructive Evaluation and Quality Control, 1989, 465–482.

2. F.J. Cichanski. Method and System for Dual Phase Scanning Acoustic Microscopy, U.S. Patent 4866986.

3. C. Bloomer, R. Mepham, and R.M. Sonnicksen. "Advanced Failure Analysis Techniques for Multilayer Ceramic Chip Capacitors." *Proc. International Symposium for Testing and Failure Analysis Microelectronics Symposium,* Los Angeles, CA, 1987, ASM International, Metals Park, Ohio, pp. 25-33.

4. American National Standards Institute (ANSI), Electronic Industries Association (EIA), JEDEC (Joint Electron Device Engineering Council) Joint Industry Standard, J-STD-020, Moisture/Reflow Sensitivity Classification for Plastic Integrated Circuit Surface Mount Devices, October 1996.

5. U.S. Military Standard—MIL STD 883D, Method 2035, Ultrasonic Inspection of TAB, Bonds.

7.7 SUGGESTED READING

Plastic Encapsulated Electronics. M.C. Pecht, L.T. Nguyen, and E.B. Hakim. New York: John Wiley & Sons, 1995.

Electronic Materials Handbook, Vol. I—Packaging. Materials Park, OH: ASM International, 1989.

Fundamentals of Acoustics. L.E. Kinsler and A.R. Frey. New York: John Wiley & Sons, 1962.

Proc. International Acoustic Micro Imaging Symposium (IAMIS), 1996, San Diego, CA. Bensenville, IL: Sonoscan, Inc.

Proc. International Acoustic Micro Imaging Symposium (IAMIS), 1997, Anaheim, CA. Bensenville, IL: Sonoscan, Inc.

Proc. International Acoustic Micro Imaging Symposium (IAMIS), 1998, Anaheim, CA. Bensenville, IL: Sonoscan, Inc.

CHAPTER 8
METALLOGRAPHY

David J. Roche

HI-REL Laboratories

8.1 INTRODUCTION

Metallography is both an art and a science and, in the hands of an accomplished individual, will result in the acquisition of information in the areas of structural and interfacial features that cannot be obtained by any other analytical method. The entire principle of metallography is to prepare a representative cross-sectional view of a sample, free from artifacts and pseudo structures. It is the ability to recognize these artifacts that separates the novice from the expert.

Metallography has gained wide acceptance due to the ease with which a variety of samples can be evaluated for structural and interfacial features. An additional, equally attractive feature is the relatively small amount of capital investment needed to set up a metallographic laboratory. The basic equipment needed for a metallographic lab includes grinding and polishing wheels, chemicals for etchants, and an metallurgical microscope. More sophisticated equipment, such as automated preparation systems, ion beam mills, scanning electron microscopes, and energy dispersive spectrometers, may augment a basic laboratory.

Throughout this chapter, procedures and materials will be discussed that have shown good, repeatable results. They are not "set in stone," and each individual is encouraged to experiment to find the most productive combination.

8.2 HISTORICAL PERSPECTIVE

The beginnings of metallography date back to when man first began abrading and chipping stones to use as tools. Progress in the field continued and, by the time of Christ, crude mirror-like finishes could be made on metal surfaces using pottery wheels.[1] However, it was not until the 1860s that a petrographer named William Chilton Sorbey began his studies in what is now modern metallography.

The history of modern metallography goes hand in hand with the history of the steel industry in the early 1900s. It was during this period that metallography saw its greatest growth. The steel industry used metallography as both a production control and research and development tool. It also allowed engineers and scientists to correlate mechanical properties with microstructures and establish the principles of physical metallurgy.

With the advent of the first transistor by Bell Laboratories in the late 1940s, it became evident that metallography would once again be a viable resource in the production and research of a fledgling microelectronics industry. It was realized that significant adaptations and refinements of the metallographic process would be required, due to the variety of hard and soft materials coupled to form a cohesive unit and the monumental reduction in the size of the samples to be cross-sectioned.

The past ten years have been marked by consistent improvements in the consumable products used in the preparation of metallographic samples. In recent years, the emphasis has been placed on fine tuning the automated systems currently available. These systems allow a large number of samples to be processed quickly and repeatably.

8.3   SETTING AN OBJECTIVE

The first step in the preparation of any metallographic sample is to determine what the desired result is going to be. This is probably the most important step because. once begun, the cross-sectioning procedure becomes a destructive test. If not thoroughly thought out, a one-of-a-kind sample can be lost in a matter of seconds.

The amount of "up front" work generally depends on the complexity and uniqueness of the sample. This can range from obtaining a general microstructure, which typically requires at most the identification of grain flow direction, to cross-sectioning into a specific 1 μm contact, which would require detailed mapping in a scanning electron microscope.

8.4   SAMPLE REMOVAL

Once the objective has been determined, the next step is to evaluate the best method of removing the sample from the assembly. This step is not always needed when dealing with microelectronic samples. However, when required, there are a wide variety of methods to choose from, and the decision generally depends on the sample. Possible extraction techniques and apparatus for different samples are listed in Table 8.1.

Table 8.1   Suggested Sample Removal Techniques

Removal technique/apparatus	Suggested sample
Cutoff wheel	Large metallurgical samples
Wafering saw, diamond blade	Ceramic substrates and packages
Wafering saw, Al_2O_3 blade	Hybrid packages without ceramic substrates
Router	Printed wiring boards
Scribe and break with a diamond tool	Dice in wafer form
Eutectic reflow	Metallurgically bonded dice
Elevated temperature shear	Epoxy and silver glass attached dice
Hand-held abrasive cutting tool	Metal packages and printed wiring boards

8.4.1 Cutoff Wheel and Wafering Saw

The cutoff wheel and the wafering saw are the most versatile of all tools. They will allow both large and small samples to be dissected into smaller, more manageable portions. Added versatility can be achieved by blade material selection as shown in Table 8.2. Stand-alone and tabletop models are available (Fig. 8.1). Both styles incorporate a coolant system that prevents overheating of the sample and minimizes the generation of dust.

Table 8.2   Cutoff/Wafering Blade Selection Chart

Application/material	Abrasive mineral	Abrasive bonding media
Extra hard/Rockwell C 60+	Al_2O_3	Rubber/resin or rubber
Hardened/Rockwell C 45–60	Al_2O_3	Rubber
Carbon steel/medium hard steels	Al_2O_3	Rubber or resin
Medium soft steels	Al_2O_3	Resin
High nickel alloys	Al_2O_3	Rubber
Tough nonferrous/titanium alloys	SiC	Rubber
Soft nonferrous/aluminum/copper alloys	SiC	Rubber/resin
Ceramics/composites	Diamond	Metal
Carbides/fragile nonmetallics	Diamond	Resin

8.4.2 Router

The use of a router has gained wide acceptance within the printed wiring board industry. This acceptance is based on its high output and the nominal amount of damage introduced to the sample. A router has no other significant application in the removal of metallographic samples. Similar results can be achieved with a wafering saw.

8.4.3 Scribe-and-Break

The scribe-and-break method of die removal is a quick and easy procedure for removing dice from a wafer form. The entire method consists of scribing a line into the surface of the wafer with a diamond tipped tool and then applying a downward bending force sufficient to break the silicon along the line. This method is not the most precise, but it does yield favorable results with a very minimal amount of tooling and fixturing.

8.4.4 Eutectic Reflow

Eutectic reflow is a method utilized in removing die which are metallurgically bonded to a package. The procedure consists of placing the base of a package on a hot plate maintained at a temperature above the die attach melting point. Once the die attach has reflowed a slight scrubbing and lifting action with tweezers will result in the removal of a die. Care

FIGURE 8.1   Overall view of a precision wafering saw. *Source:* Courtesy of Allied High Tech Products.

should be taken to minimize the amount of time at these elevated temperatures to prevent alloying between the aluminum metallization and silicon die.

8.4.5   Elevated Temperature Shear

Elevated temperature shear is a procedure designed to remove epoxy and silver glass attached dice from a package. The fillet of silver glass attach material around the die is initially removed by cutting it away from the sides of the die with a razor blade. The package is then placed base down, flush to a hot plate maintained at approximately 370° C. At this temperature, the adhesion between the die and attach or the package floor and attach degrades sufficiently to allow the die to be popped off when a force is applied to the side of the die with a pair of blunt-nose tweezers. Care must be taken to minimize the amount of time the sample remains at this elevated temperature.

8.4.6 Hand-Held Abrasive Cutting Tools

Hand-held cutting tools can be used to remove more precise locations while minimizing the amount of damage to surrounding areas. This technique is quite useful when a number of samples need to be taken from a relatively small area. The drawback to these tools is that they quickly heat the sample and produce large amounts of expelled particulate from their blades.

8.5 *ENCAPSULATION*

Encapsulation is the process of applying a molding material around the sample to be cross-sectioned. The purpose of this procedure is twofold. First, it allows the sample to be easily manipulated during the grinding, polishing, and etching stages of sample preparation. Second, it provides the added edge retention needed when the areas of interest are near the surface of the sample.

8.5.1 Molding Compounds

As with every step so far, a multitude of molding compounds are available. Each has unique properties that make it desirable in certain instances. For the most part, a molding compound that is to be used in cross-sectioning microelectronic components must exhibit a high hardness, excellent edge retention, good clarity, low shrinkage, good resistance to chemicals, and a low-temperature cure. An additional feature that is highly desirable, but not needed in every instance, is the ability to be used in either a vacuum or pressure impregnation system. Compression techniques of encapsulation *should never* be used for microelectronic components, because they will crush the delicate structures and materials used in these devices.

The three compounds that are the most versatile and fulfill the majority of requirements stated above are the epoxies, polyesters, and acrylics. There are, once again, numerous derivatives of these three compounds, and a distributor should be consulted about specific applications. The general features of each are as follows:

- *Epoxies* are two-part liquid systems consisting of a resin and a hardener. They provide excellent edge retention due to their high viscosity, which makes them the only system suitable for vacuum or pressure impregnation. They also provide the least amount of shrinkage upon polymerization. They are typically clear and have adequate hardness and chemical resistance. Their cure times are generally the longest of the three categories.

- *Acrylics* are two-part systems generally consisting of a liquid and powder that catalyze when mixed. They range from clear to milky white when cured and typically have poor adhesion to the surface of the sample. They are hard and chemically resistant. Their cure times are less than half an hour. An acrid odor is also typically associated with the curing and grinding of these compounds.

▪ *Polyesters* are also two-part systems that polymerize when mixed. They tend to exhibit better adhesion to a sample than the acrylics. They have good chemical resistance, and their cure times fall between those of epoxies and acrylics.

The encapsulation process is not completely foolproof, and invariably a poor mount will result when it can be least tolerated. In most instances, a mount can be placed in an organic solvent that will decompose the polymer encapsulant. It is recommended that the excess encapsulant be ground away from around the sample to limit the amount of time the sample is exposed to the solvent. These solvents can attack various metallic structures commonly used in microelectronic components. Solvents should never be used to extract printed wiring board samples, because they will quickly degrade the integrity of the laminants. Common problems encountered with cold mount systems are described with their causes in Table 8.3.[2]

8.5.2 Cleaning

A successful encapsulation depends upon the cleanliness of the sample. Finger oil and dust can severely degrade the adhesion of the compound to the surface of the sample. To prevent such occurrences the sample should be cleaned immediately prior to encapsulation. Typical cleaning would entail a mild detergent and water wash followed by an isopropyl alcohol rinse and a blast of dry compressed air. More aggressive cleaning may be accomplished by using an ultrasonic cleaner. This method can severely damage bond wires and therefore should not be used when such damage cannot be tolerated.

8.5.3 Vacuum Impregnation

The vacuum impregnation technique for encapsulation is indispensable when (metallographically) preparing porous samples such as compressed metal and ceramics powders or when surface features and edge retention are critical, as in the case for most microelectronic samples. Vacuum impregnation allows small pores and surface contours to come in direct contact with the encapsulant by removing the buffer zone of air. Because of its high viscosity, epoxy is the only molding compound suitable for this application.

The vacuum impregnation technique is a relatively simple process. This process requires a bell jar type of container with a bib access that allows a hose from a vacuum pump to be attached to it. An epoxy mount that has just been poured is placed into the bell jar. The contents of the jar is then evacuated until the epoxy starts to froth. At this point in the process, air is admitted back into the chamber to ensure that the epoxy has contacted the surface of the sample and allowed to fill any surface feature. This process of pulling a vacuum and admitting air is generally repeated two or three times (Fig. 8.2).

8.5.4 Techniques to Improve Edge Retention

When the standard encapsulant (to sample interface) does not provide adequate edge preservation, additional techniques may be employed to increase the edge retention. These techniques either increase the hardness of the encapsulant or provide a hard adherent overcoat to the sample. The systems that rely on hardening the encapsulant generally require the addition of a nonsoluble particulate. These additions tend to make the encapsulant opaque and as such limit their applications in the metallography of microelectronic devices.

Table 8.3 Typical Problems of Castable Mounting Materials

Problem	Cause	Solution
Acrylics		
Bubbles	Too violent agitation while blending resin and hardener.	Blend mixture gently to avoid air entrapment.
Polyesters		
Cracking	Insufficient air cure prior to oven cure, oven cure temperature too high, resin-to-hardener ratio incorrect.	Increase air cure time, decrease oven cure temperature, correct resin-to-hardener ratio.
Discoloration	Resin-to-hardener ratio incorrect, resin has oxidized.	Correct resin-to-hardener ratio, keep containers tightly sealed.
Soft mounts	Resin-to-hardener ratio incorrect, incomplete blending of resin-hardener mixture.	Correct resin-to-hardener ratio, keep containers tightly sealed.
Tacky tops	Resin-to-hardener ratio incorrect, incomplete blending of resin-hardener mixture.	Correct resin-to-hardener ratio, blend mixture completely.
Epoxies		
Cracking	Insufficient air cure prior to oven cure, oven cure temperature too high, resin-to-hardener ratio incorrect.	Increase air cure time, decrease oven cure temperature, correct resin-to-hardener ratio.
Bubbles	Too violent agitation while blending resin and hardener mixture.	Blend mixture gently to avoid air entrapment.
Discoloration	Resin-to-hardener ratio incorrect, oxidized hardener.	Correct resin-to-hardener ratio, keep containers tightly sealed.
Soft mounts	Resin-to-hardener ratio incorrect, incorrect blending of resin-hardener mixture.	Correct resin-to-hardener ratio, blend mixture completely.

FIGURE 8.2 A typical vacuum encapsulation station with an assortment of disposable molds.

The techniques that apply an overcoat to the sample to improve edge retention do so by placing the critical surfaces under the protective coat. The coat thereby replaces the surface of the sample along the weaker encapsulant interface. These procedures have gained wide acceptance in metallographic labs. The most common are either electrolytic or an electroless plating. Both nickel or copper are suitable for either application. The electroless plating systems tend to be simpler and have a lower risk of damaging the sample due to their less aggressive pretreatment procedures. These overplate systems are available from metallographic supply distributors.

These metallic plating systems are not adequate for added edge retention on deglassivated semiconductors, because they adhere poorly and obstruct viewing of the circuitry. Because of these shortcomings, a simple, reliable method of applying a glassivation layer has been developed by T. Devaney.[3] In this technique, a deglassivated die is placed on the platen of a standard metallographic wheel. The backside of the die is attached to the platen with a double-sided adhesive tape. The wheel is then turned to the highest setting, and a few drops of a commercially available spin-on glass is applied to the surface of the die. The die is allowed to rotate on the wheel so that the glass is uniformly spread across the die surface and is pushed into any microcrack by the centrifugal forces. The die is then removed from the wheel and placed on a hotplate to cure. Once the glassivation is cured, the sample can be encapsulated.

8.5.5 Optimized Procedure

For metallographic applications involving microelectronic components and materials, vacuum encapsulation procedure using an epoxy yields the best and most consistent results.

The first step in the vacuum impregnation technique is to prepare what is termed a *half-mount.* A half-mount is an epoxy puck that is approximately half the thickness of a standard mount. Half-mounts are used as a bed upon which the metallographic sample can be placed. This allows all surfaces of the sample to be encased with epoxy, thereby enabling the metallographer the versatility of viewing and potentially sectioning into any portion of the sample. This would not be the case if one were to use a standard metallographic sample clip. This added freedom becomes increasingly more important as the structures to be sectioned decrease in size (Fig. 8.2).

A half-mount can be made easily by mixing a batch of epoxy according to the manufacturer's instructions and pouring it to the desired thickness in a disposable mold. The half-mounts are then placed into a vacuum chamber (usually a bell jar connected to a vacuum pump). A vacuum is then pulled until the epoxy froths up to the top surface of the disposable mold, indicating that the majority of air bubbles have been removed. Air is admitted into the chamber and the molds removed, then the epoxy is then allowed to cure.

Once cured, the disposable molds are cut away from the epoxy. The top surfaces of the half-mounts are then ground to remove the meniscus and to roughen the surface for subsequent encapsulations. The half-mounts are placed back into disposable molds, and clean metallographic samples are placed upon their surfaces. Another batch of epoxy is mixed and poured over the samples to fill the molds. The mounts are placed back into a vacuum chamber and evacuated until the epoxy starts to froth up to the surface of the mold. Air is then readmitted into the chamber. This cycle of pulling a vacuum and admitting air is generally repeated two or three times. After the final cycle, the mounts are removed from the chamber and placed on a flat surface to cure.

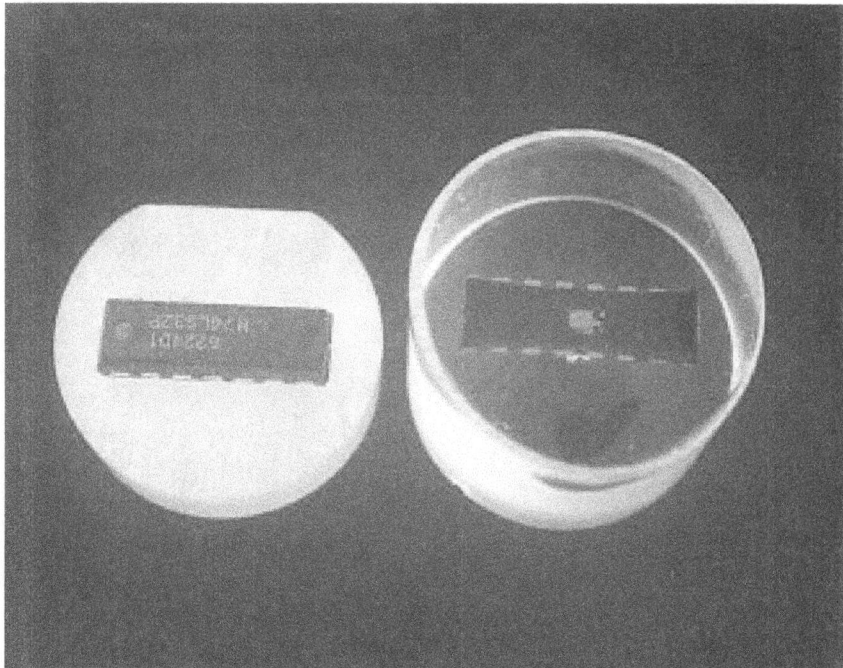

FIGURE 8.3   An example of the "half-mount" system.

The vacuum encapsulation procedure is, for the most part, a reliable process. However, Murphy's Law will invariably strike when encapsulating a critical one of kind sample. Procedures have been established that can compensate for some of the problems encountered with an epoxy mount.

Epoxy pull-away is a condition in which the epoxy pulls away from the surface of the sample. Pull-away can be the result of a dirty sample prior to encapsulation, imbalances in the ratio of hardener and resin, or an etching procedure. Preparing a mount that exhibits this condition will result in the complete loss of edge retention and the ability to delineate any critical surface features. Pull-away can generally be corrected with a backfill of epoxy (Fig. 8.4).

Backfill Procedure. The initial step in the backfill procedure is to ultrasonically clean the sample in a beaker of water. The sample is then removed from the water and blown dry with compressed air. The sample is placed in a 100° C oven to drive off any residual moisture present within the pull-away interface. A piece of transparent tape is then wrapped around the top edge of the mount to form a cup. A batch of epoxy is then mixed and a thin layer of it poured on the surface of the mount. The mount is placed in a vacuum chamber and the air evacuated thereby allowing the epoxy to fill the gap produced by the pull-away. Air is then admitted into the chamber. The cycle of pulling a vacuum and admitting air is repeated two or three times to ensure a complete backfill of epoxy. The mount is then taken out of the vacuum chamber and allowed to cure at room temperature.

FIGURE 8.4 An example of epoxy pull-away from a poor encapsulation procedure. Note that the pull-away occurs along the surface of the braze.

Soft epoxy is a condition that results from an imbalance in the hardener/resin ratio. Epoxy is considered soft if it deforms rather than being scratched when probed with tweezers or a sharp instrument. If used in this state, considerable rounding and a overall lack of edge retention will result. This lack of hardness at times can be corrected by placing the mount in an oven maintained at approximately 100° C for 30 to 60 min. The mount is then removed from the oven and allowed to cool to room temperature (Fig. 8.5).

Fire epoxy refers to a yellow/orange epoxy filled with bubbles and the pull-away that occurs from the violent exothermic reaction produced by the addition of an excessive amount of hardener. This is a common condition when pouring large volumes of epoxy as a single mount. If the mount were to be used in this condition, a lack of edge retention would result, and the transfer of material between steps would occur. This condition may be corrected by grinding into the separation surrounding the area of interest and then performing the backfill procedure described earlier (Fig. 8.6).

If the described solutions have not worked, there is only one remaining option: to decapsulate and reencapsulate the sample. This is done by grinding/cutting away as much of the epoxy surrounding the sample as possible. The remainder is then placed in a 100° C bath of Dynasolve 180™ (Dynaloy Inc.) until the last of the epoxy is removed. The sample is then rinsed off in sequence with acetone, isopropyl alcohol, and water. The sample is blown dry with compressed air. The sample can now be reencapsulated following the procedures described earlier.

8.6 INTERMEDIATE SAMPLE PREPARATION

Prior to initiating the grinding procedure, a few simple steps should be performed. The epoxy mounts should be removed from their molds and any pertinent sample identification

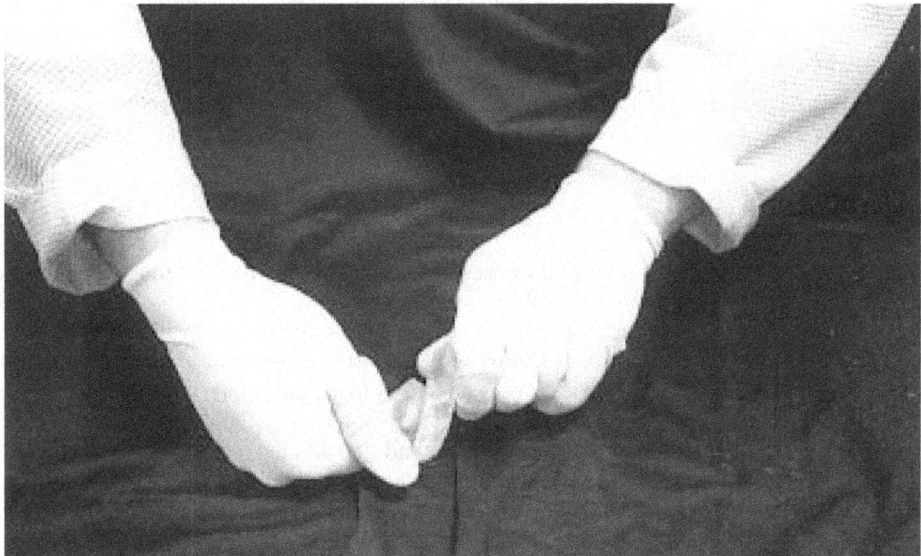

FIGURE 8.5   An example of soft epoxy. Here, the epoxy mount is being twisted.

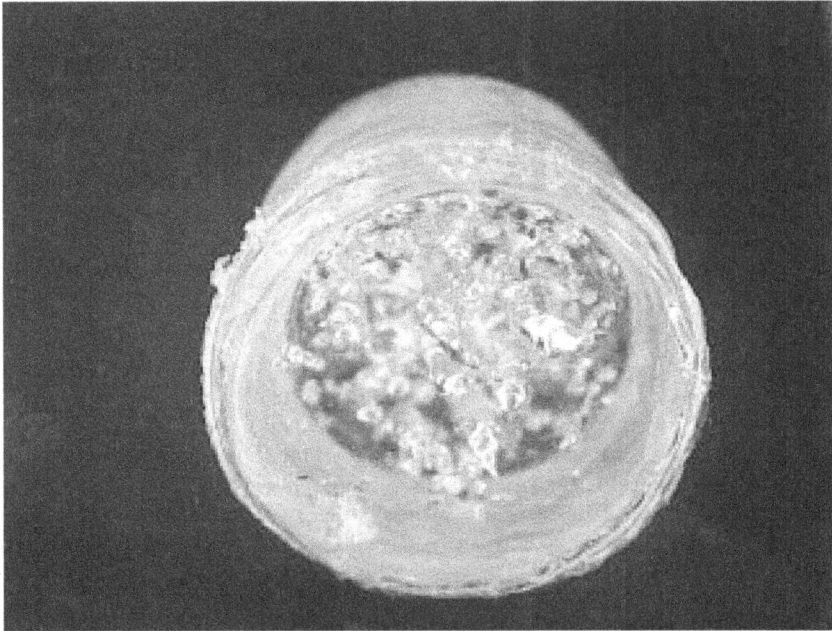

FIGURE 8.6 An example of "fire epoxy." Note the bubbles within the epoxy that can entrap grinding and polishing debris and provide poor edge retention.

scribed into the surface of the encapsulant. The sharp meniscus around the top surface of the mount then should be beveled off to prevent cutting the metallographer as well as catching and tearing the papers and cloths used in the grinding and polishing steps. Beveling of the edge can be easily accomplished by rotating the top edge of the mount on a piece of silicon carbide paper. The final and most critical of the steps is to verify the plane of sectioning. If not identified correctly, the structure of interest could be ground through within a matter of seconds.

8.7 GRINDING

Grinding has several purposes in the metallographic procedure. First, it establishes a flat plane on which each subsequent step depends. Second, it is used to remove any damage that may have been imparted into the sample from a removal technique. Finally, it is allows access to the desired location with a nominal amount of induced damage and deformation.

The grinding process uses abrasives to remove material from the surface of the sample. Contact between the surface of the sample and the abrasive can result in one of two actions. Mulhearn and Samuels found that an abrasive will scratch and cut a material if it has a favorable contact angle with the sample. They measured this contact angle to be $90° \pm 2°$ for steel. If the critical contact angle is not achieved, the abrasive will tend to produce a ploughing action. This action results in a layer of deformation.[4]

The thickness of this deformation layer depends on the abrasive size. The larger (more coarse) the abrasive, the thicker the deformation layer. Because of this relationship, all grinding procedures will incorporate numerous steps using sequentially finer abrasives. Nelson displays this relationship between abrasive size and deformation layer thickness in Fig. 8.7.[5]

The grinding process has been traditionally divided into *coarse* and *fine* grinding categories. Coarse grinding is the initial step. Abrasives in the range of 60 to 320 grit are used to grind through the encapsulant past the damage incurred during sample removal and to gain close proximity to the desired final location. This is followed by fine grinding, which uses 400 and 600 grit abrasives to prepare the sample for polishing.

For microelectronic components, coarse grinding is done only to remove the encapsulant that is in front of the sample. In most situations, any contact with an abrasive greater than 400 grit will cause severe damage to the component. Fine grinding will generally consist of a sequence of 400, 600, and 1200 grit abrasives. In some instances, additional steps of 12, 9, and 6 μm abrasives are used.

8.7.1 Grinding Abrasives

Fixed abrasive systems have gained universal acceptance for grinding due to their ease of use and compatibility with most metallographic laboratory equipment. Depending on the type of abrasive, they can be bonded to paper, metal, or a mylar type of film. These products come in sheets, disks, and belts of various sizes.

There is a greater choice of quality abrasives today than five years ago, because high-quality aluminum oxide and diamond can be economically manufactured. However, the overwhelming choice is still silicon carbide. Silicon carbide is hard enough for most applications. Papers of it are readily available in variety of shapes and sizes to meet the requirements of most laboratories. These abrasives papers can be purchased in grades ranging from 60 to 1200 grit (Figs. 8.8 through 8.10).

Aluminum oxide is a slightly softer abrasive than silicon carbide. In some applications, where a soft and hard material are coupled, it has been found to provide superior flatness as compared to silicon carbide. Aluminum oxide can be purchased as a sheet or disk. It is available as a paper product but it is more commonly found as a film. Bonded abrasive sizes range from 100 to 0.5 μm (Fig. 8.11).

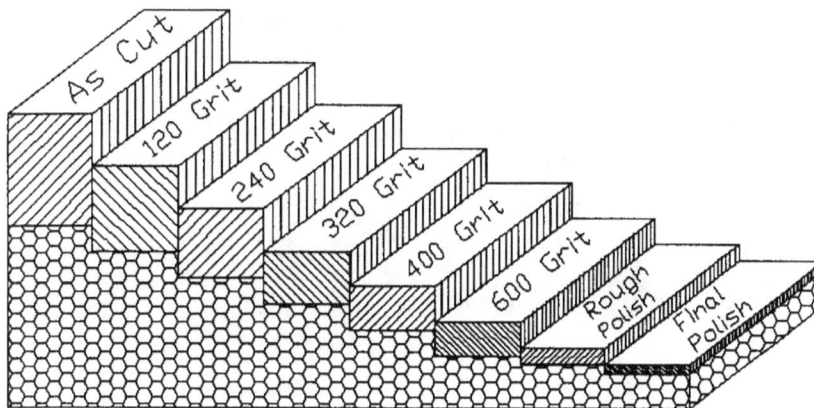

FIGURE 8.7   Correlation between abrasive size and deformation layer thickness.[5]

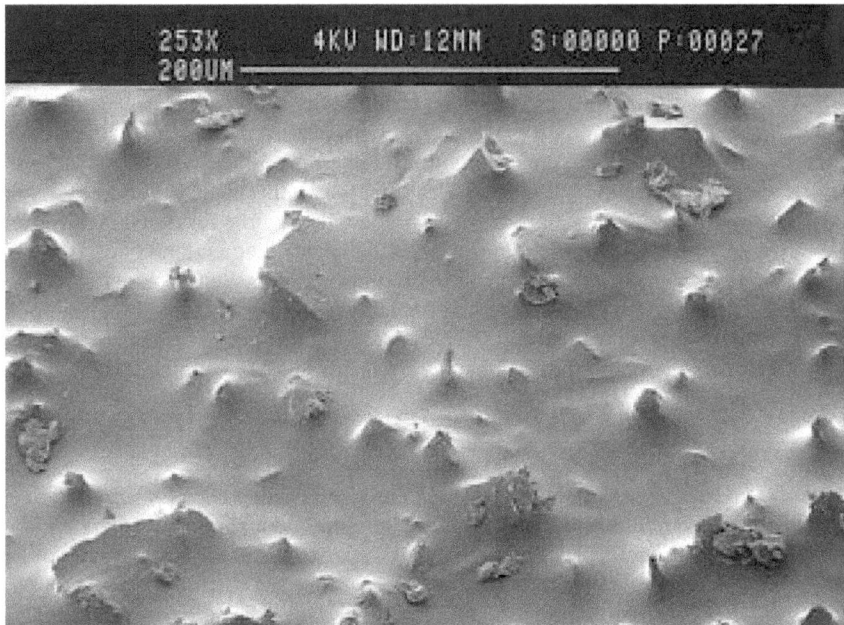

FIGURE 8.8 SEM photo showing the orientation and structure of the abrasives on a standard 400 grit silicon carbide disk.

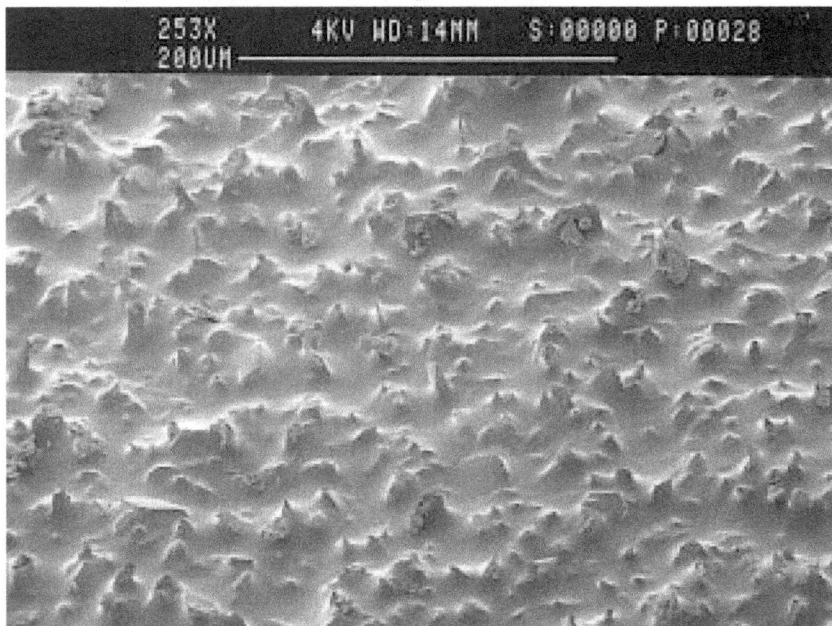

FIGURE 8.9 SEM photo showing the orientation and structure of the abrasives on a standard 600 grit silicon carbide disk.

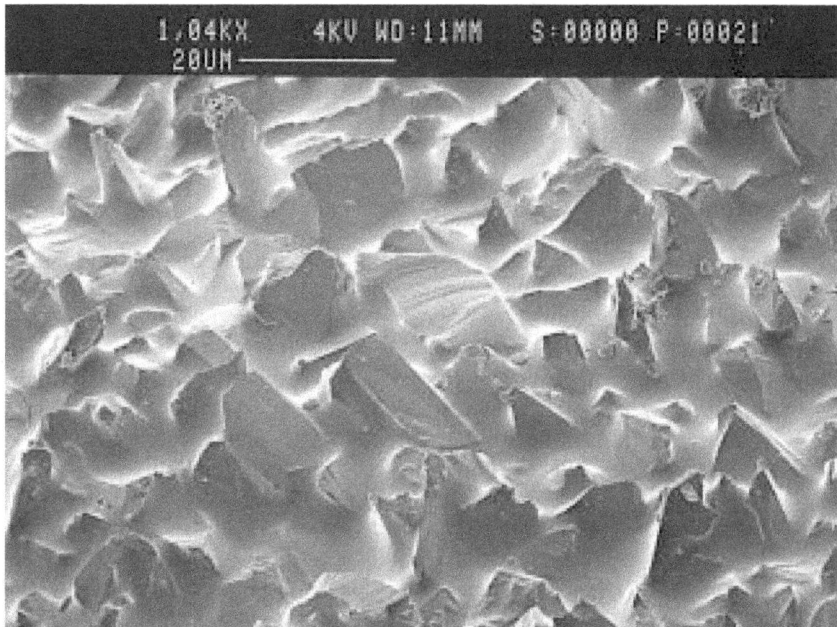

FIGURE 8.10   SEM photo showing the orientation and structure of the abrasives on a standard 1200 grit silicon carbide disk.

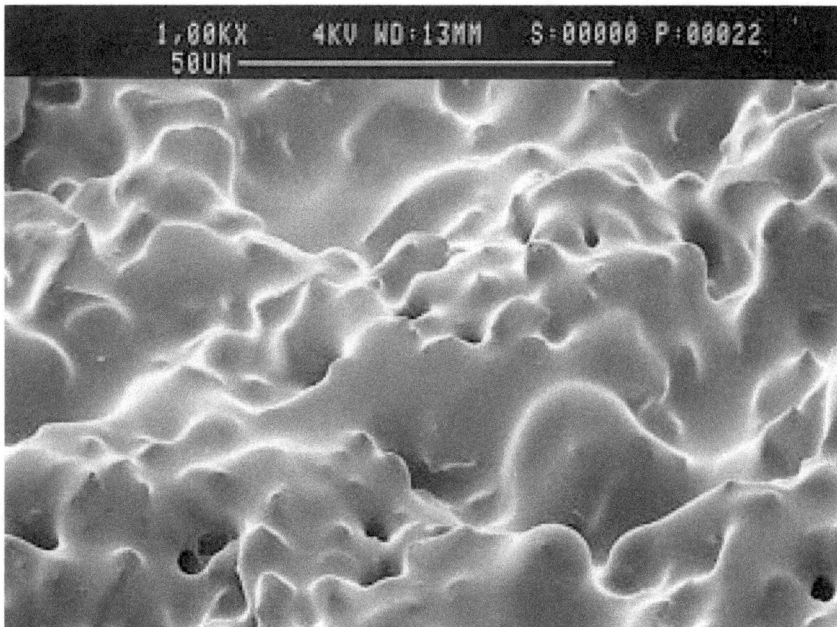

FIGURE 8.11   SEM photo showing the orientation and structure of the abrasives on a standard 10–15 μm diamond film.

Diamond is the hardest of all commercially available abrasives. Even with improvements in manufacturing it is still the most expensive. However, in certain applications such as in the preparation of ceramics and sapphire it is indispensable. Diamond is bonded to either a metal disk or to a film. Diamond is available in the largest range of sizes which extend from 50 mesh to 1 μm (Fig. 8.12).

8.7.2 Equipment and General Procedure

Grinding can be done either as a hand lapping procedure or with the aid of a motorized wheel. The latter has gained wide acceptance, because it is less tedious and allows the metallographer more control over material removal rates by adjusting the rotational speed of the wheel (Fig. 8.13).

The first step of the grinding process is to choose a suitable abrasive grade. The disk is then attached to the wheel platen with a metal ring that forms a compression fit around the circumference. The wheel is then turned to a selected speed. For coarse grinding through the encapsulant, a high setting is generally chosen to maximize the material removal rate. A lower setting is then used for grinding into the device.

A lubricant is then flowed onto the face of the abrasive. Water is the most common choice, because it is nontoxic and readily accessible in a laboratory environment. There are rare instances, however, when water should not be used, because it could react with the sample.

The flow rate of water is very important. The flow rate must be high enough to control frictional heating between the abrasive and sample so that exaggerated smearing and

FIGURE 8.12 SEM photo showing the orientation and structure of the abrasives on a standard 10–15 μm diamond film.

FIGURE 8.13   Typical grinding and polishing wheel.

deformation do not occur. It must also be sufficient to wash the grinding debris from the abrasive so that cutting can be optimized and the life of the abrasive can be extended. If the flow rate is too high, there will be a tendency for the sample to hydroplane across the surface of the abrasive. This will also result in poor cutting and an increase in the thickness of the smeared and deformed zone on the surface of the sample.

The distribution and magnitude of load applied to the metallographic mount are also important. The distribution of load will determine the plane in which a sample is prepared. A uniformly applied load will result in a flat surface, which is the desired goal. Varying the distribution allows the metallographer to remove more material where the load is highest and to adjust the plane of the section. Once the plane has been adjusted to the correct position, the load should be redistributed in a uniform manner to maintain the correct plane. The ability of the metallographer to make fine adjustments becomes more critical as the geometry of the structures being cross-sectioned decreases.

The amount of load applied to the mount should be just enough so that the metallographer can feel a slight to moderate tugging action as the sample is held against the rotating abrasive. The amount of resistance felt will decrease with the size of the abrasive. If the applied load is not high enough, the sample will ride on top of the abrasive resulting in the conditions described as hydroplaning. If the load is too great, the abrasives can become embedded in soft materials, or it can crack delicate structures.

Once the abrasive has been chosen, and the wheel speed, water flow rate, and applied load have been optimized, the sample is held in a stationary position against the abrasive. The sample is then slowly tracked back and forth across the radius of the wheel so that the disk is worn at an even rate. The sample should be inspected with a stereomicroscope and/

or metallurgical microscope frequently to ensure that the correct plane of the section is maintained and to verify the current position with respect to the final desired location.

Changing to the next finer abrasive should be done when all scratches from the previous stage are removed and a uniform scratch pattern from the current abrasive is observed. Traditionally, metallographers have been taught to rotate the sample 90° when changing between abrasive grades. This has allowed for easy recognition of scratches left from the previous step. This practice is not recommended when metallographically preparing microelectronic components, because any edge or corner that is not fully supported can be chipped or cracked, leading to misdiagnosis of quality or failure mechanism. The critical edge of the sample therefore should be kept facing into the rotation of the wheel at all times. A metallurgical microscope must then be fully utilized to determine when scratches from the previous step have been fully removed. Figures 8.7 through 8.10 show the typical scratch patterns produced in a semiconductor from the various grades of abrasives.

When changing between abrasive sizes, both the metallographers hands and sample should be thoroughly rinsed with water. This will prevent the transfer of grinding debris.

8.8 POLISHING

Polishing is the final step in the mechanical preparation of a metallographic sample. The purpose of polishing is to produce a surface that is as free of deformation as possible. Traditionally, the goal has been to achieve a mirror finish. However, due to the variety of materials joined together and the critical nature of surface features for microelectronic components, polishing to a mirror finish, if not performed correctly, can lead to rounding and smearing of critical interfaces. For these reasons, it is imperative that suitable abrasives and cloths are used.

The principles of polishing are in essence the same as those of grinding. The only difference is that the abrasives are not fixed but suspended. Because of this freedom, the abrasives tend to tumble rather than cut. As a result, material removal rates are much lower when compared to comparable times using fixed abrasive systems of similar size. Samuels has theorized that the abrasives will tumble and plough until they become embedded in the cloth. It is when the abrasives become embedded that the cutting action is optimized.[4] Figures 8.14 and 8.15, respectively, show 6 μm diamond abrasives in a napless coarse polishing cloth and 0.05 μm gamma alumina abrasives in a napless final polishing cloth.

8.8.1 Polishing Abrasives

Polishing abrasives are available in a variety of grades and forms. Each abrasive has unique properties and applications. There are three general-purpose polishing abrasives used in a metallographic lab. The abrasives are diamond, aluminum oxide, and colloidal silica.

Diamond is the hardest of the polishing abrasives. The diamond abrasive provides superior flatness and is generally used as an initial polishing step. Diamond polishing compounds can be purchased as a paste or as a suspension in grades ranging from 0.05 to 60 μm. The diamond polishing compounds come in either a mono- or polycrystalline form. As shown in Figs. 8.16 and 8.17, the polycrystalline form has more facets and therefore has a higher cutting rate.

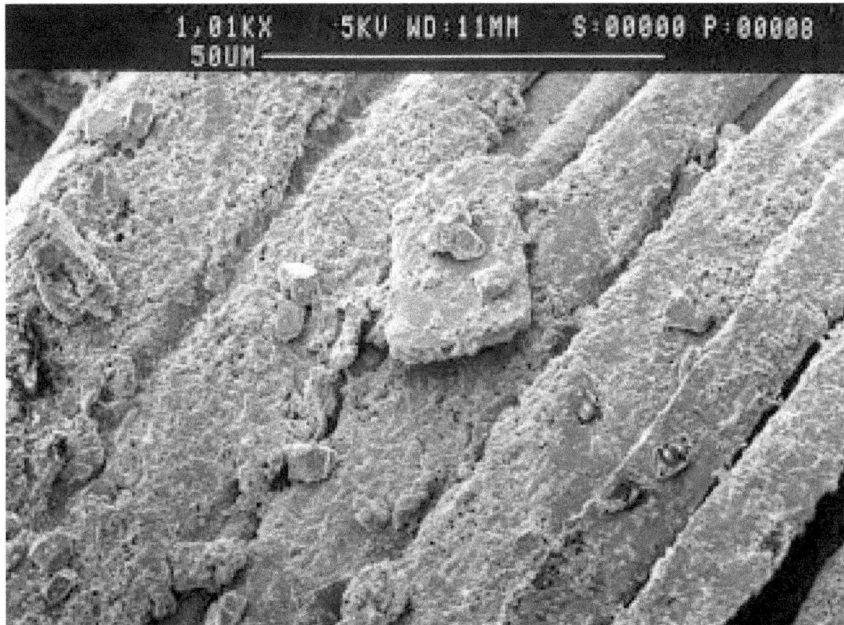

FIGURE 8.14   SEM photo showing how the 6 μm diamond abrasives (angular structures) become embedded in the fibers of a napless cloth during polishing.

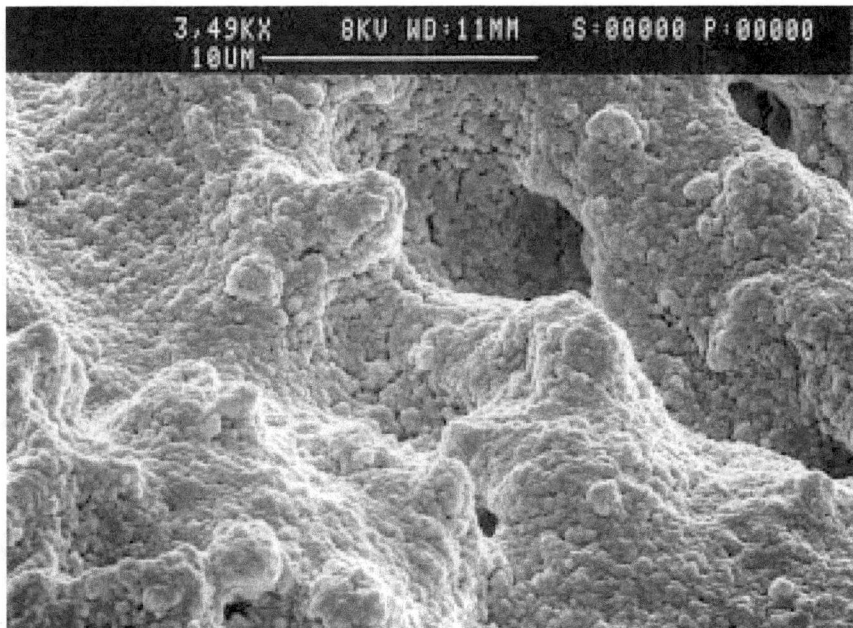

FIGURE 8.15   SEM photo showing how the 0.05 μm gamma alumina slurry attaches to the structures of a napless final polishing cloth.

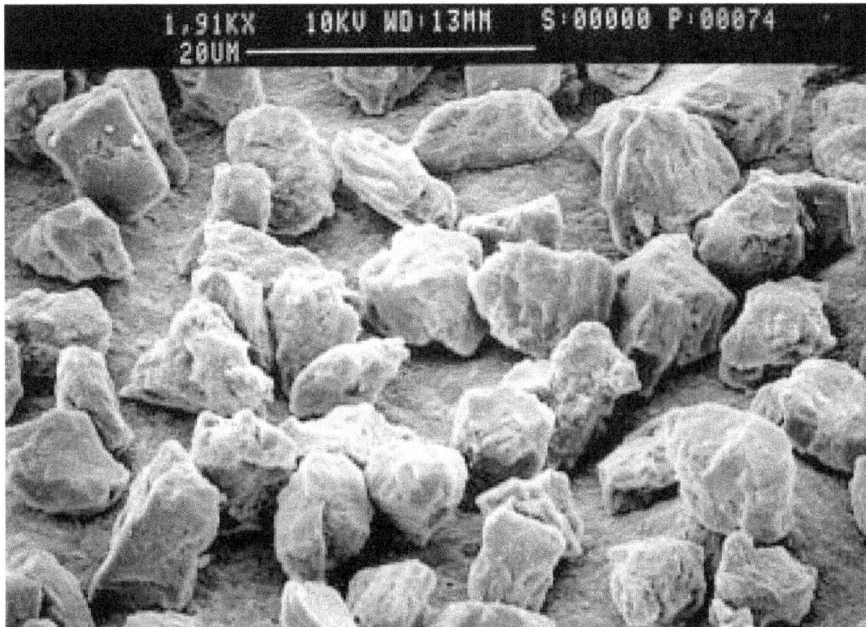

FIGURE 8.16 SEM photo of a monocrystalline form of diamond commonly found in polishing compounds.

FIGURE 8.17 SEM photo of a polycrystalline form of diamond commonly found in polishing compounds. Note the greater number of corners, which improves the cutting action.

Aluminum oxide is also available in two forms, alpha and gamma structures. The alpha aluminum variety traditionally has been used as an intermediate polishing step, because it is slightly harder than the gamma form. The use of alpha aluminum has become almost nonexistent because of the superior performance and the wide variety of diamond products that are commercially available. In contrast, the 0.05 gamma alumina is still the most widely used abrasive for final polishing. It can be purchased as either a premixed slurry or in powder form.

Colloidal silica is gaining wider acceptance as a final polishing compound. In applications dealing with high-purity metals and couples of hard and soft materials, colloidal silica has given superior results. Colloidal silica can be purchased as a premixed slurry in a grade of 0.05 μm. Care must be exercised when using colloidal silica. If the polishing wheel is allowed to dry, the silicon dioxide will crystallize. These glassy particles can severely scratch and gouge the next metallographic sample placed on the wheel. If a dry wheel is encountered, it should be thoroughly rinsed with water.

8.8.2   Polishing Cloths

The choice of a polishing cloth is as important to the outcome of a sample as is the choice of an abrasive. The purpose of the polishing cloth is to hold the abrasive against the surface of the sample. The cloth accomplishes this by entrapping the abrasive particles within its texture or weave.[4] The choice of a cloth should be based on its compatibility with an abrasive system and with the type of metallographic sample under preparation.

For coarse polishing, a napless cloth is generally used to take full advantage of the flatness achievable from using diamond abrasives. A low-nap cloth can also be used for this stage of polishing. However, a low-nap cloth tends to produce relief between hard and soft structures.

Low-, medium-, and high-nap cloths are used for fine polishing. Relief between structures increases as nap length increases. The relief produced from even the low-nap cloths is too great for most microelectronic components and assemblies. Because of the need to counter this situation, napless final polishing cloths have been developed. Table 8.4 provides a selection guide for some of the commercially available polishing cloths.

8.8.3   Equipment and General Procedures

Polishing, like grinding, is generally done on a rotating wheel. The wheel or wheels should be dedicated to polishing to prevent residual grinding debris from contacting the polishing cloths. Also, the wheels should be placed in a clean environment and kept covered when not in use. This will help prevent dirt from gathering on the cloths.

Cross contamination is of great concern while working with any sample. Because of this, the first step in the polishing procedure is to wipe any dust and/or dirt buildup from around the polishing wheel(s). The metallographer must wash both hands, and the metallographic sample, with a mild detergent solution and dry thoroughly prior to starting the process. The metallographer must then choose an appropriate cloth and attach it via its adhesive backing to the wheel platen.

The coarse polishing cloth is then charged with diamond by applying four small dots of paste around the wheel. The paste is worked into the cloth with either a clean fingertip or the side of a clean mount. The wheel is turned on, and a water- or oil-based extender is dispensed onto the cloth. The purpose of the extender is to provide lubrication between the metallographic sample and the cloth.

Table 8.4 Polishing Cloth Selection Guide

Polishing cloth	Description
Glendur	Fine stainless steel mesh on an aluminum foil polyethylene backing. Recommended when grinding and polishing metal matrix and tough-to-polish composite materials.
Gold Label	Specially woven napless nylon with plastic backing. This cloth has a long life and provides excellent flatness and edge retention. Recommended for all diamond compounds.
Nylon	Durable napless nylon weave. Provides excellent finish on most materials.
Encloth	A short pile wool cloth excellent for all final polishing stages.
Glenco	Synthetic rayon flock bound to woven cotton backing. Excellent for final polishing with diamond or alumina.
Vel-Cloth	Low-napped synthetic velvet cloth designed to produce excellent final polish with minimal edge rounding.
Kempad-Pan W	Nonwoven textile both oil and water resistant. Recommended for rough and intermediate polishing.
Billiard	100% virgin wool sheared pile recommended for intermediate polishing stages.
Blue Felt	100% woven wool material used with outstanding success for polishing hard materials.
Velvet	Made from the finest synthetic velvet available. Excellent for final polishing of soft materials.
Spec-Cloth	A finer version of Glenco specially designed for superior final polishing on a verity of materials.
Imperial Polish Film	Rayon flock coated on a polyester film backing used in conduction with fine-grade slurries and compounds.
Cotton	A tightly woven cotton cloth for use with diamond and alumina in the rough and intermediate polishing stages.
Red Felt	Plucked pile, 100% virgin wool fabric. Recommended for rough and intermediate polishing stages with diamond or alumina.
Silk	100% pure woven silk for final or skid polishing on friable materials.
Selvyt	Medium nap cloth made from high quality cotton. Excellent for intermediate and final polishing with diamond or alumina.
Final B	Specially designed for long life. Made from a rayon flock on a cotton canvas backing. Excellent for all final polishing.
Final A	Napless cloth bound to tightly woven cotton backing. Excellent for final polishing with alumina, diamond, or colloidal silica.

It is important that a proper amount of extender is on the wheel at all times. If there is not enough extender present, the diamond abrasives will become embedded in the epoxy or any soft material present in the sample. If too much extender is on the wheel, the sample will tend to hydroplane and will not polish the sample properly.

The proper amount of extender can be determined easily by removing the sample from the wheel and inspecting the surface that was in contact with the cloth. If the surface exhibits a thin, glistening film, a proper amount of extender is on the wheel. If the surface appears dry, additional extender needs to be added to the wheel. If extender builds up around the sample, the wheel should be allowed to rotate without use until the excess has been spun off.

Once the correct amount of extender is applied, the polishing wheel is set to a low speed, and the sample is placed on the wheel with a medium amount of evenly distributed pressure. If there is no concern about potentially damaging a critical surface feature, the sample is moved circumferentially around the polishing wheel. The sample should be moved in a direction opposite to the rotation of the wheel.

Throughout the coarse polishing process, it may become necessary to recharge the wheel with abrasives and extender. Diamond may be reapplied in a paste form. It is more convenient to maintain the abrasive level with a suspended solution. Extenders can be applied easily with a squirt bottle.

Between the coarse and final polishing steps, the metallographer must again thoroughly wash both hands and the sample with a mild detergent solution and dry thoroughly. An appropriate final polishing cloth may then be applied to the platen. If using an existing cloth, it should be thoroughly rinsed with water and, while the wheel is rotating, the wheel should be squeegeed off by running a clean thumb from the midpoint of the wheel to its outer edge. This procedure should remove any debris from the cloth that could scratch the surface of the sample.

The cloth should then be saturated with the abrasive slurry. The sample is then placed on the cloth with a light amount of uniform pressure distributed across the surface of the sample. The sample may then be moved in a manner consistent with the coarse polishing step.

If the polishing cloth tends to grab the sample, the slurry may be too thick and should be thinned by applying water to the wheel. During the final polishing step, water and slurry may need to be added to the wheel as it dries off. Additions of both materials can easily be made with squirt bottles.

The majority of artifacts and edge relief are produced during the polishing stage of sample preparation. For this reason, polishing times should be kept to a minimum. Polishing for extended periods of time indicates that the previous steps were not performed properly and that artifacts are likely to have been produced. Figures 8.18 through 8.23 show typical damage caused by various abrasives.

8.9 ETCHING

Etching is used to bring out contrast in an "as-polished" sample. This contrast enhancement is achieved by producing variations in the reflectivity of a sample when viewed under an optical microscope or variations in the topographical features when viewed under a scanning electron microscope. Etching is also used to remove the residual layer of deformed metal in a properly prepared metallographic sample.[5]

FIGURE 8.18 Typical damage produced in a semiconductor die from a 400 grit silicon carbide abrasive.

FIGURE 8.19 Typical damage produced in a semiconductor die from a 600 grit silicon carbide abrasive.

FIGURE 8.20 Typical damage produced in a semiconductor die from a 1200 grit silicon carbide abrasive.

FIGURE 8.21 Typical damage produced in a semiconductor die from a 12 μm aluminum oxide abrasive.

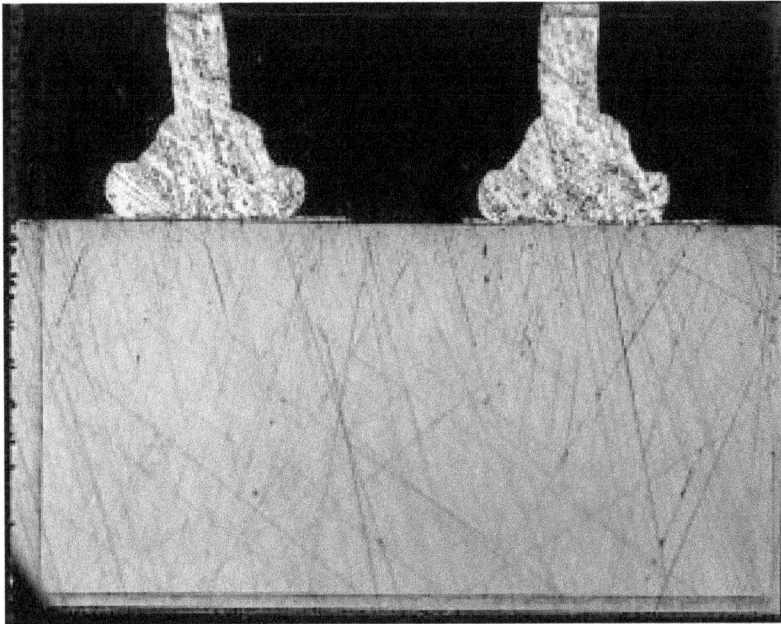

FIGURE 8.22 Typical damage produced in a semiconductor die from coarse polishing with a 6 μm diamond abrasive on a napless cloth.

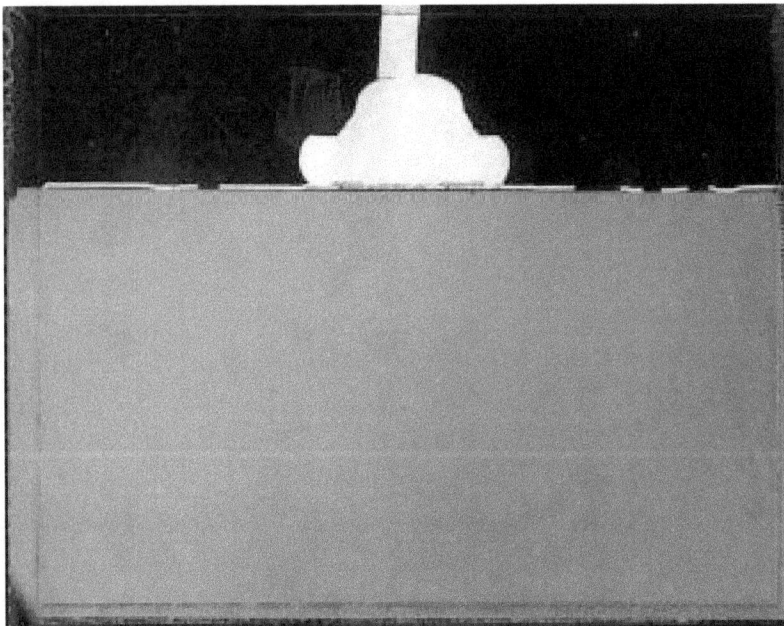

FIGURE 8.23 Typical damage produced in a semiconductor die from final polishing with a 0.05 μm gamma alumina slurry on a napless cloth.

Etching is generally required for metallurgical samples that tend to exhibit homogeneous features showing little if any contrast in the as-polished sample. However, microelectronic components and assemblies tend to be a composite of nonhomogeneous materials. As such, variations in the reflectivity of the sample tend to be greater than the 10 percent difference needed for visual contrast.[6] This does not mean that microelectronic components and assemblies are never etched but, rather, a great deal of information can be obtained by inspecting an as-polished sample.

Electrochemical etching is the oldest and most commonly employed method for producing contrast in an as-polished sample. The electrochemical technique relies on establishing a galvanic cell between the microstructural features in a metallographic sample. As with corrosion, variations in the electromotive potential will render areas either anodic or cathodic. Under corrosive action, the anodic regions will be attacked while the cathodic regions will be preserved. In a metallographic sample, anodic and cathodic regions can be established between a grain and grain boundary, grains and impurities, mechanically worked and nonworked zones, and between unlike microstructural phases.

These anodic and cathodic regions produce contrast by reflecting light at various angles due to the relief and faceting of various structural components. Optical contrast is achieved by the anodic regions reflecting the incident light off at oblique angles (producing dark indications) and the cathodic regions reflecting the light in a normal/incident direction (producing bright indications). Inspection with an electron microscope will delineate the topographical contrast produced by the anodic regions being recessed below the cathodic regions.

Etchants are typically arrived at by trial and error. However, some of the guesswork is removed by knowing that most etchants have three common features: a corrosive agent (such as an acid), a modifier (such as alcohol or glycerin) and an oxidizer (such as hydrogen peroxide).[4]

Electrochemical etching is generally performed at room temperature under a fume hood. The sample is immersed into the etchant with tongs and slightly agitated to break up any vapor barrier layer that may be produced on the surface from the corrosive action. It is not recommended that a cotton swab be used to apply etchant onto the surface of the sample, because the fibers could scratch the polished surface.

The etching time can be anywhere between a few seconds and several minutes. Even comparable samples can etch at different rates. Therefore, as a general rule, it is better to etch for several short durations and monitor the progress between steps rather than to etch for one long cycle because additional etching is easier to perform than resurfacing the sample.

It is also important to realize that electrochemical processes are not ideally suited for the majority of microelectronic components and assemblies. This is because of the variety of materials coupled together and the large potential differences that generally exist between them. As a result, artifacts such as pitting and the complete removal of the most anodic metal are common. Because of this, the metallographer must have a clearly defined goal and must be willing to play a game of "give and take." Table 8.5 shows common etchants that are used for microelectronic components and assemblies.

There are a variety of adaptations to the traditional electrochemical process. However, these techniques which include precipitation etching, electrolytic etching, anodizing, and potentiostatic etching have not been proven viable for assembled microelectronic components and assemblies.

Vapor deposited interference layers produce delineation by amplifying microstructural differences that are present on the "as polished" surface but are not visible under normal

Table 8.5   Common Etchants for Microelectronic Components and Assemblies

Etchant composition	Comments
Aluminum	
85% saturated solution of KOH	Used to etch pure Al as in the case of microcircuit metallization; swab 2–10 s.
2.5 mL HNO_3 1.5 mL HCl 1.0 mL HF 95 mL DI water	Keller's reagent; general-purpose reagent for Al and Al alloys; outlines constituents and brings out grain boundaries in some alloys.
Copper	
10 mL NH_4OH 20 mL H_2O_2	General-purpose etch for Cu and its alloys; used for PWBs because it etches Pb/Sn at a comparable rate; use fresh swab, 3–5 s.
15 mL NH_4OH 15 mL H_2O_2 15 mL DI water 4 pellets NaOH	For Cu-BE alloys; add NaOH last.
Gold	
20 mL HCl 20 mL HNO_3	Aqua regia; for Au and other noble metals.
1–5 g CrO_3 100 mL HCl	For pure Au and its alloys; swab or immerse sample for up to 60 s.
Iron	
1–10 mL HNO_3 90–99 mL methanol	Nital; general-purpose etch for iron and its alloys with the exception of stainless steel; good for Kovar® packages.
4 g $CuSO_4$ 20 mL HCl 20 mL DI water	Marbles reagent; reveals the grain structure of most stainless steels.
Lead	
84 mL glycerol 8 mL acetic acid 8 mL HNO_3	For Pb and its alloys; use fresh and for a few seconds.
15 mL acetic acid 5 mL H_2O_2 (30%)	For Pb and its alloys; etch for 6–15 min.
Silicon	
10 mL HNO_3 10 mL acetic acid 1 mL HF	Provides relief between oxides and delineates doped junctions; swab for 2–10 s.

Table 8.5   Common Etchants for Microelectronic Components and Assemblies *(continued)*

Etchant composition	Comments
Silver	
50 mL NH₄OH 20 mL H₂O₂	For Ag and Ag solders; use fresh and immerse for up to 60 s.
60 mL HCl 40 mL HNO₃	for Ag alloys; immerse for up to 60 s.
Tin	
2–10 mL HNO₃ 90–98 mL DI water	Nital; for Sn and its alloys; swab or immerse several minutes.
2–10 mL HCl 90–98 mL alcohol	For Sn and its alloys; swab or immerse several minutes.
Titanium	
1–3 mL HF 2–6 mL HNO₃ 100 mL DI water	Kroll's; for Ti alloys; swab or immerse for 3–30 s.
15 mL latic acid 5 mL HF 5 ml HNO₃ 5 mL DI water	For Ti and its alloys; swab sample.
Tungsten	
10 g KOH 10 g K₃Fe(CN)₆	Murakami's; for W and its alloys; immerse sample for up to 60 s.
H2O2 (3%)	For W; use boiling for up to 90 s.

illumination. Amplification is achieved by depositing a thin film of iron, platinum, lead, gold, copper, indium, nickel, or palladium onto the sample surface. Reflections are then obtained from both the film surface and sample to film interface producing the interference. Good contrast has been reported for aluminum alloys, high-temperature nickel and cobalt alloys, cemented carbides, plasma sprayed layers, brazed joints, and sintered metals.[6]

Ion etching has been used in a variety of analytical techniques that include depth profiling for auger analysis, the production of thin section samples for transmission electron microscopy, and as a production tool in the manufacturing of semiconductor die. It is only recently that these techniques have been used for etching metallographic samples.

Ion etching consists of placing a polished sample in a vacuum chamber. Ions of particular gas are accelerated to a high energy state and allowed to impinge upon the surface of the sample. On impact, the ionized gas will "pull-off" surface atoms from the sample. Contrast is subsequently produced by material removal rate differences of the various structures present in the sample. These removal rates depend on the type of ionized gas, the material being etched, and the angle at which impingement occurs.

8.10 INSPECTION TECHNIQUES

Today, the metallographer has a wider choice of inspection techniques from which to choose than at any other time in history. Light microscopy has always been, and still is, the first choice, due to ease of use, numerous contrast enhancement techniques, and the variety of samples that it can accommodate. However, with the general availability of electron microscopes, there is a second viable tool readily available to a metallographer. The choice of technique will depend on the sample requirements. Table 8.6 gives a comparison between light microscopy and scanning electron microscopy.[7]

Table 8.6 Light Microscopy vs. Scanning Electron Microscope[*]

Criterion	Light microscopy	Scanning electron microscope
Resolution	0.2 mm	0.002 mm
Magnification	2,000×	74,000×
Depth of focus	d	200 d
Working distance	w	20w
Sample preparation	Less	More

* *Source:* Data from J.R. Devaney, SEM Training Material.

General techniques and applications of light microscopy and scanning electron microscopy will be discussed here in summary form. More detailed discussions can be found in Chaps. 4 and 11.

8.10.1 Optical Microscopy

Optical microscopes can be used to inspect both polished and etched samples. In all adaptations, light from an external source is sent through an optical system down onto the surface of the sample and reflected back up to the eye. It is the manipulation of this light and its reflections that provides contrast.

Bright-field illumination provides contrast between surface features that are normal to the incident light and those that are oblique to it. The features normal to the incident light appear bright, while those that are oblique appear dark. In most instances, this contrast needs to be enhanced by etching to provide the variations in reflectivity. This technique is most useful when inspecting etched samples. However, as-polished samples are viewed when inspecting for separation between interfaces, porosity, and inclusions.

Dark-field illumination provides contrast by collecting the oblique light and blocking the incident. This method provides a significant improvement in contrast as compared to bright-field illumination. It is particularly useful when inspecting for fine features such as voids, cracks, and delaminations.

Polarized illumination provides contrast by converting incident light into plane polarized light with use of a polarized filter. Upon reflectance (back up) from the sample, the plane polarized light goes through an analyzer which filters out all light not parallel to the analyzer filter. This technique provides microstructural contrast of anisotropic metals and phases. It has also been reported that polished sections of transparent resin, glass, and ceramic layers when viewed under polarized light will exhibit their inherent colors and any flaws such as cracks.[6]

Differential interference contrast provides topographical delineation by sending linearly polarized light into a prism that splits it into two parallel paths. One path is slightly ahead of the other as it impinges the surface of the sample. The reflected paths are then recombined by the prism. Interference based on the topographical features of the sample are produced by the analyzer. Differential interference contrast is useful when analyzing structures that exhibit relief from polishing due to adjoining soft and hard regions.

Phase contrast is an infrequently used technique that provides structural delineation based on relief variations on an "as polished" sample. Height variations as little as 10 to 50 Å can be observed. However, optimum differences should be in the 200 to 500 Å range. Contrast is achieved by placing an angular disk at the front focal plane of a condenser lens and a transparent phase plate in the back focal plane of the objective. Phase contrast will produce light and dark features that correlate to the high and low points on the polished sample.[6] This technique can be used to bring out microstructures and slip lines in as-polished samples.

8.10.2 Scanning Electron Microscopy

Like optical microscopy, scanning electron microscopy can be used to inspect both polished and etched samples. Scanning electron microscopy relies on monitoring the reactions that occur when an highly accelerated electron beam (1 to 40 kV) strikes the surface of the sample. These reactions can provide both structural and compositional data to the metallographer.

Secondary electron imaging is accomplished by collecting the electrons that are expelled from the first 10 nm of the sample surface. As such, the electrons provide structural contrast based upon topographical features. On as-polished samples, the surface variations tend to be minimal. To enhance contrast, the metallographic sample is generally etched. This mode is indispensible when inspecting semiconductor construction features.

Backscatter electron imaging relies on the collection of electrons that are expelled from deeper within the sample (0.1 to 1.0 μm, depending on the acceleration voltage). This type of imaging provides compositional contrast based on the atomic number of the material from which the electrons are removed. Backscatter imaging is particularly effective in studying interfacial reaction zones in as-polished samples.

X-ray analysis is predominantly used for compositional identification of features found during either secondary electron imaging or backscatter electron imaging. However, these x-rays can be manipulated to provide compositional contrast in the form of a dot map.

8.11 *OPTIMIZED GRINDING AND POLISHING TECHNIQUES FOR SPECIFIC APPLICATIONS*

The following section outlines techniques that have been found to yield good, repeatable results. They are continually refined as more and better consumable products become available. These procedures should be used as a guideline from which each individual can develop a technique that works best.

8.11.1 Printed Wiring Board Cross-Sectioning Procedure

Sample Preparation

1. Identify the location to be sectioned
2. Remove it from the remainder of the board or coupon with a router, jewelers saw, or a hand held abrasive tool. Allow approximately 3 millimeters of clearance between the cut and location of interest.
3. Rinse the sample with water and dry with compressed air.

Encapsulation
Follow the procedures of Sec. 8.5.5.

Intermediate Sample Preparation
Follow the procedures of Sec. 8.6.

Coarse Grinding

1. Grind up to the edge of the board with 60 grit silicon carbide disks.
2. Rinse the sample with soapy water.

Fine Grinding

1. Grind just into the through-hole barrel with 400 grit silicon carbide disks.
2. Rinse the sample with soapy water.
3. Continue grinding with 600 grit silicon carbide to approximately 0.25 mm from the center point of the through hole.
4. Rinse the sample with soapy water.
5. Grind to approximate center point with 1200 grit silicon carbide.
6. Rinse the sample with soapy water and dry with compressed air.

Coarse Polishing
Follow the procedures of Sec.8.8.3. The wheel speed should be set low and the polishing time kept between 30 and 60 s.

Final Polishing
Follow the procedures of Sec. 8.8.3. The wheel speed should be set low and the polishing time kept under 60 s. (See Figs. 8.24 and 8.25.)

8.11.2 Encapsulated Semiconductor Cross-Sectioning Procedure

Applications
By performing precision metallography on semiconductor devices, a great deal can be learned. This information can include:

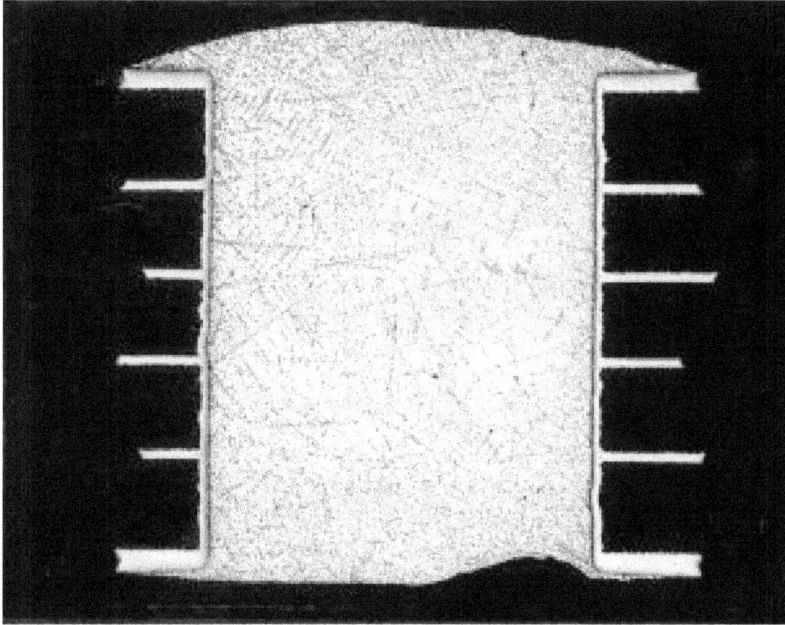

FIGURE 8.24   Cross section of a typical plated-through hole in the as-polished condition.

FIGURE 8.25   The importance of inspecting samples in the as-polished condition. Note the separation of the innerlayer connection.

1. Degree of intermetallic formation in a gold ball bond
2. Integrity of mono- or intermetallic bonding systems
3. Metallization step coverage
4. Junction depths
5. Thickness of construction layers
6. The list continues....

Sample Preparation

1. It is essential that the device to be cross-sectioned is fully mapped out prior to encapsulation. Light microscope documentation may not be adequate in all situations. Therefore, it should be complemented with documentation from a scanning electron microscope.
2. If at all possible, the die should be removed from its package following the procedures of Secs. 8.4.4 and 8.4.5. Cleaning, other than a blast of dry compressed air to remove any particulate, is generally not needed for semiconductor dice.

Encapsulation
Follow the procedures of Sec. 8.5.5

Intermediate Sample Preparation
Follow the procedures of Sec. 8.5.6.

Coarse Grinding

1. Grind up to the edge of the die with 60 or 120 grit silicon carbide while maintaining a plane as closely parallel to the final desired plane as possible. Do not contact the die with this grit of paper, because it will crack it.
2. Wash the sample with soapy water

Fine Grinding

1. With 600 grit silicon carbide, grind to approximately 3 µm of the desired site. Fine grinding is optimized by holding the face of the die against the rotation of the wheel. The wheel speed should be low with an adequate flow of water to cool, lubricate, and wash away the debris.
2. Rinse the sample with soapy water.
3. Use 1200 grit silicon carbide to get within 2 µm of the desired location.
4. Wash the sample with soapy water and dry with compressed air.
5. Place a piece of 12 µm alumina lapping film on a paper towel covering a clean glass plate. Proceed to hand lap the mount with the top surface of the die facing the cutting edge. Continue lapping until approximately 1 µm from the desired location.
6. Wash the sample with soapy water and dry with compressed air.

Coarse Polishing

1. Follow the procedures of Sec. 8.8.3. Use this step to just enter the desired location. Coarse polishing should not be used when preparing gallium arsenide, because it will shatter and chip the die.

2. Wash the sample with soapy water and dry with compressed air.

Final Polishing

1. Use the procedures of Sec. 8.8.3 to reach the final location. The polishing times can range between approximately 30 s and 5 min, depending on the amount of material that needs to be removed to reach the final location.

2. Wash the sample with soapy water and dry with compressed air.

Etch/Stain

Choose an etchant based on the desired results. Typical etchants are listed in Table 8.5. Figures 8.26 through 8.28 result from the processes described above.

8.11.3   Unencapsulated Semiconductor Cross-Section Procedure

Unencapsulated cross-sectioning was conceived by IBM in the 1970s. In their procedure, still used today, the die was held stationary in a tool on a rotating etched glass wheel. This produced a very flat representative section of the structure. The only drawback was that this was very time intensive. For this reason, glass coverslip techniques and diamond lapping films have been developed that provide for higher grinding and polishing rates.

Sample Preparation

1. Identify and map the location to be sectioned.

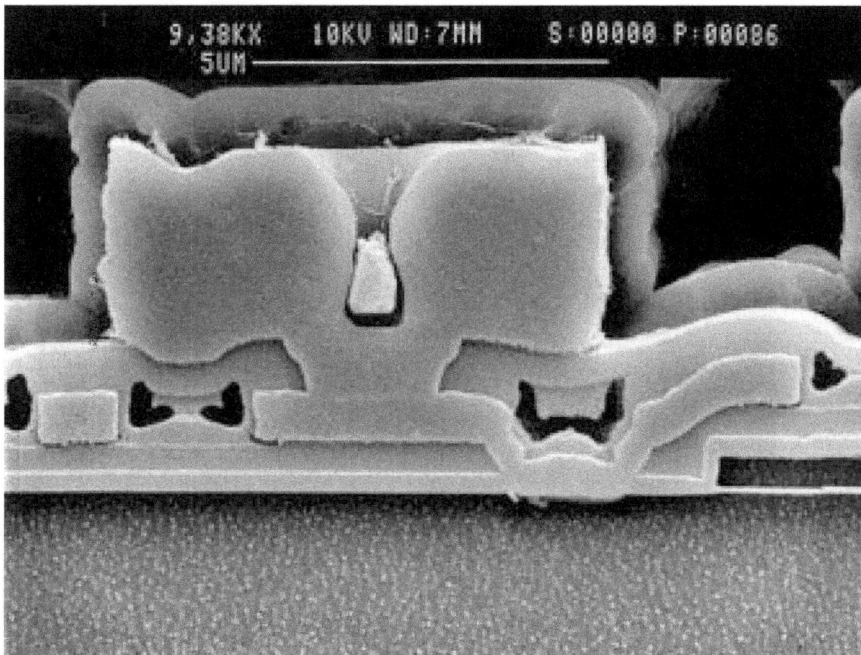

FIGURE 8.26   SEM photo of a typical semiconductor prepared following the procedures of Sec. 8.11.2. The sample was etched with ionized CF_4.

FIGURE 8.27 Cross section of a gallium arsenide die prepared using a coarse polishing procedure. Note the chipouts in the die.

FIGURE 8.28 Cross section of a gallium arsenide die prepared without a coarse polish step. Note that there are no chipouts in the die.

2. Mix 1 g of epoxy according to the manufacturer's instruction.

3. Apply a single drop of the epoxy to a glass cover slip.

4. Place the die top-side down into the drop of epoxy. Gently slide the die back and forth to allow any trapped air bubbles to escape.

5. Place the die in an oven to allow the epoxy to cure.

6. Once cured, remove the die and cover slip assembly from the oven. Place a small piece of hot wax on a cross-sectioning paddle and heat it on a hot plate until the wax flows. Place the backside of the die in the wax allowing the location to be sectioned to overhang the paddle. Allow the assembly to come to room temperature.

7. Insert the die and paddle assembly in the holder.

Grinding

1. Grind to within approximately 3 μm of the desired location with a 15 μm diamond bonded metal wheel. Grinding is optimized by holding the face of the sample away from the leading edge of the wheel. The wheel speed, water flow, and pressure applied to the holder should be low. **Omit this step if the die is silicon or gallium arsenide.**

2. Rinse the sample with soapy water.

3. Apply the glass plate to the platen followed by a sheet of 15 μm diamond film. Grind to within approximately 2 μm of the point of interest.

4. Switch to a 6 μm diamond film. Grind to approximately 1 μm of the desired location.

5. Rinse the sample with soapy water.

6. Use a 1 μm diamond film to get within 0.5 μm of the final location.

7. Rinse the sample with soapy water.

Polishing
Silicon on Sapphire

1. Polish into the desired location using a napless cloth with 0.05 μm colloidal silica. The face of the die should be held stationary away from the leading edge of the wheel and the wheel speed should be low.

2. Rinse the sample with soapy water.

Silicon

1. Follow the coarse and final polishing techniques described in the encapsulated procedure for semiconductors. The face of the die should be held stationary away from the leading edge of the wheel and the wheel speed should be low.

Note: The holder for the unencapsulated procedure should only be used on diamond films because silicon carbide would quickly wear the Teflon® feet (see Figs. 8.29 and 8.30).

8.11.4 Metallographic Procedure for the Preparation of Ceramics

Ceramics are usually encountered as a package or substrate material. Cross sections are generally performed on these assemblies to provide information on cofired circuitry, thick

FIGURE 8.29 Typical tool used for unencapsulated cross-sectioning. *Source:* courtesy of Allied High Tech Products.

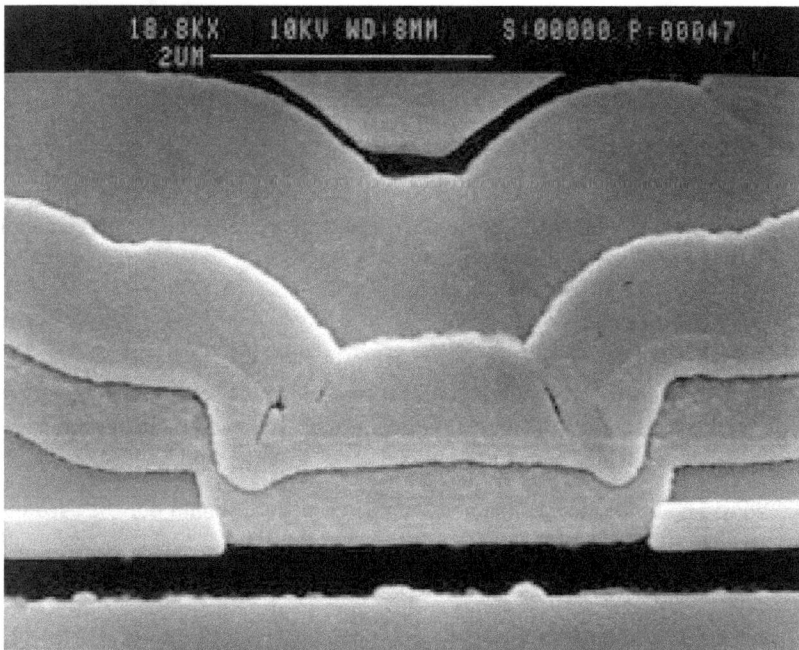

FIGURE 8.30 A silicon-on-sapphire die prepared using the unencapsulated method of Sec. 8.11.3.

film quality, and the integrity of metallurgical attachments. The preparation of ceramic capacitors is the exception to this procedure.

Sample Preparation

1. Most packages and substrates are too large to be easily manipulated during the grinding and polishing procedures. Therefore, the area of interest should be removed. This is most easily accomplished by cutting the location out with a diamond watering blade.

Cleaning
Rinse the sample with water to remove any cutting debris. Thoroughly dry with compressed air.

Encapsulation
Follow the procedures of Sec. 8.5.5.

Intermediate Sample Preparation
Follow the procedures of Sec. 8.6.

Coarse Grinding

1. Grind up to the edge of the ceramic with 60 or 120 grit silicon carbide.
2. Wash the sample with soapy water

Fine Grinding
Fine grinding of ceramics is optimized by using metal bonded diamond wheels. These wheels will maintain a flat surface on the face of the sample. When using metal bonded diamond wheels, the wheel speed should be kept low and the flow rate of water minimized to prevent hydroplaning. The pressure needed to maintain the cutting action of these wheels is generally greater than with silicon carbide abrasives.

1. Proceed into the ceramic using a 220 mesh metal bonded diamond wheel to within approximately 5 μm of the desired site.
2. Rinse the sample with soapy water.
3. Use a 30 μm metal bonded diamond wheel to get within 3 μm of the desired location.
4. Rinse the sample with soapy water and dry with compressed air.
5. Use a 15 μm metal bonded diamond wheel to remove the 30 μm damage and to enter the desired site.
6. Wash the sample with soapy water and dry with compressed air.

Coarse Polishing

1. Follow the procedures of Sec. 8.8.3. Use a napless cloth freshly charged with 6 μm polysilicon diamond. Remove the 15 μm damage but do not overpolish, because rounding will occur rapidly between the ceramic and any adjacent metal or mounting media. Approximately 30 s of polishing time is needed on a low wheel speed setting.
2. Wash the sample with soapy water and dry with compressed air.

Final Polishing

1. Follow the procedures of Sec. 8.8.3. Use a napless cloth with a 0.05 μm gamma alumina slurry. The polishing time should be kept to a minimum (approximately 15 to 30 s on a low wheel speed setting for a properly prepared sample) to reduce any potential rounding and relief of structures.

2. Wash the sample with soapy water and dry with compressed air.

Figure 8.31 shows the result of the above process.

8.11.5 Metallographic Procedures for Passive Components

Passive components are generally cross-sectioned to provide construction information. This may be in conjunction with a failure analysis or lot qualification. Passive components are generally unified collections of metals, ceramics, and polymers. As such, material removal rates are varied, and rounding becomes a problem.

Cleaning
Rinse the component with isopropyl alcohol and dry with compressed air.

Encapsulation
Follow the procedures of Sec. 8.5.5.

FIGURE 8.31 Cross section of a gold ball bond and gold vias found on a ceramic substrate mounted in a ceramic package. The procedures of Sec. 8.11.4 were used in the preparation of this sample.

Intermediate Sample Preparation
Follow the procedures of Sec. 8.6.

Coarse Grinding

1. Grind up to the edge of the component with 60 or 120 grit silicon carbide while maintaining a plane as closely parallel to the final desired plane as possible. Do not contact the component with this grit of paper—it can initiate cracks and chip outs.
2. Wash the sample with soapy water

Fine Grinding

1. Grind into the component with 400 grit silicon carbide disks.
2. Rinse the sample with soapy water.
3. Continue grinding with 600 grit silicon carbide to remove the 400 grit damage and to enter the desired location.
4. Wash the sample with soapy water and dry with compressed air.
5. Remove the 600 grit damage with a 1200 grit silicon carbide.
6. Wash the sample with soapy water and dry with compressed air.

Coarse Polishing

1. Follow the procedures of Sec. 8.8.3. Use a napless cloth with a 6 µm diamond abrasive. Polishing times should be kept below 30 s for a low wheel speed setting.
2. Wash the sample with soapy water and dry with compressed air.

Final Polishing

1. Follow the procedures of Sec. 8.8.3. Use a napless cloth with a 0.05 µm gamma alumina slurry. Polishing times should be kept below 30 s with a low wheel speed setting.
2. Wash the sample with soapy water and dry with compressed air.

Figure 8.32 shows the result of the above process.

8.11.6 Metallographic Procedures for Glass-Body Diodes

Diodes are generally cross-sectioned to provide construction information. This may be in conjunction with a failure analysis or lot qualification. Glass-body diodes are typically composed of a fragile silicon die, a glass envelope, a refractory metal slug, and a soft lead material. As such, material removal rates are varied, and rounding can easily become a problem.

Cleaning
Rinse the diode with isopropyl alcohol and dry with compressed air.

Encapsulation
Follow the procedures of Sec. 8.5.5.

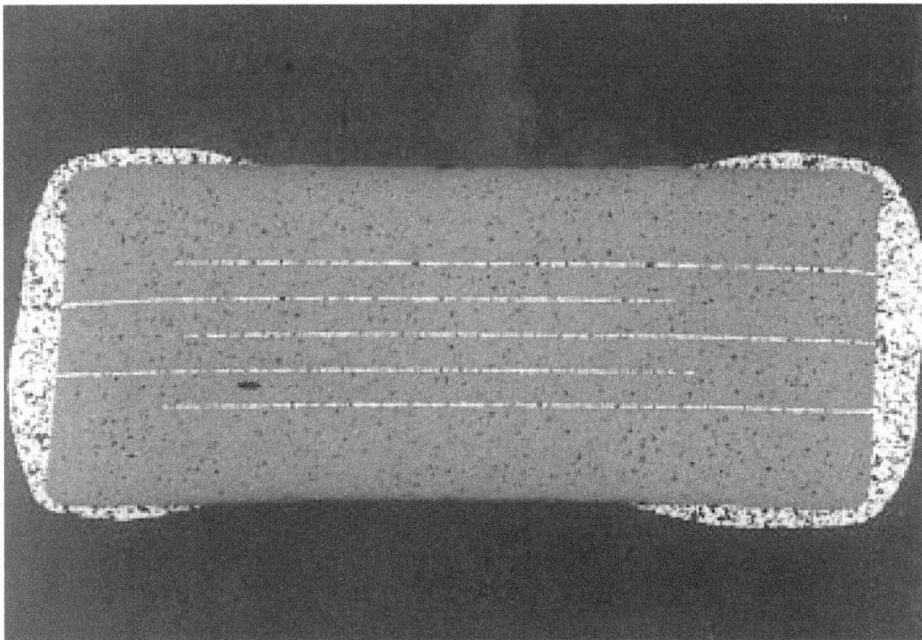

FIGURE 8.32 Cross section of a capacitor prepared following the procedures of Sec. 8.11.5.

Intermediate Sample Preparation
Follow the procedures of Sec. 8.6.

Coarse Grinding

1. Grind up to the edge of the component with 60 or 120 grit silicon carbide while maintaining a plane as closely parallel to the final desired plane as possible. Do not contact the component with this grit of paper because it can crack the die and glass envelope.

2. Wash the sample with soapy water

Fine Grinding

1. Grind into the component with 400 grit silicon carbide disks.

2. Rinse the sample with soapy water.

3. For the majority of diode configurations, there will be a pocket/cavity around the die that the glass does not fill. Therefore, to preserve the integrity of the die, the cavity should be backfilled following the procedures in Sec. 8.5.5.

4. Continue grinding into the die with 600 grit silicon carbide to remove the 400 grit damage and to enter the desired location.

5. Wash the sample with soapy water.

6. Remove the 600 grit damage with 1200 grit silicon carbide.

7. Wash the sample with soapy water.

8. Apply a disk of 12 μm aluminum oxide film to the wheel and remove the 1200 grit damage. This is followed by sequences of washing the sample with soapy water and sequential grinding with (9, 6, and 1 μm) aluminum oxide abrasive disks.

Coarse Polishing

Coarse polishing is not performed on this device type, because the refractory metal slugs will round almost instantaneously on contact with the coarse polishing wheel.

Final Polishing

1. Follow the procedures of Sec. 8.8.3. Use a napless cloth with a 0.05 μm gamma alumina slurry. Polishing times should be kept below 15 s on a low wheel speed setting. Polishing longer than this will cause severe rounding of the various materials in the diode.

2. Wash the sample with soapy water and dry with compressed air.

Figure 8.33 shows the result of the above process.

8.11.7 Scribe-and-Break

The scribe-and-break method is a quick and easy sectioning procedure that does not require encapsulation, grinding, or polishing. The entire method consists of scribing a line

FIGURE 8.33   Cross section of a glass-body diode prepared using the procedures of Sec. 8.11.6.

into the surface of the wafer with a diamond tipped tool and then applying a downward force sufficient to break the silicon along the line. This method is not the most precise, but it does yield adequate results with very little effort (see Figs. 8.34 and 8.35).

8.11.8 Spin-On Glass Procedures

When it is necessary to cross-section a metallization structure on an integrated circuit or transistor after the surface glassivation has been removed, or if the device has no glassivation layer over the metal, an application of spin-on glass will greatly assist the metallographer. Without the surface layer of glass, the aluminum or metallization tends to smear, exaggerating the original metal thickness as well as filling microcrevices. The following steps use a commercially available spin-on glass (SOG) and a standard metallographic wheel.

Steps

1. If the die is packaged, it is recommended that it be carefully removed (see Sec. 8.4.4).

2. Apply double-sided tape to the center point of a standard metallographic wheel. Place the backside of the die on the tape with tweezers.

3. Turn the wheel on and allow for it to get to high speed.

4. Apply 10 to 20 drops of high-viscosity SOG to the die surface in a time frame of 30 to 60 s.

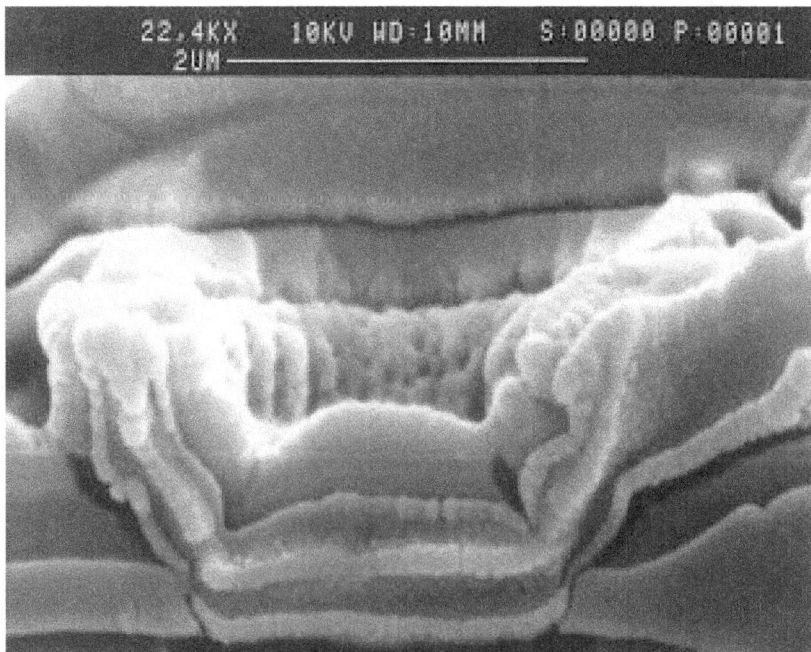

FIGURE 8.34 Cross section of an integrated circuit prepared using a room-temperature scribe-and-break procedure. Note the elongation/plastic deformation of the aluminum metallization.

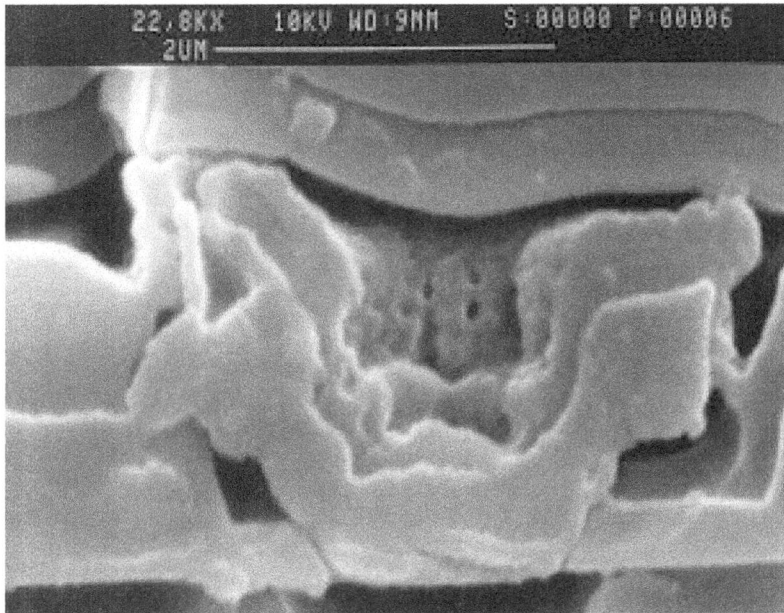

FIGURE 8.35 Cross section of an integrated circuit prepared using a scribe-and-break procedure while immersed in liquid nitrogen. Note that the aluminum exhibits less deformation than is present in the room temperature scribe and break.

5. Allow the die to continue rotating at high speed for an additional 15 s.
6. Carefully remove the die from the tape and place it on a 300° C hot plate for 60 s. The hot plate should be under a fume hood because fumes are released as the glass cures.
7. Metallographically prepare the sample using either the encapsulated or the unencapsulated technique with a coverslip.

See Fig. 8.36.

8.12 ARTIFACTS

It is the ability to recognize artifacts from true structures that separates the novice from the expert. Following the guidelines described throughout this chapter will help the metallographer establish a solid foundation from which to build. However, professional skills must be honed by the individual metallographer. Throughout a metallographer's career, questions will arise on whether a feature is representative of a particular structure or an artifact of sample preparation. The following section will familiarize the reader with common artifacts encountered throughout the metallographic procedure.

8.12.1 Artifact: Epoxy Pull-Away

Epoxy pull-away is the separation of the encapsulant from the surface of the sample.

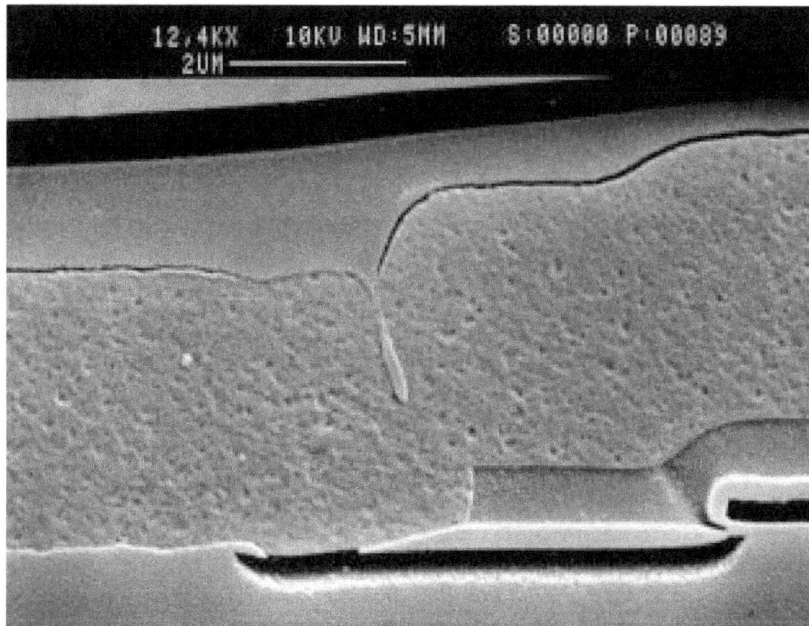

FIGURE 8.36 Cross section of a metallization step showing the ability of spin-on glass to fill microcracks and provide edge retention.

Cause. There are many causes for epoxy pull-away. The surface of the sample can be dirty prior to encapsulation, an adequate vacuum may not have been pulled during the vacuum impregnation process, or the epoxy may have shrunk back during the etching process.

Solutions. Ensure that the sample is properly cleaned prior to encapsulation. If cleanliness is not the cause, a backfill procedure should provide adequate edge retention for subsequent preparation (see Fig. 8.37).

8.12.2 Artifact: Scratches in the Polished Sample

Cause. There are two possible causes of scratches of the polished sample. First, steps prior to final polish may not have removed the damage of preceding grinding or polishing steps. Second, debris may have been transferred between steps.

Solution. Inspect the sample after each grinding or polishing step to ensure that the damage from the preceding step is completely removed. If each step has removed the damage of the prior step, thoroughly wash the sample in between steps or clean both the coarse and final polish cloths (see Fig. 8.38).

8.12.3 Artifact: Bleedout

Bleedout is an unwanted staining of the surface.

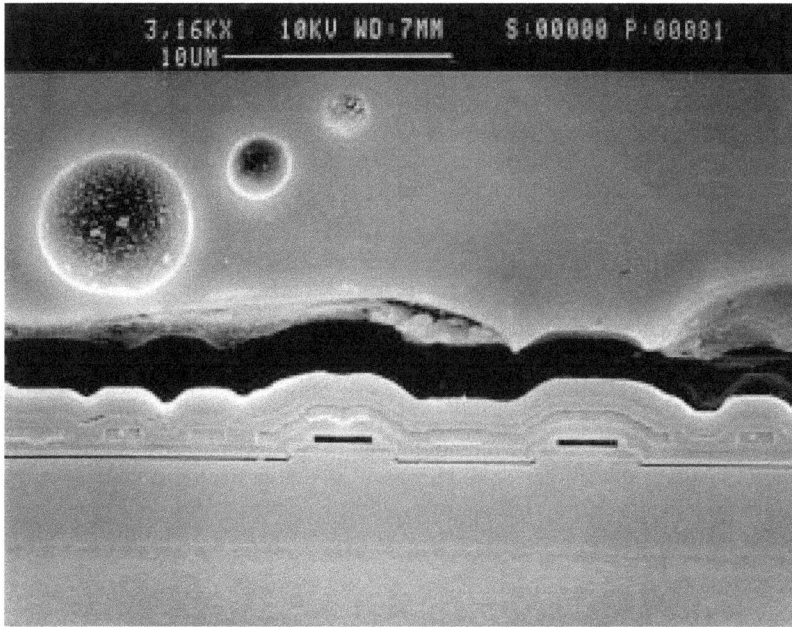

FIGURE 8.37 Typical epoxy pull-away along the surface of the die.

FIGURE 8.38 Scratches in a polished sample.

Cause.   This artifact is a result of entrapped lubricants or etchants wicking out of a separation between the epoxy and surface of the sample or from delaminations within the sample.

Solution.   Prior to drying, saturate the surface of the sample with a low-boiling-point fluid such as isopropyl alcohol. If this does not work, wash the sample with a mild detergent and water solution, soak in a bath of isopropyl alcohol for approximately 1 min, and then place in a vacuum oven maintained at 100° C for approximately 5 min (see Fig. 8.39).

8.12.4   Artifact: Smearing

Smearing is the plastic deformation of a metal over an adjacent material or into a gap.

Cause.   This artifact is a result of poor edge retention and/or overpolishing.

Solution.   Apply an edge retention overcoat to the sample or ensure that the steps leading up to polishing were performed properly so that polishing is kept to a minimum (see Fig. 8.40).

8.12.5   Artifact: Rounding

Rounding is an excessive relief between structures.

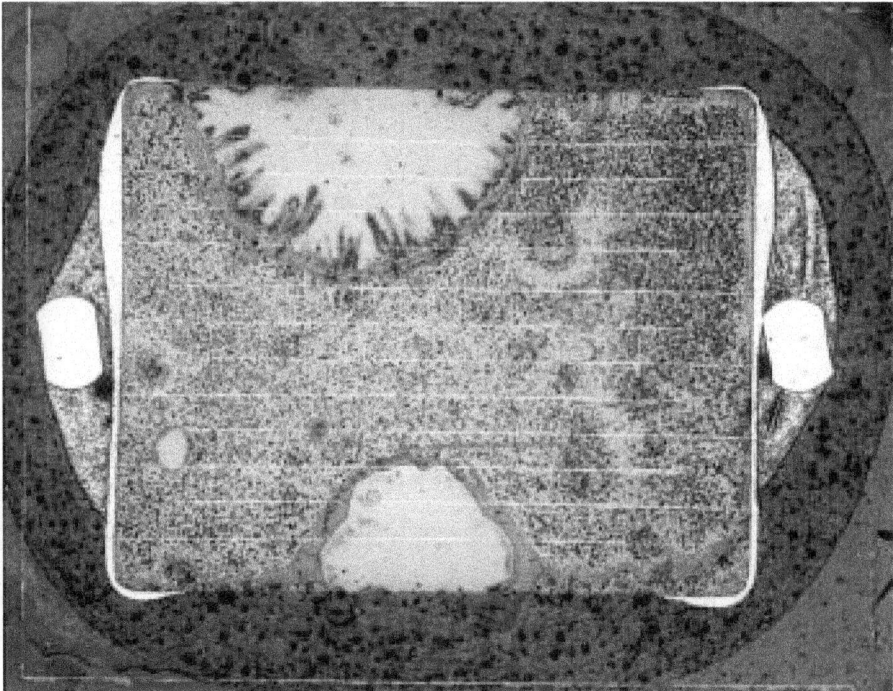

FIGURE 8.39   Typical "bleedout" found on a capacitor.

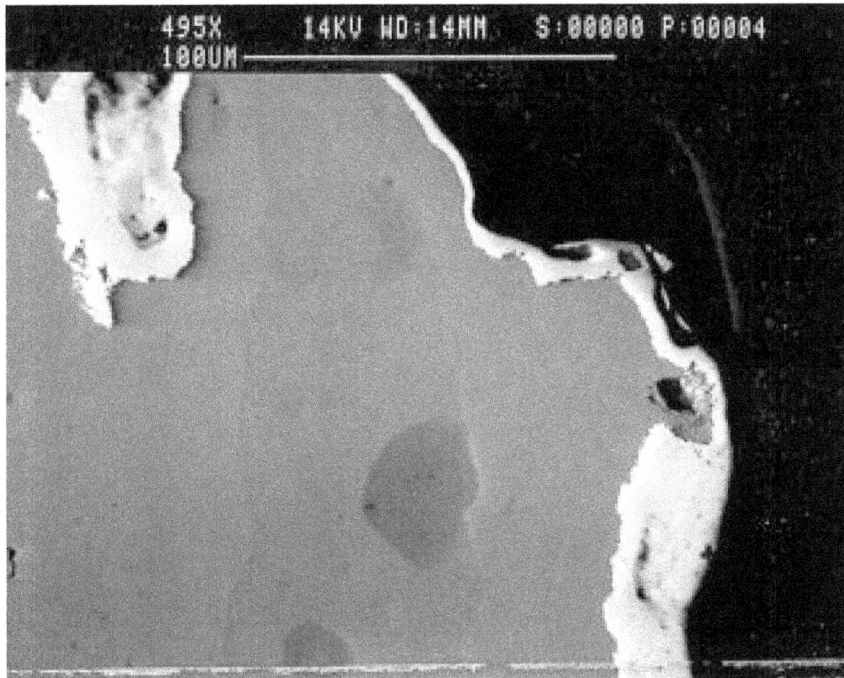

FIGURE 8.40   Smearing of a gold plate (light-colored edge) over a lead core metal (gray interior).

Cause.   The cause of rounding is the variation in material removal rate between hard and soft locations. This can be between either the sample and encapsulant or hard and soft components within the sample.

Solution.   If rounding is occurring between the sample and the encapsulant apply an overcoat to the surface of the sample to improve edge retention. If the rounding is occurring between hard and soft locations within a sample, minimize the amount of final polishing (see Fig. 8.41).

8.12.6   Artifact: Ballooning

The following discuses ballooning of a Ti-W barrier metallization.

Cause.   Titanium-tungsten barrier metals are used in the vast majority of integrated circuits. When in contact with wet chemical etchants used for structural delineation of the die, the barrier metallization thickness normally expands (balloons). This expansion can result in the true thickness of the aluminum metallization being misrepresented.

Solution.   To negate this expansion, the sample should be dry etched with ionized CF_4. A standard plasmod unit used for deglassivating dice is recommended for this procedure (see Figs. 8.42 and 8.43).

FIGURE 8.41 Rounding (dark/shadow along the transition interface between materials) found on a poorly prepared diode.

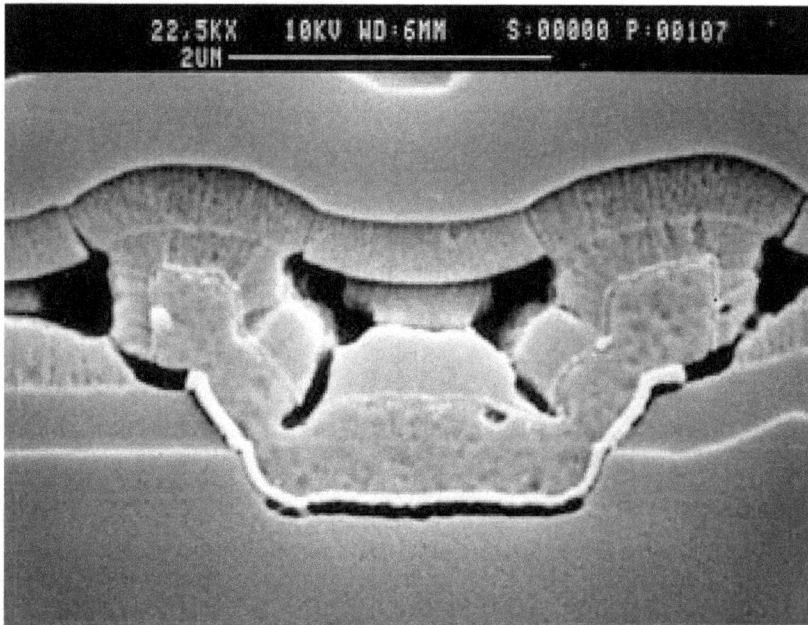

FIGURE 8.42 Semiconductor cross section delineated with a wet chemical etch. Note the thickness of the barrier metal (light colored layer).

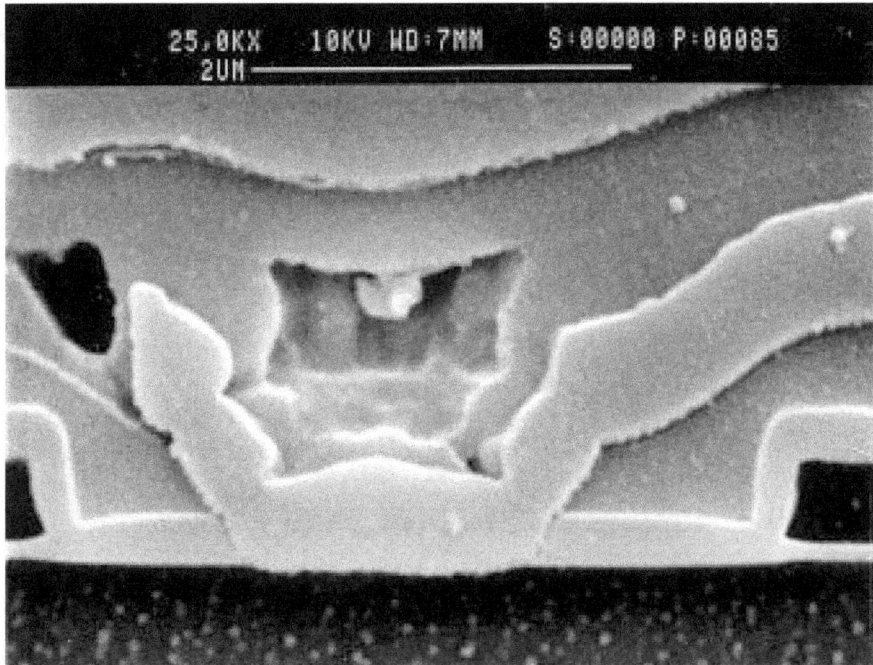

FIGURE 8.43   The semiconductor shown in Fig. 8.41 but after a dry etch with ionized CF_4. Note that the thickness of the barrier metal light-colored layer is less.

8.12.7   Artifact: Embedded Particles from Grinding and Polishing Abrasives

Cause.   Grinding and polishing abrasives are generally embedded into a material if the load applied during preparation is to great or if lubrication is insufficient.

Solution.   To eliminate this problem, reduce the pressure applied to the mount and/or increase the amount of lubrication between the sample and wheel (see Fig. 8.44).

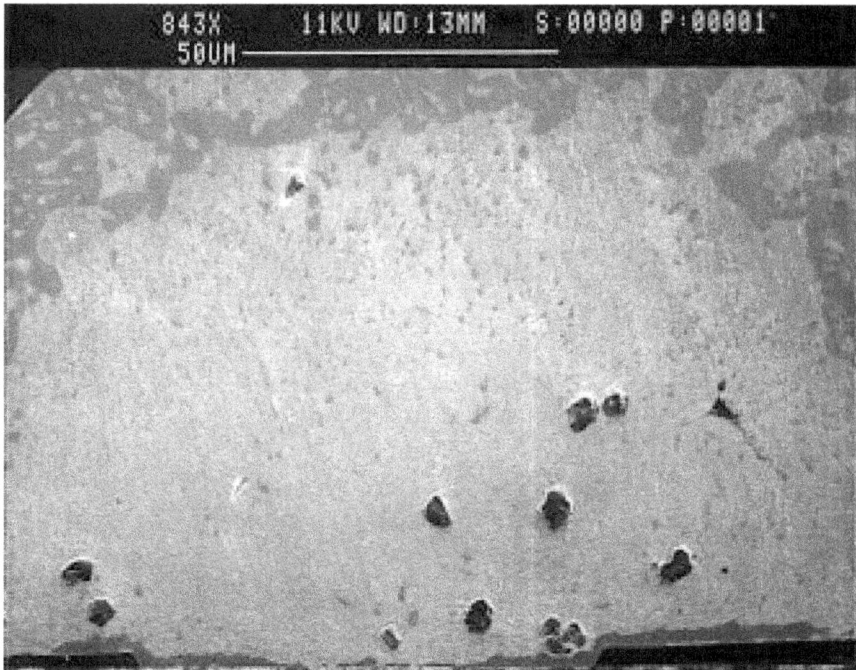

FIGURE 7.44 Close-up view of diamond abrasives (dark angular particles) that were embedded in a solder sphere of a BGA assembly.

8.13 REFERENCES

1. Mills, Tom. "Microsectioning Tutorial: Historical Development." Presented at International Reliability Physics Symposium, 2 April 1984, Las Vegas, NV.

2. *Metals Handbook,* 9th Ed., Vol. 9, Metallography and Microstructures, p. 31, Table 4. Materials Park, OH: ASM International, 1985.

3. Devaney, Trevor A. "Applications of Spin on Glass to Precision Metallography of Semiconductors." Presented at the 18th International Symposium for Testing and Failure Analysis, 19–23 October 1992, Los Angeles, CA.

4. Vander Voort, George F. *Metallography Principles and Practice.* New York: McGraw Hill, 1984.

5. Nelson, James A. "Microetching Techniques for Revealing Printed Wiring Board Microstructures," IPC Technical Paper, IPC-TP-823, 1988.

6. Petzow, Günter, and G. Elssner. *Etching, Metallography and Microstructures.* Metals Handbook, 9th Edition, Vol. 9, 57–70. Materials Park, OH: ASM International, 1985.

7. Devaney, John R. SEM Training Course.

CHAPTER 9
CHEMICAL CHARACTERIZATION

Perry Martin
National Technical Systems (NTS)

9.1 INTRODUCTION

The failure analyst will deal with four major classifications of chemical analysis:

- Bulk analysis
- Microanalysis
- Thermal analysis
- Surface analysis

Bulk analysis techniques utilize a relatively large volume of the sample and are used to identify the elements or compounds present and to verify conformance to applicable specifications. Microanalysis techniques explore a much smaller volume of the sample and are typically used to identify the elements or compounds present for studies of particles, contamination, or material segregation. Thermal analysis techniques are used to obtain thermomechanical information on sample materials to identify the coefficient of thermal expansion and other properties relevant to the failure analyst. Finally, surface analysis examines only the top few atomic layers of a material. These techniques are used in microcontamination studies, adhesion studies, and microelectronic failure analysis. Surface analysis will not be covered in this chapter, and the interested reader is referred to the materials in the recommended reading list.

Before deciding which technique(s) to use, the failure analyst must ask a number of questions:

1. What type of information is required?
   - Quantitative?
   - Qualitative?
   - A mixture of both?
2. What analytical accuracy and precision are required? *(Accuracy is the extent to which the results of a measurement approach the true values. Precision is the measure of the range of values of a set of measurements.)*
3. What is the physical state of the material?
   - Is the material a powder?
   - Is the material a pellet?
   - Is the material a paste?
   - Is the material a foam?
   - Is the material a thin film?
   - Is the material a fiber?

- Is the material a liquid?
- Is the material a rectangular bar?
- Is the material an irregular chunk?
- Is the material a tubing?
- Is the material a gel?
- Is the material a cylinder?

4. Are the events of interest minor or major?
5. What is known about the material or samples?
6. What are the important properties of the material?
7. Is the sample a single component or a complex mixture?
8. What is the material's history?
9. What is the material's future?
10. How much material is available for analysis and is there a limitation on sample size?
11. How many samples must be run?
12. What is the required analysis turnaround time?
13. Are there any safety hazards to be concerned about?

The ability to answer these questions, and the use of the answers, will depend on the experience of the failure analyst and the equipment available. Most failure analysts will have a great deal of experience in the microanalysis techniques associated with the scanning electron microscope (SEM).[*] Many failure analysts will require guidance as to the other techniques. In the past, the corporate environment employed several senior scientists working in the laboratory. Today, there may be only a single scientist overseeing that laboratory and other laboratories throughout the corporation, and expert knowledge is obtained from outsourcing. Such information can be obtained from local universities, commercial laboratories, and consultants and instrument manufacturers. It is also hoped that this handbook can provide some elementary support in the techniques discussed.

9.2   TECHNIQUES AND APPLICATIONS OF ATOMIC SPECTROSCOPY

Atomic spectroscopy[1] is actually not one technique but three: atomic absorption, atomic emission, and atomic fluorescence. Of these, atomic absorption (AA) and atomic emission are the most widely used. Our discussion will deal with them and an affiliated technique, ICP-mass spectrometry.

9.2.1   Atomic Absorption

Atomic absorption occurs when a ground-state atom absorbs energy in the form of light of a specific wavelength and is elevated to an excited state. The amount of light energy absorbed at this wavelength will increase as the number of atoms of the selected element in the light path increases. The relationship between the amount of light absorbed and the

---

[*]These are covered in the SEM chapter of this book.

concentration of analyte present in known standards can be used to determine unknown concentrations by measuring the amount of light they absorb. Instrument readouts can be calibrated to display concentrations directly.

The basic instrumentation for atomic absorption requires a primary light source, an atom source, a monochromator to isolate the specific wavelength of light to be used, a detector to measure the light accurately, electronics to treat the signal, and a data display or logging device to show the results. The light source normally used is either a hollow cathode lamp or an electrodeless discharge lamp (see Fig. 9.1).

The atom source used must produce free analyte atoms from the sample. The source of energy for free atom production is heat, most commonly in the form of an air-acetylene or nitrous oxide-acetylene flame. The sample is introduced as an aerosol into the flame. The flame burner head is aligned so that the light beam passes through the flame, where the light is absorbed.

9.2.2 Graphite Furnace Atomic Absorption

The major limitation of atomic absorption using flame sampling (flame AA) is that the burner-nebulizer system is a relatively inefficient sampling device. Only a small fraction of the sample reaches the flame, and the atomized sample passes quickly through the light path. An improved sampling device would atomize the entire sample and retain the atomized sample in the light path for an extended period to enhance the sensitivity of the technique. Electrothermal vaporization using a graphite furnace provides those features.

With graphite furnace atomic absorption (GFAA), the flame is replaced by an electrically heated graphite tube. The sample is introduced directly into the tube, which is then heated in a programmed series of steps to remove the solvent and major matrix components and then to atomize the remaining sample. All of the analyte is atomized, and the atoms are retained within the tube (and the light path, which passes through the tube) for an extended period. As a result, sensitivity and detection limits are significantly improved.

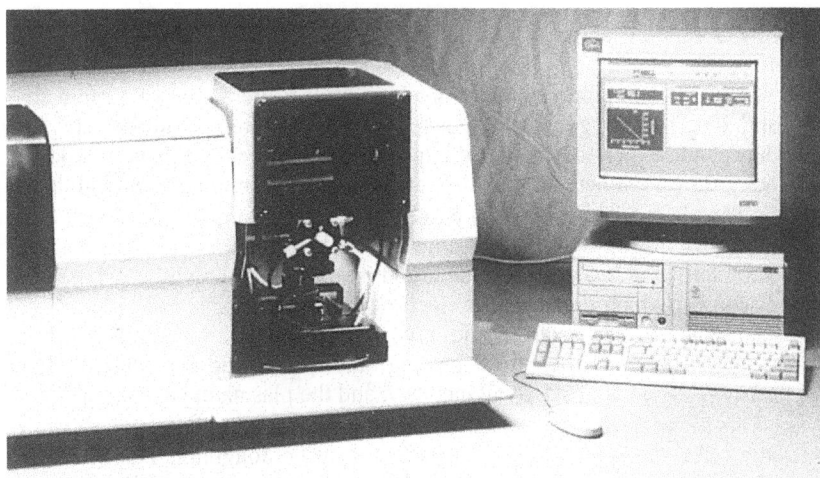

FIGURE 9.1 Typical atomic absorption (AA) spectrometer. *Source:* photo courtesy of Perkin-Elmer Corp.

Graphite furnace analysis times are longer than those for flame sampling, and fewer elements can be determined using GFAA. However, the enhanced sensitivity of GFAA and the ability of GFAA to analyze very small samples and directly analyze certain types of solid samples significantly expand the capabilities of atomic absorption.

9.2.3 Atomic Emission

Atomic emission spectroscopy is a process in which the light emitted by excited atoms or ions is measured. The emission occurs when sufficient thermal or electrical energy is available to excite a free atom or ion to an unstable energy state. Light is emitted when the atom or ion returns to a more stable configuration or the ground state. The wavelengths of light emitted are specific to elements present in the sample.

The basic instrument used for atomic emission is very similar to that used for atomic absorption, with the difference that no primary light source is used for atomic emission. One of the more critical components for atomic emission instruments is the atomization source, because it must also provide sufficient energy to excite the atoms as well as atomize them.

The earliest energy sources for excitation were simple flames, but these often lacked sufficient thermal energy to be truly effective sources. Later, electrothermal sources such as arc/spark systems were used, particularly when analyzing solid samples. These sources are useful for doing qualitative and quantitative work with solid samples but are expensive, difficult to use, and have limited applications.

Due to the limitations of the early sources, atomic emission initially did not enjoy the universal popularity of atomic absorption. This changed dramatically with the development of the *inductively coupled plasma* (ICP) as a source for atomic emission. The ICP eliminates many of the problems associated with past emission sources and has caused a dramatic increase in the utility and use of emission spectroscopy.

9.2.4 Inductively Coupled Plasma

The ICP is an argon plasma maintained by the interaction of a radio frequency (rf) field and ionized argon gas. The ICP is reported to reach temperatures as high as 10,000 K, with the sample experiencing useful temperatures between 5,500 and 8,000 K. These temperatures allow complete atomization of elements, minimizing chemical interference effects.

The plasma is formed by a tangential stream of argon gas flowing between two quartz tubes. RF power is applied through the coil, and an oscillating magnetic field is formed. The plasma is created when the argon is made conductive by exposing it to an electrical discharge, which creates seed electrons and ions. Inside the induced magnetic field, the charged particles (electrons and ions) are forced to flow in a closed annular path. As they meet resistance to their flow, heating takes place, and additional ionization occurs. The process occurs almost instantaneously, and the plasma expands to its full dimensions.

As viewed from the top, the plasma has a circular, "doughnut" shape. The sample is injected as an aerosol through the center of the doughnut. This characteristic of the ICP confines the sample to a narrow region and provides an optically thin emission source and a chemically inert atmosphere. This results in a wide dynamic range and minimal chemical interactions in an analysis. Argon is also used as a carrier gas for the sample.

9.2.5 ICP-Mass Spectrometry

As its name implies, ICP-mass spectrometry (ICP-MS) is the synergistic combination of an inductively coupled plasma with a quadrupole mass spectrometer. ICP-MS uses the ability of the argon ICP to efficiently generate singly charged ions from the elemental species within a sample. These ions are then directed into a quadrupole mass spectrometer.

The function of the mass spectrometer is similar to that of the monochromator in an AA or ICP emission system. However, rather than separating light according to its wavelength, the mass spectrometer separates the ions introduced from the ICP according to their mass to charge ratio. Ions of the selected mass/charge are directed to a detector that quantifies the number of ions present. Due to the similarity of the sample introduction and data handling techniques, using an ICP-MS is very much like using an ICP emission spectrometer.

ICP-MS combines the multielement capabilities and broad linear working range of ICP emission with the exceptional detection limits of graphite furnace AA. It is also one of the few analytical techniques that permit the quantization of elemental isotopic concentrations and ratios.

9.3 *SELECTING THE PROPER ATOMIC SPECTROSCOPY TECHNIQUE*

With the availability of a variety of atomic spectroscopy techniques (such as flame atomic absorption, graphite furnace atomic absorption, inductively coupled plasma emission, and ICP-mass spectrometry), laboratory managers must decide which technique is best suited for the analytical problems of their laboratory. Because atomic spectroscopy techniques complement each other so well, it may not always be clear which technique is optimal for a particular laboratory. A clear understanding of the analytical problem in the laboratory and the capabilities provided by the different techniques is necessary.

Important criteria for selecting an analytical technique include detection limits, analytical working range, sample throughput, interferences, ease of use, and the availability of proven methodology. These criteria are discussed below for flame AA, graphite furnace AA (GFAA), ICP emission, and ICP-mass spectrometry (ICP-MS).

9.3.1 Atomic Spectroscopy Detection Limits

The detection limits achievable for individual elements represent a significant criterion of the usefulness of an analytical technique for a given analytical problem. Without adequate detection limit capabilities, lengthy analyte concentration procedures may be required prior to analysis.

Typical detection limit ranges for the major atomic spectroscopy techniques are shown in Fig. 9.2, and Table 9.1 provides a listing of detection limits by element for six atomic spectroscopic techniques: flame AA, hydride generation AA, graphite furnace AA (GFAA), ICP emission with radial and axial torch configurations, and ICP-mass spectrometry.

Generally, the best detection limits are attained using ICP-MS or graphite furnace AA. For mercury and those elements that form hydrides, the cold vapor mercury or hydride generation techniques offer exceptional detection limits.

Detection limits should be defined very conservatively with a 98 percent confidence level, based on established conventions for the analytical technique. This means that, if a

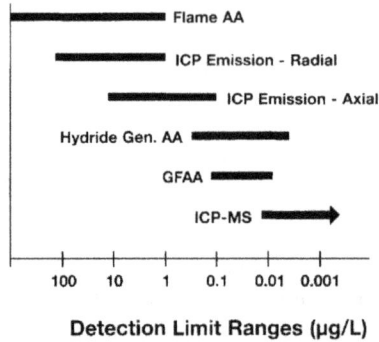

FIGURE 9.2 Typical detection ranges for the major atomic spectroscopy techniques. *Source:* courtesy of Perkin-Elmer Corp.

concentration at the detection limit were measured many times, it could be distinguished from a zero or baseline reading in 98 percent (3σ) of the determinations.

9.3.2 Analytical Working Range

The analytical working range can be viewed as the concentration range over which quantitative results can be obtained without having to recalibrate the system. Selecting a technique with an analytical working range (and detection limits) based on the expected analyte concentrations minimizes analysis times by allowing samples with varying analyte concentrations to be analyzed together. A wide analytical working range also can reduce sample handling requirements, minimizing potential errors.

Figure 9.3 shows typical analytical working ranges with a single set of instrumental conditions.

9.3.3 Sample Throughput

Sample throughput is the number of samples that can be analyzed or elements that can be determined per unit time. For most techniques, analyses performed at the limits of detection, or where the best precision is required, will be more time consuming than less demanding analyses. Where these factors are not limiting, however, the number of elements to be determined per sample and the analytical technique will determine the sample throughput.

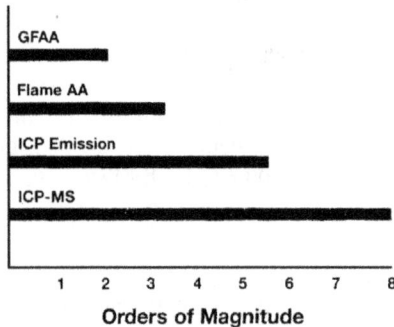

FIGURE 9.3 Typical analytical working ranges for the major atomic spectroscopy techniques. *Source:* courtesy of Perkin-Elmer Corp.

Flame AA. Flame AA provides exceptional sample throughput when analyzing a large number of samples for a limited number of elements. A typical determination of a single element requires only 3 to 10 s. However, flame AA requires specific light sources and optical parameters for each element to be determined, and it may require different flame gases for different elements. In automated multielement flame AA systems, all samples normally are analyzed for one element, the system is then automatically adjusted for the next element, and so on. As a result, even though it is frequently used for multielement analysis, flame AA is generally considered to be a single-element technique.

Table 9.1 Atomic Spectroscopy Detection Limits in Micrograms/Liter *(courtesy of Perkin-Elmer Corp.)*

Element	Flame AA	Hg/hydride	GFAA	ICT emission	ICT-MS	Element	Flame AA	Hg/hydride	GFAA	ICT emission	ICT-MS
Ag	1.5		0.02	0.9	0.003	Mo	45		0.08	3	0.003
Al	45		0.1	3	0.006	Na	0.3		0.02	3	0.003*
As	150	0.03	0.2	50	0.006	Nb	1500			10	0.0009
Au	9		0.15	8	0.001	Nd	1500			2	0.002
B	1000		20	0.8	0.09	Ni	6		0.3	5	0.005
Ba	15		0.35	0.09	0.002	Os	120			6	
Be	1.5		0.008	0.08	0.03	P	75000		130	30	0.3
Bi	30	0.03	0.25	30	0.005	Pb	15		0.06	10	0.001
Br					0.2	Pd	30		0.8	3	0.003
C				75	150	Pr	7500			2	<0.0005
Ca	1.5		0.01	0.02	0.05*	Pt	60		2.0	10	0.002
Cd	0.8		0.008	1	0.003	Rb	3		.0.3	30	0.003
Ce				5	0.0004	Re	750			5	0.0006
Cl					10	Rh	6			5	0.0008
Co	9		0.15	1	0.00089	Ru	100		1.0	6	0.002
Cr	3		0.03	2	0.02	S				30	70
Cs	15				0.0005	Sb	45	0.15	0.15	10	0.001
Cu	1.5		0.1	0.4	0.003	Sc	30			0.2	0.02
Dy	50			2	0.001	Se	100	0.03	0.3	50	0.06
Er	60			1	0.0008	Si	90		1.0	3	0.7
Eu	30			0.2	0.0007	Sm	3000			2	0.001
F					10000	Sn	150		0.2	60	0.002
Fe	5		0.1	2	0.005*	Sr	3		0.025	0.03	0.0008
Ga	75			4	0.001	Ta	1500			10	0.0006
Gd	1800			0.9	0.002	Tb	900			2	<0.0005
Ge	300			20	0.003	Te	30	0.03	0.4	10	0.01
Hf	300			4	0.006	Th					<0.0005
Hg	300	0.009	0.6	1	0.004	Ti	75		0.35	0.4	0.006
Ho	60			0.4	<0.0005	Tl	15		0.15	30	0.0005
I					0.0008	Tm	15			0.6	<0.0005
In	30			9	<0.0005	U	15000			15	<0.0005
Ir	900		3.0	5	0.0006	V	60		0.1	0.5	0.002
K	3		0.008	20	0.015*	W	1500			8	0.001
La	3000			1	0.0005	Y	75			0.3	0.0009
Li	0.8		0.06	0.3	0.0001*	Yb	8			0.3	0.001
Lu	1000			0.2	<0.0005	Zn	1.5		0.1	1	0.003
Mg	0.15		0.004	0.07	0.0007	Zr	450			0.7	0.004
Mn	1.5		0.035	0.4	0.002						

* Denotes that the ICP-MS detection limit was measured under cold plasma conditions.

All detection limits are given in micrograms per liter and were determined using elemental standards in dilute aqueous solution. All detection limits are based on a 98 percent confidence level (three standard deviations).

All atomic absorption (Model 5100) detection limits were determined using instrumental parameters optimized for the individual element, including the use of System 2 electrodeless discharge lamps were available. ICP emission (Optima 3000) detection limits were obtained under simultaneous multielement conditions with a radial plasma. Detection limits using an axial plasma (Optima 3000 XL) are typically improved by 5 to 10 times.

Cold vapor mercury detection limits were determined with a FIAS™-100 or FIAS-400 flow injection system with amalgamation accessory. The detection limit without an amalgamation accessory is 0.2 µg/L with a hollow cathode lamp, 0.05 µg/L with a System 2 electrodeless discharge lamp. (The Hg detection limit with the dedicated FIMS™-100 or FIMS-400 mercury analyzers is <0.010 µg/L without an amalgamation accessory and <0.001 µg/L with the accessory.) Hydride detection limits shown were determined using an MHS-10 mercury/hydride system.

Graphite furnace AA detection limits were determined using 50-µL sample volumes, a L'vov platform, and full STPF conditions (Model 5100 PC with 5100 ZL Zeeman Furnace Module or Model 4110 ZL). SIMAA 6000 detection limits are similar in its multisource mode and are typically 2× to 5× better in its single-source mode. Graphite furnace detection limits can be further enhanced by the use of replicate injections.

ICP-MS detection limits were determined using a three-second integration.

Graphite Furnace AA. As with flame AA, GFAA is basically a single-element technique. Because of the need to thermally program the system to remove solvent and matrix components prior to atomization, GFAA has a relatively low sample throughput. A typical graphite furnace determination normally requires 2 to 3 min.

ICP Emission. ICP emission is a true multielement technique with exceptional sample throughput. ICP emission systems typically can determine 10 to 40 elements per minute in individual samples. Where only a few elements are to be determined, however, ICP is limited by the time required for equilibration of the plasma with each new sample, typically about 15 to 30 s.

ICP-MS. ICP-MS is also a true multielement technique with the same advantages and limitations as ICP emission. The sample throughput for ICP-MS is typically 20 to 30 element determinations per minute, depending on such factors as the concentration levels and required precision.

9.3.4 Interferences

Few if any analytical techniques are free of interferences. With atomic spectroscopy techniques, however, most interferences have been studied and documented, and methods exist to correct or compensate for interferences that may occur. A summary of the interference types seen with atomic spectroscopy techniques, all of which are controllable, and the corresponding methods of compensation are shown in Table 9.2.

Table 9.2   Atomic Spectroscopy Interferences *(courtesy of Perkin-Elmer Corp.)*

Technique	Interference type	Compensation method
Flame AA	Ionization	Ionization buffer
	Chemical	Releasing agent or nitrous oxide-acetylene flame
	Physical	Dilution, matrix matching, or method of additions
GFAA	Physical and chemical	STPF conditions
	Molecular absorption	Zeeman or continuum source background correction
	Spectral	Zeeman background correction
ICP emission	Spectral	Background correction or the use of alternate analytical lines
	Matrix	Internal standardization
ICP-MS	Mass overlap	Interelement correction, use of alternate mass values, or higher mass resolution
	Matrix	Internal standardization

9.3.5 Other Comparison Criteria

Other comparison criteria for analytical techniques include the ease of use, required operator skill levels, and availability of documented methodology.

Flame AA. Flame AA is very easy to use. Extensive application information is available. Excellent precision makes it a preferred technique for the determination of major constituents and higher concentration analytes.

GFAA.   GFAA applications are well documented, although not as completely as with flame AA. GFAA has exceptional detection limit capabilities but a limited analytical working range. Sample throughput is less than that of other atomic spectroscopy techniques. Operator skill requirements are somewhat more extensive than for flame AA.

ICP Emission.   ICP emission is the best overall multielement atomic spectroscopy technique, with excellent sample throughput and very wide analytical range. Good documentation is available for applications. Operator skill requirements are intermediate between flame AA and GFAA.

ICP-MS.   ICP-MS is a relatively new technique with exceptional multielement capabilities at trace and ultra-trace concentration levels and the ability to perform isotopic analyses. Good basic documentation for interferences exists. Application documentation is limited but growing rapidly. ICP-MS requires operator skills similar to those for ICP emission and GFAA.

9.3.6   Comparison Summary

The main selection criteria for atomic spectroscopy techniques, concentration range, and analytical throughput are summarized in Fig. 9.4. Where the selection is based on analyte detection limits, flame AA and ICP emission are favored for moderate to high levels, whereas graphite furnace AA and ICP-MS are favored for lower levels. ICP emission and ICP-MS are multielement techniques, favored where large numbers of samples are to be analyzed.

9.4   OVERVIEW OF FOURIER TRANSFORM INFRARED MICROSPECTROSCOPY

Typical applications of Fourier transform infrared (FT-IR) microspectroscopy[2] include bulk composition, surface contamination, and inclusions.

9.4.1   Sample Requirements/Constraints

The overall physical size of the sample is restricted to what can be accommodated by the stage of an optical microscope. The sample thickness for transmission is generally limited to <0.5 mm for mid-IR, and a few millimeters for near IR. A typical system is shown in Fig. 9.5.

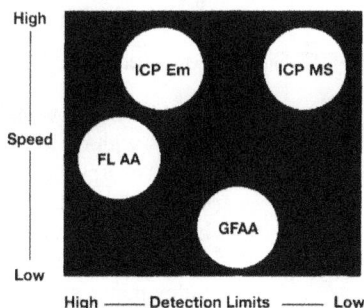

FIGURE 9.4 General selection guide for atomic spectroscopy instrumentation based on sample throughput and concentration range. *Source:* courtesy of Perkin-Elmer Corp.

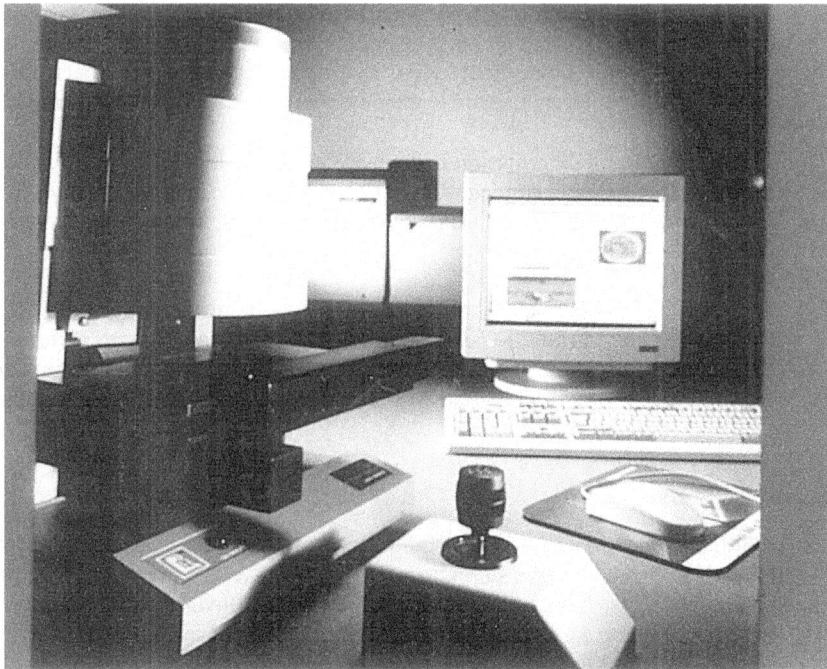

FIGURE 9.5 Typical atomic absorption (AA) spectrometer. *Source:* photo courtesy of Perkin-Elmer Corp.

9.4.2 Elements Detected

Detection is by fingerprints of individual chemical compounds or characteristic absorptions of chemical functional groups organic or inorganic.

9.4.3 Detection Limits

This is an absorption technique where the limit is determined by a concentration x-path length product. In absolute terms, this puts the smallest amount detectable in the picogram range. Monomolecular layers are readily detectable on metal surfaces.

9.4.4 Signal Detected

The measured signal is a ratio of the infrared energy detected in the presence of a sample to that with no sample present. Indirectly, this gives the amount of IR absorbed by the sample. Absorption arises from molecular vibrations and can be associated with specific chemical bonds that absorb at different frequencies (wavelengths). The measurement can be by transmission (thin samples), reflection (reflective surface), or by attenuated total reflectance (ATR). ATR is a near-field technique that uses a crystal of a high refractive index material brought into contact with the surface. It is used for opaque samples.

9.4.5 Organic Information Available

Information is available on the presence or absence of particular chemical functional groups. Identity of specific compounds is provided by comparison with libraries of spectra, the IR spectrum giving a unique finger print. Libraries with tens of thousands of spectra are available. Quantitative information can be extracted on the amount of material present and the composition of mixtures.

9.4.6 Depth Resolution

Reflection from a homogeneous material is that of the surface molecules. Layers on a reflective substrate give transmission/reflection spectra. These correspond to transmission through a layer of double the actual thickness. It is useful from monolayers to hundreds of micrometers. ATR spectra come from a thickness of less than the wavelength, the effective depth being proportional to the wavelength. In the mid-IR, this depth typically would be a few micrometers.

9.4.7 Area/Size Resolution

For transmission, this is limited by diffraction, giving a typical limit of 10×10 μm. The normal arrangement is to mask the region to be measured by using an aperture at an image of the sample rather than at the sample itself. Physical masking of the sample can give spectra from smaller regions. Spectra can also be obtained from smaller samples if they can be physically isolated.

9.4.8 Image Mapping

Using an x-y stage, samples can be mapped automatically pixel by pixel. A full spectrum is obtained at each point. Maps then can be generated based on specific spectral features. Multiple maps can be created from a single experiment to show the distribution of different species.

9.4.9 Strengths and Weaknesses

IR is very versatile, being capable of obtaining spectra from almost any surface. It is a very simple technique, operating in air. IR is the best technique for identifying specific compounds. It provides very good specificity for different chemical types. It has limited spatial resolution because of the wavelengths utilized. Because maps are generated sequentially pixel by pixel, mapping is fairly slow, requiring several seconds per pixel.

9.5 THERMAL ANALYSIS

Material physical properties that can be analyzed by thermal analysis[3] include:

- Softening point
- Effects of plasticizer

- Glass transition temperature
- Melting point
- Heat of fusion
- Specific Heat
- Purity
- Oxidative stability
- Rates of reaction, curing and cross linking
- Degree of cure
- Decomposition
- Crystallization temperature
- Pyrolysis
- Weight percent filler
- Weight loss vs. time
- Weight loss vs. temperature
- Loss of water, solvent, or plasticizer
- Impact properties
- Viscoelastic behavior as a function of stress, strain, frequency, and temperature
- Mechanical behavior
- Long-term behavior such as creep and creep recovery
- Modulus vs. temperature
- Damping vs. temperature
- Chain branching
- Molecular weight
- Molecular weight distribution
- Coefficient of thermal expansion
- Liquidus temperature
- Solidus temperature

Often, a combination of techniques is required to solve the problem as follows:

Multiple thermal analysis techniques, ASTM methods
D1525 Test Method for VICAT Softening Temperature of Plastics
D0256 Test Method for Impact Resistance of Plastics and Electrical Insulating Materials
D0789 Test Method for Determination of Relative Viscosity, Melting Point, and Moisture Content of Polyamide

Tables 9.3 through 9.5[4] summarize the techniques of choice for cured and uncured material identification and degree of cure analysis.

Table 9.3 Techniques for Degree-of-Cure Analysis[*]

Technique	Description	Use	Value
Thermomechanical analysis (TMA)	Measure probe displacement as a function of temperature	Determines glass transition temperature	Indication of degree of cure or environmental effects
Differential scanning calorimetry (DSC)	Performs enthalpy measurements	Determines glass transition temperature Determines residual heat of reaction	Indication of degree of cure or environmental effects
Dynamic mechanical analysis (DMA)	Measures mechanical response to oscillating dynamic loading	Determines glass transition temperature Observes mechanical transition due to additional cross-linking	Indication of degree of cure or environmental effects Indication of undercured condition
Infrared spectroscopy	Measures IR spectrum	Distinguishes between reacted and unreacted functional groups	Indicates amount of unreacted functional groups to determine extent of cure
Solvent extraction	Exposure to an organic solvent	Removes unreacted material leaving reacted network behind	Indication of the degree of cure

[*] *Source:* from Kar, R.J., WL-TR-91-4032, *Composite Failure Analysis Handbook,* Vol. II, Part I, pp. 6–4, Northrop Corp., Hawthorne, California, February 1992.

Table 9.4 Techniques for Uncured Material Identification[*]

Technique	Description	Use	Value
High-pressure liquid chromatography (HPLC)	Produces liquid chromatograms of any soluble liquid	Identifies individual components of differing solubilities or size	Formulation verification
Infrared spectroscopy	Measures IR spectra	Identifies functional groups attached to carbon backbone	Formulation verification
Differential scanning calorimetry (DSC)	Performs enthalpy measurements	Determines heat of reaction	Formulation verification
X-ray fluorescence	Measures X-ray fluorescence spectra	Determines sulfur content	Hardener content

[*] *Source:* from Kar, R.J., WL-TR-91-4032, *Composite Failure Analysis Handbook,* Vol. II, Part I, pp. 6–4, Northrop Corp., Hawthorne, California, February 1992.

Table 9.5 Techniques for Cured Material Identification[*]

Technique	Description	Use	Value
Pyrolysis—gas chromatography	Determines gas chromatograms formed from nonvolatile organics by thermal decomposition	Qualitative and quantitative analysis of cured epoxy	Formulation/impurity verification
Pyrolysis—gas chromatography mass spectroscopy (PGC/MS)	Allows mass spectrometer to act as a detector for the gas chromatograph	Qualitative and quantitative analysis of cured epoxy	Formulation/impurity verification
Infrared spectroscopy	Measures IR spectra	Functional group analysis	Formula verification
Thermomechanical analysis (TMA)	Measures material thermal-mechanical response	Determines glass transition	Identification of general resin system
X-ray fluorescence	Measures X-ray fluorescence spectra	Determines sulfur content	Hardener content

[*] *Source:* from Kar, R.J., WL-TR-91-4032, *Composite Failure Analysis Handbook,* Vol. II, Part I, pp. 6–4, Northrop Corp., Hawthorne, California, February 1992.

9.5.1 Differential Scanning Calorimetry

Differential scanning calorimetry (DSC) is the workhorse of the thermal analysis laboratory. DSC can examine materials between −170 and +750° C (see Fig. 9.6).

DSC is a technique with which the heat flow to or from a sample specimen is measured as a function of temperature or time as it is subjected to a controlled temperature program in a controlled atmosphere. Some of the plastics applications of DSC are to

- Identify the softening point of the material (glass transition)
- Compare additive effects on a material
- Identify the glass transition temperature
- Identify the material's minimum process temperature
- Identify the amount of energy required to melt the material
- Quantify the material's specific heat
- Perform oxidative stability testing (OST)
- Understand the reaction kinetics of a thermoset material as it cures
- Compare the degree of cure of one material to another
- Characterize a material as it cures under UV light
- Characterize a material as it is thermally cured
- Determine the crystallization temperature upon cooling

FIGURE 9.6 Typical differential scanning calorimeter (DSC) system. *Source:* photo courtesy of Perkin-Elmer Corp.

	Differential scanning calorimetry (DSC) ASTM methods
D2471	Test Method for Gel Time and Peak Exothermic Temperature of Reacting Thermosetting Resins
D5028	Test Method for Curing Properties of Pultrusion Resins by Thermal Analysis
D4816	Test Method for Determining the Specific Heat Capacity of Materials by DSC
D4565	Test Method for Determining the Physical/Environmental Performance Properties of Insulation and Jackets for Telecommunications Wire and Cable
D4591	Test Method for Determining Temperatures and Heats of Transitions of Fluoropolymers by DSC
D3012	Test Method for Thermal Oxidative Stability of Polypropylene Plastics Using a Biaxial Rotator
D4803	Test Method for Predicting Heat Buildup in PVC Building Products
D2117	Test Method for Melting of Semicrystalline Polymers by the Hot Stage Microscopy Method
D3417	Test Method for Heats of Fusion and Crystallization of Polymers by Thermal Analysis
D3418	Test Method for Transition Temperature of Polymers by Thermal Analysis
D3895	Test Method for Oxidative Induction Time (OIT) of Polyolefins by Differential Scanning Calorimetry
D4419	Test Method for Determining the Transition Temperatures of Petroleum Waxes by DSC
E698	Standard Test Method for Arrhenius Kinetic Constants (of thermally unstable materials) Using DSC
E1559	Standard Test Method for Contamination Outgassing Characteristics of Space Craft Materials by DSC
E537	Standard Test Method for Determining the Thermal Stability of Chemicals by DSC
E793	Standard Test Method for Determining the Heat of Crystallization (of solid samples in granular form) by DSC
E1269	Standard Test Method for Specific Heat Capacity by DSC
E1356	Standard Test Method for Glass Transition Temperature by DSC

9.5.2 Thermogravimetric Analysis

Thermogravimetric analysis (TGA) examines materials between ambient and $+1,500°$ C. TGA is a technique in which the mass of a substance is monitored as a function of temperature or time as the sample specimen is subjected to a controlled temperature program in a controlled atmosphere.

Some plastics applications of TGA are to

- Identify the filler content of a material by weight percent
- Identify the ash content of a material by weight percent
- Characterize the materials weight loss within a certain temperature range
- Characterize the material's weight loss vs. time at a given temperature

- Quantify the material's loss of water, solvent, or plasticizer within a certain temperature range
- Examine flame retardant properties of a material
- Examine the combustion properties of a material

Thermogravimetric analysis (TGA) ASTM methods	
D2288	Test Method for Weight Loss of Plasticizers on Heating
D4202	Test Method for Thermal Stability of PVC Resin
D 2115	Test Method for Volatile Matter (including water) of Vinyl Chloride Resins
D2126	Test Method for Response of Rigid Cellular Plastics to Thermal and Humid Aging
D3045	Recommended Practice for Heat Aging of Plastics Without Load
D1870	Practice for Elevated Temperature Aging Using a Tubular Oven
D4218	Test Method for Determination of Carbon Black Content in Polyethylene Compounds by a Muffle-Furnace
D1603	Test Method for Carbon Black in Olefin Plastics
D5510	Practice for Heat Aging of Oxidatively Degradable Plastics
E1131	Standard Test Method for Compositional Analysis by TGA
E1641	Standard Test Method for Decomposition Kinetics by TGA

9.5.3 Dynamic Mechanical Analysis

Dynamic mechanical analysis (DMA) involves examining materials between −170 and +1000° C. DMA is a technique in which a substance, while under an oscillating load, is measured as a function of temperature or time as the substance is subjected to a controlled temperature program in a controlled atmosphere. Some plastics applications of DMA are to

- Quantify the impact properties of a material
- Examine the viscoelastic behavior of a material as a function of stress, strain, frequency, time, or temperature
- Examine a material's mechanical behavior
- Examine a material's long term behavior with respect to creep or creep recovery
- Identify the material's modulus vs. temperature
- Identify the material's damping qualities vs. temperature
- Examine effects of temperature on molecular chain branching
- Compare material's molecular weight and molecular weight distribution
- Examine additive effects on a material's mechanical properties

Dynamic mechanical analysis (DMA)
ASTM methods

D4440	Practice for Rheological Measurement of Polymer Melts Using Dynamic Mechanical Procedures
D4473	Practice for Measuring the Cure Behavior of Thermosetting Resins Using Dynamic Mechanical Properties
D5023	Test Method for Measuring the Dynamic Mechanical Properties of Plastics in Three Point Bending
D5024	Test Method for Measuring the Dynamic Mechanical Properties of Plastics in Compression
D5026	Test Method for Measuring the Dynamic Mechanical Properties of Plastics in Tension
D5418	Test Method for Measuring the Dynamic Mechanical Properties of Plastics Using a Dual Cantilever Beam
D0638	Test Method for Tensile Properties of Plastics
D0695	Test Method for Compressive Properties of Rigid Plastics
D1708	Test Method for Tensile Properties of Plastics by Use of Microtensile Specimens
D5296	Test Method for Molecular Weight Averages and Molecular Weight Distribution of Polystyrene by DMA
D3029	Test Method for Impact Resistance of Flat, Rigid, Plastic Specimens by Means of a Tup (falling weight)
D4508	Test Method for Chip Impact Strength of Plastics
D4812	Test Method for Unnotched Cantilever Beam Impact Strength of Plastics
D5083	Test Method for Tensile Properties of Reinforced Thermosetting Plastics Using Straight-sided Specimens
D0790	Test Method for Flexural Properties of Plastics
D0882	Test Method for Tensile Properties of Thin Plastics
D0952	Test Method for Bond or Cohesive Strength of Sheet Plastics and Electrical Insulating Materials
D0953	Test Method for Bearing Strength of Plastics
D1043	Test Method for Stiffness Properties of Plastics as a Function of Temperature by Means of a Torsion Test
D5420	Test Method for Impact Resistance of Flat, Rigid, Plastic Specimen by Means of a Striker Impacted by a Falling Weight
D5628	Test Method for Impact Resistance of Flat, Rigid, Plastic Specimen by Means of a Falling Dart (tup or falling weight)
D0671	Test Method for Flexural Fatigue of Plastics by Constant Amplitude of Force
D0747	Test Method for Apparent Bending Modulus Plastics by Means of a Cantilever Beam
D0785	Test Method for Rockwell Hardness of Plastics and Electrical Insulating Materials
D2990	Test Method for Tensile, Compressive, and Flexural Creep and Creep Rupture of Plastics
D2765	Test Method for Determination of Gel Content and Swell Ratio of Cross-linked Ethylene Plastics
D4476	Test Method for Flexural Properties of Fiber Reinforced Pultruded Plastic Rods
D2343	Test Method for Tensile Properties of Glass Fiber Strands, Yams, and Rovings Used in Reinforced Plastics
D1939	Practice for Determining Residual Stresses in Extruded or Molded ABS
E1640	Standard Test Method for Glass Transition by DMA

9.5.4 Thermomechanical Analysis

Thermomechanical analysis (TMA) examines materials between −170 and +1000° C. TMA is a technique in which the deformation of a substance, while under a nonoscillating load, is measured as a function of temperature or time as the substance is subjected to a controlled temperature program in a controlled atmosphere. Some plastics applications of TMA are to

- Determine a material's coefficient of thermal expansion
- Determine the volumetric growth of a material, dilatometry measurements
- Identify a material's glass transition temperature
- Identify the material's VICAT softening point
- Measure a material's heat deflection temperature (HDT)

Thermomechanical analysis (TMA) ASTM methods	
D2566	Test Method for Linear Shrinkage of Cured Thermosetting Casting Resin During Cure
D0789	Test Method for Determination of Relative Viscosity, Melting Point, and Moisture Content of Polyamide
D2732	Test Method for Unrestrictive Linear Thermal Shrinkage of Plastic Film and Sheeting
D0696	Test Method for Determination of Coefficient of Linear Thermal Expansion of Plastics
D3386	Test Method for Determination of Coefficient of Linear Thermal Expansion of Electrical Insulating Materials by TMA
E1545	Test Method for Determination of Glass Transition Temperatures Under Normal Range (−100 to +600° C)
E831	Test Method for Determination of Glass Transition Temperatures of Solid Materials by TMA

9.6 ACKNOWLEDGMENTS

I would like to thank the Perkin-Elmer Corporation for providing much of the material in this chapter. Special thanks go to Joyce Gallagher of Gibbs & Soell, Inc., for tracking down much of this information, and Richard Spragg for filling in the gaps.

9.7 REFERENCES

1. Adapted with permission from "The Guide to Techniques and Applications of Atomic Spectroscopy," Perkin-Elmer Corporation, January 1997.
2. Private communication with Richard Spragg, Perkin-Elmer, Infrared Spectroscopy Business Unit, Beaconsfield, England, February 17, 1997.
3. Andrew W. Salamon, "Thermal Analysis Method Development Poster," presented at 1996 ANTEC, Society of Plastics Engineers 54th Annual Technical Conference, May 7, 1996
4. Kar, R.J., WL-TR-91-4032, *Composite Failure Analysis Handbook* Vol. II, Part I, pp. 6–4. Hawthorne, CA: Northrop Corporation, February 1992.

9.8 RECOMMENDED READING

D. Briggs and M. P. Seah, *Practical Surface Analysis.* New York: John Wiley & Sons, 1983

Materials Characterization, Metals Handbook, Vol. 10, 9th ed. American Society for Metals, 1986.

CHAPTER 10
ELECTRONIC AND ELECTRICAL CHARACTERIZATION

J.J. Erickson

Hughes Space and Communications Company

10.1   INTRODUCTION

In this chapter, electronic and electrical characterization, as it is planned and performed for failure analysis, is presented. The differences between electrical characterization or testing for failure analysis and *standard* electrical testing are discussed. A general approach is recommended for failure analysis electrical testing that will help the analyst to electrically characterize the device under analysis (DUA*) without causing additional damage that may obscure or obliterate the original cause of failure.

Failure analysis is performed on a device to determine the root cause of the failure. If, during the analysis, electrical testing causes damage to the device or alters the device's failure mechanism, then the wrong conclusions will be made concerning the root cause of failure.

The results of a failure analysis are used to determine what corrective action is needed to prevent future failures. If the wrong root cause is determined, then the wrong corrective action will be applied, and future failures will not be eliminated. In effect, the failure analysis is a failure if it induces a failure mechanism in the device that was not present at the beginning of the analysis.

This chapter is not an introduction to *standard* electrical characterization or test techniques. It will be assumed that *standard* electrical testing techniques are already known and understood by the reader.

10.2   STANDARD ELECTRICAL CHARACTERIZATION TECHNIQUES

The term *standard* electrical characterization techniques implies those techniques that are typically used to characterize electrical and electronic components. In a *standard* electrical test, the test approach is defined by standard test procedures. In addition, a specification for the device under test provides details of electrical bias conditions that are to be applied to the device during the standard test. The test parameters and approach are specifically defined, and the results of the test are compared to limits in the device specification to determine if the device "passes" or "fails." If, during the testing, the component is tested

---

*The abbreviation DUA is used to avoid using cumbersome terms such as "the device being analyzed" or "the reportedly failed device." Some references refer to the "failed" device with failed in quotes. A large percentage of failure analyses that are performed determine that the "failed" device is not actually a failure, or the failure could not be confirmed. DUA also avoids this possible contradiction.

correctly and is damaged by the tests, this is of little consequence, since the device was obviously flawed and will be placed in the "failed" device bin.

The previous *standard* test techniques and specifications are designed on the expected behavior of a device that has been built according to certain design guidelines. These tests also assume that subsequent testing or handling of the device has not altered the device from its expected configuration. In addition, the standard test is not intended to protect the device from additional damage if the device has been degraded or damaged during previous tests. The purpose of the standard test is to determine only if the device meets its intended electrical specifications. If the device has been degraded during previous tests or handling, the standard test may cause additional damage to the device during its characterization, thereby degrading it further. For the purposes of testing, this does not significantly matter. The initially degraded device would have failed the test just as the further degraded device did. Both would have been set aside as electrical test rejects.

10.3 FAILURE ANALYSIS ELECTRICAL CHARACTERIZATION

During the failure analysis of a component, electrical characterization should be planned and performed using a very different approach from that used for standard electrical characterization.

During a failure analysis, when testing the DUA, the test has to be performed on the assumption that the DUA is not the same as a standard or "normal" device. The DUA may be different from a standard device for one of the following reasons.

- The device may have been electrically overstressed.
- The device may have been damaged by ESD (electrostatic discharge).
- The device may be altered internally due to chemical reactions.
- The device may have been misassembled.
- The device may include a manufacturing defect.
- The device may be intermittently faulty.

Specifically, when performing a failure analysis, the standard test procedure or test program could cause additional damage to a device that is already degraded. This may result in the wrong conclusion being drawn from the information found during the subsequent failure analysis.

For example, suppose an input on an IC has been damaged by ESD. If the failure analyst were to proceed directly to an automated IC tester to characterize the device, the device could be damaged by the tests performed by the automated tester. The program that was generated for the automated tester was written on the basis of testing a "good" device. Assume that the good device routinely will draw no more than a few tens of nanoamperes with 5 V applied to the input. The automated test program will direct the tester to apply 5 V to the input with all other pins biased normally. The 5 V will not be current limited (except by the capability of the power supply used in the tester), since a good device should not draw any substantial current.

However, the ESD-damaged input on the IC may draw enough current that additional damage will be generated in the device's input circuitry during the automated tester evaluation. During the subsequent steps of the failure analysis, examination of the appearance of the damage could indicate that the device has been damaged by electrical overstress. The original damage induced by ESD (which is typically very small and localized) may have been modified during the additional testing to appear to be the result of electrical overstress (which is generally much grosser and easier to detect than ESD damage). Thus, not only was the device a failure, but the failure analysis was also a failure, since the wrong conclusion was drawn regarding the cause of the failure.

10.4 OTHER ASPECTS OF ELECTRICAL CHARACTERIZATION FOR FAILURE ANALYSIS

Failure analysis electrical testing may vary from standard electrical testing for other reasons. As mentioned previously, standard electrical tests are based on the expected behavior of a device. These tests are also designed in accordance with the anticipated physical layout of the device.

10.4.1 Testing for Unexpected Behavior

Sometimes a device fails because it has internal built-in flaws or internal design layout problems that are not anticipated when the standard test was planned. The standard basically tests the device to determine that it performs as it is supposed to perform. Electrical testing for failure analysis must also include tests that verify that a device does not do what it is not supposed to do. This concept is best illustrated by some examples.

Example 1   A failure analysis was performed on an IC that had four separate AND gates. Each AND gate had two inputs, A and B. As expected, when the inputs A1 and B1 of AND gate one were both "high," the output was "high." If either or both of the inputs were "low," then the output was "low." The other AND gates performed in the same manner. The device passed the functional test that verified this operation. However, when the device was used in an application, it failed to function correctly.

Electrical tests performed during failure analysis revealed that when the output of AND gate 1 was "high," the output of AND gate 2 was also "high," regardless of the inputs to AND gate 2. The reverse was also true. When the AND gate 2 output was "high," the output of AND gate 1 was also "high." regardless of the its inputs. These two AND gates were supposed to operate independently, but they did not.

Additional analysis revealed that the internal circuitry in the device used nichrome resistors in the outputs. The output circuitry for AND gate 1 was near the output circuitry for AND gate 2. When the nichrome resistors were processed, a masking error allowed the resistors for the two outputs to be connected. As a result, the two connected outputs resulted in the observed behavior.

Example 2   Failure analysis was performed on an ASIC that performed a complicated logic function for an application. As in the previous example, the device was tested using an automated tester and successfully passed the functional testing. When placed in the application, the ASIC caused a failure. During failure analysis, it was discovered that one of the ASIC's outputs was generating an extraneous pulse. However, this pulse was generated about three clock pulses after the last logic transition that it was supposed to generate. Figure 10.1 illustrates the problem.

The automated test program verified that the output stayed "low" after clock pulses 4 and 5 but did not check the output beyond that point. The automated tester, therefore, did not detect the anomalous pulse on the output after clock pulse 7. Since the ASIC was not supposed to be able to generate the pulse after clock pulse 7, the tester program did not check for it.

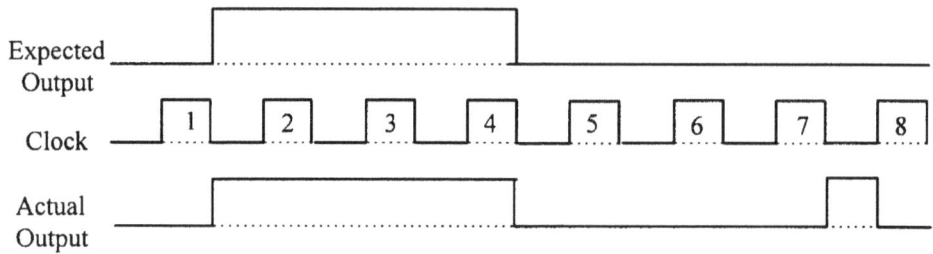

FIGURE 10.1   Logic waveforms for the ASIC device. The expected output was supposed to switch to a "high" state after clock pulse 1 and then stay "high" until the end of clock pulse 4, at which time it was supposed to switch to a "low" state, and then stay "low." The actual output from the DUA did switch "low" after clock pulse 4 but then switched "high" for a short period of time after clock pulse 7.

10.4.2   Testing for Intermittent Electrical Behavior

When standard electrical tests are performed, the device under test is usually tested only for a period of time necessary for the device to stabilize after any transients. The parameter or function being tested is documented, and then the next test is performed. Also, if the device is to be tested with an applied voltage of 50 V, no effort is made to characterize the device as the applied voltage is increased to 50 V. When the device is tested over its temperature range, the device under test is usually tested only at room temperature and at its maximum and minimum rated temperatures. There typically is no attempt made to characterize the device continuously over the full temperature range.

When failure analysis electrical testing is performed, it is important to characterize the device at all intermediate values of applied bias rather than just at the typical and maximum rated biases. Under some bias conditions, the device may be exhibiting intermittent electrical characteristics that are not detected at the normal applied bias or at the maximum rated bias.

Even more important is testing the DUA over the full temperature range while the temperature is being changed. Some device failures occur as a result of mechanical mismatches between materials that is exacerbated by the application of temperature. If one material absorbs heat faster than the other material, the mismatch may be maximized during the temperature transitions. Therefore, the device may fail only during these transitions. However, during the soak at the temperature extremes, both materials will reach the same temperature and the mismatch may be diminished. Standard testing is usually performed during the temperature soak at the extremes and could fail to detect the intermittent failure.

10.5   *REPORTED FAILURE MODE AND BACKGROUND INFORMATION*

Information is usually supplied with a device that is to be subjected to a failure analysis. This information or data can usually be separated into two categories: (1) the reported failure mode, and (2) background information. In some cases, the information may be very detailed and will be a valuable asset to the failure analyst. In other cases, there may be no information supplied or the available information may be virtually worthless. In the latter case, the failure analyst should make it a priority to obtain as much information as possible. This information will provide some guidelines to the failure analyst as to the best approach for performing electrical testing.

10.5.1 Reported Failure Mode

When the analyst receives a device to be analyzed, there is usually a failure mode reported for the device. Typical examples are shorted, open, high leakage current, functional failure, fails to switch, and so on. If the analyst is fortunate, these failure modes are further defined with actual data. For example, "The device exhibits a leakage current of 1 mA, but it should only be 100 nA maximum."

If the reported failure mode is not defined with specifics, the analyst should attempt to get additional data, since meanings of common terms vary from one person to another and also from one application to another. For example, in a power circuit, a 10 Ω impedance may be a significant load. However, in a logic circuit, a 10 Ω load would often be classified as a short. In other applications, the expected impedance may be hundreds of megohms. In these applications, a resistance of a few kilohms may be classified as a short.

Occasionally, even very detailed data that is reported may be of little use to the failure analyst. This often occurs when the device fails in a complicated circuit being tested. For example, it might be reported that the device failed during "test procedure 4.4, at step 2.6.1.4, with the gain reduced by 4 dB." If the DUA is a resistor, the analyst might wonder how to measure the gain of a resistor. However, in this case, it was probably the overall application that had a reduced gain, and the resistor was determined to be the cause of the reduced gain. It will be incumbent on the analyst to contact the circuit designer to determine the most likely failure mode of the resistor that would decrease the circuit gain.

Sometimes a device is submitted for analysis with less than useful information such as "the device fails to function," "the device is bad," or "unknown failure." In these cases, the analyst again should attempt to determine more specific information about the device failure.

Information supplied with a device often will save the analyst a tremendous amount of time and effort. For example, if a resistor is submitted with the information that it is intermittent only during temperature transitions from cold to hot at approximately 30° C, the analyst is saved the trouble of performing detailed characterizations of other aspects of the device. Also, temperature cycling can be performed early in the analysis to try to confirm the failure. Without this detailed information, it is possible that the analyst might never have confirmed the device's failure.

Therefore, it is mandatory that the analyst determine as much as possible about the failure mode that was observed for the device in its application. This information could also prevent the analyst from pursuing a problem with the component that may be unrelated to the reason that it caused an application failure. For example, suppose the analyst has tested an IC and determined that one of the inputs exhibits a leakage current that is marginally outside of specification limits. If the reported failure is that the device output failed to switch when inputs other than the leaky input were switched, the analyst might set aside the leaky input data and concentrate on other aspects of the device's electrical performance. (*Note:* The leaky input should be further investigated at a later point, but the reported failure mode decreases its probable importance in the failure analysis.)

10.5.2 Background Information

In addition to the reported failure mode, the failure analyst should obtain as much background information about the DUA and its application as possible. Again, this information will aid the analyst in developing an electrical test plan. The following background information should be obtained.

1. Test conditions at the time of failure, including
- Biases applied to the device

  DC bias
  AC signals
  Frequency of signals
  Transients

- Environmental conditions

  Temperature (room, hot, cold?)
  Temperature cycling
  Humidity testing
  Vacuum testing (or pressure testing)
  Electric or magnetic fields present
  Radiation exposure
  ESD exposure

- Mechanical stresses

  Vibration testing
  Shock
  Mechanical loading of the device (or its circuit board)

This background data will provide potentially critical information regarding the conditions under which the device should be electrically tested.

2. Previous history of the specific device
- Was it a failure at first test at the manufacturer?
- Was it a failure at incoming test?
- What was the device subjected to during installation in the application?
- Did it function correctly in the application before failing?
- What other tests were performed before the device failed?
- After the component was removed and replaced, did the original problem go away?
- What other tests were performed on the device after it was determined to be a failure?
- How was the device removed from the application?

Answers to these questions will provide information as to which electrical tests may need to be performed. Also, the last two questions may provide information to indicate whether the device could have been further damaged during subsequent testing or by excessive stresses during removal. Information gathered during electrical testing should be analyzed with respect to this information to evaluate this possibility.

3. Information about other components in the application
- What was the circuit configuration in which the device was used?
- Were other components removed along with the failure?
- Did other components fail?

This information can again provide valuable information for the analyst when planning electrical tests. The circuit configuration will provide valuable information regarding possible biases applied to the device, plus information regarding current limiting provided in the application. If other components failed or were also removed, the analyst should be aware of their condition to determine possible scenarios for testing the DUA.

If other components were found to be failures, testing of the DUA may result in "Not a failure" for the DUA. This will be easier to accept than in the case where the DUA was the only device that was removed or in the case where each of the other components was found to also be "not a failure."

4. Information about other components of the same type as the DUA

   - Have other components of this type been successfully used in the same application where the DUA reportedly failed?
   - Have there been other failures of this component type in this application?
   - Have there been other failures of this component type in other applications? (This information can be found if the analyst has access to a failure analysis database.)
   - Have other users experienced failures of this component? (This information is available in databases such as GIDEP, NASA and other databases.)
   - Information should be obtained from the component engineer for this device to determine if there is any other important information about either this specific component or other similar components from the same manufacturer.

This information could be very valuable for planning electrical tests, especially if there were other failures of this component in the same application. If these failures were analyzed, then the results of these analyses may tell the analyst exactly which electrical tests need to be performed on the DUA. Other analyses of this device from other applications could also alert the analyst to subtle or unusual failure mechanisms that the analyst might not have included in the test plan.

10.6 GENERAL APPROACH FOR FAILURE ANALYSIS ELECTRICAL TESTING

Much of the material previously presented in this chapter describes electrical testing for failure analysis in general terms that might be described as *philosophical*. This section attempts to provide the analyst with a more practical approach to failure analysis electrical testing.

10.6.1 Sequence of Electrical Tests for Failure Analysis

For all failure analysis techniques in general, the best approach is to perform the least destructive tests first and gather the data from these tests. Then proceed with destructive (or potentially destructive) tests.

The best approach for electrical testing during failure analysis is the same. Perform nondestructive electrical tests first, gather the resulting data, and then proceed with potentially destructive tests. Since a DUA is possibly different from a normal device, it is sometimes difficult to determine exactly which electrical tests may be potentially destructive.

The following is a general approach for failure analysis electrical characterization of a device, assuming that there is little or no detailed information on the DUA's failure mode. If detailed information is available, the following approach should be modified as appropriate.

The best general approach for failure analysis electrical testing is to perform initial tests with low voltages and low currents. As a voltage is applied to a device, the current should be limited to a low value. As more and more voltage is applied and significant current begins to flow, the power dissipation should be limited and the current/voltage characteristics monitored for stability.

It would be difficult to meet all of these requirements using a discrete power supply with discrete voltage and current meters. However, it is quite easy to meet all of these requirements using a curve tracer. The curve tracer itself is described Sec. 10.6.2.

10.6.1.1 Curve Tracer Testing. Some believe that the curve tracer was designed to test and evaluate only discrete semiconductor devices such as diodes and transistors. While this may have been the basic purpose of the curve tracer, this piece of equipment is an invaluable tool for the initial electrical characterization of all electrical and electronic components. For some devices, especially discrete semiconductor components, the curve tracer will be all that is needed to completely electrically characterize the device. However, for other components, the curve tracer is used only to evaluate the basic integrity of the DUA.

There may be no need to develop an extensive automated test program or to build an elaborate test setup if a device has a basic problem such as a shorted or open connection. Also, as mentioned previously, it is necessary to proceed with caution when electrically characterizing a DUA, since the device may not behave as expected. Standard approaches may damage, obliterate, or change the initial failure mode present in a device.

The curve tracer is used to verify continuity between connections in a device that should be continuous. For example, the curve tracer could be used to verify that the primary coil in a transformer is continuous. It could also verify that the primary coil is not shorted.

The curve tracer is also used to verify that there is no conduction path between portions of a device that are supposed to be isolated. For example, the curve tracer could be used to verify that there is no short between the primary and secondary coils in a transformer.

The curve tracer provides information about any intermittences in the basic electrical behavior of a device. For example, if there was in intermittent connection to the primary coil in the transformer, the curve tracer could detect this problem.

The DUA can also be subjected to a limited amount of temperature cycling during curve tracer testing using a carefully controlled heat gun and/or *freeze spray* to determine whether it shows any indications of intermittence in its electrical characteristics. Obviously, this testing is not intended to provide detailed information about the DUA's behavior over temperature but is intended as a quick screen to detect problems associated with intermittences.

10.6.1.2 Standard Electrical Characterization (with Modifications). After the curve tracer has been used to verify the basic integrity of the DUA, then the DUA is electrically characterized using standard electrical tests that may be modified for the purposes of failure analysis electrical characterization.

For example, if a capacitor were being tested for leakage current at high voltages, a series resistor should be used to limit the current to avoid a catastrophic overstress of the device. If high currents are to be applied to a device, the voltage should be limited to a safe value to prevent excessive voltages from being applied if the device should suddenly open circuit during testing.

If a device is being tested on automated test equipment (ATE), the program might be modified to stop the testing as soon as any failure is detected. Also, the test program should be modified to not only monitor test currents but to also limit test currents to safe values.

10.6.1.3 Electrical Characterization over a Temperature Range.   If the DUA passes all parametric and functional testing at room temperature, then the DUA should be characterized over its rated temperature range. Initially, it is sufficient to test the DUA at its temperature extremes only.

If the device passes the temperature tests at its rated extremes, it is generally safe to assume that the device failure is not due to temperature-related affects. However, if the device was initially reported to fail only under conditions of varying temperature, then the device should be continuously electrically characterized while varying the temperature from room temperature to its maximum rated temperature and then allowing it to equilibrate at the maximum temperature. The temperature should then be decreased to the minimum rated temperature of the device while continuously monitoring the DUA's electrical behavior. After the device has equilibrated at the minimum rated temperature, the temperature should then be raised back to room temperature, again while continuously monitoring the DUA's electrical behavior. This temperature cycle should be repeated several times to attempt to identify any intermittence in the device.

10.6.1.4 Other Electrical Characterization Testing.   In addition to the previous tests, additional testing may be performed to determine whether the DUA is a failure. These additional tests most often will be based on information that is reported about the DUA's testing conditions at the time it was initially determined to have failed. These tests may include combinations of the following test conditions:

- Varying the power applied to the DUA
- Varying the frequency of the test signals applied to the DUA
- Monitoring the DUA's electrical behavior vs. time
- Applying various environmental or mechanical stresses to the DUA including
    - humidity
    - vacuum (or pressure)
    - mechanical loading
    - shock or vibration
    - acceleration

Other tests may be performed based on known failure modes for the DUA. For example, if the DUA is a semiconductor, tests may be performed to determine if its electrical behavior may have been previously degraded due to ionic contamination induced inversion. (See Sec. 10.8.1.2.)

10.6.1.5 Evaluation of Electrical Test Data.   If the device fails at any time during the electrical characterization of the DUA, either a parametric requirement or a functional requirement, the failure data must be evaluated. The purpose of the evaluation is to deter-

mine if the failure of the device has actually been verified. Electrical testing may have detected a minor problem in the device that is unrelated to the failure that was reported in the application for the DUA.

For example, if the device is a functional failure at normal operating conditions, then it can be concluded that the device failure has been verified. However, if the device fails only at some extreme operating condition, additional testing may be necessary. Examples to illustrate this follow.

Example 3   Assume that the DUA is an IC that is supposed to function with a power supply voltage of 4.5 to 5.5 V over its entire rated temperature range. Also assume that the device fails with an applied power supply of 4.5 V but only at 125° C. If these conditions could have existed in the application where the DUA was used, then the failure could be claimed to be verified. However, if the application had a power supply that provided a solid 5.0 V over temperature, then additional testing should be performed.

Example 4   In this case, suppose that the DUA is a capacitor rated for a maximum leakage current of 1.0 nA at its maximum rated voltage of 50 V. Also assume that the measured leakage current was 2.0 nA at 50 V. If, in the application, the maximum applied voltage was 5.0 V (where the measured leakage current was only 0.01 nA), the effect of this leakage current would have to be evaluated to determine if it would affect the application. If this leakage current would have no effect, simply stating that the capacitor failure has been verified because it does not meet its specification requirements does not determine the cause of the original reported failure. Additional testing should be performed to try to identify the true failure mode of the capacitor.

In some cases, the DUA will fail a parametric requirement under standard operating conditions at room temperature, but the failure may again be unrelated to the reported failure of the DUA in the application. Again, additional electrical characterization should be performed. An example of this type of parametric failure follows.

Example 5   Assume that the DUA is a diode that failed its forward voltage requirement with a forward voltage drop of 0.85 V at 10 mA, while the specification is 0.81 V maximum. If the purpose of the diode in the application was transient suppression, the forward voltage drop probably is not significant, and additional tests need to be performed. However, it the diode was used as part of a precision voltage reference, then the parametric failure would be significant, and the failure could be stated to be verified.

10.6.1.6   Additional Diagnostic Electrical Characterization.

Suppose that the DUA has been tested, and the failure has been verified. Additional diagnostic electrical testing may still provide additional information about the failure of the device.

For example, in a complex IC, the use of specially designed test vectors might further locate the functional failure of the IC to a certain portion of the device's circuitry. Detailed evaluation of a diode's leakage current over all bias conditions could provide additional information about the probable failure mechanism affecting the device.

10.6.1.7   Failure Analysis Electrical Characterization Completion.

When is the failure analysis electrical characterization completed? It is obviously completed if the failure of the DUA is verified and all diagnostic testing has been completed. This may happen at any point in the testing of the DUA. It could happen early at curve tracer evaluation of the DUA's basic integrity, or it could happen at a later point, after many hours of testing, when a very specific set of conditions determined that the DUA was intermittent.

When is the electrical characterization completed if the failure cannot be verified? It is completed when the device has been thoroughly tested to all possible electrical and environmental conditions to which it was subjected in the application, and the failure could not be verified. The failure analyst should discuss the tests that have been completed with the application engineer and designer to confirm that the device has been adequately tested. Also, the failure analyst should review all possible failure mechanisms for the device to be sure that the tests were adequate to screen for these mechanisms.

10.6.2 Introduction to the Curve Tracer

The curve tracer is a piece of electronic test equipment used to characterize the current-voltage (I-V) characteristics of a component. The curve tracer applies a triangular voltage waveform to the DUT. The triangular voltage waveform is applied to the x-axis of a CRT display. While the voltage waveform is applied to the x-axis, the resulting current that flows through the DUT is displayed on the y-axis of the same display.

Figure 10.2 is a photograph of one of the most widely used curve tracers for failure analysis electrical testing, a Tektronix Model 576. Figure 10.3[*] is a photograph of the dis-

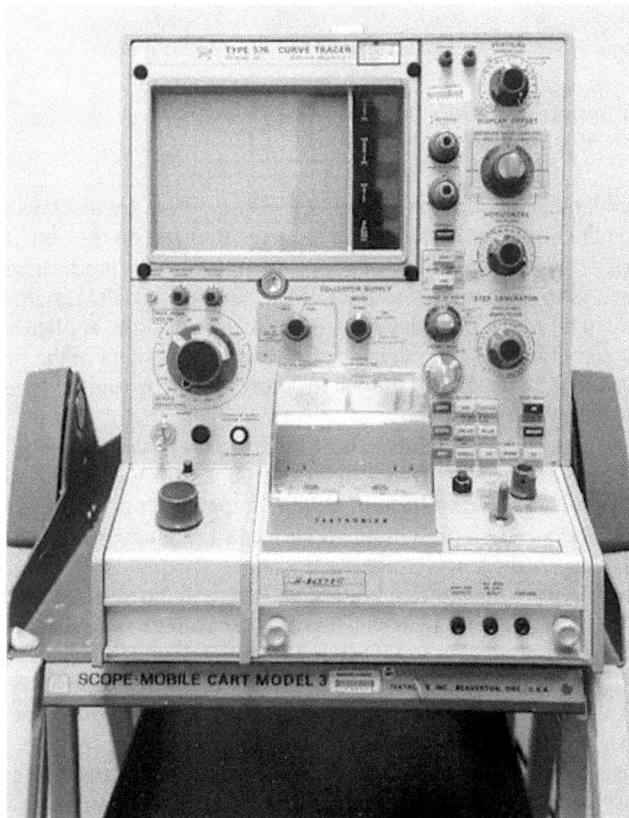

FIGURE 10.2   Tektronix 576 curve tracer.

---

[*]This is the only true photograph of a curve tracer display included in this section. All other figures showing curve tracer characteristics for devices are graphical representations of curve tracer displays.

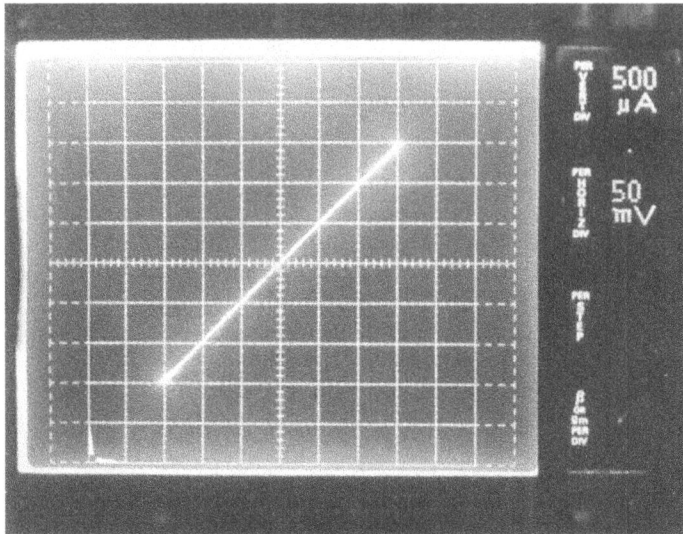

FIGURE 10.3 The displays on the curve tracer shown in Fig.10.2. The main display shows the current-voltage characteristic for a 100-Ω resistor. The numeric displays on the side show the volts and amps per division on the main display.

plays on the curve tracer. In addition to the CRT display, there are displays that indicate the volts/division on the horizontal (x) axis and amperes/division on the vertical (y) axis.

The curve tracer has the capability of applying voltages varying from zero to hundreds of volts, and zero to many amperes to the DUT. The applied voltages are controlled by a voltage-range switch (which provides a coarse adjustment to the applied voltage) and the applied voltage potentiometer (which controls the fine adjustment in the range selected by the coarse adjustment switch). The maximum current flow is limited by a series resistor that can be selected by the operator.

The controls on the curve tracer also allow the operator to vary the displayed sensitivity on the x-axis of the CRT display from 1 mV/division to 100 V/division. The y-axis sensitivity can be varied from 1 nA/division to several amperes per division.

Figure 10.4 is a diagram representing some basic I-V characteristics obtained from a curve tracer display. If the curve tracer test voltage is applied to an open circuit, the applied voltage waveform generates a horizontal line, with no vertical deflection, since no current flows (trace 1 in Fig. 10.4). If the curve tracer voltage is applied to a short circuit, the applied voltage is converted to current and a vertical line is generated (trace 2 in Fig. 10.4), with no horizontal deflection, since no voltage is sustained across a short circuit. If the curve tracer voltage is applied to a resistor, a straight line with a slope proportional to the resistance is generated (trace 3 in Fig. 10.4).

The previous examples show the basic function of the curve tracer. The curve tracer's intended purpose is demonstrated on a nonlinear device. Figure 10.5 shows the I-V characteristics of a typical diode at low power levels. The curve tracer easily generates the I-V characteristics of the device. (To generate this characteristic using a power supply and current meter, discrete points on the I-V characteristic would have to be generated and then plotted to generate the characteristic generated by the curve tracer.) The additional utility of the curve tracer is displayed in Figs. 10.6 and 10.7. By simply adjusting the voltage

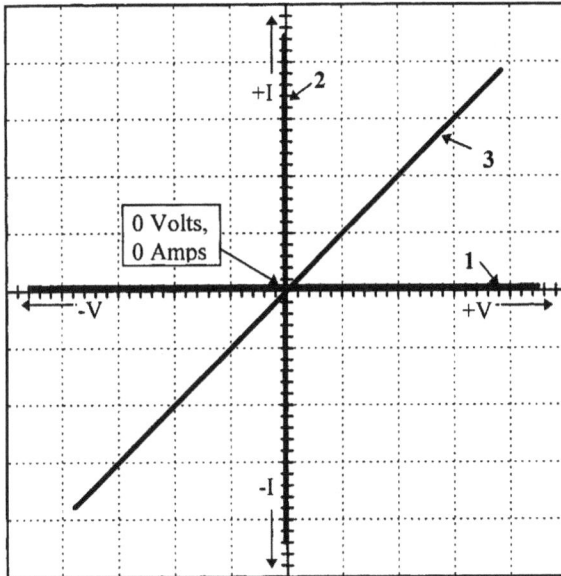

FIGURE 10.4   Three straight-line curve tracer characteristics indicating (1) an open circuit, (2) a short circuit, and (3) a resistor. (The origin [0 V, 0 A] and I, V polarities noted on this figure are common to all curve tracer figures that follow.)

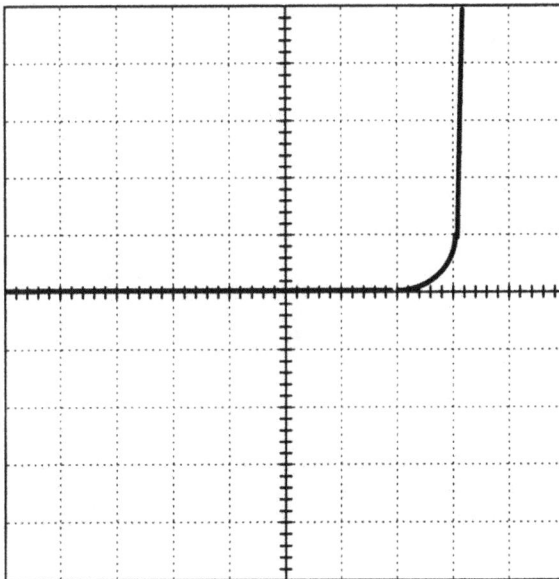

FIGURE 10.5   Curve tracer display representing a diode characteristic at low power (horizontal display of 0.2 V/division, and a vertical display of 100 μA/division). The forward voltage drop at 100 μA is about 0.61 V.

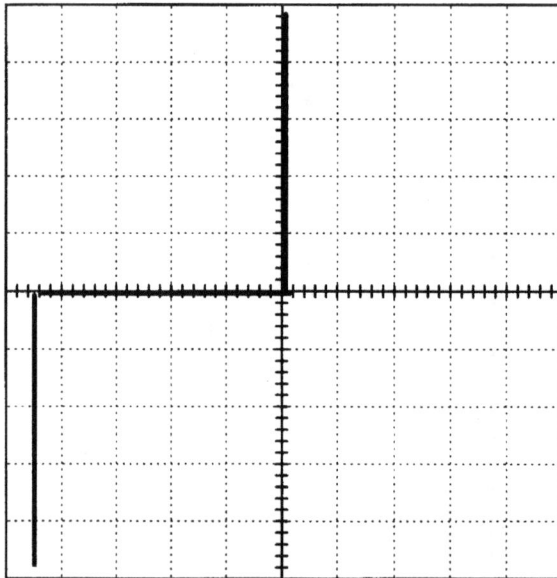

FIGURE 10.6   Curve tracer display representing a diode characteristic at higher power (horizontal display of 10 V/division and a vertical display of 1 mA/division). The reverse breakdown voltage at 1 mA is 55 V.

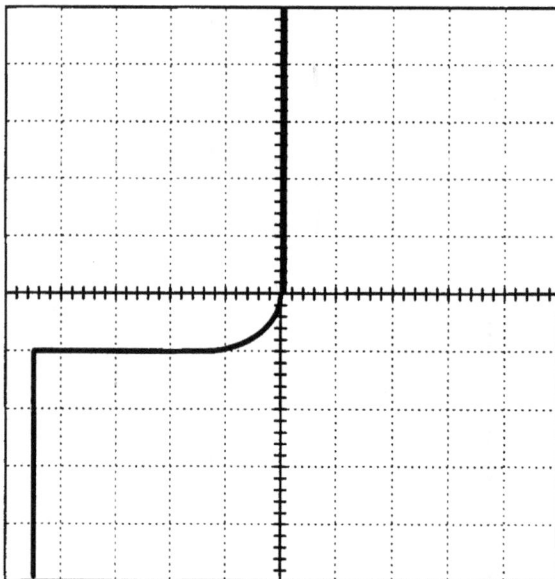

FIGURE 10.7   Curve tracer display illustrated in Figure 10.6 with the vertical scale expanded (1 μA/division) to measure the leakage current. At a reverse voltage of 50 V, the reverse leakage current is about 1 μA.

range of the curve tracer power supply and then adjusting the horizontal and vertical display sensitivities, the trace in Fig. 10.5, which represents the low power I-V characteristics of the diode, is changed to the trace in Fig. 10.6, which represents the higher power I-V characteristics of the diode. Figure 10.7 shows a different vertical scale sensitivity setting for Fig. 10.6, which enables the operator to view the reverse leakage current of the diode at high voltage settings.

The previous discussion is an introduction to the basic operation of a curve tracer. The following section discusses some of the other advantages of the curve tracer as compared to other methods of characterizing components during failure analysis electrical testing.

10.6.3 Advantages of Curve Tracer Testing

The curve tracer is a powerful tool for rapidly characterizing the electrical characteristics of a variety of components. It was designed primarily for rapidly characterizing nonlinear semiconductor components. However, during failure analysis electrical testing of a wide variety of components, the curve tracer has many advantages over other methods of testing.

1. The curve tracer provides rapid feedback on the test being performed. With the curve tracer, a small voltage can be applied to a device, and there is an instantaneous I-V characteristic generated on the display. The analyst can rapidly assess whether there is any indication of a problem and proceed to increase the applied voltage if no problems are detected. If the analyst had to rely on discrete power supplies, voltmeters, and current meters, only point-by-point data would be available.

2. The curve tracer provides automatic current limiting and instantaneously displays the amount of current that is flowing in the DUA. Failure analysis testing often includes verifying that device pins that should be isolated from one another are actually isolated. Using the curve tracer on the most sensitive current scale, it can be rapidly determined if the pins are truly isolated. If the pins are not isolated, the current limiting provided by the curve tracer prevents potentially destructive currents from flowing.

3. The curve tracer provides an indication of rapidly varying (or intermittent) electrical characteristics. For example, resistors can fail due to internal intermittently open (or highly resistive) mechanical connections. If an intermittent resistor were being measured using a digital ohmmeter, the variation in resistance might be missed if the sampling rate of the meter was slow. If a meter using a galvanometer mechanism were used, the inertia of the mechanism may again be too slow to detect the intermittence. However, if the resistor was being tested on a curve tracer, the trace would indicate an intermittent condition by rapidly moving on the display. Figure 10.8 is an illustration of the possible behavior of the curve tracer display while testing an intermittent resistor.

4. The curve tracer provides rapid feedback on excessive power dissipation or heating in a DUA. If excessive power is being applied to the DUA, the device may start to heat up. As the device gets hotter, the device's resistance may decrease resulting in more current flow and even higher power dissipation. If this continues unimpeded, the device could "self destruct" due to runaway heating. However, if the device is being characterized on the curve tracer, heating effects are evident as a gradual shift in the I-V characteristic curve (under constant applied voltage) as time passes. Figure 10.9 is

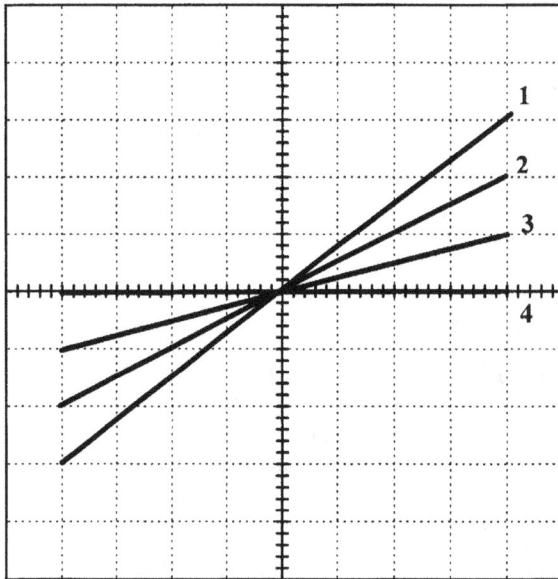

FIGURE 10.8   Curve tracer results from measurement of an intermittent resistor. Initial measurement is curve 1, with curves 2, 3, and 4 representing subsequent sequential results after different periods of time.

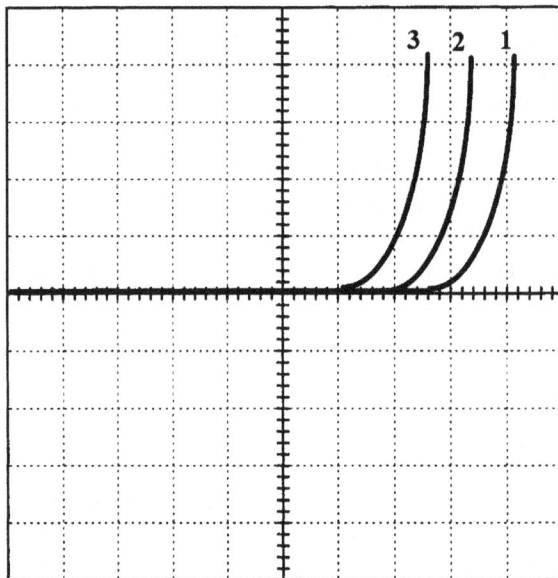

FIGURE 10.9   Curve tracer results from measurement of a semiconductor junction as it heats due to power dissipation. Curve 1 is the initial measurement with curves 2 and 3 representing the change in the characteristic as time passed and the junction temperature increased.

an illustration of the behavior of the curve tracer display when testing a diode with sufficient power applied to cause heating.

5. The curve tracer provides an overview of a device's I-V characteristic rather than just a point-by-point indication of its electrical characteristics. This enables the analyst to easily detect problems that might otherwise be overlooked during conventional testing. For example, some resistors were submitted for analysis with the complaint that they were out of tolerance. Testing them using a digital ohmmeter indicated that they were well within their specification limits. However, examination of the resistors' I-V characteristics on the curve tracer indicated that they were nonlinear. At the particular bias provided by the digital ohmmeter, the resistances were within the requirements of the specification. However, at other bias points, the resistors were not within the specification limits due to their nonlinear behavior.

6. The curve tracer can provide nondestructively provide indications of parasitic silicon controlled rectifier (SCR)-type behavior in semiconductors. Occasionally, some ICs are designed that accidentally include a combination of adjacent diffused structures that provide parasitic SCR-type behavior. Devices with this structure can be electrically overstressed simply by applying normal electrical biases and then triggering the SCR structure by either applying a negative bias to an input or by applying a pulse to a sensitive pin. If the SCR structure is triggered, the device draws extremely high currents that can either overstress portions of the device's semiconductor die or melt open bondwires to the device. Since the curve tracer limits the amount of power available to the device, these parasitic SCR structures sometimes can be detected by electrically characterizing the device's inputs on the curve tracer. If I-V characteristics are detected that indicate a negative resistance characteristic (see Fig. 10.10), then parasitic SCR behavior is a potential problem that can be investigated.

A list of suggested readings appears at the end of this chapter. The book by Devaney et al. is a good source for additional information on curve tracer testing for failure analysis. The paper by Holmes is an excellent reference for interpreting curve tracer results when testing semiconductor devices.

10.6.4 Automated Test Equipment

Automated test equipment (ATE) is used to rapidly characterize electronic components. These test devices are computer-controlled combinations of power supplies, voltmeters, current meters, and pulse generators. The test equipment is controlled by a software program written to test each of the devices that are addressed by the ATE.

ATE is most often used to test ICs, due to their complexity. The ATE can rapidly test the functionality of an IC and measure all of its electrical parameters. The trade-off is that the program required to test the IC may require several days to write and debug. However, once the program is available, it only takes a couple of minutes to completely characterize a complex IC. Also, the variation of a device's electrical behavior with temperature can be determined by programming the ATE to control the temperature of the device using a hot/cold module.

ATE is extremely valuable for testing ICs for failure analysis. The electrical characteristics of even a very complex DUA can be rapidly determined. Also, the test program for the

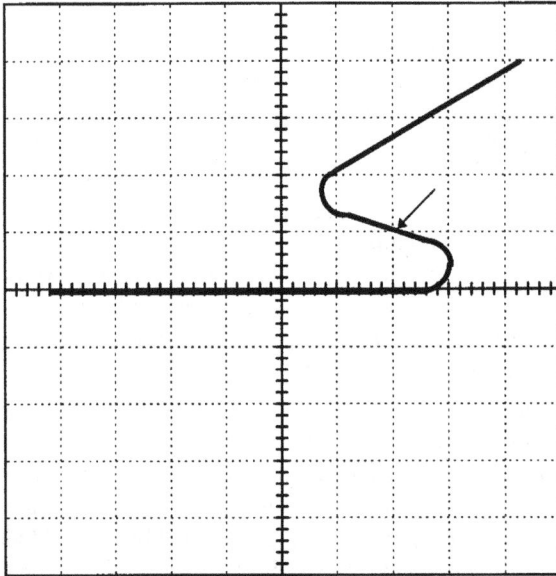

FIGURE 10.10   Curve tracer measurement indicating SCR-type behavior. The arrow on the display indicates an area of negative resistance in the electrical characteristic, which is typical of SCR-type behavior.

ATE can be modified to characterize the DUA while testing it for a variety of combinations of test conditions. These tests would require unacceptably long times to complete using discrete test equipment.

10.7   ELECTRICAL CHARACTERIZATION OF NONSEMICONDUCTOR DEVICES

Specific devices will, of course, require specific tests during failure analysis electrical testing. Some of the specific nonsemiconductor device types that will be briefly discussed in regard to failure analysis electrical test include

- PCBs and cables
- Connectors
- Resistors
- Capacitors
- Inductors
- Transformers
- Switches
- Relays

Some other device types are either specific subcategories of the above types or not included in the list. These may be devices used for very specialized applications or devices that require very specialized testing. In some cases, information for testing these other devices can be obtained by choosing a component from the above list that is closest to the device and applying the described approaches (with appropriate modifications).

The following test approaches are suggested, assuming that the background information or reported failure mode does not indicate a specific problem with the DUA. If information is available that indicates a specific problem, this information can be used to appropriately modify the following suggested approaches.

Prior to performing electrical tests on any device, make sure that the physical method of making contact to the device under test will not damage or obliterate any physical evidence that might be present on the device (see Sec. 10.11 for more details).

10.7.1 Printed Circuit Boards and Cables

Printed circuit boards (PCBs) and cables have very different physical appearances but provide the same basic electronic functions: electrical interconnection between various points or between various components. At the same time, they provide isolation between the various signals that are simultaneously transmitted within these devices.

To characterize these devices, the interconnection and isolation functions must be verified. This means that the conduction path between points that should be connected must be measured to ensure that it is intact and of the appropriate resistance. Typically, the resistance of the conductance paths is specified to be less than some maximum.

To verify the isolation function of these devices, the resistance between conduction paths must be verified to be greater than some specified minimum, or the leakage current must be determined to be less than some specified maximum at a particular voltage.

Both of these components types are susceptible to various failure mechanisms that could cause leakage paths between conductors. However, in many cases, these leakage paths may not support significant amounts of current, so the current must be limited to avoid "blowing out" any leakage paths between traces.

The suggested approach for failure analysis electrical test of PCBs and cables is to first characterize the isolation between conductors and then the continuity of each conductor using a curve tracer. (If background information is available to indicate a problem with a specific conductor, then the testing can be concentrated on that conductor.) The isolation should be performed initially using low voltages and with the instrument configured to detect very low currents. Then, the applied voltages should be increased to the maximum rated while limiting the current to very low values. If at any point, excessive current is detected, the then the power should be decreased to prevent damage to the leakage path.

Continuity should then be verified beginning with very low voltages while monitoring the current. The applied current should be increased to the maximum rated value while monitoring for any intermittence or instability in the resistance of the current path. If any anomalies are noted, the power should be decreased to avoid damaging the failure site.

If initial curve tracer tests verify the continuity of conduction paths and the isolation between them, then additional tests should be performed on the curve tracer to attempt to verify that there are no intermittent anomalies. The electrical tests should be repeated while applying thermal cycling and/or mechanical stresses to the DUA.

If the curve tracer tests indicate no problems, then the DUA's electrical behavior may have to be characterized using more sensitive test equipment to detect extremely low leakage currents or marginally high conductor resistance. (Note that four-point resistance measurements are usually necessary when accurately measuring conductor resistance due to the very low resistances in these devices—see Sec. 10.9.1.) While performing these tests, it is important to limit the current while applying high voltages and to limit the available power when applying high currents. These precautions will tend to limit the potential for changing or altering the original failure mechanism. In some cases, other tests may also have to be performed to measure specific attributes of the DUA based on information received with the DUA.

Once an anomaly has been verified, additional electrical tests can be performed (often in conjunction with other analysis techniques) to isolate the failure location. In the case of opens, probing is usually used to narrow down the location of the open in the PCB or cable. In the case of shorts or leakage paths, various techniques can be used to isolate portions of the conductor to locate the failure.

Resistive shorts might be located by applying power to the short and then using thermal imaging techniques to locate the failure. (Low-resistance shorts, or *dead* shorts, will not dissipate sufficient power. They will not be distinguished from the conductor, which will also heat up under high-current conditions.) The curve tracer is a good power source in cases such as this, since the I-V characteristic of the short can be monitored while power is applied. If the I-V characteristic starts to drift excessively, this is an indication that the short may be dissipating too much power. The power can then be reduced to prevent damage to the failure site.

10.7.2   Connectors

Failure analysis electrical test for connectors is very similar to that for PCBs and cables, because connectors provide a similar electrical function. Connectors vary from these other devices in that there are often problems associated with interfaces to the pins and/or sockets in the connectors. Therefore, care has to be taken in the manner that electrical connections are established with these devices.

Visual examination should be performed very carefully connector pins and sockets before attempting to make contact with them for electrical test. If this is not done, the failure mechanism could be altered or obliterated while trying to make electrical connections.

Connectors are sometimes also required to conduct very high currents or handle very high voltages without shorting. Therefore, they may require special test equipment capable of providing these higher currents or voltages.

Some connectors include special provisions for filtering the signals that they conduct. They may include built-in capacitors and inductors to provide this filtering function. Other connectors include diodes for blocking signals of the wrong polarity or for protecting against signals of too high a voltage. These specialized connectors should be electrically tested to include test procedures that would be used for the other components present in the connectors. For example, connectors with internal diodes should be tested as a combination of a connector and a diode to avoid damaging either device.

10.7.3   Resistors

Electrical testing for resistors is usually very simple. However, as in all other failure analysis electrical testing, care must be taken to avoid altering or obliterating the original failure mechanism. It is again recommended that these devices be initially characterized on the curve tracer at low power. If no intermittences or instabilities are noted, then they can be tested at higher power-up to the rated maximum power. Note that, at higher power, there will be some self-heating of the device, and the I-V characteristic may start to drift. This is normal. However, if the characteristic is drifting very rapidly, the power should be reduced to avoid damaging the device.

Additional tests should be performed on the curve tracer to attempt to verify that there are no intermittent anomalies. The electrical tests should be repeated while applying thermal cycling and/or mechanical stresses to the DUA.

The curve tracer can also be used to give an estimate of the linearity of the resistor. That is, the resistor should provide a straight I-V characteristic (neglecting heating effects) over all biases for which it is rated. If the I-V characteristic is not linear on the curve tracer display, this may be an indication of an internal interface problem in the device's construction.

Following the curve tracer evaluation, the device is then measured using typical resistance characterization equipment to accurately determine the resistance value. Also, the device can be characterized to determine if it is within specification limits with respect to its change in resistance with temperature.

Carbon composition resistors are subject to a failure mechanism not encountered in other resistor types. These resistors will absorb moisture when exposed to normal room atmospheric conditions. The resistance of carbon composition resistors can increase by as much as 12 percent (depending on the power rating and the nominal resistance value) due to moisture absorption. The devices can be restored to their nominal resistance values by baking them at high temperatures for a sufficient amount of time. This drives the absorbed moisture out.

Variable resistors (potentiometers) can also be evaluated using a similar approach. Curve tracer testing can be especially useful when testing these devices. If the curve tracer is used to monitor the variable resistance as the resistor is varied, any intermittence or intermittent open is immediately detected when the trace on the curve tracer jumps or appears erratic. (Similar tests using a digital meter may miss the intermittent behavior due to the time delay between changing digits. If the same test was performed with a meter having a galvanometer driven display, this meter may have too slow a response to detect dropouts in the resistance as it is varied.)

10.7.4 Capacitors

Capacitor electrical testing for failure analysis should also begin with curve tracer tests. Note the polarity of voltages being applied to polarized capacitors. Application of ac voltages or reverse voltages to polarized capacitors could result in electrical degradation and, in some cases, even overt physical damage to the device.

Even though a good capacitor will only generate a *balloon* on the curve tracer display (see Fig. 10.11), this balloon characteristic is an initial indication to the failure analyst that the DUA is not shorted or electrically open.

The curve tracer can also be used in the dc mode to measure the leakage current of the DUA and monitor this leakage current for intermittence or drift. As in all tests, low voltages should be applied initially, with voltage slowly increased with the current limited at all times.

After the initial curve tracer tests, the DUA can be thermally cycled or mechanically stressed to determine if there are any indications of intermittence or drift in its electrical behavior. Following curve tracer tests, the typical electrical parameters for the capacitor can be measured using standard test equipment. These parameters include capacitance, leakage current, dissipation factor (DF), and equivalent series resistance (ESR). These

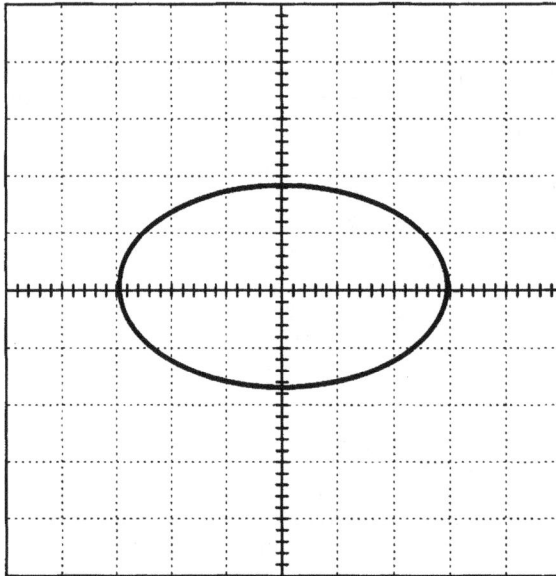

FIGURE 10.11 Curve tracer measurement results from a capacitor. If the capacitor were shorted or open, this characteristic would not appear.

parameters can then be characterized over temperature to determine if the device meets its electrical specification requirements.

Unencapsulated ceramic chip capacitors are somewhat fragile devices and are susceptible to cracking. Sometimes cracks induced in these devices are very fine and not easily found by visual inspection. If a ceramic chip capacitor exhibits slightly high leakage currents, it can be tested for fine cracks using the following procedure. The ceramic chip capacitor is dunked in isopropyl alcohol for about a minute. It is then removed from the alcohol and externally dried off using a warm heat gun. (Do not dry the device excessively.) The leakage current is then remeasured. If the device contains fine cracks, the isopropyl alcohol will penetrate the cracks and significantly increase the device's leakage current.

10.7.5 Inductors

An inductor can be simplistically modeled by a coil of wire wrapped around a core. (Some inductors, used mainly at high frequencies, are chip-type inductors using films on ceramic substrates. These are best tested as if they were resistors, since most failures are either shorts or opens rather than parametric failures.) A curve tracer is the starting point for failure analysis electrical testing of these devices also. Again, the curve tracer is used to verify that the coil of wire is continuous using low currents and voltages initially, then proceeding to maximum rated currents. As in the case of capacitors, the trace for an inductor will not be a line but rather a distorted loop. If the resulting trace is a straight line, then the inductor is either open or shorted. The characteristic trace can be monitored while thermally cycling the coil to test for intermittent opens or shorts in the coil winding.

Using the dc mode of the curve tracer, the analyst can determine if the coil resistance is approximately correct. As the voltage is increased while measuring the dc resistance of the coil, the resulting collection of points should create a straight line. If there are spikes superimposed on the straight line, this is an indication of possible breakdowns in the windings of the coil.

After the curve tracer has verified that the coil winding is intact and is of the approximate correct resistance, then the standard electrical test equipment can be used to verify the actual inductance of the coil. Also, the resistance of the coil can be accurately measured.

If the coil winding is the correct resistance, but the inductance is too low, this is an indication that the core of the inductor may be faulty or damaged.

10.7.6 Transformers

A transformer is basically two or more coils of wire wound on a common core. Curve testing of this device should be performed in the same manner as an inductor. In addition, the insulation resistance between coils should also be checked using the curve tracer. This test should be performed by starting with low voltages and then proceeding to the maximum rated voltage of the transformer, with the current limited at all times.

After the initial curve tracer tests have verified that the various transformer coils are intact and not shorted to one another (unless they are designed to be connected together), the coils can be monitored for intermittent shorts or opens while thermally cycling the transformer.

Once curve tracer testing has been completed, standard electrical tests can be performed to verify the transformer's correct operation. As in the case of the inductor, if the transformer's coils are not shorted and are of the proper resistance, but the transformer's output voltage is too low, the transformer's core has probably been damaged.

10.7.7 Switches

Electromechanical switches also should be initially characterized using a curve tracer. These devices are also subject to some failure mechanisms (such as internal chemical corrosion) that could be modified if excessive power was applied. Contacts that are expected to be open should be verified to be open, and the leakage current between them should be below the specification requirement. Contacts that are expected to be closed should be verified to be closed, and there should be no indications of intermittence or variation in the contact resistance. After the device has been switched, the expected open and closed contacts should be verified as in the preceding discussion.

After curve tracer testing has confirmed that the switch's basic functionality is confirmed, additional test should be performed using standard test equipment to verify that leakage currents between open contacts is below the specification requirements. Also, tests should be performed using four-point resistance measurements (see Sec. 10.9.1) to determine the resistance of the closed contacts.

10.7.8 Relays

Relays should also be initially characterized using a curve tracer. The curve tracer can be used to verify that the coils in the relay are intact and not open or shorted. Also, the curve tracer can be used to verify that there are no anomalous conductive paths between pins of the relay that should be isolated from each other.

Note that some relays have diodes in series with their coils to ensure that the correct polarity of voltage is applied to the coil. This diode will be evident in the curve tracer analysis of the coil. Other relays have diodes in parallel with the coil to suppress transients when the voltage is removed from the coil and it is deenergized. Again, this diode will be evident in the curve tracer analysis of the coil. In relays with diodes in series or parallel with the coil, the diode may the cause of the relay's failure. In some cases, destructive analysis of the relay will have to be performed to separate the diode from the coil to determine if the diode or the coil has failed.

Once the basic integrity of the relay's coils has been established, the device can be electrically characterized using standard relay tests. For failure analysis, the pickup (or actuation) voltage and the dropout (or deactuation voltage) voltage of the relay should be carefully tested several times. Large variations in these parameters indicate possible mechanical problems within the relays. Excessive switching times and large variations in switching times are also indications of possible mechanical problems in the relay.

Also, the contact resistances of the relay should be measured several times. If there are large variations in the contact resistance, this is a possible indication that contact surfaces are contaminated.

10.8   ELECTRICAL CHARACTERIZATION OF SEMICONDUCTOR DEVICES

This section will discuss failure analysis electrical characterization of semiconductor devices. The following semiconductor devices will be specifically addressed:

- Diodes
- Transistors
- Integrated circuits
- Hybrids

Semiconductor devices are continuously being improved and made more complex. Failure analysis techniques for the analysis and characterization of these devices are constantly being generated and improved. In addition, equipment for the analysis of these devices is constantly being invented and also improved.

Since so much current literature is being generated that is devoted to techniques for analyzing and electrically characterizing these devices, this chapter will provide only an overview of the general failure analysis electrical characterization techniques used for these devices. Two additional sources for information on specific techniques for characterizing these devices are the yearly *Proceedings of the International Reliability Physics Symposium (IRPS)* and *Proceedings of the International Symposium for Testing and Failure Analysis (ISTFA)*. These references also include papers on electrical characterization of nonsemiconductor devices, but the vast majority of the papers deal with semiconductor device analysis and characterization.

10.8.1   Semiconductor Devices

Semiconductors are considered by many to be the more "modern" of the electronic devices. These devices continue to be studied and characterized in an attempt to make them even more reliable and robust.

However, these devices are subject to failure mechanisms not encountered in other electronic devices. This is due to the fact that many of these devices utilize extremely small structures or layers in their construction. The small dimensions that are encountered give rise to new and unique problems. Some of the unique failure mechanisms encountered in semiconductor devices are briefly described below. Consult additional reference materials for additional information on detecting and characterizing these mechanisms.

10.8.1.1 Semiconductor Failure Mechanisms

1. *Time-dependent dielectric breakdown (TDDB).* Most semiconductors use very thin layers of dielectric (glass) to provide insulation between other layers of the semiconductor. The fields across these dielectric layers can be several million volts per centimeter. Depending on the type of glass, the included impurities in the glass and other factors, the dielectric may fail due to this mechanism. Typically, this mechanism is not a concern to failure analysts since devices in the field have been designed to mitigate this mechanism.

2. *Electromigration.* Semiconductors use small thin traces of metal (most often aluminum with traces of other elements) to conduct current. Due to the very small size of these conductors, the current densities can be very high. Devices are designed to limit the current density to a maximum of about 5×10^5 A/cm^2. At this current density, the effects of electromigration are negligible. However, at higher current densities, the momentum of the electrons moves the atoms of the conductor and can cause voids in the conductor, which eventually can lead to open circuits. Unintended flaws introduced during device processing can cause electromigration.

3. *Hot carrier injection.* Due to the small structures involved in semiconductors, high electric fields are often generated. These electric fields can impart high energies to electrons (or holes) in these devices. These very energetic (or *hot*) carriers can be injected into the glass dielectrics used in these devices. The charges generated by these trapped carriers can degrade the electrical parameters of the devices to the point where they fail.

4. *Ionic contamination-induced inversion.* This mechanism is often encountered in semiconductor devices. There are specific electrical characterization approaches for detecting this mechanism in semiconductors. This mechanism is detailed in the next section.

10.8.1.2 Ionic Contamination-Induced Inversion.   Semiconductor devices are subject to a particular failure mechanism that can result in significant degradation of a device's electrical performance at one point in time but have no effect at another. This failure mechanism is called *ionic contamination-induced inversion.*

If ionic contamination is present in the glass on the surface of a semiconductor die, it can cause degradation of the device's electrical performance if present in sufficient quantity in the device's glass. Even if the ionic contamination is not initially present in sufficient quantity, electrical biases in the device can cause the ionic contamination to move and concentrate in vulnerable areas.

The most common contaminant causing this problem is sodium, which, when ionized, has a positive charge. (There are other positively and negatively charged ions that can cause problems, but this discussion will be limited to sodium.) Sodium can be introduced into the surface passivation of a device during processing of the semiconductor. Other contaminants are capable of causing problems, but sodium is especially insidious because it can diffuse through the silicon dioxide glass that is used as a passivation layer on the surface of semiconductors. (Sodium also diffuses through silicon nitride, which is also used as a passivation layer, but at a slower rate than through silicon dioxide.) Manufacturers are well aware of this problem, but since it sometimes requires less than 0.1 ppm of sodium in the glass to cause a problem, devices still occasionally fail due to this mechanism.

Figure 10.12 is a diagram of the cross section of a portion of a p-n junction. In this diagram, sodium is shown randomly dispersed in the glass over the junction.

If the previous junction is reverse biased, as shown in Figure 10.13, there will be fields at the surface of the junction just below the glass layer. These fields result from the depletion layer induced in the junction by the reverse bias. The depletion region on the p-type side of the junction will be biased negatively with respect to the n-type side of the junction. The negative bias on the depletion layer in the p-type silicon will attract the positively charged sodium ions. The ions will diffuse through the glass layer over the p-n junction and accumulate at the area above the p-type silicon adjacent to the junction as illustrated in Figure 10.13.

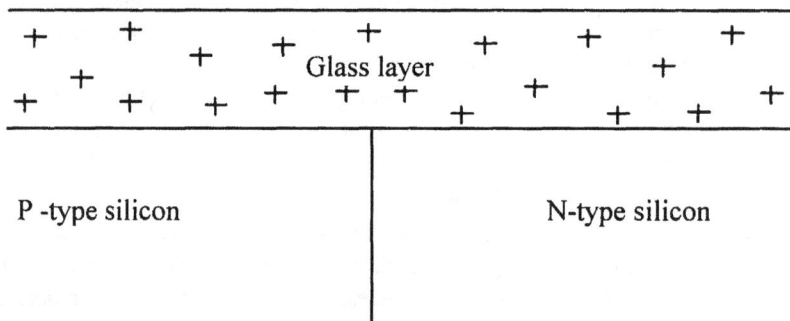

FIGURE 10.12 Diagram of a cross section of a portion of a p-n junction in a silicon semiconductor device. The + symbols in the glass layer on the surface represent randomly distributed sodium ions in the glass layer.

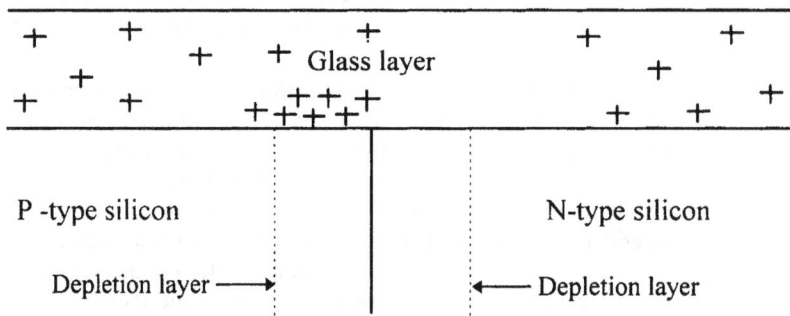

FIGURE 10.13 With the p-n junction reverse biased, a depletion layer forms across the junction. The depletion layer in the p-type silicon will be biased negatively with respect to the bias on the depletion layer in the n-type silicon. This will cause the sodium ions to diffuse toward the p-type silicon and away from the n-type silicon.

The accumulated positive charge above the p-type silicon attracts negative carriers (minority carriers) in the p-type silicon just below the accumulation of sodium ions. If enough minority carriers are attracted, the silicon in this area acts like n-type silicon due to the presence of so many negative carriers. Thus, the silicon in this area can become "inverted" from the expected p-type silicon to n-type silicon. The area of inverted silicon is referred to as the inversion region (see Fig. 10.14).

The inversion region generated by the sodium contamination will significantly degrade the electrical characteristics of the original p-n junction. This area acts as an extension of the n-type silicon that is intentionally formed on the n-type side of the junction. However, this portion of the junction is obviously not part of the carefully processed original p-n junction. The electrical characteristics of this portion of the junction will generate a large leakage current when the junction is reverse biased. Even the forward characteristic of this junction can be degraded due to the inversion layer.

Ionic contamination can degrade semiconductor structures other than p-n junctions. Figure 10.15 is a diagram of a portion of a metal-oxide semiconductor field-effect transistor (MOSFET). The MOS transistor functions on the basis of intentionally creating an inversion layer. When a voltage is applied to the metallization (in Fig. 10.15) that is positive (with respect to the bias on the p-type silicon), minority carriers (negative charges) in

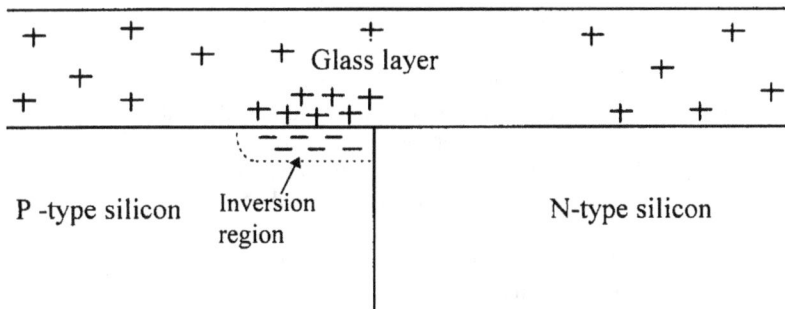

FIGURE 10.14 After the reverse bias is removed, there is an accumulation of sodium ions in the glass over the p-type material. This attracts negative minority carriers in the surface of the p-type material. If sufficient minority carriers are attracted, the material in this area may be inverted and act as n-type material with no bias on the gate metallization. (Thus, the FET is anomalously *on* with no gate bias.)

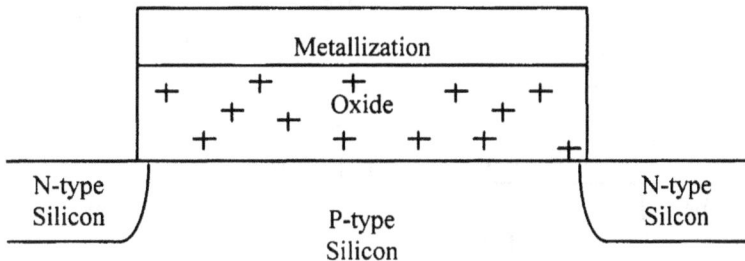

FIGURE 10.15 Diagram of a cross section of a portion of a MOS gate region on a silicon semiconductor device. The + symbols in the oxide layer represent randomly distributed sodium ions in the oxide layer.

the p-type silicon are attracted to the surface of the silicon just below the oxide beneath the metallization. If the positive voltage on the metallization is large, there are enough negative carriers attracted to invert the p-type silicon just below the oxide and cause it to act like n-type silicon. This n-type inversion layer will connect the previously separated n-type areas at the edges of the oxide layer. If a voltage potential exists between the separated n-type areas, the linking inversion layer will result in current flow between the two areas. (Prior to the presence of the inversion layer, the reverse bias from the n-type areas to the p-type area prevented current flow.)

The preceding discussion is the basis for the operation of a MOS transistor. The metallization is referred to as the gate, the oxide is called the gate dielectric, the n-type areas are the source and drain, and the inverted layer that is formed is the channel. When the gate voltage is sufficient to cause the inversion layer resulting in conduction between the n-type areas, the MOS transistor is "on." Without the inversion region, the transistor is off. In this example, an NMOS transistor has been described, since the channel needed for the transistor to be *on* is a n-type channel. (A PMOS transistor is described by reversing the n-type materials to p-type, the p-type materials to n-type, and the changing the *on* gate voltage to negative.)

If sodium contamination is present (as indicated by the + symbols in Fig. 10.15), positive voltage on the metallization will drive the sodium ions away from the metallization toward the surface of the p-type silicon, as shown in Figure 10.16. If sufficient sodium ions are accumulated near the surface of the p-type silicon, an inversion layer could be generated without any bias on the metallization. Thus, current could flow from one n-type region to the other with no voltage applied to the metallization and the transistor would be *on* with no bias applied to the (gate) metallization as illustrated in Fig. 10.17. (For a PMOS transistor, the positive sodium ions would require a more negative voltage than normal on the gate metallization to cause the transistor to be "on.")

In summary, sodium ion inversion may degrade NMOS transistors to the point at which they will be anomalously *on* with no potential applied to the gate. (PMOS transistors will require more gate voltage than normal to turn "on.")

In both of the previous types of ionic contamination-induced inversion (p-n junction inversion and MOS transistor degradation), the sodium was affected by a bias that caused the sodium that diffused through the silicon dioxide to accumulate in a specific area. In both cases, the effect of the ionic contamination will decrease with time if no bias is applied to maintain the accumulation of ionic contamination. Without a bias, the ionic

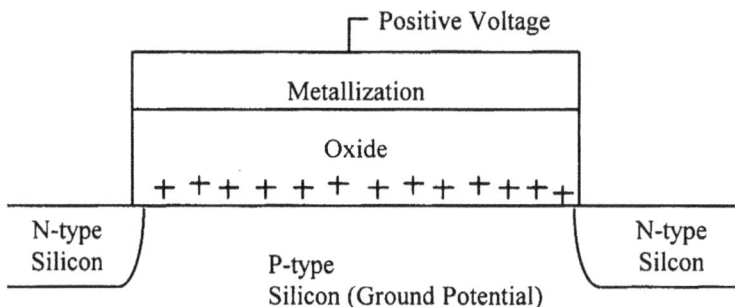

FIGURE 10.16   With a positive bias on the metallization with respect to the p-type silicon, the sodium ions are repelled from the metallization and diffuse toward the silicon.

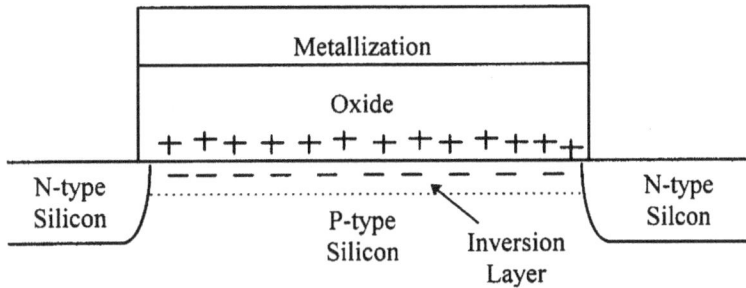

FIGURE 10.17 With a positive bias on the metallization with respect to the p-type silicon, the sodium ions are repelled from the metallization and diffuse toward the silicon.

contamination will tend to randomly diffuse throughout the glass, and its effect will diminish as it moves away from the silicon surface.

The rate of the diffusion is highly dependent on temperature. Raising the temperature of the device will increase the rate of diffusion. If the temperature is raised while the reverse bias is applied to the p-n junction, the sodium will cause the inversion layer faster than at reduced temperatures. This is the basis for high-temperature reverse bias (HTRB) testing used to test for inversion problems. Also, with no bias applied, the accumulation of ionic contamination will disperse faster at higher temperatures. Therefore, both types of ionic contamination induced inversion will be "cured" faster at higher temperatures.

At the beginning of this section, it was noted that ionic contamination-induced inversion may be present and cause a failure in a device. If the unbiased device is heated for a short time or simply sits at room temperature for a longer period of time, the ionic contamination may randomly diffuse, and the failure may disappear. The device will continue to function normally until certain biases are applied for the critical time/temperature combination to reinduce the inversion.

For failure analysis electrical testing, to detect the presence of ionic contamination, HTRB should be performed on p-n junctions. For MOS structures, to detect the presence of positive ionic contamination, PMOS transistors should be biased with a positive voltage on the gate. (Note that the degradation on the p-n junction will be exhibited in the form of increased leakage current with reverse bias applied. However, the degradation in the MOS transistor will be not be noted while the inversion inducing voltage is applied to the gate metallization.)

10.8.1.3 P-N Junction Breakdown Voltage vs. High Leakage Current.

All semiconductor devices include p-n junctions in their circuitry. When characterizing a p-n junction, two parameters are measured to characterize the reverse characteristics of the p-n junction: *reverse leakage current* and *reverse breakdown voltage*. For the purposes of failure analysis electrical characterization, the reverse avalanche breakdown should be measured rather than the reverse breakdown voltage as measured per the device specification.

Reverse leakage current is measured by applying a reverse voltage to the junction that is less than the reverse breakdown voltage of the junction, and then measuring the amount of current that flows. The reverse breakdown voltage of a p-n junction is measured per the specification by increasing the reverse voltage until a specified amount of current flows. The voltage required to obtain the specified current flow is then (by the device specification) the breakdown voltage of the junction.

However, for failure analysis electrical characterization, it is important to determine the actual reverse avalanche breakdown voltage of the junction. This may be different from the breakdown voltage measured according to the specification if the p-n junction is very leaky. Figure 10.18 is a curve tracer display diagram that illustrates the problem. Trace number 1 in Fig. 10.18 shows a "good" p-n junction. Assume that the specification for this device requires the breakdown voltage to be measured at a reverse current of 1 μA. The "good" p-n junction shows a reverse breakdown of about 20 V, per the specification requirements. In this case, the reverse avalanche breakdown voltage is also 20 V.

Trace 2 in Fig. 10.18 shows a p-n junction with very high reverse leakage current. Note that the leakage current causes the reverse breakdown voltage per the specification measurement requirements to be only about 16 V, but the actual avalanche breakdown voltage is still at 20 V.

Trace 3 in Figure 10.18 illustrates another p-n junction that has been severely overstressed and now has an actual decrease in the avalanche breakdown voltage to only about 8 V. (The breakdown voltage per the specification measurement requirements would also be 8 V.)

For the purposes of failure analysis electrical characterization, it is important to distinguish between high reverse leakage current and degraded avalanche breakdown. Different failure mechanisms can be implied, depending on whether the change (or the degradation) is associated with the leakage current or the breakdown voltage.

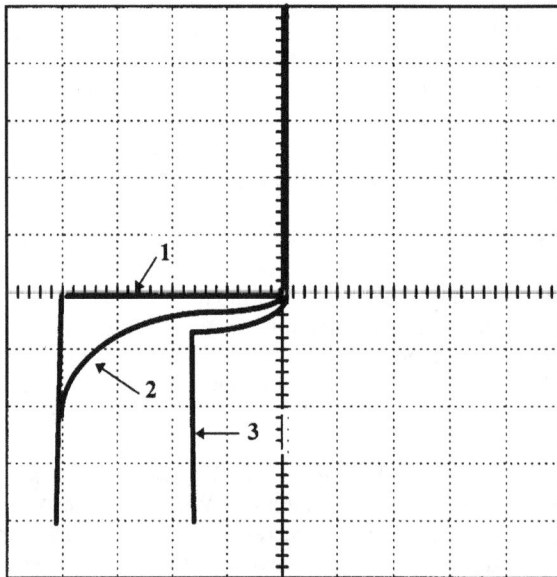

FIGURE 10.18 Diagram of a curve tracer display of a p-n junction's I-V characteristic. (Horizontal = 5 V/division; vertical = 1 μA/division). Curve trace 1 illustrates a "good" p-n junction with a breakdown voltage of about 20 V at a reverse current of 1 μA. (The avalanche breakdown voltage is also 20 V.) Curve trace 2 illustrates the same p-n junction with high reverse leakage current. The specification breakdown voltage at 1 μA is now only about 16 V, but the avalanche breakdown voltage is still 20 V. Trace 3 illustrates a severely degraded junction where the avalanche breakdown voltage has decreased to about 8 V.

10.8.2   Diodes

Diodes can be almost completely electrically characterized using a curve tracer. As in all devices, curve tracer testing of diodes starts using low voltages and currents and then proceeds to higher currents.

The basic parameters to be measured for a diode include the forward voltage drop at various specified currents, the reverse leakage current at a specified reverse voltage, and the reverse avalanche breakdown voltage (see Sec. 10.8.1.3).

If these electrical parameters for the DUA are within specification limits, then the device should be temperature cycled to ensure that it does not exhibit any intermittent electrical behavior. If the DUA still exhibits no anomalous electrical behavior, then it could be subjected to HTRB testing to determine if it is susceptible to ionic contamination induced inversion (see Sec. 10.8.1.2).

If the DUA still exhibits no electrical anomalies, then it should be characterized to determine if it exhibits sufficient capability to transmit any dissipated power to its case or leads. This is done by measuring either the θ_{JC} (thermal impedance from junction to case) or θ_{JL} (thermal impedance from junction to leads) of the device. If these parameters are too high, the device may be failing due to excessive heating of the junction when it dissipates power.

10.8.3   Transistors

Transistors can also be almost completely electrically characterized using a curve tracer. The previous discussion of the curve tracer (Sec. 10.6.2) was a basic introduction to two-terminal device characterization. The curve tracer actually has a third connection so that transistors can be completely characterized. However, failure analysis electrical characterization should be initiated by testing pairs of terminals.

It should be noted that there are two basic types of transistor: the bipolar junction transistor (BJT) and the field-effect transistor (FET). BJTs include NPN and PNP transistors. FETs include two subcategories of devices including the junction FET (JFET) and the metal-oxide-semiconductor FET (MOSFET). All of these transistor types can be electrically characterized using a curve tracer.

Numerous failure mechanisms are associated with these devices, and each has its own combination of *signatures* during curve tracing. The paper by Holmes (see the suggested readings at the end of this chapter) is an excellent source of information for the analysis of curve tracer results from various semiconductor devices.

Once the basic parameters of the DUA have been tested and shown to be acceptable, the DUA should be temperature cycled to determine if it exhibits any intermittent electrical behavior. Tests should then be performed on the DUA to determine if it is susceptible to ionic contamination induced inversion (see Sec. 10.8.1.2).

As in the case of the diode analysis, if the DUA still exhibits no electrical anomalies, it should be characterized to determine if it exhibits sufficient capability to transmit any dissipated power to its case or leads. This is done by measuring either the θ_{JC} or θ_{JL} of the device. If these parameters are too high, the device may be failing due to excessive heating of the junction when it dissipates power.

10.8.4   Integrated Circuits

Failure analysis electrical characterization on ICs also should be begun by characterizing the device using a curve tracer. Some people will be surprised by the previous statement, since ICs obviously have more than three leads, which is the maximum number of connections on the curve tracer. However, much information can be obtained from ICs by using the curve tracer to characterize the I-V characteristics of two leads at a time.

Typically, the approach used for an IC is to characterize the I-V characteristics of all of the input pins with respect to each of the power supply connections and also with respect to the ground connection. The same is done for the output pins with respect to the same pins. In addition, low-power curve tracer measurements should be performed on adjacent pairs of pins to check for anomalous leakage paths caused by either internal or external contamination of the insulation between pins.

As in all initial curve tracer evaluations of a DUA, low voltages are applied first, and then higher voltages if no anomalies are noted. The curve tracer characteristic obtained for an input pin with respect to a power supply pin can be evaluated in one of two ways. If the complete schematic for the IC is available, the curve tracer characteristic can be evaluated to determine if it is consistent with the schematic. (Sometimes the resulting curve tracer results do not seem to be compatible with the schematic for the device. There often are input protection devices that are not shown on the schematic, and these structures can cause the curve trace results to be different from what is expected.) If the complete schematic is not available, then the curve trace result for one input can be compared to the curve trace result for another similar type of input to see if they are similar. Output pin I-V characteristics from the curve tracer are analyzed in the same manner.

Another approach to evaluating curve tracer results from pairs of IC pins is to compare the results from the DUA to the results from a known-good device. In this way, direct comparisons can be made from the suspect device to a known-good result.

There are many advantages of initially electrically characterizing the inputs and outputs of an IC on the curve tracer, as enumerated below.

1. The main advantage is that degraded or shorted inputs or outputs can be detected. If degraded circuitry is detected on these pins, they should not be subjected to the stresses that would be applied during typical ATE testing. Typical ATE testing applies full rated power to the inputs and forces the outputs to deliver full rated power. This could result in additional damage to circuitry that has already been damaged causing the original failure mechanism to be changed or obliterated.

2. Curve tracer testing also detects open input or output pins, which would certainly cause the DUA to fail on the ATE.

3. Obvious anomalies at the inputs or outputs of the DUA will cause no ATE testing to be required. This could save the cost and time of both writing a test program for this device and possibly building a test fixture compatible with the ATE.

If no obvious anomalies are detected during curve tracer testing, then the DUA is tested using ATE that can fully characterize the device over a large range of test variables (see Sec. 10.6.4).

ICs are also susceptible to ionic contamination-induced inversion problems. These devices also can be tested for this failure mechanism (see Sec. 10.8.1.2), but the biasing schemes for detecting this mechanism in these complex devices can be correspondingly complex and may require more than one set of test conditions.

10.8.5 Hybrid Microelectronics

Hybrid microelectronics can include semiconductors of all types plus other electronic devices including resistors, capacitors, and inductors. However, the initial approach to failure analysis electrical characterization is the same as for ICs.

Using the curve tracer, characterize the I-V characteristics of all of the input pins with respect to each of the power supply connections and also with respect to the ground connection. Subsequently, the same is done for the output pins with respect to the same pins. In addition, low-power curve tracer measurements should be performed on adjacent pairs of pins to check for anomalous leakage paths caused by either internal or external contamination of the insulation between pins.

As in all initial curve tracer evaluations of a DUA, low voltages are applied first, followed by higher voltages if no anomalies are noted. The curve tracer characteristic obtained for an input pin with respect to a power supply pin can be evaluated in one of two ways. If the complete schematic for the hybrid is available, the curve tracer characteristic can be evaluated to determine if it is consistent with the schematic. If the complete schematic is not available, then the curve trace result for an input on the DUA can be compared to the curve trace results for the same input on another known-good hybrid. (Since hybrids are such complex devices, different inputs on the same device cannot be compared, since it is possible that they will be different). Output pin I-V characteristics from the curve tracer are analyzed in the same manner.

If the DUA passes the initial curve tracer characterization, then the DUA is tested using ATE to fully characterize its electrical behavior.

Since hybrids are such complex devices with such a mix of technologies, the next step in the analysis would have to be determined based on a case-by-case evaluation. For these devices, reported failure modes and background information are critical in determining the best approach for failure analysis electrical characterization.

10.9   SPECIAL TEST METHODS FOR FAILURE ANALYSIS ELECTRICAL CHARACTERIZATION

Some test methods are especially useful for failure analysis electrical characterization. This section includes two test procedures that can be applied to many devices, and two other test methods that are limited in their application to one semiconductor technology type.

10.9.1   Four-Point Probe Measurements

When measuring a large resistance (greater than $100\ \Omega$), the resistance of the contacts or the probes used for the measurement are often neglected. However, when measuring smaller resistances, the contact resistance of the measuring device can be larger than the resistance that is being measured or add significant resistance to the measurement. Therefore, when small resistances are measured, four-point resistance measurements are used.

In Fig. 10.19, suppose R_x is the unknown resistance. R_{m1} and R_{m2} are the resistances of the probes connected to points B and C to measure R_x. (These resistances include both the internal resistances of the probes plus the contact resistance between the probes and R_x.) As already stated, R_{m1} and R_{m2} are included in the measurement of R_x for large values of R_x since these other resistances are typically small (on the order of less than $1\ \Omega$). When R_x is measured, current is forced from point A to point D through the probes and the voltage drop is measured from A to D. The resulting resistance is then determined.

In the four-point resistance measurement, two additional probes are connected to R_x at points B and C. These additional probes will have resistances R_{m3} and R_{m4} as also shown

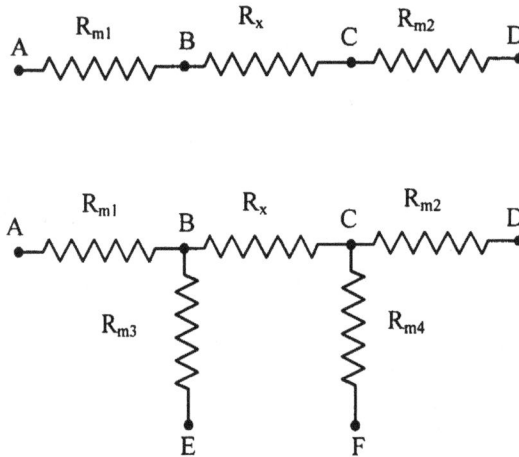

FIGURE 10.19   The upper circuit diagram represents a typical resistance measurement. R_x represents the unknown resistance being measured. R_{m1} and R_{m2} represent the resistances of the leads and probes attached to points B and C to perform the measurement. The lower circuit diagram represents the four-point probe approach for measuring an unknown resistance. R_{m3} and R_{m4} represent the resistances of two additional leads connected to points B and C.

in Fig. 10.19. Again, a current is forced from A to D, but the voltage is measured at E and F using a voltmeter. (Since the impedance of the voltmeter is typically a megohm or greater, the current through the voltmeter path is negligible and the voltage between E and F is virtually the same as the voltage between B and C.) Thus, the four-point resistance measurement technique eliminates the resistances that are introduced by the contact probes, and small resistances are measured accurately.

This technique is also used when measuring voltage drops across semiconductor junctions when very high currents are applied. For example, suppose the forward voltage drop of a power diode is rated at 1.10 V maximum with 1 A of forward current. If clip leads were used to connect the power supply to the diode's leads, the voltage drop measured at the power supply would include the voltage drops through the leads. However, if a second set of clip leads is connected to the diode's leads and used to measure the voltage drop across the diode, then the true diode forward voltage drop would be measured.

10.9.2   "Cold Starts" for Detecting Moisture

Many devices have an internal sealed cavity. Most of these types of devices are semiconductors that are very susceptible to any moisture that may be in the cavity atmosphere. The moisture can be most deleterious when the temperature of the package is below the dew point of the water vapor in the package. Below the dew point, water vapor can condense on the surface of the semiconductor die that is in the package and cause leakage currents or corrosion of the materials on the die.

When such packages are sealed, attempts are made to seal them with an atmosphere containing <5000 ppm of water. The dew point for this water vapor concentration is below $0° C$. Therefore, if the water vapor condenses on the die below $0° C$, it instantly freezes and causes no problems.

However, some packages are either sealed with more than 5000 ppm water in their atmosphere, or the packages are damaged and additional water vapor leaks into the package. However, sometimes this excess water vapor is not detected during normal electrical testing. This is because devices often are tested at room temperature first and then tested at cold temperatures. The initial testing at room temperature can cause the semiconductor die in the package to be warmer than the rest of the interior of the package. When the device is cooled, the water vapor condenses on the cold package but not on the warmer die.

This can be avoided by performing a "cold start." The device is cooled down to below $0°$ C with no power applied. Water vapor condenses on the package interior and the die. Then power is applied, and the package is warmed to room temperature. If moisture-related problems are causing electrical degradation, this sequence of testing is more likely to detect the problem.

10.9.3 Threshold Voltage Measurements on CMOS ICs

Threshold voltage measurements on MOSFETs can provide information on the amount of ionic contamination present in the oxide of these devices, especially if measurements are performed before and after performing biased tests at high temperature (see Sec. 10.8.1.2 for a discussion of ionic contamination effects on MOS devices).

However, many people are not aware that threshold voltage measurements can sometimes be performed on individual MOSFETs on each of the inputs of a CMOS (complementary MOS) IC. Figure 10.20 shows the typical input circuitry on a CMOS IC. Note

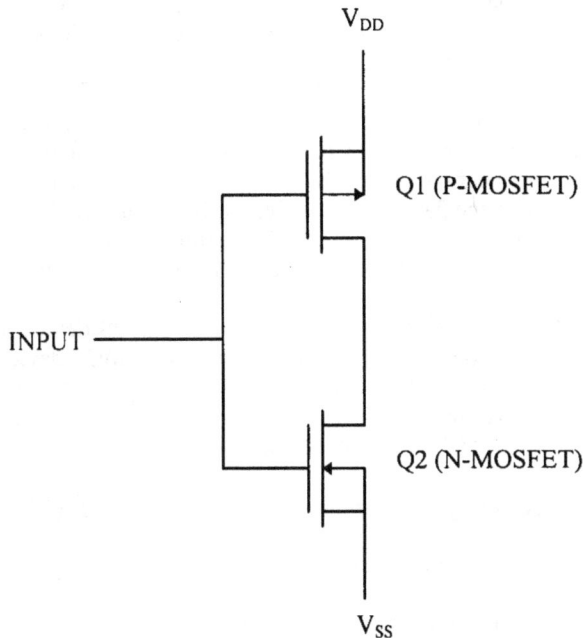

FIGURE 10.20   Typical CMOS inverter circuit.

that there is an N-MOSFET and a P-MOSFET connected together to form the input buffer circuit. In the normal operation of a CMOS device, one of these devices will always be in the *off* condition while the other one is "on."

To measure the threshold voltage of one of the two series devices, the other device must be *on* while the device that is being measured is being turned "on." Current will flow through the *on* device and the device that is just at its threshold voltage where it starts to conduct (or turn *on*) will determine the current that flows. This is accomplished by biasing the IC in a manner that is not typical for normal operation, but is not detrimental to the device as long as the test currents are limited.

To measure the threshold voltage of the N-MOSFET (Q2 in Fig. 10.20), the following procedure is followed. V_{dd} is biased with +5 V while the input is grounded. Then the V_{ss} terminal is initially biased at ground. This turns on the P-MOSFET (Q1 in Fig. 10.20) in series with the N-MOSFET.

Then the voltage on the V_{ss} terminal is slowly decreased to a negative voltage until the N-MOSFET turns *on* and a predetermined amount of current flows from V_{dd} to V_{ss}. The voltage at which this current flows is the threshold voltage for the N-MOSFET.

The threshold voltage for the P-MOSFET is similarly determined by biasing the V_{ss} terminal with –5 V with the input grounded. V_{dd} is then biased at ground and slowly raised in voltage until current flows at the P-MOSFET threshold voltage.

10.9.4 Power Supply Current Testing for CMOS ICs

CMOS ICs draw little current (compared to the similar logic function implemented using bipolar technology), since there is always an *off* transistor between the V_{dd} power supply and the V_{ss} power supply that blocks any substantial current flow. However, when there is a flaw in a CMOS IC, the blocking transistor conducts current through the flaw, or the transistor is not turned *off* as a result of the flaw.

By monitoring the power supply current, I_{dd}, the location of a flaw in a failed CMOS IC can often be detected by an increase in the power supply current (I_{ddq}) that results when the logic in the IC appropriately biases the flaw to cause current flow. Using a cleverly designed logic string, the location of the flaw can be inferred by the location of the logic sequence needed to activate the flaw. For example, if a 16-bit shift register contains a flaw, a 1 can be shifted through the shift register, one bit at a time. If the current for the shift register suddenly increases after the 12th clock pulse, the flaw in the circuitry is located in the 12th shift register in the circuit.

Figure 10.21 illustrates the basic circuitry in a simple CMOS device. In this case, an inverter is shown connected to a second inverter. (Obviously, CMOS devices are not composed entirely of inverters, but the following description of flaw detection holds for more complex logic functions). Various defects that might be present in the circuitry associated with the first inverter (Q1 and Q2) are listed as follows:

- Defect 1: Short between the gate and drain in the PMOS transistor (Q1)

- Defect 2: Short between the gate and source in the PMOS transistor

- Defect 3: Open gate connection to the PMOS transistor gate

- Defect 4: Short between the gate and drain in the NMOS transistor (Q2)

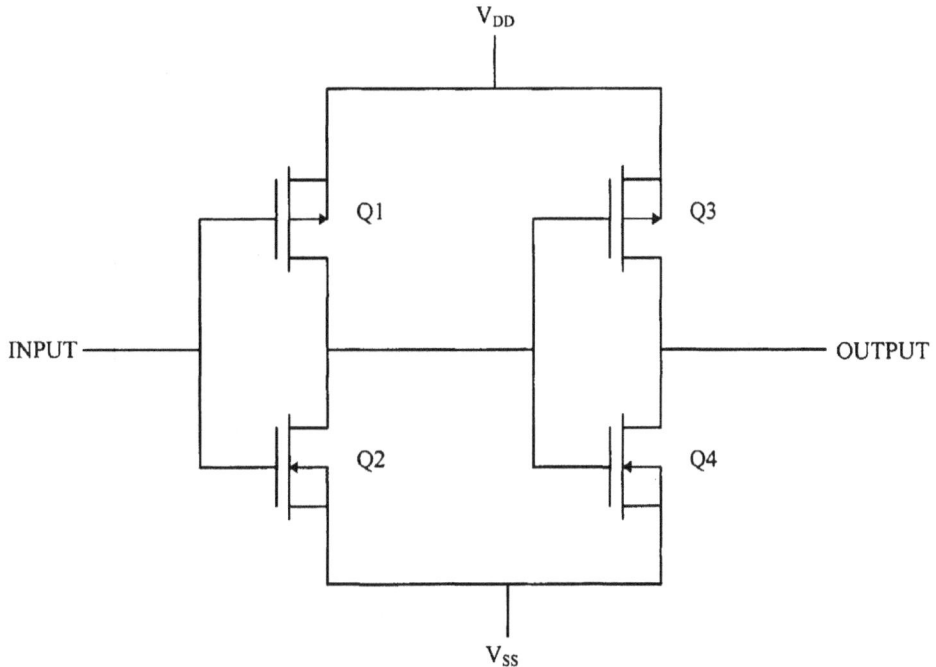

FIGURE 10.21   Simple CMOS circuit made up of two inverters in series. Q1 and Q2 form the first inverter, whose output is tied to the input of the second inverter formed by Q3 and Q4.

- Defect 5: Short between the gate and source in the NMOS transistor

- Defect 6: Open connection to the NMOS transistor gate

   When the input to the first inverter is a logic "0" (or a voltage close to V_{ss}), defects 1 and 2 would be detected by excess current flow from V_{dd} through the flaws to the input (at bias V_{ss}). The current paths are shown as I_1 and I_2 in Fig. 10.22. Defect 6 could also be detected, since the gate of transistor Q2 would be floating. This transistor may not be turned *off* as it should be if the gate were intact. With defect 6, the current could flow directly from V_{dd} through *on* transistor Q1 then through Q2 to V_{ss}.

   When the input to the first inverter is a logic 1 (or a voltage close to V_{dd}), defects 4 and 5 would be detected by excess current flow from V_{ss} through the flaws to the input (at bias V_{dd}). The current paths are shown as I_3 and I_4 in Fig. 10.22. Defect 3 could also be detected, since the gate of transistor Q1 would be floating. This transistor may not be turned *off* as it should be if the gate were intact. With defect 3, the current could flow directly from V_{dd} through transistor Q1 then through *on* transistor Q2 to V_{ss}.

   The same analysis applies to these defects in transistors Q3 and Q4, except the current would flow through the defective device and then through Q1 or Q2, whichever device is *on,* to provide the logic 1 or 0 that was the input for the first inverter. The current paths for the flaws in the second inverter are illustrated as I_1 and I_2 for defects 1 and 2; and I_3 and I_4 for defects 4 and 5 in Fig. 10.23. Defects 3 and 6 would cause current to flow through Q3 and Q4 from V_{dd} to V_{ss}.

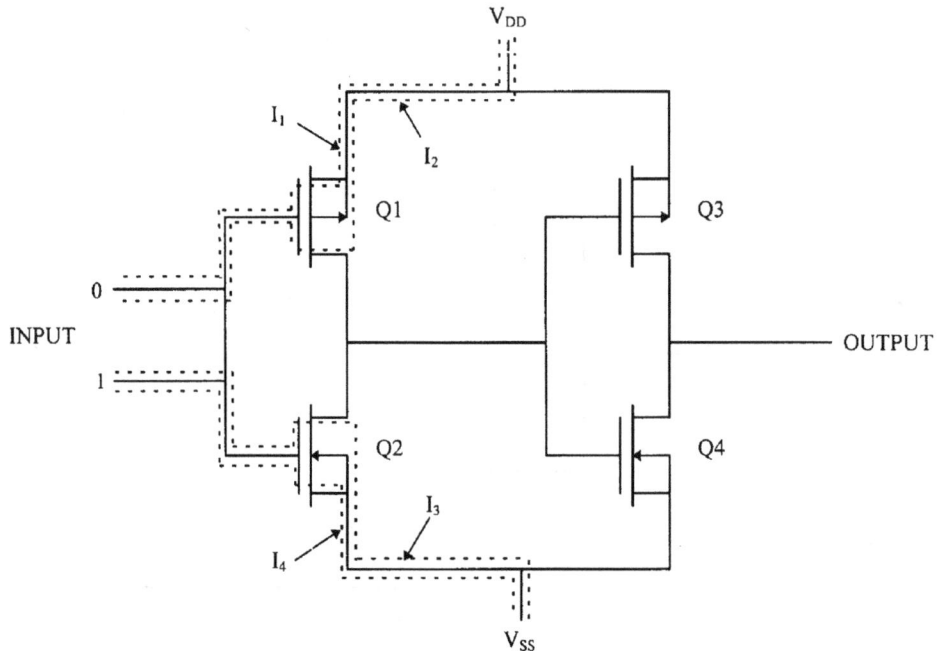

FIGURE 10.22   Circuit shown in Fig. 10.21 with current paths shown for various defects in the first inverter. (Two inputs are shown to separate the current paths for a 0 on the input and a 1 on the input.)

10.10 PRECAUTIONS FOR FAILURE ANALYSIS ELECTRICAL CHARACTERIZATION

When performing electrical characterization for failure analysis, there are many problems that the analyst should be aware of and precautions to avoid these problems.

1. *False opens.* Verify that electrical opens are not electrical connection problems between the DUA and the test equipment. Sometimes, when devices are installed on circuit boards, the entire board may be coated with a thin layer of conformal coating. After the device is removed from the board, the conformal coating may still be coating the leads of the device. This insulating layer may cause problems in establishing electrical connection to the leads of the device.

    Other causes for opens in electrical connections to the DUA are broken conductors in test leads, broken pins in sockets, DUA leads that are too short to make contact with test socket connectors, or worn springs in test lead clips. If opens are detected in the DUA, confirm these opens by appropriate additional measurements to verify the test setup.

2. *False indications during temperature testing.* Often, the DUA must be tested while its temperature is varied. Regardless of the method used to cause the temperature change

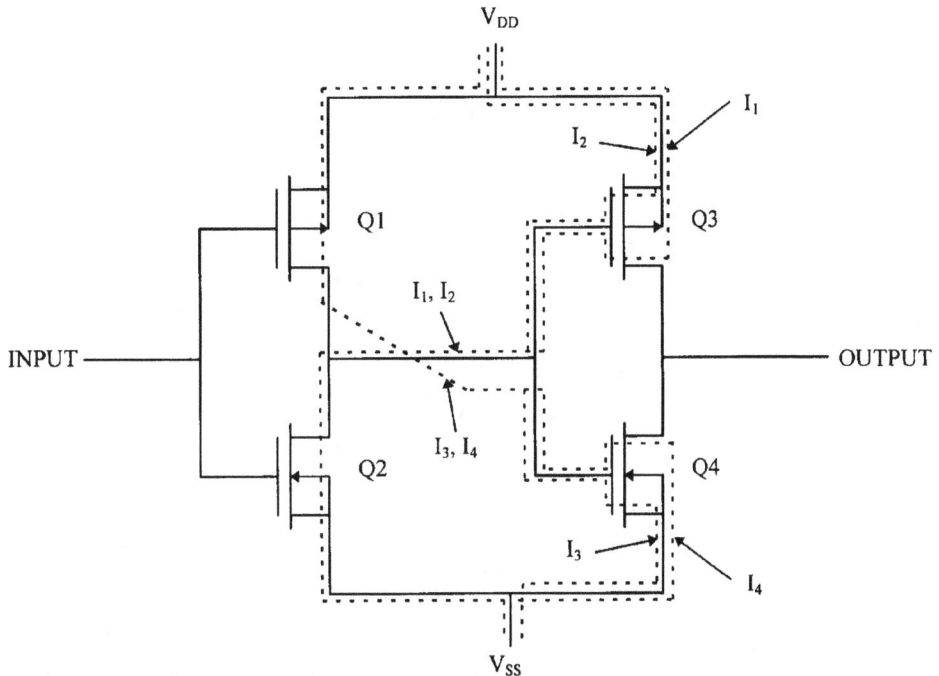

FIGURE 10.23   Circuit shown in Fig. 10.21 with current paths shown for various defects in the second inverter.

(temperature chambers, heat guns, or cold spray), confirm that the failure indication is not due to a problem in the electrical connections to the device caused by the temperature exposure. Also, when testing at cold temperatures, moisture condensation on the DUA must be avoided to prevent false indications of increased leakage currents.

If using a temperature chamber, be aware that there may be glitches introduced in the power line when the chamber heating elements are switched on. These glitches may be seen as instabilities in a curve tracer display or other test equipment. The glitches can be coupled directly to the test gear through the ac power line, or they may be picked up by test leads acting as antennae. Be careful to not interpret these glitches as DUA instabilities.

3. *ESD precautions.* When testing a static sensitive device, always use ESD precautions to prevent inducing ESD damage into a device, which could lead to a completely incorrect conclusion in a failure analysis. (*Note:* If in doubt about the static sensitivity of a device, always use ESD precautions.)

4. *Semiconductor photosensitivity.* Virtually all semiconductor devices are photosensitive when the semiconductor element is exposed to light. Most devices are packaged in such a manner that ambient light cannot impinge on the semiconductor element and cause photoelectric effects. However, when performing a failure analysis, devices are usually delidded or deencapsulated, and the semiconductor element may be exposed to ambient light. The photosensitivity of the semiconductor element can cause increased leakage or power supply currents, voltage offsets, and other electrical problems. Shield open semiconductor devices from any light to avoid these problems. Note that the inside of some "dark" ovens or temperature chambers may be illuminated by the visible and IR glow from the heating elements.

5. *Thermocouple effects.* When testing devices at temperature, especially when making very low-voltage measurements, be careful to avoid voltage offsets caused by thermocouple effects. If two different metals are connected and then subjected to temperature, a small voltage can be produced. The magnitude of the voltage depends on the metals and the magnitude of the temperature change. For most tests, this effect will not be a problem. Typical connections to the DUA tend to offset these effects. However, when it is necessary to add nontraditional connections to devices using dissimilar materials, this problem might have an effect.

6. *VOM testing.* Do not use a volt-ohmmeter (VOM) for testing semiconductor devices. A VOM, even if battery powered, can damage semiconductor devices. In the resistance measuring mode, these devices are capable of delivering high currents that can cause damage to small semiconductor junctions. In the past, battery-powered VOMs were often used by untrained personnel to troubleshoot problems, which induced damage in semiconductors. This damage, of course, was unrelated to the original failure mechanism. This problem is currently much less prevalent but is noted to prevent recurrence.

7. *Soldering to a DUA.* Prior to soldering a connection to a DUA, carefully evaluate the possibility of the soldering heat causing damage to or a change in the unit. If the DUA has been previously soldered to a board (and also possibly desoldered), the probability of damage due to soldering is diminished considerably. However, as much testing should be performed as possible prior to soldering to the DUA. If the DUA has an internal solder joint, attempt to minimize the amount of soldering heat applied to the DUA to avoid reflowing or damaging the internal solder joint.

8. *DUA with broken leads.* Many devices are removed from boards and then are submitted for analysis. Often the removed devices have had their leads cut or damaged during removal from the board. (This is usually done to prevent damage to the board so that it can be reworked.) If a DUA's leads have been obviously cut, there is probably no evidence that might be lost either by probing or connecting clip leads to the remainder of the leads.

However, if the leads on a DUA appear to have been fractured, it is best to thoroughly document the appearance of the leads prior to probing them or attaching clips leads to them. (Of course, if information is available to explain why the leads are fractured, this is not necessary.) Documentation of the leads may include both optical photography and scanning electron microscope (SEM) examination. Fractured leads should be thoroughly documented prior to probing to avoid causing damage to any of the features on the fracture face of the lead. This information could be the evidence that is needed to determine the actual cause of failure of the device.

9. *Retest of the DUA.* This section has discussed the approaches to be used for the electrical characterization of a device for failure analysis. These techniques have covered only the initial electrical tests performed on a device prior to any disassembly or destructive operations being performed on the device. Following any disassembly or

destructive operation, the device should be retested, if possible, unless the operation that was performed is known to have changed the electrical performance of the device.

For example, following the delidding of an transistor, the device should be retested to be sure that the delidding procedure has not caused damage. (*Note:* The transistor's electrical behavior may affected by light impinging on the semiconductor die following delidding.) The device does not have to completely electrically recharacterized, but sufficient testing should be performed to show there have been no significant changes in the DUA. Occasionally, a procedure that would seem to have no effect on the device's electrical behavior will actually produce a change. If a series of additional nondestructive tests are performed with no retesting in between the tests, an electrical behavior change may be discovered after all tests are completed. Without intermediate tests, the test that caused the electrical behavior change will not be known. Without this knowledge, the cause of the change may never be determined.

10.11 SUGGESTED READINGS

Devaney, John R., Hill, Gerald L., and Seippel, Robert G. *Failure Analysis Mechanisms, Techniques, and Photo Atlas.* Monrovia, CA: Failure Recognition & Training Services, Inc., 1988, 3-27 through 3-32 and 3-38 through 3-81.

Holmes, P.J., "Transistor Abnormalities as Revealed by Current-Voltage Characteristics," *The Radio and Electronic Engineer* 38(5), November 1969.

Proceedings of International Reliability Physics Symposium (annual publication). New York: IEEE.

Proceedings of International Symposium for Testing and Failure Analysis (annual publication). Materials Park, OH: ASM International, yearly publication.

CHAPTER 11
SCANNING ELECTRON MICROSCOPY AND X-RAY ANALYSIS

John R. Devaney

HI-REL Laboratories

11.1 *HISTORY OF MICROSCOPY*

Scanning electron microscopy (SEM) has been heralded as one of the most significant inspection and analytical tools to evolve since mankind first aided eyesight with globes of glass filled with water. The telescope allowed observation of the heavenly bodies, moon, planets, stars while the microscope allowed the interested and curious to explore the fine structures of the natural world.

The optical microscope (OM) with which most of us are familiar dates back to the early years of the seventeenth century. In fact the use of the term *microscope* was first coined in 1625 by the Italian Giovanni Faber for the "new occhiale for viewing minute objects." By the middle of the seventeenth century, a physician, Henry Power, a doctor of medicine, wrote the first extended publication regarding the application of the microscope, *Experimental Philosophy, in Three Books.*[1] This predated the work of the great practitioners in microscopy with whom we are all familiar: Malpighi, Swammerdam, Leeuwenhoek, Grew, and Hooke.

The subsequent history of classical optical microscopy follows the same lines as do many other scientific advances and instruments. It progressed from crude instrument, to parlor curiosity, to increasingly useful and sophisticated scientific tool. As the scientific community increased its understanding of the basic principles involved, it put increasing pressure on the equipment manufacturers to improve their instruments. Thus, optical microscopy, defined as microscopy using visible light and combinations of glass lenses, continued to improve in resolution and image quality due to the correction of spherical and chromatic aberration through the end of the nineteenth century, and on into the first decades of the twentieth century. In the following years the instrumentation became more refined, as would be expected of a mature discipline. The outward appearances changed dramatically (see Fig. 11.1).

The changes became more subtle, and phase contrast, polarized light, and interference contrast were added to the techniques available to attack a problem. The illuminating medium and signal were still the same—a beam of light was projected on or through the specimen, and some portion of that illuminating beam was either transmitted or reflected into the viewers' eye or onto a piece of film.

By 1873, knowledge about the properties of light had progressed to the point that the great optician, Ernst Abbé, had determined the theoretical limits of light optics to be about 0.1 μm (100 nm, 1000 Å). A path for a supermicroscope to overcome these limitations was opened in the 1924–1926 time frame by three major discoveries. In 1924, de Broglie

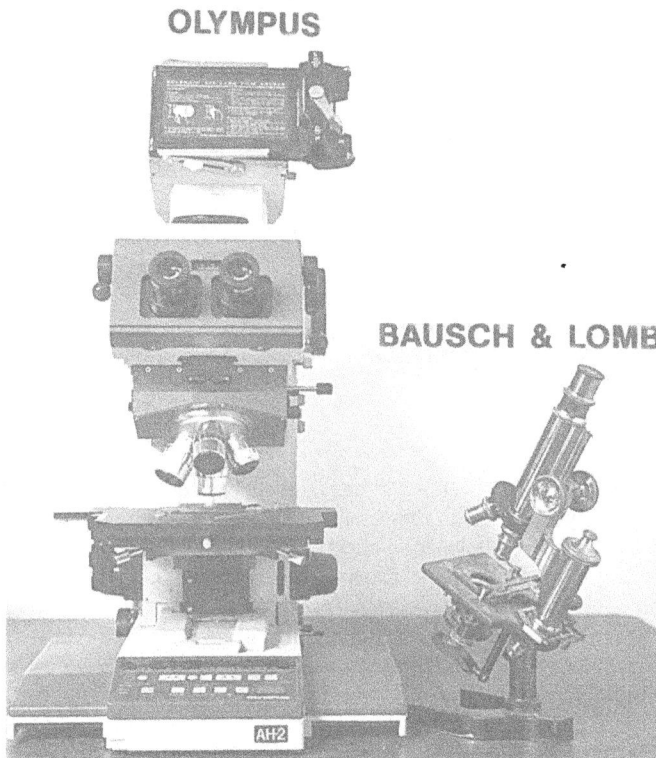

FIGURE 11.1   Circa 1910 Bausch & Lomb microscope next to a 1990 Olympus AH2 metallurgical microscope.

hypothesized the wave nature of electrons; in 1925, Schröeder discovered wave mechanics; and finally, in 1926, H. Busch discovered the lens properties of axially symmetric electric and magnetic fields and laid the foundations of geometrical electron optics.[2] What is surprising in hindsight is that the "concentrating coil" used in cathode ray oscilloscopes since 1899 was not recognized as a means to focus a beam of electrons until Busch presented his work.

These discoveries broke the deadlock that had hampered microscopists and, by 1927, the scientific community in Berlin was abuzz with the possibility of an electron supermicroscope. Work was initiated immediately in several locations (by Knoll and Ruska at the Technische Hochschule, Berlin, and by Bruche and his collaborators at the A.E.G Laboratory, Berlin). By 1932, publications showed results comparable to a good hand magnifying glass! But within four years, results were comparable in resolution to the best optical microscopes and, by 1940, resolution 100 times better than any optical microscope in the world was demonstrated. The electron "supermicroscope" was on its way. This microscope is known today as the *transmission electron microscope* (TEM).[3]

The *scanning electron microscope* (SEM) (that is, the scanning of a focused beam of electrons across a sample surface instead of transmission through the sample) was first theorized by Knoll in 1935. The concept of creating an image with a time-position based generation and detection of a current was followed within three years by V. Ardenne, who built the first SEM. This first instrument used two magnetic lenses to provide a small electron spot at the specimen surface. However, this first SEM utilized the currents absorbed by the specimen to generate an image. Not until 1942, seven years later, did Zworykin, Hillier, and Snyder of RCA incorporate a detector that collected electrons emitted from the surface. This collected electron signal was used to create a time-location dependent image. This technique is the basis for all modern scanning electron microscopes.[4]

Research on scanning electron microscopy was suspended during the war years and was resumed only in 1948 when Prof. Oately of Cambridge University initiated further work as a series of graduate research projects for his doctoral candidates in the Electrical Engineering department.[5] These graduate students became the leading names of scanning electron microscopy: Oately himself, Nixon (his successor), McMullan, Smith, Everhart, Wells, Thornley, Ahmed, George, Spreadbury, Stewart, Broers, Plows, Chang, and Banbury, to name a few. As mentioned in Everhart's history, Atack and Smith built a functional SEM and delivered it to the Canadian Pulp and Paper Research Institute. Images from this instrument, when published, were the first to exhibit micrographs of present-day quality. K.C.A. Smith left the University and was later instrumental in the efforts at Cambridge Instruments to bring into the market the first commercial scanning electron microscope in 1965.

The first commercially available SEMs did not burst forth on the scientific world as a complete surprise. By the 1963–1964 time frame, rumors began to filter out of England of a new, wonderful supermicroscope that could perform miracles. In my own experience, I was in the Failure Analysis Lab at the Texas Instruments facility at Dallas in 1964. We heard the rumors and sent some sample transistors to Dr. Patrick Thornton at the University of Bangor, Wales. Several months later, we received back some Polaroid® photographs that confirmed the exaggerated claims of the rumors. By 1965, Cambridge Instruments, of Great Britain, and JOEL, of Japan, were taking orders for scanning electron microscopes. These first instruments contained all the basic elements of today's most modern instruments (see Fig. 11.2). Their resulting photographs delivered the desired results: obtainable resolution of 20 nm (200Å), depth of field 200 to 300 times that of an optical microscope, working distances of 1 to 20 mm at high magnifications, no illumination problems, and little sample preparation required. Equally exciting to microelectronics/semiconductor specialists were the possibilities of voltage contrast and beam-induced current imaging.

11.2   COMPARISON OF OPTICAL TRANSMISSION SCANNING MICROSCOPY

In the modern industrial/academic and scientific environments, instruments are used on a routine basis by personnel not necessarily trained as experts in their use. Rarely is a technician given formal training in the use of an optical microscope. Instrument design and manufacture has progressed to the point at which the modern optical microscope is almost foolproof. These optical instruments are designed as complete systems, with the novice in mind. There is no need to worry about which objective goes with what combination of eyepiece or body tube. Many of the applications of optical microscopy today are in industry, where components are assembled or inspected with stereo (binocular) microscopes. In fact, the widespread use of microscopes in the semiconductor manufacturing industry resulted in a boom in microscope sales. However, it is not this type of optical microscope that will be compared to the scanning or transmission electron microscope—the supermicroscopes. The light optical microscope of comparison is the reflected-light metallurgical

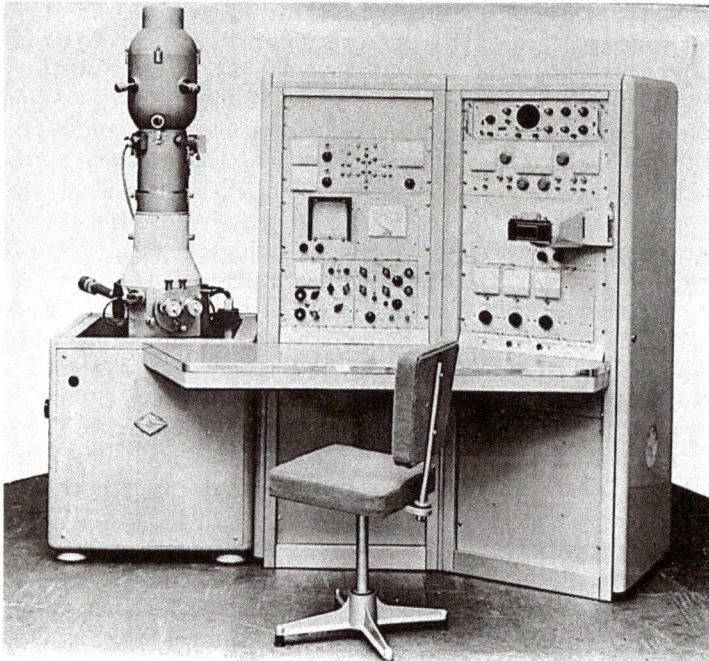

FIGURE 11.2 Cambridge Instruments 1965 Model II scanning electron microscope.

type used to examine opaque specimens in a research or testing environment. The real advantages of the SEM can be appreciated only if the limitations and capabilities of the other two microscopes are understood.

11.2.1 Comparison Criteria

Five basic criteria need to be compared among which the three types of microscopes.

- resolution
- working distance
- depth of field
- sample constraints
- magnification

Magnification is listed because it is usually stated to be the reason for use of an SEM. Sample preparation requirements, or the lack thereof, are also important.

Resolution. The resolution of a research-grade metallurgical microscope is accepted to be about 0.2 μm, whereas a good TEM should be capable of 0.002 μm of resolution. The typical SEM working in a routine laboratory environment is capable of resolving about 0.02 μm on a typical specimen.

Working Distance.   This the distance between the face of the objective in an optical microscope and the sample surface, and this comparison will be for the highest possible magnifications. For an optical microscope, this distance is about 0.5 mm, and for a scanning electron microscope it is about 1 to 20 mm. This does not apply to a TEM, because the beam goes through the specimen.

Depth of Field.   This parameter is defined as that portion of a sample surface containing significant relief that is in focus. Usually, this is defined as being within ±10 percent of the exact plane of focus. The depth of field of an optical microscope will be chosen as a standard of 1 unit. Based on this, an SEM has a usable depth of field of 200 to 300 units. A TEM has no depth of field, since the beam goes through the specimen.

Sample Constraints.   An optical microscope is very versatile in that the only constraints on sample properties are that the sample must be solid, fit on the stage, allow enough travel of the focusing mechanism, and not be affected by the illuminating radiation. At high magnifications, the surface to be examined must be relatively flat, due to the restricted depth of field. Scanning electron microscopists enjoy almost as much latitude in sample variability as does the optical microscopist. For the SEM, the sample must withstand the rigors of a modest chamber vacuum (10^{-5} torr) and be small enough to fit into the vacuum chamber. With modern instruments, this can be a sample up to 200 mm dia. if the sample is thin. If the sample is bulky, depending on the particular instrument, it can be up to $50 \times 50 \times 50$ mm (see Fig. 11.3). Samples for a TEM are very restricted. Since the electron must go through the specimen, it must be thin (≈ 0.1 μm) and small enough to fit onto a TEM sam-

FIGURE 11.3   Optical photograph of the interior of a Cambridge Instruments Model 250 SEM showing the stage with a large bulky specimen in place.

ple holder, which is about 5 mm dia. An alternative possibility in some instances is the formation of a replica of the sample surface, which can be coated and shadowed to delineate surface detail (see Fig. 11.4).

Magnification. The high-magnification capabilities of an SEM are stated as its primary attribute, but this is a meaningless parameter unless tied to an actual attainable resolution value for any specific application. Ignorance of the resolution/magnification obtainable in a TEM has led many potential users into the trap of desiring an SEM in lieu of, or in addition to, an optical microscope. A high-quality light optical metallurgical microscope has a usable magnification range of 50× to 1500×, an SEM's useful range is 10× to 100,000×, and a TEM's range is 10,000× to 1,000,000×. But (useful) magnification depends on resolution. All SEMs have a control that will allow a scanned magnification of 100,000× but, if the point-to-point resolution is not is the 50 nm range, the image will be blurred, soft, and useless (Fig. 11.5).

Sample Preparation. A sample to be examined in an optical microscope only needs to be solid, clean, and (for high-power inspection) relatively flat. As noted previously, the major disadvantage of the TEM is the need to prepare either a thin section of the sample or a replica of its surface. The strength of the SEM lies in the fact that *no* sample preparation is required for most specimens, as long as they can withstand the rigors of the vacuum and fit in the chamber. There is a widespread myth that the samples, if not conductive, must be overcoated with a thin conductive metallic film to prevent charging, but this is not the case. Depending of the actual sample, it is possible to examine and photograph samples in the SEM that are not at all conductive—aluminum oxide, for instance. This requires some experimentation with the instrument accelerating voltage, but nonconducting specimens can be examined *as is* (see Fig. 11.6). As with all rules of thumb, there are exceptions, and

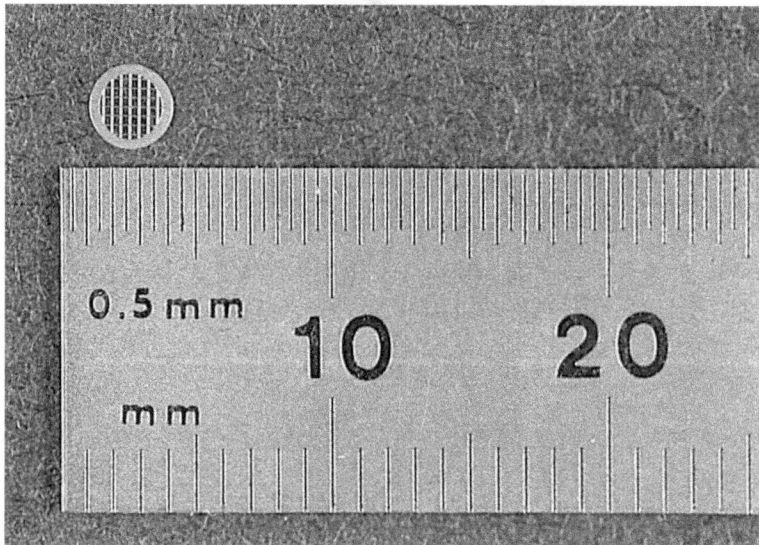

FIGURE 11.4 Optical photograph of a TEM sample holder with scale to show relative size. The area to be imaged is smaller than the grid openings.

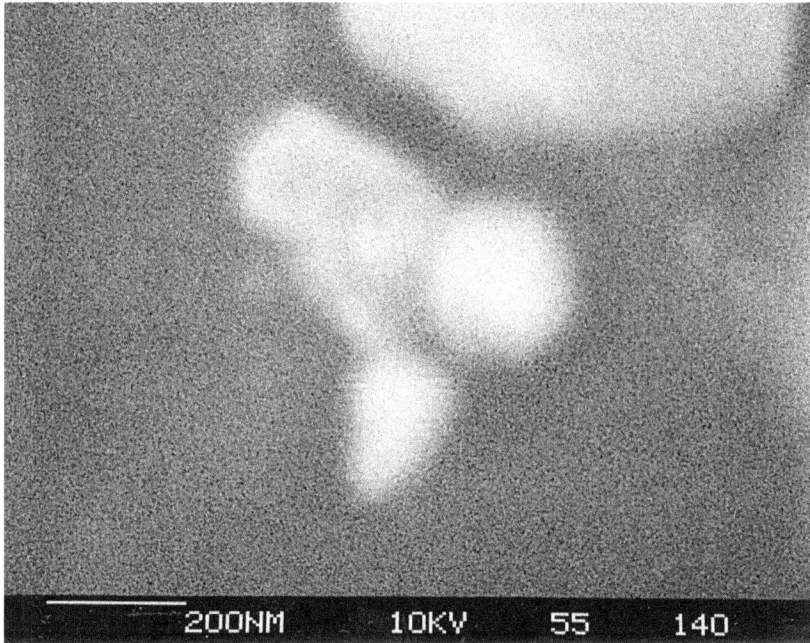

FIGURE 11.5   SEM photograph of a sample obtained at a magnification of 90,000×. The enlargement is meaningless, since there is inadequate resolution; spot size is too large.

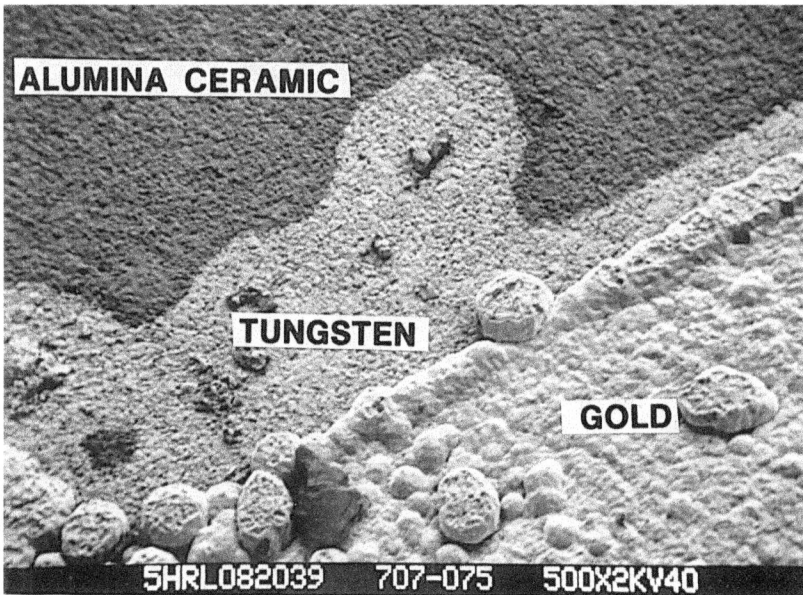

FIGURE 11.6   SEM micrograph of a ceramic substrate with a fired gold conductor. Notice that the exposed alumina (Al_2O_3) adjacent to the gold is not "charging."

the prohibition against coating is the usual exception. Often, when there is no pressing need to maintain the electrical integrity of the sample being viewed, and the complexity of the sample warrants it, coating may be the easiest way out. A typical electronic assembly may contain various types of insulating and nonconducting regions, each with its own optimum beam voltage. In this case, the operator may not be able to find a beam voltage that does not charge one of the materials. An example would be a transformer or coil where the winding wire is varnished and there are other materials close by such as an epoxy core or mylar or teflon wrapping tape. Each of these will charge at a different voltage. If no further electrical diagnosis is anticipated, a thin conductive overcoat would facilitate examination in the SEM (Fig. 11.7).

The coating procedure itself has become routine in most labs; with the advent of the tabletop sputtering systems, it is almost as simple as microwaving a donut. The coating materials in common use are gold, gold-palladium, and carbon. Carbon is preferred if EDS is anticipated, since it adds only one very low-level peak to the spectrum. However, if emissive imaging is required, carbon is a poor choice. Because it has a very low atomic number and therefore is inefficient in producing a strong signal, carbon-coated samples "look" noisy. Gold and gold-palladium are the materials of choice for imaging; since they are "heavy," they produce an excellent noise-free emissive signal.

A word of caution is in order regarding coating thickness. Most operators tend to apply too thick a coating layer, which will obscure subtle surface details and degrade true resolution. As a rough order of thickness, gold becomes gold in color only above about 100 nm thickness (1000 Å). As an example, the SEM specification for examination of integrated circuit step coverage only allows a 20 nm (200Å) coating. To get a rough idea of thickness, the operator can coat a glass slide. At about 70 nm(700Å), a gold layer will look purple/violet when held up to a fluorescent light source. A 20 nm (200Å) coating will only appear as a slight neutral density filter to the same light.

11.3   *BASICS OF SCANNING ELECTRON MICROSCOPY*

The modern scanning electron microscope has evolved significantly since delivery of the first commercial models in the mid 1960s, which filled an entire room. Today, they vary in size from large research models down to desktop personal units that are mouse controlled via a computer. All, however, contain three basic functional units.

- a vacuum-generating system
- an electron-beam generating and scanning system
- the signal detection image generating system (see Fig. 11.8)

11.3.1   Vacuum System

This portion of the scanning electron microscope has benefited greatly from the advances of vacuum technology in the 1970s. Early units required massive, intricate vacuum plumbing. They usually employed water cooling for one or two diffusion pumps, which were backed by a high-capacity roughing pump. Most modern microscopes have replaced the diffusion-pumped system with a simple turbomolecular pump, which may or may not need water cooling. This turbo pump, in turn, is backed by a high-capacity roughing pump. Lower and cleaner vacuums are obtained with this technique—easily 10^{-6} torr in the spec-

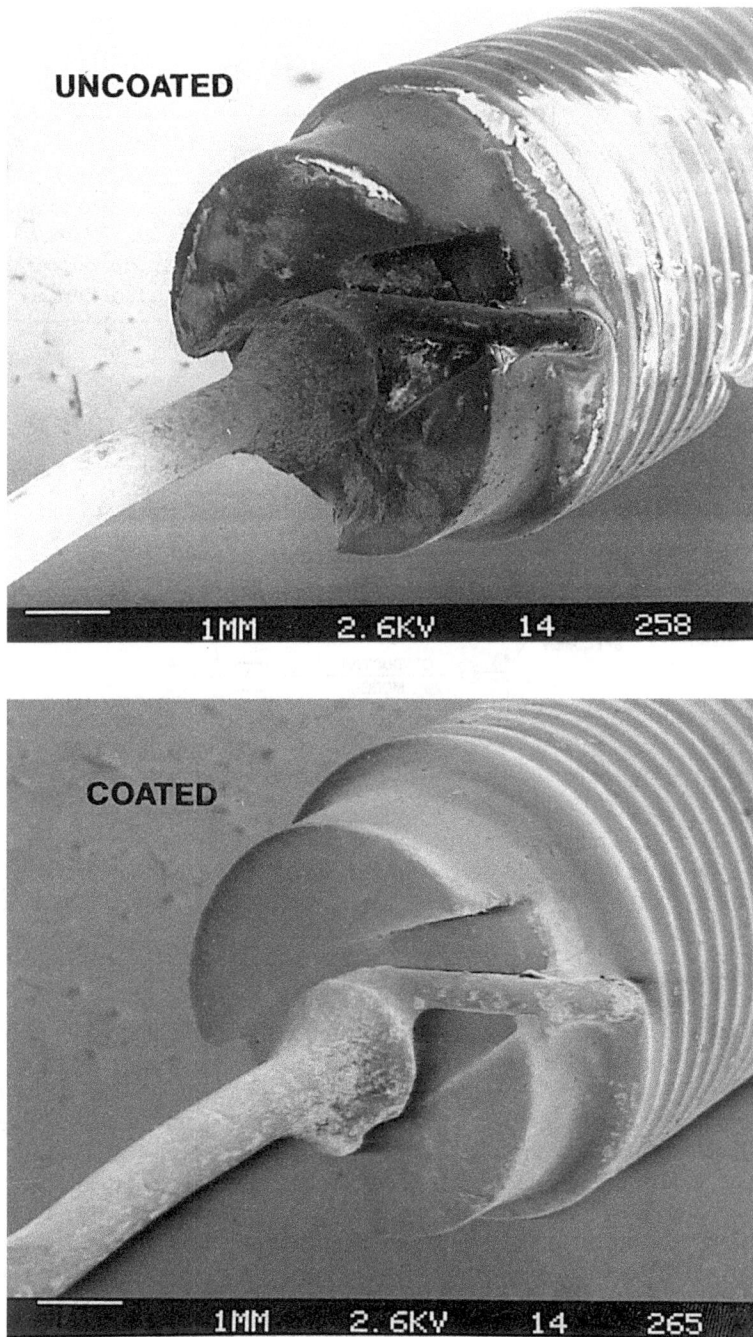

FIGURE 11.7   SEM micrographs of a depackaged power resistor obtained in the as-is and coated modes. The uncoated sample is charging, but contaminants and insulating varnish are not visible in the coated sample, because they are not charging.

ELECTRON GUN

POWER SUPPLY
I TO 30 kV

SPECIFICATIONS:

BEAM CURRENT.................10^{-12} TO 10^{7} A

BEAM VOLTAGE ....I TO 30 kV

MINIMUM BEAM DIA..................~100 Å

MAGNIFICATION.........20x TO 100,000x

DEPTH OF FOCUS...........300x OPTICAL

MAGNETIC LENSES

DEFLECTION COILS

FINAL LENS

SPECIMEN BIAS

SPECIMEN

hv

e⁻

CATHODOLUMINESCENCE MODE

EMISSIVE MODE

CONDUCTIVE MODE

RECORD CRT

VIEWING CRT

VIEWING CRT

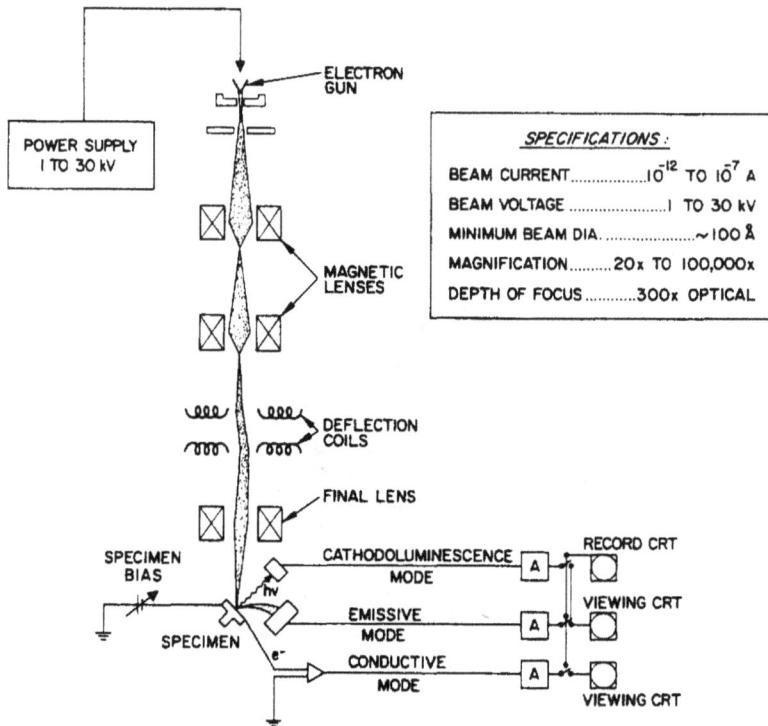

FIGURE 11.8 Basic schematic of the SEM and its principal functional elements.

imen chamber. Some systems require the addition of an ion pump on the gun if the electron source is to be field emission or lanthanum hexaboride. These advancements in vacuum generation have resulted in simpler more compact and reliable systems.

11.3.2 Electron Optics Column

The electron column consists of an electron source (gun) that emits a focused beam of electrons that are attracted toward and accelerated by the anode. The anode is at the top of the optical column and contains a central aperture through which the accelerated beam of electrons is transmitted down into the series of electromagnetic lenses. The lenses further collimate and finally focus the beam on the specimen surface. Contained within the final lens is another set of electromagnetic coils used to sweep (scan) the beam of electrons over the sample surface. Advancements in electronic technology have greatly reduced the size and power requirements of the various components that provide the signals and high voltage required to generate and focus the beam and scan it over the sample surface.

11.3.3 Signal-Detection/Image-Generation System

When an energetic, focused beam of electrons strikes the surface of a sample (be it an electronic component, fractured axle shaft, or ancient Roman coin), it penetrates the sample surface and produces a spectrum of signals. These signals exist regardless of whether the means available to detect and display them in a particular instrument. These signals consist of

- secondary electrons
- backscattered electrons
- Auger electrons
- absorbed electrons
- X-rays
- light (Fig. 11.9)

The primary signal used in most SEM applications is defined as an *emissive* or secondary electron image. Secondary electrons emitted from the surface at the point of beam impact are collected by a positive low-voltage field, converted into light by a scintillator (which channels them into a photomultiplier where they are reconverted into electrons), and amplified, resulting in a time-varying electric current. The magnitude of the current at any given location $x_1y_1t_1$ is proportional to the number of secondary electrons collected from that location on the sample. The intensity of electron emission is a function of the surface topography and, to a lesser degree, the atomic number of the material or the elec-

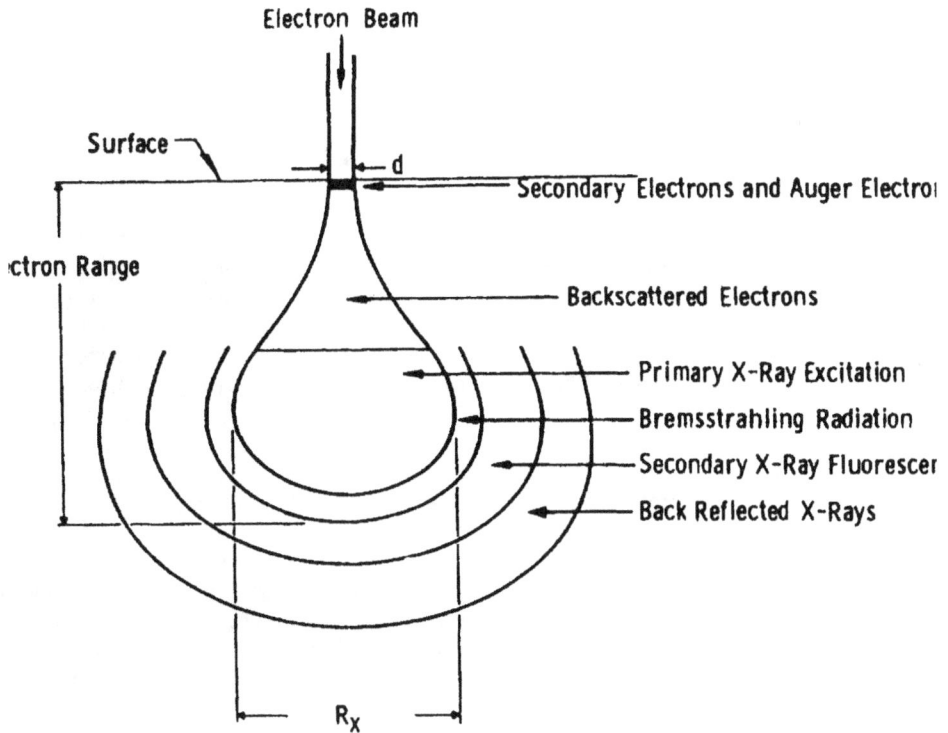

FIGURE 11.9 Schematic of interaction region between beam and specimen with the various types of signals produced from the sample.

tric potential on the surface. This amplified electric signal is fed to a display cathode ray tube (CRT), which is swept in synchronization with the electron beam on the sample surface. The intensity of the CRT display is modulated and is proportional to the signal collected from each corresponding location of the sample surface. One unique aspect of an SEM is that it is the first instrument to create enlarged images of a specimen without using some portion of the illuminating radiation in the formation of the image.

The enlarged image of the scanned portion of the specimen surface, as displayed on a TV monitor or on film, is independent of the resolution obtained in the image. This magnifying of a region of the raster-scanned specimen surface is achieved by a very simple mechanism, similar to the pantograph used by draftsmen. The displayed image on the visual viewing screen or the photographic screen has a fixed size, e.g. 100×100 mm. As the length of scanned line on the specimen surface is varied, the displayed image will exhibit varying degrees of enlargement. If a magnification of 1000× is desired and the displayed image is 100×100 mm, the region scanned on the specimen surface would be 0.1×0.1 mm. This is accomplished by changing the magnification control on the instrument, which in turn adjusts the current being fed to the scan coils. The signal to the scan coils determines the angle of deflection of the beam, which adjusts the length of line scanned at the focal plane.

11.4 IMAGING MODES

11.4.1 Emissive-Mode Imaging

Emissive imaging is the mode with which the public is most familiar, and it most exemplifies the SEM image. This is the mode in which most high-magnification, high -resolution images are obtained, and it also displays the great depth of field and uniform (nonspecular) illumination typical of SEM images (Fig. 11.10).

The principle advantages of scanning electron microscopy *vis a vis* optical microscopy are resolution, depth of field, nonspecular illumination, and working distance. As noted in the previous section, resolution is a function of the instrument and how well it is set up, e.g., the quality of optical column and apertures and the specimen itself. An emissive image is formed on a point-by-point basis by collection of secondary electrons emitted and collected from the sample as the beam is rastered over the surface.

The interaction between the electron beam and specimen surface is very complex and has been described in great detail in several books.[6-8] For the application of the instrument on a routine basis, the operator needs to be familiar with just a few basic concepts. The first and foremost is that the signal generated is *not* the reflection or bouncing of the incident beam electrons from the sample surface. The incoming beam electrons, usually accelerated to 10 or 20 kV, penetrate the specimen surface and are diffused based on the density of the specimen itself. Penetration will be greater in aluminum (atomic no. 13) than in gold (atomic no. 79). In bulk samples, this may not be a crucial factor but, in samples consisting of thin films of material, this could affect the image information. For instance, in aluminum, beam penetration and reaction with subsequent signal generation at 20 kV is on the order of 2 μm whereas, for gold under the same conditions, it will be only about 0.2 μm (see Fig. 11.11). The instrument beam or accelerating voltage is a key parameter that the operator needs to choose wisely. Early SEMs, and even most in use today, tend to utilize a current-heated tungsten wire as a source of electrons for the beam—a hot filament gun.

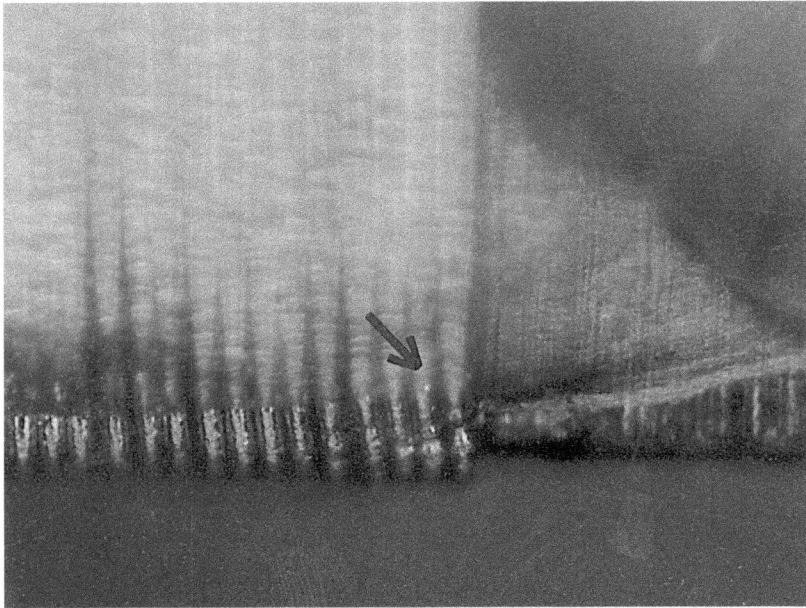

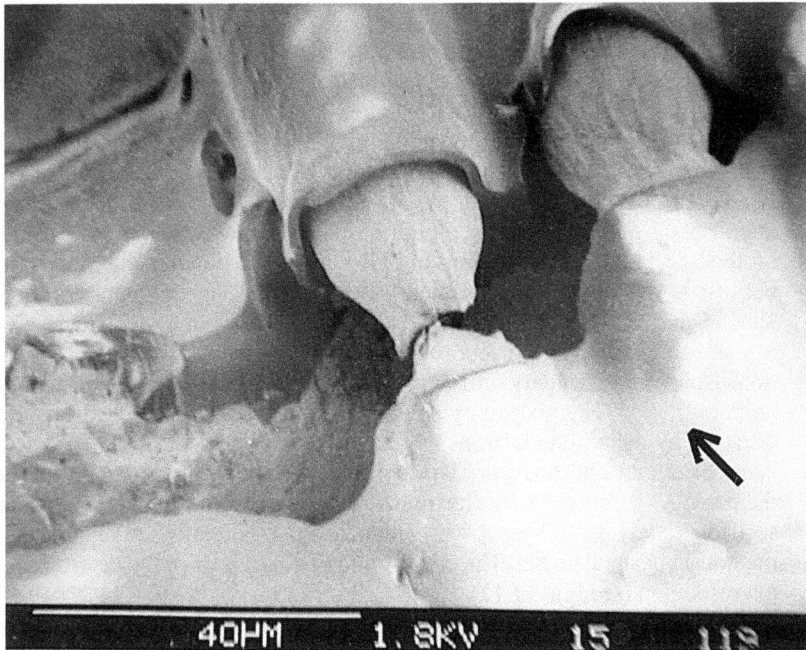

FIGURE 11.10   An important feature of the SEM, depth of field, is shown in these micrographs of a coil wire fracture. The optical microscope (top) has inadequate depth of field to produce a higher-magnification image of any value. The SEM image exhibits both higher magnification and better depth of field.

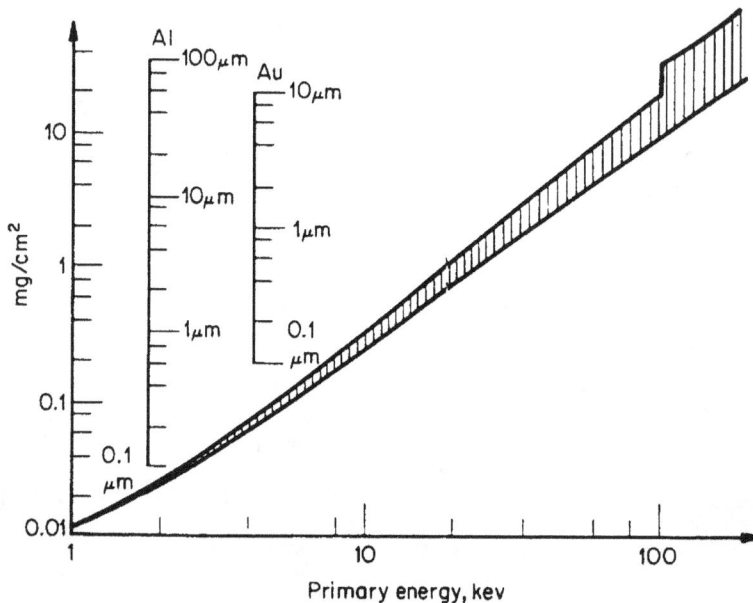

FIGURE 11.11 Comparison of electron penetration in a sample (aluminum vs. gold) at various beam voltages. *Source:* Data from Ref. 9.

Due to interactions between the beam electrons and the electromagnetic fields in the column (lenses), it is possible to generate and control a smaller, more tightly defined electron beam at the specimen surface with higher accelerating voltages. Chromatic and spherical aberrations, similar to that encountered in light optics, are minimized at the higher accelerating voltages. The ability to focus a sharply defined electron spot at the specimen surface is a critical factor determining the actual point-to-point resolution on a particular specimen. But, as noted previously, a high beam voltage results in greater depth of penetration of the specimen by the beam. Initially, concerns about penetration were minimized, because it was thought that secondary electron emission could only occur from a depth of about 5 nm in the immediate region of beam impact. Since secondary electrons are very low in energy, the large reaction volume at high beam voltages was irrelevant to resolution. However, secondary emission does occur over a large region and affects point-to-point resolution. This secondary emission is due to backscatter electrons. Backscatter electrons are by definition electrons that have 95 to 99 percent of the energy of the incoming beam electrons, and these are emitted from a region much larger than the beam impact diameter. As these energetic electrons reach the surface over a large area they can produce secondary electrons. The resulting secondary electron emission region can easily be twice the impact beam diameter. The emission region then will be the determining factor in sample resolution (see Fig. 11.12).

On a uniform specimen, or one that has been coated and thus is elementally uniform, the number of electrons emitted and collected from any specific location is primarily dependent on surface topography. Depressions on the surface, crevices, hillocks, and protrusions facing away from the collector result in regions from which fewer secondary electrons are collected. This gives rise to darker regions in the image as contrasted to regions

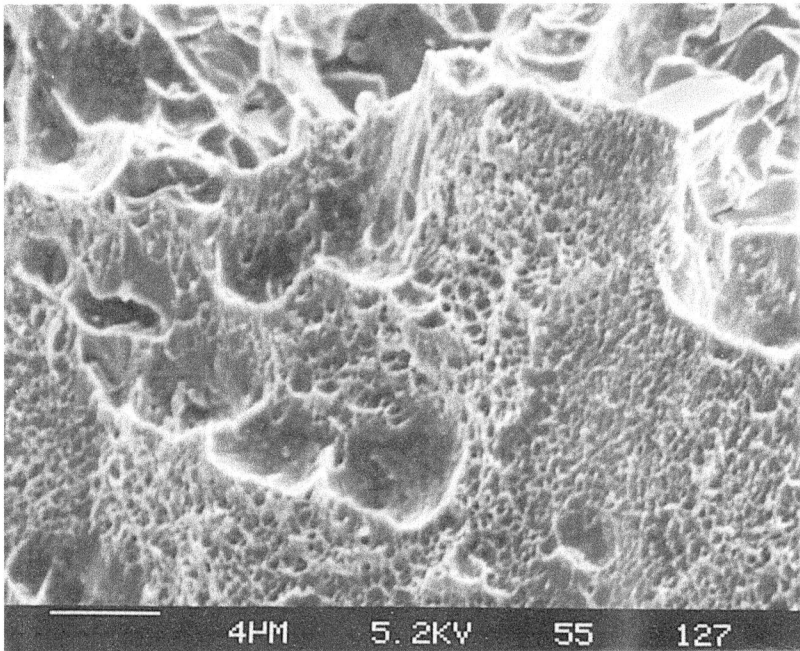

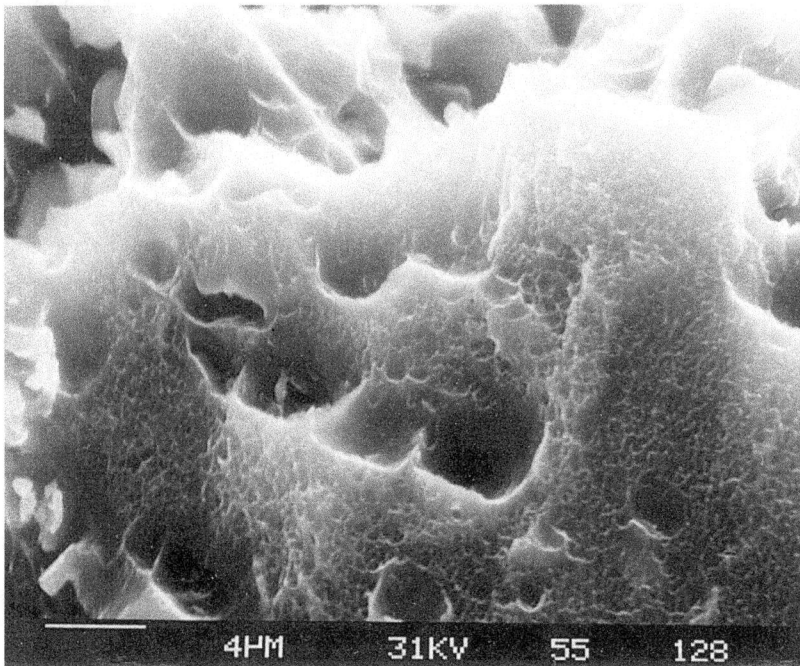

FIGURE 11.12 SEM image of fractured aluminum shaft at 5 kV and 30 kV beam voltages. Note that, at the higher voltage, fine surface structure is lost even though the instrument "should" exhibit better resolution.

from which more electrons are collected. Assuming a perfect specimen (i.e., one that has a conductive surface, is clean, and contains fine detail), the resolution will be determined by the accelerating voltage, incident spot size, and working distance.

Working distance, the distance between the surface of the final lens, polepiece, and the focal plane on the sample surface is, as noted, a critical factor in achieving high resolution. At magnifications above 10,000×, this distance must be minimized. Once the electron beam exits the final lens, the operator has lost all ability to control and shape the spot. At this time, it is beginning to disperse slightly, and the critical concentricity of the spot is degrading. The greater the distance the beam travels from the focusing lens, the more diffuse it will become, and this spot size and shape determine achievable resolution on the sample surface.

Resolution depends on the size and shape of the electron beam as it impacts the sample surface, the reaction volume, density, and the homogeneity of the sample itself. This being the case, why then do all instruments have a control that allows the size of the spot to be changed, i.e. from larger to smaller? Why not work with the smallest spot at all times? This would ensure the best possible resolution on all samples. The point-to-point resolution in the image is a direct function of spot size, and thus a smaller spot is required to resolve smaller features. A feature can be resolved only if it is larger than the electron spot impinging on it; resolution is actually about 1.5 to 2 times the spot size. As the operator decreases the spot size to improve resolution, a new variable comes into play: signal-to-noise ratio.

The *signal-to-noise ratio* (SNR) is the ratio between the signal generated from the specimen surface and the *noise* generated by the signal amplification electronics. This noise is similar to that in a radio when the volume is increased to pull in a weak station. Turning up the volume does generate more sound, but the information may be drowned out by background static. The same phenomenon occurs in an SEM image because of a limitation in the number of electrons in the beam that impact the specimen surface.

For any given electron source, (e.g., hot tungsten filament, lanthanum hexaboride, or tungsten field emitter), there are a fixed maximum number of electrons per unit area emitted from the source. This is defined as gun *brightness*. Beam current density is a fixed parameter, and as smaller and smaller beam spots sizes are chosen by the operator to improve point-to-point resolution, fewer and fewer beam electrons strike the specimen surface to generate signal electrons. Therefore, at smaller and smaller spot sizes, the operator is forced to drive the electronic signal amplification electronics system harder and harder. At some point, the signal from the specimen is swamped by the amplifier noise, and the resulting image is useless. The operator must then compromise by working with a less-than-optimal spot size to improve the SNR, or the specimen can be scanned at a slower rate to integrate the noise out of the image (Fig. 11.13).

As a caution, the operator should ensure that the secondary electron detector system is working properly at all times. The scintillating material degrades with time and use, especially if large spots sizes are used routinely for X-ray analysis. Thus, the scintillator should be replaced on a regular basis.

The spot size can also be adjusted by the use of larger or smaller final apertures, which are used to define the spot as it exits the final lens. This final aperture will affect not only the spot size, SNR, and resolution, but to a certain extent the depth of field as well. Actually, this is the only mechanism the operator can use to change the depth of field. The great depth of field exhibited by all SEM images is primarily a function of the wavelength of the illuminating radiation, in this case a beam of electrons. Electrons have a much shorter wavelength than visible light, so diffraction effects as the beam passes through the defining final aperture are negligible. The final aperture results in the formation of a very small,

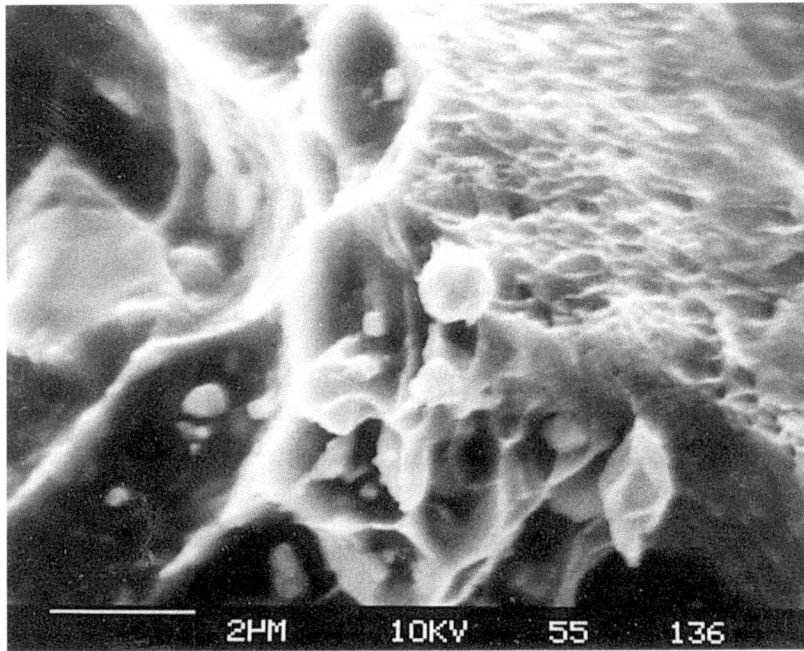

2HM   10KV   55   136

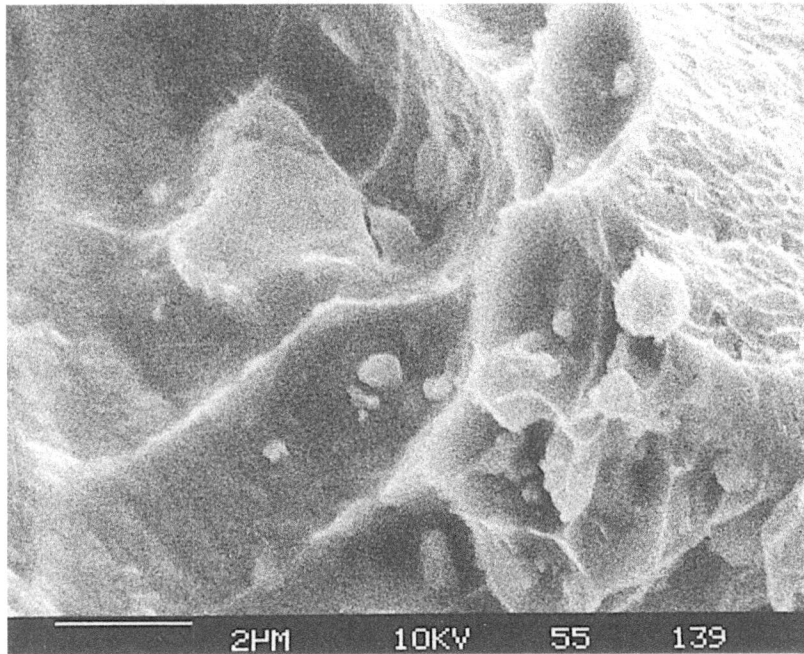

2HM   10KV   55   139

FIGURE 11.13 Noisy vs. clean SEM micrographs showing that additional features are images with a smaller spot but with some increase in noise in the micrograph.

slowly converging beam to its minimum focused spot diameter on the sample surface. The distance along the beam axis which lies within ±10 percent of the minimum spot size is essentially in focus. This determines how much of the surface topography is in focus, and this depth, as noted previously, is 200 to 300 times greater than that of an optical microscope at the same magnification (Fig. 11.14).

The great depth of field of all images taken with an SEM, although recognized as a desirable property, can also result in misinterpretation of the structures in the image. Thus, since large variations of relief will be in focus in the image, the true extent of the surface relief will have no visible frame of reference, especially if a fractured surface is being viewed directly at no-tilt angle (Fig. 11.15).

When working with optical light microscopes at higher magnifications (greater than 50×), two major constraints are a short working distance and a shallow depth of field. These two factors prevent the microscopist from examining complex specimens to inspect a feature that is below surrounding portions of the specimen, which might require the specimen be tilted.

Information acquisition is the goal and intent of examining any sample with any type of microscope. In this, the SEM excels, because the sample can be tilted to very high angles, and large portions of the surface will be in focus. Why is this important? Because the microscopist can easily interpret the information: this is the way we view the real world around us. We've all experienced the problem of flying over our neighborhood and having difficulty in recognizing where we live. We ordinarily view our world at some slight angle

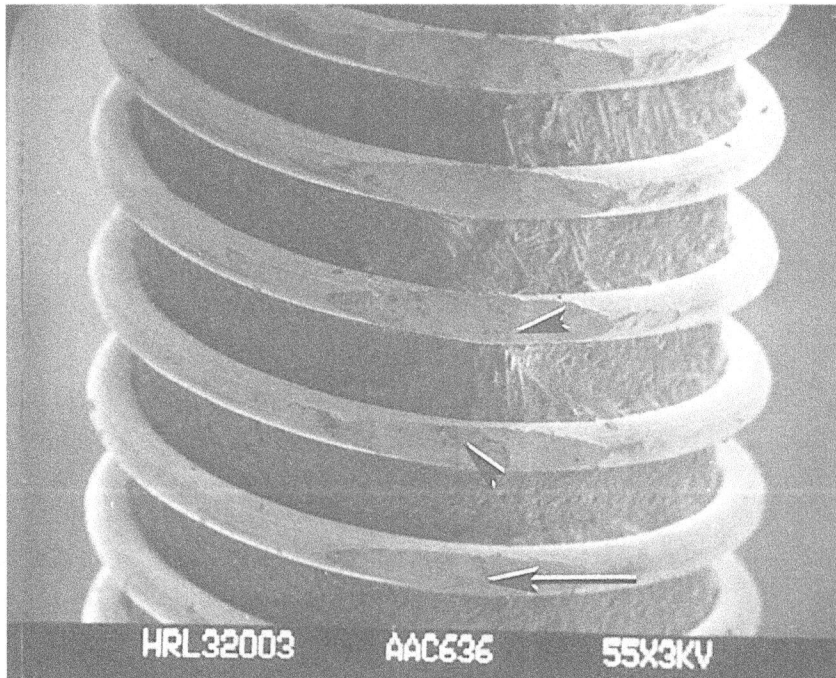

FIGURE 11.14 SEM micrograph of a depackaged coil. Note the great depth of field typical of SEM images.

FIGURE 11.15   Fracture surface viewed at 0° tilt vs. 60° of tilt. In the 0° tilt image, no information is conveyed regarding the overall protuberance of the fracture surface above the plane of fracture.

from the horizontal, the surface on which we stand. We do not typically view the real world in a plan view. So it is with the ability to tilt the specimen and, as it were, to view it at a slight angle to the specimen surface. We generate an image familiar to us, and the information is easily assimilated (Fig. 11.16).

This 3–D-like image gave rise to the name of the first commercial SEM, the Cambridge Instruments Stereoscan. Based on this feature, the microscopist can obtain true three-dimensional images of the surface by photographing the surface at two slightly different angles of tilt—about 6° apart for most specimens. When viewed in a stereoviewer, the image does indeed provide more information than either of the two-dimensional images.

However, a caveat must be attached to the use of tilt angles in the SEM. The microscopist must be aware that, as the specimen tilt angle increases, a keystoning effect occurs. The magnification (micron marker) indicated on the screen or photograph is correct only along the centerline of the tilt axis. If the tilt axis lies along the x-axis of the image, the magnification at the top of the image will be more than that indicated, and the magnification at the bottom of the image will be less than that indicated. If structural details are to be precisely measured, they can only be measured along the center line of the tilt axis.

This keystoning effect is inherent in the formation of the SEM image. As noted previously, magnification of the image of the specimen surface features is achieved by scanning the electron beam over the specimen surface along a length of line proportional to the length of the line on the on the display monitor. The length of scanned line is determined by the angle through which the beam is deflected by the scanning coils. At any given deflection angle, the length of scanned line will depend on the distance of the sample from the final lens surface. All scanning electron microscopes have a specific working distance at which the magnification indication and micron marker are calibrated. On the Cambridge units, this is usually 10 mm. As the sample is brought nearer to the final lens, the line subtended by any given deflection angle will be shorter, thus resulting in a higher magnification for that particular magnification setting. The same is true for images obtained for working distances greater than at the calibration distance. Thus, if a sample is focused at, for example, 20 mm instead of 10 mm, the indicated magnification will be less than that indicated for that magnification setting. With these concepts in mind, it is then obvious that, at high angles of specimen tilt and low magnifications, the portion of the sample nearest the final lens (short working distance) will be at a higher magnification than that portion of the sample farthest away from the final lens (long working distance). This makes it very difficult to measure feature sizes that are not coincident with the tilt axis.

Contrast in the image is a crucial factor is detecting and displaying surface features in an SEM image. As noted, contrast arises from the variations in collection of the emitted signal. These secondary electron are collected by the low-voltage field of the collector, +300 V. A small percentage of the image information is contributed by the energetic backscattered electrons that emerge from the sample surface along a vector directly into the collector. Since backscattered electrons are very energetic (90 to 99 percent of the energy of the incoming beam electrons), they are not affected by the positive field of the collector (the +300 V). They contribute to the picture information and are important since they contain chemical information of the sample surface—if it is not overcoated or the overcoating is kept very thin (<50 nm). This chemical information arises from the fact that the number of backscattered electrons is proportional to the density of the sample surface material. The intensity of the backscattered signal is roughly dependent on the atomic number of the specimen (Fig. 11.18). This relationship between the atomic number of the specimen and the backscattered signal strength is the basis for Robinson backscatter imaging, which will be discussed. This backscatter effect produces additional information in emissive images.

FIGURE 11.16   Sequence of micrographs of a flat surface with features. The optical and 0° tilt SEM image convey about the same amount of information. Only in the 50°-tilt SEM micrograph is it obvious that the features are pits in the surface. *(continues)*

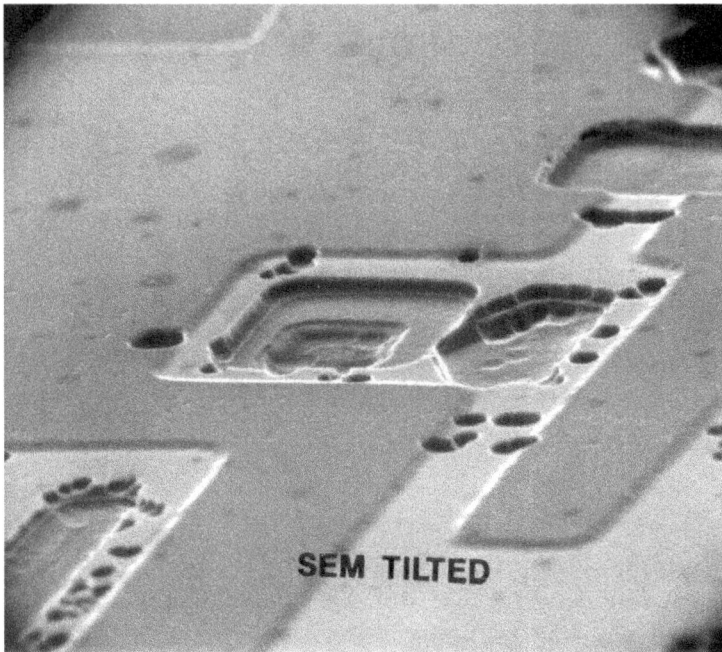

FIGURE 7.16   (*continued*)

KEYSTONE EFFECT

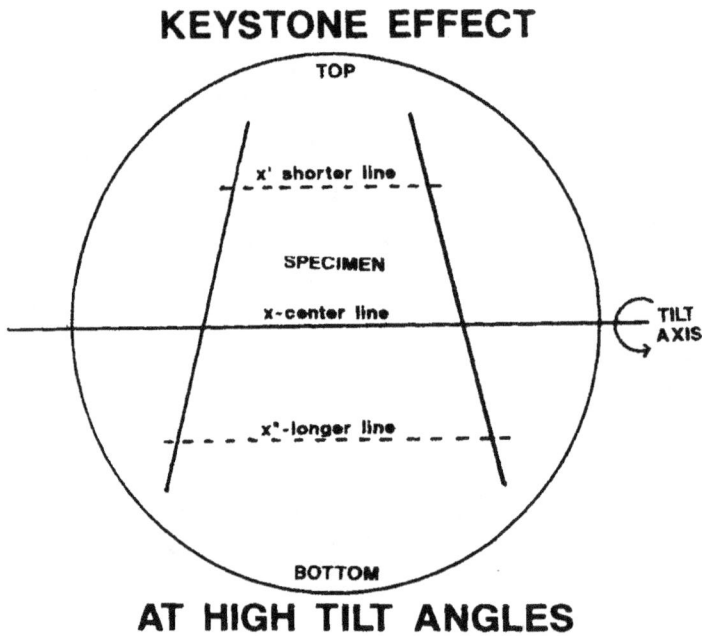

FIGURE 11.17   Line drawing of keystone effect. Accurate dimensions can be obtained only along the centerline axis of tilt.

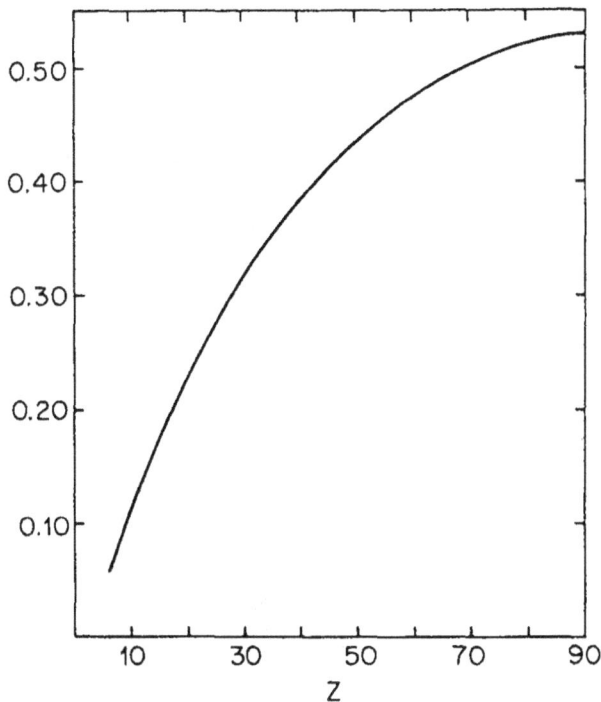

FIGURE 11.18 Electron backscattering coefficient n as a function of the sample atomic number, 0° tilt at 30 kV. *Source:* Data from Ref. 10.

For instance, if a typical aluminum specimen holder is polished and then masked in such a fashion that one-half can be overcoated with a thick layer of gold, subsequent examination of the surface will result in a bright region and a dark region. The bright region will be the gold coated area (atomic no. 79), and the darker region the aluminum surface (atomic no. 14). Contained within these regions will be the normal topographic structures. The effect occurs in all samples to a greater or lesser degree, depending on the relative density differences of the surface materials (Fig. 11.19).

Specimen charging is a topic of routine discussion among SEM operators and is listed as one of the major sample preparation requirements for nonconducting specimens. As discussed previously, higher accelerating voltages are used to "improve" resolution, but higher accelerating voltages can result in *charging* of nonconducting specimens. For bulk metallic specimens, which are not composed of layers of various types of material, and some of which may or may not be conductive, sample preparation for SEM inspections is indeed minimal. But the types of specimens encountered in electronic systems failure analysis usually are not of this simplicity. Specimens will have nonconductive regions such as glass insulators, standoffs, insulating varnishes, or contaminating films. To overcome the problem of such insulating materials (i.e., charging under the high-voltage electron beam), the operator is forced to coat the specimen.

The option of coating a sample should be a last desperate resort for the failure analyst. In electronics, many failures are obviously of an electrical nature, so electrical testing and retesting will be performed numerous times during the analysis. Once the sample is coated

FIGURE 11.19   Example of backscatter signal dependence on sample density. Left side of sample (bright) is gold (atomic no. 79), and the right side (dark) is aluminum (atomic no. 13).

for SEM inspection, the sample is a *dead short*. If further diagnostic testing is required, the analyst is then faced with the unwelcome task of removing the overcoating layer without introducing some new extraneous variable into the analysis.

The SEM operator has an option that can be exercised to bypass some of these difficulties. This option is to examine the specimen under less-than-optimum instrument operating conditions. By reducing the beam accelerating voltage to some low value (1 kv to 3 kV, depending on the material), an operating regime can be obtained that results in a secondary electron emission current greater than unity; in other words, more electrons are being emitted by the sample than are in the incoming beam, and the sample is discharging.[11–12] (See Fig. 11.20.)

The best example to demonstrate this property is to examine a sample with nonconducting regions at normal operating beam voltages, 10 to 20 kV, long enough for the sample to display normal charging characteristics. The operator then lowers the beam voltage to the 1 to 3 kV range and scans the specimen. The higher secondary emission at the lower beam voltage will gradually "erase" the charged regions of the sample. The drawback to the method is the need to operate the instrument at the lower beam voltage. For most instruments, resolution will be poor at magnifications above 2 to 3,000×, but this may not be relevant to the task at hand. The operators' task is the acquisition of information about the sample under conditions least likely to alter the sample (Fig. 11.21).

In addition to inspecting the specimen at low beam voltages to prevent charging, the operator has another option to diminish charging effects: examination of the sample in the backscatter mode.

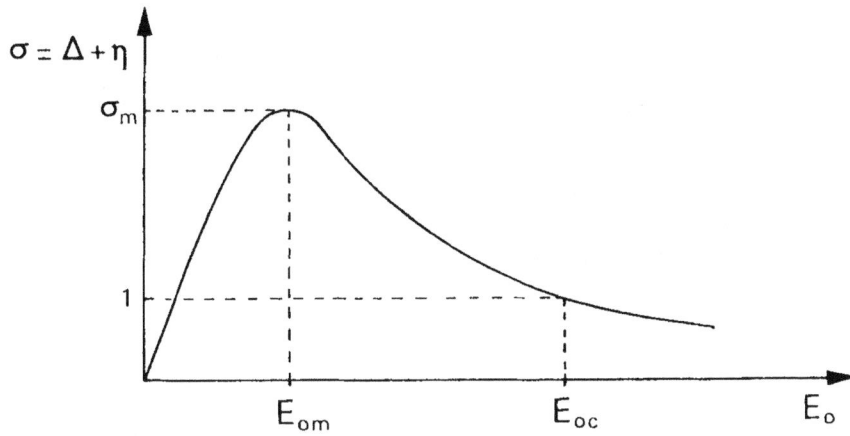

FIGURE 11.20 Graph depicting dependence of total signal emission from an insulator as a function of beam voltage. The task for the operator is to determine the crossover voltage for a particular specimen to be above the net discharge line of 1.0 to prevent the sample from charging.

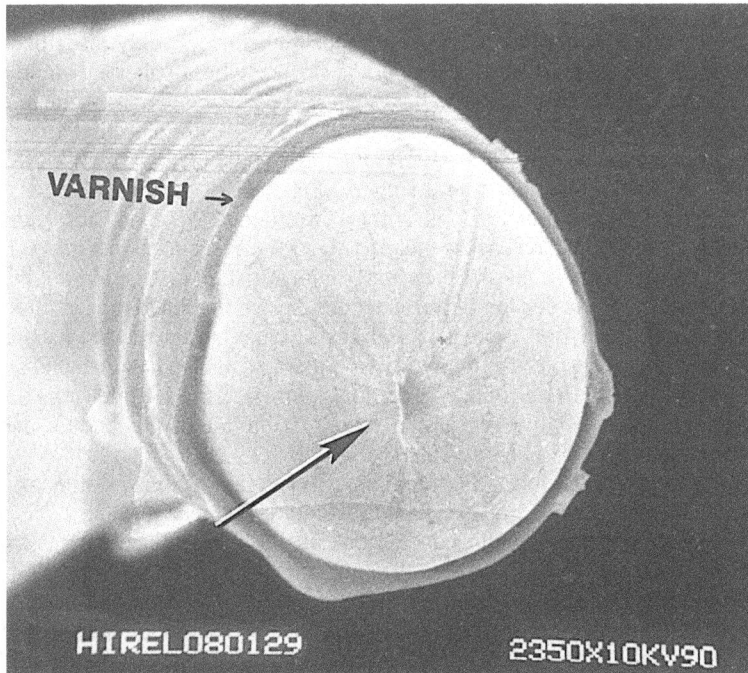

FIGURE 11.21 SEM image of a fractured wire, uncoated, illustrating the ability to image the insulating varnish layer without charging.

11.4.2   Backscatter Mode

Backscattered electrons are defined as electrons emitted from the specimen with energies of 90 to 99 percent of the energy of the incoming beam electrons. These arise from elastic collisions between the beam electrons and atoms in the sample. Backscattered electrons can escape from deep within the specimen and are generally emitted over an area much greater than the focused beam diameter striking the sample surface. Collection and amplification of these backscattered electrons give rise to two imaging modes distinctly different from each other or emissive images.

Surface backscatter imaging is obtained by turning off the collector grid voltage on the detector. This positive voltage (+300 V) can be switched off (0 V) or, on some SEMs, can actually be continuously decreased with a variable potentiometer control from the +300 V to –50 V. As the positive collector voltage is reduced, fewer secondary electrons are collected from the specimen to form an image. As the secondary electron signal decreases, the remaining signal will contain proportionally more backscattered electrons. This remaining signal will consist of backscattered electrons whose vector, when emerging from the sample surface, is directly toward the detector aperture. Since these backscattered electrons follow a line-of-sight path from the sample surface to the detector, the resulting image will exhibit drastic shadowing contrast, and subtle surface features that might be lost in a normal emissive image will be accentuated. The backscattered electrons are not affected by surface charging effects, and an image that would be unacceptable in the emissive mode would be quite usable in the backscatter mode (Fig. 11.22).

A disadvantage to the use of this backscatter mode is the requirement to use larger spot sizes to generate adequate signal strengths for image formation. If the operator visualizes both the secondary and backscattered electrons as being emitted from a hemispherical region centered around the beam on the specimen surface, only those backscattered electrons are collected from this hemispherical surface whose spherical angle is subtended by the aperture of the collector. The backscattered emission signal is much smaller than the secondary emission signal and thus requires either greater amplification or a much larger incoming beam current. This mode is useful only at relatively low magnifications (<3000×) because the resolution is so poor (Fig. 11.23).

Elemental backscatter imaging is the better-known type of backscatter image and is typically called *Robinson* backscatter after one its earliest proponents. As mentioned in the section on elemental contrast in emissive images, the backscattered signal strength is a direct function of the density of the specimen, which is related to its atomic number. Thus, higher atomic numbered elements will have a higher backscattering coefficient than lower numbered elements. This is graphically illustrated in Robinson images of a cross-sectioned solder joint. In this image, the solder joint is mottled in appearance with lighter and darker regions. The lighter regions are lead (Pb), atomic no. 82, and the darker regions are tin (Sn), atomic no. 50 (Fig. 11.24).

Many analysts are now exploiting this effect in lieu of X-ray dot maps since it is much faster with greater more precise spatial resolution. Typically, the analyst captures a backscatter image and then identifies the various regions by localized point X-ray spectrometry, EDS.

There are several constraints for operating in the Robinson backscatter mode, the first being similar to that for surface backscatter: use of large spot sizes to generate adequate signal strength. Also, Robinson backscatter images (RBIs) are obtained of relatively flat specimens (cross sections are ideal) with low or 0° tilt angles and short working distances.

The typical elemental backscatter detector consists of a flat circular plate composed of a large silicon diode with a central aperture. This plate can be either attached to the surface

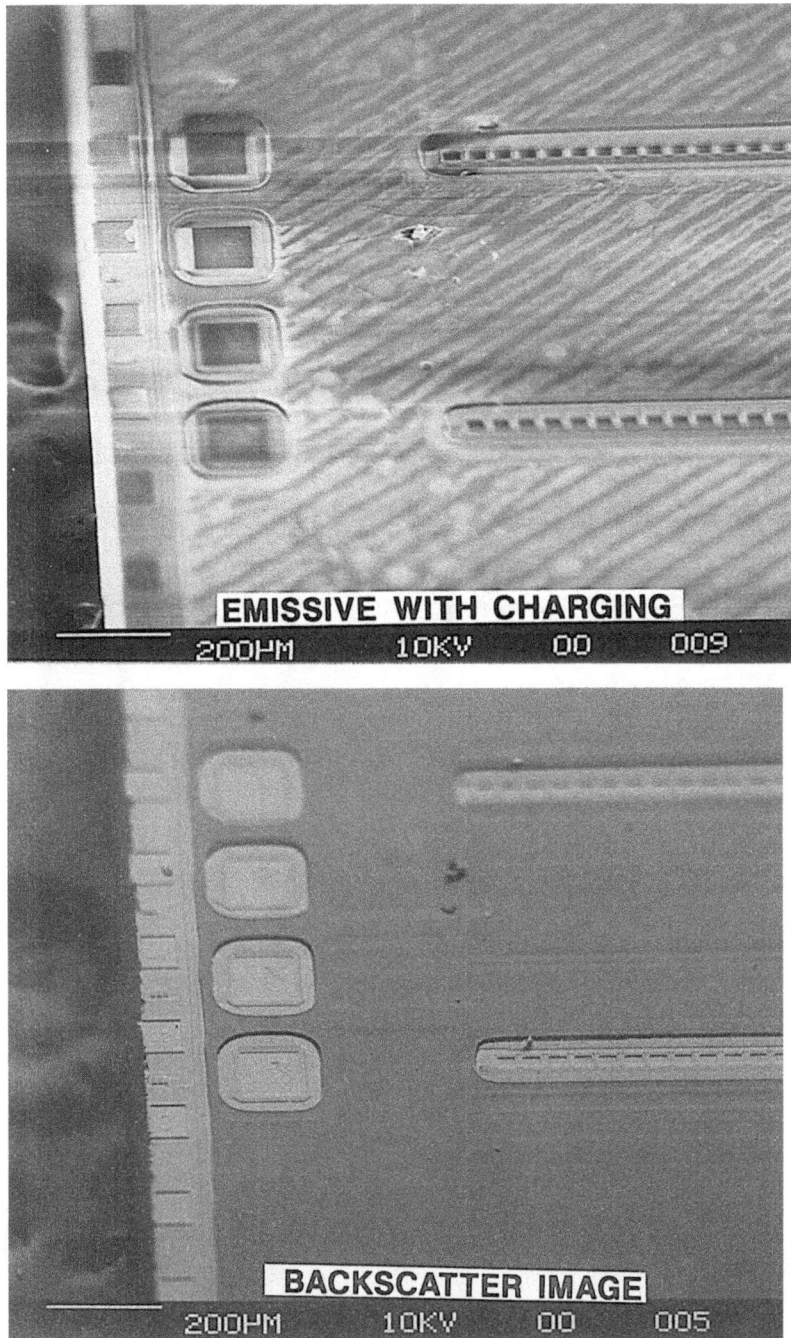

FIGURE 11.22   The advantage of basic backscatter imaging is shown in this pair of micrographs of a sample with a thick (4 mils) layer of polyimide. In the top photo, the layer is charging, as shown by the ripple pattern. With the detector turned off and imaging only backscatter electrons, the surface

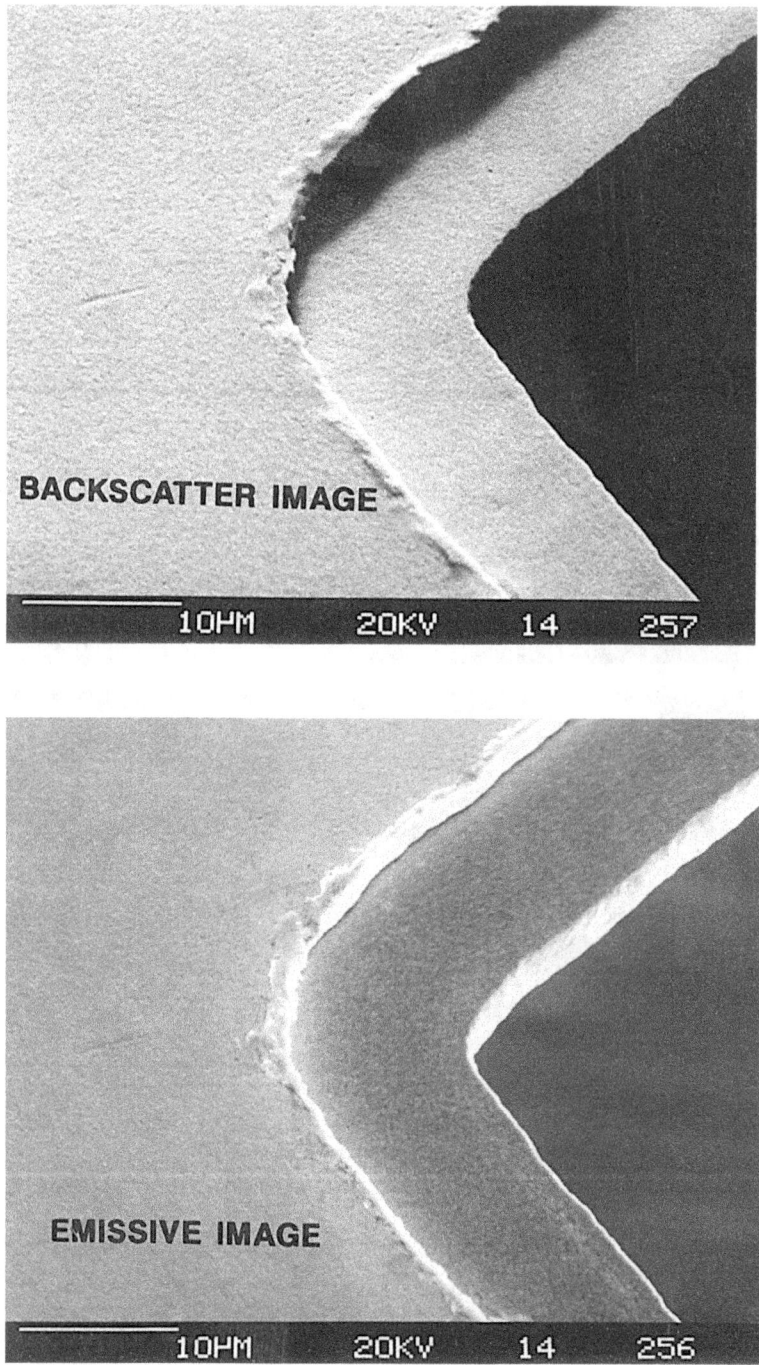

FIGURE 11.23 Pair of SEM micrographs showing enhanced surface detail in the backscattered image (top) as compared to a typical emissive image (bottom).

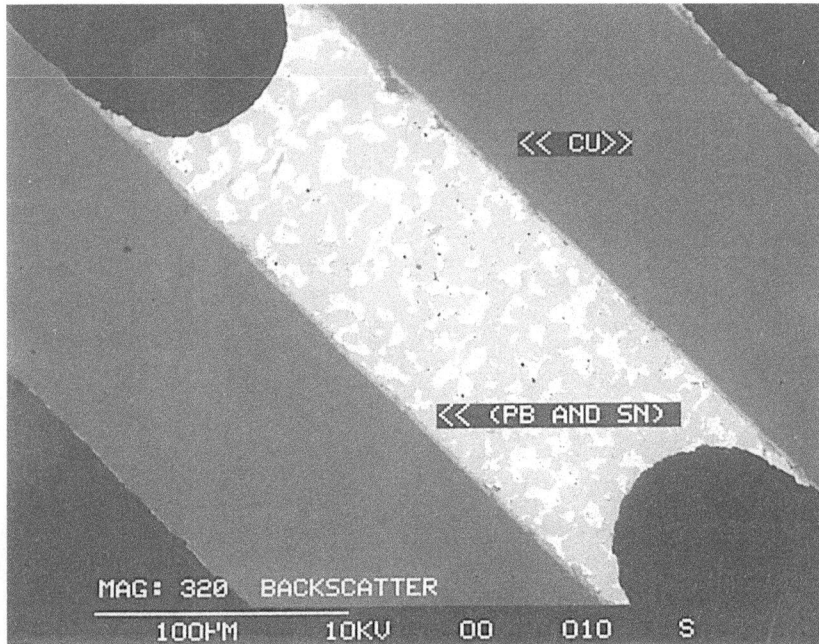

FIGURE 11.24 Robinson backscatter image of a cross-sectioned joint. Note the significant difference between the copper (atomic no. 29) and the brighter regions in the solder, lead (atomic no. 82) and tin (atomic no. 50).

of the final pole piece or on an articulated swing arm that can move the detector out of position when not in use. This plate also has a central aperture aligned with the beam exit from the final lens (Fig. 11.25).

The beam is scanned over the specimen surface, and both backscattered and emissive signals are generated simultaneously. For effective formation of RBI beam, voltages in excess of 7 kV are needed to produce backscattered electrons with sufficient energy to generate a signal when impacting the detector. This is a very useful technique for a sample that has been cross-sectioned and cannot be etched to delineate various features, or the sample has little topographical relief but is composed of various types of material. This is also a very effective mode to image thin organic films on an otherwise conductive surface (Fig. 11.26).

Specimen absorbed current is the imaging mode in which the signal is affected by the variations in conductivity of the sample surface. This can be used in applications where the analyst is attempting to establish the existence of a conducting path across the sample or the conductivity of a layer of material on the sample surface without an external biasing circuit.

This imaging mode takes advantage of the reactions occurring in the sample as it is scanned by the beam. As the beam scans the sample secondary electrons are emitted, as are backscattered electrons. Some electrons are absorbed (secondaries and backscattered generated deep within the sample that cannot escape) by recombination and thus become an absorbed current signal, which is normally conducted to stage ground. If this stage current is used to provide a signal to a current amplifier, then this amplified signal can be used

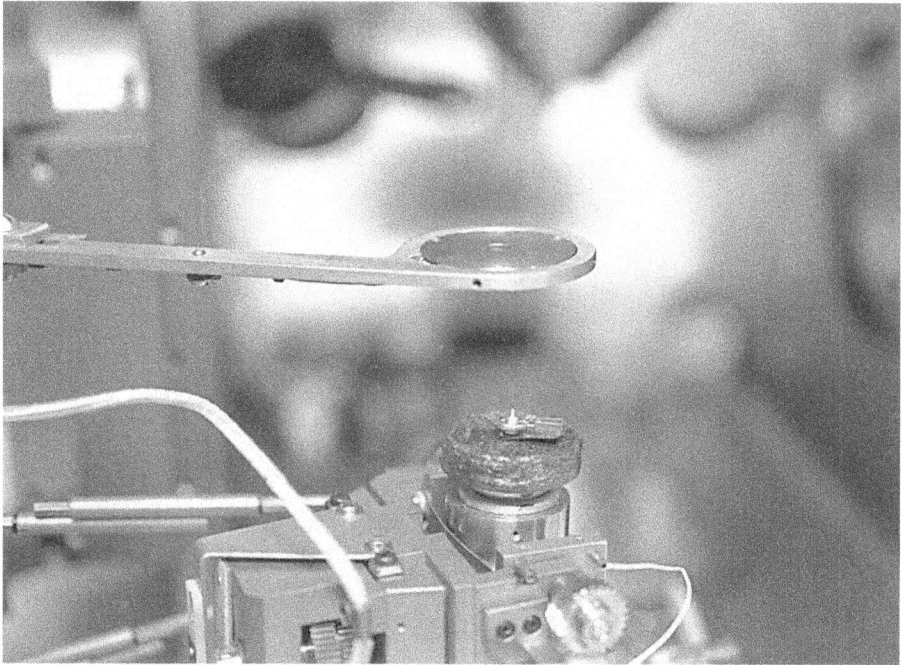

FIGURE 11.25   Photo of a typical backscatter detector that can be moved way from the specimen when not in use.

FIGURE 11.26   Robinson backscatter image of a screw with a thin film of organic (grease) on its surface.

to generate a signal that is roughly dependent on the conductivity of the region being scanned. For instance, if one portion of a complex printed wiring board (PWB) is isolated from the rest of the pattern, no absorbed current image of that conductor will be generated. To implement this technique, the analyst merely needs to isolate those portions of the sample from stage ground that are of no interest. The remaining areas connected to ground will then be current imaged based on their conductivity and connection to the amplifier circuit (Fig. 11.27).

Voltage contrast and electron beam induced current (EBIC) are two imaging modes unique to analysis of semiconductors and some types of electronic components such as thin-film resistors. They are specialized cases of the emissive mode and the specimen current mode. These will be discussed only briefly, because they have been extensively described in journals pertaining to semiconductor failure analysis.

Voltage Contrast.   Voltage contrast imaging is a special case of emissive-mode imaging. As described in the previous section, emissive-mode images are composed primarily of secondary electrons emitted and collected from the sample surface at the point of beam impact on the surface. This signal is affected by surface topography and, to a small degree, by surface composition. An unwelcome effect in emissive mode imaging is charging of the sample wherein localized regions, being nonconductive, acquire a large negative charge from the beam electrons that cannot be drained off to ground. This field introduces severe image distortions. A more subtle effect occurs when the surface of the sample has regions of varying potential distributed spatially over its surface, usually in the form of a patterned layer of conducting material. In most integrated circuits, transistors, and diodes, this con-

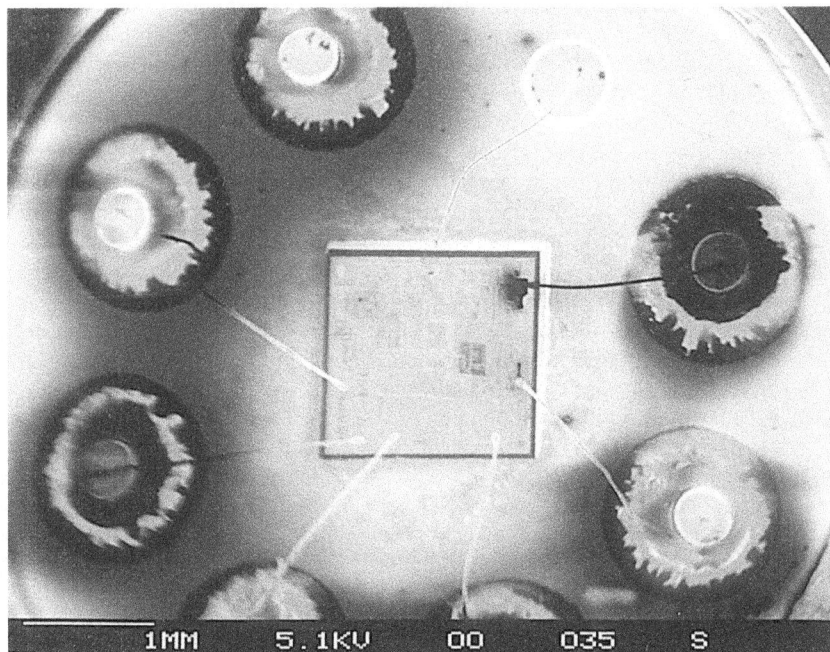

FIGURE 11.27 Specimen current image of a circuit. Only the black wire is connected to the stage ground, resulting in signal flowing into the detector/amplifier.

ducting layer is an aluminum film of about 1 to 3 μm in thickness. The actual material composition of the conductor is irrelevant for most purposes. This conductor may be in contact with and/or lying on the surface of an insulator, i.e., silicon dioxide. If the analyst can "wire up" the component in the chamber and bring these connections to the outside world, then various voltages supplied by power supplies or batteries can be impressed on the sample. Typically, this would be some arrangement similar to the actual application biasing of the device or some unique condition that mirrors the mode of application during failure. These impressed voltages on the sample surface will affect the emission and collection of secondary electrons from the surface. On regions of positive voltage, the emission of secondary electron will be decreased; it is very difficult for a low-energy negatively charged electron to escape from a positively charged surface. The converse is true for films with an impressed negative charge. Here, the negative potential on the conductor will enhance secondary electron emission from its surface. The end result will be a normal emissive image containing topographical information but, in addition, the conductors will vary in brightness depending on the impressed voltage, and positive regions will be darker than negative regions. In a normal, unmodified SEM using a standard Everhart detector, the range (gray scale) will extend from black (no signal) to white (maximum signal). The voltage variations that can be displayed and detected as differences and noted by eye are at a minimum about 0.5 V, a forward diode drop. Complicating factors in achieving voltage contrast will be the presence of insulating films over the metal layers, glasses on the surface of integrated circuits, and insulating portions of packages in both semiconductors and other types of electronic components.

For implementation of voltage contrast imaging, there are somewhat different operating conditions from those for routine high-resolution emissive images. First and foremost, the Everhart secondary electron detector must be in excellent working condition. As previously noted, the scintillating material degrades with use, and this is a key factor in initial signal amplification which, in turn, affects the SNR in the image. Larger beam spots are used to maximize the signal generated from the specimen surface. Since spatial resolution is not the primary goal, longer working distances are used to maximize the collection of signal from the sample surface. The most critical variable is the beam accelerating voltage. This must be adjusted to maximize the voltage contrast on a specimen and will vary from specimen to specimen. This is sensitive to the degree, for example, that no voltage contrast will be visible on a sample at 6.5 kV beam voltage but is obvious at 6.8 kV (Fig. 11.28).

Electron Beam Induced Current. Electron beam induced current imaging is a special case of specimen-absorbed current imaging specific to semiconductors. When formed, a *p-n* junction, of which all semiconductors are composed, instantaneously forms a depletion layer at the interface between the two differently doped regions. This region, without external bias, is a region that contains no free carriers—either holes or electrons. To initiate conduction (current flow) across the junction, some small external bias must be applied. For a silicon diode, this is about 0.6 V. In a circuit, this is supplied by the power supply or battery, but another mechanism can achieve similar results. If an electron beam is scanned across the surface of the device and interacts with the exposed depletion layer, the beam electrons ionize the atoms in the depleted region. The built-in field in the depletion region sweeps this ionization current across the junction—a current flow. If the SEM operator feeds this current to the high-gain specimen current amplifier, then a time-location varying signal will be produced that can be used as an image forming signal. This image will display variations in signal that mirror the current flow across the junction, a technique used to image localized defects such as ESD damage (Fig. 1.29).

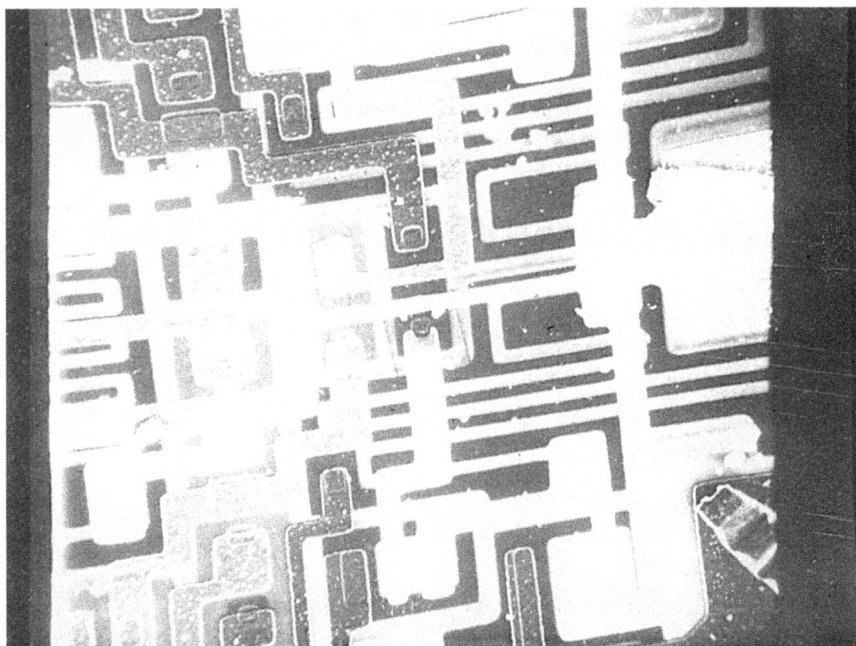

FIGURE 11.28   Voltage contrast micrograph of an open metal run on an integrated circuit surface. The bright region is negative, and the dark region is positive.

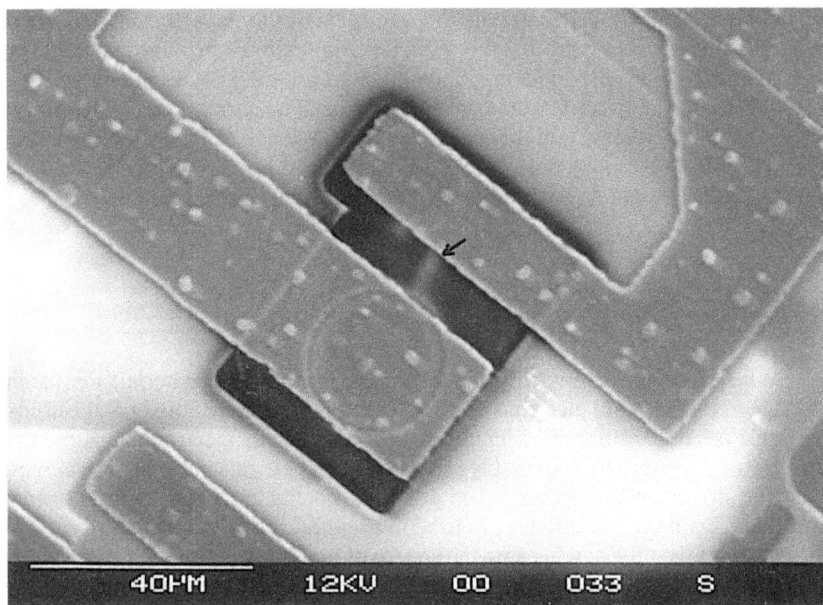

40µM   12KV   00   033   S

FIGURE 11.29   Superposition of emissive image over an EBIC image. The arrow notes the leakage site across the junction, visible only in the EBIC mode.

The implementation of the EBIC mode is similar to that for specimen current imaging in that large spot sizes are used to create an adequate signal. Unlike voltage contrast, high beam voltages are used to ensure penetration of the surface layers with sufficient energy remaining in the beam electrons to generate carriers in the sample. Most high gain amplifiers do not work well at visual TV rates and the sample must be examined at the slower scan modes, one to two seconds per frame.

11.4.3 Summary

The electronics component failure analyst will use an SEM for qualities that include

- better resolution
- higher magnifications
- little or no sample preparation
- great depth of field
- ability to tilt the sample (three-dimensional information)
- long working distance at higher magnifications
- electrical information
- chemical versus spatial information in the sample

There are, however, other advantages that are not as easily recognized. Most modern SEMs have versatile stage mechanisms with which the sample can be rotated, tilted, and traversed in the *x, y,* and *z* coordinates while being examined. Another more esoteric feature of SEM imaging of surfaces lies in the area of illumination effects. Many samples contain features and materials that are optically transparent or translucent, or specularly reflective to the extent that it is difficult to actually inspect the surface. The SEM does not image in the visible spectrum, and optical transparency does not extend to the SEM image. Thus, subtle surface features on a glass or ceramic surface are easily imaged in an SEM, as are smooth, rounded, highly reflective surfaces such as metallic spheres, solder joints, and ball bonds. It is also possible to "see" into deep holes and pits, since the positive field of the collector can draw the signal up out of the cavities (Fig. 11.30).

11.5 X-RAY SPECTROSCOPY WITH THE SEM

When an electron beam impinges the surface of any material, the resulting interaction between the beam electrons and atoms in the sample results in a spectrum of emitted signals, secondary and backscattered electrons, Auger electrons, absorbed currents, and X-rays. The production of X-rays from an object when struck by fast-moving electrons was first discovered by Roentgen in 1895. In Roentgen's experiment, the energetic electrons were supplied by cathode rays used in a cathode-ray tube. These electrons struck the glass walls of the tube and became the source of the X-rays. The inefficiency of this technique to produce X-rays was quickly recognized, and a source whose primary purpose was the generation of X-rays was devised. This source is known as the Coolidge-type X-ray tube wherein an intense beam of electrons is directed to a target (often copper or chromium), and the resulting X-ray flux is used to illuminate a sample (Fig. 11.31).

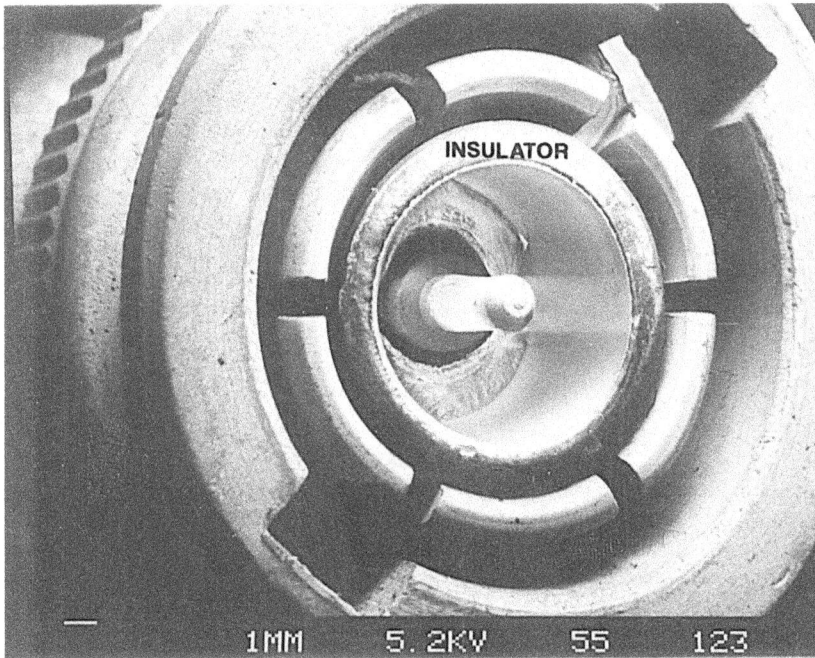

FIGURE 11.30 SEM micrograph displaying all the advantages of the method. A BNC connector is imaged with both the surface and structures inside the receptacle in focus. In addition, the insulator sleeve is not charging at this low voltage.

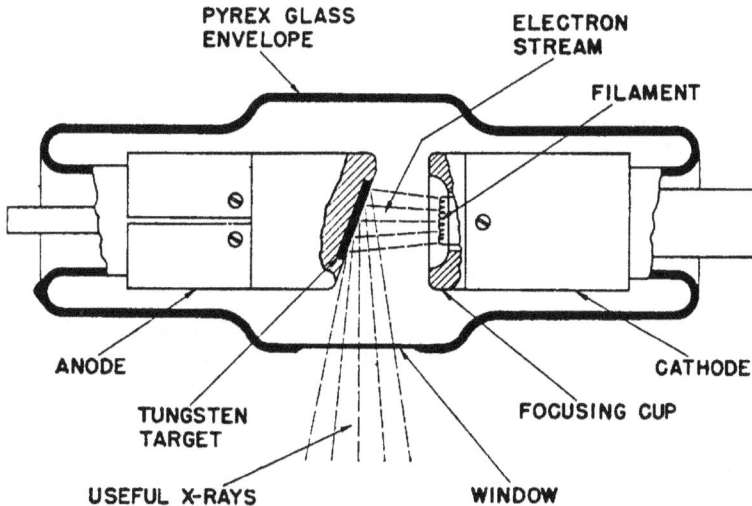

FIGURE 11.31 Schematic of a typical modern X-ray tube, an evolution from the original Coolidge design.

The transmission of X-rays through the sample depends on its atomic density. These are the radiographs used by doctors, dentists, and component analysts to "look" inside a device.

When an incident electron strikes an atom with sufficient energy, either it can be scattered or it can eject an orbital electron. If an orbital electron is ejected, the atom is now in an excited state, and the orbital vacancy must be filled by electronic relaxation. This relaxation can result in the emission of an X-ray photon. Another possibility would be for the X-ray photon energy being transferred to another orbital electron and the ejection of an Auger electron. The emission of X-ray photons is statistically more probable than the emission of Auger electrons. The emitted X-rays will be characteristic of a given element, as noted by Moseley in 1913.[13] This relationship is described by

$$Z \cong \lambda_x^{-1/2} \tag{11.1}$$

By 1912, M. von Laue theorized that the lattice spacings within a crystal would be small enough to act as a diffraction grating for the short wavelength of X-rays. This concept was proved out shortly by Friedrich and Knipping. W. L. Bragg studied these complex diffraction patterns and, in his studies, derived the equation that bears his name, the Bragg equation, for the reflection of X-rays from a series of atomic planes in a crystal.

$$n\lambda = 2d \sin\theta \tag{11.2}$$

Based on this concept, Bragg set up his system to have the crystal reflect the incoming X-rays to a photographic plate. With knowledge of the grating space of a crystal and the use of the Bragg equation, it is possible to measure the wavelengths of the X-rays emitted by the target of the X-ray source. When resolved by a crystal spectrometer, the heterogeneous beam of x-rays from a target is found to consist of two distinct types of spectra: a continuous spectrum and a sharp line spectrum superimposed on the continuous spectrum. The sharp line spectra will be unique to the material of the tube target or sample struck by the electron beam (Fig. 11.32). The logical evolution of the principle led to the work by Casting and his coworkers in France in the late 1950s to create the *electron microprobe* or *E-beam microanalyzer*. This instrument utilized a stationary electron beam to generate X-rays from a small spot on the sample surface. The resulting X-rays were analyzed by their wavelengths through the use of a complex reflection grating setup wherein the angle between X-ray source and crystal surface could be varied over a wide range, thus reflecting different wavelength X-rays into the counting source (ionization gage). This was the first true location-specific microanalyzer. These instruments became commercially available in the early 1960s.

This X-ray wavelength dispersive spectrometer consists of a crystal that is curved or bent slightly to achieve better sensitivity and accuracy. The X-rays from the sample are reflected to a gas-filled detector whose output is fed to a multichannel analyzer. In the large E-beam microanalyzers that preceded the energy-dispersive spectrometer, all three of the critical elements of the system (sample, crystal, and detector) were mounted on the same focusing circle. (See Fig. 11.33.) This need for precise alignment of all three of the elements gave rise to an intense debate over sample location and, more specifically, the *takeoff* angle between the sample (source of the X-rays) and the crystal. The takeoff angle would be the critical factor in determining the circumference of the Rowland circle. The

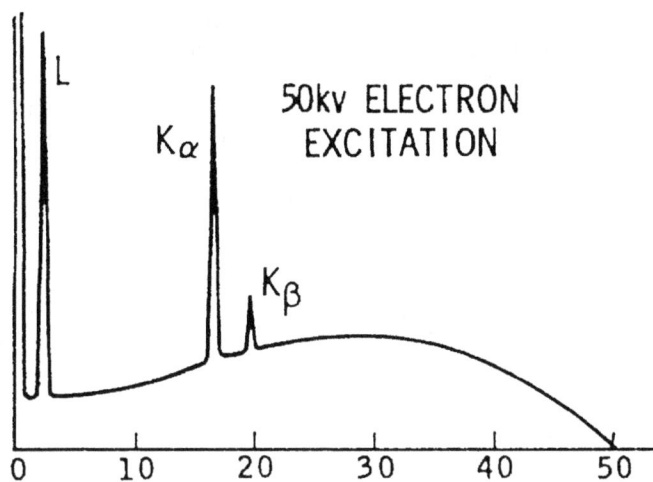

FIGURE 11.32 Schematic characterization of the X-ray emission from a tungsten target excited by electrons. This shows the tungsten lines superimposed on the continuous spectrum.

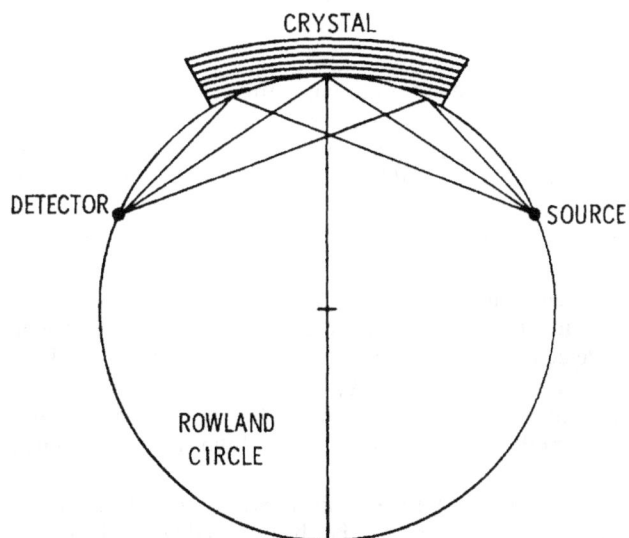

FIGURE 11.33 Schematic of Johansson focusing in a wavelength spectrometer. The crystal is bent to an arc generated by a radius equal to the diameter of the Rowland circle.

debate had as its basis the possible self-absorption of X-rays by the sample before emission. An important factor in microprobe analysis was then the positioning of the sample height in the instrument to ensure the correct takeoff angle. In actual practice, the accuracy of the instrument was a function of how well all the mechanical movements could be controlled and the angle measured.

The SEM debuted in 1965, and it also was an electron-beam instrument. The scanning electron microscope, however, scanned the electron beam over the sample surface with the intent of generating an image of the surface topography. It was not long before the scanning electron microscopist wanted to "know" the composition of various regions on the sample surface. Initial efforts included the attachment of crystal spectrometers to the SEM, but the smaller beam currents used in scanning microscopy resulted in count rates that were woefully inadequate, leading to impractically long counting times. For unknowns, this factor resulted in analysis times in *hours* or *days*.

Fitzgerald, Keil, and Heinrich[14] were the first to describe the addition of a solid state X-ray detector to an E-beam microanalyzer. Although it suffered from several deficiencies (poor resolution, poor sensitivity, and peak overlap), it was immediately recognized as a possibility for use on an SEM. Even in its earliest and crudest form, it overcame some of the limitations of a wavelength spectrometer, providing simultaneous detection and analysis of all X-ray energies, higher count rates, real-time display of the spectrum as it was acquired, and simpler manufacture and use than a crystal spectrometer.

The solid-state detector does not determine the composition of the specimen by analyzing the wavelength spectrum emitted. Rather, the solid-state detector determines the characteristic energies of the emitted X-rays, which corresponds to specific elements. This is based on the familiar equation:

$$E = \frac{hc}{\lambda} \tag{11.3}$$

The solid-state detector consists of a lithium-drifted (doped) silicon wafer, a PIN diode with an external bias. When an X-ray strikes the diode surface, it penetrates and ionizes atoms in the depletion region. The production of each free electron requires 3.8 eV of energy. Thus, a 3800 eV X-ray would produce, at most, about 1000 free carriers. These are collected as a charge pulse of a given quantity, analyzed by pulse height, and displayed on a chart with energy as its abscissa and number of collected counts as the ordinate.

The chief advantage of the solid-state detector, the energy-dispersive X-ray spectrometer (EDS), is the simultaneous analysis of X-rays of all energies entering the detector (Fig. 11.34). Thus, the detector does not need to be adjusted to detect iron (Fe) X-rays and then be changed to detect chromium (Cr) X-rays as in a wavelength spectrometer. As the X-rays are detected and analyzed, the display shows a real-time formation of spectrum peaks, which are related to both the elements present and the relative amount of each element present.

The complexities involved in EDS analysis are discussed in great detail in several excellent publications,[10,11] and these will not be repeated here. The intent of this material is to broadly sketch an outline for the everyday use of EDS to solve failure analysis problems, with some of the more common pitfalls clearly noted. These will be classed under headings such as beam voltage, beam specimen interactions, spectrum analysis, and the mechanics of the instrument itself that may lead to erroneous results.

11.5.1 Beam Voltage

The energy of the electrons in the exciting beam (beam voltage) must be greater than the energy of the X-ray line to be excited. To efficiently produce an X-ray from a sample, the

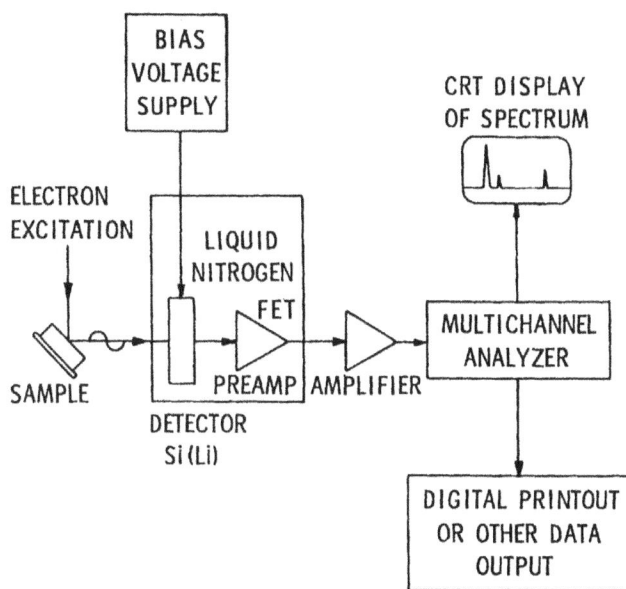

FIGURE 11.34   Block diagram of an energy-dispersive spectrometer. The sample and detector are within the SEM chamber, at vacuum.

beam should be at least 2× the energy of the line to be excited for quantitative analysis. For this reason, it is standard practice for SEM-EDS be performed at high beam voltages, 20 kV and higher. This is a good standard operating methodology unless the sample is not a homogeneous bulk specimen. All of the cautions discussed regarding beam penetration and reaction volumes for image signal emission are equally pertinent. The point, in fact, is that beam voltage, sample penetration, and signal-producing reaction volume are much larger for X-rays (because of their much greater escape depth) than for secondary or back-scattered electrons. This being recognized, any sample that has features smaller than several micrometers in size, either spatially or as layers of dissimilar materials, may need to be analyzed at less than the optimum beam voltage (Fig. 11.35). This is especially true for low-density materials such as aluminum, silicon, magnesium, and carbon compounds. The analyst must recognize that, when analyzing a 1 µm-thick film of aluminum on a dissimilar material, the spectra will contain significant data about the underlying material. To overcome some of these difficulties, it may be necessary to work at lower beam voltages.

The first objection usually voiced is that the primary lines for the heavier elements lie at higher energy levels. This is often true, but *not* for many of the heavier elements of interest in electronic components. Copper (29), a common material in many electronic components and systems, is typically analyzed with a 20 kV beam to excite the Ka and Kb lines at 8.040 kV and 8.904 kV, respectively. Based on standard practice, this would require use of an 18 kV beam. However, if the analyst is concerned about penetration and fluorescence, then the copper La line at 0.930 kV can be easily excited by a 2 to 3 kV beam (Fig. 11.36).

This is possible only because of the nature of atomic structure. The nucleus of an atom is composed of protons and neutrons with the number of protons equal to the atomic number, Z, of the element. In its rest or un-ionized state, the positive charge of the nucleus is balanced by an equal number of negative charges (electrons) in orbit around the nucleus. Based

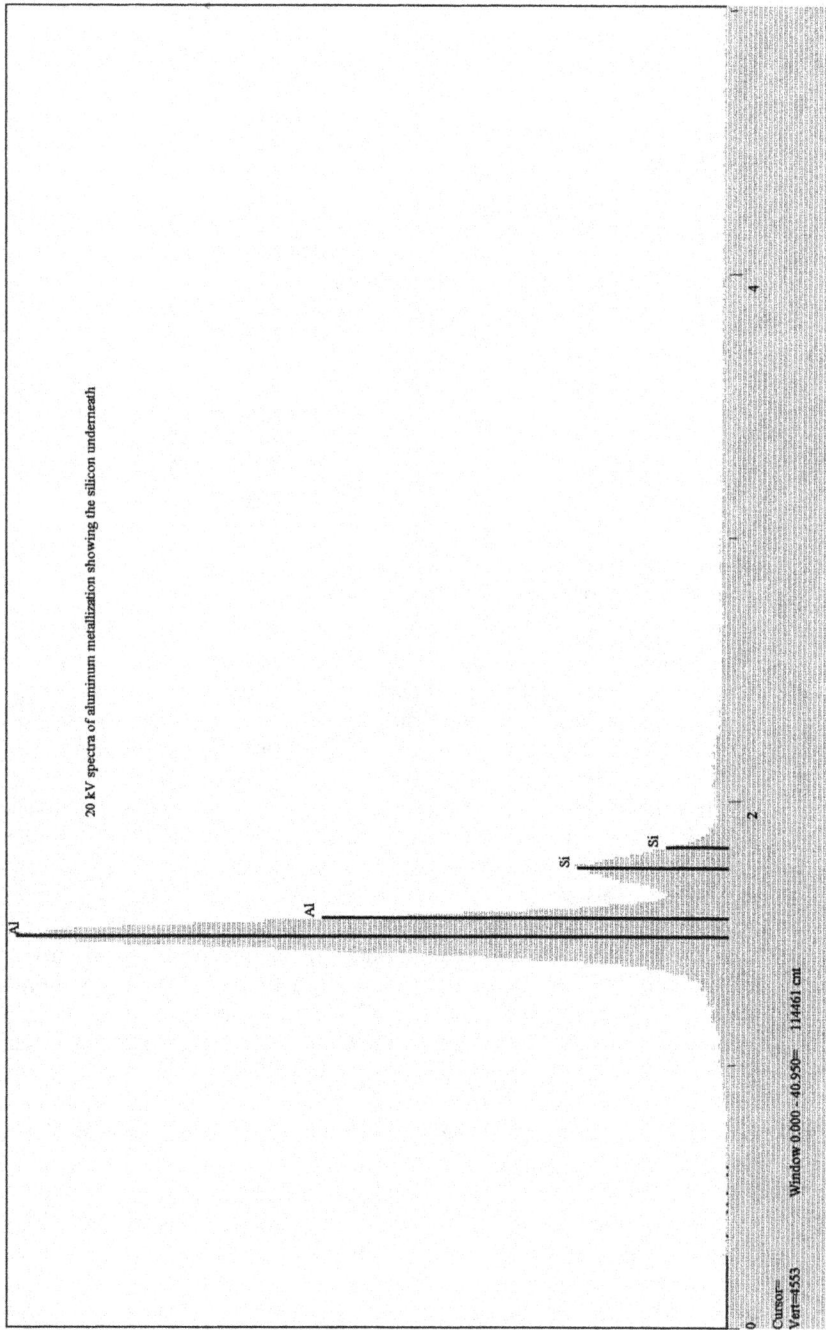

FIGURE 11.35 Spectrum of a 1 μm thick aluminum film over insulating silicon dioxide. At a beam voltage of 10 kV, the beam is penetrating the aluminum and producing X-rays from the underlying silicon.

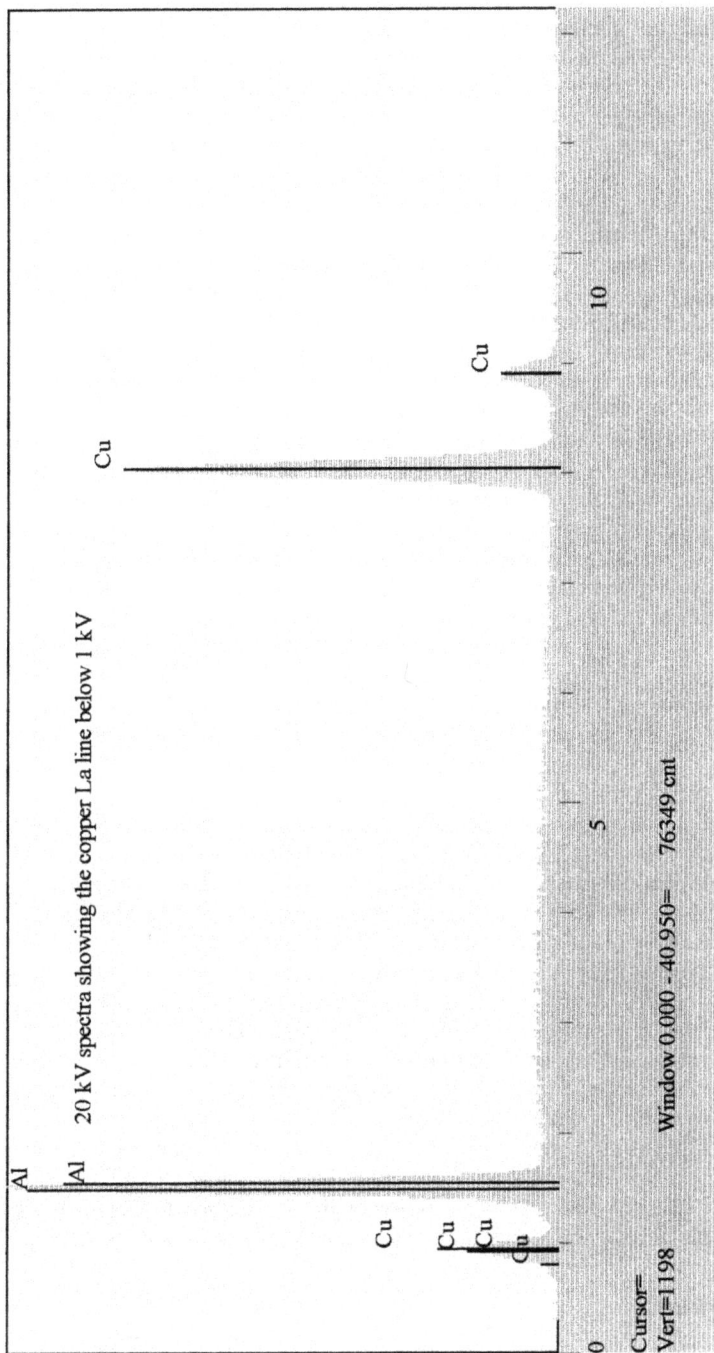

FIGURE 11.36 Spectrum of copper and aluminum obtained at 20 kV. Note that the copper La line at 929 eV is quite obvious.

on Bohr's theory, these electrons must occupy specific orbits around the nucleus. As the number of protons increases with the atomic number Z, the number of electrons surrounding the nucleus must also increase. But these orbital electrons are restricted to certain energy states defined on the basis of the three quantum numbers, n, 1, m. The first quantum number, n, defines the shell of the electron ($n = 1$ corresponds to the K shell, $n = 2$ to the L shell, and $n = 3$ to the M shell). Based on Bohr's theory of the atom, the number of electrons in a given shell is fixed and, as the atomic number increases, the number of shells filled will increase. The K shell can only contain 2 electrons in a single orbit, whereas the L shell can contain up to 8 electrons, 2 in each of the first 2 sub-shells and 4 in the third sub-shell. The M shell can contain up to 18 electrons arranged in 5 sub-shells in the order 2:2:4:4:6. It is the interaction between the incoming beam electron and the electrons that are ejected from a given shell, which gives rise to the multiplicity of X-ray lines for the heavier elements. The elements hydrogen through phosphorous (1 through 15) are characterized only by K series X-ray lines. The elements sulfur through cesium (16 through 55) are characterized by multiple lines in the K and L series. The heavier elements barium through uranium (56 through 92) are usually characterized by their L series and M series lines.

This multiplicity of lines in the X-ray spectra can be both a confusing factor and a method to positively identify a given peak out of the jumble that may be present for a complex specimen. If high beam voltages are possible due to the relative homogeneity of the specimen, confirmation of both high- and low-energy lines will aid in confirmation of a given element. Often, this can be accomplished with a single set of lines that are unique for a given element. For example, tin (Sn) Z = 50, exhibits a triplet of lines from the L series at 3.443, 3.662, and 3.904 kV. The presence of these three peaks in the correct ratio is a unique identifier for tin. The 3.443 Lal peak is the major, with the Lbl peak at 3.662 being 75 percent as intense and the Lb2 peak at 3.904 only 17 percent as intense as the principal. On most newer EDS systems, when the K, L, M markers are called up for display, the marker lines will be scaled to show this relative intensity.

What other heavy elements have lines at the lower beam voltages? Nearly all of them. Iron, chromium, nickel, tin have lines under a kilovolt, while the truly heavy elements like tungsten, platinum, gold, and lead have lines in the 1 to 3 kV range. The author has found the Ma gold (79) line at 2.121 keV to be extremely useful for examining thin gold-plated layers. Tungsten's (W) Ma line at 1.774 unfortunately coincides with silicon, which can be a problem with many semiconductor components. Lead (Pb) emits a strong Ma line at 2.342 kV, but it overlaps the Ka line of sulfur (16), which poses quite a problem in many systems (Fig. 11.37). Beam spread and X-ray emission from a large area at higher beam voltages can affect the accuracy of results when analyzing materials in cross section. If the analyst is attempting to analyze a thin plated layer with other layers over or under it, beam spread can generate a spectra including the materials of the adjacent layers. The analyst is then confronted with the trade-offs of resolution, spot size, beam current, volume of X-ray emission, and adequate energy to excite the desired X-ray lines. The analysis of small particles is quite similar, wherein the compromise is between the possibility of exciting all necessary lines without X-ray emission from the surrounding regions.

Beam voltage in conjunction with the atomic number of the element determines the penetration/reaction emission volume of X-rays from the specimen.

11.5.2   Beam Current

Beam current density (gun brightness) is a function of the electron gun source itself, whereas beam current is a function of the gun brightness and the spot size. For a given gun

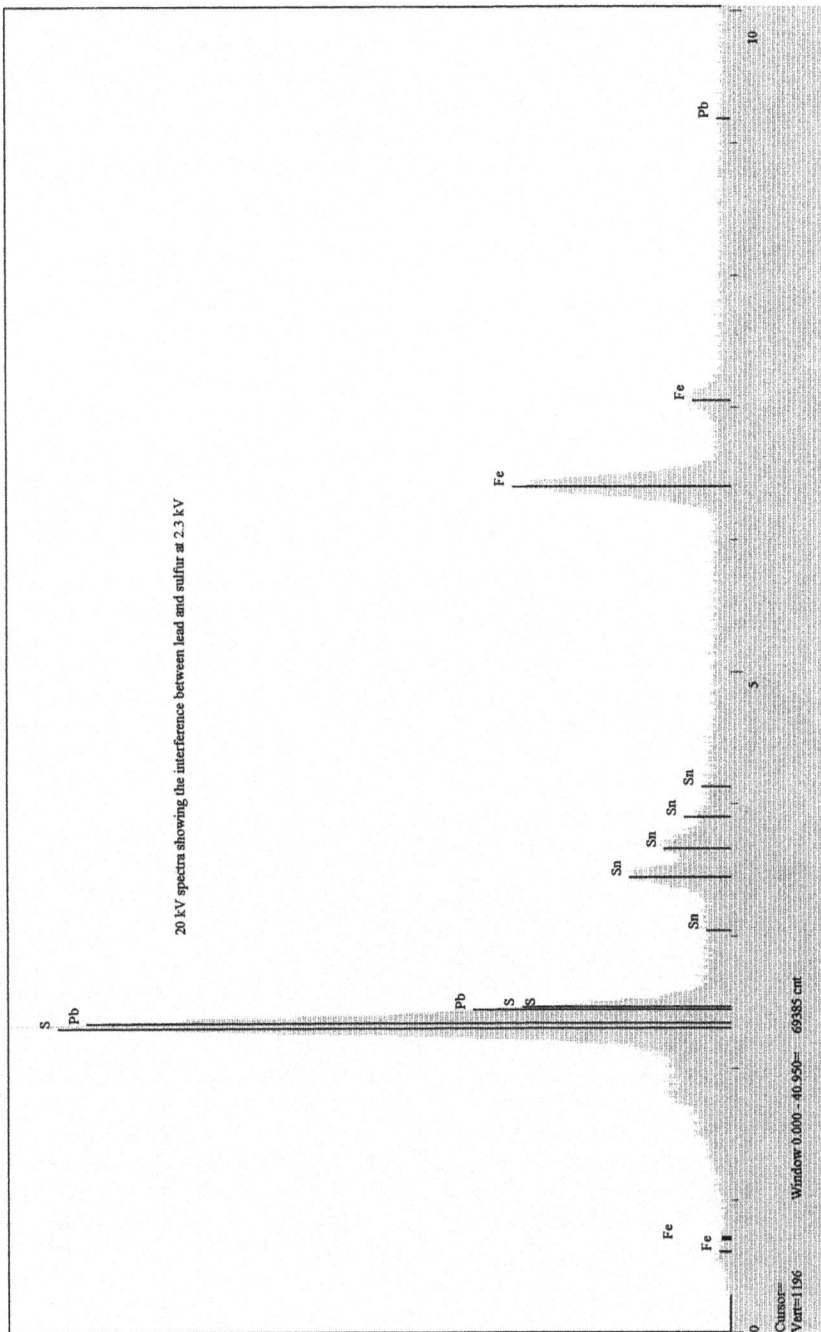

FIGURE 11.37   Spectrum of sample containing lead (Pb), sulfur (S), tin (Sn), and iron (Fe). The lead Ma line obscures the primary Ka sulfur line.

brightness, the actual beam current striking the specimen is determined by the diameter of the spot itself. The requirements for the generation of a high-resolution emissive micrograph are quite dissimilar to those for the generation of an X-ray spectra. Typical high-resolution imaging requires a small spot (10 nm) with beam currents in the 10 to 50 pA range. This is a flux of only 8×10^8 electrons per second. X-ray generation usually requires beam currents in the 10 to 50 nA range for acceptable counting rates. The actual optimum counting rates will vary with each individual system design. Total count rate is a function of the beam current and beam voltage and the particular specimen.

11.5.3   Windows

There are three basic types of detector systems in common usage today:

- windowless
- thin window
- beryllium window

Beryllium window systems were the first and are still the most common. The window is actually the opening between the specimen-chamber sample (under vacuum) and the X-ray detector element itself. In nearly all systems, the silicon drifted detector element is kept at a liquid nitrogen temperature, and this accounts for the presence of the LN_2 dewar. The low temperature is needed to prevent migration of the lithium doping in the silicon diode from drifting under bias. In addition, it minimizes electronic noise in the front-end high-gain portion of the preamplifier attached directly to the detector. At these temperatures, exposure of the detector element to air when venting the chamber would result in condensation of moisture on the element, degrading its performance. It would also result in excessive use of LN_2 to keep the element cold. In use, in the chamber under vacuum, the LN_2-cooled surfaces act as a cold trap for chamber contaminants and pump oil. The beryllium window is placed between the detector and the chamber and acts as a vacuum plate. Beryllium was initially chosen for this purpose because it has adequate strength as a thin film to withstand the pressure differential when the chamber is vented to air. At atomic no. 4, it is the lightest of useful metals and, as a very thin film, will transmit X-rays with energies greater than 0.68 keV, the Ka line for fluorine. Weaker X-ray lines are absorbed by the window material, so beryllium window systems cannot detect the lighter elements such as carbon, oxygen, and nitrogen. Because carbon, oxygen, and nitrogen are of great interest to failure analysts, this is a serious deficiency in the system. To overcome this weakness, there are two other types of systems available. The windowless system uses a turret mechanism with which the beryllium window normally in place can be rotated out of the detector path, and the region between detector and sample can be left open. This allows X-rays of all energies to strike the detector and be analyzed. This technique allows detection and analysis of X-rays of the light elements including carbon, oxygen, and nitrogen. Thus, all hydrocarbon compounds can be confirmed, although not to the degree that other analytical techniques provide. All organics are the same except for trace elements, and these may provide the subtle clue to their origin. Oxygen and nitrogen are critical elements, because they occur in many samples as compounds of other elements as oxides or nitrides. The windowless system suffers from several defects, the turret window changing system tends to develop vacuum leaks, and the LN_2-cooled detector surface still acts as a cold trap for contaminants and oil.

Thin-window systems are now produced that overcome the difficulties of the other systems. With recently devised manufacturing techniques, it is possible to manufacture a window out of thin films of oxides, such as silicon. These very thin, light-element oxide windows overcome most of the drawbacks of the other systems, but they must be treated carefully to ensure that the window material is not damaged.

11.5.4 Mechanical Concerns

The electron probe microanalyzer that preceded the energy-dispersive spectrometer was a combination electron beam and mechanical system, with the quality of its mechanical mechanism determining its analytical accuracy. Location of the sample in relation to detector crystal on the Rowland circle was critical. Due to the restrictions of *takeoff* angle, it was difficult to analyze samples with extreme surface topography. The EDS as used with an SEM overcomes some of these difficulties. Takeoff angle becomes irrelevant for qualitative EDS, as does surface topography. As long as the analyst can ensure a clear line-of-sight path from the point on the sample to the EDS detector window, X-rays can be detected and analyzed. Location of the sample in relation to the detector is critical insofar as height within the chamber is concerned. The sample X-ray emitting region must be positioned to maximize the surface area of the detector pointed toward the specimen. Since, in most systems, the detector motions are limited compared to sample movement, it is the sample that must be positioned to increase collection/detection efficiency. For any given detector-chamber specimen combination, there will be an optimum height and tilt arrangement in the chamber (Fig. 11.38).

A common mistake in X-ray spectroscopy is the use of the emissive image to determine the point on the specimen surface for X-ray analysis. The X-ray detector does not "see" the specimen in the same way as the secondary electron collector. X-rays are line-of-sight only and cannot be collected by the detector—only intercepted. There may be obstructions (portions of the sample) between the X-ray detector and area of analysis that block off the direct line between X-ray source and X-ray detector. Another common mistake is attempting analysis of a curved surface, such as a wire. The +300-V field of the secondary detec-

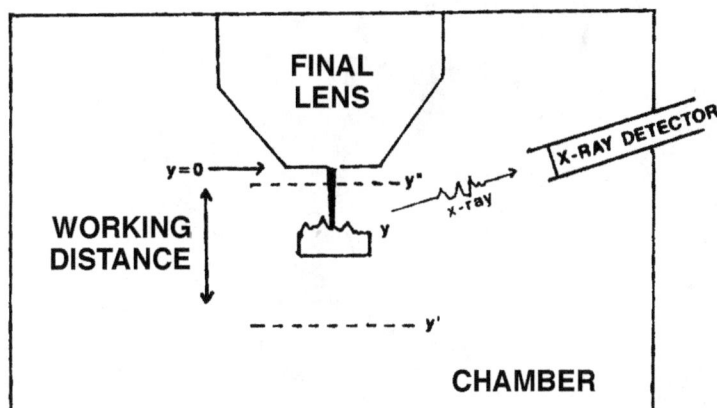

FIGURE 11.38 Schematic of sample chamber showing location of final lens, sample, and X-ray detector. There is an optimum sample location for each SEM configuration to acquire X-rays for spectral analysis.

tor allows imaging of the wire surface away from the secondary detector, but there is no similar collecting field for X-rays. Therefore, although the side of the wire away from the secondary detector can be imaged, it cannot be analyzed Fig. 11.39).

11.5.5   Detectability, Accuracy, and Sensitivity

Typical questions regarding EDS are its sensitivity, quantitative accuracy, and detectability limit. For detectability-sensitivity limits, there is no single number that can be quoted for any system. The ability of the instrument to detect the presence of any given particular element depends on several factors, amount of unknown present, atomic number of unknown, atomic number of matrix in which unknown is present, and conditions regarding unknown distribution such as thin film, particle, or regions of concentration in the bulk [e.g., as lead (Pb) in solder]. In most discussions of analytical instrumentation, an EDS appears to be crude and clumsy. With its lower detectability limit of 0.1 to 1 percent, it cannot be compared to atomic absorption spectrometry with its part-per-billion capability. Its accuracy at lower concentrations is often ±100 percent. What must be kept in mind is the actual task set for the instrument, nondestructive analysis of a specimen or portion of a specimen that cannot even be seen with the naked eye. Often, the analytical area will be small portions of a micrometer in size.

When this is kept in mind, the analytical capabilities begin to take on a new light. The capability to analyze a region 1 µm in diameter and the ability to determine material

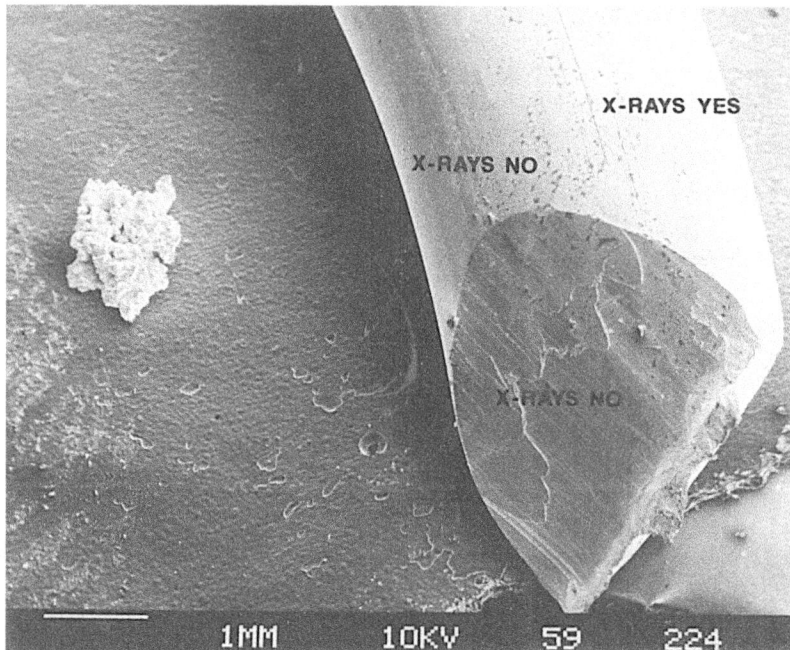

FIGURE 11.39   This SEM micrograph of a wire shows the locations where X-rays may be detected for this particular detector location. They do not necessarily agree with the location from which secondary electrons can be captured for imaging.

present down to the 1 percent range is an astounding feat that is repeated routinely by the SEM-EDS operator. For failure analysis applications, the ability to analyze small samples overrides the need for quantitative accuracy. Often, there is no need to quantize the data merely to determine the elements present or in what relative concentrations. A coil failure mechanism in miniature relays is an open or high resistance coil winding. After location of the failure site by electronic diagnosis the cause of the anomalous high resistance must be determined. Examination and analysis by SEM may determine the mechanism to be corrosion of the copper wire through a pinhole in the enameling. The detection of chlorine by EDS would point toward activated solder flux, a fingerprint, or possibly exposure to chlorinated washing solutions. The amount of chlorine present at the failure site is irrelevant, since it can vary by orders of magnitude throughout the corroded region. The key factor is the presence of chlorine, or in another case it could be sulfur. The analyst is looking for the reaction initiator and quantitative analysis has limited value in the failure analysis.

11.6 REFERENCES

1. Power, H. *Experimental Philosophy, in Three Books.* McCormick, 1987.
2. Gabor, D. *The Electron Microscope.* Chemical Publishing, 1948
3. Gabor, D., ibid.
4. Thornton, P.R., *Scanning Electron Microscopy.* New York: Chapman & Hall, 1968.
5. Everhart, T. Reflections on scanning electron microscopy, *Proc. Symposium on Scanning Electron Microscopy,* IITRI Research Institute, 1968.
6. Thornton, P.R., ibid.
7. Wells, O. *Scanning Electron Microscopy.* New York: McGraw-Hill, 1974.
8. Meny, F.M., and R. Tixier. *Microanalysis-Scanning Electron Microscopy.* Les Editions de Physique, 1978.
9. Holliday, J.E., and E.J. Sternglass. Backscattering of 5–20 keV electrons from insulators and metals, *J. Appl. Phys. 28,* 1957.
10. Bishop, H.E. Some electron backscattering measurements for solid targets, *Proc. Fourth Cong. Intl. X-Ray Opt. Mic.,* Orsay, 1966.
11. Thornley, R.F.M. New Applications of the Scanning Electron Microscope (PhD diss.). Cambridge Univ., 1960.
12. Thornley, R.F.M., and L. Cartz. Direct examination of ceramic surfaces with the scanning electron microscope, *J. Amer. Ceram. Soc. 45,* 1962.
13. Moseley, H. The high frequency spectra of the elements, *Phil. Mag. 26,* 1913.
14. Fitzgerald, R., K. Keil, and K.F.J. Heinrich, *Science 159,* 1968.

CHAPTER 12
MISCELLANEOUS TECHNIQUES

Dennis H. Van Westerhuyzen

Raytheon System Company, Component and Materials Laboratory

12.1 *PACKAGE OPENING TECHNIQUES*

12.1.1 General Considerations

Package opening techniques refers to the opening of metal, ceramic, or metal-ceramic type packages. *Decapsulation* and *depotting* refer to opening of plastic encapsulated type packages. The package is opened to expose the component only after all pertinent electrical tests and package evaluations are complete. This includes radiographic examination, surface contamination analysis, hermeticity, and residual gas analysis. Every effort must be made to preserve the interior of the package without introducing foreign material or losing contaminate material that may be in the package cavity.

12.1.2 Metal Packages with Seam-Sealed or Brazed Lids

In opening metal packages with seam-sealed or brazed lids, the weld bead or seam is ground thin with a hand-held miniature electric grinder or a file. One edge or corner is ground through to allow the blade of a sharp knife to be inserted and the lid pried off. Again, great care must be exercised to prevent damage to the component or the introduction of foreign material.

A motor-driven grinding wheel with 400- or 600-grit silicon carbide paper may also be used to thin the weld seam on some package configurations.

In removing metal lids from ceramic packages, the whole lid may be ground until only a thin layer remains, as evidenced by the *oil canning* of the lid. Place tape over the bulging lid and quickly remove the tape; the lid should come off with the tape.

Round metal cans can be opened with a miniature can opener designed specifically for the job. It operates like a pipe cutter. A cut is made just above the seam, and the is lid carefully removed. Care must be taken not to cut into the base—only into the lid.

Grinding the weld seam using a hand-held miniature grinder could open larger or irregularly shaped metal cans. The seam is nearly ground through, then the lid is peeled away from the body using the sharp blade of a miniature knife.

It is important to check the lids frequently for the observance of the oil-canning effect so as not to grind through into the package cavity. It is also very important to clean the package before applying the adhesive tape to minimize the possibility of introducing contaminants into the cavity. To hold the packages during the grinding operation, it will be necessary to fabricate a holding fixture (to save the ends of your fingers) either from a zero-insertion force (ZIF) socket or a specially modified epoxy mount with predrilled holes for the package leads.

12.1.3 Ceramic Packages

The same technique also works for ceramic packages. Again, grind the surface of the ceramic package on a motor-driven grinding wheel with 400- or 600-grit silicon carbide paper. Carefully grind the ceramic until the material over the cavity begins to *oil can.* Apply adhesive tape over the surface and quickly pull it off. The thinned ceramic will adhere to the tape and break away from the package.

As with the process described in the previous section for metal devices, it is important to check frequently for the observance of the oil-canning effect so as not to grind through to the package cavity.

Safety note: Some ceramic packages are made from beryllium compounds to take advantage of these compounds' thermal properties. Beryllium is extremely toxic in its powdered form. Only wet grinding should be used (to minimize the dust), and all papers and water effluent must be handled as hazardous waste.

12.2 DECAPSULATION/DEPOTTING

12.2.1 General Considerations

Chemical decapsulation/depotting refers to chemical techniques or to a combination of mechanical and chemical techniques used for gaining access to the component in question for failure analysis. The chemicals used for depotting/decapsulation are strong acids, bases, or solvents to attack and remove portions of packages. These chemicals pose a serious risk to the operator and to the device being examined. Therefore, the use of these chemicals require utmost care.

Device manufacturers are constantly improving the density and formulations of their potting and encapsulating compounds to increase the protection for the components from the external environment, including improved hermetic properties and thermal properties. This makes the task for the analyst increasingly more and more difficult. The task of the analyst is to remove these "impervious" materials and expose the component without damaging the failure site.

Compounding the problem for the analyst is the wide variety encapsulants and potting materials used today. These include pourable castings, injection moldings, and dipped/ sprayed coatings of epoxies, silicones, phenolics, and urethanes. Fillers are often used in the encapsulants and potting materials, and these alter their properties with respect to the etchants.

12.2.2 Chemical Removal of Encapsulants/Potting Materials

In general, mechanical methods should be used to remove as much of the encapsulant/potting materials as possible without inducing damage to the device being analyzed. This minimizes the time the device will be exposed to the harsh chemicals. These mechanical techniques include the use of a spot facer in a mill or a hand grinder. After as much of the material as possible is mechanically removed, the appropriate chemical, or a combination of chemicals, is used to remove the remaining material and expose the device for analysis.

Table 12.1 is a listing of the commonly used solvents and the material types that they remove.

Table 12.1 Solvent Selection Chart[*]

Solvent	Epoxy	Silicone	Urethane	Photoresist
Alcohol base (alkaline activator)		✔	✔	✔
Methylene chloride base (acid activator)	✔	✔		
Propylene glycol ether base (alkaline activator)	✔	✔	✔	✔
Methy-lpyrrolidone	✔		✔	✔
Concentrate of sulfuric or nitric acid	✔			

[*] *Source:* Data from *Electronic Manufacturing* Vol. 35, No. 13, December 1989, p. 31, used by permission.

An example of the technique is the opening of a plastic encapsulated microcircuit. A well is milled into the package over the area of interest. The package must be well supported during the milling operation or the package may fracture. The cavity should extend to just above the wire loops (as determined by radiographic examination). The purpose of the well is to contain the acid etchant. The process is as follows:

1. Heat the sample on a hotplate in a fume hood to 60 to 100° C.
2. Place a drop of the appropriate acid (fuming sulfuric or nitric) into the well.
3. Allow the reaction to proceed until it begins to subside (approximately 15 s).
4. Remove from the hotplate.
5. Rinse with glacial acetic acid to remove chemically attached packaging material.
6. Rinse the sample gently with a stream of acetone to rinse away the glacial acetic acid.
7. Allow the excess acetone to evaporate from the part.
8. Examine the device under low-power magnification.

Repeat the above steps as required to completely expose the area of interest. Limit the exposure of the part to the acids to a minimum. Glacial acetic acid will attack aluminum after extended exposure.

Modifications of this procedure can be used to decapsulate or depot most electronic components. In some cases, the device must be immersed in the chemical solvent or stripping agent. In some cases, to prevent stress on internal components as the encapsulant is removed, it may be necessary to solder a wire frame to the component leads for support.

Many chemical strippers are commercially available that are highly selective. These have been designed to remove silicones, elastomers, and epoxies. Caution must be exercised in using some these commercial products, as they cause swelling of the material being removed, which may cause internal damage. When using these commercial products, the technique often used is to allow the solvent to work for a short while, then remove and rinse the part, and very carefully, under a low-power microscope, mechanically remove the residual material. The part is then place backed in the solvent, which allowed to work again.

Commercial equipment is currently available that helps to automate the stripping process for plastic semiconductors and minimizes the potential for damage to the die. This equipment pumps a small jet of the etchant (fuming nitric or sulfuric acid) against the surface of the package in the area of interest. The unit is self-contained, which minimizes the exposure of these chemicals to the operator.

12.3 CLEANING ISSUES

12.3.1 General Considerations

Occasionally, during the initial stages of a failure analysis, it is determined that low-level leakage currents are present between different pins of a component. These leakage currents may be attributed to contamination on surface of the device. Technological changes are accelerating the move to ever smaller dimensions, higher densities, and higher complexities. This causes the physical package to shrink, placing greater burdens on the insulating materials and on the external cleanliness of the packages. Because of this decrease in geometric package size, many modern electronic components are quite sensitive to ionic contamination. These ionic contaminations may come from the component manufacturing process (i.e., contaminated processing solutions or contaminated final rinses) or from the use environment (e.g., salt spray, handling, or humidity testing). The insidious problem with low-level current leakages is that they may not manifest themselves until quite some time after the unit is placed in use, because moisture, bias, and temperature exposures may be required before the leakage paths are formed. The identification of external package leakage is not normally a straightforward process but is usually accomplished indirectly by secondary evidence. Direct methods such as *electron spectroscopy for chemical analysis* (ESCA), *energy dispersive X-ray* (EDX), and *auger electron spectroscopy* (AES) may be used by comparing a known-good device with the unit in question. This can be quite time consuming and expensive.

The indirect methods include the following techniques:

- Ambient vacuum bake
- High temperature vacuum bake
- External wash, followed by vacuum bake

12.3.2 Ambient Vacuum Bake

An ambient or low-temperature vacuum bake for 16 hr, or more, with a resulting shift in device electrical parameters is indicative of absorbed or trapped moisture or other volatile material on the surface of the package. The ambient or low-temperature vacuum bake is used when the device being evaluated cannot be exposed to high temperatures.

If residual gas analysis cannot be performed on the device, for any of a variety of reasons, a hole punctured in the package allows removal of moisture and volatile gases from the package interior during this vacuum bake. Change or recovery in the electrical performance of the device suggests the presence of trapped moisture in the package.

12.3.3 High-Temperature Vacuum Bake

A high-temperature vacuum bake (100 to 200° C), with a resulting shift in electrical parameters, is again indicative of contaminants on the package surface. The high-temperature vacuum bake will allow a shorter processing time for changes in electrical leakage to develop.

In both cases of vacuum bake, the device should be placed in a nitrogen-purged cabinet or desiccator until ready for testing. In the case of the high-temperature vacuum bake, this will allow the device to cool and stabilize prior to testing without absorbing any additional water. The devices must also be handled very carefully after the vacuum bake to prevent recontamination.

12.3.4   External Washes

External package washes should be considered when any one of the following have occurred or been observed:

1. The external vacuum bake has shown a positive change in the electrical performance of the device.

2. The external package surface shows signs of degradation.

3. Chemical deposits or corrosion are evident on the package surface.

Based on the previous observation, several types of chemical washes are available. The primary and most often used is the deionized or distilled water wash. When salts or ioniz-able contaminates are suspected, a 5 to 10 min wash in hot deionized or distilled water is performed. After blow drying, the device is given a 16-hr vacuum bake to remove the residual moisture from the surface. The device is then retested, and any improvement in electrical parameters indicates the presence of ionic contamination.

If organic contamination is suspected, the device is washed in boiling acetic acid for 5 min, followed by a double boiling wash in deionized or distilled water. A vacuum bake for 16 hr follows to remove residual moisture. Any improvement in the electrical parameters indicates the presence of organic contamination.

If deposits of metals are suspected, a wash of dilute nitric or hydrochloric acid is rec-ommended. The analyst, based on a knowledge of the manufacturing and use environ-ments and the materials of construction, determines the choice of acid. As with the acetic acid wash, the dilute acid wash is followed by double boiling in a deionized or distilled water rinse, followed by a 16-hr vacuum bake. Any change in subsequent retest parameters is indicative of metallic contamination.

Caution should be exercised in that acetic acid should not be used on devices with high lead (Pb) content glasses or frits. The boiling washes should not be used on packages if the temperature will damage the device. The washes should not be used on nonhermetic devices, as they may be permanently damaged.

12.4   PARTICLE IMPACT NOISE DETECTION TESTING

12.4.1   General Considerations

Particle impact noise detection (PIND) testing is used to detect loose particles in the sealed cavity of a package. The technique can be used only on devices that have an internal cavity. If conductive particles are trapped inside a package cavity, it may contribute to device failure (e.g., shorts). However, the test cannot differentiate between conductive and nonconductive particles. It detects only loose particles of a sufficient mass that results in the excitement of a transducer. The test is currently defined by MIL-STD-833, Method 2020.

12.4.2 Basic Equipment

Several commercial PIND testers are available, but the overall theory and principle of operation are the same. The basic equipment consists of a vibration system that shakes the device under test (DUT) at various frequencies. The system is equipped with a mechanism that can impart a series of mechanical shocks to the device. A transducer, on which the part is mounted, detects the noise generated by a loose particle striking the cavity walls. The signal from the transducer is fed to the filter/amplifier, which provides an audio output and a signal to a visual display.

12.4.3 Test Procedure

The part to be tested is placed on the transducer head using either a grease-like couplant or a double-face adhesive tape. This attachment medium must minimize particle noise signal attenuation and any extraneous noise. The transducer is attached to the vibration system and is made to vibrate over a range of frequencies, and the part is struck with a high g-force three times during the vibration. Any loose particles in the package that strike the cavity walls create acoustic waves in the 100 to 300 kHz regions. The transducer coverts these into electrical signals that are used to power an audio speaker and are simultaneously displayed on an oscilloscope.

If a grease-like couplant is used to attach the device to the transducer, the grease-like couplant must be cleaned off (typically by degreasing) after the test. If this grease-like couplant is not removed, it can give erroneous readings in subsequent hermeticity tests.

The shock portion of the test is important to dislodge any particles that are loosely attached to internal surfaces, or that might be held in place by electrostatic forces. If the detection of the particle is not repeated, the device should be considered suspect, because the particle may have again become lodged so that it cannot be vibrated free.

The vibration level required depends on the cavity size. The frequency must cause the particle to maximum travel in the package cavity and to give the maximum number of "strikes" with the package walls. The optimum frequency is normally in the range of 30 to 90 Hz.

12.5 HERMETICITY TESTING

12.5.1 General Considerations

Hermetic seals on electronic devices serve one of two purposes. The first is to exclude the external gases and fluids from entering the package, and the second is to prevent gases or fluids sealed in the package from leaking out. Examples of the first are the egress of moisture, halogenated organic molecules, or other vapors into a package from the external atmosphere. For many electronic devices, moisture or halogenated organic molecules in the package cavity, even in small quantities, can cause a wide range of failures from electrical leakage to corrosion. An example of the second case is to prevent a corrosive material, (e.g., sulfuric acid) from leaking out of an electrolytic capacitor, resulting in failure of the device and corroding the electronic substrate on which the device is mounted.

Every hermetically sealed package leaks at some defined rate. This leak will cause the internal package ambient to change over time through a diffusion process until the internal and external package chemical compositions are nearly identical and are near equilibrium. If the leak rate of the package is low enough so that the internal atmosphere is maintained for the expected life of the device, then the package is considered to be *hermetic*. The larger the internal package volume, the higher the leak rate can be while maintaining the required package atmosphere for the life of the device.

The primary source for moisture in the package is ingress because of a poor or failed seal. However, moisture may also be sealed in a package as a result of processing or assembly, which can come from two areas:

1. It can be adsorbed moisture on internal surfaces.

2. It can be internally generated after package sealing from material decomposition or outgassing.

The detection of trapped internal moisture is addressed in Sec. 12.6.

The spectrum of leak rates is divided into two different regimes: gross and fine leaks. The standard leak rate is defined as

...the quantity of dry air at 25° C in atmosphere cubic centimeters flowing through a leak or multiple leak paths per second when the high pressure side is at one atmosphere (760 mm Hg absolute) and the low-pressure side is at a pressure of not greater than 1mm Hg absolute.[1]

Units are expressed as atm cm^3/s. Figure 12.1 shows the resolution limit for the various leak-testing methods.[2] Table 12.2 also defines the range for gross leaks is 10^{-1} to 10^{-4} atm cm^3/s, and the range for fine leaks is 10^{-5} to 10^{-12} atm cm^3/s. A gray area exists between the leak rates of 10^{-4} to 10^{-5} atm cm^3/s, where leaks are not easily characterized by either the gross- or fine-leak test techniques. Table 12.2 shows the operational range, the test method, tracer, and the MIL-STD-883D specification requirement.

Two different test methods are required, because gross leaks cannot be detected with the fine-leak methods. When a device with a gross leak is placed in test equipment used for fine-leak testing, all of the tracer gas would be removed by the vacuum system before the mass spectrometer could be operational.

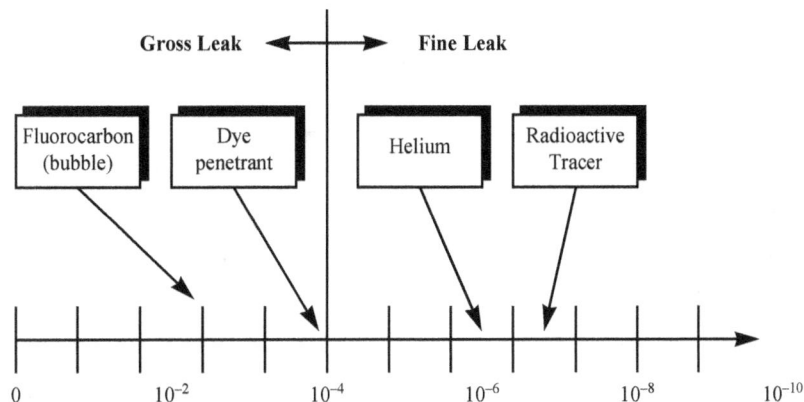

FIGURE 12.1   Resolution limit for various leak-test methods.

Table 12.2 MIL-STD-883D Specifications for Fine- and Gross-Leak Tests

Leak test	Operational range (atm-cm^3/s)	Test method	Tracer	MIL-STD-833 test condition
Fine leak	10^{-6} to 10^{-10}	Helium leak detection	Helium	A_1, A_2, A_3
		Radioisotope	Krypton-85	B
Gross leak	10^{-1} to 10^{-5}	Bubble	Fluorocarbon	C
		Dye penetrant	Fluorescent dye	D
		Weight gain	Fluorocarbon	E

12.5.2 Fine-Leak Tests

Under normal conditions, fine-leak testing is nondestructive and precedes gross-leak testing. If gross-leak testing is performed first, two possible problems can arise. First, it is possible to have the gross-leak tracer fluids plug the capillary leak paths, which would prevent the tracer gasses used in the fine-leak testing from entering the package. Second, gross-leak testing can be potentially destructive if the tracer fluid enters the package.

The most commonly used tracer gas used for fine-leak testing is helium. The device is placed in a pressurizing bomb, which is evacuated and then pressurized with helium at a specific pressure for a specified amount of time, depending on the volume of the package. After the pressurizing cycle is complete, the devices is removed from the bomb, blown off with dry nitrogen, and placed in a mass spectrometer, where the rate of helium leakage from the device is measured.

Table 12.3 is from MIL-STD-883, Method 1014[1] and is used to determine the appropriate time and pressure values, based on the internal volume of the DUT using the fixed volume method (A_1) of that specification. These test conditions ensure the test sensitivity for the required measured leak rate, R_1. The flexible method, A_2 allows a variance in the test method conditions to detect the leak rate, and the details are in Ref. 1, test condition A_2.

Table 12.3 Fixed Condition for Testing Condition A_1

Volume of package in cm^2	PSIA (±2)	Bomb condition		Reject limit (atm cc/s He)
		Minimum exposure time in hours (T_1)	Maximum dwell hours (T_2)	
<0.05	75	2	1	5×10^{-8}
$\geq$0.05 to $\leq$0.5	75	4	1	5×10^{-8}
$\geq$0.5 to $\leq$1.0	45	2	1	1×10^{-7}
$\geq$1.0 to $\leq$10.0	45	5	1	5×10^{-8}
$\geq$10 to <20.0	45	10	1	5×10^{-8}

The other fine-leak test method is the radioisotope test (see Ref 1, Condition B). It is essentially the same as the helium mass spectrometer test but uses radioactive krypton, Kr^{85}, as the tracer gas. Here the device is subjected to pressurization in dry nitrogen gas containing approximately 1 percent of the radioactive krypton. After the pressurization, the package is placed in a scintillation counter, which detects the gamma emissions from the beta decay of the Kr^{85}. Since the gamma radiation is powerful enough to penetrate the package walls, the scintillation count is a direct measure of the amount of Kr^{85} in the package. The advantage of the Kr^{85} testing over the helium mass spectrometer is that of much higher sampling rates. The radioisotope is also more sensitive that the helium method. The disadvantage is the potential hazard of the high radioactivity. In addition, the equipment is more costly and special permits usually are required by local agencies.

A note of caution is in order with any of the fine-leak testing techniques. Any contamination, such as polymeric materials or highly modeled surfaces on the device, may lead to the absorption of the tracer gas, which will be detected in the final measurement and will provide a false reading. If possible, all foreign material should be removed from the device before the pressurization cycle.

12.5.3 Gross-Leak Testing

One of three gross leak tests may be used to find leaks in the range of 10^{-1} to 10^{-4} atm cm^3/s. They are the fluorocarbon bubble test, the fluorocarbon weight gain test, and the dye penetrant test. In the fluorocarbon bubble test, the devices are subjected to a vacuum evacuation of 5 torr for 1 hr. While in vacuum, the device is covered with a high vapor pressure fluorocarbon fluid (typically 3M Electronic Product Divisions' FC-84), which is forced into the device through any leak paths when the pressure is returned to 1 atm. The device is then placed in a heated (125° C) fluorocarbon bath (typically 3M Electronic Product Divisions' FC-40), which is higher than FC-84's boiling point. Any FC-84 that had leaked into the package during the evacuation-pressurization step will now vaporize and appear as bubbles that, with the aid of a magnifying glass, are observable against a black background. The greater the leak rate, the greater the generation and effluence of bubbles. As in the helium leak test, time and pressure values are based on the internal volume of the device and are detailed in Ref. 1.

The weight gain test is similar to the bubble test. Prior to the pressurization step, the device is cleaned, baked, and weighed to 0.1 mg. As in the bubble test, the devices are evacuated and then covered with the low-viscosity, low-vapor-pressure fluorocarbon fluid. After the devices are removed from the tracer fluid and dried, the devices are weighed again to determine the weight gain from the entrapped tracer fluid. Depending on the devices' internal volume, weight gains of 1 to 2 mg typically are cause for rejection. The disadvantage of this technique is that it does not provide the location of the leak as does the bubble test. As with the fine-leak testing techniques, the packages should be free of foreign materials to ensure accurate results.

The dye penetrant test uses a fluorescent dye as the penetrant liquid and is typically used on devices with transparent walls. The devices are pressurized in a fluorescent dye, removed, washed with a suitable solvent, and then visually inspected using an ultraviolet light as the light source. Any evidence of the dye penetrant inside the package is cause for rejection. The disadvantage of this method, besides requiring transparent walls on the package, is the fact that gross leakers many be damaged by the penetrant dye, preventing any further analysis.

12.6 *RESIDUAL GAS ANALYSIS*

12.6.1 General

Residual gas analysis (RGA) is used to determine the stoichemical composition of the atmosphere inside package cavities. These could be integrated circuits, hybrids, relays, or other cavity-type packages.

As described in Sec. 12.5, the gases inside a sealed package can have a significant effect on the device function, with moisture being the main concern, because of its ability to contribute to many different failure mechanisms. Moisture in a hermetic package can come from the moisture in the sealing atmosphere, from decomposition materials used inside the package, or from outgassing of materials inside the package.

12.6.2 Procedure

The RGA procedure is covered in MIL-STD-883D, Method 1018 and, while the description of the RGA process appears relatively simple, it is indeed a complex process with subtle technical difficulties. A mass spectrophotometer, typically a quadrupole, is used to identify the atmosphere in a package. The package is placed in a holding fixture in a small vacuum chamber. The chamber is evacuated and heated to 100° C for 12 to 24 hr to reduce the background gas levels before the test. The package is then punctured with a sharp-pointed tool, and the gas is released into the mass spectrometer, where it is analyzed. The quadrupole mass spectrometer is normally used because of its ability to be tuned to unit resolution across the mass range of interest (1 to 100 atomic mass units) and its rapid scan feature, which allows it to characterize flowing gases or rapidly changing gas concentrations.

Gas flow dynamics plays an important part in the accuracy of the analysis. The puncture pin will make a mechanical seal with the metal lid of the DUT. The pin must then instantaneously retract to completely break this seal and introduce the cavity gas into the system. If this does not happen, the gas dynamics will be considerably different from those produced during the calibration sequence. Low atomic mass gases will escape first, followed by the heavier gases. This will introduce erroneous results.

Ceramic-lidded packages must have their lids thinned by mechanical grinding prior to being placed in the RGA chamber. This will allow the lid to be easily fractured and the gas introduced to the analytical system. If the lid is not thinned, the package could fracture, introducing contamination into the system and producing false results. It the lid is thinned excessively, the lower atomic mass gases in the package could diffuse through the thinned section, also resulting in erroneous readings. This diffusion would be accelerated during the vacuum-bake cycle prior to the puncturing operation.

System calibration is the key to the accuracy in measuring such a small volume of gas from one of these devices. The calibration system must contain a number of different calibrators that will simulate the sample bursts using different gases. The volume of these different calibrators can range is size from 0.01 to 20.0 cm^3.

12.6.3 Interpretation of RGA Data

A good understanding of processes and materials used in device manufacture is necessary to interpret the results of RGA.

Moisture. Moisture in a package can come from four different sources: a hermetic seal failure, thermal decomposition of materials sealed in the package, internal outgassing of material in the package, or a poor sealing atmosphere. The loss of hermetically is the easiest to determine. By evaluating the analysis spectrum for oxygen and argon peaks, and if they are present, then the moisture in the package is probably due to a hermetic seal failure.

The thermal decomposition of materials sealed in the package is usually accompanied by the generation of carbon dioxide. Desorbed moisture outgassing of material in the package or from package walls is not typically accompanied by other gases. Moisture entering the package from the sealing gas system typically is a rare occurrence, and more difficult to determine. Packages from the same lot or date code can be analyzed and, if gas stoichemistry is similar, the sealing gas system may be the cause.

Nitrogen.   Many hermetic devices are sealed in nitrogen, and this gas is normally the major constituent found during RGA. Other sealing gas mixtures are dry air (78 percent nitrogen) or nitrogen-oxygen mixtures.

Oxygen.   Oxygen may be used in the sealing gas when sealing ceramic packages with a glass frit. However, because of the high sealing temperatures, the oxygen may react with the carbonous materials in the glass frit and produce carbon dioxide and water.

Argon.   Argon may be detected in a packaged sealed with dry air (approximately 9000 ppm by volume), or dry nitrogen (less than 500 ppm by volume). If the device is suspected as being a leaker, the oxygen-to-argon ratio would be approximately 20:1, and other gases, such as carbon dioxide and traces of carbon monoxide and helium, would also be present in the spectrum.

Carbon Dioxide.   As explained under the paragraph on oxygen, carbon dioxide may be found in packages sealed with a glass frit where a partial pressure of oxygen is used. Carbon dioxide may also be generated as a result of the thermal decomposition of organic die and/or substrate attaches or coating materials used in the package.

Hydrogen.   Hydrogen is used when the package must be sealed in a reducing atmosphere, and concentrations are normally reported at greater than 10 percent. Hydrogen may be used either to prevent the oxidation of sealing surfaces or to prevent performance degradation of certain types of devices.

Other Gases.   The presence of helium may indicate that gas was used as component in the sealing gas. If helium is used in the sealing gas, the unit can be immediately fine-leak checked after the seal operation and would not have to go through the bombing procedure described in the fine-leak sections. The device manufacturer would have to be consulted to determine if this was the case. If helium was not used in the sealing gas, and it and/or fluorocarbons were detected during the RGA, this indicates a fine- or gross-leak failure. The presence of organic gases or vapors is typically outgassing of the die and/or substrate attach materials. Vapors from isopropyl alcohol, 1,1,1-tricholoethane, and methylethylketone may be residuals from the cleaning processes, indicating the device was not adequately baked before sealing.

12.7   *CROSS-SECTIONAL ANALYSIS*

Cross-sectional analysis of a component is typically used as a "last resort" in performing a failure analysis on an electronic component. Cross-sectioning results in irretrievable damage to the component, and additional testing and evaluation cannot be performed. Many times, in performing the cross-sectional analysis, the analyst is looking for defects of 1 μm or less in size. Because of the destructive nature of the cross-sectional analysis, a thorough knowledge of the component is required. This includes the construction, materials used, and the area of interest. Once these are understood, the analysis plan can be formulated.

Because of the small size and precision required, most electronic components must be vacuum encapsulated in a clear medium. Epoxy is normally used because of its transparency and hardness. After the epoxy has completely cured, a window is polished perpendicular to the plane of the section on the side of the mount to give the analyst a window to view the component during the cross-sectioning. In performing vacuum encapsulation, several cautions must be exercised. First, the epoxy must not be outgassed for too long, or some of the hardener will evaporate, resulting in an incorrect ratio of hardener to resin. Epoxy pull-away and shrinkage will also result from excessive outgassing. Outgassing of the freshly mixed epoxy typically requires a 1 to 2 min exposure to 0.1 torr. Second, the epoxy cure is an exothermic reaction and can generate enough heat to damage some components. Placing the mount in a shallow pan of water will provide mass to absorb the heat.

If the sample being prepared is a very hard or brittle material (e.g. ceramic capacitors, semiconductors), barium titanate or glass beads may be added to the epoxy to increase the hardness of the mounting medium. These, however, may obscure the sample in the mount and make finding the area of concern more difficult.

The sectioning procedure usually starts with a course cut or grind using 240- to 320-grit paper on a flat grinding wheel. When the area of interest in near, 320- to 400-grit silicon carbide wet grinding is then used. Final grinding is performed using 600-grit silicon carbide. This is followed by a course polish using 6-μm diamond on an artificial cloth. The final polish is a slurry of 0.05-μm alumina on a short-nap cloth. Polishing time must be kept very short, or microscopic rounding of the specimen will occur.

As with the metallurgists of old, with their lists of numerous etches, today's analyst must now select an etch that will remove smeared material and preferentially enhance certain features of the sectioned component to make them more visible to the optical or scanning electron microscope. In working with semiconductors, the analyst must delineate regions that have no physical or chemical differences in the experience of the metallurgist of old. The region on either side of a *p-n* junction is still silicon.

Most of the etches for silicon contain hydrofluoric and nitric acids. The ratios of these two, along with various buffering agents such as acetic acid, determine the final solution. Some examples are given in the table below.

		Parts hydrofluoric acid (49%)	Parts acetic acid (99.7%)	Other
Stain	4			
Dash etch	3	1	12	
3-1	3	1	10	
10-1	10	1	7	
20-1	20	1		
3-1 with copper sulfate	3	1	10	*

* 4 to 6 drops copper sulfate solution: 10 ml hydrofluoric acid, 200 g $CuSO_4$, 1 L deionized water.

12.8 REFERENCES

1. MIL-STD-883D, Test Method 1014.10. Washington, DC: U.S. Government Printing Office.

2. *Microcircuit Manufacturing Control Handbook, A Guide to Failure Analysis and Process Control.* Scottsdale, AZ: Integrated Circuit Engineering, 1974.

Electronic Failure Analysis for Specific Technologies

CHAPTER 13
SOLDER JOINTS

L.G. Vettraino

Electronic Failure Analysis Laboratory
National Technical Systems (NTS)

13.1   INTRODUCTION

Over the past several decades, the attachment of components to printed wiring boards (PWBs) has been accomplished primarily by solder joining technologies. Electronic assembly designs have incorporated various types of technology configurations to form mechanical, electrical, and thermal interconnections. These configurations have developed from through-hole technologies that incorporate, single-sided, double-sided, and plated-through hole (PTH) architectures, to surface mount technology (SMT) designs, which includes standard, fine, and very fine pitch devices for leaded and leadless component mounting. Recently, new SMT designs based on ball and column grid arrays, located under the component, have received wide interest and application. This technology maximizes the input/output-to-component size ratio without excessive use of board acreage. Other soldered interconnection technologies focus on direct chip attachment; for example, flip-chip applications. With this type of interconnection, solder bumps located on the chip are directly mounted to the board. These various types of solder joining technologies provide for durable, long-lasting, and inexpensive methods of mass or isolated interconnect production. Due to the low temperature melting aspects of these types of solders, the soldered joints can be reworked or repaired easily and selectively.

Depending on the operational environment(s), electronic assemblies may incur use for months or years and may experience extensive power and thermal cycling, vibration and other mechanical degradation mechanisms, and exposure to hostile environments. The failure of soldered interconnections is not only directly dependent on the environmental conditions experienced, but is also inherently associated with the prior manufacturing, rework, and repair history. This history includes joint design and geometry, fabrication techniques and parameters, thermal history, solder type and chemistry, and conductive circuitry and lead material type and condition.

The solder joint, regardless of configuration, is a complex composite system incorporating substrate materials, the bulk solder, component (leadless or leaded) metallization, and metallurgical bonding to both the substrate and component metallizations. For failure analysis purposes, the solder joint properties are also largely dependent on both the mechanical and physical properties (which are dependent on chemical composition) intrinsic to the solder itself. The bulk solder and substrate compositions (and prior thermomechanical history) defines the condition, state, and properties of the joint and directly determines the microstructure of the joint. Subsequently, these factors largely contribute to joint properties such as creep and fatigue strengths, ductility, electrical and thermal conductivity, diffusivity, coefficient of thermal expansion, resistance to corrosion, and other environmental effects.

This chapter focuses on the identification, characterization, and analysis of mechanically, thermally, and electrically defective and "failed" low melting temperature soldered interconnections.

13.2 SOLDER ALLOYS

Solder alloys are alloys that melt below 454° C (850° F) and which are used to join other metals and metal containing materials by metallurgical bonding. For electronic joining applications, typically the low-temperature soft solders (melting temperature below 260° C) are employed to provide strong and highly reliable interconnections without damaging heat-sensitive substrates or components during normal processing and assembly. Two major categories of solders are currently utilized: lead-containing and lead-free solders.

The chemical composition determines the melting and, hence, application temperatures of the solder used. In turn, other important properties such as creep strength, which is highly dependent on temperature, also rely on the solder type. Thermal cycling, which can lead to thermal fatigue, depends not only on the number of thermal cycles but also on the range of temperatures over which the joints cycle.

The processing temperatures for soldering are also governed by the melting temperatures of the solder used. Processing temperatures are normally selected as 30 to 50° C above the respective solder liquidus temperature for soldering in air environments. When soldering in inert gas environments, the processing temperatures are generally lower, since higher temperatures are not required to achieve the same degree of wetting.

Table 13.1 lists many of the common lead containing solder alloys used for electronic interconnection. Many solder alloy specifications exist, and each specification gives specific composition ranges for each of the major and minor alloying elements. Residual elemental impurity levels are also given. During failure analysis, the solder alloy specification that was originally used during manufacture should be used, and this is especially critical when determining alloy type and residual contamination. For many commercial uses, ASTM specification B 32 is commonly specified. For military and governmental use, Federal Specification QQ-S-571 has been typically mandated. It should be noted that QQ-S-571 has been recently replaced by IPC/ANSI-J-005/6. Currently, many government specifications are being rescinded and, as a result, new government contracts are required to substitute and specify DoD approved commercial specifications. Other specifications such as ISO/DIS 9453 have received international approval, especially in the European communities. A listing of common solder alloy specifications is given as Table 13.2.

13.2.1 Lead-Bearing Alloys

Tin-lead Alloys. Tin-lead alloys have formed the basis for joining components to circuit board substrates. Both new and mature electronic assembly designs, whether of through-hole or SMT configurations, have been based on the use of the eutectic (63Sn–37Pb wt.%) or near-eutectic (60Sn–40Pb wt.%) tin-lead alloys for mechanical and electrical connections. Figure 13.1[22] is the tin-lead binary phase diagram and includes all tin-lead compositions. As shown, the tin-lead eutectic composition is 61.2 wt.% Sn–38.8 wt.% Pb. Under equilibrium conditions, this alloy isothermally undergoes solidification from a liquid to a two-phase solid. The commonly referenced *eutectic alloy* (63 wt.% Sn–37 wt.% Pb)

Table 13.1 Compositional Ranges of Common Lead-Containing Solders

Alloy grade	Composition, weight percent[*]											Melting range[†]			
	Sn	Pb	Sb	Ag	Cu	Cd	Al	Bi	As	Fe	Zn	Solidus		Liquidus	
												°C	°F	°C	°F
Sn63	62.5–63.5	Rem	0.50	0.015	0.08	0.001	0.005	0.25	0.03	0.02	0.005	183	361	183	361
Sn62	61.5–62.5	Rem	0.50	1.75–2.25	0.08	0.001	0.005	0.25	0.03	0.02	0.005	179	354	189	372
Sn60	59.5–61.6	Rem	0.50	0.015	0.08	0.001	0.005	0.25	0.03	0.02	0.005	183	361	190	374
Sn10A	9.0–11.0	Rem	0.50	0.015	0.08	0.001	0.005	0.25	0.02	0.02	0.005	268	514	302	576
Sn10B	9.0–11.0	Rem	0.20	1.7–2.4	0.08	0.001	0.005	0.03	0.02	0.02	0.005	268	514	299	570
Sn5	4.5–5.5	Rem	0.50	0.015	0.08	0.001	0.005	0.25	0.02	0.02	0.005	308	586	312	594
Ag5.5	0.25	Rem	0.40	5.0–6.0	0.30	0.001	0.005	0.25	0.02	0.02	0.005	304	580	380	716

* Limits are maximum percentages unless shown as a range or stated otherwise. For the purpose of determining conformance to these limits, an observed value of calculated value obtained from analysis shall be rounded to the nearest unit in the last right-hand place of figures used in expressing the specified limit, in accordance with the rounding method of ASTM Recommended Practice E 29.
† Temperatures given are approximations and for information only.
©ASTM, reprinted with permission.

Table 13.2 Commercial and Federal Solder Specifications

Standard	Title	Comments
ASTM B-32	Standard Specification for Solder Metal	Current revision, 1996
QQ-S-571	Federal Specification, Solder; Tin Alloy, Tin-Lead Alloy, and Lead Alloy	Replaced by IPC/ANSI-J-005/6
ISO/DIS 9453	Soft Solder Alloys—Chemical Composition and Forms	Current edition, 1993
IPC/ANSI-J-005	Requirements for Soldering Pastes	DoD adopted in 1994
IPC/ANSI-J-006	Requirements for Electronic Grade Solder Alloys and Fluxed and Non-Fluxed Solid Solders for Electronic Soldering Applications	DoD adopted in 1994

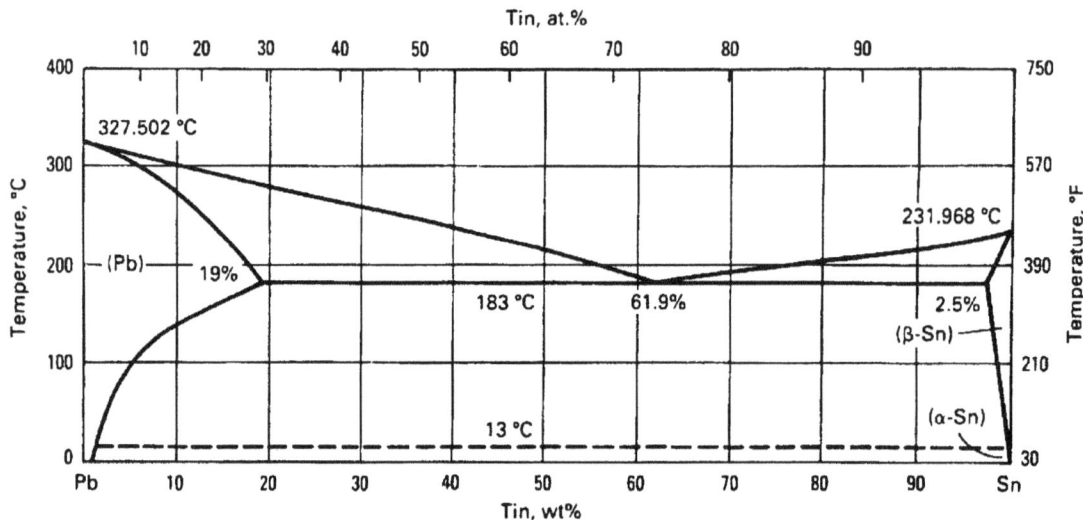

FIGURE 13.1   Tin-lead equilibrium phase diagram.

approaches isothermal solidification into two distinct phases under nonequilibrium cooling, as would normally be experienced during soldering. The solder alloys to either side of the eutectic composition solidify over a temperature range. This solidification range comprises both a solid and liquid phase, and the solder is said to be *pasty* in these regions. These off-eutectic alloys with large volume fractions of the proeutectic phases are generally large grained and exhibit dendritic growth. The proeutectic phases tend to make the solder joint appear grainy under low-power magnification. The solidification ranges vary on the composition of the solder, with the high-lead and high-tin solders exhibiting higher liquidus temperatures. The noneutectic alloys melt and solidify more slowly than eutectic alloys, since the off-eutectic solders melt (and solidify) over a temperature range. The melting temperatures and ranges for many tin-lead solders are given by both Fig. 13.1 and Table 13.1. Tables 13.3 and 13.4, respectively, give bulk physical and mechanical properties for some common lead-containing solders.

Table 13.3   Bulk Physical Properties of Lead-Containing Solders

Alloy designation	Density, g/cm^3	Electrical conductivity of Cu, %IACs*	Thermal conductivity, W/m • °K (BTU • in/s • f^2 • °F)	Poisson's ratio[†]
Sn63Pb37	8.34	11.8	50.0 (0.098)	Data n/a
Sn62Pb36Ag2	8.50	6.8	Data n/a	Data n/a
Sn60Pb40	8.52	11.5	49.8 (0.096)	0.4
Sn10Pb90	10.50	8.2	35.8 (0.069)	Data n/a
Sn5Pb95	10.80	8.1	35.3 (0.068)	Data n/a
50Pb50In	9.14	6.0	Data n/a	Data n/a

* International Annealed Copper Standard (IACS)
† at 25° C (77° F)

Table 13.4 Bulk Mechanical Properties of Lead-Containing Solders

Alloy designation	Tensile strength, KPa (Ksi)	0.2% yield strength, KPa (Ksi)	Shear strength, KPa (Ksi)	Elongation, %	Modulus of elasticity, GNm^2 ($\times 10^6$ psi)	Hardness (HB)
Sn63Pb37	54 (7.8)	Data n/a	37 (5.4)	37	31 (4.5)	17
Sn62Pb36Ag2	65 (9.4)	Data n/a	Data n/a	Data n/a	23 (3.3)	Data n/a
Sn60Pb40	52 (7.6)	19.8 (2.89)	39 (5.6)	40	30 (4.35)	16
Sn10Pb90	30 (4.4)	13.8 (2.02)	17 (2.4)	30	19 (2.76)	10
Sn10Pb88Ag2	28 (3.9)	Data n/a	Data n/a	15.9	Data n/a	Data n/a
Sn5Pb95	28 (4.0)	13.2 (1.93)	14 (2.1)	45	Data n/a	8
50Pb50In	3.2 (4.7)	Data n/a	18.5 (2.7)	55	Data n/a	9.6

Sn63Pb37. This is commonly referred to as the *eutectic* tin-lead alloy. The most common application for this alloy is the attachment of electronic components to PWBs. This solder, for many applications, is preferred over the Sn60Pb40 alloy due to its very narrow solidification range. This solder is also used is many step-soldering applications, where a high-temperature solder is initially utilized. Subsequently, other joints are made by soldering with this lower-melting-temperature alloy, thereby not remelting the initial high-temperature interconnections. Recently, this alloy has also found wide application in the attachment of ball grid array (BGA) or column grid array (CGA) packages.

Sn62Pb36Ag2. The addition of approximately 2 percent silver (Ag) limits the solubility of Ag into the soldered joint during assembly or processing. When soldering to silver-coated or silver-containing substrates, leaching (dissolution) of the silver into the soldered joint is greatly reduced. The addition of silver also increases the monotonic and thermal-fatigue resistance without significantly reducing ductility.

Sn60Pb40. As with the eutectic tin-lead alloy, the primary use of this alloy is the assembly of circuit boards. Since this near-eutectic alloy exhibits a *pasty range,* this characteristic allows for unique applications not associated with the Sn63Pb37 material. These include the filling of large holes and vias on printed wiring boards and the prevention of solder wicking for some leaded surface mounted devices (SMDs). This off-eutectic alloy also finds use in SMT applications where component leads and PWB pads heat at differing rates. Since this solder melts over a temperature range, it allows both the leads and the pads to reach the same temperature, subsequently wetting both surfaces.

High Lead-Tin Alloys. The Pb90Sn10 and Pb95Sn5 alloys are used for high temperature attachment and for step soldering applications as already mentioned. These solders are largely found in *flip-chip* or controlled-collapse chip component (C^4) attachment of silicon chips to substrates. Recent development of BGA and CGA packaging technologies utilizes the 90Pb10Sn alloy for the solder ball and column material. Attachment of the balls and columns to the respective component or board substrate is accomplished by use of a lower-melting-temperature solder, typically Sn63Pb37.

Lead-Silver Alloys. These alloys find use in high-temperature applications. The Pb94.5Ag5.5 (Ag5.5) material is commonly used in the soldering of thermocouples in aircraft engines for aerospace applications.

Lead-Indium Alloys. The 5 to 50 wt.% indium alloys have been extensively used in C^4 chip attachment applications. In addition, these alloys are used to solder to precious metal (Au, Pt, Ag, and so on) coatings or substrates and for attaching components to hybrid thick-film networks. The advantage is that lead-indium solders, relative to tin-lead alloys, have a much lower precious metal dissolution rate. The lead-indium solders also fill a niche in cryogenic applications due to their low-temperature ductility. These alloys exhibit better fatigue properties than their tin-lead alternatives. In addition, these alloys are more ductile than their tin-lead alternatives, which minimizes loading in more fragile thick-film ceramic bonding applications. The Pb-In solders, however, are very susceptible to corrosion, and the 50Pb-50In alloy is particularly susceptible to excessive intermetallic formation. The fatigue properties of the these solders varies parabolically with indium content, with the minima at approximately 15 wt.% indium. Corrosion severity decreases rapidly with decreasing In concentration, and below 25 wt.% is considered negligible; however, 50Pb-50In has very good resistance to alkaline corrosion.

13.2.2 Lead-Free Alloys

The vast majority of lead-free solder alloys are based on elemental tin as the major component and utilize secondary metals such as silver, bismuth, antimony, copper, indium, zinc, selenium, and tellurium as alloying additions to form binary, ternary, and complex alloys. Lead-free alloys are gaining wider acceptance due primarily to their reduced toxicity and environmental concerns avoided by limiting the concentration of lead (typically < 0.2 wt.%). In addition, many lead-free solders can provide better mechanical properties than tin-lead alloys, such as improved fatigue strength. Lead-free solders can also be used for different applications from those of tin-lead solders, since the melting (and hence processing) temperatures vary significantly, as shown by Table 13.5. Typical composition requirements for lead-free solders are given in Table 13.6. Bulk physical and mechanical properties for some lead-free solders are given by Tables 13.7 and 13.8, respectively.

Tin-Silver and Tin-Silver-X(-X) Alloys. These materials have received wide use as high-temperature solders and, in particular, step-soldering operations where the near-eutectic tin-lead solders are used as the lower-melting-temperature alloy. The eutectic, 96.5Sn3.5Ag alloy, is the most commonly used composition. Relative to the tin-lead solders, these materials maintain superior wetting, strength, and fatigue properties. It has been reported that tin and silver form stable Sn-Ag intermetallic compounds; subsequently, these alloys are not prone to electromigration reactions. In soldered joints fabricated with 96Sn4Ag alloy, no appreciable silver migration has been observed.

The 91.8Sn-3.4Ag-4.8Bi ternary eutectic alloy has a melting temperature of 211° C. This alloy has similar wetting times relative to the tin-lead eutectic alloy; however, the surface tension of this alloy is greater than the Sn3Pb37 alloy, and reduced spreading is observed.

The 96.2Sn-2.5Ag-0.8Cu-0.5Sb quandary alloy has a melting range of 210 to 216° C and has been largely tested as a drop-in replacement for the tin-lead near-eutectic alloys. This particular alloy is reported to have monotonic mechanical properties superior to that of the eutectic tin-lead alloy. Mechanical fatigue properties of this alloy are similar to those of the Sn63Pb37 alloy; however, the results displayed very little if any strain dependence.

Table 13.5 Melting and Processing Temperatures for Various Solders

Solder alloy	Melting temperature or range, °C	Estimated process range, °C	Typical soldering iron tip or (reflow) temperature, °C	References
52In/48Sn	118	148–168	Data n/a	26
58Bi/42Sn	138	168–220	(190)	3, 32, 53, 56
77.2Sn/20.0In/2.8Ag	114[*], ~175–187	217–250	(250)	3, 30
50In/50Pb	180–209	239–259	Data n/a	32
63Sn/37Pb	183	213–260	315–371 (250-260)	3, 54, 57
60Sn/40Pb	183–189	219–260	315–371 (250-260)	3, 55
91Sn/9Zn	199	229–250	Data n/a	21
91.8Sn/4.8Bi/3.4Ag	211	230–261	371	3, 54, 57
95.5Sn/4.0Cu/0.5Ag	216–222	252–272	(250)	57
95Sn/3.5Ag/1.5In	218	248–268	Data n/a	44
93.5Sn/3.0Sb/2.0Bi/1.5Cu	218	248–268	Data n/a	44
96.5Sn/3.5Ag	221	235–271	371 (50)	53,57
95.5Sn/3.5Ag/1.0Zn	221	251–271	Data n/a	44
99.3Sn/0.7Cu	227	250–277	(250)	44
95Sn/5Sb	232–240	270–290	Data n/a	2
10Sn/88Pb/2Ag	268–290	320–340	Data n/a	32
90Pb/10Sn	268–302	332–352	Data n/a	22
5Sn/95Pb	308–312	342–362	Data n/a	45

* Indicates limited melting.

Tin-Antimony and Tin-Antimony-X-X Alloys. The Sn95Sb5 alloy is the most commonly used material in this category for electronic interconnections. Due to its very high melting temperature range, it has primarily been used as a high-temperature solder in step processing. The very high melting temperature severely limits the use of this material in printed wiring assembly (PWA) processing, because the high processing temperatures and subsequent high thermal input could damage heat-sensitive electronic devices as well as PWBs. These materials, however, maintain excellent wetting and creep strength, similar to the tin-silver solder alloys.

Tin-antimony-bismuth alloys have an increased melting range over the eutectic Sn-Bi alloy. The 75Sn-6Sb-19Bi alloy exhibits a melting range of 208 to 212° C. At room temperature, the solid solubility of antimony and bismuth is limited to 2 wt.% each. Additions of 1 to 2 wt.% Ag and 1 wt.% Indium lowers the solidus temperatures. The silver content, however, increases the dendritic characteristics of this alloy.

Table 13.6 Compositional Ranges of Some Lead-Free Solders

| Alloy name | Short name | Former name** | Composition, weight percent[†,‡] | | | | | | Melting range* | | | |
			Sn	Pb	Au	In	Ag	Other component elements	Solidus °C	Solidus °F	Liquidus °C	Liquidus °F
Sn96Ag4	Sn96	Sn96	Rem–96.3	–	–	–	3.7	–	221	430	221	430
Sn95Sb5	Sb05	Sb5	Rem–95.0	–	–	–	–	Sb = 5.0 ± 1.0%	233	450	240	464
In52Sn48	In52	In52	Rem–48.0	–	–	52.0	–	–	118	244	188	244
Sn42Bi58	Bi58	Bi58	Rem–42.0	–	–	–	–	Bi = 58.0	138	280	138	280

* The solidus and liquidus temperature values are provided for information only and are not intended to be a requirement in the formulation of the alloys. Although efforts have been made to document the correct solidus and liquidus temperatures for each alloy, users of this document are advised to verify these temperature values before use.
† Except where other wise indicated, the component elements in each alloy shall not vary from their tabulated percentage by more than ± 0.20 when their tabulated percentage is equal to or less than 5.0 or by more than ± 0.50 when their tabulated percentage is greater than 5.0. (E.g., the actual percentage of a component element having a tabulated percentage of more than 5.0 must fall within the following limits inclusive: {tabulated percent – 0.50} to {tabulated percentage + 0.50}.) The letter "R" appearing with a *number* for an element of an alloy (e.g., R-10.0) denotes that the element makes up the remainder of that alloy, and the *number* indicates the approximate percentage of that element in the alloy.
‡ Alloys are listed in these tables as purity grade A (standard) alloys. The four gold alloys can be obtained as purity grade B (gold non-barrier die attachment) alloys. Purity Grade A alloys are identified by a "B" suffix on the alloy name and short name; purity grade A alloys may be identified as such by an "A" suffix.
**The presence of a former alloy name indicates that the current alloy is substantially the same as the indicated QQ-S-571E alloy. An asterisk (*) following a former alloy name indicates that the former (QQ-S-571E) alloy was required to have 0.20 to 0.50% antimony, but the required antimony content in the current alloy is 0.50% maximum (0.0 to 0.50%).
Source: ANSI/IPC J-STD-006.

Table 13.7 Bulk Physical Properties of Lead-Free Solders

Alloy designation	Density, g/cm³	Electrical conductivity of Cu, %IACS*	Thermal conductivity, W/m • °K (BTU • in/s • ft² • °F)	Poisson's ratio[†]
52In48Sn	Data n/a	11.7	Data n/a	Data n/a
58Bi42Sn	Data n/a	4.5	Data n/a	Data n/a
77.2Sn20In2.8Ag	7.25	10.1	Data n/a 53.5 (0.103)	0.40
91Sn9Zn	7.27	Data n/a	Data n/a	Data n/a
96.5Sn3.5Ag	7.36	14.0	Data n/a	Data n/a
95Sn5Sb	7.25	11.9	Data n/a	Data n/a

* International Annealed Copper Standard (IACS)
† at 25° C (77° F)

Table 13.8 Bulk Mechanical Properties of Lead-Free Solders

Alloy designation	Tensile strength, KPa (Ksi)	0.2% yield strength, KPa (Ksi)	Shear strength, KPa (Ksi)	Elongation, %	Modulus of elasticity, GNm^2 ($\times 10^6$ psi)	Hardness (HB)
52In48Sn	11.8 (9.71)	41.3 (6.03)	Data n/a	Data n/a	Data n/a	Data n/a
58Bi42Sn	54.9 (7.98)	Data n/a	63 (9.2)	Data n/a	Data n/a	Data n/a
77.2Sn20In2.8Ag	46.6 (6.80)	Data n/a	32.9 (4.8)	47	38.6 (5.60)	Data n/a
96.5Sn3.5Ag	44.1 (6.40)	48.5 (7.08)	55 (8.0)	Data n/a	Data n/a	14.8 HV
95Sn5Sb	36.3 (5.30)	17.0 (2.45)	28 (4.1)	38	50.0 (7.26)	15

The Sn-Sb-Bi-Cu quandary eutectic alloy melts at 218° C. This alloy exhibits high strength and has excellent thermal fatigue properties.

Indium-Tin and Tin-Indium-X Alloys.   For the binary alloys, the eutectic and near-eutectic solders are the most commonly used. These are, respectively, 52In48Sn and 50In50Sn which, because of their low melting temperatures, have applications in attaching heat-sensitive devices or in step-soldering applications where tin-lead alloys are used as the higher-melting-temperature materials. Indium-rich alloys are also finding use in cryogenic applications due to their low-temperature properties. Indium-based tin alloys have poor mechanical properties and especially exhibit poor creep properties. Indium-based alloys, however, can wet nonmetallic surfaces such as various glasses and ceramics. These materials, however, are very susceptible to corrosion. The tin-indium-X ternary alloy(s) have been developed in response to finding suitable lead-free replacements for the near-eutectic tin-lead alloys. The melting ranges are similar to those of the near-eutectic tin-lead alloys; however, the Sn-In-Ag alloy also exhibits limited melting at approximately 113.5° C, most probably due to the formation of the low-melting temperature 52In48Sn phase. Although the wetting times for this solder are significantly greater than those of the tin-lead eutectic alloy, the mechanical properties, including creep properties, of this alloy were reported superior.

Tin-Bismuth and Tin-Bismuth-X Alloys.   The tin-bismuth eutectic alloy (42Sn-58Bi) is the most common binary alloy in this class. Because of their low melting points, these alloys find use in step soldering and processes that require attachment to heat-sensitive substrates or devices. Alloys containing in excess of 47 wt.% bismuth are reported to expand after solidification, but the amount of expansion varies, depending on the composition.

The Sn-Bi-Zn alloy metals exhibit low melting temperatures and ranges. Alloys with greater than 10 wt.% Bi and 5.5 wt.% Zn have solidus temperatures of 127 to 140° C. Other alloys in this class have wider melting ranges. A ternary eutectic melting point for all these alloys was observed 127° C, probably due to the formation of 41.7Sn-57Bi-1.3Zn. Compositions containing small amounts of bismuth maintained microstructures of tin-rich or zinc-rich dendrites.

Ternary Sn-Bi-In alloys in compositions ranging from 80Sn-10Bi-10In to 70Sn-20Bi-10In were found to have melting ranges from 163 to 209° C and 143 to 193° C, respectively. These materials have solid solubility and exhibit a single phase structure. It was reported that the single phase structure should allow for good thermal fatigue properties, but intergranular failure may be expected.

Additions of up to 0.3 wt.% Ag have no effect on the eutectic melting temperature of this alloy (138° C). Additions of up to 3 wt.% Ag have little effect on this melting temperature. When the silver concentration was increased to over 2.0 wt.%, the formation of the SnAg3 intermetallic compound was observed. An alloy of composition 97.7Sn2Bi0.3Ag has a narrow melting range of 10° C; however, an alloy with a composition of 83.5Sn14Bi2.5Ag maintains a larger melting range of 142 to 218° C. With increasing Bi content, the liquidus temperature decreases approaching that of the Sn-Bi eutectic.

Tin-Copper and Tin-Copper-X Alloys.   The tin-copper eutectic (99.3Sn0.7Cu) has a melting point of 227° C, whereas the Sn-Cu-Ag alloy (95.5Sn4.0Cu0.5Ag) exhibits a melting range of 216 to 222° C. These alloys have found limited application, in part due to their high-melting-temperature characteristics.

Tin-Zinc Alloys.   The 91Sn9Zn eutectic alloy maintains a melting point of 199° C. Although this solder has adequate wetting and strength properties, zinc, being a very reactive metal, readily oxides and reacts with other materials. As a result, this alloy is highly prone to corrosion and has found limited use.

13.3   SOLDERING PROCESSES AND STANDARDS

The assembly of manufactured hardware typically involves the use of mass soldering technologies, although manual soldering has been used for specific manufacturing applications.

While, for plated-through hole (PTH) applications, wave soldering is the joining method of choice, for SMT applications mass reflow assembly (e.g., infrared, convection, vapor phase, and so on) and wave soldering are employed. These methods of mass assembly highly differ from rework/repair technology, which largely relies on the use of manual soldering equipment in which desoldering (the removal of solder from a plated-through hole, surface pad, and so forth) and resoldering (the addition of solder to form a previously formed joint) is conducted by localized heating and material application.

13.3.1   Conduction Processes

Soldering Iron.   Soldering with soldering irons is generally limited to localized component attachment and rework/repair. This type of soldering technology is mandatory to the rework/repair of through-hole devices, since the solder is fed to the joint by wire form. Other methods of heat transfer are not conducive to this type of localized solder application. Soldering iron temperatures can be set (typically 38 to 482° C) as appropriate to the soldering application. Soldering tips can come in various shapes (e.g., conical and chisel) for PTH applications and larger forms that contact the leads on a multileaded SMD. Extremely fast cooling rates are normal to joints manufactured by this technique. Cooling rates of greater than 100° C/s are not uncommon. Resultant microstructures tend to be fine grained and tend to improve the solder joints mechanical properties.

Since heating is localized, soldering is highly dependent on the operator's skill and experience. Insufficient heating can result in cold solder or poorly wetted joints. Excessive heating or prolonged contact of the iron with the joint can cause overheated solder joints. Overheating and mechanical pressure will also cause *measling* of the board substrate material (separation of the woven fibers in the board caused by the heating and expansion of volatiles compounds or by the localized application of heat and pressure damaging the individual board matrix fibers). Since this type of soldering is generally a manual operation, these soldered joints are more prone to disturbed joints than are automated mass-soldering processes. Disturbed joints appear crystalline under low magnification optical microscopy. With this type of solder joint defect, the molten solder is disturbed prior to final solidification, leaving a jagged crystalline appearing joint. Solders with narrow or no melting (solidification) ranges, which therefore solidify faster, are used to help alleviate this problem.

Hot Bar. A resistive heating bar made of a non-solder adhering material (e.g., titanium) locally heats and reflows solder to one or several component leads. This heating method is limited to SMT, where the component leads are flat and accessible. The heat can be supplied continuously to the bar, maintaining a constant temperature, or it can be pulsed. The pulse method is activated at a preset pressure that is exerted on the leads during soldering. The major advantage of pulsed heating is that the solder joints can cool and solidify before the tool pressure is removed. The solder material can be preplaced by paste-dispensing or printing techniques prior to joint formation. The solder can also be placed directly onto the pads or components leads via soldering iron prior to component attachment. Since the soldering bar makes contact with the component leads, pressure is applied during the process, which forces the leads into position during joint formation. With this type of process, it is important to have all leads correctly positioned planar to the board. Leads that are bent up and then forced down by the hot bar during soldering can induce large residual stresses into the joint. The presence of residual stress greatly reduces the functional lifetime of the joint. Soldered joints containing large residual stresses and subjected to thermal fatigue environments are very susceptible to premature failure.

13.3.2 Convection Processes

Hot Gas. This method of soldering is generally localized to soldering or removal of a specific component. Typically applications are therefore limited to rework/repair operations. As the name implies, this technology uses hot gas as the medium to reflow or melt solder for SMT applications. Nozzles are utilized to limit and distribute the gas flow around a specific component. Both leaded and leadless components can be removed or soldered with this method.

Gas flow is usually controlled by the nozzle design, which may incorporate baffles. The nozzle distributes the flow of the gas around one or more interconnections, while the baffle plates tend to minimize the flow of the hot gas to adjacent component areas or direct impingement to the component under processing. The hot gas temperature and gas flow rates directly determine the heating efficiency. On some systems, the gas flow velocity and volume can be adjusted for optimized processing. Insufficient gas flow will minimize heating efficiency and solder reflow and, in oxidizing environments (i.e., air), may increase the oxidation of the components leads and PWB metallization, thereby reducing wetting and hence solderability. The use of nitrogen or a nitrogen-hydrogen gas mixture ($75N_2$-$25H_2$) can minimize the oxidation and improve the solderability and reliability of the soldered joints. Excessive gas flow can force hot gas across the assembly, which can overheat or thermally age adjacent areas.

Since gases and air in particular are poor heat transfer media, the gas temperature will quickly decline as the distance form the nozzle increases. Since component bodies tend to absorb heat faster than the interconnect under processing, heat damage to the component or PWB is likely where high temperature and high gas flow rates are used.

The thermal profile of the PWA itself (and not only at the component being processed) is critical to the quality and reliability of all interconnections and components on the assembly. The component density, dimensional characteristics, and heat transfer properties of the circuit board and component materials also influence the heating effects throughout the assembly. Figure 13.2[18] demonstrates the time-temperature relationship (thermal profile) to a component and associated regions on an assembly during hot gas soldering.

Multileaded component interconnects are susceptible to premature failure due to the melting or reflowing to a selective portion of these joints. Partial reflowing of soldered interconnections can induce residual stresses into the soldered joints, severely reducing joint integrity. When the joints are thermally cycled during power cycling and/or under operating conditions, fractures can result.

Gas Convection.   Hot gas or air convection heating is limited to in-line mass reflow assembly. This heating process is relatively slower than other reflow methods and thus requires more time to reach equilibrium soldering conditions. Various atmospheres (gas or air) can be employed easily with this reflow technology. The process can use forced convection, which increases and optimizes air or gas flow around the assemblies, thereby enhancing the heat transfer and uniformity of heat distribution.

Vapor Phase (Condensation).   Vapor phase soldering can be employed for batch or in-line processing. Due to the precise nature of this soldering method, fine-pitch components as small as 0.38 mm can be accommodated. Because this technology is an oxygen-free process, pitch size is not a limiting factor. This reflow method uses the condensation of a heated liquid to thermally transfer heat to the relatively cooler PWA. During reflow, the PWA is either passed through or into a layer of the saturated vapor. Uniform and rapid heating is accomplished by this method, and thermal equilibrium is normally achieved in

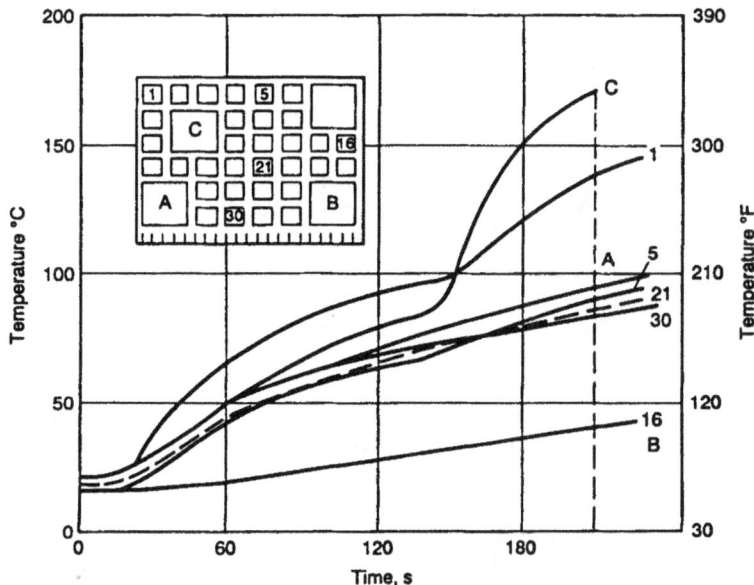

FIGURE 13.2   Hot gas thermal profiles at various locations on a PWA.

only a few seconds. The reflow temperature is limited only by the boiling point of the fluid used for the specific process.

This type of soldering offers very precise temperature control and lends itself to multi-level (step) soldering. Up to three different solders, and hence three different reflow temperatures, can be accomplished during a single reflow process. Fluid temperatures range from 100 to 265° C (212 to 510° F), depending on the type of fluid(s) used in the processing. For reflow processing of solder interconnections, a temperature 20 to 30° C above the liquidus is common.

A typical thermal profile for vapor phase soldering is shown as Figure 13.3.[31] The initial or preheating stage is set to bring the PWB to a temperature between 125 to 150° C (255 to 300° F) in 50 to 60 s. A desired heating rate is 2° C/s (4° F/s) to 100° C (180° F) of the reflow temperature is typically employed to avoid thermal shock. The thermal profile, including the time at maximum temperature, depends on the thermal mass of the load, and these profiles must be adjusted to accommodate the unique thermal mass and component type parameters for each type of assembly.

The flux/reflow fluid profiles should be characterized and established for each assembly and associated soldering materials. For example, if the flux dissolves at a rate greater than the increase in assembly temperature to the solder reflow temperature, the flux volume may be deficient and may not facilitate wetting during reflow. High heating rates associated with some thermal profiles may induce solder balling. Other defects, such as wicking, can occur when a differential heating rate between the component lead and the substrate pad is established. The solder will wick to the smaller thermal mass (typically the lead) due to the higher lead temperature.

Wave Soldering.   Wave soldering is limited to through-hole mass soldering and SMT applications incorporating leadless or standard-pitch devices containing few leads, or a combination of these two technologies. The circuit boards with attached components are carried by conveyor through each of the processing steps. During the first step (fluxing), the liquid solder flux is applied by brush, foaming, or spray application. The amount of flux can be regulated, depending on the specific type of fluxer. Deterioration of the flux during long-term processing occurs from contamination with the circuit board itself. Also, during prolonged use, the loss of the flux vehicle occurs, and monitoring must be instituted regularly to reduce solder defects in the assembly. The PWB is then preheated to

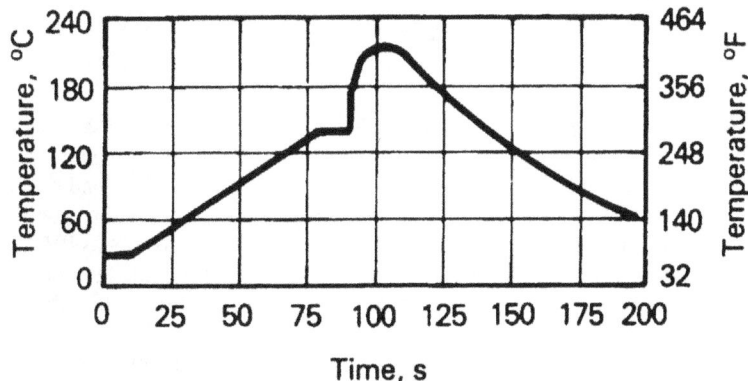

FIGURE 13.3   Thermal profile of a vapor phase soldering cycle.

thermally condition the board and components against thermal shock and to activate the flux prior to soldering. The preheat stage should bring the board into a temperature range of 100 to 150° C (210 to 300° F). To avoid cracking of surface mounted chip capacitors, a maximum preheating rate of 4° C/s (7° F/s) to within 100° C (212° F) of the solder wave temperature as measured on the bottom side of the board has been suggested. The board is next conveyed to the solder wave. Soldering is accomplished by direct contact of the board with the molten solder wave. The joint or interconnect is in contact with the solder wave for an average of 3 to 5 s, but this duration can be longer for dual-wave systems. The temperature of the molten solder is typically 50 to 80° C above the liquidus temperature of the solder, which is typically higher that those used in reflow processes. The higher temperatures are required to provide adequate heat transfer to the substrate and to allow for sufficient wetting during soldering. The formation of dross in solder pots tends to be very high in wave soldering applications. As a result, dross particles can come embedded into the joints during soldering. Boards with larger thermal mass (i.e., thicker, multilayer, or controlled-expansion substrates) require higher solder pot temperatures. With increasing solder temperature, the potential for laminate damage, including warpage and distortion, also increases. Contamination from dissolution of PWA metals accumulates in the solder pot over extended use. This buildup of contaminants can have detrimental effects on the interconnects (see Table 13.9[27]).

The last stage allows for cooldown of the soldered assembly. As the board comes out of the solder wave, it is especially crucial that the assembly not be bumped or disturbed, which can cause fractured or defected joints. An adequate cool-down period is recommended prior to moving the newly soldered assembly.

Defect trends have been identified and established for wave soldering applications. For example, fast conveyor speeds and high solder pot temperature facilitate the production of voids and blowholes. Under these processing conditions, extremely fast heating rates are generated, causing the rapid volatilization of flux components responsible for void formation. Rapid conveyor speeds cause other defects as follows:

- Insufficient dwell times in the solder wave, which produce poorly filled through holes or vias
- Formation of solder icicles and bridges from improper conveyor speed and solder wave flow combinations
- Cracking of ceramic components due to thermal shock

13.3.3 Radiation Processes

Infrared Processing.   This method of solder reflow for SMT is an in-line process conducted in a chamber in which heating and thermal transfer is accomplished by emitted infrared (IR) energy. The emitted radiant energy is partially absorbed by the portion of the energy contacting the PWA. The amount of radiant energy absorbed by the assembly, and hence the heating rate, depends on the absorptivity of the individual PWA materials, mass, specific heat, reflectivity, thermal conductivity, diffusivity, and geometry of the components and assembly. Each different PWA part maintains unique IR reflection and absorption characteristics.

IR reflow systems fall into three categories. The first is lamp IR systems, which typically use quartz-encapsulated tungsten filament emitters in the near-to-medium wavelength range (~1 to 3 μm) to provide radiant thermal energy. These systems use large

Table 13.9   Overview of Process-Related Solder Joint Defects

Category	Parameter	Potential Problems
Design	Location, size, and spacing of features	Soldering defects
	Differential thermal masses	Uneven heating in reflow, cracking of joints
	Selection of parts	Inconsistent soldering
	Performance specification	Lack of concern for manufacture and test
	Design rules	Different for through hole, difficult to optimize
Components	Dimensions	Variations make consistency impossible
	Vendor performance	Poor parts caused by lack of AQL programs
	Parts standardization	Dimensional and solderability variations
	Coplanarity	Noncoplanar leads do not solder
	Solderability	Age or condition prevents good connection
	Volatiles	Voiding and solder balls
Handling	Testing	Contamination, bent leads, internal damage
	Storage	Oxidation, contamination, damage
	Kitting	Bent leads, wrong parts, damage
	Transport	Bents leads, contamination
Solder paste	Composition	Problems created in solder flow
	Alloy mixture	Problems in solder reflow time
	Activity, volatility, and contamination	Variable reflow times and joint quality
	Paste location	No connection or bridge, or poor joint quality
	Paste quantity	Lack of consistency in soldering, tombstoning
	Paste and mesh	Improper selection
Solder mask	Consistency	Improper flow cause by voiding
	Registration	Poor soldering
	Adhesion	Solder wicks away from pad
	Quantity	Insufficient protection or smearing
Assembly	Handling	Bent or broken leads
	Placement	Misalignment, missing parts, part damage
	Transport	Lost components

Table 13.9 Overview of Process-Related Solder Joint Defects *(continued)*

Category	Parameter	Potential Problems
Solder wave	Flux density, activity, and viscosity	Wettability
	Contaminant content	Inclusions and wettability
	Solder pot temperature	Cold or overheated solder
	Wave temperature	Cold or overheated solder
	Wave heights	Insufficient, excess, or bridging solder
	Conveyor speed	Insufficient, excess, or bridging solder
	Board angle	Insufficient, excess, or bridging solder
	Air knife and temperature	Bridging
	Oil intermix specification	Wettability
Reflow	Paste dry time and temperature	Wetting, voids, graininess, solder balls, tombstoning
	Heating uniformity	Wetting, graininess, tombstoning, wicking
	Reflow time	Wetting, graininess, tombstoning
	Preheat time	Wetting, graininess, tombstoning, wicking
	Batch time	Wetting, graininess, tombstoning
	Belt speed	Wetting, graininess, tombstoning
	Heating profile	Wetting, graininess, tombstoning
Laser	Wavelength	Cold or overheated solder
	Time	Cold or overheated solder
	Alignment	Cold or overheated solder, burned boards
	Incident angle	Cold or overheated solder, burned boards
General	Preheat time and temperature	Oxidation, remaining volatiles, intermetallics
	Cooling consistency	Intermetallics and grain structures
	Handling	Movement during soldering
	Time in reflow	Excess formation of intermetallics
Inspection/test	Programming	Cost may discourage use
	Fixturing	Cost, alignment, and board damage
	Time	Slow throughput for 100% inspection/test
	Repeatability and reliability	May be inadequate for process control

amounts of electrical power and characteristically produce high emitter temperatures. This IR technology is more prone to the charring of corners on PWAs, overheating of small components, and shadowing effects of tightly spaced components, especially at higher lamp temperatures and shorter wavelength energies.

The second category is panel/natural convection IR systems. These employ both convection provided by the natural circulating air movement within the chamber and by direct absorption of the radiant energy in the mid-to-far IR (~2.5 to 15 μm) wavelength range and secondary emission from panel surfaces. Since these systems operate at lower temperatures, repeatable and uniform thermal profiles are normally achieved. In addition, the use of natural convection aids in the reduction of thermal shadowing effects. The controlled heating capability allows for very specific component heat ramping rates, so cracking of chip capacitors is greatly reduced as compared to the other IR reflow technologies.

The last IR system type, forced convection IR, utilizes a constant inflow of temperature-controlled air, which minimizes temperature gradients across the assembly. This technology is very similar to the panel/natural convection IR system, but forced air is used instead of the natural convection. The forced air can be preset to a temperature lower than that of the hottest region(s) of the PWA, which cools these hot areas and subsequently heats the cooler regions providing for consistent and uniform temperatures. Hot spots, such as those located at the corners of circuit boards, can be minimized with this type of reflow system.

When IR systems are used, the specific thermal profile for an assembly must be developed prior to reflow soldering. A temperature profile showing the interconnection temperature during an IR reflow process is presented as Figure 13.4.[59] The reflow profile consists of four distinct regions: preheat, preflow, reflow, and cooldown. During the preheat stage, the interconnection temperature is raised from ambient to approximately 100 to 120° C. The preheat minimizes the possibility of PWB delamination and thermal shock to the components. It is normally recommended that the heating rate in this region be limited to 4° C/s or less (especially for ceramic components) and should be established based on the assembly and component materials and PWA design, including component density. In this stage, some solder pastes begin to activate, and some solvents start to volatilize and evaporate.

In the second stage, or preflow, the assembly is brought to a temperature below the reflow temperature to reduce thermal gradients. The remaining portion of the nondesireable solvent inherent to the solder paste is "boiled off" at this stage. As the preformed interconnections approach the reflow temperature, the flux activators reduce the component lead and PWB pad surface oxidation. It should be noted that the temperature and dwell times used will vary, depending on the type of solder paste. *Rosin mildly activated* (RMA) fluxes are normally fully activated between 150 and 170° C, and some formulations require holding times of less than 1 min to optimize processing results. Organic acid (OA) fluxes can have preflow stages similar to the RMA fluxes; however, for many OA fluxes, activation is optimized by gradual thermal ramping from approximately 100 to 170 ±15° C (212 to 340 ±30° F). For OA fluxes, the preflow stage is very critical and is generally limited to 1 min, depending on the flux formulation. Excessive preflow duration can result in solder paste oxidation and flux degradation. The reflow segment of the thermal cycle is characterized by the melting and reflow of the solder material. At this stage, all solderable areas should reach the desired peak temperatures. For the eutectic tin-lead alloy (Figure 13.4), the solder reflows starts at 183° C. For most applications, the interconnect temperature is recommended to be approximately 15 to 30° C (27 to 54° F) above the solder alloy solidus temperature to ensure optimal wetting and fillet formation (represented as

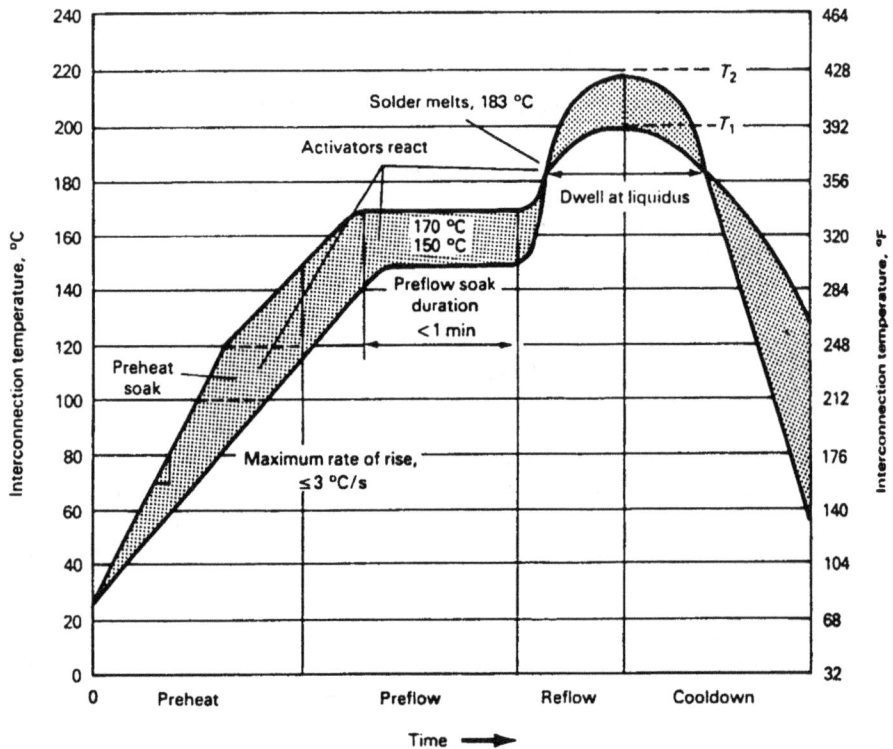

FIGURE 13.4 Thermal profile of an infrared soldering process.

T_1 in Fig. 13.4). In this figure, T_2 represents the lowest temperature during reflow at which a component can be thermally damaged and thus is considered to be the maximum profile temperature. Normally, a buffer of 5° C (7° F) is subtracted from this temperature (T_2) to establish a maximum working temperature. During cooldown, the solder interconnection temperatures are reduced, allowing for solidification of the molten solder and cooling of the PWA to stable temperatures. The cooling rate in this region is normally recommended not too exceed –4° C/s (–7° F/s) to avoid thermal shock to components such as ceramic chip capacitors.

Laser Processing.   This soldering process uses a focused, controlled beam to deliver thermal energy to an interconnection site. The energy is generated by either CO_2 or yttrium-aluminum-garnet (Nd:YAG) laser sources. The continuous wave of low-power energy is applied to a specific location for a precise length of time. Soldering times with this method are precisely controlled and generally are extremely short in duration (typically 0.01 s) Scanning reflow process times vary from 200 to 800 µs to achieve higher output levels. One advantage to this process is that it can be accomplished without the use of a flux. In this situation, the preformed joints must be properly prepared with small clearances between all connecting surfaces, and these surfaces must exhibit an appropriate degree of solderability prior to processing. Laser soldering offers several other advantages. Localized heating allows for minimal induced thermal stresses. The rapid heating and solidification rates produce fine grain sizes, which in turn results in better fatigue properties. The quick joint heating and formation rates also lends to reduced intermetallic layer thickness and improved joint ductility.

13.3.4 Processing Defects

Due to specific operational parameters for each soldering process, unique solder joint defects can be associated with a specific process or processes. A defect itself may constitute a "failure," both mechanically and/or electrically. Other defects may be categorized as failures depending on the specifications and requirements to which they are designed or processed. Most defects can and will eventually lead to premature mechanical and/or electrical failure. Table 13.9 gives a general overview of process related solder joint defects.

13.4 JOINT TYPES AND DESIGN

The design requirements for both through-hole and SMT interconnections have been well established. A listing of many of the design standards is given in Table 13.10.

13.4.1 Through-Hole Joints

Through-hole joint configurations (see Fig. 13.5[50]) refer to package types in which the component leads are inserted and soldered into predrilled holes in the PWB. Metallization may be located on one side of the board only. This type of architecture is considered to be single-sided. Double-sided and multilayer constructions use holes that are plated through such that the metallization is not only on both the top and bottom sides of the board but

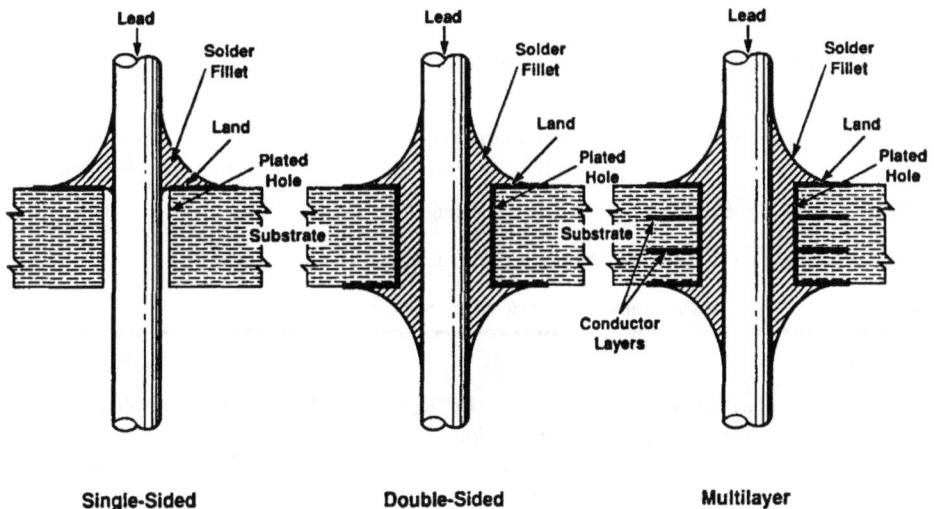

FIGURE 13.5 Various through-hole configurations.

Table 13.10 Commercial and Military Design Standards

Standard	Title	Comment(s)
IPC-D-249	Design Standard for Flexible Single and Double-Sided Printed boards	
IPC-D-275	Design Standard for Rigid Printed Boards and Rigid Printed Board Assemblies	Replaces D-319 and D-949
IPC-D-300G	Printed Board Dimensions and Tolerances	
IPC-330	Design Guide	(Section One: Specifications and Standards)
IPC-FA-251	Assembly Guidelines and Requirements for Single-Sided and Double-Sided Flexible Printed Circuits	
IPC/ANSI-J-001	Requirements for Soldered Electrical and Electronic Assemblies	Current Revision: A, January 1995, supersedes IPC-S-815
IPC/ANSI-J-002	Solderability Tests for Component Leads, Terminations, Lugs, Terminals and Wires	
IPC/ANSI-J-003	Solderability Tests for Printed Boards	
IPC/ANSI-J-012	Joint Industry Standard Implementation of Flip Chip and Chip Scale Technology	EIA J-STD-102
IPC-PD-335	Electronic Packaging Handbook	
IPC-S-816	SMT Process Guideline and Checklist	
IPC-SM-780	Component Packaging and Interconnecting with Emphasis on Surface Mounting	
IPC-SM-782A	Surface Mount Design and Land Pattern Standard	
MIL-STD-2000	Standard Requirements for Soldered Electrical and Electronic Assemblies	Rescinded January 1996
MIL-STD-275	Printed Wiring for Electronic Equipment	Rescinded 21 April 1995, use IPC-D-275
MIL-P-55110	General Specifications for Printed Wiring Board, Rigid	
MIL-P-50884	Printed Wiring, Flexible, Printed Wiring, flexible and Rigid Assembly	
EIA-PDP-100	Mechanical Outlines for Registered and Standard Electronic Parts	

also forms a barrel on interior hole wall. In multilayer applications, conductive paths within the laminate are developed.

The device leads are generally copper or one of several iron-based alloys that can be coated with a solderable metallized surface. Furthermore, the component leads are typically coated in a protective tin-lead solder finish, which is either plated, hot-dipped, or plated and fused. These coatings are placed to maintain solderability over long-term storage.

Substrates for through-hole technology include organic laminates, metal-clad materials, and ceramics typically used for hybrid microcircuits. The metallization at the substrate hole is generally solder-coated copper. These lands and conductive circuitry are constructed of copper foil (0.017, 0.035, or 0.071 mm thick and, respectively, designated as 0.5-, 1.0-, and 2.0-oz copper), or they can be plated with electroplated or electroless copper to meet the thickness requirements (0.013 to 0.076 mm). The addition of plated copper also provides for enhanced solderability.

The role of through-hole technology, as well as any interconnection, is to provide both long-term mechanical and electrical attachment of components to their respective substrates. For double-sided and multilayer architectures, the role has been expanded to include signal conduction between the two substrate surfaces and internal layers. Holes solely used for interloper signal transmission and not component attachment are termed *vias*. In this situation, the hole walls are coated with 0.0003 to 0.0025 mm thick electroless copper and then with 0.025 to 0.076 mm thick electroplated copper.

The hole design must optimize both the mechanical and electrical connection aspects, plus it should incorporate enough room to accommodate the component lead and still provide for capillary flow during the soldering operation. A gap of 0.15 to 0.20 mm between the lead and hole wall is normally specified. Smaller hole sizes can cause the formation of voids due incomplete venting of gases generated during soldering. The protrusion of the lead is typically kept small (0.8 to 2.0 mm) to minimize drainage of the solder fillet. The through-hole pad or land should be round and approximately three times the lead diameter. Additionally, to maintain appropriate fillet formation, the minimum height of the lead should equal the pad width.

13.4.2 Surface Mount Joints

SMT incorporates both leaded and leadless component types, in addition to chip-on-board technologies such as tape automated bonding (TAB) and C^4 (flip-chip) interconnections. In this technology, the component lead or leadless metallization is directly attached to a pad, which depends on the type and size of the package I/O configurations. For leaded and leadless components mounted to PWB substrates, the copper pad (land) thickness is normally between 0.018 and 0.071 mm. As with through-hole technology, the protective finishes are plated, hot-dipped solder (which is generally hot-air leveled to maintain uniformity), or organic coatings applied to the bare copper.

The conductive circuitry on ceramic substrate boards is applied by using one of the thick-film inks. Conductive lines, approximately 0.01 mm thick, are applied by a single printed layer. The pads to which components will be soldered can receive a double printing to reduce the film porosity content, improving solderability, and to increase the metallization thickness, which undergoes dissolution during soldering and subsequent intermetallic formation and growth. The thickness of the double-printed layers is approximately 0.02 to 0.025 mm and largely depends on the amount of porosity in the first layer.

Leadless Surface Mount Joints.   These joint configurations are generally limited to the attachment of chip-type resistors and capacitors. The normally rectangular and cylindrical packages have metal-capped terminations that are coated with an additional metallization to promote and maintain solderability. The chip-type resistors and capacitors incorporate a silver-bearing thick-film ink, which is fired into regions of the ceramic substrate. The addition of 35 percent palladium to the ink reduces the amount of dissolution or scavenging of the silver into the solder during assembly. In most instances, the thick film is coated with copper or a solderable nickel surface to prevent dissolution. Lastly, a protective layer of electroplated tin or tin-lead, or a hot-dipped tin-lead solder, is applied to preserve solderability.

Other leadless devices for integrated circuits have multiple I/O terminations emanating from the package exterior. The terminations are coated with a refractory layer such as tungsten or molybdenum, which provides an electrical feed through the component body. This refractory layer is then coated with a solderable nickel layer 1.3 to 8.9 μm thick. A protective 1.3 to 2.5 μm thick gold layer is applied over the nickel layer. The gold finish is removed by hot dipping prior to solder attachment.

Leaded Surface Mount Joints.   The terminations come in various configurations (gull wing, J-leaded, S-leaded, and so on) and are fabricated from copper or from one of the low-CTE iron-base alloys. For iron-base alloy leads, a solderable layer of electroplated or electroless nickel is applied, followed by a coating layer of gold. The component leads are hot dipped in a tin-lead solder to remove the gold layer or to protect the solderability of copper-base leads. For fine-pitch applications, electroplated tin or tin-lead coatings are preferred, since the plating processes do not build up and deform the leads, which can cause lead misalignment, solder joint distortion, or electrical failure.

Ball Grid and Column Grid Array Joints.   These package types can be considered a form of leaded SMT; however, for ceramic packages, the "leads" are either balls or columns of 90Pb-10Sn solder, subsequently attached by the use of eutectic tin-lead solder. The most common attachment method is the use of the eutectic tin-lead solder to bond the balls or columns onto the component. The balls or columns are seated into recesses formed directly into the PWB pad. Upon solder reflow of the eutectic solder, the joint is formed. Since both the solder balls and columns, for ceramic substrates, are joined with eutectic tin-lead solder, the joint to the substrate reflows a second time during the PWA assembly process. The high-temperature solder balls and columns do not reflow during package or PWA assembly. For plastic packages, the solder balls and the bonding solder are commonly the eutectic tin-lead alloy.

The BGA laminates are formed from various epoxy or ester plastic materials or can be manufactured from ceramic, particularly alumina or low-temperature cofired ceramics (LTCCs). In most applications, the balls or columns are solder bonded to the packages with eutectic tin-lead solder. For ceramic substrate components, an alternative solder column joins the high-temperature melting point solder directly to the ceramic chip carrier.

The large standoff heights created by the hierarchical solder structure (between the ceramic package and the PWB) accommodate the large mechanical strains generated during normal machine power cycling, caused by the thermal mismatch between the ceramic carrier and the PWB. The JEDEC standard for standard ceramic ball (CBGA) and column grid arrays (CCGA) are 1.0, 1.27, and 1.5 mm. For the 1.27 pitch, the nominal ball diameter is 0.89 mm, and CCGA column dimensions are 2.2 mm or 2.7 mm in height and 0.5 mm in diameter. The nominal pad diameter for attachment of CBGA technology is 0.86 mm. The pads are nickel and gold plated.

In plastic package applications, the PWB is generally FR-4 epoxy/fiberglass material, with a nominal copper pad size of 0.635 mm. Two common surface finishes are employed for preserving pad solderability: the finishes are either 0.17 mm thick hot-air-leveled Sn/Pb solder or an organic coating.

For TAB BGA applications, 90Pb-10Sn solder balls are mounted to copper-padded FR-4 PWBs through the use of eutectic tin-lead solder attachment. The balls are typically 1.27 mm pitch or 0.65 mm in diameter.

13.4.3 Controlled-Collapse Chip Connection (C⁴) Joints

In this method of attachment, the silicon chip is bonded directly to a ceramic substrate by small solder bumps. A film of chromium is vacuum deposited onto the chip surface to provide an adhesion layer for the adjacent copper layer. The surface is vacuum deposited with a layer of copper for solderability. To preserve the solderability of the copper layer, a final vacuum-deposited layer of gold is applied.

In situations where the solder bumps are attached to nickel surfaces, gold is the protective applied surface. When the solder bumps are adhered to molybdenum vias, the via is coated with vacuum-deposited nickel and gold films.

The solder bumps can be one of the high-lead, lead-tin alloys (90Pb-10Sn or 95Pb-5Sn) or one of the lead-indium alloys such as 50In-50Pb. When lead-tin alloy solder bumps are used, attachment is conducted using the Sn63Pb37 eutectic alloy. The solder is vacuum-deposited as a 0.100 to 0.123 mm thick film onto the silicon chip and then reflowed in a protective-atmosphere belt furnace.

C^4 solder interconnects must accommodate the thermal strain mismatch between the silicon IC chip and the metallized substrate. By maximizing and controlling the standoff distance above the substrate, limiting the size of the chip, and using substrates with CTEs similar to silicon, fatigue damage can be minimized.

13.4.4 Tape Automated Bonding (TAB) Joints

This packaging technology uses a metal lead frame to secure the silicon IC chip. The lead frame is then soldered to the PWB, forming the interconnection. The silicon chip is attached to the lead frame by a $AuxSn1-x$ solder bond formed by the intermixing and diffusing of gold and tin from metallizations deposited on both the lead and chip at elevated temperature and pressure. The metal frame lead is soldered to the substrate with a tin-lead alloy using a conventional assembly process.

The lead minimizes the thermal strain mismatch between the chip and the substrate, thereby reducing the thermal fatigue experienced by the solder joints. Jogs or bends formed in the lead frame geometry can also accommodate induced thermal strain, further reducing the thermal strain mismatch from differences in CTE properties.

13.5 INSPECTION, TEST, AND CHARACTERIZATION METHODS

13.5.1 Visual Inspection

Since visual inspection is easy to use, requires limited or little resources, and can be conducted virtually anywhere by anyone with appropriate examination skills, it is the most wildly used and applied nondestructive technique for the examination of soldered joints. For this method, inspection is limited to external surfaces or exposed internal surfaces, e.g., large open cracks or voids. These include anomalies such as cracks, voids, wetting or dewetting, filet size, surface characteristics (e.g., roughness, pitting, graininess, and so on), delamination of copper lands, charring or overheating, and cold and disturbed solder joints.

The visual inspection of soldered joints is normally limited to low-power optical microscopy (1.75 to 50× magnification). This method can be limited to the use of magnifiers or microscopes of 2 to 4 magnification; however, for referee purposes, a stereomicroscope of 10 to 30× minimum magnification is required. The smaller the pitch or conductor size, the higher the required magnification. Based on the minimum land width (pitch) size, appropriate corresponding viewing and referee magnifications should be selected for optimal examination. As presented in Table 13.11, MIL-STD-2000 lists guidelines for optical inspection and referee magnification based on conductor land size.

Table 13.11   Magnification Guidelines for Optical Solder Joint Inspection

Land width	Inspection	Referee
>0.5 mm (0.020 in)	4×	10×
0.25 to 0.5 mm (0.010 to 0.020 in)	10×	20×
<0.25 mm (0.010 in)	20×	30×

Visual inspection is limited to the examination of the exterior surfaces of soldered interconnections, except where one or more open surfaces exist (such as a hole, crack, and so on), allowing some degree of internal inspection. Criteria for inspection have been based on visual characteristics, i.e., contact angle and joint appearance (dull, grainy, shiny, and so forth). A list of solder joint defects that are detectable by visual inspection is presented in Table 13.12. The joint appearance to a large extent describes the thermomechanical processing experienced by that interconnection. For example, visual characteristics relate the degree and quality of wetting by the size of the contact angle, or dewetting or nonwetting conditions. A granular-appearing joint may describe a disturbed and/or overheated condition. Lack of wetting, and thus poor joint formation, can indicate a cold interconnection in which insufficient heat prevented wetting and metallurgical bonding. Joint overheating and/or extensive pressure can promote and cause adhesively bonded copper lands to "lift" or become permanently detached. Overheating and/or excessive pressure may also cause individual fibers in the PWB matrix to break, or the same effect may be experienced from the overheating and subsequent expansion of volatiles in the PWB damaging the matrix/fiber interface.

Generally, eutectic or near-eutectic tin-lead alloys exhibit fine-grained microstructures that appear shiny and smooth. Off-eutectic alloys, with large volume fractions of the proeutectic phase, are normally large grained and exhibit dendritic growth. The proeutectic phases tend to make the solder joint appear grainy under low-power magnification. Although some lead-free solder alloys have multiphase structures similar to tin-lead alloys, the phases tend to be larger and, as a result, display a grainy or frosty appearance that could be mistaken for a defect. Typically, the tin-silver, tin-bismuth, and lead-indium solders have a grainy-appearing solder joint. Also, the fillets of tin-silver and lead-indium alloys will be less concave relative to the tin-lead solders, due to their higher surface tensions.

13.5.2   Reflectance Inspection

Reflectance-type systems include optical systems that use conventional lenses, as mentioned in Sec. 13.2.1. Two-dimensional reflectance systems capture a reflected image on a

Table 13.12   Various Defects Detectable by Optical Examination

Dewetting
Insufficient solder
Charring, burning or other damage to insulation, PWB or conformal coating
Spattering of solder on adjacent connections or components
Icicles or bridging
Voids or holes
Cold-solder interconnections
Rosin-solder (entrapped flux) connections
Soldered joints contaminated with residue, lint, splash, flux
Insufficient solder
Insufficient strain relief
Solder that reduces electrical spacing below end product requirements
Splicing
Loose leads, terminals, or wires
Fractured (or cracked) solder joints
Clinched-lead overhang that reduces spacing between conductors below allowable minimum values
Solder balls
Dendrites
Cuts, nicks, stretching, scraping
Heat (including fire) damage and melting of interconnection
Corrosion and/or oxidation
Arcing and evidence of carbon deposits
Delamination of the PWB or adhesive failure of the pad or land
Cracked or delaminated or separations in through-hole barrels
Warping of PWB
Measling of PWB substrate

CRT monitor. The image results from incident light reflected off of the PWA and then collected by the video camera. The image can either be black and white or color, depending on the type of camera used. The captured image can be digitized into a pixel array with a frame grabber. Each pixel in the array has an assigned brightness level between 0 and 225 relative to eight bits of grayscale data for black-and-white imaging. Once the information has been digitized, it can be evaluated with previously stored light data collected by the camera. Image quality is a function of the camera resolution (expressed by the number of inherent pixels), surface illumination, and digitizing surface hardware. The processed image features can be compared to defect data in the computer's memory. The solder joints under examination then can be compared against known-defective joints. This second type of reflectance system has definite advantages over optical microscopy. Since a computer can assess the relative quality of a joint, subjective analysis, particularly from one evaluator to the next, is reduced. Also, the image can be stored and reexamined or simultaneously viewed by multiple analysts. A major disadvantage to this type of examination system is that height (and thus solder joint volume) is inferred from variations in the gray-scale data, which in turn rely on both the software algorithms and illumination techniques.

The last reflectance system generally uses laser technology to scan the solder joint surface. The reflected light is detected by one or more sensors and is then computer processed by geometrical triangulation, which produces a three-dimensional image of the joint. These systems are not as illumination dependent as the other reflectance systems and can provide detailed images of solder joint geometries by incorporation of the height dimensions. Quantitative solder joint geometry and volume data in relation to pad and component leads, is acquired allowing for full three-dimensional imaging. Subsequently, surface contours and anomalies can be fully examined with high resolution. This type of imaging is finding use in other than SMT applications, such as TAB technology applications.

13.5.3 Thermal Imaging Inspection

This type of inspection technology operates on the premise that differing materials exhibit unique thermal properties and, as a result, absorb and release heat energy differently. Heat in the form of IR energy is applied to the assembly. On cooling, the subsequent incident heat energy from the surface of the specimen is monitored. The detected thermal energy is characterized from the individual thermal signatures produced by the inspection system. For instance, a larger volume of a specific material will have a larger heat capacity and therefore will cool at a slower rate. A material of a given volume that is strongly attached to its substrate will likewise demonstrate an accelerated cooling rate over a similar amount of the same material that has not been firmly bonded to its respective substrate.

With the first technique, the entire assembly is exposed to an IR heat source. During cooling, the thermal energy released is monitored by an IR camera, and the recorded video data is digitized. In this situation, the thermal energy is represented as variations in gray-scale data. The images are then compared to previously stored images to assess variations, which then can be related to specific defects.

The second type of thermal imaging technology uses a continuous-wave laser, such as an Nd-YAG laser. In this situation, an individual solder joint is heated by the thermal laser source, and the temperature rise and decay occur by IR emission. For a YAG-type, laser the wavelength of the radiation is 1.06 μm; however, the pulse length can be varied from 10 to 200 ms, with 30 ms being a typical pulse length. The assembly can be mounted to an microprocessor-controlled x-y table that allows controlled inspection of the individual joints at an average inspection rate of 10 joints per second. The thermal signatures are then compared with ideal signatures from previously stored data, and any differences in thermal mass or surface absorption can be related to specific defects. This type of system is very sensitive to surface effects, such as the presence of a blowhole, but anomalies within the solder bulk are extremely difficult to detect.

13.5.4 X-Ray Imaging Inspection

X-ray inspection systems can be generalized into two specific types. The first, *transmissive radiography,* fixes the source and the detecting screen relative to the PWA under examination. X-rays generated by a high-voltage source are transmitted through the assembly, which in turn strikes the fluorescent screen. The amount of X-rays transmitted through the joint is used to describe the solder joint and its associated boundaries. Voids, holes, and gaps within the solder bulk allow for greater penetration of the X-rays producing an image differing from transmission through a denser volume. A video image of the detector (screen) is then digitized and processed. The image focus can be varied by changing the voltage of the X-ray source, thus allowing detailed examination at various locations within the joint.

One disadvantage to this technique is that it is difficult to detect cracks. The cracks must be properly positioned within the joint relative to the incident radiation to promote sufficient contrast. Another disadvantage is the system's inability to distinguish between devices located on differing layers.

The second type of X-ray examination, *laminographic radiography,* allows for cross-sectional viewing of the solder joint volume, including the respective pad and component lead. This nondestructive application has found significant use in BGA applications, since virtually all interconnects are sandwiched between the component body and the PWB. Under this situation, only the exposed joints located on the outer periphery can be viewed by optical techniques. Cross-sectional methods provide for three-dimensional quantitative data gathering at different positions within the joint. Measurements for BGA-type joints include ball and joint shape and solder thickness, including annular rings concentric with the pad or ball centroid. A listing of BGA solder joint defects is presented in Table 13.13.

Table 13.13   Common BGA Solder Joint Defects

Defect type	Description and cause
Short	Excess solder forms electrical connections (bridging) between adjacent traces and vias.
Void	Trapped pockets of gas within the eutectic solder joint. These voids may vary in size; however, small voids can be found at the component–ball or the ball–pad connection. Multiple voids are possible at a single joint, and large voids are experienced in larger-radius joints.
Open	Lack of eutectic solder between the component and ball or the ball and pad. It can be caused by insufficient eutectic solder (paste), bowing of the PWB, or wicking of the solder from a specific site.
Insufficient solder	Caused by insufficient solder paste volume or solder migration during reflow. Joints of this type are susceptible to fatigue. Migration can be due to poor wetting conditions, heat dissipation characteristics of the board, or reflow oven profile characteristics.
Excess solder	Caused by excess application of solder paste. During reflow, the solder moves to the location of lowest surface tension.
Misalignment	Poor placement of the BGA component before reflow.
Solder balls	Particles of solder paste ejected from the joint during processing and reflow. It can be caused by oxidized solder balls and/or volatile compounds expanding during heating.

In laminographic radiography, cross-sectional viewing is produced by a prepositioned source/detector combination specifying a specific *x* and *y* area. By translating the PWA in the *z* dimension, different layers or cross sections can be observed. This method, compared to transmission radiography, offers easy viewing of two-sided assemblies without requiring complex image processing software.

It should be noted that some of these methods, including X-ray laminography, scanning laser (IR) systems, and three-dimensional reflectance imaging systems, typically exist as automated inspection equipment. Automated equipment inspection is generally conducive only to high-volume, low-mix production applications. This equipment can incur significant capital, setup, and processing costs and is not normally seen in failure analysis laboratories unless associated with mass production environments. In addition, these detection systems can be used as process or quality tools, but particular examination technologies must be appropriately applied to assess specific defects inherent to the corresponding interconnection technology. Table 13.14[39] lists different defects types relative to the potential detection capability of three major nondestructive inspection (NDI) technologies.

Table 13.14   Detection Capability of Various NDI Methods

Defect category	Reflectance	Thermal	X-ray
PWB defect group			
Damaged PWB	P	U	P
Solder on PWB	R	U	N
PWB delamination	R	R	R
Artwork defects	R	P	R
Etching defects	R	P	R
Resist defects	R	P	P
Misaligned layers	R	P	P
Incomplete cleaning	R	P	P
Solder mask defects	R	P	P
Solder defect group			
Excess solder	P	P	P
Insufficient solder	P	P	P
Unsoldered connection	P	P	P
Improper wetting	R	P	P
Dull solder	R	P	P
Disturbed solder	R	P	P
Porosity/voids	R	P	P
Solder inclusion	R	P	P
Incomplete flow	R	P	P
Bridging	R	P	P
PTH defect group			
Wrong thickness	R	P	P
Missing holes	R	P	P
Location tolerance	R	P	P
Component defect group			
Component preparation defects	R	P	P
Component registration	R	P	P
Damaged component	R	P	P
Missing component	R	P	P
Wrong component	P	P	P
Part marking	P	N	P
Lead not through hole	P	N	P
Lead trimmed to wrong length	R	N	P
Lead improperly clinched	R	N	P
Component height	R	N	P

Note: R = reliably, P = potentially, U = unsure, N = none.

13.5.5   Electrical Testing

Electrical testing consists of two primary methods: functional and in-circuit testing. The type of anticipated defects will determine the method of electrical testing used.

Functional testing is characterized by electrically stimulating the entire PWA through an edge connector. The subsequent response is a pass/fail indication of the electrical performance of the assembly. One type of functional testing is hot mock-up in which the assembly is modified for ease of interchangability. This test method duplicates actual operating conditions very well and identifies defective assemblies to a very high rate of confidence; however, each individual board type may require a separate system. Other commercially available systems perform more generic tests, test a variety of PWAs, and require little setup time. With this test equipment, one disadvantage is that operational conditions may not be replicated.

Another form of functional testing is the self-test. With this test method, additional circuitry is built into the assembly, which allows the use of simpler test systems and results in reduced testing times. One major advantage is that field failures can be diagnosed by technicians with relatively unsophisticated equipment. The largest disadvantage with any functional testing method is the limited ability to diagnose failures. The problem typically can be traced only to one area on the assembly and not down to a specific component. It should be noted that some functional testers have no diagnosis capability but indicate only whether the assembly has passed or failed the electrical testing.

The second major type of electrical testing is in-circuit testing. In contrast to functional testing, this method tests the performance of each component and not the entire assembly. Probes make contact with nodes located on the PWA surface. The series of probes are mounted in a *bed-of-nails* fixture and can establish contact with one or two sides of the assembly. Tests signal are applied to various combinations of nodes, and the responses are measured. With this method, defects typically can be isolated to a single component or to a small component group. In general, the test probes are positioned on 2.54 mm (0.10 in) centers. In many SMT applications, pitch sizes of 1.27 and 0.63 mm (0.050 and 0.025 in) are not uncommon. With fine- and ultrafine-pitch technologies, developing fixtures of the corresponding pitch size may be prohibitive. The test probes should not directly contact solder joints where the solder and specifically metallizations can be damaged. When testing assemblies that have been coated with hard solder fluxes (such as no-clean types) and residues, establishing electrical contact with the nodes may be very difficult. The solder flux residues can act as insulating paths between the probes and the respective nodes.

The specific test type should be determined based on the anticipated faults, the PWA volume and mix, and the anticipated level of throughput versus test capability of the specific test equipment. Detection of solder joint defects during electrical testing is generally limited to opens and shorts. As a result, the solder defects that can be isolated are limited to solder bridging, solder balls, and unsoldered joints.

13.5.6   Metallographic Characterization[*]

This method of destructive physical analysis (DPA) relies on cross-sectional techniques to prepare and expose predetermined planes for optical and microstructural observation and further testing.

---

*See also Chap. 8.

Joints can be selectively ground or sectioned at differing depth levels to obtain a chemical profile throughout the joint. The soldered joint can be sectioned at any of various angles to expose specific joint details as needed. Multiple slices can be made through a single joint, provided the joint is sufficiently large and the cutting wheel is appropriately thin and rigid to accommodate this. Generally, a fine-blade cutting wheel, such as a diamond wheel, is used. The cutting equipment must come with appropriate mounting fixturing to provide a stable location before and during cutting. Care should be taken when mechanically cutting into and through the joint, since PWA materials (especially the soft materials) can easily deform and smear. The soldered joint may be cast in a mount before sectioning to encapsulate and protect the joint from external damage that can alter the shape or structural details. Cold mounting materials should be used to prevent heating and cooling damage.

The individual slices or the desired surface to be viewed should be ground and polished to produce a flat, optically reflective surface. Again, care must be taken not to apply excessive pressure that can cause smearing of the solder or copper, form multiple planes on the joint surface, or deform existing structures inherent to originally formed joint. Since many differing material combinations are commonly used in soldered interconnections, the harder materials typically will polish differently from the solder materials. Due to the relative hardness differences, it can be extremely difficult to remove all scratches from the entire polished surface. The corresponding hardness differences will also produce different metal removal rates. Generally, the softer materials will be removed faster than the harder materials, and subsequent variations in surface depths will form, causing relief between nonsimilar materials. Another form of damage that can be caused by overpressure during polishing is separation between differing materials—particularly, interfacial separation.

The polished surface can be viewed under optical or microscopic examination. In addition, the surface under examination may be etched to reveal features not otherwise detectable. Enchants normally attack a specific material only, therefore multiple enchants may be required to expose features throughout the joint. For failure analysis purposes, metallographic preparation is specifically used for the following:

- Thickness measurements of conductive circuitry, plated coatings, and intermetallic layers
- Microstructural observation, including grain coarsening, pitting, voids, and other features not exposed to an external surface, cavities, inclusions, internal defects, dendritic growth, and so forth
- Contact angle, fillet angle, and wetting information
- Assessment of bonding to the conductive pad or to the lead (metallization)
- Determining excessive or limited melting
- Preparation prior to SEM analysis
- Preparation prior to chemical analysis, including bulk and localized
- Detection of internal cracking or determining the extent and location of cracking within the joint

13.5.7 Scanning Electronic Microscope Characterization

Depending on the sample and sample preparation, this imaging tool is capable of viewing surface features from 10× to 100,000× magnification with resolution as fine as 3 to

100 nm. Aside from the greatly increased magnification capability, SEM imaging is not as significantly limited with regard to depth of field as is optical microscopy, so the region under examination can remain entirety focused. In general, the depth of field of an SEM is at least 100× greater than that achieved by optical microscopy. Scanning electron microscopy, especially at high magnification, requires extremely flat surfaces that have been metallographically prepared. A flat, polished surface is necessary for quantitative EDS or cross-sectional metallography. Applications include the evaluation of microstructural features, e.g., crack and fracture surfaces, grain size, dendrites, precipitates, dispersoids, micropores, and so forth. When coupled with analytical chemical equipment, identification of elemental species down to micron size is possible, and characterization of elemental gradients on bulk surface samples at 1 μm distances is possible.

The size of the sample to be viewed is limited by the respective sizes of the SEM chamber and associated fixtures used for holding the specimen during examination. Some SEMs are capable of holding entire PWAs, so the individual solder joints can be viewed without destructive preparation. With this type of examination, though, specimen travel can be very limited. Contamination on a failed board may also cause problems with viewing specimens. Conformal coatings, plastics, and other organic materials may also outgas during vacuum depressurization of the SEM chamber. Special attention must be made when contamination is present and, in many situations, the contamination may be removed and analyzed either at another time or with other analytical equipment. If the contamination is not limited to a specific site, a portion of the PWA can be cleaned and prepared with nondamaging and inert materials such as methyl ethyl ketone (MEK) or alcohols. The cleaned portion then can be viewed or metallographically prepared and inspected. Nonconducting surfaces, typically polymers, various adhesives, and organic materials, may cause charge buildup and not allow examination of contaminated sites. If the specimen is small enough to fit into a sputter coater of a vacuum evaporator, the part can be coated with a conductive material. This should be considered an electrically destructive technique that can complicate EDS analysis. Alternately, low sum voltages (300 V to 3 kV) can reduce or eliminate charging at a cost of reduced resolution.

Metallurgically prepared mounts can be coated with conducting materials such as gold-palladium and carbon, or conductive silver paint or carbon colloidal suspensions lines can be painted onto the joint surface to provide a path for electrical conduction and subsequent discharge of conducting surfaces to ground.

Since soldered joints for electronic applications typically are small-volume interconnections, elemental analysis generally is conducted by spot or localized chemical analysis.

Secondary Electron Characterization.   The information from secondary electrons provides the major signal for image information in the SEM. This type of imaging provides for the highest resolution and thus the most "real looking" image.

Backscattering Characterization.   This method is particularly useful in contrasting different phases within a solder or materials of differing chemical compositions, due to atomic number contrast. This method is also less susceptible to charging but requires higher operating voltages.

Energy Dispersive X-Ray (EDS) Characterization.   In conjunction with an SEM, EDS chemical analysis is used to provide an X-ray spectrum of elements inherent to the specimen analyzed. Qualitative assessment of major and minor constituents can be made relative to size and shape of the individual peaks. The minimum size sample is that which can be obtained and placed within the SEM chamber. The lower limit of sample size depends on the ability to handle the specimen with gloves or tweezers. Detection of elements of atomic number 11 or greater is normal. By using special detectors, qualitative analysis of light elements—atomic numbers 5 (boron) to 10 (neon)—is possible. Minimum thickness for quantitative analysis varies with density, but careful selection of gun voltage can allow analysis of thin samples.

X-Ray mapping by EDS analysis is especially useful in assessing the chemical makeup of the components forming the solder joint. Mapping presents a standard secondary electron image of the area analyzed, and individual X-ray maps of the same area are generated simultaneously showing distribution of a specific element. The individual regions can be qualitatively (and hence quantitatively) assessed to determine the specific chemical content of each part. Chemical dissolution and migration can also be characterized by this method. Dissolution of base material metallizations or plating and intermetallic characterization are similar analyses that can be conducted by this method.

Wavelength Dispersive Spectroscopy Characterization.   Relative to EDS chemical analysis, wavelength dispersive spectroscopy (WDS) characterizations is slower, but the elemental peaks are much more easily resolved. Masking effects, interference between adjacent peaks, and especially between elements of small concentration are reduced.

13.6 RELIABILITY CONCERNS

13.6.1 Geometry and Design Effects

The joint design, including lead shape and height, and volume of solder, has a great effect on the long-term performance of a joint. Lead length has been shown to bee a critical factor for quad flat packs (QFPs). The lead length is measured from the contact point with the component body horizontal to the lead toe. When the length of the lead was increased from 21.3 to 2.82 mm (0.085 to 0.1125 in), the fatigue life of the interconnection was increased by 67 percent. The fatigue life of a soldered joint is also a function of the lead height, as measured from the contact point with the component body vertically to the solder pad. As the lead height increases, the fatigue life increases in an almost linear manner. Conversely, the soldered joint fatigue life decreases with increasing width; however, the lead thickness is shown to have a more pronounced effect on fatigue life than lead width. Fatigue life varies inversely with lead thickness. Lead coplanarity is often related to processing problems during production but also can be attributed to rework/repair operations. Large gaps between the pad and lead can result in open joints. For fine-pitch SMT devices, a range of 0.05 mm (0.002 in) has been recommended. The industry norm for standard-pitch components is in the range of 0.1 mm (0.004 in), and for fine-pitch components the range is typically 0.075 mm (0.003 in).

Solder volume associated with a joint affects the stress distribution on the joint and can also affect crack propagation rates once a crack has been initiated. Poorly formed joints can have built-in stress concentration sites that provide locations for premature crack initiation.

Large joint volumes in leadless chip carriers have demonstrated better fatigue resistance than smaller-volume joints. It is believed that the larger solder volumes distribute the applied stress over a larger cross-sectional area and thus reduce the overall applies stress.

Secondly, the large solder volume provides additional area for the crack to propagate through, which results in reducing the crack propagation rate.

For fine-pitch QFPs, with gull wing leads and a foot length of 0.5 mm (0.020 in), the standoff height, as well as the heel fillet, were found to affect fatigue life. As the standoff height increased form 0.05 to 0.10 mm (0.0002 to 0.004 in), the fatigue life also increased. At standoff heights of 0.045 mm (0.018 in) or less, the fatigue life was determined not to be affected by the height of the heel. But with a standoff height of 0.10 mm (0.004 in), the fatigue life increased as the height of the heel fillet increased from 0.10 to 0.25 mm (0.004 to 0.010 in). It is suspected that, as the foot length is increased, most of the strain is accommodated at the heel, thereby reducing the effect of standoff height. It is assumed that, with the short foot length of 0.5 mm, the entire dimension would accommodate the deformation resulting from the plastic strain. This phenomenon has also been shown to occur in 128-lead gull wing QFPs. The magnitude of the equivalent strain was shown to diminish as the distance from the heel increased.

13.6.2   Dissolution Effects

Dissolution of base metal substrates occurs during manufacturing, processing, and rework/repair. The metal substrates used in soldered interconnections are typically precious metal coatings (e.g., Ag, Au, Pd) on component leads or metallizations, diffusion barrier materials (such as nickel), or the conductive copper circuitry itself. The rate of dissolution depends on factors such as base metal composition, solder composition, temperature, and flow velocity of the solder. Dissolution can cause embrittling effects within the joint, removal of precious metal coatings (resulting in reduced or nonsolderable components), and formation and/or growth of intermetallic layers. The rate of dissolution varies exponentially with temperature. Dissolution rates for various precious metals in molten Sn60-Pb40 solder are given by Figure 13.6.[50]

It has been shown that leaching of AgPd substrate metallizations can cause strong alloying of the solder material. When the alloying effects are associated with small cross-sectional area solder joints and rigid substrates, cracking at stress concentration sites and subsequent failure often occurs.

Small additions of various alloying elements into the solder has reduced dissolution rates. Leaching of silver and gold from silver- or gold-plated wires or component metallizations into the joint can be reduced by the use of a 1 to 3 wt.% silver-containing solder. The silver in the solder limits the leaching of silver from the metallizations into the joint. Dissolution of both silver and gold metallizations can be significantly decreased by using palladium-silver or platinum-gold metallizations, respectively. The dissolution of copper can be reduced by additions of Ag or Cd added to the solder. This effect is attributed to the formation of coherent Ag-Cd-Sn intermetallic at the solder base metal interface, which acts as a diffusion barrier to the copper. The use of tin-free solders (Pb-In) was found to be effective in reducing the dissolution of gold.

13.6.3   Intermetallic Formation and Growth Effects

The formation and growth of intermetallic compounds has a strong effect on the physical and mechanical properties of soldered joints. Intermetallic formation is a requirement for solder joint strength but, as the intermetallic layer(s) are usually brittle, these compounds can have a detrimental effect on mechanical properties, particularly fatigue strength.

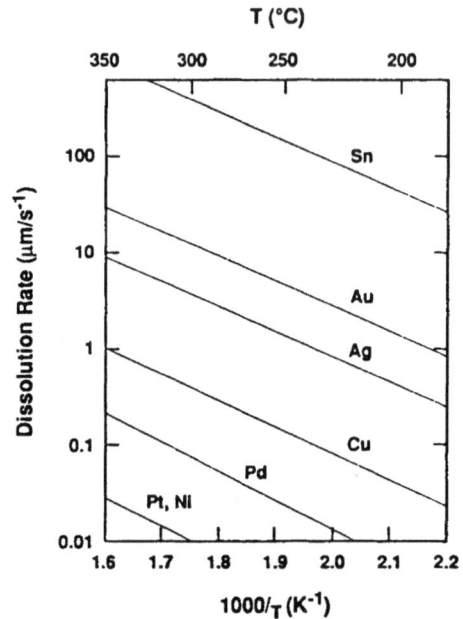

FIGURE 13.6 Dissolution rates for various metals in molten Sn60Pb40 solder.

The growth of these compounds is an exponential function of temperature and time raised to an exponent n (where $n = 1$ or $1/2$). Given the particular solder/substrate system and service environment (time and temperature), the intermetallic layers can continue to grow over the life of the assembly. These layers can become significantly thick (in excess of 20 μm). Figure 13.7[50] demonstrates the growth of copper-tin intermetallics for various coatings and temperatures.

Electronic interconnection technology has largely been based on joining component terminations to copper substrates with tin-lead alloys. During soldering, subsequent processing and annealing, formation, and growth of two intermetallic compounds (Cu_6Sn_5 [η] and Cu_3Sn [ε]) will occur. Both are observed to grow in the solid state. A mechanism for dissolution of the copper substrate, and hence formation of the intermetallic compounds, is provided by the diffusion of the tin into the solid copper and by the diffusion of the copper into the liquid solder. The Cu_6Sn_5 [η] phase forms adjacent to the solder and the Cu_3Sn [ε] phase forms adjacent to the copper substrate. In situations where the combined intermetallic thickness is approximately 10 to 15 μm or greater, a lead-rich layer of solder adjoining the Cu_6Sn_5 intermetallic can be formed. The growth of these compounds occurs during normal processing and in application temperature ranges; that is, from room temperature to soldering temperatures (approximately 250° C) for both soldered joints and pretinned components.

In virtually all cases using tin-lead solders for the attachment of PWA components, a fusible coating of tin or tin and lead solder is applied via plating or hot dipping to component terminations, and to the copper conductive circuitry on PWB substrates. These coatings have been applied to preserve solderability. Under these situations, intermetallic compounds are formed. Excessive solid-state growth of these layers on pretinned parts can lead to poor or nonsolderable conditions. The intermetallic film can consume very thin layers of tin or tin-lead solder. The copper/tin intermetallics are wettable only by a strongly

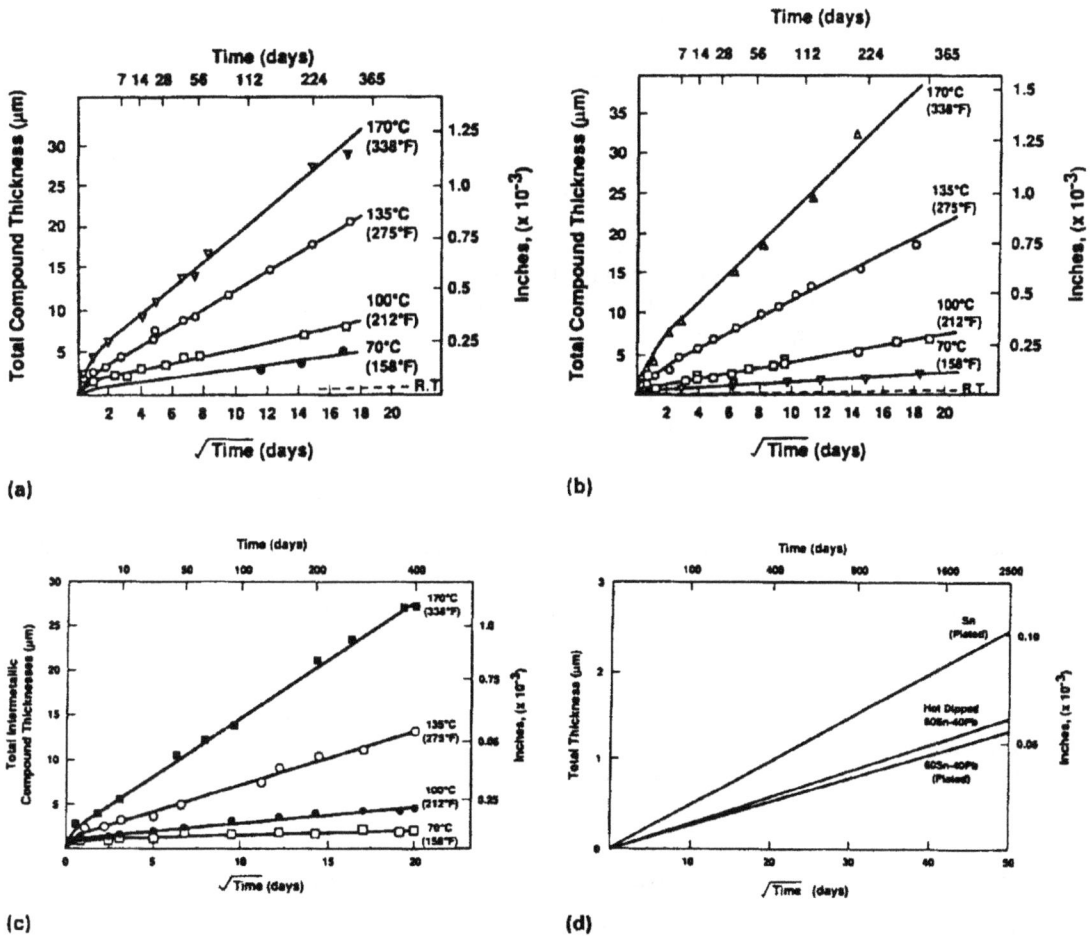

FIGURE 13.7   Total copper-tin (Cu_6Sn_5 and Cu_3Sn) intermetallic compound thickness growth rates for various coatings and temperatures. (a) Electroplated tin coating, (b) electroplated 60Sn-40Pb coating, (c) hot-dipped 63Sn-37Pb coating, and (d) various tin-lead coatings at room temperature.

activated flux. If the solder or fusible coating is sufficiently thin on the copper substrate, the oxidation of either intermetallic can occur. The oxidation of the intermetallic layer may occur as isolated patches surrounded by solder or, in a worst case scenario, a uniformly oxidized surface with extremely poor wettability.

Molten solder-substrate reactions can form intermetallic compounds that can embrittle the solder joint, such as the compound $AuSn_4$. For instance, in tin-rich solders, the molten tin, reacts with the Au (normally, from 1.3 to 3.8 µm thick plated component leads) to form either $AuSn_2$ or $AuSn_4$. The $AuSn_4$ intermetallic is the predominant compound formed. This intermetallic readily embrittles tin-lead solders. The maximum limit of gold in a tin-lead solder has been given as 3 to 4 percent. Above this content limit, embrittlement is likely. Given sufficient loss of substrate materials, embrittlement of the joint can result in lack of impact strength or brittle tensile behavior.

13.6.4 Coefficient of Thermal Expansion

Both local and global effects result from the differences in coefficient of thermal expansion (CTE) between solder, component bodies, component leads, metallization, and PWB materials, and they can promote and cause failure and reliability problems in soldered interconnections. Mismatch between different materials used to produce soldered interconnections, including substrate materials (see Table 13.15), with thermal cycling, can lead to premature thermal fatigue failure of the solder joint. Local effects are generally limited to mismatch at a specific soldered joint or to the interconnections of a specific component. Global effects include thermal expansion mismatches over portions, regions, or the entire PWA itself. Large effects between the circuit board and component materials account for most of the stresses and strains produced by this differential. The effects of CTE mismatch generally are not experienced in through-hole technologies where lead compliance is large. SMT, and in particular leadless SMT, devices are very susceptible to CTE effects if the mismatch is considerable.

Table 13.15 Coefficient of Thermal Expansion (CTE) Data for Various Solders, Component Bodies, Component Leads, and Substrate Materials

Solder alloy	CTE ppm/°C	Component body or component lead material	CTE ppm/°C	PWB or substrate material	CTE ppm/°C
52In/48Sn	20[32]	Plastic leadless chip carriers (PLCCs)	19[2]	Cu PWB lands	15[2]
58Bi/42Sn	14[53]	Ceramic leadless chip carriers (CLCCs)	6[2]	Phenolic paper laminate (FR-2)*	19–30[58]
77.2Sn/20.0In/2.8Ag	28[30]	Cu leads	15[2]	FR-4® boards	15[2]
63Sn/37Pb*	24.5[58]	Alloy 42 leads	6[2]	Kevlar® boards	7[2]
60Sn/40Pb	25[58]	Kovar™ leads	6.5[53]	Alumina*	6[58]
91.8Sn/4.8Bi/3.4Ag	23[53]	Invar™ leads*	1.6[58]	Polyimide quartz	6–8[40]
95Sn/5Sb	28.2[32]			Polyimide fiberglass	12–16[40]
96.5Sn/3.5Ag	30.2[32]			Epoxy aramid fiber	6–8[40]
				Fiberglass Teflon®	20[40]
				Thermoplastic resin	25–30[40]
				Alumina-beryllia	5–7[40]
				Beryllia (BeO)	6.4[58]
				Kevlar/cyanate ester	15[26]

* All CTE data measured at 20° C except where noted with asterisk, in which case data measured at 25° C.
Data from references shown in superscript.

Heating and subsequent cooling operations during soldering can result in extremely large changes in temperatures (ΔT), and complicated states of stress and strain may result. Very rapid solidification and cooling are typical of hand-soldered interconnections. The conditions of stress and strain may be further compounded by poorly formed soldered joints, especially due to differences in solder volume on the same component (particularly leadless types).

Through-Hole Technology.   Because most of the stress and strain is accommodated by the typically long and relatively compliant leads, mismatch between the component body and PWA have little effect on this solder joint life. However, thermal fatigue in through-hole technology is normally due to mismatch between the lead and through thickness expansion value of the PWB laminate. The solder lessens this effect for the joint, and strain relief bends or loops placed in the leads minimize this mismatch. Repeated heating of the joint such as is incurred by rework and/or repair can cause degradation of multilayer PTH printed wiring boards. This has been shown to be a function of the number of board layers. PWAs with larger number of layers incur greater thermally induced stresses acting on the PTHs, resulting in an estimated higher failure rate. The stresses have been attributed to the direct thermal mismatch between the copper barrel and the conductive circuitry and the dielectric material. For this reason, it has been recommended that repair/rework operations be limited to a maximum of three excursions, to reduce the risk of introducing failure mechanisms into the PWA.

Surface Mount Technology.   Soldered interconnections of SMT components such as leadless chip carriers and ceramic capacitors are particularly susceptible to low-cycle fatigue failures caused by this mechanism. Leaded devices can accommodate some of the stress and strain produced during thermal cycling. The presence of brittle intermetallics at joint interfaces and residing at stress concentration sites also accentuates failure. Cracks on leadless packages begin under the component body and propagate out toward the open surface. As a result, this type of cracking typically grows to a significant size before being visually delectable.

Localized CTE mismatch between a bulk solder and an intermetallic layer adjacent to the copper substrate can produce cracking. The cracking occurs at the interface between the solder and the intermetallic layer. The high bulk solder strength, in conjunction with the rapid solidification and cooling of the soldered joints, contributes to "impact loading" experienced at the solder/intermetallic layer during thermal contraction. The failure mode is typical of overloading, especially during high loading rates.

Fatigue resistance can be improved by increasing the standoff height between the component and the substrate. Quickly solidified solder joints that subsequently produce a fine microstructure (e.g., hand conduction methods, hot-gas, and hot-bar processes) improve fatigue properties. The use of lead-free or alloyed solders can also improve fatigue properties over those of the eutectic or near-eutectic tin-lead solders.

13.7   FAILURE MECHANISMS

13.7.1   Wetting

Wetting of a base metal substrate by the solder can qualitatively be deduced by the contact angle θ. Excellent wetting is said to be achieved when θ is positive and small; that is, approaching $0°$ but less than $\approx 60°$. Optimal wetting contributes to optimal solder joint properties (e.g., fatigue strength and shear strength).

Nonwetting conditions typically arise form the lack of formation of the intermetallic compound. Under such situations, a film or layer is typically present that does not allow for the formation of metallurgical bonding. Such films can be oxide layers or preexisting and exposed (oxidized) intermetallic layers. Contaminant films such as silicones, epoxies, and other residues such as oils can also cause this problem. Entrapped dirt at the base metal or at the interface between metal layers will prevent the formation of metallurgical bonding and result in nonwetting conditions. Lack of wetting produces defects in the adhering areas, which act as stress risers and are particularly deleterious to solder joint life in fatigue environments. In SMT soldering applications, nonwetting can occur from non-reducible films or localized regions of insufficient heating (i.e., too low to facilitate bonding).

Dewetting (localized wetting) is a mechanism in which wetting is not complete such that the solder will not wet specific sites and, as a result, will pull back from these areas. This form of inhibited wetting can be barely visible or can be extremely evident, depending on the situation, and may require magnification aids to fully determine the degree of wetting. A source of dewetting has been linked to the evolution of gases during the sudden exposure of the part to the molten solder. The volatile expansion of water, and degraded or hydrated inorganics, produces this effect. In addition, these materials can be entrapped or codeposited in a coating. In many situations, the solder surface exhibits fine porosity. Vaporization of water can oxidize the molten solder film or a subsurface interface, such as an intermetallic layer. The degree of dewetting will depend on volume of gas evolved, the type of gas, and area location of the gaseous release. The larger the volume and depth of water vapor relative to the solder thickness, the greater the affected dewetted region. Contamination of the base metal by embedded particles or inorganics will enhance the dewetting effect, since the intermetallic and solder film can be oxidized. Dewetting can also be caused by gas-releasing impurities in any substrate metal layer.

Finishes such as tin, solder, gold, and silver are used to preserve solderability and can themselves be contaminated and/or entrap materials that in turn promote poor wetting. Since these coatings are typically removed during soldering or presolder preparation (e.g., hot dipping), the solder and/or the copper-intermetallic interface region can be affected.

The degree of dewetting can be extremely small, giving rise to little effect on the joint properties, or it can be similar in effect to nonwetting conditions. The effect to solder joint properties is normally to reduce fatigue resistance, since void formation creates stress risers, which can reduce fatigue life by 50 percent.

Magnification aids are extremely useful when discerning the type and degree of wetting problem. Characterization by scanning electron microscopy may be required to appropriately assess wetting phenomena limited by optical methods. The use of surface analytical techniques like EDS, auger, or ESCA spectroscopy can be crucial to film and oxide characterization or particle or residue identification.

13.7.2   Mechanical and Thermomechanical

Overload.   This failure mode is rarely experienced in soldered interconnections. Designs have accommodated for this failure mode and, if failure does occur by overload, this mechanism is generally associated with shear failure due to shock environments. Rarely are tensile failures experienced.

Low-Cycle Fatigue.   Low-cycle fatigue is commonly related to thermal expansion differences (see Sec. 13.6.4). The sources of temperature change (ΔT) can be attributed to the following:

- Powering on/off cycles
- Variations in heating/cooling of different components/interrelated assemblies
- Environmental temperature changes
- Differences in electrical loading relative to different assemblies/equipment

Through-Hole Technology.   Thermal fatigue in through-hole technologies is normally due to mismatch between the lead and through-thickness expansion of the PWB laminate. Fatigue begins as a "matted ring" around the component lead and then propagates through the solder joint as shown in Figure 13.8.[50] A common appearance of fatigue initiation in plated-through hole (PTH) interconnections is severe roughening and cracking of the solder surface. This has been stated as being only superficial in nature. At a much later stage, the fatigue cracks can propagate from either end and proceed inward.

Surface Mount Technology.   In leadless SMT component applications, the fatigue cycle usually starts with grain coarsening of the solder and may exhibit roughening on the exposed surfaces. Typically, the fatigue will start under the component and progress to the outer surface. The cracks usually propagate to the corners of the components (stress risers) and then through the bulk of the joint. It should be noted that cracking can also occur at the component/solder interface. This is dependent on the thermal expansion coefficients (TCEs) of the board, solder, and components and the subsequent resulting strains.

In leaded SMT component applications, thermal fatigue begins as grain coarsening and possible roughening of the exposed solder surface (generally at the component heels). Depending on the lead-frame material, fatigue cracking can start at the toe and/or heel. The cracks will then propagate into the joint.

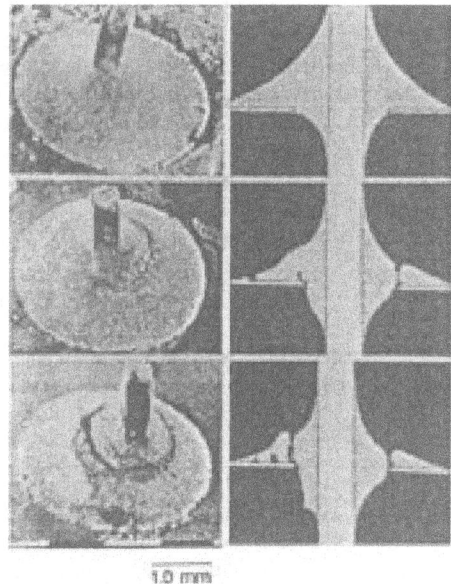

FIGURE 13.8 Progression of low-cycle fatigue through a single-sided through-hole soldered joint.

1.0 mm

Optimizing Reliability. There are many methods available to improve the reliability and hence increase the life of soldered joints. Some common ways are listed as follows:

- *Providing strain relief.* This is commonly accomplished in through-hole applications where the leads are long (compliant) and can be bent to allow for additional relief. SMT leaded components can have long or shaped leads, depending on the configuration. However, SMT devices tend not to be as compliant as through-hole devices.

- *Matching CTE.* This is especially important in leadless SMT applications where there is virtually no strain relief and the solder joints accommodate the vast majority of the resultant strain. CTE matching should be accomplished not only between the board and component body material but, if possible, incorporate the component lead and solder materials.

- *Decreased solder joint thickness.* This is applicable where the strain is relatively small. By decreasing the joint thickness such that the intermetallic compound is continuous from the substrate to the lead, the intermetallic layer will not fracture under small strains.

- *Increased solder joint thickness.* It has been proposed that the total shear strain across a joint is inversely proportional to solder joint thickness. Therefore, by increasing the joint thickness, the strain is reduced. However, experimental results have shown this effect to be minor.

High-Cycle Fatigue. For most practical applications, cracking or failure by high-cycle fatigue is highly unlikely. Component leads can crack when exposed to high-frequency energy (i.e., ultrasonic).

Creep and Stress Rupture. Creep is the plastic deformation associated with constant load (stress) at temperatures usually greater than or equal to half the melting temperature (T_m). For many applications, creep resistance can be the most important design parameter. Since electronic solders are by definition low-melting-temperature metals and are used at high homologous temperatures (T/T_m), creep deformation influences nearly all aspects of their mechanical behavior. In general, solders have low creep strengths such that Pb-30-Sn50 alloy at room temperature produces a creep rate of 0.0001 mm/mm/day at a constant stress of 830 kPa (120 psi).

Creep, in general, consists of three main regions. During the first (*primary* or *transient* creep), the creep rate (strain rate) decreases with time. During the second region, or *steady-state* creep, the creep rate is constant. Finally, *tertiary* creep is that region in which the strain rate increases with time. Cavitation is the primary cause for tertiary creep in bulk solder samples loaded in tension. This process is facilitated by a large grain size. The triple or quadrary grain junctions provide locations for cavitation and stress concentration. Solder creep tested in shear often exhibits tertiary creep as plastic instability. Relative to tertiary creep by cavitation, tertiary creep due to plastic instability does not necessarily result in rapid failure and can experience very large tertiary strains. For example, soft solders, such as the Sn-Bi eutectic, can experience more than 50 percent of its creep strain to failure at 20° C.

Intergranular Creep. In situations where the solder microstructure is fine grained, the stress is low to intermediate, and the temperature is high, creep can be exhibited intergranularly. This mechanism leads to very stable and nondamaging plastic deformation. When the grain size is small, nearly equiaxed, and remains stable under deformation, intergranular deformation occurs by grain boundary sliding.

Stress Rupture. Stress rupture, as related to solder joints, is the time to failure (fracture) under constant loading.

13.7.3 Corrosion

The presence of corrosion and residues promoting corrosion at an interconnection can increase the resistivity of the conductive path, cause shorting, facilitate dendritic growth, cause decohesion of conformal coatings and, in extreme cases, produce mechanical as well as electrical failure. Additionally, corrosion products affect the thermal fatigue life of chip joints, and selective corrosion attack can produce stress concentration sites.

Chloride, Carbonate, and Sulfate Corrosion. Lead-containing solders, and in particular tin-lead versions, are very susceptible to this form of corrosion. Under normal conditions, lead in the solder is protected against environmental attack by the formation of tenacious, strongly adhering lead oxides. In the presence of water, the chloride ions can form with the disassociated water hydrogen ions to form hydrochloric acid. The hydrochloric acid reacts with the lead oxide (PbO) to from $PbCl_2$ and water. The lead chloride is a very loosely adhering material. The lead chloride combines with water and carbon dioxide to form lead carbonate ($PbCO_3$). The final corrosion product is formed when the sulfate reacts with the lead complex to form lead sulfate ($PbSO_4$), which is a white porous layer. In the presence of water, carbon dioxide, and sulfates, this reaction is self-generating such that, as the very porous layer of lead sulfate does not protect the solder, the exposed lead-oxide surface can continue to be converted into the lead carbonate and finally into the lead sulfate.

Aqueous and Gaseous Mildly Acidic Electrolytic Solutions. In air-saturated aqueous environments, the dissolution rate for lead and lead alloys containing 1 percent or less Sn was an order of magnitude greater than that for alloys containing 3.5 to 80 percent Sn and two orders of magnitude greater than pure Sn. In nitrogen-saturated aqueous environments, the dissolution levels for all concentrations of Sn were typically reduced by an order of two; however, the corrosion rates were considered negligible.

In an aqueous solution with a pH of 4, the dissolution rate is relatively constant for alloys with Sn levels below 80 percent. From 80 percent concentration to pure tin, the corrosion rate linearly decreased. At pH levels of 4 to 7 the Sn was shown to passivate relative to the Pb which demonstrated negligible corrosion rates.

In a gaseous environment simulating office environmental conditions (containing H_2S, SO_2, NO_2, and Cl_2), the lead oxide readily reacts to form a stable thick corrosion product consisting largely of $PbSO_4$ and small amount of $PbCO_3$ and $Pb(OH)_2$.

It was determined that the maximum thickness of the corrosion product on a low-Sn solder (≤ 2 wt.%) was 12 nm, which would form a C^4 joint of acceptable quality and strength. The lead oxides were found to be thicker but were readily reducible.

Soldering Fluxes. The fluxes used for soldering in electronic applications are many and varied, and some are more prone to corrosive effects than others. Pure rosin flux (type R) requires a temperature of 160° F to become activated. At ambient temperatures, the rosin is inert an can act as an insulating barrier to the solder joint. Activated rosin fluxes do contain activators, which can be halides such as fluorides, chlorides or bromides, organic acids, or amines or amides that must be removed shortly after soldering. All visible indications of flux residues should be removed and ionic contamination testing conducted to ensure appropriate removal of ionic contamination. Water-soluble fluxes are typically citrus acid based, and aqueous or semiaqueous cleaning techniques should be employed to remove and dilute these residues. No-clean fluxes, which contain organic acids, may remain in an activated state if improper thermal profiling or hand-soldering techniques are used.

Flux residues can and do leave ionic and acidic residues on the PWA. If sufficient residue or corrosion products are available, chemical analyses can detect the various species present.

Corrosion of Sn-Pb and Pb-Sn Alloys. Both tin-based and lead-based solders are susceptible to attack by acids and alkalies. Lead-rich phases form many insoluble compounds, relative to tin-rich materials, in near-neutral aqueous media. Alloying additions to the Sn-Pb solders do not readily affect the corrosion resistance of these alloys.

Lead-rich alloys are particularly susceptible to attack by volatile organic acids. Wood or insulating materials can give off acetic acid (CH_3COOH) and/or formic acid ($HCOOH$) or other acids that can produce a "white encrustation" that is detrimental to the lead-rich material.

In many atmospheric conditions, a layer of lead sulfate ($PbSO_4$) forms, which protects the solder. In chloride and sea environments, the solder becomes anodic to nickel, copper, and copper alloys.

Corrosion of In-Pb Alloys. Testing at 70° C and 70 percent RH for 6000 hr was conducted on various alloys containing from 5 to 50 percent In. After 3000 hours, the 50In-50Pb showed extensive corrosion. At concentrations of 25 to 50 percent In, the severity and frequency of corrosion decreased rapidly. Below 25 percent In, the corrosion was considered negligible and, below 20 percent, the corrosion was considered nonexistent.

Galvanic Corrosion. The soldered joint is a composite system made up of many different materials. Within the joint or between joints and other conductive circuitry, dc circuits can be established that will corrode the most anodic material. For this type corrosion, a cathode, anode, and electrolyte are required. Humid environments with conductive species (such as flux residues) can form electrolytes. Since the PWA is electrically powered, sufficient electrical energy exists to form a voltaic cell.

In multiphase solders, such as a two-phase tin-lead solder (see Fig. 13.1), the electromotive potential between the two or more phases can with the presence of an electrolyte establish a galvanic cell.

13.7.4 Ionic Contamination

Conductive ionic contamination left on the solder joint or PWA can cause current leakage, electrical malfunction, dendritic growth, shorting, and corrosion. Ionic contamination can be detected and measured by solvent extract methods measuring the conductivity or resistivity of the extracted solution. The PWA is placed in a solvent wash consisting of 75 percent deionized water and 25 percent isopropyl alcohol by volume. Normally, the solution is heated and agitated. This procedure is described in detail by both MIL-STD-2000 and IPC-TM-650. The amount of contamination is usually expressed in micrograms NaCl equivalent per square area (of the PWA). It should be noted that this method of ionic contamination testing does not differentiate between ionic species and only measures those contaminates that are removed from the PWA and placed into solution.

Other chemical test methods such as ion chromatography and high-performance liquid chromatography (HPLC) can identify and quantify ionic and organic species, respectively.

Ionic contaminates can result from the PWB manufacturing process, soldering processes, handling, or airborne contamination. The following are common sources and types of ionic contaminates:

- Underetched particles of copper
- Metallic smearing of copper, tin, and lead
- Flux residues, typical species being fluorides, chlorides, and bromides
- Adsorption of chlorinated and other (organic) cleaning solvents via the PWB surface
- Entrapped chemicals from PWB manufacture forming rough PWB surface or surface tension effects, such as PTH barrels

13.7.5 Elemental Contamination

Contamination (Lead-Containing Solders).   Contamination by residual metallic elements can have detrimental effects on the solder joint. Excessive additions of these elements can result in poor or nonwetting, graininess, reduced resistance to corrosion, solder joint brittleness, and the formation of insoluble compounds. Table 13.16 lists maximum contamination limits for both preconditioning and assembly soldering. This table provides guidelines for maximum allowable contamination in weight percent for military and commercial applications governed by MIL-STD-2000 and IPC/ANSI-J-001. Purity levels established by other solder standards will be similar.

In addition, nonmetallic species can cause detrimental effects in tin-lead solders. For example, 0.01 wt.% phosphorus causes dewetting/grittiness, and 0.0015 wt% sulfur causes grittiness in Sn60/Pb40 solder.[22] Excessive sulfur is expected to appear as particles of tin and lead sulfides. Similarly, excess phosphorous can produce dewetting and result in the formation of tin or lead phosphides.

Contamination Effects (Lead-Free Solders).   Lead-free solder alloys are extremely susceptible to lead contamination. With this type of contamination, materials compatibility, including subsequent reliability, is of great concern.

The effects on melting temperature by the addition of lead from a lead-containing source, such as a substrate previously contaminated with a tin-lead solder, will vary depending on the concentration of lead relative to the chemical composition of soldered joint. The effects of lead will also depend on chemical composition of the solder used during manufacture, rework/repair, miscibility, diffusion properties, and compatibility within each joint. For instance, the melting temperature of a contaminated joint may vary little from that of the applied lead-free solder, or it could result in significant lowering of the melting temperature of the lead-free solder, such as the formation of ternary eutectic phases (Table 13.17).

Very small additions of lead (0.1 wt.% or more) to tin-silver, tin-antimony, or tin-copper solders can be very deleterious. The presence of dissolved lead in high-temperature lead-free solders has the potential to result in hot tears (cracking) during solidification or

Table 13.16 Contamination Limits for Preconditioning and Assembly of Eutectic and Near-Eutectic Tin–Lead-Bearing Solders

Contaminant	Recommended maximum contamination limits in weight percent*		Possible solder joint condition if contaminated
	Preconditioning (lead/wire tinning)	Assembly soldering (pot, wave, etc.)	
Copper	0.750	0.300	Sluggish solder flow, solder hard and brittle
Gold	0.500	0.200	Solder grainy and brittle
Cadmium	0.010	0.005	Porous and brittle solder joint, sluggish solder flow
Zinc	0.008	0.005	Solder rough and grainy, frosty, and porous; large dendritic structure
Aluminum	0.008	0.006	Solder sluggish, frosty, and porous
Antimony	0.500	0.500	Insufficient: formation of tin pest/white powder from low-temperature aging ($\leq 13°$ C) Excess: solder brittle
Iron	0.020	0.020	Formation of $FeSn_2$, which does not wet—presents resoldering problems
Arsenic	0.030	0.030	Small, blister-like spots
Bismuth	0.250	0.250	Reduction in working temperature
Silver[†]	0.750	0.100	Dull appearance—retards natural solvent action
Nickel	0.025	0.010	Blisters, formation of hard insoluble nickel/tin compounds

* The total of copper, gold, cadmium, zinc, and aluminum shall not exceed 0.4% for assembly soldering.
† Not applicable for Sn62Pb36Ag2 (limits to be 1.75 to 2.25 for both operations).

Table 13.17 Melting Temperature Changes for Various Binary Eutectic Alloys Alloyed with Lead

Binary alloy	MP for binary eutectic (°C)	MP as ternary eutectic with lead (°C)	ΔT (°C)	Reference
Sn-Ag	221	178	−43	2
Sn-Bi	138	95	−43	2
Sn-Cd	176	142	−34	4
Sn-Cu	227	162	−65	59
Sn-Sb	230	182	−48	2
Sn-Zn	199	183	−16	2

hot shortness and subsequent joint failure. This type of contamination will also negatively affect the mechanical properties of the soldered joints.

Low-melting-temperature solders have also exhibited a low tolerance for lead contamination. In many instances, after joining surface mount devices (SMDs) with Sn-Bi solder, pores were observed due to the formation of a low-melting (96° C) ternary Sn-Pb-Bi phase.

The dissolution of metal substrates or coatings into soldered interconnections can also affect the solder joint chemistry and metallurgy (see Sec. 13.6.2). The dissolved metals may react with the solder and/or the residual lead in solution. This effect may result in the formation of low-melting-temperature phases or other uncontrolled species, which could lower the physical and mechanical properties of the joint.

13.7.6 Dendritic Growth

Dendritic growth of metals, resulting in a tree- or fern-like appearance, occurs under the influence of a combination of moisture and a dc potential. The migration of silver is most common, but other metals common to PWA technology, such as copper, tin, and lead, are also susceptible to dendritic growth. Copper, tin, and lead dendritic growth is slower than silver dendritic growth. With silver, the migration may extend along a surface for millimeters but also tends to follow paths through and along fibrous materials surfaces. This mechanism, as with other metallic growths, can lead to shorting, current leakage, and electrical malfunction. The addition of as little as 2 to 3 wt.% lead is sufficient to inhibit the dendritic growth of tin.

Mechanisms for growth other than electrical potential and moisture can promote migration. Indium, commonly used as a constituent in solder material for GaAs lasers, exhibits growth by thermomigration; that is, the metal is transported from hot surfaces to cold surfaces.

13.7.7 Whiskers

Pure tin, and especially electrodeposited tin, are prone to whisker formation. Metals such as silver, gold, nickel, and palladium are less susceptible. Whiskers can grow as the result of stress in metal layers. Generally, whiskers are single crystals that have a constant cross section along their length. Whisker diameter are on the order of millimeters, whereas the length of the whiskers can be many millimeters. The whiskers can be identified as low-impedance shorts or as high-impedance paths but are usually identified as current leakage between leads. The current-carrying capacity at times can be in the tens of milliamperes. Tin-lead solders containing less than 70 percent tin are not considered susceptible to whisker formation.

The velocity of whisker growth has been given as 0.001 to 0.3 nm/s or ≈3 mm/year. Increasing the temperature also increases the mobility of atoms and subsequently the growth of the whisker. However, if the temperature is sufficiently raised to stress-relieving conditions, whisker growth can be curtailed. If the temperature is increased beyond where the annealing mechanism predominates, each metal exhibits a particular temperature at which maximum whisker growth occurs. For tin, this temperature is between 60 and 70° C.

Moist and hot environments promote whisker growth, and this growth is approximately proportional to the compression stress exerted on the material.

13.7.8 Other Mechanisms

Tin Pest.   Tin pest is the decomposition of metallic tin (β-tin) to α-tin that occurs at temperatures below 13° C (see Fig. 13.1). This transformation rarely occurs at temperatures near (just below) 13° C, as implied by the equilibrium diagram, since extremely long periods of time are required to facilitate this transformation. The maximum rate experienced has been stated to be at –30° C. This transformation changes the metallic tin into a grey powder, which is accompanied by a 26 percent increase in volume, causing disintegration of the tin. The application of a tensile stress accelerates this process. Electroplated pure tin coating at long, low-temperature exposure is susceptible to this transformation. Tin hot-dipped coatings have not exhibited this transformation.

Until recently, the U.S. military required the addition of 0.20 to 0.50 wt.% antimony to tin-containing electronic solders to prevent the possibility of this type of solder disintegration. The requirement now stands at ≤0.50 wt.% antimony (Sb). Solder containing 40 wt.% lead is generally not affected, provided that contamination with previously transformed α-tin (tin-pest) does not occur. It has been stated that this transformation can be eliminated by the addition of 0.1 wt.% antimony or 0.5 wt.% bismuth to the solder.

Conductive Anodic Filament Growth.   Ionic transport across the insulation and between conductors produces metal filament growth over the insulation. This condition can result in shorting of adjacent electrical conductive circuitry. This growth becomes very important with the use of closely spaced conductors, high applied voltages, and moist environments. The growth generally occurs from cathode to anode but, under specific circumstances, may occur from the anode or both the anode and the cathode.

CAF growth can also take place inside moistened solder resist or within the PWB along glass reinforcing fibers between oppositely biased plated holes.

13.8   CASE STUDIES

13.8.1   Gold Embrittlement on Component Leads

Problem.   Gold embrittlement of the hot-dipped tin/lead solder on component leads generally occurs when the leads of the components are not properly tinned. A correct solder coating, allows for the leads to remain in the solder pot for sufficient time to dissolve the gold (plating on the leads) in the molten solder. During this process, the leads are subsequently coated with tin/lead solder. This coating should contain less than 3 percent gold by weight. Above this approximate amount, the solder can become brittle and will tend to flake off the solder coat, especially in regions experiencing tensile loads (i.e., locations were bending is performed). The bends on the leads are intentionally formed during the lead-forming operation to allow the component to be properly seated on the printed wiring board. Figure 13.9 shows the flakes of embrittled solder coat flaking off the first bend on the component lead.

Corrective Action.   To correct this defect prior to installing the component on the board, the leads need to be retinned. The leads must remain in the molten solder bath for sufficient time to allow the original contaminated solder coat and residual gold to dissolve into the molten solder. If the component is already installed in on a printed wiring assembly, then the component should be removed form the PWB and manually retinned with a soldering iron or solder pot. It is also highly recommended that attachment sites on the PWB, which may contain contaminated solder, be reworked. The solder should be wicked or vacuumed off the PWB through hole or pad and then replenished with fresh solder. This process should be repeated at least two to three times to ensure that the concentration of gold is sufficiently below that which would cause embrittlement.

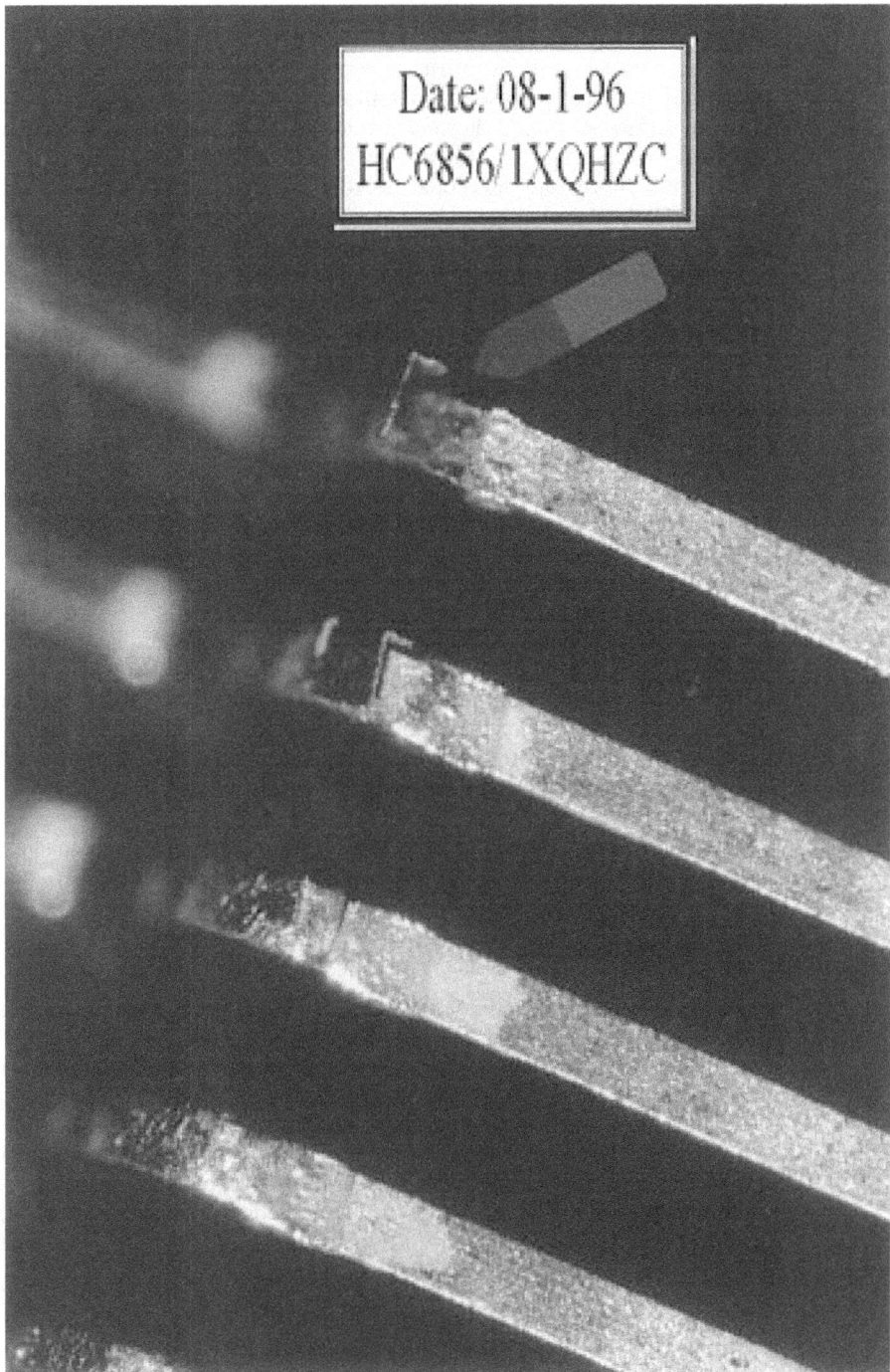

FIGURE 13.9 Embrittled tin-lead solder on tinned gold-plated lead.

13.8.2   Solder Bridging

Problem.   Solder bridging most commonly occurs during the tinning of component leads. This is typically caused by not allowing the leads enough time in the bath to minimize the surface tension of the solder onto the component leads. Subsequently, the solder remains wicked, connecting the adjacent leads as shown in Figure 13.10. This typically results when the operator has not acquired the appropriate skills to assess the duration and speed at which the leads should be exposed to and removed from the solder bath.

Corrective Action.   Periodic operator training may be required to maintain the desired level of precision required to properly tin component leads. If feasible and cost-effective, automated/robotic equipment may be used to ensure repeatable tinning results.

13.8.3   Nicked Wires

Problem.   The nicking of wires is a common defect experienced in electronic assembly. Multiple stranded wires that require tinning are very susceptible to this form of defect, which commonly occurs during insulation stripping. The wire stripping tools provide for the concentric removal of the wire insulation. However, the wire strands, to some degree, usually will be out of exact alignment. The extent of damage will depend on the degree to which the wires are out of position relative to the centroid of the wire bundle, and on operator skill. In most instances, the nicks will be very shallow and may be allowable, depending on the application and the standards used. Figure 13.11 illustrates two large "nicks" formed during stripping of the insulation. These particular defects were caused by defective stripping tool jaws that were off center relative to each other. In addition, the wire itself was slightly off center with respect to its insulation. The "nicks" were deemed excessive and unacceptable due to their depth of penetration.

Corrective Action.   New wire-stripping jaws were correctly positioned and subsequently used on other insulated wires. Periodic operator training may also be required.

13.8.4   Measling of Printed Wiring Boards

Problem.   Measling usually occurs when the PWB fibers expand and detach themselves from their original positions, causing whitish spots internal to the board. This can be due to moisture within the board being quickly heated and volatilized during soldering with a soldering iron. Measling is usually visible as shown in Fig. 13.12, depending on the degree of damage caused during the soldering operation. Measling may be considered a defect or, depending on the standard and the amount of the damage, can be a process indicator. MIL-STD-2000 and MIL-P-55110, which reference IPC-A-610, use measling as a process indicator. Measling of PWBs occurs when a combination of factors are present. These include but are not limited to:

- Medium to high amount of moisture absorbed by the PWB substrate. Organic materials tend to be hydroscopic and will absorb moisture from the surrounding environment at different rates, depending on the storage conditions, ambient temperature, time of storage, and moisture content in the air.
- Temperature of the soldering iron tip.

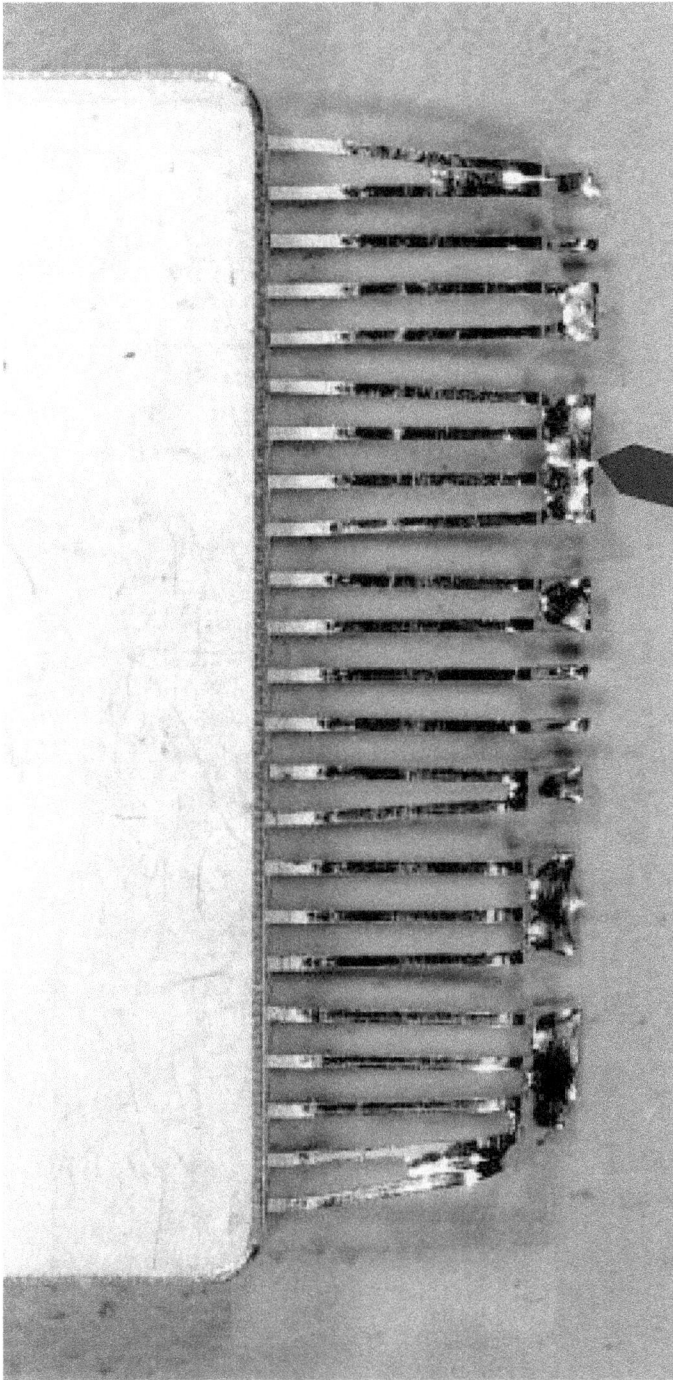

FIGURE 13.10 Solder bridging on SMD leads.

FIGURE 13.11 "Nicking" of stranded wires during insulation removal.

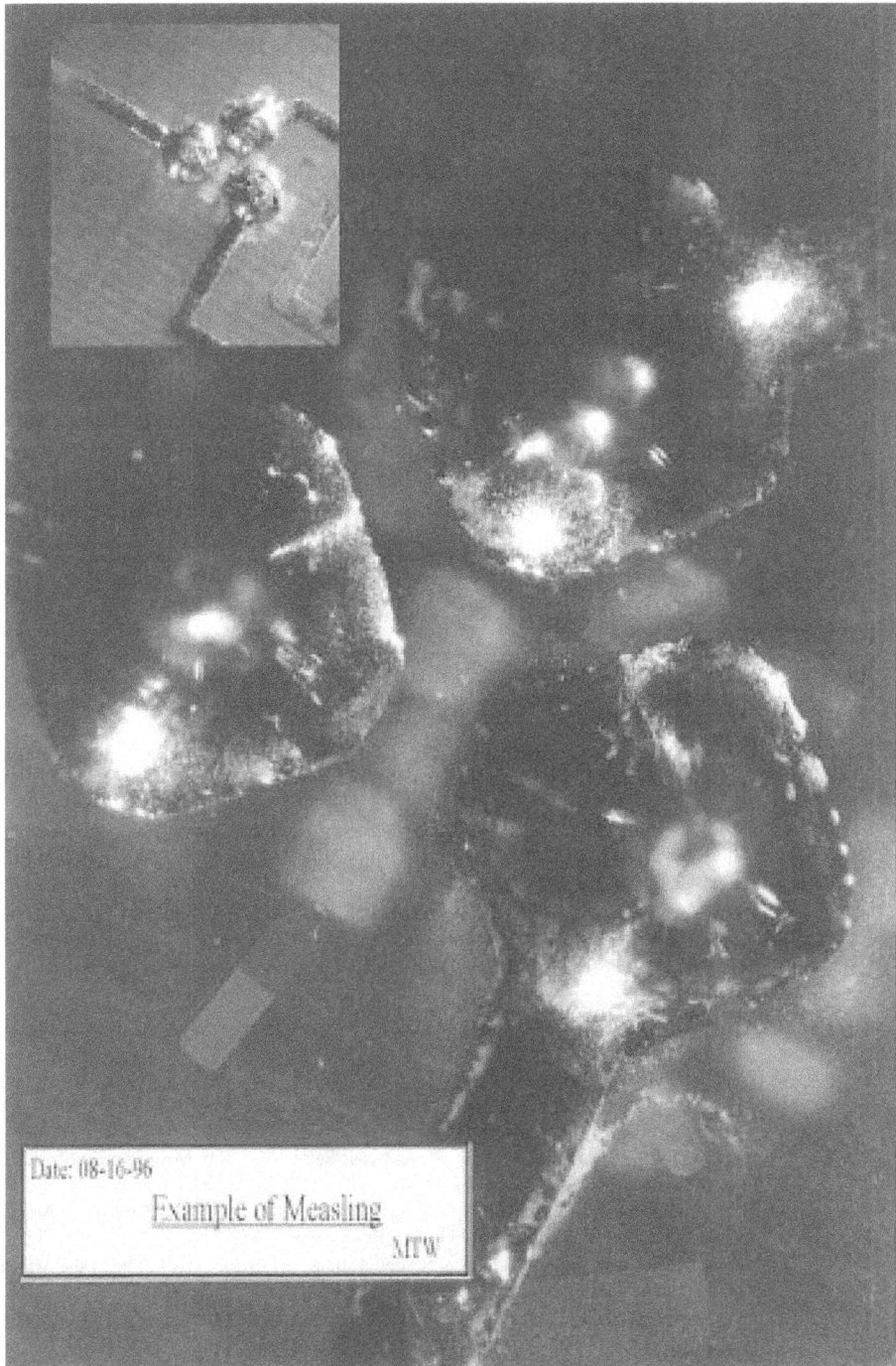

FIGURE 13.12 PWB measling surrounding three soldered joints.

- Low operator skill level. The operator skill level will determine the time (and temperature of the soldering iron tip) and pressure the PWB experiences by the soldering iron. The longer the exposure time and greater the applied pressure, the greater the chance for PWB substrate damage.

Corrective Action.   To avoid differences in operator skills and abilities, a common solution is to precondition the printed wiring board prior to soldering. Typically, the board is oven baked at a temperature and time sufficient to remove the absorbed moisture.

13.9   ACKNOWLEDGMENTS

The author would like to thank Mr. Hernando Arauco of TRW Electronic Systems Group for providing the case studies and associated photographs, Mr. Joseph Boyd of Electronic Failure Analysis Laboratory, National Technical Systems for his careful review and suggestions on SEM/EDS analysis, and Mr. Perry Martin and Mr. Sheldon Elicker, also of the Electronic Failure Analysis Laboratory, National Technical Systems, for their continual support of this effort.

13.10   REFERENCES

1. Adams, John A. Inspection of Ball Grid Array Assembly, from *Ball Grid Array Technology,* John H. Lau, ed. New York: McGraw-Hill, 1995, 465–489.

2. Allenby, B.R., and Ciccarelli, J.P., et al. An Assessment of the use of Lead in Electronic Assembly, Proc. Surface Mount International Conference, San Jose, CA, 1992, 1–28.

3. Artaki, I., Jackson, A.M., and Vianco, P.T. Fine Pitch Surface Mount Assembly with Lead-Free, Low residue Solder Paste, *Proc. Surface Mount International Conference,* San Jose, CA, 1994, 449–459.

4. AWS Committee on Brazing and Soldering, *Soldering Manual (*2d ed., revised). Miami, FL: American Welding Society, 1978,131–132.

5. Bader, W.G., Dissolution of Au, Ag, Pd, Pt, Cu and Ni in a Molten Tin-Lead Solder, *Welding Journal* 48, Welding Research Supplement (Dec. 1969), 551-s–557-s.

6. Belle, Rick. Private Communication. Utica, NY: Indium Corporation of America, Dec. 12, 1995.

7. Ibid., July 2, 1996.

8. Brusic, V., DiMilia, D.D., and MacInnes, R. "*The Corrosion of Lead, Tin and Their Alloys,*" Corrosion (July 1991), vol. 7, no. 19, pp. 509–518

9. Caufield, Thomas, Cole, Marie S., Cappo, Frank, Zitz, Jeff, and Benenai, Joseph. An Overview of Ceramic Ball and Columb Grid Array Packaging, from *Ball Grid Array Technology,* John H. Lau, ed. New York: McGraw-Hill, 1995 465–489.

10. Crenshaw, Richard, and Rooks, Steve. Cross-Sectional X-Ray Measurements Test Ball Grid Array Connections, *Proc. NEPCON West,* Anaheim, CA, 1995, 1035–1047.

11. de Kluizenaar, E.E. Reliability of Soldered Joints: A Description of State of the Art (Part 1), *Soldering & Surface Mount Technology* 4, February 1990, 27–38.

12. de Kluizenaar, E.E. Reliability of Soldered Joints: A Description of State of the Art (Part 2), *Soldering & Surface Mount Technology* 5, June 1990, 56–66.

13. DeVore, John A., Failure Mechanisms in Soldering, from *ASM Electronic Handbook, Volume 1, Packaging.* Materials Park, OH: ASM International, 1993, 1031–1040.

14. Long, Joseph B., ed. Fusible Alloys, from *Metals Handbook, 9th ed., Vol. 3, Properties and Selection: Stainless Steels, Tool Materials and Special-Purpose Metals.* Materials Park, OH: ASM International, 1980, 799.

15. Frear, D.R. Thermomechanical Fatigue in Solder Materials, from *Solder Mechanics: A State of the Art Assessment*, D.R. Frear, W.B. Jones, and K.R. Kinsman, eds. Warrendale, PA: The Minerals, Metals & Materials Society, 1991, 223.

16. Frear, D.R., and Vianco, P.T. Intermetallic Growth and Mechanical Behavior of Low and High Melting Temperature Solder Alloys, *Metallurgical and Materials Transactions* 25A (July 1994), 1509–1523.

17. Frear, D.R., and Yost, F.G. Reliability of Solder Joints, *MRS Bulletin* (Dec. 1993) 49–54.

18. Frederickson, Michael D. Hot Gas Soldering, from *ASM Handbook, Volume 6, Welding, Brazing and Soldering.* Materials Park, OH: ASM International, 1993, 361–362.

19. Frick, John P., ed. *Woldman's Engineering Alloys, 7th Edition.* Materials Park, OH: ASM International, 1990, 1085.

20. Garrison, A., Lee, M., Park, H.S., and Todd, N.L. How Much Rework is Too Much?: The Effects of Solder Joint Rework on Plated-Through Holes in Multilayer Printed Wiring Boards, *Soldering & Surface Mount Technology* 19 (Feb. 1995) 48–51.

21. Hampshire W.B. The Search for Lead-Free Solders, *Soldering & Surface Mount Technology* 14 (June 1993), 49-52.

22. Hampshire, William B. Solders, from *ASM Electronic Handbook, Volume 1, Packaging.* Materials Park, OH: ASM International, 1993, 633–642.

23. Hinch, Stephen W. *Handbook of Surface Mount Technology.* Essex, U.K.: Longman Scientific & Technical, 1988, 426–438.

24. Hosking, F.M., Vianco, P.T., et al. Wetting Behavior of Alternate Solder Alloys, *Proc. Surface Mount International Conference,* San Jose, CA, 1993. 476–483.

25. Howard, Robert T. Optimization of Indium-Lead Alloys for Controlled Collapse Chip Connection Application, *IBM Journal of Research and Development* 26(3) (May 1982) 372–378.

26. Hwang, Jennie S. *Modern Solder Technology for Competitive Electronics Manufacturing.* New York: McGraw-Hill, 1996, 75, 509–541.

27. Hymes, Les. Surface Mount Soldering, from *ASM Electronic Materials Handbook, Volume 1, Packaging.* Materials Park, OH: ASM International, 1993, 697–709.

28. Lea, C. *A Scientific Guide to Surface Mount Technology.* Ayr, Scotland: Electrochemical Publications Ltd., 1988, 329–338.

29. Ibid., 38–40, 496–499.

30. Lee, N., Slattery, J., Sovinnsky, J., et al. A Novel Lead Free Solder Replacement, *Proc. Surface Mount International Conference,* San Jose, CA, 1994, 463–472.

31. Linman, Dale L. Vapor-Phase Soldering, from *ASM Handbook, Volume 6, Welding, Brazing and Soldering.* Materials Park, OH: ASM International, 1993, 369–370.

32. Manko, Howard H. *Solders and Soldering,* 3d ed. New York: McGraw-Hill, 1992, 70–75, 163–164, 167–171.

33. Marshal, James L. Scanning Electron Microscopy and Energy Dispersive X-ray (SEM/EDX) Characterization of Solder-Solderability and Reliability, from *Solder Joint Reliability: Theory and Applications,* J. H. Lau, ed. New York: Van Nostrand Reinhold, 1991, 194–212.

34. Marshall, J.L., Foster, L.A., and Sees, J.A. Interfaces and Intermetallics, from *The Mechanics of Solder Alloy Interconnects*D.R. Frear, S.N. Burchett, H.S. Morgan, and J.H. Lau, eds. New York: Van Nostrand Reinhold, 1994, 42–86.

35. Maykuth, Daniel J., and Hampshire, William B. Corrosion of Tin and Tin Alloys, from *Metals Handbook, 9th ed.,Vol. 13, Corrosion.* Materials Park, OH: ASM International, 1987, 770–774.

36. McCormack, M., and Jin, Sungho. Progress in the Design of New Lead-Free Solders, *Journal of Metals* (July 1993), 36–40.

37. Millard, Don Lewis. Reliability-Related Solder Joint Inspection, from *The Mechanics of Solder Alloy Wetting and Spreading*, F.G. Yost, F.M. Hosking, and D.R. Frear, eds. New York: Van Nostrand Reinhold, 1993, 267–297.

38. Millard, Don L. Solder Joint Inspection, from *ASM Electronic Materials Handbook, Volume 1, Packaging.* Materials Park, OH: ASM International, 1993, 735–739.

39. Morris, J.W., Jr., Freer Goldstein, J.L., and Mei, Z. Microstructural Influences on the Mechanical Properties of Solder, from *The Mechanics of Solder Alloy Interconnects*, D.R. Frear, S.N. Burchett, H.S. Morgan, and J.H. Lau, eds. New York: Van Nostrand Reinhold, 1994, 21–28.

40. Prasad, Ray P. *Surface Mount Technology: Principles and Practice.* New York: Van Nostrand Reinhold, 1989, 125.

41. Romig, A.D., Chang, Y.A., Stephens, J.J., Marcotte, V., Lea, C, and Frear, D.R. Physical Metallurgy of Solder-Substrate Reactions, from *Solder Mechanics: A State of the Art Assessment,* D.R. Frear, W.B. Jones, and K.R. Kinsman, eds. Warrendale, PA: The Minerals, Metals & Materials Society 1991, 34–78.

42. *Requirements for Electronic Grade Solder Alloys and Fluxed and Non-Fluxed Solid Solders for Electronic Soldering Applications,* J-STD-006. Northbrook, IL: The Institute for Interconnecting and Packaging Electronics Circuits (IPC), 1994.

43. Roeder, J.F., Notis, M.R., and Frost, H.J. Physical Metallurgy of Solder Alloys, from *Solder Mechanics: A State of the Art Assessment,* D.R. Frear, W.B. Jones, and K.R. Kinsman, eds. Warrendale, PA: The Minerals, Metals & Materials Society, 1991, 2–4.

44. Socolowski, Norbert. Lead-Free Alloys and Limitations for Surface Mount Assembly, *Proc. Surface Mount International Conference,* San Jose, CA, 1995, 477–480.

45. Standard Specification for Solder Metal, B 32, *Annual Book of ASTM Standards.* ASTM (1996).

46. *Solder Alloy Data,* (I.T.R.I Publication No. 656), International Tin Research Institute, 1986, 7–83.

47. Vanzetti, Richardo. Laser Soldering, from *ASM Handbook, Volume 6, Welding, Brazing and Soldering.* Materials Park, OH: ASM International, 1993, 359–360.

48. Vianco, Paul T. Embrittlement of Surface Mount Solder Joints by Hot Dipped, Gold-Plated Leads, *Proc. Surface Mount International Conference,* San Jose, CA, 1993, 337–355.

49. Vianco, Paul T. General Soldering, from *ASM Handbook, Volume 6, Welding, Brazing and Soldering.* Materials Park, OH: ASM International, 1993, 964–984.

50. Vianco, Paul T. Soldering in Electronic Applications, from *ASM Handbook, Volume 6, Welding, Brazing and Soldering.* Materials Park, OH: ASM International, 1993, 985–1000.

51. Vianco, Paul T. Wave Soldering, from *ASM Handbook, Volume 6, Welding, Brazing and Soldering.* Materials Park, OH: ASM International, 1993, 366–368.

52. Vianco, Paul. Private communication, Sandia National Laboratories, Albuquerque, NM, August 30, 1995.

53. Vianco, P.T., Artaki, I., and Jackson, A.M. Reliability Studies of Surface Mount Circuit Boards Manufactured with Lead-Free Solders, *Proc. Surface Mount International Conference,* San Jose, CA, 1994, 437–438.

54. Vianco, P.T., and Dal Porto, J.F. Use of High Temperature Solders for Edge Clip Attachment on Polyimide-Quartz Circuit Board, *Proc. ASM International's 3rd Electronic Materials & Processing Congress,* San Francisco, CA, August 1990, 25–30.

55. Vianco, P.T., and Frear, D.R. Issues in the Replacement of Lead-Bearing Solders, *Journal of Metals* (July 1993), 14–19.

56. Vianco, P.T., and May, C. An Evaluation of Prototype Surface Mount Circuit Boards Assembled with Three Non-Lead Bearing Solders, *Proc. Surface Mount International Conference,* San Jose, CA, 1995, 481–494.

57. Vianco, P.T., and Mizik, P.M. Prototyping Lead-Free Solders on Hand-Soldered, Through-Hole Circuit Boards, *Proc. SAMPE Electronics Conference,* Parsippany, N.J. (June 20–23, 1994), vol. 7, 366–380.

58. Wassink, R.J. Klien. *Soldering in Electronics* 2nd ed., Ayr, Scotland: Electrochemical Publications Ltd., 1989, 68–73, 148–159, 165, 293–299, 373, 645–647.

59. Zarrow, Phil. Furnace and Infrared Soldering, from *ASM Handbook, Volume 6, Welding, Brazing and Soldering.* Materials Park, OH: ASM International, 1993, 353–355.

CHAPTER 14
FAILURE ANALYSIS OF PRINTED WIRING ASSEMBLIES

Richard A. Blanchard
E^{x}ponent

Donald Galler
Department of Materials Science and Engineering
Massachusetts Institute of Technology

Duncan Glover
Alexander Kusko
John D. Loud
Noshirwan K. Medora
Gregory J. Mimmack
E^{x}ponent

14.1 INTRODUCTION

Printed wiring assembly (PWA) fabrication is a complex procedure involving a variety of material processing steps. From a failure analysis perspective, a fully populated PWA is one of the most complex structures that an investigator may encounter. In this chapter, the term *printed circuit board* or *PCB* describes a bare board consisting of layers of conductors and insulators, while the term *printed wiring assembly* or *PWA* refers to a PCB with components attached to it. PCBs are fabricated from composite substrate materials and use laminated and plated copper for high-density electrical interconnections. Soldering processes are used to attach components ranging in complexity from resistors to sophisticated microelectronic devices. In addition, PWAs must often withstand significant mechanical and environmental stress in normal operation. Because of the complex nature of a modern, fully populated PWA, an investigator has a variety of areas to examine when analyzing a PWA failure.

In this chapter, some of the fundamental processes used in printed wiring assembly fabrication are reviewed, the materials used in those processes are discussed, and procedures for examining PWAs during the course of an investigation are covered.

14.2 CONSTRUCTION AND FABRICATION OF A PRINTED CIRCUIT BOARD

14.2.1 Introduction

Printed circuit board fabrication is a complex process consisting of a number of steps. The fabrication process uses copper-clad laminates that are etched to leave behind circuit traces. Multiple layers of these copper clad laminates may be stacked to arrive at the final board configuration. The boards are drilled using automatic machinery, the resulting holes are coated with a conductor to provide a conductive path between copper layers, and additional conductive material may be added to complete the PCB fabrication process. Although a variety of chemical processes and board structures have evolved over the years, the final result of these processes is a product similar to the one just described.

A variety of PCB construction methods are used today. The three basic types of PCBs use traditional construction methods. In order of increasing complexity, these types of PCBs are single-sided, double-sided, and multilayer boards. A simplified diagram of a multilayer board is show below (Fig. 14.1). This diagram illustrates features common to all three types of printed circuit boards.

The board material is an insulating substrate that provides both mechanical support for components and electrical insulation between the circuit traces. A glass-fiber and epoxy resin composite (glass-epoxy) is the most common material. The circuit traces, or the *interconnect,* which is made of patterned copper, connects the electronic components on the board. The connections are usually made by traces that connect the component mounting pads or by plated-through holes (PTHs). Plated-through holes perform two major functions. First, the holes provide a receptacle for component leads during the assembly process before soldering. Second, they provide a means for interconnecting traces on different layers. Vias are plated-through holes on two-sided and multilayer boards that are used to interconnect layers or traces, but they are not used for component mounting. Vias are usually smaller in diameter than component holes. There are two special types of vias:

1. *Blind vias.* Blind vias are visible from one exterior side of the board. The other end of blind vias terminate on interior layers.

2. *Buried vias.* Buried vias are not visible from an exterior layer of the board. They connect conductive layers in the interior of the PCB.

Solder is a eutectic tin-lead or other alloy used to attach components to the PCB, to improve its mechanical and electrical properties, and to prevent oxidation of the copper

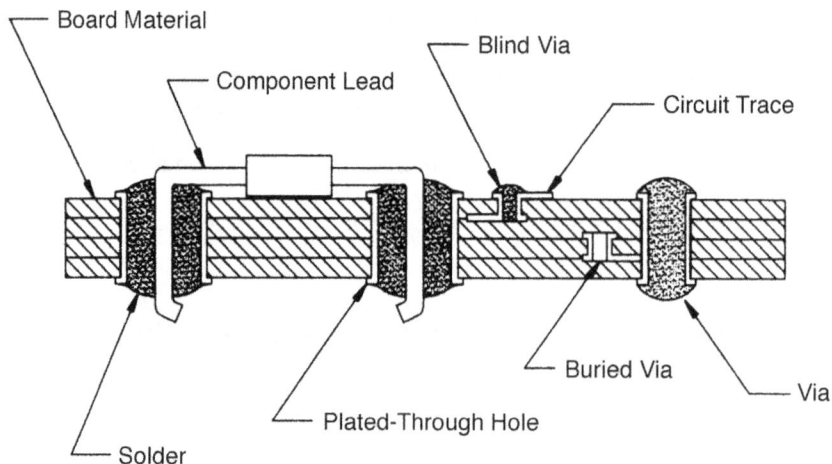

FIGURE 14.1   Simplified diagram of a multilayer printed wiring board.

traces. Solder can be applied manually or by automated machinery. When plated-through holes are soldered, capillary action causes the solder to wet both sides of the board. The solder improves the mechanical strength of the hole and provides the means for component attachment.

14.2.2 Fabrication Processes

Several fabrication process sequences are used to manufacture PC boards. A typical fabrication sequence for a two-sided PCB is shown in Fig. 14.2.

1. The process begins with a copper-clad substance. Glass fiber with epoxy resin (glass-epoxy) or polyamide resins are common materials.
2. The PCB design is used to generate a drilling pattern for vias and through holes. Numerically controlled (NC) machines drill the board using bits with sizes appropriate for the different hole types.
3. The entire surface of the PC board is chemically activated.
4. The board is immersed in a chemical solution that coats the bare surfaces of the drilled holes with copper. This process step is called *electroless copper* because electric power is not used for the plating.
5. Additional copper is plated onto the board by electroplating. This process step increases the thickness of the plating on traces and in through-hole areas. In the plating process, copper surfaces of the boards are connected to the cathode (negative potential) of the electroplating power source.

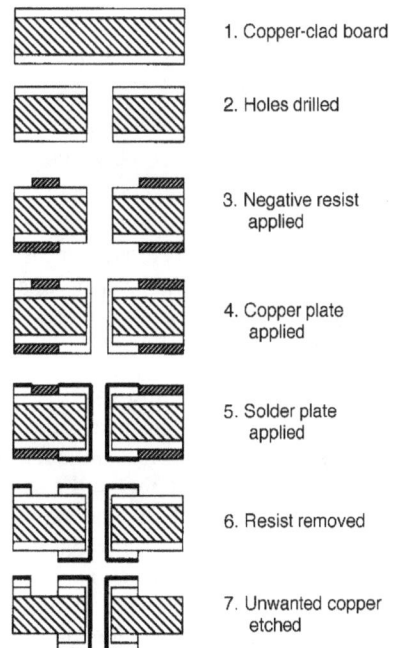

1. Copper-clad board

2. Holes drilled

3. Negative resist applied

4. Copper plate applied

5. Solder plate applied

6. Resist removed

7. Unwanted copper etched

FIGURE 14.2 Steps in a typical two-sided PC board manufacturing sequence. *From Ref. 1.*

6. The drilled board is coated with a photosensitive emulsion. The solubility of the emulsion is modified by exposing the board to ultraviolet light using a photographic image of the circuit pattern as a mask. After the board is rinsed in a solvent, the pattern in the resist leaves the board exposed where copper traces are to be etched.

7. Acid etching removes the unwanted copper from the board, leaving behind circuit traces and plated-through holes, after which the resist is removed.

8. A photosensitive *solder mask* material is applied. The solder mask is aligned and exposed to modify its solubility. The soluble solder mask material is dissolved in a solvent, and the solder mask coat is baked. Solder is applied to the exposed copper, the plated-through holes, and the vias, protecting the copper from corrosion and preparing the PCB for the subsequent attachment process during which components are attached using solder.

14.2.3 Types of Printed Wiring Assemblies

Many types of PWA construction can be found in products. Most of the common rigid PWAs fall into one of the three categories described below. Traditional PWA construction uses copper-clad board materials to construct single-sided, double-sided, or multilayer boards with copper circuit traces. Components are mounted with their leads passing through holes in the board, (through-hole technology) or to pads on the surface(s) of the board (surface mount technology).

▪ Single-sided boards have traces on one side only. The board material is one piece with holes for component leads. The inside surfaces of holes are not plated.

▪ Double-sided boards have traces on both sides. The board material is one piece with holes for component leads. Vias are often present, which interconnect the traces on the two sides of the board. If the board has vias, the vias and the holes for component leads are plated. If there are no vias, the holes for components are not always plated.

▪ Multilayer boards have traces on both sides and on interior layers. The multilayer board (MLB) can be visualized as several double-sided boards laminated together with intervening layers of unclad board material. The inside surfaces of component mounting holes and vias are plated. Multilayer boards often include internal ground and power planes to provide a distributed capacitance, which increases the immunity to noise on power and ground voltages.

Surface mount components may have conventional leads, or they may have contact regions connected by solder balls or other techniques to provide mechanical and electrical connections to the PCB.[2] They are soldered onto pads on exterior surfaces of the board without using plated-through holes. PWAs using surface mount technology (SMT) are almost always multilayer PCBs. Surface mount and through-hole mounting technologies are often found on the same board, with the through holes used to attach relatively large components such as capacitors, inductors, and transformers, as well as connectors.

Flexible printed circuit boards, or *flex circuits,* are used for flexible circuit connections and for construction of PWAs with unusual shapes. They are typically used in lightweight applications where a rigid glass-epoxy PWA does not fit or might fail prematurely due to flexing. Polyamide film is used almost exclusively as the substrate for flex circuits. In most cases, copper circuit traces are laminated between two sheets of the film. Flex circuit construction can accommodate the mounting of both through-hole and surface mount components. The adhesive between the layers is typically a polyester material.

14.2.4 Solder

Overview. Electronic solder is often a eutectic alloy, usually containing tin and lead and sometimes containing one or more other metals. A primary characteristic of a eutectic alloy is that it has a lower melting point than either of its two constituents.

Tin-Lead Alloy. Eutectic tin-lead solder is 63% tin, 37% lead by weight and has a melting point of 182° C (361° F). All other alloys of lead and tin including pure tin or pure lead have higher melting points. Electronic solders are usually composed of tin and lead in ratios close to that of the eutectic point. When noneutectic mixtures cool, the proeutectic portion solidifies first, leaving a eutectic alloy to cool at the reduced temperature. This circumstance can result in increasing internal stress and cold solder joints. The phase diagram of tin-lead is shown in Fig. 14.3.

Other Solders. Solders often contain other elements in addition to (or in some cases instead of) tin and lead. The composition and melting point of other solders used in PCB manufacture are shown in Table 14.1.

Metallic and nonmetallic contaminants are of concern in solder alloys. Contamination occurs during the solder process while the solder is in a molten state. Under normal condi-

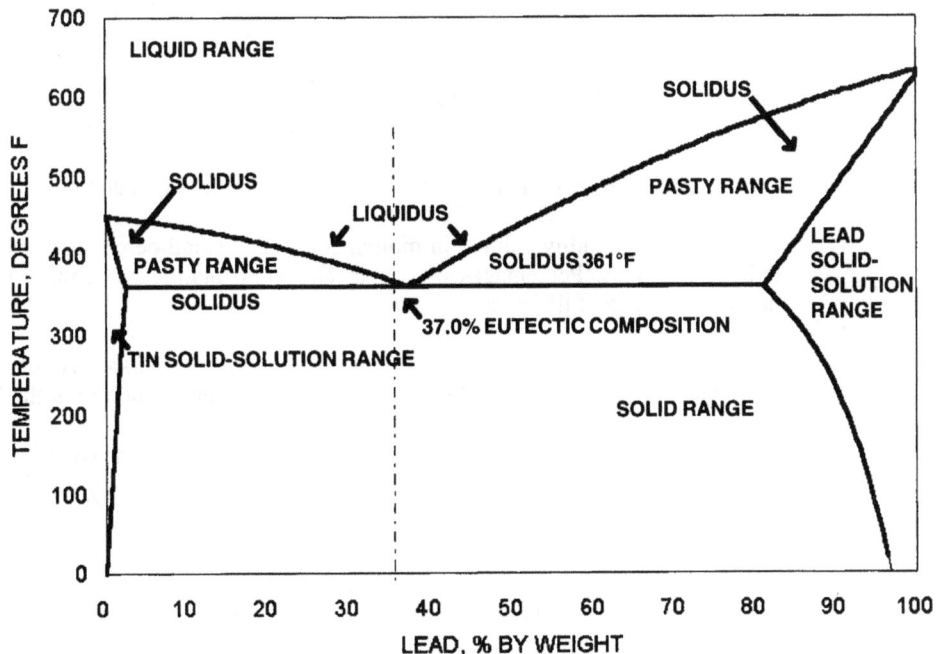

FIGURE 14.3 Phase diagram of tin-lead.

Table 14.1 Composition and Melting Temperature of Solders Used in PWA Manufacturing.[3]

Composition	Melting Point, °C
62%Sn- -36%Pb-2%Ag	179
63%Sn-37%Pb	183
60%Sn-40%Pb	191
50%Pb-50%In	209
96.5%Sn-3.5%Ag	221
65%Sn-25%Ag-10%Sb	240
75%Pb-25%In	264
80%Au-20%Sn	280
90%Pb-10%Sn	302

tions, solder joints do not become contaminated after they are formed. However, the effect of initial contamination can significantly reduce solder joint life. Nonmetallic contaminants such as sulfides and oxides are formed by the reaction of the solder with sulfur and oxygen. These contaminants affect the flow characteristics of the solder and metallurgical characteristics of the joint. The properties of solder are affected by small amounts of certain metallic contaminants. Some of these are discussed below:[1]

Copper. Copper forms two intermetallic compounds with tin: Cu_3Sn and Cu_6Sn_5. The compounds weaken the solder joint and cause the solder to become sluggish and gritty.

Iron. Iron forms two intermetallic compounds with tin: $FeSn$ and $FeSn_2$.

Gold. Gold is readily soluble in molten solder and small percentages can cause brittle, dull solder joints. Percentages of copper and gold totaling more than 0.3 percent will degrade solder joint lifetime.

Zinc. Zinc is one of the most detrimental of solder contaminants. As little as 0.005 percent zinc will cause grittiness, lack of adhesion, and eventual failure of the joint.

Other. Small amounts of nickel and aluminum can also degrade solder joints.

14.2.5 Copper-Clad Board Material

Copper-clad board material is the raw material from which almost all PCBs are made. The most common board material is a glass-epoxy composite. The manufacturing of glass-epoxy composite boards start with the glass fiber fabric. The fabric is impregnated or coated with resin, which is then polymerized to a point suitable for storage. The partially cured material is called *prepreg*. Copper foil is supplied in rolls for lamination onto the prepreg. The foil and prepreg are pressed together in large, heated presses at pressures exceeding 1000 psi to produce copper-clad board stock. The thickness of copper cladding is specified in ounces per square foot. When the fabrication process is complete, circuit traces with 1 oz of copper have a thickness of 0.0014 in or 1.4 mils.

14.2.6 Standards for PCBs, PWAs, and Their Assembly

Military. Military standards and specifications cover materials, test methods, and standard board layout procedures. Applicable standards include the following:

MIL-STD-275	Printed Wiring for Electronic Equipment
MIL-P-55110	Military Specification for Printed Wiring Boards
MIL-STD-202	Military Standard Test Methods for Electronic and Electrical Component Parts.
MIL-P-13949	General Specifications for Plastic Sheet, Laminated, Metal Clad (For Printed Wiring Boards)

Federal. The composition of solder alloys is specified in the following standard.

QQ-S-571	Solder, Tin alloy, Tin-Lead alloy, and Lead Alloy

American Society for Testing and Materials. The American Society for Testing and Materials (ASTM) (http://www.astm.com) publishes the following test procedures for materials used in PCBs.

ASTM-E-53	Standard Method for Chemical Analysis of Copper (Electrolytic Determination of Copper)
ASTM-B-487	Measuring Metal and Oxide Coating Thickness by Microscopic Examination of a Cross-Section

The Institute for Interconnecting and Packaging Electronic Circuits. The Institute for Interconnecting and Packaging Electronic Circuits (IPC) (http://www.ipc.org) publishes many standards covering PCBs, which are referenced in both military and commercial specifications. The primary IPC standards are listed below.

IPC-A-28/29	Single Sided Test Specimen
IPC-A-28	Double Sided Test Specimen
IPC-A-29	Multilayer Test Specimen
IPC-A-600	Guidelines for Acceptability of Printed Boards
IPC-A-804	Solderability Test Methods for Printed Wiring Boards
IPC-CF-150	Copper Foil for Printed Wiring Applications
IPC-S-815	General Requirements for Soldering Electronic Interconnections
IPC-SM-840	Qualification and Performance of Permanent Polymer Coating (Solder Mask) for Printed Boards
IPC-TM-650	Test Methods Manual

The Electronic Industries Alliance.   The Electronic Industries Alliance (EIA, formerly Electronic Industries Association) (http://www.eia.org) publishes the following relevant documents:

EIA/IS-46	*Test Procedure for Resistance to Soldering (Vapor Phase Technique) for Surface Mount Devices* This standard establishes a technique to evaluate plastic materials used for connector housings which will be exposed to a vapor phase reflow process.
EIA/IS-47	*Contact Termination Finish Standard for Surface Mount Devices* This document is a guide in establishing finishes for SMT terminations.
EIA/IS-86	*Surface Mount Solderability Test* This document provides a standard procedure for solderability testing of surface mount devices and simulates actual board mount performance in a reflow process.

14.3   *FAILURE CHARACTERISTICS OF PWAS*

14.3.1   Mechanical Damage

Component damage can occur when the PWA is subjected to bending, twisting, or shock loads. Shock loads on SMT boards can cause components to shear off the substrate at the solder joint. Component packages may crack, or leads may separate from the body of the component on through-hole boards, while the lead to the component may remain attached to the PWA.

PWA damage can be observed in several fundamental forms:

- Damage to a component that may be visible on its exterior
- Shearing or tearing of the leads of a component
- Cracking of the PCB substrate

An examination to determine whether such damage has occurred sometimes can be conducted visually (with the unaided eye) or with a low-power optical microscope. Inflexible, fragile components such as SMT ceramic capacitors are particularly susceptible to PWA flexing. Cracked components or leads often result in the insidious *intermittent* electronic failure.

14.3.2   Thermal Damage

The type and extent of thermal damage present on a PWA can provide significant information about the heat source and its possible role in the failure. Thermal damage to a PCB and its components can occur in several ways:

- Excessive voltage applied to the PWA, resulting in high current flow
- Traces or components on the PWA that are too small or have a power dissipation capability that is too low for the current flow
- An external heat source outside the electronic enclosure

▪ Direct exposure of the PWA to heat when a board is not enclosed

▪ Electrical failure on the PWA or of a component on the PWA, causing thermal damage

Epoxy is a thermosetting material. Thermal exposure above the temperature rating will cause chemical breakdown of the substrate material and eventual carbonization. Chemical breakdown is a time-temperature process, and the rate of breakdown increases with temperature. Most substrate materials can tolerate high temperatures for up to several seconds without suffering noticeable damage, as is required for soldering processes. The thermal mass of the substrate and copper cladding also absorbs some heat. Once thermally degraded to the point of complete decomposition, the epoxy will behave as a resistor due to local carbonization of the epoxy on the fiberglass material. Thermal damage to PWAs proceeds in several stages, as the exposure becomes more severe. The process usually occurs in the following steps:

▪ Discoloration of the PWA occurs.

▪ Solder melts and components may become detached.

▪ Cracking and charring of plastic component packages occurs.

▪ Separation of circuit traces on exterior layers of the PWA occurs.

▪ Decomposition of the PWA substrate occurs, leaving carbon-coated glass fibers that form a parasitic resistive connection between copper leads on the PWA.

▪ Separation of laminated layers of the PWA occurs.

▪ Blackening and decomposition of the substrate material occurs.

Since the chemical breakdown of the glass-epoxy substrate or other materials depends on both time and temperature, it is not possible to judge temperature accurately from the appearance of the PWA. Even when two PWAs are made from the same materials, they may vary greatly in thermal characteristics because of their size, shape, and number of circuit layers. As a consequence, any conclusion about thermal exposure should be verified by comparative tests on sample PWAs with the same configuration.

Thermal damage from electrical component failures on a PWA may be localized to the component and the circuit traces associated with it. This type of localized damage can usually be distinguished from the more uniform damage caused by external fire. In glass-epoxy boards, the epoxy resin will begin to break down thermally above approximately 250° C (482° F). Visual examination is usually sufficient when investigating a PWA with regions that have been affected by heat. Low-power optical microscopy may be useful in some cases.

14.3.3 Contamination

Contamination on the surface of PWAs can cause insulation degradation, local corrosion, and component failures. Frequently, the symptoms and underlying source of contamination can be observed on the PWA.[3] Corrosion in the form of various oxides can be observed on severely contaminated PWAs. Contamination may be caused by problems in the manufacturing process or by unexpected environmental exposure while in operation after the manufacturing process. Contamination during the fabrication process can degrade both the mechanical and electrical properties of the PWA, ultimately leading to failure. Contaminants may be introduced from the following sources:

- Incomplete removal of flux used in the solder process
- Fingerprints, dirt, or cleaning fluids from processing
- Metal slivers or solder bridges from assembly
- Contaminated atmosphere during storage, equipment assembly, or operation
- Moisture or salts from the environment

Figure 14.4 shows a fingerprint left on a PCB due to incomplete cleaning during fabrication.

Ionic contaminants become active in the presence of moisture or other fluids (particularly water) and may lead to conduction at higher humidity levels. Leakage current on PWAs can occur on the surface or in the bulk material of the laminate. (If leakage occurs in the bulk material, it is most often the result of a manufacturing problem.)

Environmental contamination can lead to corrosion of copper circuit traces, damage to components, and a rapid loss of the substrate electrical insulating properties. If contamination is suspected, the PWA should be handled only with gloves and stored in an electronic-grade container. Over long periods of time, contamination can lead to a phenomenon called *metal migration*. There are two basic forms of metal migration:

1. Whisker growth is a single crystal growth a few microns in diameter.
2. Dendritic growth occurs when a dc potential electrically transfers metallic ions between conductors. A metal dendrite forms on the cathode when the mobile ions are reduced. The dendrite grows toward the anode.[5,6]

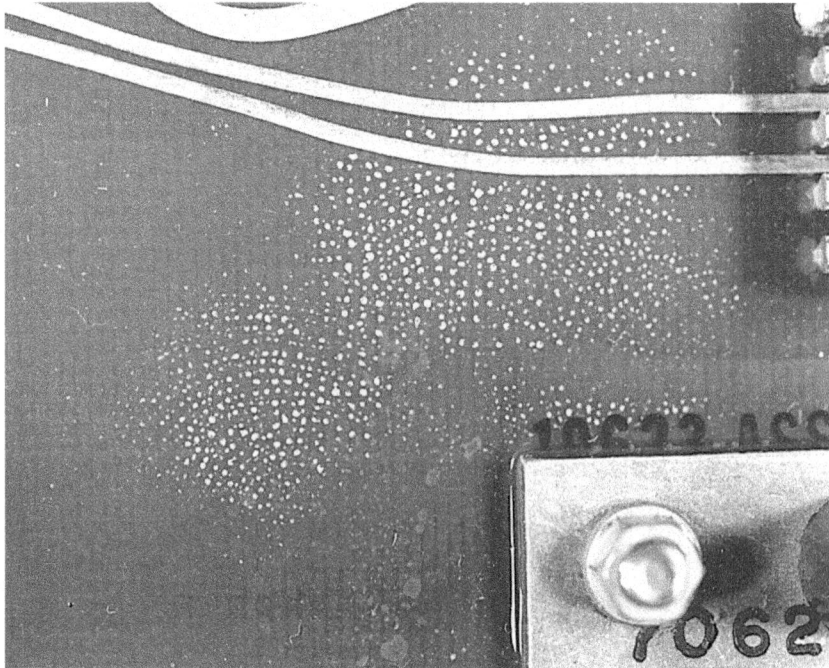

FIGURE 14.4   Contamination in the form of a fingerprint on a PCB (3×).

Whisker growth can reduce the amount of material available to conduct the current that is flowing, ultimately resulting in an open circuit if the process is allowed to go to completion. In addition, a short circuit can occur if the whisker encounters another conductor at a different potential.

Dendritic growth is usually of greater concern, since a conductive filament forms between the two conductors. A formula for the time to failure, t_f, for conductive filament formation is given below:[6]

$$t_f = \begin{cases} \dfrac{a \cdot f \cdot (10^3 L_{eff})^n}{V^m \cdot (M - M_t)}, & \text{for } M > M_t \\ \infty, & \text{otherwise} \end{cases} \qquad (14.1)$$

where t_f = time to failure
a = filament formation acceleration factor
f = multilayer correction factor
n = geometry acceleration factor
m = voltage acceleration factor
M = fraction moisture content
M_t = threshold fraction moisture content
$L_{eff} = k \cdot L$ = effective length between conductors
K = shape factor
L = spacing (inches) between the conductors

The a, f, k, n are empirical constants; they depend on geometry of the test pair and the type of surface coating on laminate.

The time to failure for traces at different distances between conductors with a voltage of 12 V between them is shown in Fig. 14.5. Figure 14.5 is valid with no contamination present on the surface of the PWA. If contamination is present, the time to failure can be reduced by several orders of magnitude.

Spacing recommendations for conductive traces as a function of voltage, altitude, and the presence of a conformal coating have been developed[7] as shown in Table 14.2.

Optical microscopy and SEM are the tools most often used to identify the presence of contamination. Electrical leakage measurements may also be used to detect contamination, if the contamination is electrically conductive. Contaminants usually can be identified with SEM/EDS, which identifies the elements that are present, or with a technique such as Fourier transform infrared (FTIR), time-of-flight secondary ion mass spectroscopy (SIMS), or X-ray photoelectron spectroscopy (XPS), which can identify the compound that is present.

14.3.4 Thermal Expansion Mismatch

Coefficient of thermal expansion mismatch results in mechanical stress on electronic components, circuit traces, and the PWA substrate. Stresses can lead to fractures in solder joints, plated-through holes, internal traces, and component leads or bodies. The mechanical failure of electronic components can be caused by exposure to extreme or rapid temperature variations. Such exposure can be caused by fire or by repeated temperature variations not anticipated in the design of the PWA.

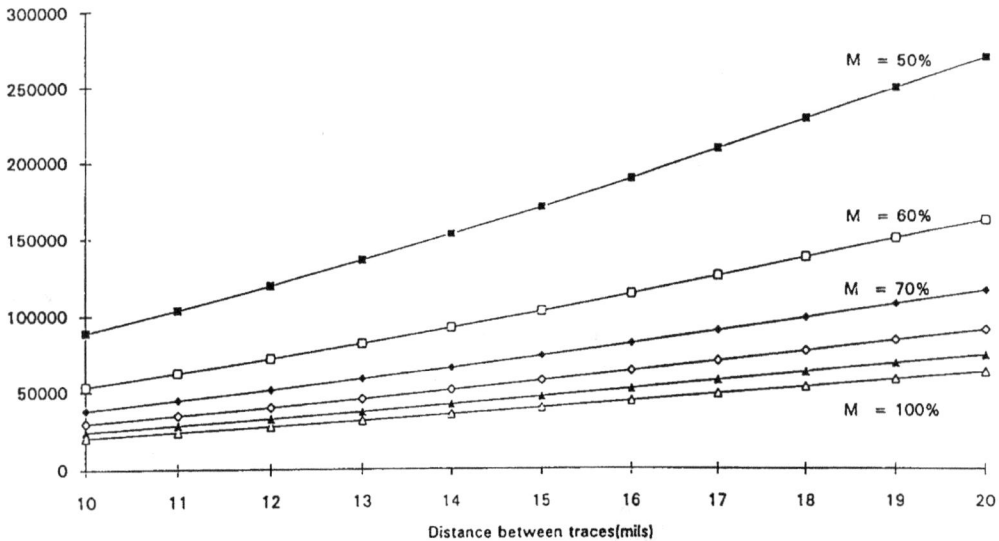

FIGURE 14.5 Time to failure vs. distance between traces for different moisture levels and constant applied voltage of 12 V.

Table 14.2 Spacing Requirements for Conductors on PWAs.[7]

Conductor condition and altitude	Peak voltage (V), dc or ac	Minimum spacing (mils)	Minimum spacing (mm)
Uncoated PWA below 10,000 ft	0–150	25	(0.636)
	151–300	50	(1.27)
	301–500	100	(2.54)
	Above 500	0.2/V	(0.0051/V)
Uncoated PWA above 10,000 ft	0–50	25	(0.635)
	51–100	60	(1.52)
	101–170	125	(3.18)
	171–250	250	(6.35)
	251–500	500	(12.7)
	Above 500	1.0/V	(0.025/V)
Conformal coated PWA, all altitudes	0–30	10	(0.25)
	31–50	15	(0.38)
	51–100	20	(0.51)
	101–300	30	(0.76)
	301–500	60	(1.52)
	Above 500	0.12/V	(0.03/V)

All materials expand and contract with changes in temperature. The rate or sensitivity to temperature is called the *coefficient of thermal expansion* or CTE. The CTE is usually expressed as a percentage or in parts per million (ppm) per degree celsius of temperature change. The CTE varies between most materials. When materials with different CTE values are physically joined, mechanical stresses arise from the dimensional change that occurs with temperature. Unless the joint fails, the materials will be elongated or compressed to accommodate the stress. The forces increase in proportion to several factors:

- The size of the assembly
- The difference between the CTE values
- The temperature variation

The CTE of organic materials can vary significantly over the temperature range of –55 to +200° C (the maximum temperature range that most PCBs are designed to withstand). This range includes temperatures encountered in manufacturing as well as those encountered in the operational environment. (It should be noted that the CTE of most metals does not significantly change over this temperature range.) CTE mismatch problems can be a life-limiting factor in some types of leadless surface mount devices. However, in PTH technology, component leads can flex and reduce stress at the solder joint. Component or PCB cracking are both symptoms that appear occasionally in failures where extreme temperature variations have occurred. It is useful for the investigator to be aware of the basic mechanism involved so the mechanism can be evaluated as a possible failure mode. Solder joint fatigue from CTE mismatch is one area of concern in the industry. Table 14.3 lists the coefficients of thermal expansion for materials found on printed circuit boards.

Table 14.3   Coefficient of Thermal Expansion for Materials Found on PCBs

Material	CTE (ppm/°C)	Material	CTE (ppm/°C)
Alumina	5–7	Copper	16.5
Epoxy-glass	12–16	Epoxy	45–65
Polyamide-glass	11–14	Steel	10–13
Copper-invar-copper	5–6	Gold	14.2
Copper-clad molybdenum	5–6	Silver	19.6
Epoxy-kevlar	6–7	Iron	12.2
Polyamide-kevlar	5–7	Kovar	5.86
Ceramic	5–7	Solder 50-50 (Pb-Sn)	13.1
Aluminum	22.9		

Cracking due to CTE mismatch may be visible with using low-power optical microscopy or SEM examination. However, the use of leadless surface mount packages such as leadless ceramic chip carriers (LCCCs), leadless chip resistors (LCRs), leadless chip capacitors (LCCs), and ball grid array (BGA) IC packages has increased the difficulty of identifying solder-joint failures.

14.3.5 PWB Interconnection Failures

Interconnection failures occur at solder joints, at the interface between layers on multilayer boards, and at the joints between traces and component mounting pads. These failures can appear as cracks in solder joints, as voids between layers and plated-through holes, and as cracks in circuit traces. Contaminants in solder, thermal stress, mechanical stress, and process problems can cause interconnection failures.

There are several basic mechanisms associated with the failure of interconnections. They include:

1. Contamination in solder can result in mechanically weak solder joints. These contaminants can cause cracking or stress-related failures at stress levels below normal operating limits.

2. Thermal excursions beyond the design limits of a joint can cause mechanical stress due to CTE mismatch.

3. Residual or applied mechanical stress can cause cracking in component leads, solder joints, and copper circuit traces.

4. Processing problems can lead to improperly filled vias or plated-through holes. This defect can lead to mechanical failure of the interconnection.

5. Volatile organic contaminants (VOCs) may be present that expand when heated, leading to voids or opens.

The highest thermal stress seen by a PWA occurs in the production soldering process. In extreme cases, the high thermal stress can crack the copper present at plated-through holes or cause separation of the laminate. A poorly controlled soldering process may damage PWAs in a fashion that will take years to show up.

The repeated heating and cooling of a package can also cause solder used to attach the package to a PCB to fail. This type of failure is often referred to as a *thermal fatigue failure*.[8,9] This type of failure can be initiated by a surface flaw in the solder but is most often a result of CTE mismatches, the properties of the materials, and the geometries. The location of a typical thermal fatigue failure is shown in Fig. 14.6.

Examples of interconnections in PWA assemblies are shown below. A plated-through hole incompletely filled with solder is shown in the microfocus X-ray radiograph of Fig. 14.7. This is usually an indication of a wetting problem with the solder process and can cause the joint to be mechanically weak.[10] Figure 14.8 shows incomplete attachment of the

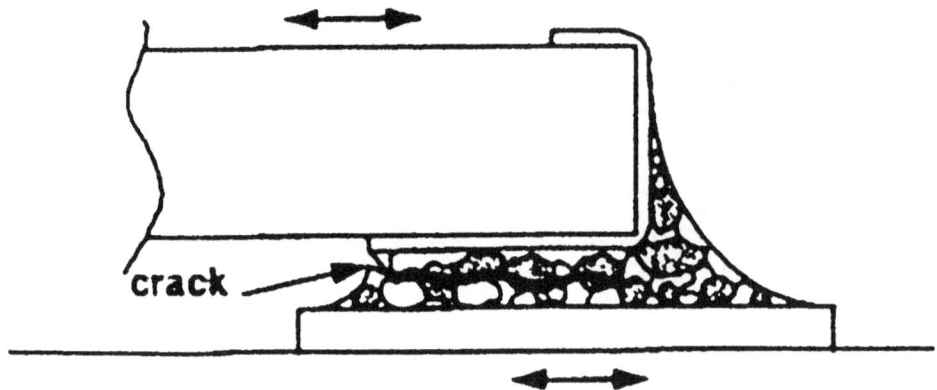

FIGURE 14.6 A crack in the solder of a leadless chip capacitor caused by thermal fatigue.

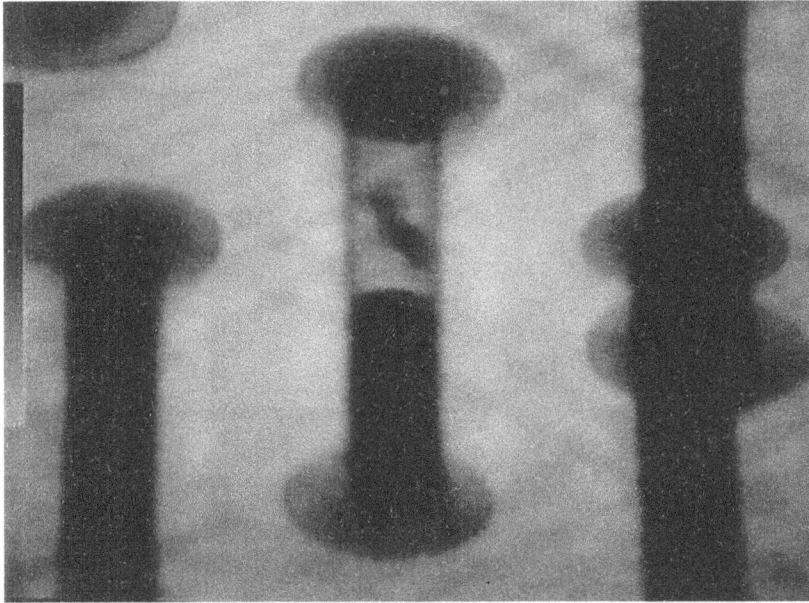

FIGURE 14.7   Microfocus X-ray radiograph showing incomplete solder filling in a plated-through hole (≈100×).

FIGURE 14.8   Microphotograph of a plated-through hole showing incomplete attachment to internal layer of multilayer board, as shown by the arrow (≈750×).

hole wall to the inner layers on a plated-through hole. This results in poor or missing connections to the layer and can cause intermittent failure of the electrical connection. Figure 14.9 shows a cracked solder joint where the component lead enters the solder fillet. This results in intermittent electrical connection and circuit malfunction.

A number of techniques may be used to investigate interconnection problems. Optical microscopy and SEM are useful for problems on external surfaces and solder joints. Real-time microfocus X-ray is a useful tool for general examination of the PWA, especially for problems within internal layers. Acoustic microscopy is useful in identifying defects such as cracks if they are perpendicular to the same beam. Metallurgical sectioning and optical microscopy are usually used to examine layer registration and through-hole problems.

14.4 FAILURE ANALYSIS PROCEDURES

14.4.1 Overview

This section outlines procedures for both the field and laboratory examination of PWAs and includes procedures for some of the specific analysis techniques. Guidelines and precautions for the field examination and handling of PWAs are provided. Accurate documentation and minimum additional damage to the PWA are of primary importance. (Special handling should be observed when the PWA includes nonvolatile memory circuits.) A list

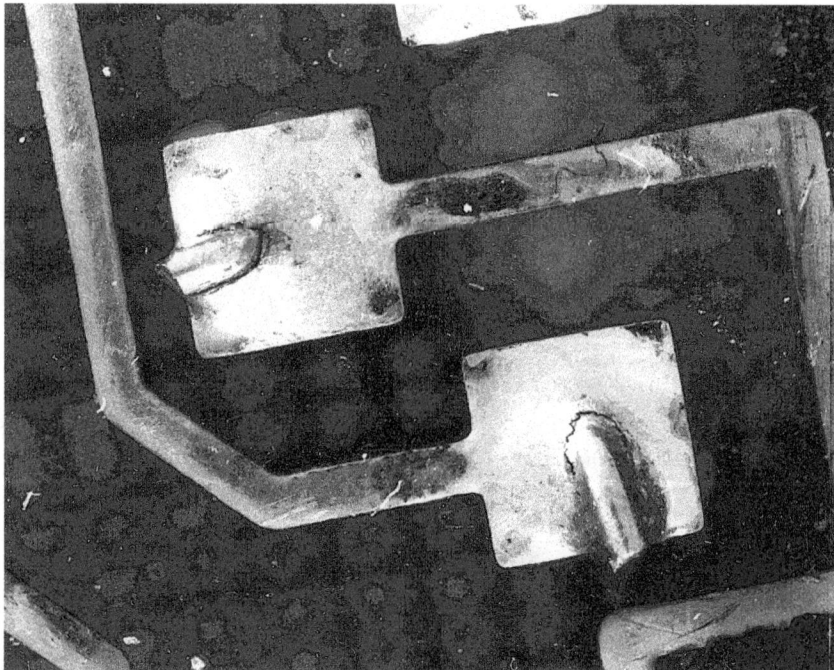

FIGURE 14.9   Optical microscope photograph of cracked solder joints on a PWB ($\approx$10$\times$). The box identifies one of the cracks.

of data to collect for PWAs is included with the discussion of field examination. An outline of the basic steps for laboratory examination is provided. Detailed procedures for analysis techniques specific to PWA are also included.

The goal of each failure analysis is to determine the root cause of the plated wiring assembly failure and to reconstruct the sequence of events that lead to the failure. In some instances, it may be necessary to duplicate the failure on an sample PWA to confirm that the conclusions from the analysis are correct.

14.4.2 Field Examination

Results from a field examination of a failed PWA are usually extremely limited because of the complex structure and the density of electronic components. However, since in-formation vital for understanding a plated wiring assembly failure may be available only in the field, it is always advisable to perform a field examination whenever possible. For that reason, the laboratory examination of a failed PWA generally should be conducted whenever the PWA may provide clues useful to the overall investigation of the failure. The recording of any identifying numbers on a PWA, and any associated electronic chassis, are essential steps in the documentation process. PWAs are marked with an assembly number and a serial number, which are both critical to record if the PWA is separated from the equipment. PWAs should not be separated from electronic equipment unless absolutely necessary. Removal of the PWA may also destroy information essential to determining the root cause of the failure and the sequence of events that resulted in the failure. Separation can cause damage to electronic components on the PWA and obscure the condition of the circuits. If separation is necessary, tag the PWA and any associated electronic chassis for identification. A sample identification form is shown in Table 14.4. PWAs should be wrapped in antistatic plastic bags before packaging for shipment to the laboratory. Under no circumstances should a PWA be wrapped in a nonconductive material.

Table 14.4   Printed Circuit Board Identification Sample Form

Equipment _____	
Sample ID# _____	PWA 101
Date of incident _____	04/30/90
Date removed from site _____	04/30/90, adjacent to power supply
Assy # _____	Assy-39765-39A
Serial # _____	313
LRU or chassis function _____	Motor control
Type of PCB _____	Multilayer with int. ground plane
Overall size _____	5×11 in
System model # _____	
System serial # _____	
System manufacture date _____	Burned circuit traces to power transistors
Comments _____	

The following general precautions should be followed for any type of PWA. These precautions are special for preserving data on PWAs that contain volatile memory devices:

- Enclose the PWA in antistatic bag prior to handling or shipping.

- Do not remove or disconnect on-board batteries.

- Do not expose to X-rays or intense UV radiation.
- Do not short or connect to card-edge connectors.

14.4.3 Laboratory Examination

The general steps to be followed during the laboratory examination of PWAs are outlined below. An examination usually occurs after some type of electrical failure has been identified. Testing should progress from noninvasive photographic or X-ray documentation to nondestructive functional testing. Destructive testing such as cross-sectional analysis should only occur after all nondestructive testing has been completed and reviewed. Tests that are destructive and should be initially conducted on exemplars of the same PWA. Complete photographic documentation is very important prior to and during any laboratory examination.

Perform a careful visual examination of the PCB. Specific features to look for are:

- *Cracking of the substrate.* Cracking indicates mechanical flexure and stress.
- *Overheated or discolored circuit traces.* They are often an indicator of overcurrent.
- *Cracked solder joints.* Their presence suggests solderability problems or solder contamination.
- *Solder joints that have a dull surface or have too much or too little solder.* These characteristics are often an indicator of poor soldering technique, contamination, or overheating or may indicate the presence of contamination, which caused staining.
- *Discolored components or regions of the substrate.* Any discoloration may indicate overheating.

Once a specific area of the PWA has been isolated, X-ray examination may be used to identify internal problems that may not be otherwise visible. Magnifications up to 1600× can be obtained. Real-time microfocus X-ray allows the board to be rotated to improve the viewing angle while the examination is being conducted. Some features to look for during an X-ray examination are:

- Any discontinuity or damage to traces
- Incompletely filled vias or through-holes
- Misregistered traces or component pads

Optical microscopy can help identify external defects or problems. Dendritic growth or metal migration can be examined on single- and double-sided PCBs by backlighting the PWB and using the light transmitted through the substrate to perform the examination.

Electrical measurements are useful in identifying open traces and fractured solder joints. Caution should be used to keep voltage and current to a minimum to avoid damage to the component being measured. Excessive leakage current due to contamination can sometimes be measured by placing a voltage between the points of concern and measuring the resulting current flow. (Limiting the maximum voltage to 0.5 V guarantees that no conduction will occur across silicon *p-n* junctions, which may result in high current flow.) Gentle flexing of the PWA may uncover an intermittent fault.

Problems with solder joints and internal defects on multilayer boards can be examined by sectioning and preparing metallurgical samples. The mounting and polishing procedures are covered later in this section. Solder failures, voids, cracks, and internal trace connections are best examined after their presence has been identified by other means due to the destructive nature of this analysis.

In some cases, contamination on the surface of the substrate can be identified with SEM/ EDS. The contamination can occur on the surface of the board or below the conformal coating. In the latter case, the coating must be removed first without affecting the contaminant, and the EDS analysis must be performed on a section of the board where the contaminated surface is exposed. Chlorine, fluorine, sulfur, sodium, and bromine are elements of interest, and they usually can be detected with the SEM. Bromine is used as a flame retardant in some materials. Solder joints can also be examined for contamination using EDS. Sulfur, oxygen, copper, gold, aluminum, and zinc are contaminants that may cause problems in solder joints. The exact location of contaminants in the solder joint is critical. Solder joint fractures caused by contamination frequently occur in the intermetallic phase at the interface between the component lead and solder interface.

14.4.4 Removing Conformal Coatings

The removal of conformal coatings may present a significant problem in mishap investigation of a PWB failure. Four methods are typically used to remove conformal coatings.

- Dissolving with solvents
- Thermal parting
- Abrasion
- Plasma etching

Solvents such as xylene, trichlorethane, methyl ethyl ketone, and methylene chloride may be used to remove coatings. However, care must be taken not to damage the board or components or to remove the contaminant of interest. Environmental restrictions may apply to the use of these solvents, so alternatives may be required. Commercial compounds are also available that are designed specifically for the removal of conformal coatings.

Thermal parting is a method that uses controlled, low-temperature heating. This technique is best for the removal of thick coatings. Heat is applied directly to the coating, which causes it to separate from the base material. If the coating is sufficiently thick, it can be peeled away from the base material.

Abrasive jet equipment, similar to sand blasting, can be used to remove coatings that cannot be removed by solvents. However, care must be taken, since charge can build up in this type of equipment, resulting in ESD damage.

The board can be placed in a vacuum chamber, and a low-temperature plasma can be used to remove the coating. The plasma is created by exciting a gas with an RF (radio frequency) field. This method is useful for removing parylene.[1]

14.4.5 Solder Joint Examination

Solder joints can be examined nondestructively using real-time microfocus X-ray equipment. There are practical limitations to the resolution of this equipment, and cracking and other defects with finer details may not be visible. Sectioning and optical microscopy may be used successfully for diagnosing manufacturing problems. These techniques will work equally well in failure investigations if it is acceptable to section the board. Contaminants in the solder can be analyzed by SEM/EDS analysis.

As a consequence of the low melting point of 63Sn-37Pb solder (182° C or 361° F), cracked or broken solder joints will not survive even a modest fire. However, if the solder reflows during a fire, the appearance should be distinguishable from that of an undamaged PCB. Visually, reflow will cause some redistribution of the solder. Broken solder joints that have reflowed due to external heat may have additional voids. Additionally, SEM/EDS analysis may reveal contaminates trapped in the voids.

14.4.6   Insulation Resistance

Degraded insulation resistance can be the result of dendritic growth, contamination, and other problems. The military test procedures for moisture and insulation resistance are defined in MIL-STD-55110 and are based on the environmental test procedures in MIL-STD-202. Preconditioning and thermal cycling are required prior to the electrical measurements, which are then made at 100 Vdc. A minimum resistance for 500 MΩ is required. The IPC has additional procedures that make use of two types of special PCB patterns designed specifically for test purposes.

Insulation resistance measurements are extremely sensitive to environmental conditions. This behavior is primarily due to the fact that polymeric materials absorb moisture over time. This absorption can activate conductive contaminants on the substrate surface or in the base material of the substrate.

Humidity is the largest single factor responsible for most insulation-related problems. Insulating properties of PCB materials are extremely sensitive to relative humidity, especially if the PWA is already contaminated. On new boards, the surface resistivity can change from 1015 to 1011 Ω/sq when relative humidity changes from 30 to 90 percent.[1] This extreme sensitivity to moisture requires that laboratory tests be conducted under conditions simulating the actual field failure conditions.

The measurement of resistances above 100 MΩ are extremely difficult to make on PWAs from field equipment, especially if there has been any exposure to fire-fighting materials, water, or smoke. In addition, high-voltage methods are not practical, since they can cause additional component failures and measurement interactions. Fortunately, insulation resistance failures are usually well below 1MΩ, thus simplifying the measurement process.

Some guidelines for verifying insulation resistance problems are provided below.

1. The PWA should be exposed to humidity levels experienced in service before the measurements are made.

2. Circuit traces may need to be physically cut to isolate circuitry. This procedure allows a single trace to be measured more accurately. Complex components or circuit areas may need to be removed to isolate the area of interest.

3. The PWA should be examined carefully using an optical microscope or SEM to verify the physical source of the leakage or contamination.

4. Measurements should be made using low-voltage techniques to avoid damage to components.

High-resistance tests on PWAs removed from field equipment may need to be supplemented with tests on sample boards having the same trace patterns and components. High series resistances must be used to meter circuits and limit applied voltages to avoid damaging electronic components.

14.4.7 Metallurgical Sample Preparation

Sample preparation consists of the five basic steps described in Tables 14.5 and 14.6. The parameters and materials for automatic polishing equipment are also provided.[11] (Most of this information pertains to manual preparation as well.) The principles of metallography and sample methods are covered in Ref. 12.

PWA sections with plated-through holes are often mounted with carbide stops to ensure that grinding will terminate before damaging the sample. The stops are removed prior to polishing. Other special accessories may also be used.

Sectioning using a *cutoff wheel* and coolant is described below. Mounting of the sample utilizes a castable media. The procedure to use when preparing a sample with plated-through holes is described in Tables 14.5 and 14.6.

Table 14.5 Procedures To Use for Grinding PWA Sections

SiC grit size	Time (s)	Wheel speed (rpm)	Pressure (psi)
180	Until all carbide stops are hit	300	35
600	60	300	35

Remove ring with carbide stops.

Table 14.6 Procedures To Use for Polishing PWA Sections

SiC grit size	Time (s)	Wheel speed (rpm)	Pressure (psi)
3 μm diamond compound/silk cloth/oil	240	250	30
1 μm diamond compound/red felt cloth/oil	60	250	30
Optional: Colloidal silica/wetted Imperial cloth	30	150	10

14.5 DETERMINING THE ROOT CAUSE OF A PLATED WIRING ASSEMBLY FAILURE AND THE SEQUENCE OF EVENTS THAT OCCURRED

14.5.1 Overview and Introduction

The techniques and methods discussed in Sec. 14.3 provide the information required to determine the root cause of a PWA failure and the sequence of events that occurred. However, this information must be analyzed, often in conjunction with information on the PCB as well as data about components that were in the damaged area of the PWA and field failure information. This section describes steps to be undertaken to determine both the root cause of a PWA failure and the sequence of events that lead to the failure.

This section assumes that the results of the work completed in Sec. 14.3 has not identified the cause of the failure since, in many failures, the damage extends over an area that includes many components and the PCB. The term *propagating PWA failure* is often used to describe such a failure.

Trends in PWA design and manufacturing technology include the following:

- A decrease in the size of the components assembled on PC boards
- A decrease in the width of the conductive traces and the spacing between traces
- An increase in the number of conductive layers in PC boards
- An increase in the size of the largest PWAs
- An increase in the complexity of PWA assembly technology including
  - More PWAs with a high density of components on both surfaces
  - An increase in the number of PWAs that use surface mount technology (SMT) instead of through-hole technology
  - An increase in overall PWA complexity
  - An increase in the power density on PWAs

These trends make it more difficult to identify the root cause of a PWA failure for at least two reasons.

1. The failure of one component on a PWA often causes the failure of nearby components, making it difficult to determine which component initially failed.
2. A component failure can be followed by the propagation of this failure across the PWA, which makes it extremely difficult to identify the location of the initial failure.

For these reasons, a general approach is required for identifying the root cause of a PWA failure and the sequence of events that led to the failure. Such an approach is provided in the remainder of this section.

14.5.2   General Types of PWA Failures

A failed PWA may range greatly in appearance, depending on the type of failure and whether the failure propagated. Types of PWA failures often encounter are listed below.

1. A failure of the PC board itself resulting from a problem in the design or the manufacture of the board
2. A failure of the PWA caused by unwanted conduction between conductive regions of the board that were designed to be electrically isolated
3. The failure of a component on the PWA that only affects the operation of the board
4. The failure of a component on a PWA that affects surrounding components
5. The failure of a component on a PWA that results in the propagation of the failure into the PC board, usually resulting in a burned PC board

These five general types of failures are covered in the next five subsections. A general approach to detecting the root cause of a PWA failure and determining the sequence of events that ultimately resulted in the PWA failure is covered in the next section.

14.5.3   Determination of the Sequence of Events that Resulted in the PWA Failure

The construction of an event flow can be extremely helpful in determining the specific sequence of events that occurred. An event flow shows all possible sequences of events that might have resulted in the observed failure. This event flow is used to determine the testing that is performed to determine the most likely sequence of events. A typical event tree is shown in Fig. 14.10.

The event flow is used to determine that tests or analysis should be performed. The results of the tests and analysis are used to eliminate or "prune" sequences events that could not have, or were unlikely to have, caused the failure. Testing and evaluation can then focus on the remaining event sequences, ultimately identifying the most likely event flow(s).

14.5.4   A Methodology for Investigating Propagating PWA Failures

The steps described below should be followed when investigating a propagating PWA failure after the PWA has been documented and photographed.

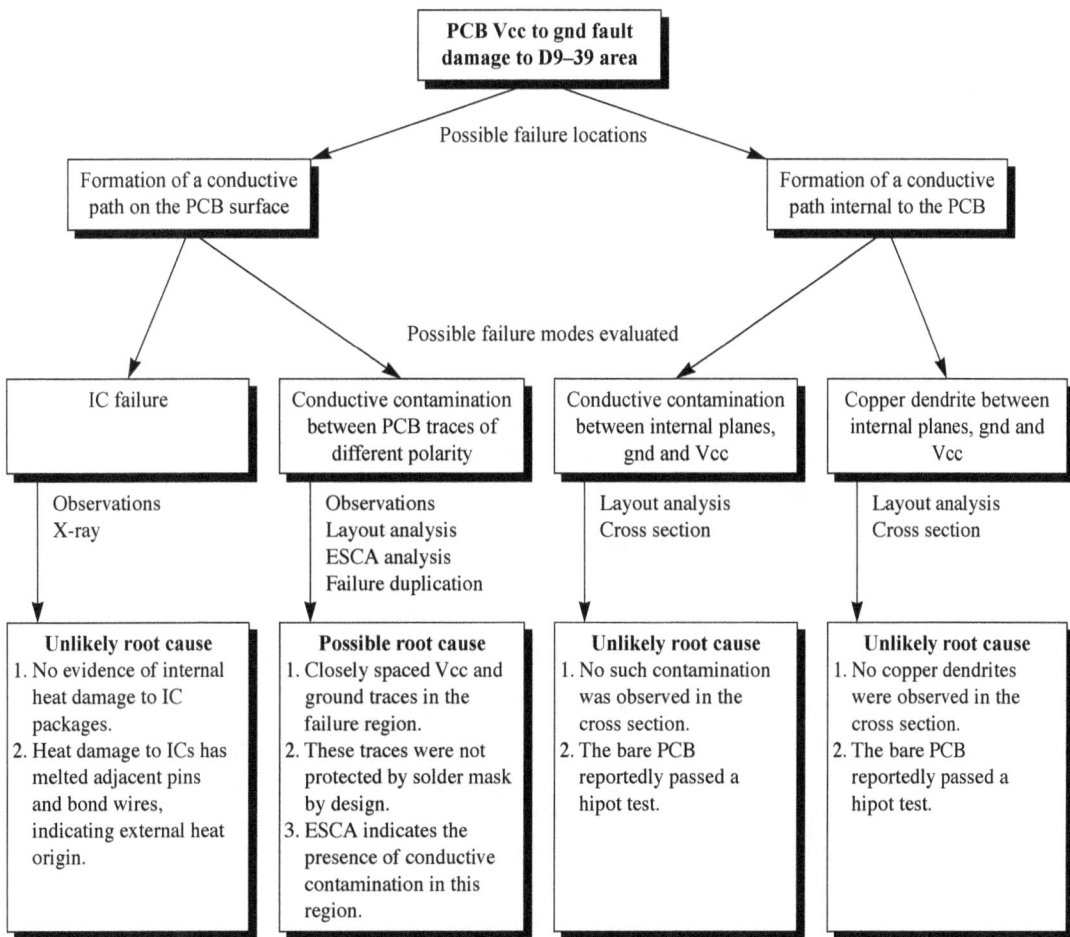

FIGURE 14.10   The event tree and analysis performed to identify the cause of the failure.

1. Locate the area of the PCB that includes the failure region.
2. Use foils of surface and internal layers, schematics, and PCBs to map power and ground traces or planes in this region.
3. Locate the likely area of origin.
   a. If the region includes continuous power and ground planes, the origin is in the geometric center of the burn.
   b. If the region includes only traces, the origin is at the furthest burned area from the power source. (The fault migrates toward the power source.)
   c. Physical orientation and airflow at the time of the failure may affect the direction of propagation. The "V" burn pattern may help locate the origin.
4. Investigate the following items in the area of origin.
   a. PCB-mounted components
       (i) Failure modes and minimum stress required to initiate failure
       (ii) Electrical, physical, environmental stress on component in the circuit; also circuit failures that could occur in operation.
       (iii) Remove components in the failure region and determine their failure mode, if any.
   b. Areas of minimum power-to-ground spacing
       (i) Between PCB traces, PTHs, internal or surface planes
       (ii) To other metals such as heatsinks
   c. Connectors
       (i) Investigate contact resistance.
       (ii) Check PTH-to-trace or PTH-to-plane connection.
       (iii) Check PTH-to-plane spacing.
       (iv) Check force exerted on connector.
   d. PCB flex points

When these steps are completed, the specific cause of the PWA failure may be fully understood.

14.5.5 Case Study: The Analysis of a Failure PWA

The following observations were made concerning the PWA failure:

1. The PCB was observed to be charred in the region of an integrated circuit. This charred region extended from pin 1 of the integrated circuit to pin 10. The board did not appear to be charred beneath the integrated circuit. No damage to the integrated circuit package was observed. These observations indicated that a failure of the integrated circuit was not a likely cause of the incident failure.
2. The damaged region including four plated-through holes (PTHs). On the solder side of the board, one of these PTHs connects to a surface mount capacitor via a metal trace approximately 0.1 in long. This trace is normally protected by a solder mask. However,

on the incident board, the copper was visible and appeared to have been overheated by the failure. This trace provided a ground connection to a surface mount capacitor. The other terminal of the capacitor was connected to Vcc. This capacitor was removed from the board and was measured to have a dc resistance greater than 40 MΩ. No damage was observed on this capacitor or the PCB beneath the capacitor. These observations indicate that the damage to the PCB ground trace was not likely due to excessive current. The most likely cause of this damage was heat from a Vcc to ground fault at another location which was conducted by the metal trace and the PTH.

3. The integrated circuit was removed from an exemplar board, and the Vcc and ground traces in the region of this component were mapped on both sides of the PCB. This map revealed that no Vcc traces were routed on the PCB surface in the failure region. However, three of the four PTHs in this region were connected to ground. These observations indicate that the incident PCB fault was not likely due to a failure between Vcc and ground that occurred on the surface of the PCB.

4. The internal layout of the PCB was analyzed to determine whether low-impedance traces were routed through the failure region internal to the board. This analysis revealed five signal traces that were routed near the failure region inside the board. All of these traces were routed between two nearby ICs and the first IC. The manufacturer provided the information that the output impedance driving these traces was 100 Ω to Vcc and 10 Ω to ground. Using these values, the maximum power, which could be supplied, was calculated to be 0.625 W. This power would be dissipated in a region internal to the PCB between the Vcc plane on layer 5 and the signal plane on layer 6.

5. X-ray films were taken on the incident board in the region of the failure. These films were examined under a microscope to characterize the damage, which has occurred internal to the board. The X-ray showed that the damage to the board was most extreme in the region of pins 1 through 3, all of which connect to the ground plane. The failure region internal to the board appeared to be centered between pins 1 and 2. Pin 1 was observed to be misaligned on its pad, possibly due to heat generated by the failure.

6. A cross-sectional analysis was performed on the incident board. The region of the failure was observed to consist of charred PCB material and melted copper. The surrounding region was examined for defects in the fiberglass weave, inclusion of particles, and the formation copper dendrites on plated-through holes. None of these defects was observed.

7. X-rays of the first integrated circuit revealed no likely failure of this device. The bond wires to the die were all intact. No regions of varying density typically associated with a failed IC were observed.

Analysis and testing of the incident board have revealed that the most likely cause of the PWA failure was the formation of a conductive path between Vcc and ground or between a signal trace and Vcc internal to the printed circuit board. These signals provide a low-impedance path to ground and are routed on layer 6 of the PCB, which is adjacent to the Vcc plane on layer 5. The hipot test, which was performed on this board, would not likely have identified a fault between a signal trace and a power plane. The event flow and analysis is summarized in Fig. 14.11.

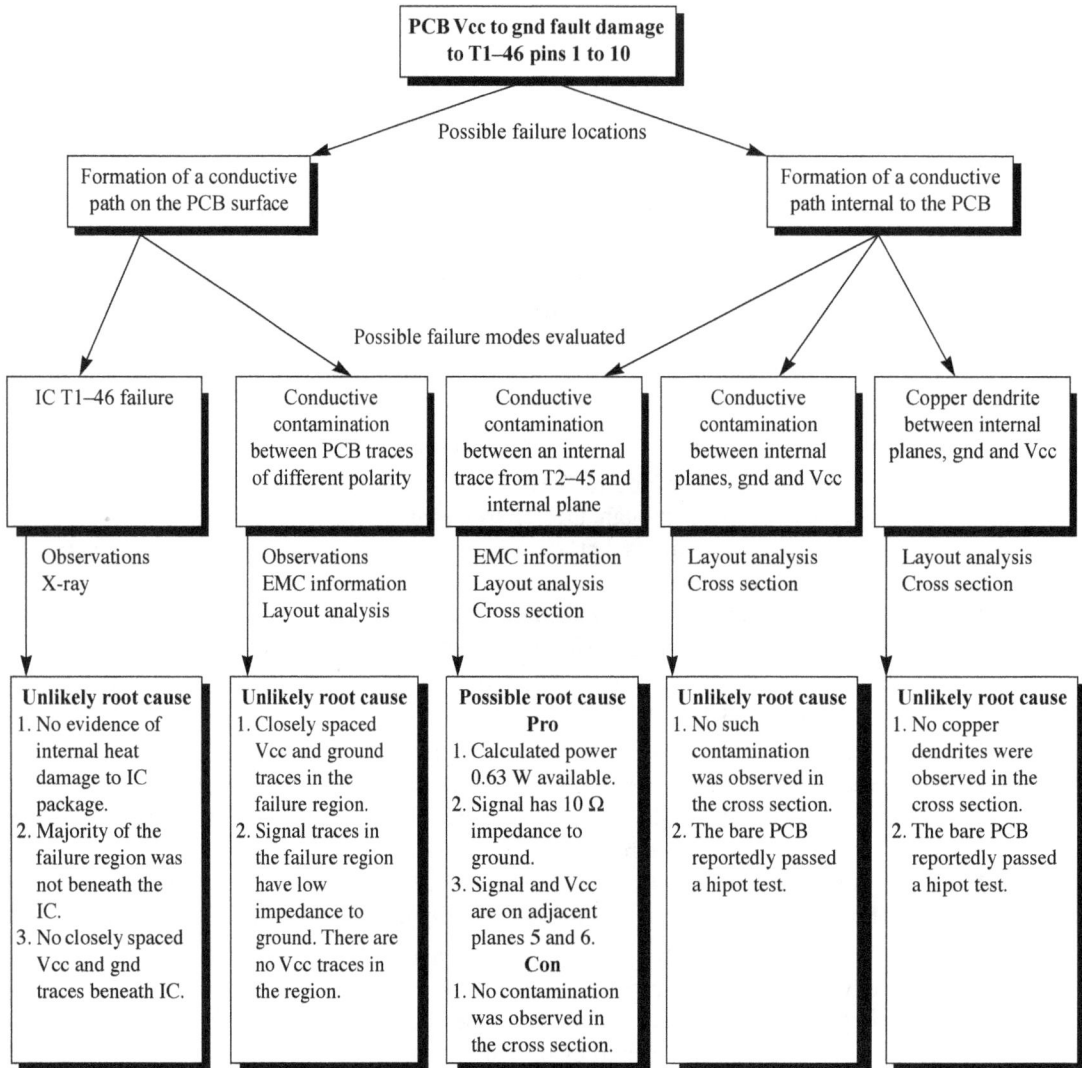

FIGURE 14.11   The event flow and the analysis performed to identify the cause of the PWA failure.

14.6   REFERENCES

1. Coombs, Jr., C.F., ed. *Printed Circuit Handbook* 4th ed. New York: McGraw Hill, 1995.

2. Hinch, S.W. *Handbook of Surface Mount Technology.* New York: John Wiley & Sons.

3. Dobbs, B. Corrosion of Electronic Components, *Proc. of the 62nd AGARD Conference,* April 1986, DTIC No. AD-A194 868.

4. Trapp, O.D., Larry J. Lopp, Richard A. Blanchard. *Semiconductor Technology Handbook,* 6th ed. Portola Valley, CA: Technology Associates, 1992.

5. Krumbein, S.J. Electrolytic Models for Metallic Electromigration Failure Mechanisms, *IEEE Transactions on Reliability* 44(4), 539–549.

6. Rudra, B., and D. Jennings. Failure-Mechanism Models for Conductive-Filament Formation, *IEEE Transactions on Reliability* 43(3), 354–360.

7. Fink, D.G., and D. Christiansen. *Electronics Engineers' Handbook,* 3d ed. New York: McGraw-Hill, 1987, 7–25.

8. Yu, Qiang, and Masaki Shiratori. Fatigue-Strength Prediction of Microelectronics Solder Joints Under Thermal Cyclic Loading, *IEEE Transactions on Components, Packaging, and Manufacturing Technology—Part A* 20(3), 266–273.

9. Pau, Y.-H., E. Jih, V. Siddapureddy, X. Song, R Liu, R. McMillan, and J.M. Hu. Thermal Fatigue of Surface Mount Leadless Solder Joints, *Advancing Microelectronics* (July/August 1997), 33–39.

10. Colangelo, J. Advanced Radiographic Techniques in Failure Analysis. Texas Instruments, DSEG Failure Analysis Laboratory, Plano Texas. (Seminar conducted at ISTFA, 1991, in *Proc. 17th International Symposium for Testing and Failure Analysis*).

11. Dillinger, L. *Suggested Procedures from Grinding and Polishing Various Alloy Systems Using Automatic Polishers.* St Joseph, MI: LECO Corporation.

12. *Metallographic Principles and Procedures.* St Joseph, MI: LECO Corporation.

CHAPTER 15
WIRES AND CABLES

George Slenski
Air Force Wright Laboratory
Wright-Patterson Air Force Base

Donald Galler
Department of Materials Science and Engineering
Massachusetts Institute of Technology

15.1 INTRODUCTION

The increased emphasis and reliance on electronic systems for modern aircraft have resulted in wiring becoming a critical safety-of-flight system. Aircraft now routinely use fly-by-wire systems with minimal or no mechanical backup systems. A recent study initiated by the Materials Directorate reported that 34 percent of all aircraft mishaps related to electrical systems were caused by interconnection failures involving wiring and connectors.[1] Wiring failures have been found to initiate hydraulic and fuel fires by electrical arcing or cause malfunctions in flight control systems and in other critical areas. At high operating temperatures, some insulations can soften or crack and become susceptible to chafing damage that normally would not occur at room temperature. Examples where wire chafing led to arcing, a fire, and an aircraft mishap are shown in Figs. 15.1 and 15.2.

This chapter discusses current wire constructions, basic materials properties, wiring failure mechanisms, analysis techniques, and case studies. A variety of wire insulations and conductors are used in electronic applications. The discussions in this chapter focus on high-performance products typically associated with aerospace and high-reliability applications. However, many of the topics are applicable to all wire products.

Aircraft wiring continues to be a high-maintenance item and a major contributor to electrically related aircraft mishaps. A review of maintenance records on fighter aircraft show that chafed or damaged insulation is the most common mode of wire failure.[2] This is shown in Table 15.1. Improved wiring constructions are expected to increase aircraft performance and decrease costs by reducing maintenance actions.[3]

Table 15.1 Wire Problem Areas According to Aircraft Maintenance Records during June 1993 through May 1994 (*from Ref. 15.2*)

Failure description	Percentage of problems	Failure description	Percentage of problems
Chaffed/damaged insulation	44	Missing wire	2.5
Broken wire	12	No defect found	24.0
Open conductor	3	No determination	9.5
Shorted conductors	5		

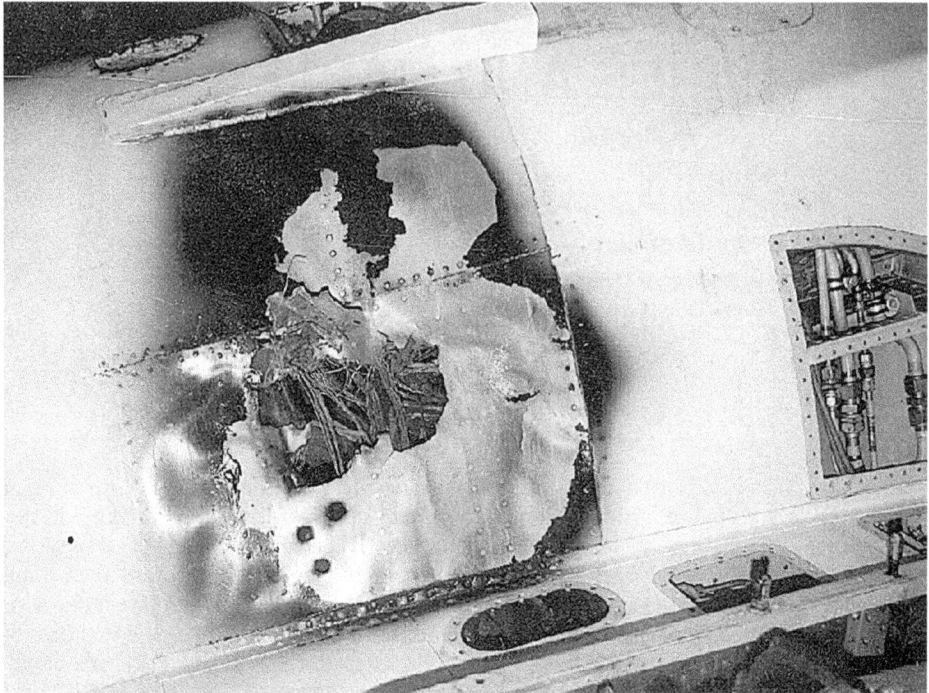

FIGURE 15.1 Aircraft damage from an electrical fire. Part of the aft aluminum structure was consumed in the fire.

15.2 *WIRING TYPES, CONSTRUCTIONS, AND PROPERTIES*

This chapter deals with the type of wire classified as *hookup* wire, which is the largest class of wire used in aircraft. Only copper-based hookup wire is considered. Hookup wire is used for interconnect cabling between pieces of equipment, harnesses attached to the airframe, and for interconnect wiring inside equipment. In addition to hookup wire, other types of wiring used include:

- Multiconductor cabling
- High-voltage cables
- Special-purpose data cables
- Shielded coaxial or multiaxial cables

Most commercial and military aircraft wiring are based on two military specifications, MIL-W-81381 and MIL-W-22759. The insulation materials used are principally aromatic polyimide, polytetrafluoroethylene (PTFE), and cross-linked ethylene tetrafluoroethylene (XLETFE). Examples of insulation properties of interest typically include flammability, high-temperature mechanical and electrical performance, fluid immersion, and susceptibility to arc propagation under applied power and chafing conditions.

FIGURE 15.2   Close-up of the aircraft's internal structure. Note exclusive thermal damage to the wiring and surrounding components.

15.3   COPPER PROPERTIES

The properties of finished copper wire are the result of a number of processing steps that begin with the final steps of processing the copper ore. The last step in processing the ore results in a product that is at least 99.5 percent copper. The processed ore is cast into ingots. These are used as anodes and electrolytically refined. In the process, the copper is transferred onto cathodes, which results in a product that is at least 99.95 percent pure copper. These copper cathodes are the end product of the copper-producing companies.

The copper cathodes are melted and cast into wire bars followed by hot rolling into wire rod. Continuous casting is also used to produce wire rod directly from the cathodes. Wire rod is cold drawn through a series of dies until it is reduced to the proper diameter. Cold working during the drawing process introduces stresses and an elongation of the crystal grain structure. Table 15.2 lists the properties of pure copper.[4] Copper-wire grain structure is shown in Fig. 15.3.

Table 15.2   Data on Copper

Atomic ID	Cu	Density	8.96 g/cm
Atomic #	29	Resistivity	$1.72 \times 10{-}6\ \Omega\text{-cm}$
Melting point	1083° C	Temperature coefficient of resistance	0.0039/ C
Boiling point	2595° C	Temperature coefficient of expansion	$16.5 \times 10{-}6/°C$

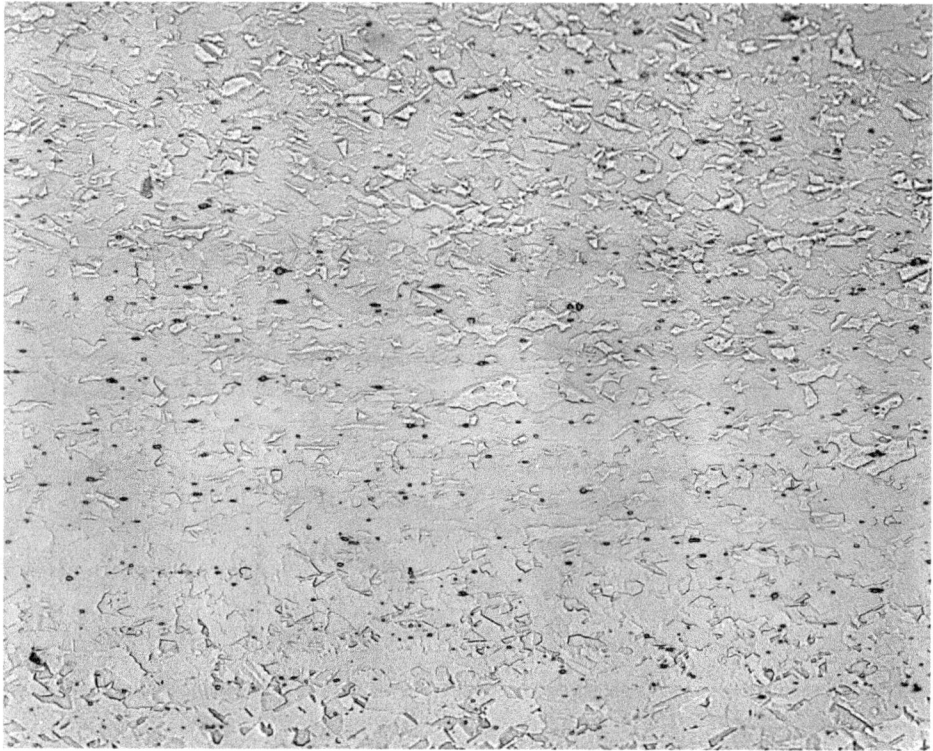

FIGURE 15.3 Cross section of normal OFHC copper wire showing grain structure (500×).

Two copper alloys are used in aircraft, as described below.

▪ *Oxygen-free high-conductivity* (OFHC) copper is made by induction melting prime quality copper cathodes. The heating is done in a nonoxidizing environment. It is produced by using a granulated graphite bath covering the copper and a protective reducing atmosphere that is low in hydrogen. OFHC is at least 99.99 percent pure copper and has an annealing temperature of 700 to 1200° F (370 to 650° C).

▪ *High-strength copper alloy* (HSCA) is usually a cadmium-copper or cadmium-chromium-copper alloy. The alloy has improved mechanical properties including increased tensile strength and resistance to annealing at elevated temperatures. HSCA has an annealing temperature of 1000 to 1400° F (535 to 760° C).

Copper wire used in aircraft is coated to provide environmental protection and promote better connections. The coatings are applied to the individual strands of the conductor, usually by electroplating. The required minimum thicknesses for the most common coating materials are listed below:

Silver	40 μin
Nickel	50 μin
Tin	20 to 50 μin

Aircraft wire is stranded to provide flexibility. The stranding arrangements are similar to those used for general-purpose electronic hookup wire. Data on stranded conductors appear in Table 15.3.

Table 15.3   Data on Stranded Conductors from MIL-W-81381

Wire size	Nominal area (cir. mils)	Stranding	Strand dia. (inch)	Conductor dia. (inch)	Resistance (Ω/1000 ft.)
30	112	7 × 38	0.0040	0.012	100.7
28	175	7 × 36	0.0050	0.015	63.8
26	304	19 × 38	0.0040	0.020	38.4
24	475	19 × 36	0.0050	0.025	24.3
22	754	19 × 34	0.0063	0.032	15.10
20	1216	19 × 32	0.0080	0.040	9.19
18	1900	19 × 30	0.0100	0.050	5.79
16	2426	19 × 29	0.0113	0.057	4.52
14	3831	19 × 27	0.0142	0.072	2.88
12	5874	37 × 28	0.0126	0.089	1.90
10	9354	37 × 26	0.0159	0.112	1.19
8	16983	133 × 29	0.0113	0.169	0.658
6	26818	133 × 27	0.0142	0.213	0.418
4	42615	133 × 25	0.0179	0.268	0.264
2	66500	665 × 30	0.0100	0.340	0.170
1	81700	817 × 30	0.0100	0.380	0.139
0	104500	1045 × 30	0.0100	0.425	0.108
00	133000	1330 × 30	0.0100	0.475	0.085
000	166500	1665 × 30	0.0100	0.540	0.068
0000	210900	2109 × 30	0.0100	0.605	0.054

- *Concentric lay* stranding is used for the smaller wire sizes, usually below AWG 10. In this arrangement, a central strand is surrounded by one or more layers of helically wound strands.
- *Rope lay* stranding is used for larger wire sizes. In this arrangement, a stranded central wire is surrounded by one or more layers of helically wound stranded wires.

Stranded conductors are provided with insulated coverings that provide environmental and mechanical protection as well as electrical insulation. Almost all aircraft wire uses polymeric insulation materials. Glass braid, glass tape, and other materials are used to a lesser extent.

Insulation materials can be applied by two processes. They are:

- *Extrusion* is a process in which the plastic insulation material is melted and flowed over the finished conductor under pressure as the conductor is pulled through a die.
- *Tape-wound* insulation uses two layers of tape spirally wound around the bare conductor in opposite directions. A resin coating may be applied to aid in wire marking.

Jacket materials are applied over the insulation for additional mechanical protection of the insulation and conductor. Some insulation and jacket materials are shown in Tables 15.4 and 15.5. Information on other insulation materials and their properties can be found in Ref. 5. Properties of polyimide can be found in Ref. 6.

Table 15.4 Insulation Materials for Military Aircraft Wiring

Material	Trade name	Military standard
Extruded PVC		MIL-W-16878, 5086
Fluorocarbon polyimide tape	Kapton®	MIL-W-81381
FTFE tape	Teflon®	MIL-W-22759
Extruded PTFE tape	Teflon®	MIL-W-22759
Extruded ETFE tape	Teflon®	MIL-W-22759
Extruded PTFE	Tefzel®	MIL-W-22759
Polyalkene	Teflon®	MIL-W-81044

Table 15.5 Jacket Materials for Military Aircraft Wiring

Material	Trade name	Military standard
Glass braid		MIL-W-81381, 5086
TFE coated glass braid		MIL-W-22759
Extruded PVC		MIL-W-5086
Extruded Polyimide	Nylon	MIL-W-16878, 5086
Extruded polyvinylidene	Kynar	MIL-W-22759
TFE tape	Teflon®	MIL-W-22759
Polyimide coating		MIL-W-22759
Aromatic polyimide resin	"Liquid H"	MIL-W-81381

15.4 WIRING FAILURE MECHANISMS AND ANALYSIS TECHNIQUES

Wire is damaged in different ways by a variety of fundamental mechanisms. In a broad sense, damage to wire is caused by one of the following three failure conditions:

- Electrical
- Mechanical
- Thermal

In this chapter, we differentiate failure characteristics from failure conditions.

- *Failure characteristics* are the physical features tested or examined by the investigator. This physical evidence may be used to infer the conditions present before, during, and after an event. Each failure characteristic will have associated with it one or more sets of possible conditions called *failure conditions*.

- *Failure conditions* are the mechanisms that caused the damage. The investigator will use the evidence to try to determine which failure conditions were actually present during the failure.

The failure characteristics should be differentiated from the failure condition, since there is not a one-to-one relationship between the conditions and the resulting characteristics. For example, recrystallization could be the result of either electrical overcurrent or fire exposure.

Several failure characteristics can appear in localized areas. For example, less than one inch of wire may be damaged. In other cases, the damage is uniform over a larger area—sometimes up to several feet in length. The extent to which the damage is localized can help verify the failure mechanism at work. The failure characteristics covered in this chapter are listed below.

- Recrystallization
- Beaded wire ends
- Metal transfer
- Cup-and-cone fracture
- Insulation failure
- Thermal damage to insulation
- Conductor discoloration

15.4.1 Recrystallization

The process of cold drawing copper wire results in a reduction in grain size and an elongation of the crystal grain structure along the axis of the wire. If the wire is subsequently heated to the recrystallization temperature, the grain structure undergoes a transformation. Three temperature-dependent transitions are observed during this transformation. These are recovery, recrystallization, and growth. This phenomenon is time and temperature dependent and is illustrated in Fig. 15.4.[7]

During cold working, dislocations are introduced into the crystal structure due to localized breaks in the atomic structure. These dislocations prevent slip along crystal planes, strengthening the material and lowering ductility. As the temperature increases toward temperature *a* in the figure, the material undergoes recovery. During recovery, these dislocations coalesce to form pseudo-grain boundaries internal to the original grain structure. Internal stresses are relieved, and electrical conductivity increases. At temperature *a*, recrystallization begins. During recrystallization, these pseudo-grain boundaries fully form into grain boundaries, forming new, smaller, more uniform grains. If the material is further exposed to elevated temperature *b*, these grains will grow larger, with more favorably oriented grains consuming smaller, less favorably oriented grains. During this growth phase, internal stress is further relieved, material strength decreases, and ductility increases (annealing). As growth continues and grain size approaches the diameter of the wire, ductility will decrease.

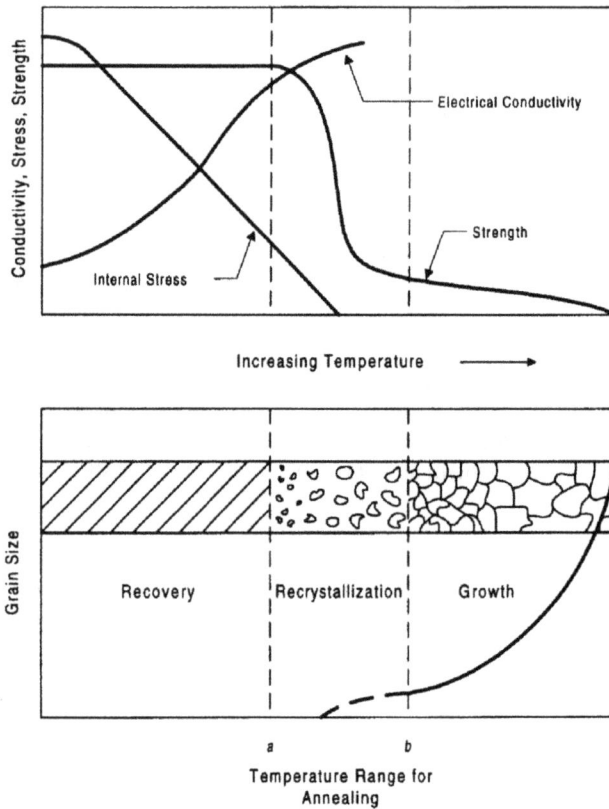

FIGURE 15.4 Recrystallization process.

The grain structure changes during recrystallization, and growth may be observed in a properly prepared microscopic examination. This may provide evidence that the material was exposed to elevated temperatures for a period of time. Recrystallization in copper wire occurs at temperatures as low as 700° F (370° C). Care is required in the preparation of samples for analysis to ensure that the wire is not recrystallized by the experimental procedures.

The appearance of equiaxed grains is the basis for one of the traditional procedures for metallurgical examination of conductors that have been involved in building fires.[7–10] The amount of recrystallization depends on the temperature and the length of time the conductor is exposed. The rate of recrystallization is governed by the same principles as a chemical reaction. As a general rule, a 10° C increase in temperature requires only one-half the exposure time to achieve the same level of recrystallization. Temperatures for recrystallization are listed below:

Conductor type	Annealing temperature
OFHC	700 to 1200° F (370 to 650° C)
HSCA	1000 to 1400° F (535 to 760° C)

Procedures for recrystallization analysis are discussed later in this chapter under *Sample Preparation.* Testing OFHC wire illustrates the recrystallization mechanism. Samples of new insulated wire were heated in an oven to produce the results shown in Fig. 15.5.

15.4.2 Beaded Wire Ends

Extreme localized heating results when wiring is involved in electrical arcing. The heating causes the copper to melt and form spherical globules, or beads, on the ends of wire strands. The bead shape is characteristic of arc phenomenon and is an indicator that temperatures well in excess of the melting point were achieved. Beaded wire ends are good indicators that the wires were energized when the damage took place, even if nonelectrical mechanisms were the original cause of the damage.

Electrical arcing frequently occurs when two conductors of opposite polarity come in contact or close proximity. Arcing from conductors to the aluminum airframe is also common, since the airframe is generally at the ground potential of the electrical system. The voltage and current of the electric arc cause heating that melts the wire near the arc. The voltage-current product associated with the arc represents the thermal power available to cause conductor melting. For example, a 20 A short-circuit on a 28 V system will result in 560 W of thermal power. Most of this power will be dissipated in a relatively small area near the arc.

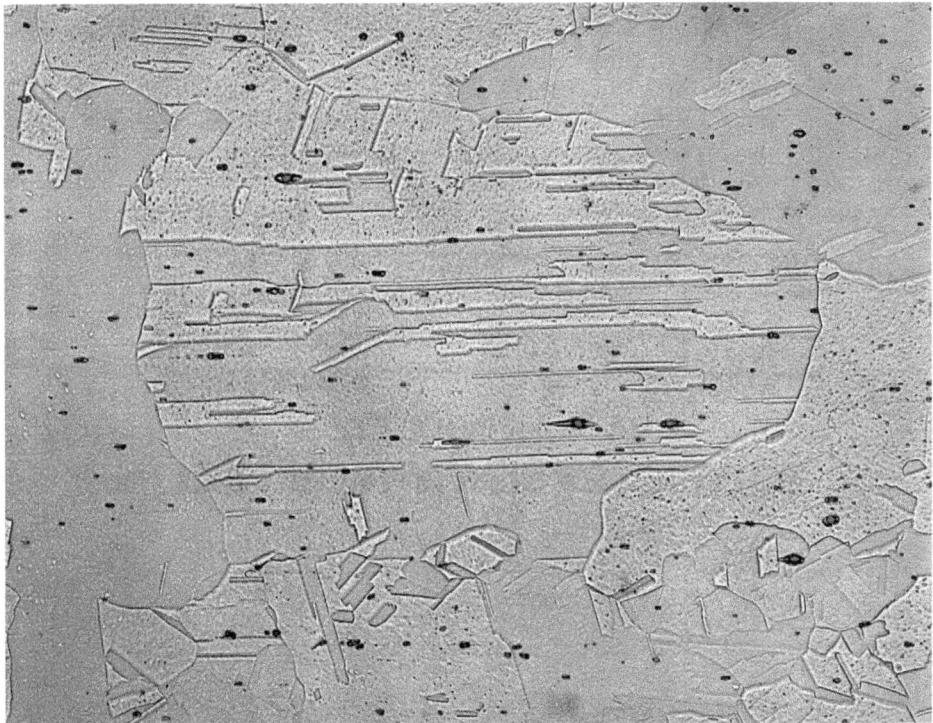

FIGURE 15.5 Cross section showing grain structure of OFHC wire after 5 min at 600° C, 500×. Note large grains compared to Fig. 15.3.

Arcing can occur in dc or ac circuits. Arcing in ac circuits is frequently aided by intermittent contact, chafing, or contamination. Sustained arcing is difficult to achieve without these factors because the arc will tend to extinguish itself during the zero crossing of the current. Current flow, resulting from direct contact between conductors on either ac or dc systems, will frequently be interrupted by overcurrent protection devices before significant melting can occur. This is another reason why these additional factors are needed to promote sustained arcing.

Recrystallization will tend to occur near the point of arcing and along the current-carrying portion of the conductor. The amount of recrystallization will depend on the current level and duration of the fault. Beaded wire ends are localized in nature. Fig. 15.6 shows the appearance of a wire that was involved in arcing in a 28 Vdc system.

15.4.3 Metal Transfer

Metal transfer occurs when an arc forms between two conductors. The arc is usually accompanied by a transfer of metal droplets from one material to the other. The direction and amount of material transferred depends on several factors, including the materials themselves, their orientation, the electric circuit, and specific mechanical structure in the area of the arcing. One experiment conducted between a copper conductor and aluminum panel connected to a 28 Vdc circuit (panel at ground) showed the prevailing metal transfer was from the wire to the panel. More damage will usually be found in the material with the lowest melting point (aluminum in this case).

Examination of the arcing site is commonly done with a scanning electron microscope (SEM) with energy dispersive spectroscopy (EDS) to determine the extent of material transfer. This testing should always be done on a comparative basis using samples of the undamaged materials as a baseline. One reason for this precaution is that several aircraft aluminum alloys contain enough copper to be detected by EDS systems. This analysis requires that the suspect conductors be of different materials. Aluminum structural members and copper-based conductors are a typical example.

15.4.4 Cup-and-Cone Fracture Surface

A cup-and-cone fracture is an indication of tensile overload that frequently occurs in wires during a mishap. The wire will elongate during the initial period of tensile stress. The elon-

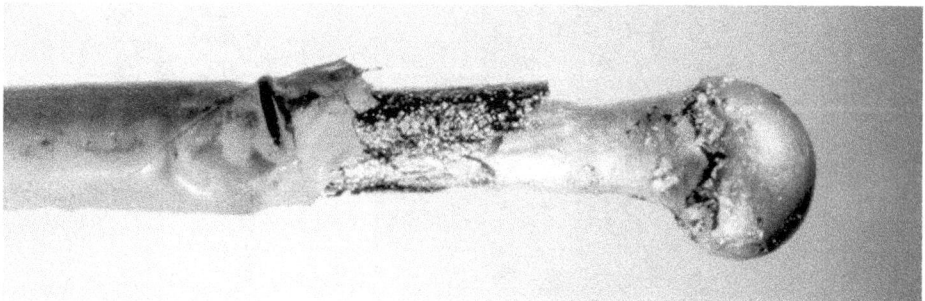

FIGURE 15.6 Wire sample exhibiting arcing damage from a 28 Vdc power system, 10×. Note balled end and degraded insulation.

gation is accompanied by reduction in a cross-sectional area called *necking*. Once necking occurs, voids form in the center and are edged by radially forming cracks. The spreading cracks form a shear lip near the surface at a 45° angle. The characteristic cup consists of a flat center region with *dimples* and an outer 45° shear lip; the opposite fracture surface forms the cone. Tensile failure may occur while the conductors are carrying load current. Fracture appearances under this condition have not been well characterized. Many types of force and effects occur that can cause mechanical failure in different conditions. The classical ductile failure, shown in Fig. 15.7, is just one type of mechanical failure that can occur. Other mechanical failure mechanisms include fatigue, stress corrosion cracking, and torsion. For stranded wire, breaks may occur initially in the outer strands on the tensile side. Therefore, it is important to examine the center strands. The approximate breaking force of some stranded conductors is listed in Table 15.6.[11]

Table 15.6 Approximate Breaking Forces for Selected Copper Conductors *(from Ref. 8)*

Gauge	Stranding	Approximate breaking force
8	133/29	480
10	137/26	295
12	19/25	172
14	19/27	108.5
16	19/29	68.7
18	19/30	53.8
20	19/32	34.4
22	19/34	21.3

15.4.5 Insulation Failure

Insulation failure is the catastrophic electrical breakdown of the insulation construction. It is typically associated with poor arc-tracking resistance and can be propagated by carbonization of the insulation materials. The carbonized insulation is conductive and provides a current path between the conductor and other powered electrical components. Insulation failure may occur over time or may happen very suddenly, as when initiated by mechanical abrasion of the insulation material.

Flashover is the sudden catastrophic failure and carbonization of the insulation material. Insulation failure is an electrical failure condition, but it may be initiated by mechanical abrasion or chafing. Arc-tracking resistance is a fundamental property related to insulation failure and flashover.

In the early stages of insulation breakdown, low-current discharges occur between the conductor and other energized components. These may occur at pinhole flaws in the insulation or at sites where chemical contamination has compromised the insulation. Tracks develop along the discharge path on the surface of the insulation. The tracks are generally more conductive than the virgin insulation. The material is said to have poor arc-tracking resistance if these tracks carbonize quickly into significant conducting paths.

Poor arc-tracking resistance has been reported to be a property of highly phenylated polymers such as aromatic epoxies, phenolics, and polymers with the para-phenylene group in the polymer chain. Aromatic polyimide films have this structure and are susceptible to flashover.[12]

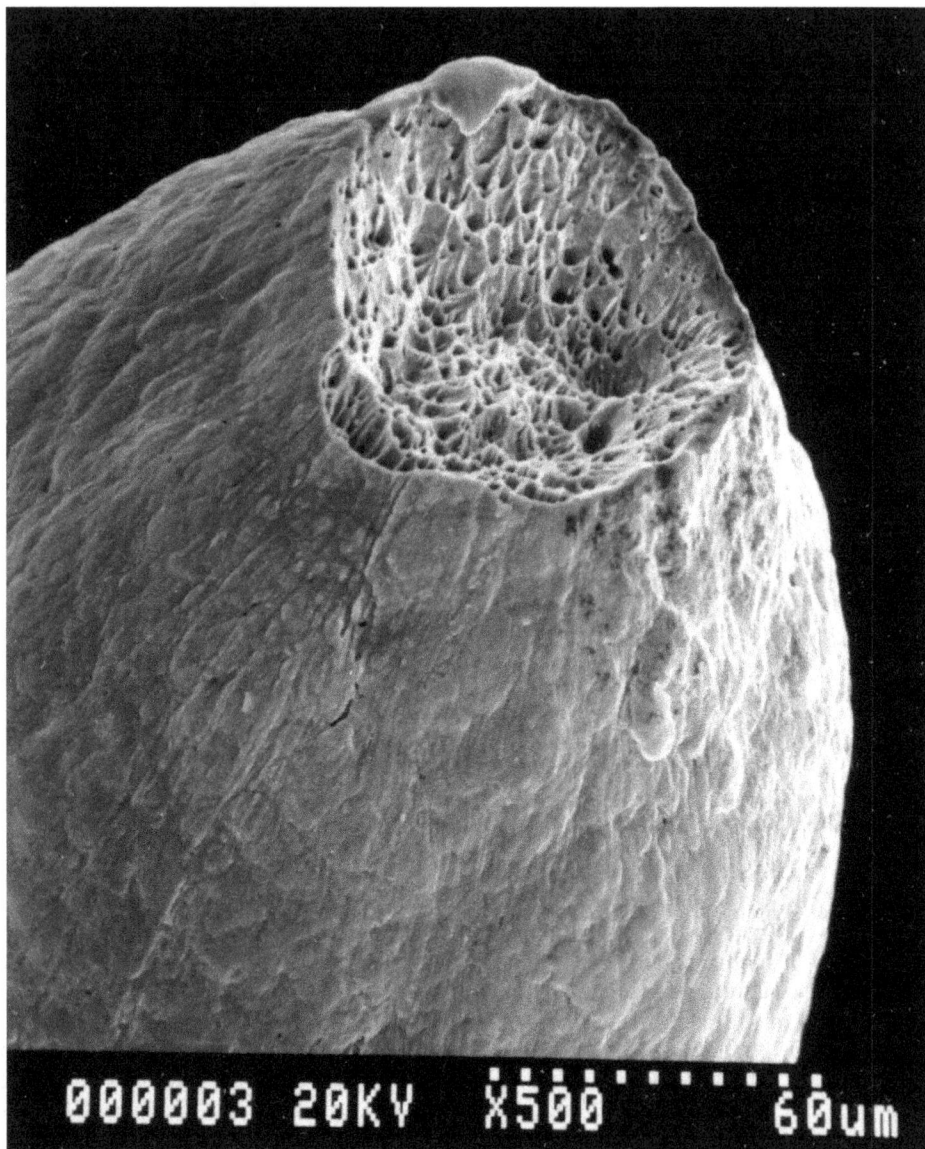

FIGURE 15.7 SEM micrograph showing the cup-and-cone fracture surface and elongation typically associated with mechanical failure of a copper conductor strand, 420×.

At bends in the wire, radial cracks may develop from manufacturing defects, installation damage, or chafing and then propagate completely through the insulation material to the conductor. The bare conductor is thereby exposed rather quickly, catastrophically accelerating the breakdown. Environmental factors can greatly accelerate insulation mechanical property degradation. Examples are sunlight (ultraviolet radiation), moisture, and various aircraft fluids.

Insulation failures in polyimide insulated wire are difficult to diagnose because of the rapid, catastrophic failures that can occur. The problem can be especially severe in moist environments. Initially, the conduction occurs by scintillating arcs in small, localized areas that eventually link together and create a carbonized path. Other materials, such as polyethylene and polytetrafluoroethylene, degrade into gaseous products rather than solid, carbonized products. These materials will only flash over when arc tracking materials come in contact with their surfaces. This can occur in wet environments. SEM analysis can be conducted on polymeric insulation in an attempt to identify metallic contaminants or deposits from preliminary arc tracking events. In specific cases, it is possible to reconstruct the failure scenario by building up a portion of a wire bundle in the laboratory. It is possible to confirm hypotheses about arc tracking as a failure mechanism in a specific failure by energizing the bundle and mechanically damaging the insulation.

15.4.6   Thermal Damage to Insulation

Wire insulations can be damaged easily at elevated temperatures. Since insulation has poor thermal conductivity, damage is usually restricted to portions of the insulation that are close to the heat source. Thermal damage is caused by exposure to any heat source and is not specific to one failure condition.

Specific characteristics will be discussed that may indicate which condition caused the damage. Polymeric materials undergo chemical breakdown at elevated temperatures. Although each material is different, some common changes in properties are:

1. Gradual discoloration, usually darkening
2. Loss of flexibility, melting, or cracking
3. Loss of electrical resistance properties

The maximum temperature to which the wire or insulation was exposed helps determine the conditions that were present during a failure event. The maximum temperature is difficult to judge from a visual examination of the insulation alone, since the thermal damage depends on the temperature and exposure time. Nonetheless, it is customary to get some indication of temperatures that might have been attained by noting the damage relative to the known properties of the insulation material. The location of the damage is sometimes more useful than the extent. A few notable cases are listed below.

Damage on Inside Surface of Insulation.   When the insulation is peeled away from the conductor, the inside surface appears discolored, while the outside surface does not. As in the case of electrical overcurrent, this suggests the conductor itself was the heat source.

Damage on Outside Surface of Insulation.   The insulation is damaged on the outside surface only. This suggests involvement of an external heat source.

Insulation Resistance Changes.   Under ideal conditions, typical resistance values for undamaged insulation exceed well over 100 MΩ when measured between the outside surface of the insulation and the conductor. After significant thermal damage, resistance values below 1 MΩ may be measured. This can be used to verify thermal damage. Polyimide is an example of a material that exhibits this behavior.

Glazing.   Some materials become shiny or glazed when exposed to a radiant heat source. Again, polyimide is an example. Melting of polymeric insulation is a common indicator. Insulation made of thermoplastic polymers will exhibit melting before significant chemical breakdown occurs. The melting temperatures for some materials of this type are listed below and provide a rough guide during examination:

1. Polyethylene   212 to 300° F (100 to 150° C)
2. PVC   221° F (105° C)
3. Nylon   392 to 482° F (200 to 250° C)
4. PTFE   621° F (327° C)
5. ETFE   512° F (267° C)

Thermosetting polymers will not melt at elevated temperatures. The first signs of damage may be cracking or discoloration. Typical maximum service temperatures for some thermosetting materials are listed below:

1. Silicones   500° F (260° C)
2. Polyimides   500 to 600° F (260 to 315° C)

Carbonization of aromatic polyimide occurs in the range of 650° C. Additional data and properties of polymers can be found in Ref. 13. Insulation damage can be determined by visual inspection and using a multimeter to determine insulation resistance. Thermal damage to Teflon® insulation at several temperatures for a five-minute duration is shown in Fig. 15.8.

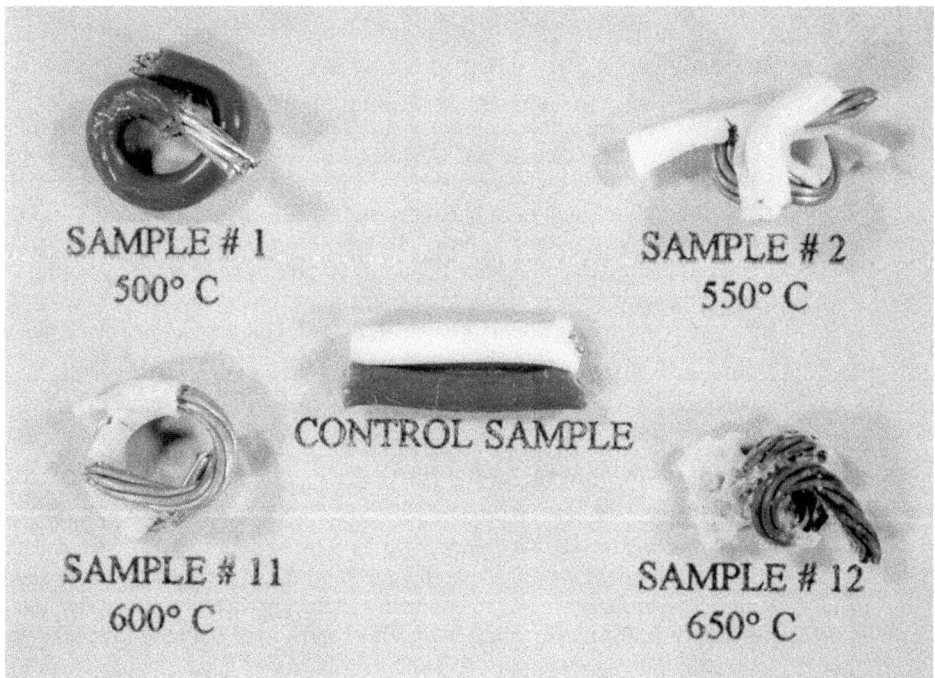

SAMPLE # 1
500° C

SAMPLE # 2
550° C

CONTROL SAMPLE

SAMPLE # 11
600° C

SAMPLE # 12
650° C

FIGURE 15.8 Samples of 20-gauge Teflon® insulation wire exposed to several temperatures for a 5-min duration (2.5×). Note that, under bending stresses, insulation appears to pull away from the conductor (550°), begins to melt (600°), and finally forms into a small globule (650°).

15.4.7 Conductor Discoloration

Discoloration of the conductor is caused by thermal exposure and is a result of oxidation and absorption of coating materials into the copper. Characteristic color changes have been noted on wires at slightly elevated temperatures, even before significant thermal damage occurs. Although the color changes themselves are not important, they can provide some useful clues to the investigator during initial visual examinations. In the case of aircraft mishaps, a variety of chemical compounds present in post-impact fire can cause conductor discoloration. An example of a discolored conductor is shown in Fig. 15.9. In this case, nickel plated copper conductors were exposed to excessive heating over an extended period of time. The top sample is shown as received (note shiny appearance), and the bottom sample was exposed to 1004° F (540° C) for 24 hr (note dull oxidized appearance). The loss of the nickel plating caused an increased resistance and corresponding temperature rise at a crimp connection. Over time, the heat generated by the high resistance and high current passing through the connection ignited the surrounding organic materials. Some known color changes are listed below.

1. *Nickel-plated wire.* Between 500° F and 519° F (260° C and 270° C), the nickel and copper diffuse into each other. The wire loses its shiny appearance and turns black.

2. *Silver-plated wire.* Between 392° F and 482° F (200° C and 250° C), the silver and copper diffuse into each other. The wire loses its shiny appearance and takes on a dull

FIGURE 15.9 Discolored conductor (2×). Note bright and shiny appearance of the Teflon®-insulated nickel-plated copper wire (top) compared to the darkened appearance of the wire after being exposed to 540° C for 24 hr (bottom).

brown color. Silver coated wires frequently take on a dull yellow color at relatively low temperatures. This may be caused by the formation of sulfides or other compounds of silver. Visual changes in a silver plated copper wire at specific temperatures are shown in Fig. 15.9. Brown, red, and black copper oxides form on the surface of the conductor after the plating has disappeared. The color may depend on the thickness of the oxide layer. Green copper-based compounds form when the base metal is exposed to high humidity, standing water, or other liquids.

3. *Tin-plated wire.* A similar mechanism as described in the above paragraph occurs at lower temperatures, in the range of 302° F to 338° F (150° C to 170° C).

15.5   SAMPLE PREPARATION: GRAIN STRUCTURE

While metallographic examination is covered elsewhere, the following is a procedure for obtaining high-quality cross sections when examining conductor grain structure. Sample preparation consists of the following four basic steps:

1. Sectioning
2. Mounting
3. Grinding
4. Polishing
5. Etching

The parameters and materials for automatic polishing equipment are provided in Tables 15.7 and 15.8.[14] Most of this information pertains to manual preparation as well. The principles of metallography and sample preparation methods are covered in Ref. 15.

Table 15.7   Rough Polishing Parameters

SiC grit size	Time (s)	Wheel speed (rpm)	Pressure (psi)
240	60	300	40
600	60	300	40

Table 15.8   Final Polishing Parameters

SiC grit size	Time (s)	Wheel speed (rpm)	Pressure (psi)
1 μm diamond compound/red felt/oil	180	250	30
Ferric oxide slurry/Lecloth	60	150	20

Sectioning should be done with an Al_2O_3 cutoff wheel.

Mounting can be done with bakelite, epoxide, or castable mounting media. In some cases, the epoxide or other-cold mounting materials should be used to avoid thermal or mechanical damage to the sample.

Grinding can be done with SiC grits as follows:

- 240 grit for 60 s on a 300-rpm wheel
- 600 grit for 60 s on a 300-rpm wheel

Polishing can be done with SiC grits as follows:

- 1 μm diamond compound for 180 s at 250 rpm wheel speed
- Ferric oxide slurry for 60 s

The ferric oxide final polish is recommended for microscopic examination in the as-polished condition; however, it leaves a passive film that is inert to etching. A few turns on an alumina polishing cloth will remove the passivity for etching purposes. Gamma alumina (0.05 μ) can be used as the final polishing medium. The addition of a few drops of a solution composed of 50 ml NH_4OH and 5 ml H_2O_2 will facilitate polishing.

Etching can be done with the following solution: 50 ml ammonium hydroxide (NA_4OH), 5 ml hydrogen peroxide (30 percent) H_2O. If this etchant is too fast, add 50 ml H_2O. To differentiate between cuprous oxide and copper sulfide inclusions, examine in the as-polished condition under polarized light. Cuprous oxide will be red, whereas copper sulfide will remain dark. Both are medium gray under bright-field illumination.

15.6   CASE STUDIES

Wiring can be affected by exposure to electrical, mechanical, chemical, or thermal events. Many times, due to secondary failure events, it is difficult to determine how a wire initially failed. The following case studies discuss specific failure mechanisms and their effects on system performance. They illustrate how primary failure mechanisms can be identified when significant secondary damage is present.

15.6.1   Red Plague

The corrosion phenomenon known as *red plague* occurs in silver-plated copper wire, and similar phenomena can occur in tin-plated copper wire when the copper comes into contact with water and oxygen.[16] The electrolytic cell that is formed produces cuprous oxide (Cu_2O), which is red, and can produce cupric oxide (CuO), which is black. Carbonates and halides may also form if their constituents are present. In the case of tin plating, a green corrosion product is formed.

Corrosion occurs when there is a break in the plating or if there is porosity in the plating. Copper is anodic to the silver plating and is oxidized when there is an electrolyte and moisture between the silver plating and copper substrate. The oxidation of the copper results in loss of metallic copper and flaking of the silver plating. A water bath is typically used during the manufacture of extruded insulation and can provide the source of the moisture and electrolyte. Exposure to solder fluxes, cleaning processes, and the "use environment" can also produce red plague under certain conditions. An example of red plague on a silver plated copper wire is shown in Fig. 15.10. The cross section shows the conductor damage (dark area). Red plague can cause poor solderability and, in serious cases, destroy the copper conductor. In the case study example, the silver plating was damaged

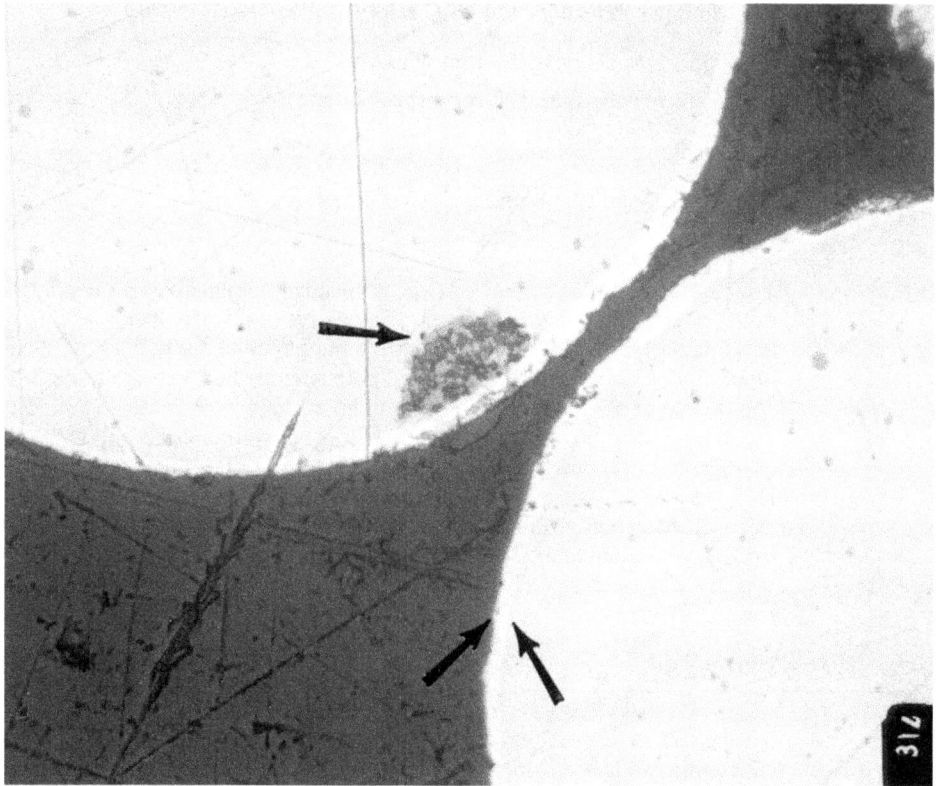

FIGURE 15.10   Cross section of two wire strands showing corrosion damage attributed to red plague (680×). The copper strands are plated with silver (two lower center arrows). Note dark area (arrow near center) representing corrosion damage.

when the insulation was applied to the conductor. Moisture was introduced from a water bath used in the insulation extrusion process.

15.6.2   Insulation Cracking

Insulation cracking is a common failure seen in wiring. Exposure of the conductor can cause system malfunctions or fires initiated by electrical arcing. Wire insulations are typically under a tensile stress as a result of stretching during a tape wrap operation or being drawn down during extrusion. The wiring is then typically subjected to bending when installed. The continuous application of stress, chemical attack, or extreme temperatures can accelerate insulation cracking. Radial and longitudinal cracking can expose conductors with can lead to electrical arcing under worst-case conditions. Insulation cracking, shown in Fig. 15.11, was due to improper manufacturing during the insulation extrusion process. This failure did not become evident until after several years in the field, where the insulation was under a bending stress. The insulation of interest is a mineral-filled fluorocarbon manufactured to MIL-W-22759/6. A tensile value of 8400 psi and elongation of 58 percent were obtained for samples removed from uncracked areas of the failed wire

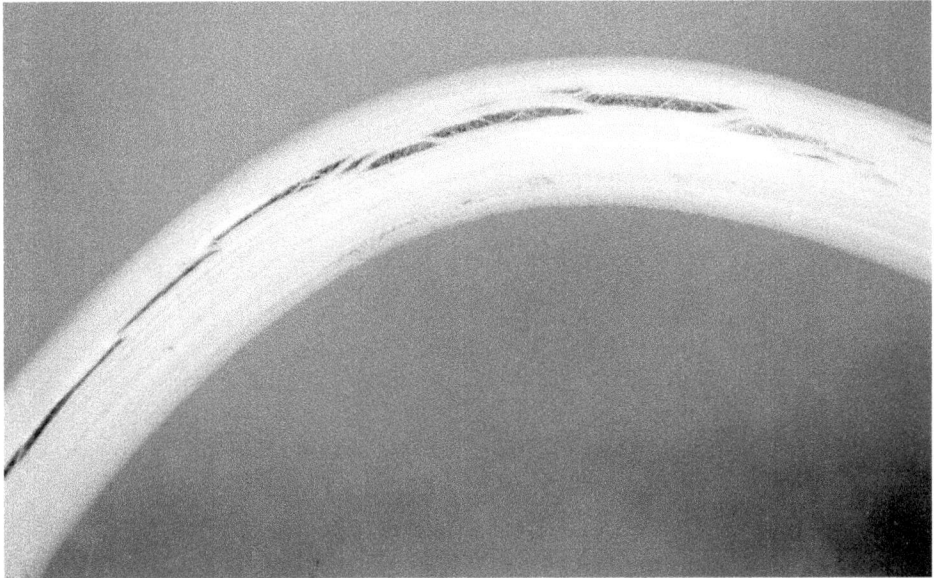

FIGURE 15.11   Cracking found in Teflon®-insulated wiring (1.7×). Note that cracks are longitudinal and are associated with bending stresses.

insulation. Typical elongation values should be in the range 150 percent for this type of insulation.

15.6.3 Melted Conductors

The temperature of a fire usually does not exceed 1500° F (815° C). Since copper melts at 2000° F, melted copper is frequently considered an indication of electric arcing. This is especially true if the melting appears in a small section of the wire—for example, one inch or less. However, there are other conditions that result in a similar appearance but do not involve arcing.

Copper can form alloys with other metals and result in materials with much lower melting points than copper alone. Two common alloying elements are aluminum and zinc. Consider the case of an aluminum panel and copper wiring, both involved in a fire. By the time the temperature is high enough to melt aluminum, the insulation on the copper conductor has decomposed. This exposes the copper, and the aluminum may melt directly onto the exposed conductor.

A simplified version of the copper-aluminum alloy phase diagram in Fig. 15.12[16] shows states of the alloys as a function of aluminum concentration. The upper line of the diagram, called the *liquidus,* depicts the alloy compositions that will be a liquid at or above the temperature on the line. Two features of the liquidus are important. First, the alloy melting point can be less than that of either aluminum or copper. Second, only 35 percent of aluminum by weight is required to reduce the alloy melting point to 1500° F (815° C).

If the temperature is 1292° F (700° C), the aluminum will be melted and in direct contact with the copper conductor. The molten aluminum will diffuse into the copper at the interface and create an alloy that is rich in aluminum. This small alloy region will become

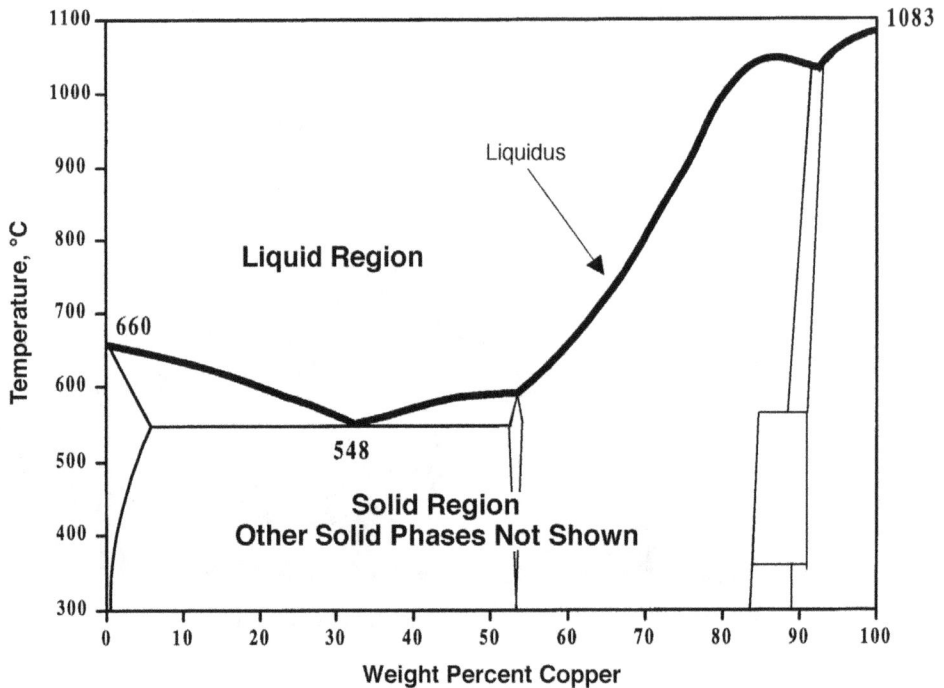

FIGURE 15.12   Copper-aluminum phase diagram.

a liquid, since it is above the liquidus on the far right of the phase diagram. Aluminum will continue to diffuse into the copper, and the melted alloy region will expand. This process will continue until the diffusion results in a region whose composition and temperature describe a point below the liquidus. This composition is still a solid at the temperature of 1292° F (700° C).

The diffusion produces a highly localized region where the aluminum contacts the copper and the result is similar in appearance to arcing. The two mechanisms can be difficult to distinguish since arcing can produce metal transfer and erosion of the conductors. The alloying process generally requires more time to develop and produces softer features than arcing. The depth of the diffusion is another indicator, usually being deeper from alloying than arcing. In summary, the appearance of a localized melt region on a copper wire may be deceiving. If the copper wire is adjacent to aluminum, what appears to be arcing evidence may actually be molten aluminum that has dripped on and alloyed with a hot copper wire.

15.6.4 Arc Tracking Failures

Wire insulations are susceptible to both wet and dry arc (flashover) tracking. The failure mechanism was described in an earlier section. An example of a dry arc flashover failure in polyimide insulation is shown in Figs. 15.13 and 15.14.

Note the carbonized insulation, highly localized damage, and melted conductor ends. Insulation resistance between the damaged wires is typically in the 1 kΩ to 1 MΩ range.

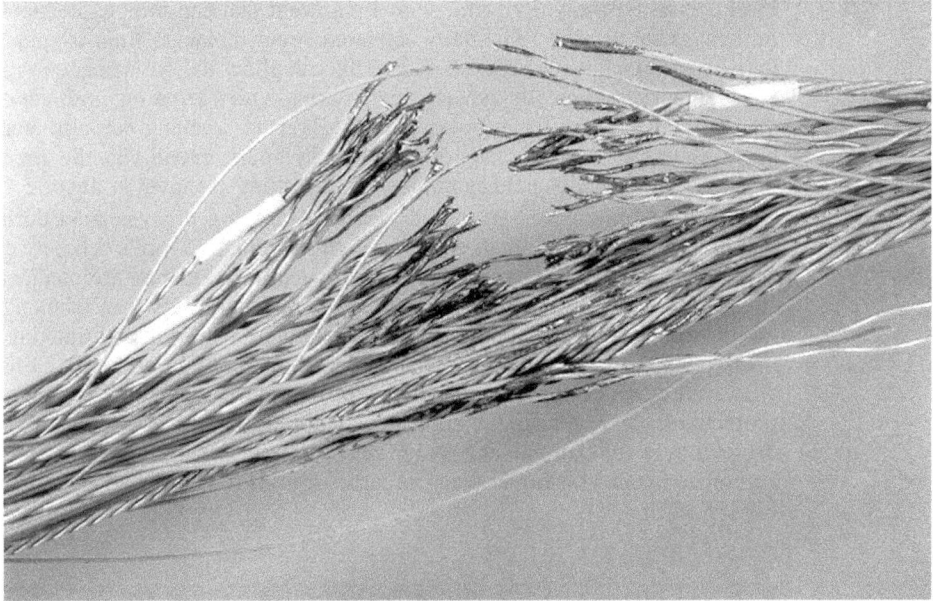

FIGURE 15.13   The arc track damage to this polyimide insulated wire bundle (0.5×) was initiated by chafing. Shorting of the 115 Vac three-phase, 400 Hz power provided the thermal energy required to carbonize the insulation.

FIGURE 15.14   Close-up of the damaged wiring of Fig. 15.13 showing carbonized insulation and melted conductor ends (4×).

Flashover is typically a violent, short-time event that can initiate serious fires. The intermittent arcing can keep thermally activated circuit breakers from tripping and the arcing may be sustained for several seconds. This can allow the arc damage to rapidly spread to adjacent wiring. Typically, power supply wiring with alternating or direct current provides the energy necessary for a flashover event. Wet arc tracking can also lead to a flashover event. In this case, arcing may occur for many hours; eventually, the circuit will open, or flashover will occur. A field example of wet tracking is shown in Figs. 15.15 and 15.16.

The field failure was attributed to a hot-stamping process used during the original manufacturing of the polyalkene insulation. Conductive fluids fell between the exposed conductors, setting up an arcing path that eventually opened the conductors. While hot stamping is an effective and common method of marking wires, when used it should be monitored carefully. A laboratory experiment demonstrating wet tracking in a polyimide insulation is shown in Figs. 15.17 and 15.18. In Fig. 15.17, a polyimide insulated seven-wire bundle has been predamaged by exposing two conductors. A 115 Vac power source protected by a 7.5 A circuit breaker was applied, and a 5 percent NaCl solution was dripped over the damage site. After seven minutes of arcing, insulation surfaces were carbonized, and a flashover event occurred that damaged the adjacent wires as shown in Fig. 15.18.

FIGURE 15.15 Polyalkene wire insulation with improperly controlled hot-stamp printing that exposed the conductor (6×). Conductive fluids between the 115 Vac three-phase wires led to arcing.

FIGURE 15.16   Close-up of the arcing damage from Fig. 15.15 showing conductor and insulation damage (22×).

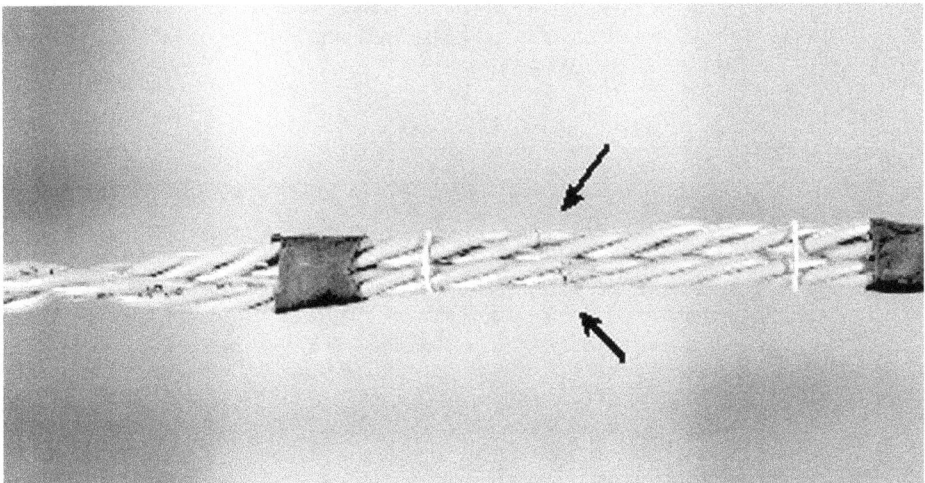

FIGURE 15.17   The polyimide insulated wet arc tracking bundle (1×) was damaged before testing (see arrows). A 115 Vac power source was connected between the damaged and undamaged wires and a 5 percent NaCl solution was dripped onto the damaged area.

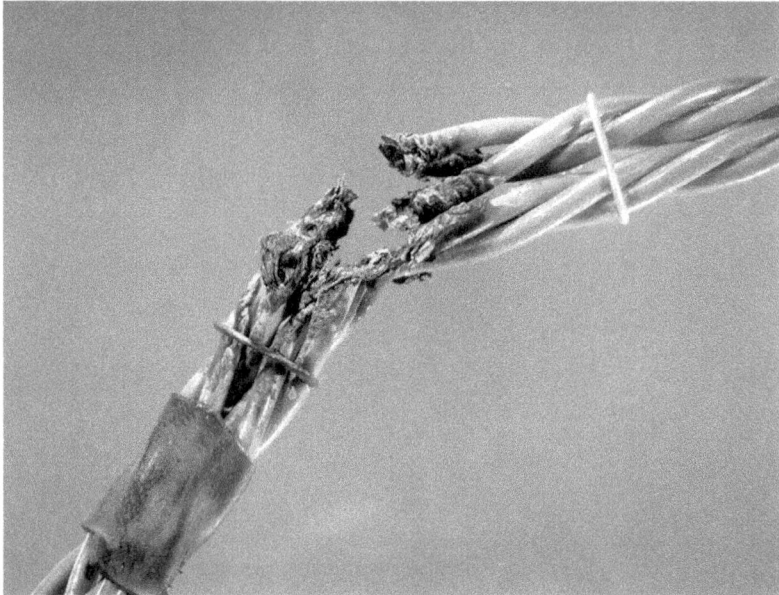

FIGURE 15.18 After seven minutes of active arcing, the wire bundle was severely damaged by a flashover event. Note the collateral damage that occurred to the initially undamaged wires.

15.7 ACKNOWLEDGMENTS

The authors would like to thank Dr. David Berkebile of CommScope Inc. for his technical review of this manuscript, and Ms. Marianne Ramsey of WL/MLSA for manuscript preparation and editorial review.

15.8 REFERENCES

1. Galler, D., and Slenski G. Causes of Aircraft Electrical Failures, *Proc. National Aerospace and Electronics Conference,* 1991.

2. Mehrotra, Yogesh. *Life Prediction of Aging Aircraft Wiring Systems, Technical Report WL-TR-96-4027,* Materials Directorate, Wright Laboratory, Wright-Patterson AFB, OH 45433-7734, 1995.

3. Soloman, R., Woodford, L., and Domalewski, S. *New Insulation Constructions For Aerospace Wiring Applications, Vol. 1, Technical Report WL-TR-91-4066,* Materials Directorate, Wright-Patterson Air Force Base, OH, 1991.

4. *Metals Handbook,* 9th ed., Vol. 2, *Properties and Selection: Non-ferrous Alloys and Pure Metals.* American Society for Metals, 1979.

5. Harper, Charles A., ed. *Electronic Packaging and Interconnection Handbook.* New York: McGraw-Hill, 1991.

6. Kapton®, Summary of Properties. Wilmington, DE: DuPont Company, Electronics Department.

7. Clark, D.S. and W.R. Varney. *Physical Metallurgy for Engineers.* D. Van Nostrand, New York.

8. Levinson, D.W. *Copper Metallurgy as a Diagnostic Tool for Analysis of the Origin of Building Fires.* National Fire Protection Association, 1977.

9. Beland, B. *Examination of Electrical Conductors Following a Fire.* National Fire Protection Association, 1980.

10. Singh, R.P. Scanning Electron Microscopy of Burnt Electric Wires, *Scanning Microscopy* 1(4) (December 1987).

11. *Tensolite Product & Technical Handbook.* Buchanan, NY: Tensolite Division of Carlisle Corp., 1974.

12. Campbell, F.J. *Flashover Failure from Wet-Wire Arcing and Tracking.* Washington, DC: Naval Research Laboratory, December 17, 1984. NRL Memorandum Report 5508.

13. *Engineered Materials Handbook,* Vol. 2, *Engineering Plastics.* ASM International, 1988.

14. Dillinger, L. *Suggested procedures for Grinding and Polishing Various Alloy Systems using Automatic Polishers.* St. Joseph, MI: LECO Corporation.

15. *Metallographic Principles and Procedures.* St. Joseph, MI: LECO Corporation.

16. *ASM Handbook,* Vol. 3, *Alloy Phase Diagrams.* ASM International, 1992.

CHAPTER 16
SWITCHES AND RELAYS

Perry Martin
National Technical Systems (NTS)

Richard Blanchard
E^{x}ponent

William Denson
IIT Research Institute (IITRI)

Duncan Glover
E^{x}ponent

Alexander Kusko
E^{x}ponent

Donald Galler
Department of Materials Science and Engineering
Massachusetts Institute of Technology

16.1 INTRODUCTION

Switches and relays electrically transfer power or functionality from one circuit to another, resulting in the making or breaking of an electrical circuit. Except for the special class of switches called circuit breakers, the actuation of switches is usually manually applied which differentiates them from relays. This chapter outlines the various types, construction, and operation of switches and relays. This chapter also provides general guidance as to the utilization of these devices in electrical circuits. General information about prevalent failure modes is presented for each device type, and an example of specific analysis procedures is presented for aircraft circuit breakers. Because solid-state devices are covered in Chap. 19, this chapter concentrates on presenting failures specific to contact technology.

Some of the military specifications relevant to switches, circuit breakers, and relays are listed in Table 16.1.[1]

16.2 SWITCHES[2]

The transfer function performed by switches is attained in two different ways:

1. Mechanically, by contact mating
2. Electronically, using solid-state or inductive means

Solid-state switching devices do not use mechanical contacts to perform their function. These devices transfer power by way of transistor-like saturation of a semiconductor

Table 16.1 Relevant Military Specifications

Relays	Switches	
MIL-R-27745	MIL-S-1743	MS-25068
MIL-R-28750	MIL-S-22885	MS-25098
MIL-R-39016	MIL-S-24236	MS-25100
MIL-R-5757	MIL-S-24263	MS-25201
MIL-R-6016	MIL-S-24523	MS-25253
MIL-R-83726	MIL-S-24524	MS-25306
MS-24143	MIL-S-24525	MS-25307
MS-24166	MIL-S-3950	MS-25308
MS-24168	MIL-S-55433	MS-27406
MS-24192	MIL-S-83731	MS-27716
MS-24376	MIL-S-8805	MS-27719
MS-24568	MIL-S-8834	MS-27753
MS-25269	MS-16106	MS-27903
MS-25271	MS-21350	MS-2885
MS-25323	MS-21352	MS-35058
MS-25327	MS-21354	MS-35059
MS-27222	MS-24524	MS-3508
MS-27400	MS-24525	MS-35258
MS-27401	MS-24547	MS-75038
MS-27418	MS-24655	MS-90311
MS-27997	MS-24656	

Rotary Switches	Circuit Breakers	
MIL-S-3786	MIL-C-39019	MS-24510
	MIL-C-55629	MS-25244

device and are primarily utilized in low-power applications where clean transfer and isolation properties are required. These characteristics also make solid-state switching devices popular in microwave electronic circuits.

Mechanical switches usually mate contacts through three primary actuation constructions:

1. Snap action
2. Wiping
3. Cross bar

A typical switch assembly drawing is shown in Fig. 16.1.[3]

Each type of switch construction is utilized on the basis of the particular application for the switch. For example, lamp or inductive loads will require snap-action configurations, since inductive loads will produce arcing, which accelerates contact degradation. Applications that encounter humid conditions may require the cross-bar arrangement to exclude any corrosion buildup problem that could create a high-resistance condition. The majority

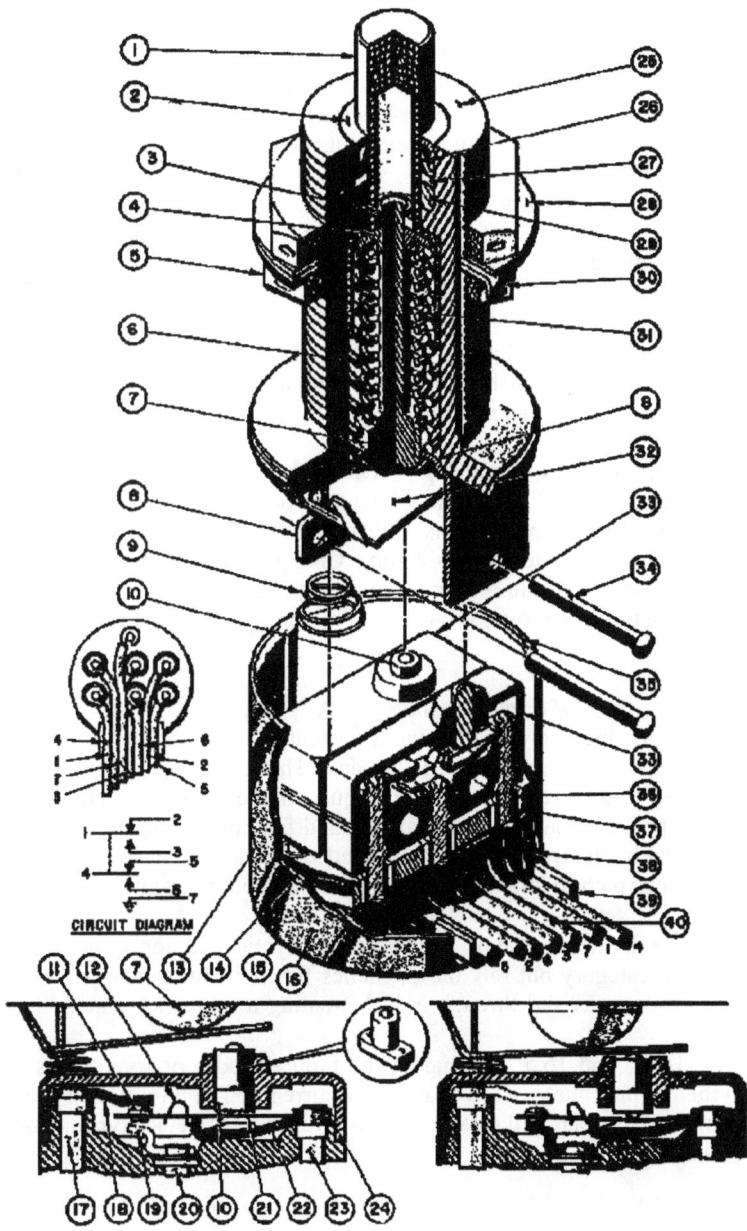

1. plunger actuator
2. guard
3. plunger washer
4. plunger spring retainer
5. nut—special
6. spring—compression
7. plunger
8. bracket
9. spring—compression
10. plunger
11. bracket
12. spring—compression
13. housing
14. header assembly
15. guard
16. paint
17. normally closed stationary contact terminal
18. normally closed stationary contact
19. normally open stationary contact
20. normally open stationary contact terminal
21. movable contact spring
22. anchor—stationary
23. common terminal
24. washer—plain
25. bushing—actuator
26. seal—ring
27. washer—plain
28. washer—lock
29. seal—ring
30. washer—special
31. spring—actuator
32. lever—switch actuation
33. case—switch
34. rivet—oval head
35. inert gas arc weld
36. glass seal
37. header tubes for terminal solder joint
38. lead wire embedment
39. insulation electrical
40. contact and ground wires

Figure 16.1 Typical switch assembly drawing.

of switch failure modes and mechanisms relate to the contacts. This topic will be covered in detail in a later section.

Some specific switch types are as follows:

Centrifugal Switches.   A centrifugal switch is actuated by rotational velocity. The simplest type consists of a speed-sensing unit that mounts directly on a rotating shaft, and a stationary contact switch assembly. The basic control element is a conical-spring steel disc that has centrifugal weights fastened to the outer edge of its circular base.

Capacitive Touch Switches.   A capacitive touch switch consists of two conductive layers on opposite sides of an insulating material such as glass or a printed circuit board. The conductive layers create a capacitance that decreases when a layer is touched. Interface circuitry converts the capacitance change into a usable switching action.

Membrane Switches.   Membrane switches are devices in which conductive leads on the underside of a flexible membrane are pushed through a hole in a spacer to make contact with conductive leads on a base. Optional overlays are provided for user interface.

Circuit Breakers with Hydraulic-Magnetic Trip Mechanism.   The hydraulic-magnetic construction circuit breaker consists of a solenoid with a dash pot time-delay element (i.e., iron core). The dash pot time-delay tube contains a silicone fluid and a return core spring. Operation depends on changes in the magnetic flux. Changes in flux are caused by changes in coil current, which in turn cause changes in the position of the iron core within the coil. The speed at which the core moves is controlled by the damping effect of the silicone liquid in the tube.

Ground-Fault Interrupters (part of circuit breakers).   A ground-fault interrupter is composed of many elements, including a differential current transformer, opamps, synchronous demodulator, resistors, capacitors, and diodes. The ground-fault interrupter removes power when it senses a current imbalance (not just an overload) between the hot and neutral conductors supplying operating power. A ground fault occurs when a current-carrying part of a circuit inadvertently contacts any grounded conducting material, regardless of whether the resistance path to ground is high (e.g., human body) or low.

Slide Switches.   Inductive switches, mainly used for high cyclic rate applications, are classified in the electronic category but rely on magnetics for their functionality. As the switch is actuated, an iron core is slid through a coil, creating a frequency change that results in a signal transfer.

Tables 16.2,[4] 16.3,[5] 16.4,[6] and 16.5[7] summarize the failure modes of some types of switches, along with their approximate relative rates of occurrence. The information listed in the tables is the best available and will clearly vary as a function of device type, manufacturer, application, and so on.

16.3   RELAYS[8]

The two main categories of relays are *electromechanical* relays and *solid-state* relays (SSRs). Electromechanical relays are electromagnetically operated devices obtainable in

Table 16.2 General Failure Modes for Switches

Failure mode	Percent occurrence	Failure mode	Percent occurrence
Open	15	Other	18
Shorted	8	Unstable	10
Intermittent	19	Drift	9
Out of spec.	14	Leaking	7

Table 16.3 Float Switch Failure Modes

Failure mode	Percent occurrence	Failure mode	Percent occurrence
Cracked/fractured	8	Out of adjustment	15
False response	23	Seized	8
Leaking	8	Stuck closed	8
No operation	23	Stuck open	8

Table 16.4 Reed Switch Failure Modes

Failure mode	Percent occurrence	Failure mode	Percent occurrence
Intermittent	10	Out of spec.	30
No operation	30	False response	20
Open	10		

Table 16.5 Toggle Switch Failure Modes

Failure mode	Percent occurrence	Failure mode	Percent occurrence
Open	24	Intermittent	25
Short	16	Mechanical	35

many different styles, each having unique mechanical construction and electrical characteristics. Solid-state relays control load currents through solid-state switches such as TRIACS, SCRs, or power transistors. Unlike electromechanical relays, solid-state relays have no moving parts and are often used in applications where rapid on–off cycling would lead to wear out of conventional electromechanical relays.

Failure modes in solid-state relays are primarily associated with the TRIAC or SSR switching characteristics. A common SSR failure takes the form of SSR false turn-on with no turn-on signal. For example, turn-on can occur if operating temperatures exceed the thyristor rating or transients from the switched load or ac line momentarily exceed the thyristor breakover voltage. Other failure modes/mechanisms include thermomechanical fatigue caused by cyclic temperature surges, chemical reactions such as channeling, and physical changes such as crystallization of materials. When an SSR fails catastrophically, it most often becomes permanently "on," preventing the relay from being controlled.

The major failure modes/mechanisms for electromechanical relays consist of contact sticking, contact material transfer, contact welding, high contact resistance, mechanical failure, and coil opening or shorting. For some applications, contact sticking and high contact pressure may be intermittent and difficult to diagnose. Coil failures are usually attributed to excessive voltage, electrolysis, or other chemical reactions or harsh environments. Excessive temperature, especially if prolonged, may deteriorate the insulation, causing the coil to fail. Many electromechanical relay failure modes are fairly easily discovered by visual inspection (see Fig. 16.2).[9]

The physical design of an electromechanical relay can be described by the contact combination or form and the construction type. The current-carrying parts of a relay, used for making and breaking the electrical circuits, are available in various combinations of contact forms. The choice of contact material and the shape of contacts affect relay failure rate. Contact reliability concerns for relays are very similar to those for switches, so the contact reliability discussion presented in the section that follows is applicable.

Relay failure rate is significantly influenced by application variables including ambient temperature, shock and vibration, contact material, shape of contacts, the amount of contact force, and the wiping or sliding of contacts. As an example, the failure modes and mechanisms for armature relays are summarized in Table 16.6.

Table 16.6   Armature Relay Failure Mechanisms

Failure mechanism	Accelerating factors	Distribution (%)
Contact contamination	Moisture, temperature	18
Poor contact alignment	Actuations, vibration	8
Contact corrosion	Actuations, voltage, humidity	6.5
Opened coil	Current, vibration	8.5
Unstable coil	Humidity, voltage, temperature	15
Contact welding	Current	7
Spring fatigue	Actuations	9
Contact corrosion	Humidity, temperature	19
Binding, jamming	Actuations, contaminants	9

The selection of a relay for a particular application is based on user requirements including:

- Class of application (e.g., military, commercial, industrial, machine tool control)
- Environmental requirements (e.g., high temperature, corrosion, shock, sand)
- Enclosure (e.g., open, sealed)
- Coil specification (e.g., resistance or impedance, voltage or current, temperature rise)
- Contact specification (e.g., form, current, voltage, ac/dc, frequency)
- Mechanical life expectancy
- Electrical life expectancy
- Electrical characteristic specifications (e.g., contact resistance, insulation resistance, dielectric strength)
- Operational specifications

Figure 16.2 Visual inspection found the travel stop set screw on one side of this power contactor to be "backed out." This eventually caused uneven wear and chatter during operation.

A number of test methods have been standardized to ensure reliable performance of relays. Several of the more important tests are listed in Table 16.7. External and X-ray views of the contactor of Fig. 16.2 are shown in Fig. 16.3.

Table 16.7   Tests Performed to Ensure Relay Reliability

Test type	Description/purpose
Contact resistance	Determines the resistance offered by electrically contacting surfaces to a flow of current. For practical reasons, leads and terminal resistance within the unit or test may be included in the measurement. In many applications, contact resistance is required to be low and stable to avoid voltage drop across the contacts, which adversely affects the accuracy of circuit conditions, and to prevent overheating at high currents.
Insulation resistance test	Measures the resistance between mutually insulated members of a relay. Values of insulation resistance can be important in the design of high-impedance circuits. Low insulation resistance may permit excessive leakage current that can affect isolation of independent circuits. Excessive leakage current can also be indicative of the presence of corrosive impurities that can cause deterioration by electrolysis or heating.
Dielectric withstand voltage test	Detects flaws in materials, design, or construction of the unit that might result in failure to withstand the specified test potential. It is a static test, conducted without contact switching and in the absence of contact arcing.
Winding resistance test	Measures the dc resistance of a relay coil winding.
Winding inductance test	Measures the inductance of the coil winding. In relays, coil inductance is a function of the number of turns of wire and the geometry and reluctance of the magnetic circuit.
Winding impedance test	Measures the impedance of relay windings, designed for use on alternating current.
Contact bounce test	Measures the duration of the intermittent opening and closing of contacts caused by contact bounce.
Contact chatter test	Monitors contact chatter when relays are subjected to vibration, shock, and acceleration tests.
Functioning time test	Measures the operate and release times of relays.
Leak test for hermetically sealed relays	Determines the seal effectiveness of a hermetically sealed relay, which is either evacuated or contains air or gas. A defect in any portion of the surface area of a seal part can permit the entry of damaging contaminants that could reduce effective relay life.

16.4   CIRCUIT BREAKERS

The construction and operation of a typical thermal circuit breaker is illustrated through the specific case of a particular aircraft circuit breaker. The general requirements for aircraft circuit breakers are found in military specification MIL-C-5809.[10] The specification covers thermal (Type I) and magnetic (Type II) circuit breakers. The circuit breakers are rated from 28 Vdc to 208 Vac and 0.5 to 100 A. Requirements for the construction, performance, and testing of circuit breakers are also included. The construction of one type of aircraft circuit breaker is shown in Fig. 16.4.[11] It is shown in the open circuit position.

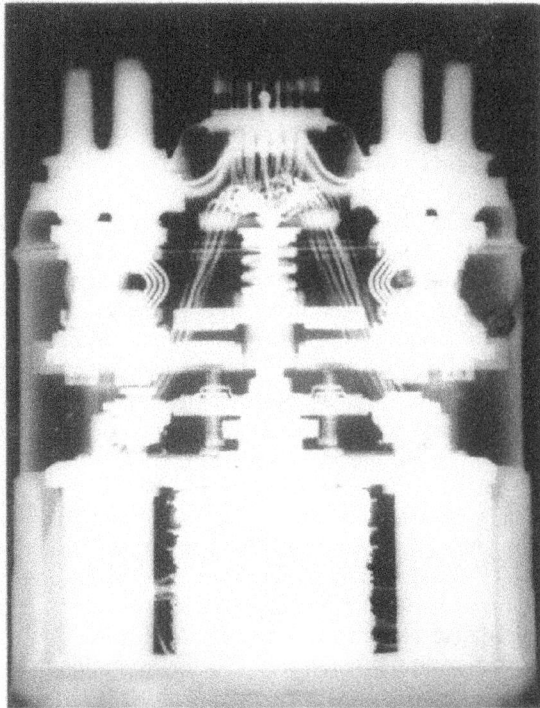

Figure 16.3 External and X-ray views of the contactor shown in Fig. 16.2.

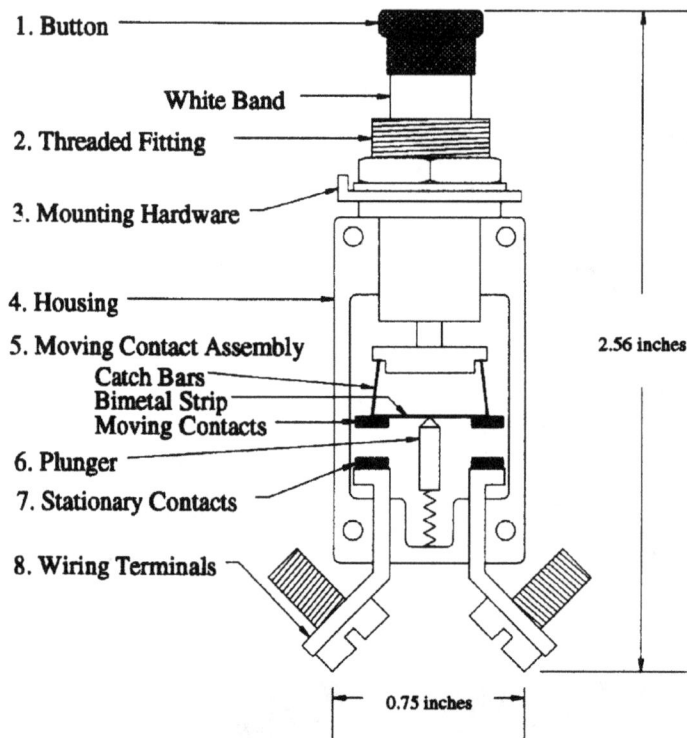

Figure 16.4   Typical aircraft circuit breaker construction.

The button allows the circuit breaker to be operated manually. The position of the button also indicates the state of the circuit breaker. Pushing the button closes the circuit and latches the button mechanism. The button can be pulled out to open the circuit. The button shaft has a white band that is visible only when the button is in the extended (open) position. This white band is an additional indicator that the circuit breaker is open. If an overcurrent trip condition occurs when the breaker is in the closed position, the electrical contacts open, and the button pops out. The circuit breaker must be manually reset after tripping. The circuit breaker is reset by pushing in the button until it latches.

The moving contact assembly is an essential element in the overcurrent trip mechanism of the circuit breaker. The assembly contains the moving contacts that mate with the stationary contacts when the circuit breaker is closed. Load current flows between the moving contacts through a bimetal strip in the center of the moving contact assembly. Two metal catch bars extend from the bimetal strip and rest against the button assembly. As the temperature of the bimetal strip rises, the metal component with a higher coefficient of thermal expansion (CTE) expands at a greater rate than the lower-CTE component. This differential expansion rate causes the strip to bend, separating the ends of the catch bars. When the catch bars separate a specific distance due to a overcurrent load, a spring-loaded mechanism in the button assembly is released. The entire button assembly moves toward the front of the circuit breaker and away from the stationary contacts as the button pops out. Many circuit breakers have an adjustable screw for calibrating the overload trip current. A spring-loaded plunger forces the sliding contact frame away from the stationary contacts and opens the circuit. The plunger pushes the sliding contact frame against a stop near the button assembly. The circuit breaker cannot be reclosed until the bimetal strip and catch bars return to their normal condition.

Aircraft circuit breaker operating mechanisms are reasonably reliable and fail-safe in their design. In addition, the condition of the operating mechanism is relatively unaffected by the electrical parameters of the load circuit. In contrast, the contacts can show signs of wear, arcing, and other features that relate to the number, type, and current level of electrical operations. The contact surfaces, therefore, contain information about the condition of the electrical equipment connected to the circuit breaker. For this reason, contacts are the focus of the following section.

16.5 CONTACTS

Most electrical contacts used in high-reliability applications are made either of pure silver or silver alloys. Alloy materials include cadmium oxide (CdO), tin oxide (SnO), and tungsten (W). The most common alloy is silver-cadmium oxide. This alloy is usually 10 to 15 percent CdO by weight and made by sintering powdered mixtures of the components. The alloy can be made using either cadmium oxide or cadmium. Preoxidized contacts are made with CdO. Postoxidized contacts are made using cadmium and then oxidizing them after the sintering process. Postoxidized contacts may contain some residual unoxidized cadmium. The addition of cadmium oxide to silver reduces the susceptibility of the contacts to welding, improves erosion characteristics, and increases current handling capability.[12]

The contacts are made separately then brazed or soldered onto contact assemblies. Common base materials for the contact assemblies are copper, brass, and phosphor-bronze.

The predominant accelerating stresses in mechanical contacts are temperature and load during switching. Temperature is generated by the Joule heating that occurs from the resistance associated with the mating of contacts. The resulting effects of this increased temperature include contact material fatigue, oxidation, and contact contamination. All of the above conditions cause increased contact resistance, which results in even higher temperature increases.

Typical failure mechanisms associated with contacts are contact pitting due to arcing on break, contact material transfer, contact weldment on make (resulting from excessive resistance and heat generation), and mechanical failures resulting from the construction or packaging of the particular device. Application and design selection factors affecting the failure rate of contacts include:[13]

1. *Peak make and break current and voltage.* For source voltages less than 14 V, arcing typically does not cause serious problems, but for source voltage greater than 14 V, arcing can occur, causing contact pitting.

2. *Contact materials.*

3. *Actuation frequency.* Contacts wear when exposed to more frequent actuation.

4. *Surrounding atmosphere.* Altitude can be a factor, since the dielectric strength of air is less at higher altitudes, causing arcing to occur for longer duration.

5. *Type of load—ac, dc, resistive, inductive or capacitive.*

6. *Chatter or bounce at make.*

7. *Arc suppressive methods used.*

8. *Contact closing force.* The higher the contact pressure, the greater the contact area due to the yielding of contact asperities (microscopic peaks and valleys). It also can degrade the contact faster due to wear.

9. *Contact opening force.*

10. *Wiping or rotary action.*

11. *Snap action on break.*

12. *Maximum contact separation.*

13. *Contact backing material.*

14. *Method of contact assembly.*

15. *Mechanical wear.*

16. *Heat generated and its dissipation during operation.*

17. *Ambient temperature.*

18. *Dimensions of contacts and backing.*

19. *Foreign material—dust, oil, etc.*

Table 16.8[14] shows various physical properties for different contact materials.

Table 16.8   Contact Material Properties

Contact material	Melting point (°C)	Resistivity (µΩ-cm)	Temp. coefficient of resistivity (per °C)	Thermal conductivity (cal-cm sec-°C-cm²)	Oxidation resistance	Arcing effects
Gold	1063	2.42	0.0034	0.71	excellent	Pits and transfer at high current and voltage
Molybdenum	2625	5.7	0.0033	0.35	good	Pits and transfer at high current and voltage
Palladium	1552	11	0.0038	0.11	fair	Resists arc erosion
Platinum	1773	10.60	0.0030	1.17	very good	Resists arc erosion
Silver	960	1.63	0.0038	1.01	excellent	Pits and transfer at high current and voltage
Tungsten	3410	5.52	0.0045	0.48	good	Resists arc erosion

16.6   FAILURE CHARACTERISTICS

16.6.1   Contact Wear

Contact wear is the result of arcing and/or mechanical abrasion. It results in a visible change in the appearance of the contacts, which can provide clues about the number of operations and types of loading conditions. Excessive contact wear suggests repeated overcurrent tripping and may indicate abnormal electrical conditions in the loads supplied by the circuit breaker. Mechanical operation below the current rating does not cause significant contact wear.

Circuit breaker contacts wear under normal use due to arcing and mechanical abrasion. Material is generally not lost from wear but simply redistributed, changing the surface shape of the contacts. Wear is accelerated when the circuit breaker is frequently required to interrupt high load currents. Contact wear resulting from dc loads differs dramatically from that observed as a result of ac service; dc inductive circuits will cause more severe wear on contacts than dc resistive circuits.

The ac aging results shown in Figs. 16.5 through 16.9 were conducted at 60 Hz. Arcing between contacts as they begin to open under load is the primary wear mechanism. The arc usually extinguishes when the sinusoidal current waveform passes through zero. At higher power frequencies, the time required to reach a zero crossing and the average arcing time are both reduced. As a result, wear is much less severe at 400 Hz than at 60 Hz.

Usually, circuit breaker contacts may be operated up to 5 or 10 thousand cycles without showing abnormal wear. Contact wear at 120 Vac is much more severe than dc operation. Contact wear can be examined by optical microscopy or SEM. The photographs in this section show new circuit breakers aged in an apparatus that mechanically opened and closed them using their operating button. The circuit breakers were each operated in 28-Vdc or 120-Vac 60 Hz circuits, with current applied and resistive loads selected to draw the rated circuit breaker current.

16.6.2 Material Transfer

Material is transferred between mating contacts of a device due to the arcing that occurs as the contacts part under load. Their appearance can provide clues about the direction of cur-

Figure 16.5 Contact from a new 10 A circuit breaker (25×).

Figure 16.6 Contact from a 10 A circuit breaker cycled 10,000 times at rated dc load current at 28 Vdc (30×).

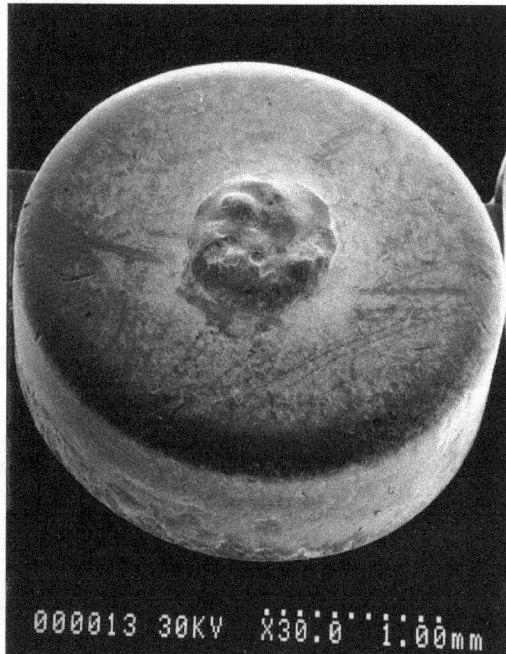

Figure 16.7 Contact from a 10 A circuit breaker cycled 20,000 times at rated dc load current at 28 Vdc (30×).

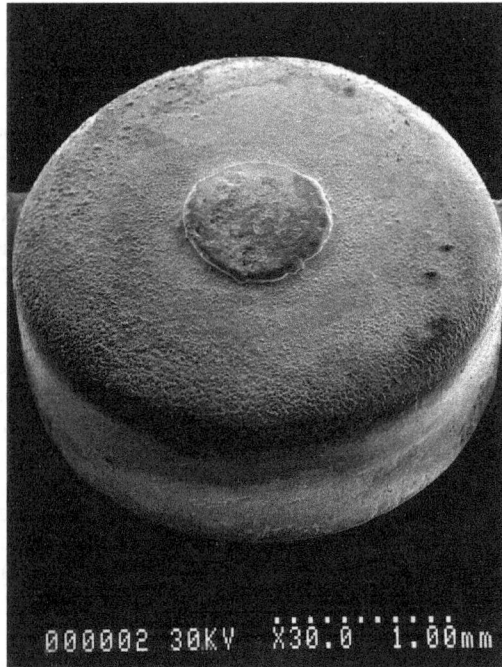

Figure 16.8 Contact from a 10 A circuit breaker cycled 10,000 times at rated ac load current at 120 Vac (30×).

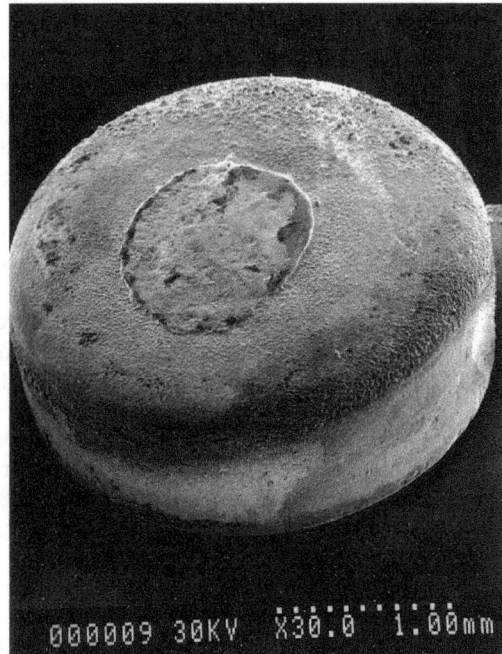

Figure 16.9 Contact from a 10 A circuit breaker cycled 20,000 times at rated ac load current at 120 Vac (30×).

rent flow in dc circuits and information about certain types of abnormal conditions. Material transfer is caused by repetitive operation of contacts in dc circuits. As contacts begin to part under load, the current is confined to a reduced area at the mating surface between the contacts. The resistance of the contact material and the high current density cause local heating that melts the contacts as they begin to separate. At the same time, the circuit inductance attempts to maintain the current by forcing an additional voltage to be developed across the contacts. An arc is created as a result of this inductance.

As the arc begins to bridge the gap between the contacts, electrons flow from the cathode to anode. Electrons leaving the cathode absorb energy from the surface as they enter the arc. This results in a slight cooling effect at the cathode. There is still a net heating effect in the contacts because of the I^2R loss in the bulk material of the contacts. Heating occurs at the other end of the arc as electrons deposit energy into the anode. This adds cathode heating to the I^2R loss at the anode. The net effect of the electron energy in the arc makes the positive contact hotter than the negative contact.[15]

Material transfer is generally in the direction of conventional current flow. The terms *cathode* and *anode* as used above relate to conventional current flow and must not be confused with markings on electric circuit components. It is also possible for current to flow backward through devices during abnormal conditions. These situations must not be overlooked when examining the device's contacts.

The appearance of material transfer may be obscured when a circuit breaker interrupts high dc currents. Material transfer is from the hotter contact to the cooler and thus occurs in the direction of conventional current flow (in the opposite direction of electron current flow). In dc circuits, the anode will be marked with depleted areas called *craters*. The cathode will be marked with mounds or cone-shaped deposits called *pips*. These characteristics are displayed in Figs. 16.10 through 16.13.

16.6.3 Overcurrent

Circuit breaker contacts can be damaged by the interruption of currents in excess of approximately ten times their rating. Their appearance may indicate that overcurrent interruption may have occurred and provide clues about abnormal electrical events. Overcurrent damage can be caused by electrical failure of equipment, wiring shorts, and accidental electrical connections. Electrical system component and wiring failures can lead to overcurrent conditions that cause circuit breaker tripping. The basic mechanism resulting from overcurrent conditions is prolonged arcing. This causes melting and redistribution of the contact materials. Inductive loads will cause more damage to contacts than resistive loads because of the stored energy available to maintain the arc as the contacts part. Contacts take on a *splattered* appearance as displayed in Fig. 16.14. The investigator should always examine all the contacts from a device to increase the confidence level of any findings. Mating contacts should be properly identified as to their location and purpose in the existing circuitry.

Most contacts will have completed a number of normal cycles prior to an overcurrent event. This aging effect will mask the damage caused by more recent overcurrent events and complicate the investigation. When aging is an issue, the investigator may need to perform a test to re-create the damage patterns of the device in question.

Severe arcing due to overcurrent events causes melting of the contact materials. Silver melts at 960° C (1760° F) and cadmium oxide decomposes at 1010° C (1850° F). Contact wear can be examined by optical microscopy or SEM. A redistribution of the Ag and CdO components may be detected with SEM/EDS.

Figure 16.10 Positive contact (anode) from a 10 A circuit breaker cycled 10,000 times with a dc resistive load at rated current (30×). Notice crater formation in center of contact.

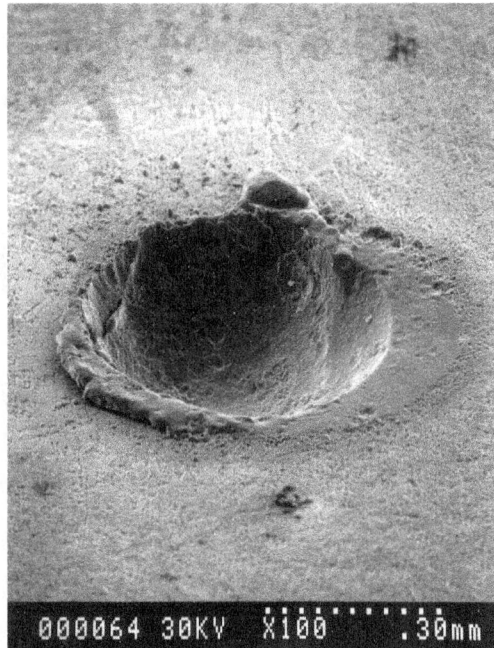

Figure 16.11 Close-up of crater formation on positive contact (100×).

Figure 16.12 Negative contact (cathode) from a 10 A circuit breaker cycled 10,000 times with a dc resistive load at rated current (30×). Notice pip in center of contact.

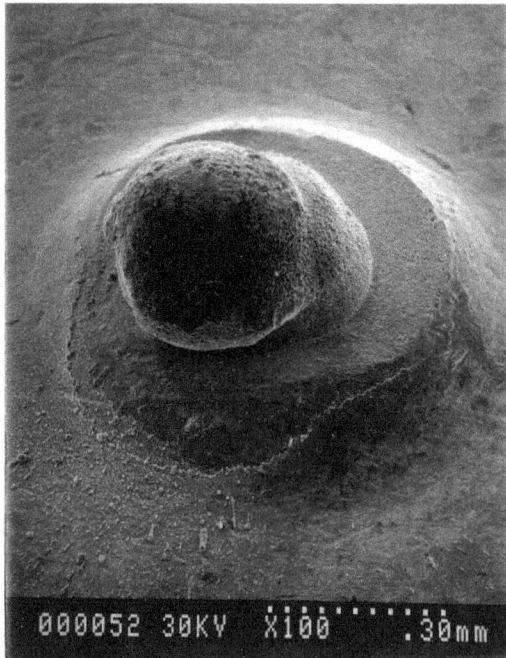

Figure 16.13 Close-up of pip on negative contact (100×).

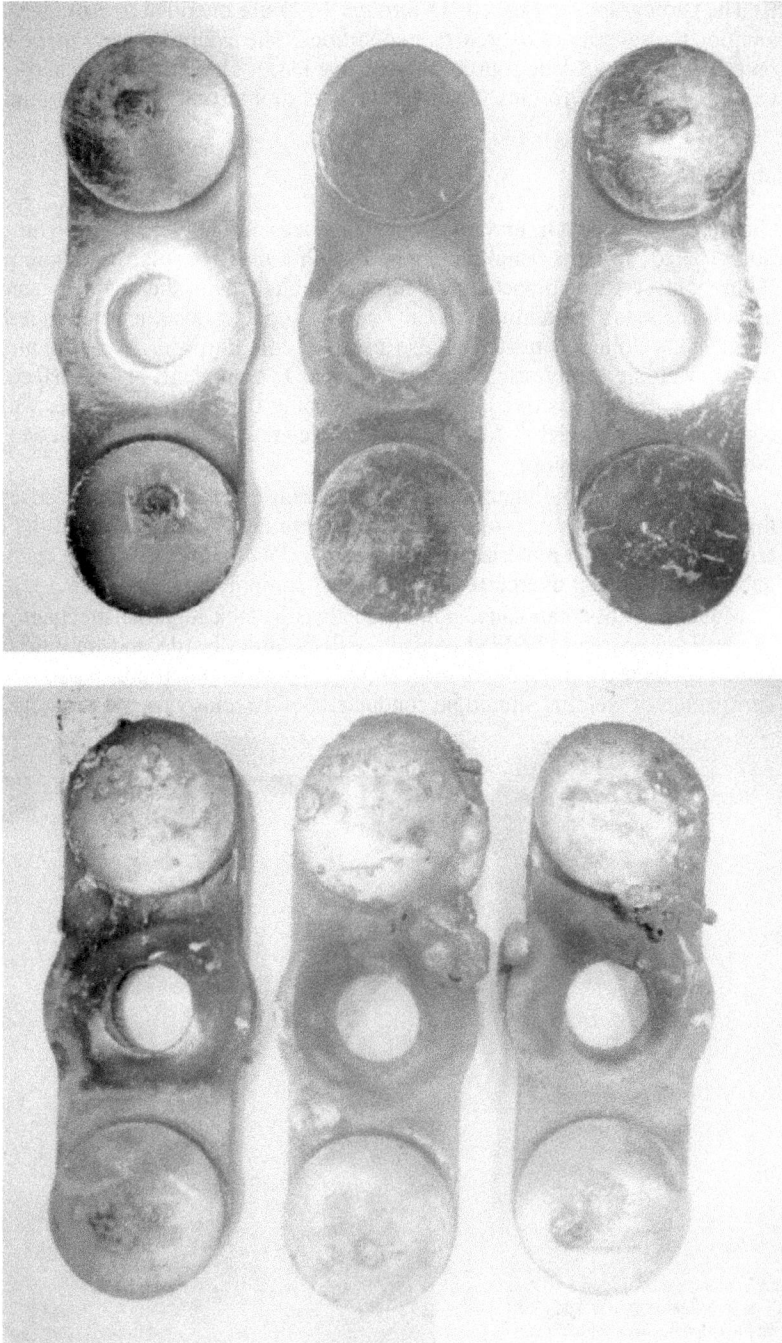

Figure 16.14 Upper photo shows a layer of soot on a set of power contactor contacts. This indicates that the contacts were open and away from the arcing. The lower photo shows an example of the melting that can occur when the contacts are involved in the arcing.

The photographs in Figs. 16.15 through 16.18 are intended to provide some guidance in gauging the severity of overcurrent conditions. The examples were made with new circuit breakers using a 28-Vdc source and resistive loads. The circuit was energized with the circuit breaker closed, forcing the circuit breaker to interrupt the load current.

16.6.4 Contact Welding

Circuit breaker, switch, and relay contacts occasionally weld together in the closed position. The condition is usually associated with a massive internal failure of the device. A failure with welded contacts may indicate problems with the device or unusual surge currents in the load. The addition of CdO in silver contacts was adopted to reduce the susceptibility to welding. Tungsten is also used for this purpose. A condition associated with contact welding is migration of the Ag and CdO components at the surface of the contacts. In dc circuits, the positive contacts will become CdO rich over time, while the negative contacts will be Ag rich.[16] Migration of the contact material alloys can be examined with a metallurgical microscope.

Mechanical misalignment may also be a cause of welding. This causes a reduction in the mating area of the contacts. A current surge melts the contacts in the reduced mating area, and the contact material solidifies before the circuit breaker can open. Failure occurs during a subsequent overcurrent condition or equipment failure.

Metal migration can cause mating contacts to lock together mechanically and behave like welded contacts. Welded contacts will be quite brittle and may be damaged by the standard mounting procedures for metallographic specimen preparation. X-ray and visual verification of welding should be conducted before removing the contacts from the circuit

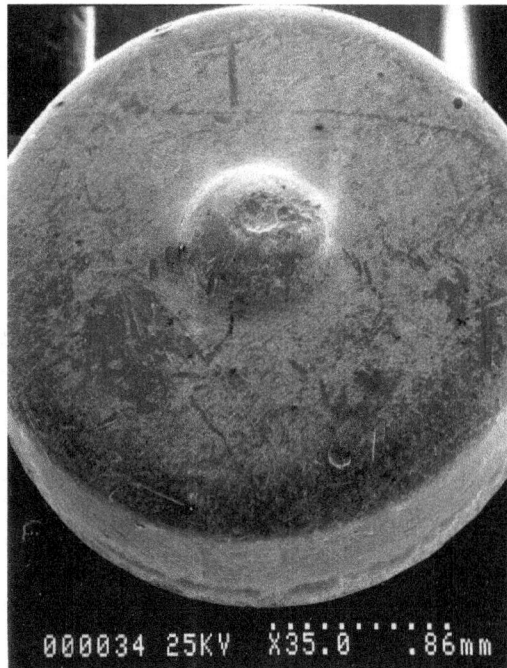

Figure 16.15 Positive contact (anode) from a 10 A aircraft circuit breaker after one interruption at 100 Adc (35×). The duration of the current interruption was about 75 ms.

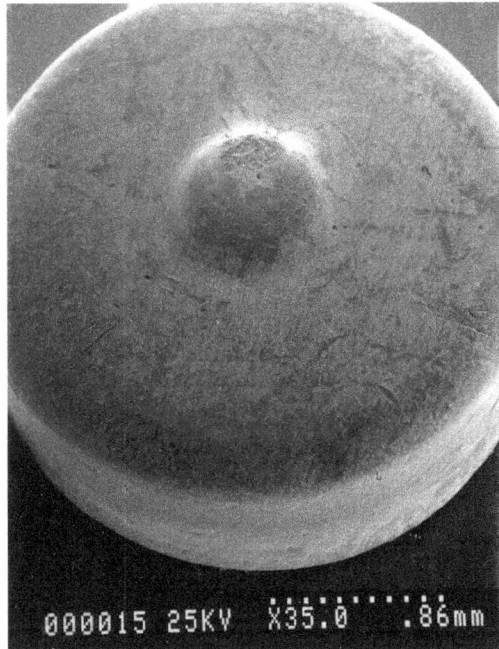

Figure 16.16 Positive contact (anode) from a 10 A aircraft circuit breaker after 10 interruptions at 100 Adc (35×). The duration of the current interruption was about 75 ms.

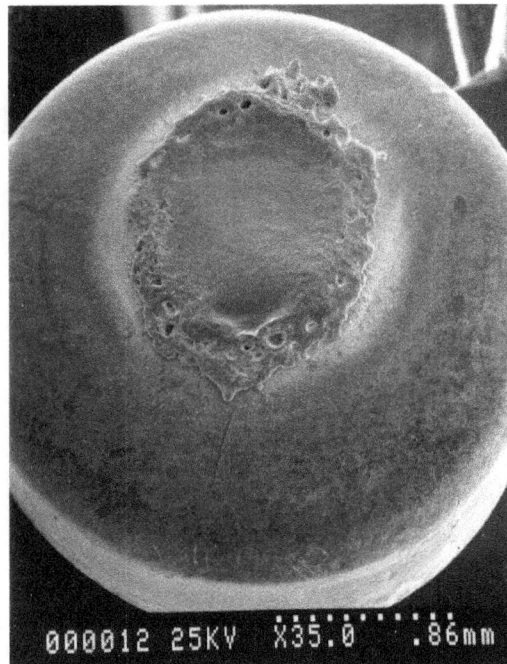

Figure 16.17 Positive contact (anode) from a 10 A aircraft circuit breaker after one interruption at 1000 Adc (35×). The duration of the current interruption was about 2 ms.

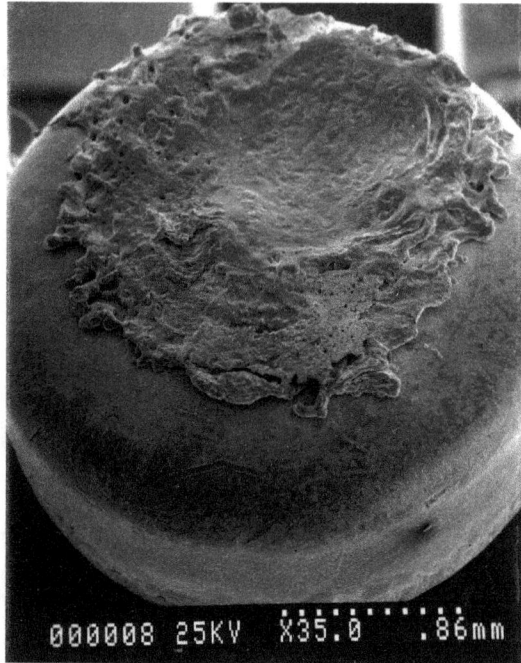

Figure 16.18 Positive contact (anode) from a 10 A aircraft circuit breaker after 10 interruptions at 1000 Adc (30×). The duration of the current interruption was about 2 ms.

breaker. X-ray radiography may also be conducted to verify that welding has occurred. Figure 16.19 shows the normal appearance of the AgCdO alloy on the top surface of the circuit breaker contacts.

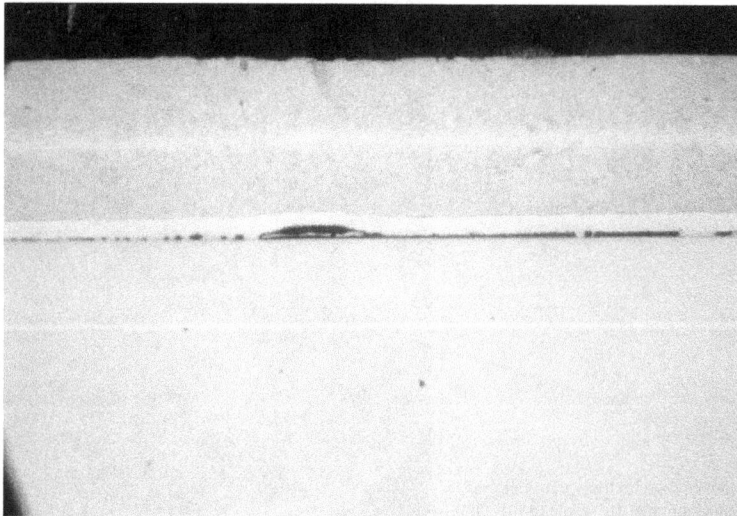

Figure 16.19 Metallurgical cross section of an electrical contact. The upper layer is the AgCdO "button," which is brazed to a copper bus bar. Note the separation that has occurred.

16.6.5 Contamination

Contaminants inside a switch, circuit breaker, or relay can cause the contacts to have a high resistance in the closed state. Contaminants can also cause arcing between the contacts and can foul the mechanical operation of the device. Contaminants can be observed visually or with the SEM. Identification may be done with the SEM or with an infrared spectrograph.

Contaminants can enter the housing of the apparatus if it is damaged or improperly sealed. Pieces of metal, plastic, clothing fibers, and other organic material may be trapped in the device during manufacturing. While initially harmless, these materials may eventually relocate and interfere with the contacts. Contamination can also be caused by severe arcing, which can burn and dislodge housing materials. If conductive, the foreign matter may cause an electrical short between the contacts when the device appears open. If nonconductive, the material may prevent proper electrical contact when the mechanism is closed. The materials may also prevent or impede proper mechanical operation.

Environmental contaminants such as dirt, moisture, and cleaning agents may enter the circuit breaker if the housing is not sealed properly or if it is damaged after installation. These materials will degrade the insulating properties of the housing. When the contacts open under load the contaminated surfaces become an alternate path for arcing, as the load current is being interrupted. The arcing can bum off small fragments of the housing material and vaporize volatile compounds in insulating materials. Arcing can both introduce additional loose solid particles and coat the contacts with condensed volatile insulating compounds.

Note: Moisture, fibers, and small particles can be lost when the device housing is opened. Disassembly and examination must be conducted with care to avoid losing this evidence.

Infrared (IR) spectrographic analysis can be used to identify specific materials. In one case, clothing fibers were found inside a circuit breaker. IR analysis showed that the fibers were an ester-based polyacrylonitrite, which is an acrylic fiber used in clothing. Other examples of materials to notice are tungsten-oxide, which can form in high-temperature and high-humidity environments, and silver tungstate, which may be found on tungsten alloy contacts.

16.6.6 Mechanical Failure

Although electrical failures are more common, mechanical failures of electromechanical devices can occur. The failures can be from abnormalities in the manufacturing process, wear, corrosion, or contamination obstructing the mechanism.

Mechanical failures can be caused by failure of internal components or foreign materials trapped in the housing. Internal component failures can result from changes in the construction materials of the device. This sometimes happens when alterations are made to the manufacturing process. Mechanical component failure is not common and should be verified by testing additional units from the same manufacturing lot. A fractured reset button spring on an aircraft circuit breaker is one example of mechanical failure that has been observed. The pieces of the spring fell into the contacts, causing an internal short.

X-ray examination of an apparatus before disassembly may reveal possible mechanical problems. Radiography is one of the best diagnostic tools, since most electromechanical devices will not work properly after disassembly. Real-time microfocus X-ray is the ideal tool for this type of investigation.

16.7 *FAILURE ANALYSIS PROCEDURES*

16.7.1 Field Examination

An aircraft circuit breaker is used as an example of a typical device throughout the failure analysis procedures section. Substitute other applicable specifications as required for other electromagnetic devices.

Field examination must be limited to what the investigator can see or measure with little analytical assistance. A multimeter is the best field tool for this work. The purpose of this examination is to identify any electromagnetic device that should be removed for laboratory analysis. There are three basic reasons why an electromagnetic device can become important during an investigation:

1. It was the ignition source of a fire.

2. It contributed to the failure of a critical system.

3. It provides information about possible failure scenarios.

Recording the function, rating, and location of the exhibit is essential. Hand sketches and photographic records are recommended. Include both the front and rear of the panel in the sketches and photographs. This is especially true if the device will be removed from a panel at the site. A circuit breaker identification form is presented as an example in Table 16.9. Be sure to record this information and tag each device with an ID number when it is removed. Mark the wires with tape or wire markers and cut them several inches away from the device. The wires should be marked so that conductors to each terminal of the exhibit can be identified later. Without changing the position of the exhibit, carefully loosen the nut on the front panel and remove it from the panel. Reassemble mounting hardware so that it remains with the exhibit.

Note: Circuit breaker buttons should be examined carefully during investigations of mishaps involving fires. The white band on the button may become covered with soot if the circuit breaker tripped before the fire. This may provide a clue as to the sequence of electrical events.

Precaution: Aircraft circuit breakers have what is called a *trip-free* feature. This ensures that the breaker will trip in response to an overload, even if the button is jammed in the closed position. This provides a fail-safe mechanism that allows the breaker to perform its function regardless of external forces. Therefore, even though a breaker may appear to be closed, it may have tripped. It should also be noted that a breaker may trip as a result of the shock of an impact. Therefore, the appearance of a trip may not indicate an overload.

16.7.2 Laboratory Examination

The general steps for laboratory examination are outlined below. This list should be considered as a guide only. Only the appropriate steps should be performed, depending on the device's condition.

1. Verify that all necessary identifying data have been recorded.

2. Make a preliminary measurement of the contact resistance to verify the electrical state of the device.

3. Obtain an X-ray radiograph to confirm the internal state of the device.

Table 16.9 Sample Circuit Breaker Identification Form

Aircraft	*F111*
Sample ID#	*CB 101*
Date of mishap	*04/03/90*
Date removed from site	*04/03/90*
Breaker type	*Push-pull 1-phase*
Breaker rating	*10 A*
Location found at site	*In cockpit panel*
Function labeled on aircraft panel	*LH fuel pump*
Location in panel	*Middle of top row*
Panel or instrument name	*Center panel between seats*
Manufacturer's name	*Mechanical products*
Manufacturer's part number	*700-801-705-10*
MS number	*MS25244*
Date code	*7024*
Wiring configuration: line side	*2 #16 white*
Wiring configuration: load side	*1 #18 white*
Position of button	*Closed*
Contact or leakage resistance	*Closed, <0.1 Ω*
Measuring instrument	*Fluke 75 DMM*
Comments	*Button melted*

4. Measure the contact resistance after radiography has been completed to verify that the exhibit has not been inadvertently operated.

5. Perform contact resistance measurements in accordance with applicable specifications for the device.

6. Perform insulation resistance measurements in accordance with applicable specifications for the device.

Once again, an aircraft circuit breaker is used here as an example of a typical device inspection.

Most circuit breaker housings consist of two halves joined together with rivets. Terminal assemblies are sometimes cemented together with an epoxy-like material, which may also be used to seal the two halves of the housing. The exterior of the terminals may be covered with an elastomeric potting compound. The steps below outline the procedure for disassembly and inspection of the internal components.

1. Remove potting compound using a razor blade or sharp knife. Examine the terminal area for arcing or leakage. Check for loose screws, corrosion, improperly terminated wires, and so on.

2. Drill out the heads of the rivets and remove the rivets from the housing.

3. Separate the housing halves with a sharp knife.

4. Photograph the internal components of the circuit breaker.

The steps below offer another procedure for inspecting the internal components. This procedure allows for operation of the circuit breaker and measurement of the temperature of the contacts and other components.

1. Grind away an area of the housing at the contact level until the plastic housing is very thin.

2. Cut out the remaining plastic to form a window exposing the contacts.

A through visual examination may be made when the circuit breaker has been disassembled. The condition of the contacts may then be determined by optical microscopy and the SEM where applicable. The overall steps are outlined below:

1. Visually examine the contacts and internal surfaces of the circuit breaker. Note any pertinent damage or discoloration of components, and make a sketch or photograph the details before removing the contacts.

2. Label the contact assemblies before removing them from the circuit breaker. Identify any specific terminology used in the labeling on a sketch or photograph. Military circuit breakers are marked on the line side, and this notation can be useful in identifying contacts and test results. Also note whether the contacts are moving or stationary.

3. Optical and SEM micrographs similar to those shown in this chapter can reveal information about the condition of the circuit breaker.

4. Cross-sectioning of the contacts and metallurgical examination can sometimes provide clues about the composition and possible causes of welding.

16.7.3   Contact Resistance

Preliminary contact resistance measurements can be made in the field or laboratory using a conventional multimeter. This will confirm that the electrical and mechanical states of the circuit breaker are consistent. Contact resistance measurements will be possible only on circuit breakers that are found in the closed position. Do not close the circuit breaker in the field to make the measurement, since this could cause internal damage and alter the condition of the circuit breaker.

Formal measurements should be made in accordance with MIL-C-5809 for consistency. The procedure passes current through the circuit breaker and measures the voltage drop across the contacts. The current should be 200 mA or one-half the current rating of the circuit breaker, whichever is less. MIL-C-5809 does not specify the contact resistance necessary for qualification. Instead, it refers to the change in contact resistance from the original condition of the circuit breaker after certain environmental and storage tests. The change should not be more than 250 mV above the original value. This corresponds to a resistance increase of no more than 1.25 Ω. The contact resistance of a new 10 A circuit breaker should be less than 0.1 Ω. Contact resistance can also be related to the power dissipation in the breaker while operating at rated current. MIL-C-5809 requires that this power not exceed 15 W.

Precaution: Caution should be taken in handling the circuit breaker if the contacts are apparently closed but have a high resistance. The concern here is to prevent any accidental movement of the contacts that could obliterate the cause of the high resistance and restore electrical continuity.

16.7.4 Insulation Resistance

Preliminary insulation resistance measurements can be made in the field with a multimeter if the circuit breaker is found in the open position. Do not open a closed circuit breaker in the field to make the measurement. The opening operation could cause internal damage and change its condition. These measurements are useful to have on record as part of the documentation with a circuit breaker, or any device, that is removed for laboratory analysis. The measurements may provide an indication of insulation degradation if contamination, especially moisture, has been a problem.

Formal measurements should be made in accordance with MIL-C-5809 for consistency. The procedure is to apply 500 Vdc to the breaker and measure the leakage current. The test has two parts:

1. The voltage is applied between the line and load terminals with the breaker in the open position.
2. The voltage is applied between the terminals and parts normally grounded, such as the mounting hardware and frame. This part is done with the breaker both opened and closed. In the field, take only data with the circuit breaker in the position in which it was found.

Note. The insulation resistance of a new 10 A circuit breaker should be more than 20 $M\Omega$ when measured with a multimeter. Also, MIL-C-5809 requires the insulation resistance to be greater than 100 $M\Omega$. The leakage current of a new 10 A circuit breaker at 1500 Vac, 60 Hz should be less than 500 μA. If the insulation resistance measured below 1 $M\Omega$ with a multimeter, it may be an indication of severe contamination and/or moisture trapped in the circuit breaker. Since moisture could evaporate during shipment from the mishap site, the laboratory should be notified of this condition.

16.7.5 Dielectric Withstand Voltage

The procedure from MIL-C-5809 involves applying a high voltage to the circuit breaker terminals to verify that no damage results. The test has two parts:

1. The voltage is applied between the line and load terminals with the breaker opened.
2. The voltage is applied between the terminals and parts normally grounded, such as the mounting hardware and frame.

The voltage is applied at 60 Hz ac and must be equal to twice the rated voltage of the circuit breaker plus 1000 V, but not less than 1500 Vac. For the test to be successful, there should be no evidence of breakdown, flashover, or leakage current in excess of 1.0 mA. The leakage current of a new 10 A breaker at 1500 Vac, 60 Hz should be less than 500 ΩA.

Precautions: These tests are generally considered destructive and should be used only on noncritical samples. In particular, the test can break down insulating contaminants on contacts, thereby obliterating any useful information about the condition of the circuit breaker. This test should be conducted only on circuit breakers that have successfully passed the insulation resistance test. The dielectric withstand test is usually reserved for production samples.

16.7.6   X-Ray Radiography

X-ray examination of circuit breakers can be a valuable procedure for determining the internal condition of the circuit breaker before opening the housing. Most circuit breakers have plastic housings that are reasonably transparent to X-rays. The internal metal components will be rendered clearly visible in the radiographs. Microfocus X-ray equipment is the best choice for this type of examination, because of the ability to view the small internal components of the circuit breaker with a high resolution.[17,18] Conventional X-ray radiographs can be photographed and enlarged using a light table and camera stand. Radiographs may also be placed in a photographic enlarger to magnify details. X-ray radiography should be performed before the circuit breaker is disassembled or operated to verify the condition and position of internal components.

16.7.7   Contact Preparation for Metallurgical Examination

Sample preparation consists of six basic steps: sectioning, mounting, grinding, polishing, etching, and cleaning. The parameters and materials for automatic polishing equipment are provided in Tables 16.10 and 16.11.[19] Most of this information pertains to manual preparation as well. The principles of metallography and sample preparation methods are covered in Chap. 8.

Table 16.10   Leveling and Grinding Process Parameters

SiC grit size	Time (s)	Wheel speed (rpm)	Pressure (psi)
240	40	300	25
600	60	300	25

Table 16.11   Final Polishing Process Parameters

SiC grit size	Time (s)	Wheel speed (rpm)	Pressure (psi)
3 μm diamond compound/silk cloth/extender (oil)	60	250	25
1 μm diamond compound/red felt cloth/extender (oil)	120	250	30
0.05 μm gamma alumina/Imperial cloth/water	30	150	20

Contact Etchant: 50 ml H_2O, 25 mL NH_4OH, 3 mL H_2O_2 immerse
Cleaning: Rinse and dry sample before mounting
   Note: Rinse with soap and water between grit sizes to avoid contamination with a larger size grit.

16.8 REFERENCES

1. Denson, W. RL-TR-92-197, Reliability Assessment of Critical Electronic Components, IITRI for USAF Rome Laboratory, July, 1992.
2. Ibid.
3. Taken from MIL-STD-1580A.
4. Denson, W., RL-TR-92-197.
5. Ibid.
6. Ibid.
7. Ibid.
8. Ibid.
9. Ibid.
10. MIL-C-5809G, Circuit Breakers, Trip-Free, Aircraft, General Specification For.
11. Galler, D., Glover, D., Kusko, A. *WL-TR-95-4004, Aircraft Mishap Investigation Handbook for Electronic Hardware.* Failure Analysis Associates, for USAF Wright Laboratories, 1995.
12. Holm, R. *Electric Contacts,* 4th ed. New York: Springer-Verlag, 1967.
13. Green, H., Kleis, J., and Pitney, K. Private notes taken during seminar, "Design and Application of Electrical Contacts." U of W Madison, 23–25 August, 1993.
14. Denson, W., RL-TR-92-197, Reliability Assessment of Critical Electronic Components, IITRI for USAF Rome Laboratory, July, 1992.
15. *The Theory and Practice of Overcurrent Protection.* Jackson, MI: Mechanical Products Inc., 1987.
16. Wingert, et al. Developments of the Arc-Induced Erosion Surface in Silver-Cadmium Oxide, *IEEE Trans. on Components, Hybrids and Manufacturing Technology,* CHMT-8(1), March 1985.
17. FeinFocus X-Ray Technology brochure. Agoura Hills, CA: FeinFocus USA Inc.
18. Colangelo, J. Advanced Radiographic Techniques In Failure Analysis. Seminar conducted at ISTFA, 1991, in *Proc. 17th International Symposium for Testing and Failure Analysis.*
19. Dillinger, L. *Suggested Procedures for Grinding and Polishing Various Alloy Systems Using Automatic Polishers.* St. Joseph Michigan: LECO Corp.

16.9 RECOMMENDED READING

Buechler, D.W. Real-Time Radiography for Electronics Reliability Assessment, *Materials Evaluation* 45(1), November 1987.

Jones, F.L. *The Physics of Electrical Contacts.* Oxford, England: The Clarendon Press, 1957.

Winquist, N.H. Using Scanning Electron Microscopes to Uncover Contact Problems, *Bell Laboratories Record* 49(5), May 1971.

Trent, D.M. Failure Analysis of Silver-Cadmium Oxide Contacts, *Proc. 12th Electrical/Electronics Insulation Conference,* 1975, IEEE Publication 75CH1014-0-EI-70.

Mason, John R. *Switch Engineer Handbook.* New York: McGraw Hill, ISBN 0-07-040769-X.

CHAPTER 17
CONNECTION TECHNOLOGY

Noshirwan K. Medora

E^{x}ponent

17.1 INTRODUCTION

Just a few years ago, packaging technologies used low power density systems, which permitted hard-wired interconnections between components. More recently, submicron, multidisciplinary technologies have produced high-speed, "smart" information-processing ICs with improved power densities, contributing to the rapid growth of the electronics industry. The proliferation of different families of intelligent devices and miniaturized components in parallel processing subsystems has created major packaging issues that need to be resolved for the ever-shrinking circuits.

As opposed to hard-wired systems, a well designed system of connectors provides a ready means for quick connect/disconnect options during installation or replacement. Connectors permit data communication between two systems or subsystems and can afford a power flow path between a power-driver and the corresponding receiver. Connectors may also be used to interface between components or subsystems. Moreover, connectors permit the rapid addition or removal of modules and subsystems, allowing repair, replacement, and upgrade in the field.

The increased demand for high-performance, high-density, interconnecting systems has created design problems that must be addressed at the initial stages of design rather than after completion. Connector design issues such as power levels, operating voltages, overload conditions, and ambient temperatures are factors that need to be uniquely addressed and resolved. Connectors, the workhorse of component packaging and assembly, are sometimes used as an afterthought. Oftentimes, this results in misapplication, misoperation, reduced or compromised reliability and, in some cases, failure of the connector. This can, in turn, create a possible fire or electrocution hazard.

This chapter examines several connector-related topics, including connector fundamentals and the different types of connectors available. The chapter will include connector electrical specifications such as contact resistance, current-carrying capacity, and contact design. Mechanical specifications, including contact material, contact block insulating material, effect of contaminants, and durability, are discussed. The chapter also includes a section on industry standards and guidelines for connectors.

Failure analysis of connectors is addressed, along with several topics including root cause investigation and events leading to connector failure. Crimping problems, whisker formation, fretting corrosion, and contact operation under overload and transient fault conditions are reviewed.

Since connectors are sometimes associated with electric shocks, an entire section has been dedicated to this issue. Due to the inherent mechanical nature of connectors, they are often a prime cause of system failure. Section 17.8 (p. 17.47) outlines case studies involving connector failures due to design defects, installation defects, and harsh environmental conditions that have resulted in misoperation, fire, and electrocution. A summary section comments on current connection technology.

17.1.1   Definition

Connectors provide a convenient means for interconnecting and disconnecting components, modules, and systems. Several definitions are listed for connectors, including those found in electrical engineering handbooks, technical articles, manufacturers' application notes, the *IEEE Standard Dictionary,* and the *IEC Multilingual Dictionary.*

The new *IEEE Standard Dictionary of Electrical and Electronics Terms*[1] defines connector as

A coupling device employed to connect conductors of one circuit or transmission element with those of another circuit or transmission element.

17.1.2   Connector Fundamentals

In its simplified form, a typical connector assembly consists of several functional components such as the following:

1. *Connector Housing.* This is a plastic, electrically insulating, thermally stable material providing mechanical integrity and electrical insulation for the contact pins. The contact pins are inserted into the contact cavity of the housing.

2. *Contact Pins.* These are metallic, fixed and movable mating surfaces for the connector, which provide a path for the electrical power flow.

3. *Contact Plating.* Male and female contact pin surfaces are routinely coated with selected contact metals to provide specialized features based on specific applications, including resistance to corrosion, reduction of wear of the contact surfaces, and reliable operation throughout the specified insertion and extraction cycles.

4. *Contact Spring.* Contact springs supply the normal mechanical force between the male and female contacts to achieve consistent, minimal contact resistance.

5. *Contact Retainer.* This refers to a mechanical insert or detent, usually in the connector housing, that provides a mechanical locking mechanism to prevent the withdrawal of the connector by accidental means.

6. *Connector Seals.* Contacts of connectors used in damp or wet environments are protected from the elements by thermoplastic O-rings or gaskets. Hermetically sealed connectors used for high- and low-pressure applications have an outer steel shell with seals for withstanding the pressure differentials.

A sectional view of a typical plug shell connector with sealing grommets and a basketweave strain relief is presented in Fig. 17.1.

17.1.3   Family of Connectors

In the extremely competitive environment of connection technology, the strong industry trend toward miniaturization has forced the development and proliferation of a large family of connectors to fulfill the unique design requirements of a wide variety of applications.

A comprehensive, market-driven design of connection topology permits electrical and electronic systems to be manufactured in modular form and to be integrated using banks of connectors to lower manufacturing costs and enhance interchangeability.

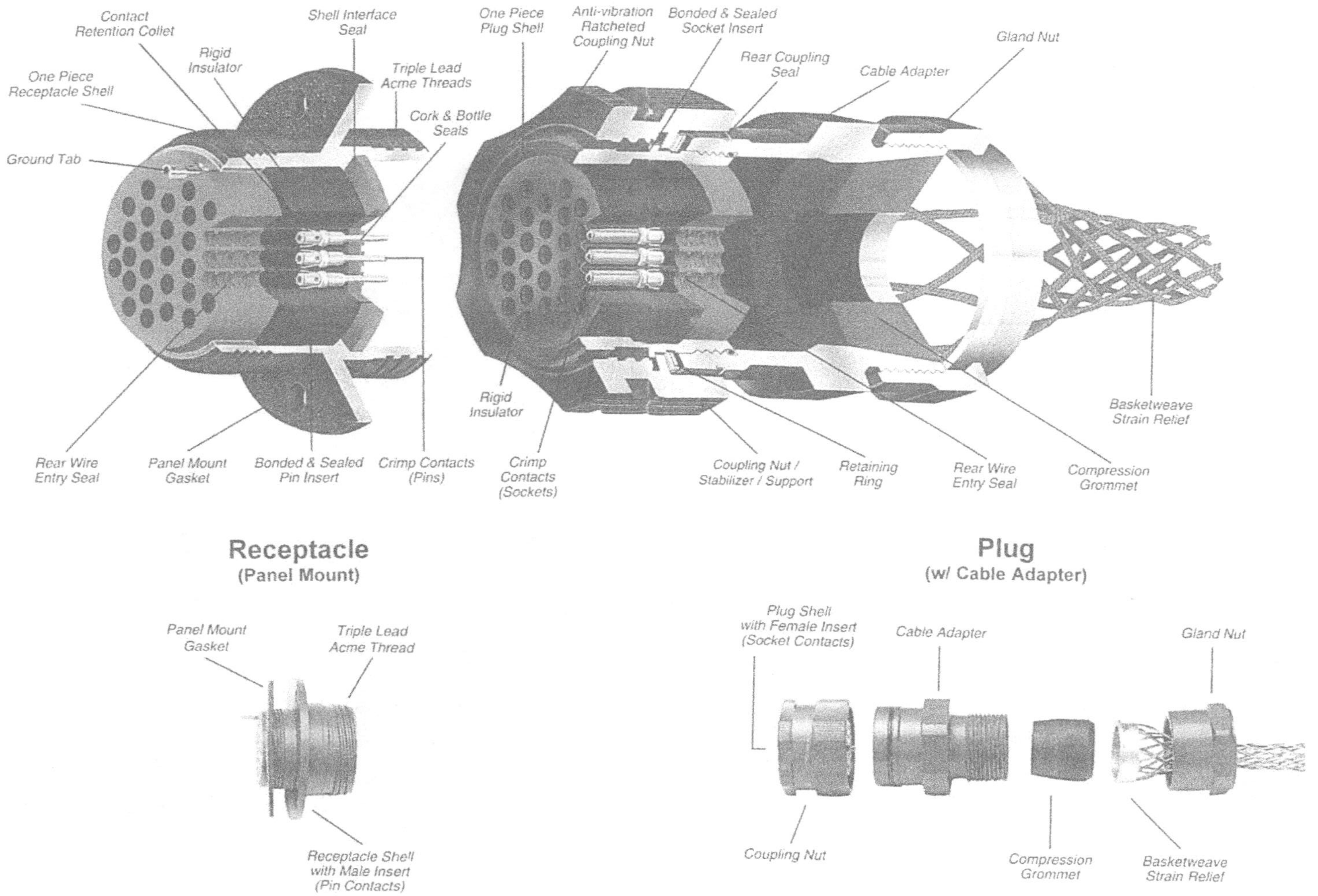

Figure 17.1 Sectional view of a typical plug shell connector with sealing grommets and basketweave strain relief *(reprinted with permission of The Sine Companies, Inc.).*

17.1.4 Types of Connectors

Connectors are available in a variety of shapes and sizes with a wide range of voltage and current-carrying capacities intended for design applications including

- Power connections
- Plug-in
- Splash-proof installations
- Automotive environments
- Military applications
- Built-in filtering
- Hermetic sealing
- High pressure
- High voltage
- Surface-mount
- Terminal blocks
- ESD attenuation
- EMI/RFI suppression
- Pluggable bus-bar connections

17.2 CONNECTOR SPECIFICATIONS

The large family of connectors encompasses a wide spectrum of electrical and mechanical specifications that may be unique to each type as outlined below. Specifications, which will be discussed in greater detail according to their electrical, mechanical, and thermal properties, include the following:

1. Contact resistance
2. Operating voltage
3. Current-carrying capacity
4. Capacitance between contacts
5. Inductance
6. Propagation delay
7. Crosstalk
8. Ambient temperature
9. Maximum operating temperature
10. Conductor size
11. Number of connector circuits
12. Contact geometry
13. Contact block insulating material
14. Contact material

15. Contact finish

16. Effect of lubricants

17. Operation in harsh environments

18. Mechanical durability

17.2.1 Electrical Specifications

The performance and reliability of connectors relies on their electrical specifications.

Contact Resistance.   Contact resistance in connectors is primarily the result of two factors: constriction resistance, and film resistance. Constriction resistance is the electrical resistance due to multiple single-point contacts at the interface surface. Film resistance is due to microscopic film formation on the mating surfaces. Films, primarily oxides and halides, resist the flow of currents. For the contact mating pair to provide useful electrical conductivity, the oxide film has to be removed, allowing contact with the exposed metal.

Contact resistance between mating surfaces is one of the critical parameters in a connector and is typically measured in milliohms (mΩ), or fractions of a milliohm, at specified current and ambient temperatures. Contact resistance varies as a function of time and depends on the size and shape of the mating contacts, contact material, number of prior mating cycles, and environmental conditions such as vibration, temperature, moisture, contaminants, and so forth.

Contact resistance can be adversely affected in a situation where the contact is only partially inserted into the mating surface. A partially inserted contact, under maximum electrical and environmental stress conditions, may eventually overheat, possibly resulting in an ignition. Typical contact resistance and calculated contact power dissipation as a function of contact current is presented in Table 17.1. Figure 17.2, derived from Table 17.1, shows various connector parameters as a function of rated contact current for a family of connectors with increased current-carrying capacity. Parameters include (1) contact pin resistance, (2) calculated contact power at rated current, (3) specified contact diameter, and (4) connector extraction force. It is observed that the parameters follow an approximately logarithmic function, and that the contact pin resistance decreases with increasing current, whereas the other parameters increase as a function of contact current. Calculated contact power has an approximately straight-line relationship on a log-log plot. Dissipated power level is approximately 0.20 W at 15 A, increasing to 3.7 W at 100 A, and further increasing to 10 W at 200 A. Above 200 A, the contact power remains constant at approximately 10 W. Experience indicates that a typical off-the-shelf connector, operating at its maximum specified rating of 15 A and dissipating over 0.4 to 0.5 W of power, may start to overheat. At power levels of 0.8 to 1 W, this connector will rapidly overheat. If the dissipated power level exceeds 1 W, the connector will go into thermal runaway, with contact pin temperatures exceeding several hundred degrees and subsequent melting and ignition of the plastic insulating material. Thermal runaway is discussed further Sec. 17.8, which describes case studies (p. 17.47).

Operating Voltage.   Connectors have a specified operating voltage rating at rated load and rated ambient temperature. The nominal voltage rating is dependent on several factors, including the type of plastic insulating material, the creep distance, and environmental conditions.

Table 17.1   Contact Resistance vs. Contact Size, Current Rating, and Contact Power *(reprinted by permission of Hypertronics Corp.)*

General Specifications										
Contact dia., inches (mm)	0.018 (0.45)	0.030 (0.76)	0.040 (1.02)	0.059 (1.50)	0.098 (2.50)	0.138 (3.50)	0.169 (4.30)	0.241 (6.12)	0.500 (12.70)	0.650 (16.50)
Contact resistance, mΩ	<8	<5	<2.5	<1.5	<0.8	<0.5	<0.37	<0.25	<0.08	<0.04
Current rating, A	2.5	5	7.5	10	15	35	100	200	350	500
Extraction force, oz (lb)	0.5–1.6	0.5–3.2	1.0–3.9	1.2–5	6–25	7–32	15–90	80–160	(9–20)	(20–37)
Contact power dissipation at rated current, W[*]	<0.05	<0.13	<0.14	<0.15	<0.18	<0.61	<3.7	<10.0	<9.8	<10
Material	Beryllium copper wires and brass body (socket), brass (pin)									
Plating	Gold over nickel									
Contact life cycles	>100,000									

[*] Calculated by Failure Analysis Associates, Inc.

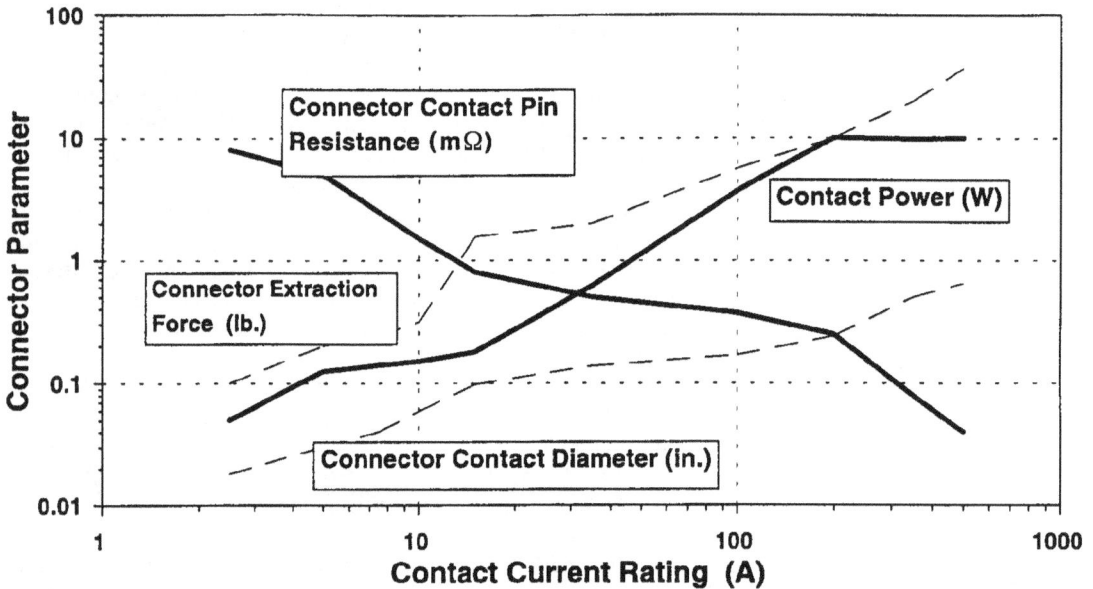

Figure 17.2   Connector parameters as a function of contact current using the information in Table 17.1 (*source: courtesy of Failure Analysis Associates).*

To verify the continuous voltage rating, connectors are subjected to dielectric withstand voltage tests performed according to established safety standards and guidelines, including IEC 512, IEC 130, MILSTD202, and MIL-STD-1344. For a typical connector, the dielectric withstand voltage rating may be several times the nominal voltage rating. As an example, a 15-A multiple-lead connector rated at an operating voltage of 600 Vac may have a dielectric withstand voltage rating of 5 kVac for 1 min between adjacent contacts.

Current-Carrying Capacity.   One of the most significant parameters of a connector pin, and one that is often implicated in the failure of a connector, is its current-carrying capacity. Maximum current-carrying capacity is determined by several factors including ambient temperature, conductor size, and number of circuits in the connector housing.

The current-carrying capacity of connectors varies with the size of the mating surface, contact material, and contact pressure, and can range from a few amperes for miniature connectors to several hundred amperes for high-power, plug-in type connectors with high current-carrying capacity. High-power connectors are typically used in hot pluggable switch-mode power supplies (SMPSs). Often, an SMPS multipin power connector includes several different contact pin formats for carrying high-power and low-power signals. This type of multipin connector may include contact pins for control signals, auxiliary power, and main power all on one connector, incorporating a multimode pin layout with two or more pin sizes. Power connectors utilize pins with unique configurations and typically have high-quality surface finishes with gold or another type of metal electroplating. Large contact surface areas result in very low submilliohm contact resistance, with the consequence of low voltage drop, even at high current levels and multiple insertion/withdrawal mechanical cycles.

Plastic insulating materials have a range of operating temperatures and are commonly specified for a particular operating condition. However, when an insulating material is exposed to high electrical, mechanical, and environmental stress conditions, the recommended maximum operating temperature may be significantly lower due to the added stresses. Connector current-carrying capacity is related to the operating temperature of the connector and is discussed below. Under certain operating conditions, a connector may overheat and melt.

Capacitance between Contacts.   Multiple-pin connectors create an inherent parasitic capacitance between adjacent pins, resulting in capacitive coupling between two or more galvanically isolated circuits in the connector. The magnitude of the parasite capacitance is governed by several factors including contact pin geometry, contact spacing, dielectric constant of the molded plastic insulating material, and the number of adjacent contact pins encircling the subject pin. Capacitance between contacts is generally measured in accordance with the Electronic Industries Alliance (EIA) Standard EIA-364-30, Capacitance Test Procedure for Electrical Connectors. Typical capacitance for a low-current, 5-A connector is approximately 1 to 2 pF between one connector pin and an adjacent pin, and 2 to 5 pF between one connector pin and all other surrounding pins.

Inductance.   A connector pin has self-inductance and mutual inductance. Mutual inductance is usually measured between the subject pin with all other pins used as the return. This parasitic inductance is generally measured in accordance with EIA Standard EIA-364-33, Inductance of Electrical Connectors. For a low-current, 5-A connector, typical inductance is approximately 15 to 30 nH.

Propagation Delay. Parasitic inductance and capacitance, inherent in multipin connectors, have a transmission line effect that results in propagation delays for high-frequency signals. Propagation delays in a connector may be simulated using lumped or distributed modeling techniques.

Crosstalk. At high system frequencies, parasitic elements (inductance and capacitance) may generate crosstalk between circuits in a multipin connector. High levels of crosstalk in a connector may result in erratic behavior, misoperation, or even failure of the overall system.

17.2.2 Thermal Specifications

Connector contacts carrying high currents can dissipate appreciable power along the length of the connector-conductor system. Power dissipation is due primarily to I^2R losses in the male-female contact interface, the contact-to-conductor interface, and conductor resistance losses. High current values may also cause an appreciable voltage drop in the connector. In a high-power, low-voltage system, millivolt drops may result in unacceptable loss of voltage due to the connector resistances. Increased operating temperatures will further exacerbate the problem. Thus, temperature rise in a power connector may be one of the principal parameters considered in determining its use in a particular application.

Ambient Temperature. Ambient temperature determines the maximum acceptable rise in temperature. Any increase in ambient temperature above a predetermined value reduces the current-carrying capacity of the connector. Consequently, connector pins have to be derated accordingly when operating at a higher ambient temperature. Figure 17.3 shows the decreased current rating of a 4-circuit and a 12-circuit connector for an ambient temperature above 75° C.

Maximum Operating Temperature. Maximum operating temperature is usually determined by the maximum specified operating temperature of the plastic insulating housing. For a polyamide connector with a melting temperature of approximately 250° C, the typical, specified maximum operating temperature is 105° C. Exposing contacts to long-term high temperatures may increase contact resistance due to stress relaxation, resulting in loss of spring force. Higher operating temperatures will also result in the buildup of an oxide layer, increasing contact resistance and power dissipation and further increasing the temperature. Connectors may occasionally be exposed to short-term overload conditions with a small rise in temperature. A large, high-current connector may have a thermal time constant of several tens of minutes and, consequently, short duration overload conditions are acceptable. However, connector failure usually occurs when connectors are exposed to abnormally high currents, resulting in operating temperatures above the specified maximum operating temperature. Figure 17.4 presents the temperature rise curve of a high-power connector as a function of time. It is observed that, with a contact current of 340 A, the connector requires almost 40 min of operation to achieve steady-state temperature.

Conductor Size. Maximum current-carrying capacity is determined by conductor size. Wires connected to the terminal end of each contact pin act as heat sinks, conducting and dissipating the heat generated at the mating surfaces. The cross-sectional area of the connecting wires is critical in determining the amount of heat that can be conducted from each

4 Circuit Connector (Wire-to-Wire)

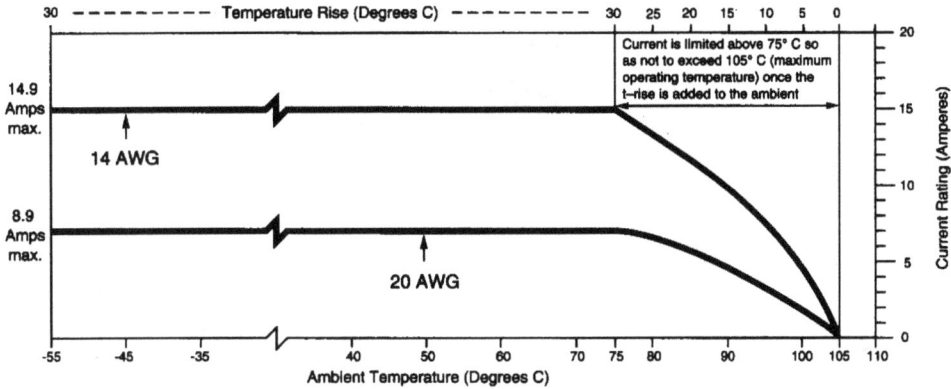

12 Circuit Connector (Wire-to-Wire)

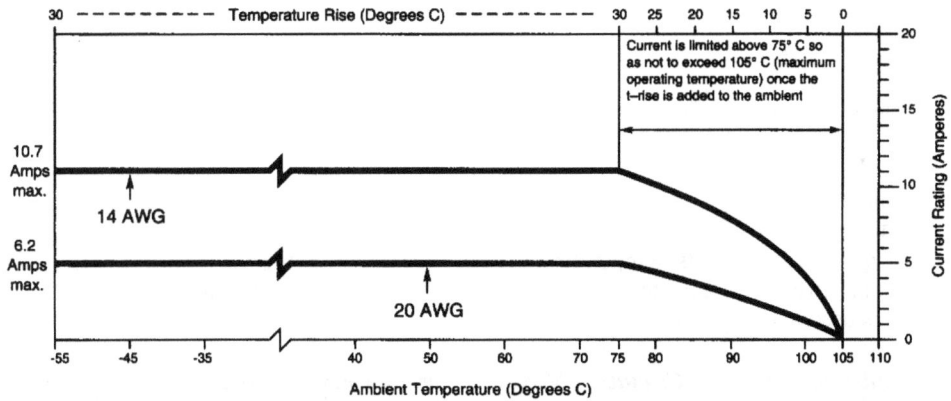

Notes: 1. Data for these curves based on initial T-Rise vs. Current Testing.
2. Current is limited above 75°C so as not to exceed 105°C (maximum operating temperature) once the T-Rise is added to the ambient.
3. Housings are fully loaded with all circuits 100% energized. Current rating is per circuit.

Figure 17.3   Decreased current rating of 4- and 12-circuit connectors above 75° C ambient *(reprinted with permission of AMP).*

contact pin. A larger cross section allows a higher current-carrying capacity for the same maximum temperature rise. With a larger wire size, contacts can carry up to 50 percent more current. Operating connectors at their maximum current rating, but with a wire size smaller than specified, compromises the current-carrying capacity and results in higher temperatures on the contact block molding material. This causes insulation degradation, melting, and possible failure of the connector. Figure 17.5 presents a comparison of contact current-carrying capacity for a #20 AWG and a #14 AWG conductor, for a 4-circuit and a 12-circuit housing. It is observed that, for a 30° C rise in temperature, the 12-circuit

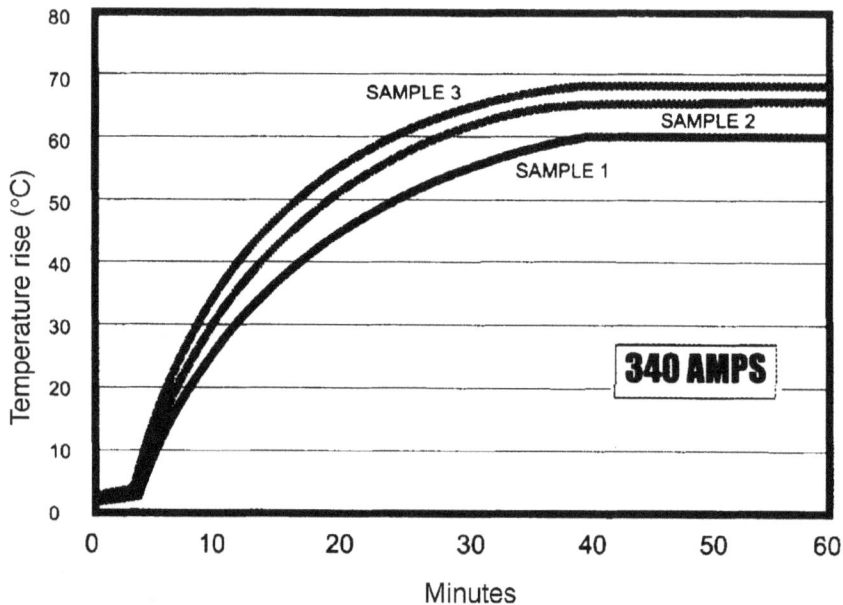

Figure 17.4 Temperature rise of a high-power connector as a function of time *(reprinted with permission of KonneKtech).*

connector can carry 11 A when connected to a 14 AWG wire, but only 5 A when connected to a 20 AWG wire.

Figure 17.6 presents temperature-current curves for a high-current connector. The 95 mm^2, two-pole connector is rated at 630 V, 330 A, for an ambient temperature of 20° C, and is used in electric vehicles.

Number of Connector Circuits. Maximum current-carrying capacity of the connector is also determined by the number of connector pins. As the number of circuits increase, the current-carrying capacity of each circuit decreases. This is observed in Fig. 17.3, where the contact current under identical operating conditions is decreased by approximately 30 percent when the number of circuits is increased from 4 to 12, with all circuits carrying the specified current.

17.2.3 Mechanical Specifications

Under normal conditions, connectors may be subjected to a variety of mechanical abuses including multiple insertion/extraction operating cycles. In addition, they may be exposed to environmental factors such as humidity, temperature, and vibration. Mechanical specifications for the connector and its housing play a vital role and are often indicative of the operating life and reliability of the connector.

Contact Geometry. The mechanical design of the contact is a critical connector parameter, contributing significantly to the operation and reliability of the contact. Geometric design of the contact, including size and shape, must take into account several factors,

A comparison of Universal MATE-N-LOK contacts on #20 AWG wire and #14 AWG wire used in 4-circuit housings. All circuits carrying indicated current.

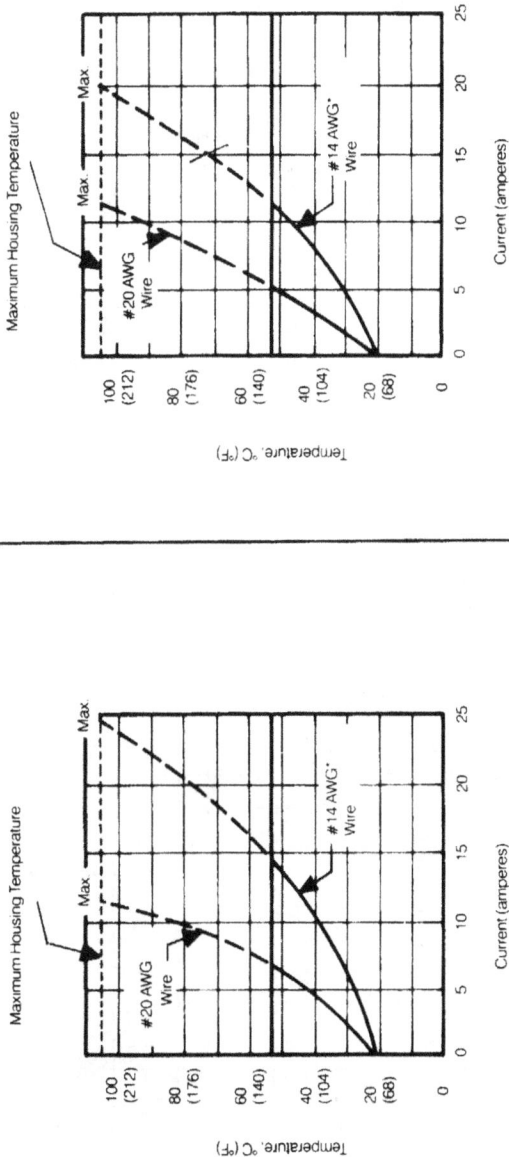

A comparison of Universal MATE-N-LOK II contacts on #20 AWG wire and #14 AWG wire used in 12-circuit housing. All circuits carrying indicated current.

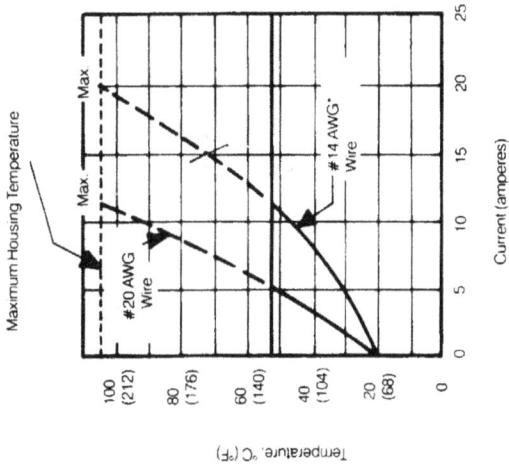

Figure 17.5 Comparison of contact current-carrying capacity for #20 AWG and #14 AWG conductors *(reprinted with permission of AMP)*.

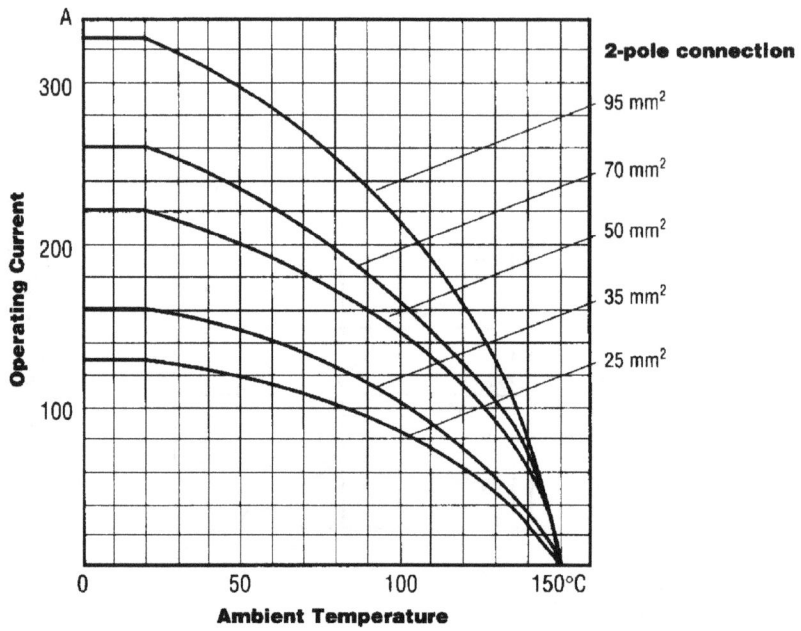

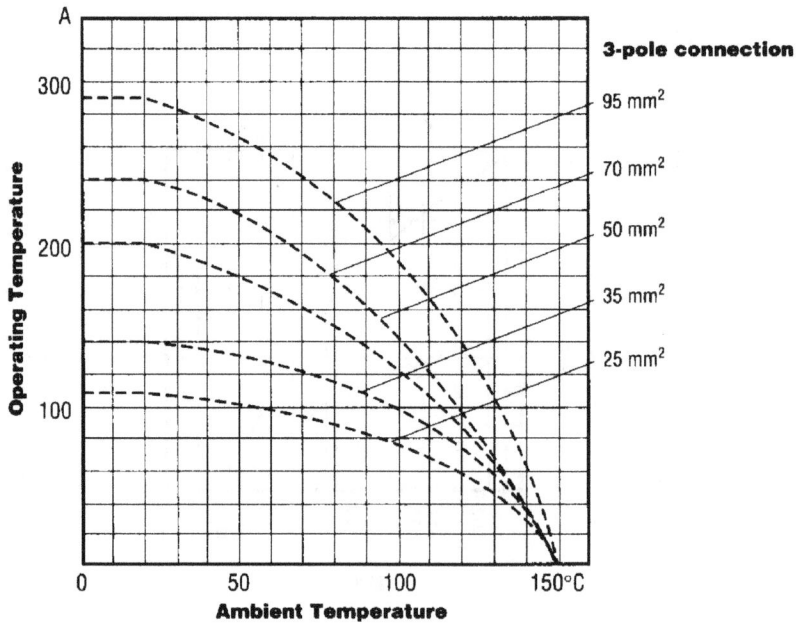

Figure 17.6 Temperature-current curves for a high-current connector *(reprinted with permission of Multi-Contact).*

including stress relaxation, number of insertion/removal cycles, magnitude of the spring force, and resistance to permanent set. Insertion force may be reduced by changing the geometrical shape of the pins, and by the application of various coatings to the contact surfaces. Horn and Egenolf[2] present concepts and formulas for shape optimization of the contact pin and socket for reduced insertion force and reduced wear.

Figure 17.7 illustrates three contact pin shapes, including an unmodified (original) pin and the pin with optimized shape. Figure 17.8 displays the measured insertion force versus distance for the optimized and original pins. It is observed that the steady-state force is the same for the optimized and original pins; however, the peak insertion force is reduced from 88 cN to 65 cN, a reduction of almost 25 percent. The article concludes by stating that, in one case, the contact insertion force prior to optimization was about 50 N. The revised shape with the geometrically optimized contact pin resulted in a reduced insertion force of about 20 N.

Contact Block Insulating Material.   Contact block insulating material is manufactured from high-quality, complex plastic materials with unique electrical and mechanical prop-

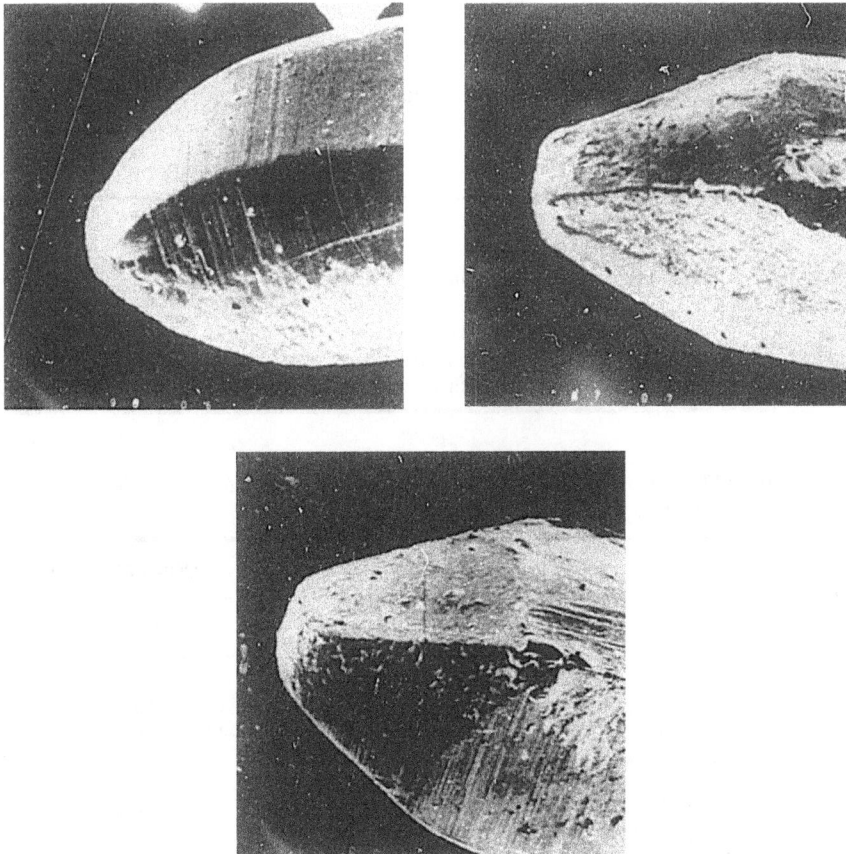

Figure 17.7 Three illustrations of contact pin shapes, including original and pins with optimized shapes *(reprinted with permission of AMP).*

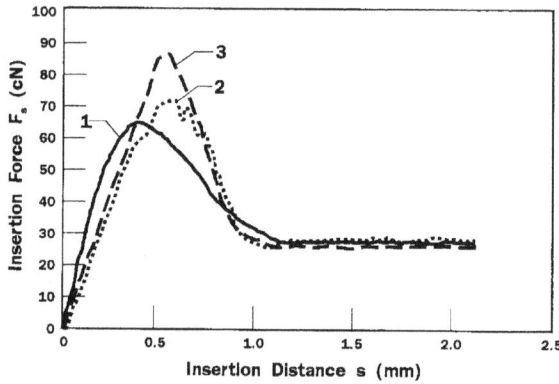

Figure 17.8   Measured insertion force vs. distance for optimized and original wedge-shaped pins *(reprinted with permission of AMP)*.

erties. Table 17.2 presents plastic insulating materials used by one company in the manufacture of connectors. Table 17.3 lists the operating temperatures of the connector materials. Table 17.4 displays the electrical and mechanical properties of the plastic insulating materials produced by one manufacturer.

Table 17.2   Plastic Insulating Materials *(courtesy of W.W. Fischer Electronic Connectors, Inc.)*

International symbol	Chemical name	Commercial name	General performance
PAI	Polyamide-imide	Torlon	Physical/chemical/electrical
PBT	Polybutylene-terephthalate	Celanex, Tenite, Valox	Physical/electrical
PEEK	Polyether-etherketone	Victrex PEEK	Physical/chemical/electrical
PTFE	Polytetra-fluorethylene	Tevlon PTFE, Halon	Temperature/chemical/HF
POM	Polyoxy-methylene	Delrin, Amcel, Kematal	Mechanical/chemical

Table 17.3   Operating Temperatures of the Connector Materials *(courtesy of W.W. Fischer Electronic Connectors, Inc.)*

	Operating temperature limits	
	Long-term	Short-term
Standard sealed connectors without sealing rings		
Contact block of PAI and PEEK	−65 to +200° C	+220° C
Contact block of PTFE	−65 to +160° C	+175° C
Contact block of PBT	−65 to +135° C	+150° C
Standard sealed connectors	−65 to +135° C	+175° C
Free connectors, e.g. SE, SVE, SOVE, WSE		
Fixed connectors, e.g., DEE, DEU, DBEE, DBEU, DBPE, DBPU, WDE	−25 to +135° C	+175° C
Delrin (POM) covers and collets	−40 to +100° C	−50 to +150° C

Table 17.4 Electrical and Mechanical Properties of Insulating Materials *(courtesy of W.W. Fisher Electronic Connectors, Inc.)*

Characteristics*	ASTM	International standards	DIN	Measurement units	PAI	PBT	PEEK	PTFE
Volume resistivity	D-257	IEC 93	53 482	Ω cm	10^{17}	$>10^{14}$	$> 10^{16}$	10^{18}
Dielectric strength (3 mm sample)	D-149	IEC 243	53 481	kV/mm	23	19	19	20
Dielectric constant at 10 kHz	D-150	IEC 250	53 483		3.5	3.9	3.3	2.1
Dielectric loss tangent at 1 MHz	D-150	IEC 250	53 483		0.009	0.016	0.003	0.0003
Tensile strength	D-638	ISO R527	53 455	N/mm^2	190	120	130	27
Flexural strength at 23° C	D-790	ISO R178	53 452	N/mm^2	215	170	200	no breakage
Flexural modulus at 23° C	D-790	ISO R178	53 457	kN/mm^2	4.6	8.0	6.5	0.54
Maximum continuous temperature	UL-746			°C	260	140	250	(250)
Flammability rating (2 mm sample)	D-635	UL – 94			V-0	V-0	V-0	†
Limiting oxygen index, LOI	D-2863	ISO 4589		% 0	43	31	35	95
Moisture absorption, 24 hr at 23° C	D-570	ISO R62A	53 495	%	0.28	0.2	0.3	0.01
Steam sterilization‡					limited	limited	very good	very good
Ionizing radiation resistance Co60				Rad	10^9	10^8	10^9	10^4
Natural color/other color					greenish	(black)	gray-beige	white

* Typical values based on manufacturers' data. The use of connectors under certain environmental conditions requires specific tests that take into consideration all of the specifications of all the materials used.
† Not UL listed.
‡ W.W. Fischer's own experimental values, to be taken as a guide only.

Plastic materials used to manufacture the connector housing are selected for their unique properties and tested accordingly. Materials are tested to U.S. and international standards to determine their characteristics under specified operating conditions. On comparison, each insulating material has certain properties that may prove to be superior or inferior. Properties examined include dielectric strength, dielectric constant, dielectric loss, volume resistivity, surface resistivity, tensile strength, flexural strength, maximum continuous operating temperature, flammability rating, moisture absorption, and arc resistance.

Contact Material.   Because the connector pin is exposed to high mechanical, electrical, and thermal stresses, the contact material must have good properties in these three areas. Due to contact resistance, the contact mating surfaces generate heat. It is essential that the electrical conductivity be as high as possible to have minimum heat generation. Furthermore, the contact material is called upon to have high thermal conductivity to rapidly conduct the heat to the adjoining cables and surroundings.

Table 17.5 presents the electrical and thermal conductivity of several contact materials. Beryllium copper alloys are known to have high electrical and thermal conductivities with

Table 17.5 Electrical and Thermal Conductivity *(reprinted with permission of Brush Wellman, Inc.)*

Alloy	Temper	Electrical (% IACS min)	Thermal BUT/ft · hr · F	W/m · K	Conductivity product % IACS × BTU/ft-hr/F
25	AT 1/4 HT 1/2 HT HT	22	60	105	1320
190, 290	All	17	60	105	1020
165	AT 1/4 HT 1/2 HT HT	22	60	105	1320
	AM 1/4 HM 1/2 HM HM SHM XHM	18	60		1080
3	AT HT HTR HTC	45 48 48 60	140	240	6300
174	1/2 HT HT	50 48	135	230	6750 6480
260	H	28	70	120	
194	S	60	150	260	
510	1/2 H H S	15 15 15	40	70	600
521	1/2 H H S	13 13 13	36 36 36	62	468
654	1/2 H, H X, XS	7	21	36	147
688	1/2 H, H S XS	17	40	69	680
725	1/2 H, H S	11	31	54	341 310
7025	TM02 TM03	40 35	98		3920 3430

excellent mechanical properties. They are used extensively in current-carrying mechanical contact applications such as high-power connectors, fuse holders, and so on where the high power generated in a small, active area requires very rapid dissipation of heat to ensure that the temperature rise remains within acceptable limits. Figure 17.9 presents the current-temperature curves for brass crimp contacts and tellurium copper contacts. It is observed that, for a rated current of 20 A, the temperature rise varies from 40° C for tellurium copper crimp contacts to 61° C for brass crimp contacts, a 21° C difference due to contact material.

Contact Finish. A contact finish provides corrosion protection for the base metal contact spring and optimizes contact surface characteristics. Mroczkowski[3] presents a fairly detailed analysis of contact finish. The reference states that durability of a contact is dependent on several factors including:

1. Contact finish
2. Normal force
3. Contact geometry
4. Mating distance
5. Mating cycles

Table 17.6 presents a summary of contact finish properties for connectors.

Table 17.6 Summary of Contact Finish Considerations *(reprinted with permission of AMP)*

Finish	Contact resistance (mΩ @ 100 g)	Hardness Knoop	Coeff. of friction	Durability
Gold cobalt	1.1	130/240	0.2/0.5	Good–very good
Inlay	0.6	25	1	Fair–good
Palladium	1.4	150/350	0.3/0.5	Good–very good
Palladium/silver 60/40 (inlay)	1.7	100	0.7	Very good
Palladium/nickel 80/20 (plate)	8.0	400/500	0.3/0.5	Very good
Silver	0.5	50/125	0.5/0.8	Fair
Tin–matte	0.8	10	0.6/1.0	Poor–fair
Tin–bright	1.0	20	0.4/0.6	Poor–fair
Tin–hot dip*	1.2	10	0.3/0.6	Poor–fair
Tin/Lead 93/7	0.7	10	0.5/0.8	Poor–fair

* Values are quite process dependent due to intermetallic compound effects.

Effect of Lubricants. During connector mating, the contact pin shape determines the insertion force required. High-density, multiple-circuit connectors may call for low inser-

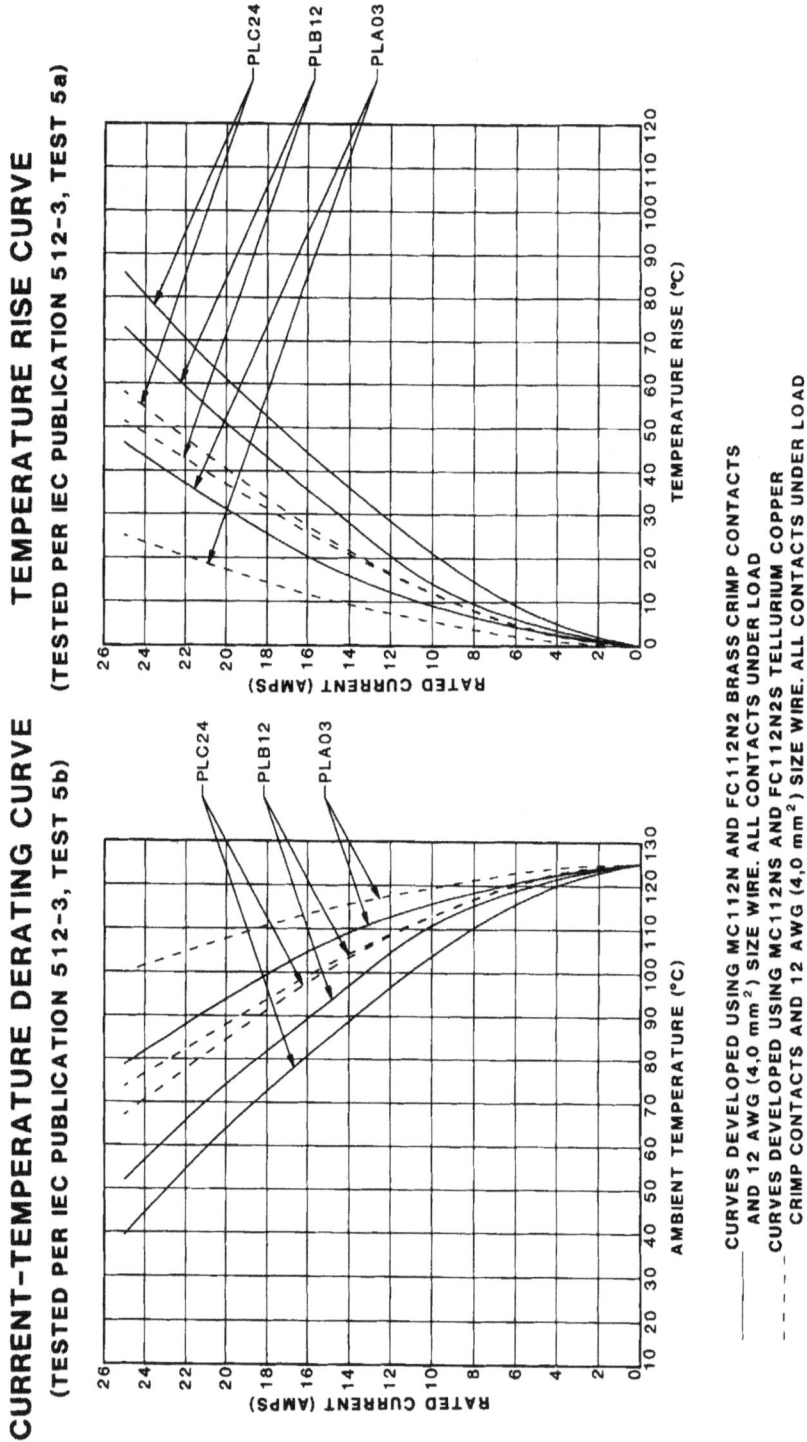

Figure 17.9 Current-temperature curves for brass crimp contacts and tellurium copper contacts *(reprinted with permission of Positronic Industries, Inc.)*.

tion force, requiring optimized geometrical shape and the use of lubricants. P. van Dijk[4] provides information on the effects of lubricants on electrical contacts. Certain problems are associated with lubricants, including cost, long-term reliability, collection of electrically conductive and nonconductive contaminants, and so on. The paper states that the force required to compress the lubricant into a thin film to permit conduction at 50 mV is about 10 N.

Figure 17.10 presents the load-wipe curves for gold-over-nickel surfaces when clean and after being lubricated with petroleum jelly. Note that at low force (< 10 N), the lubricated surface has a contact resistance exceeding 1 Ω, whereas the clean surface has a contact resistance of less than 10 mΩ. With an applied force of 10 N, the lubricated surface has a contact resistance of about 2 mΩ, and the clean surface has a contact resistance of approximately 1 mΩ. Above 10 N, the contact resistance of the lubricated surface is just slightly higher (approximately 0.5 mΩ) as compared to the clean surface.

Operation in Harsh Environments.   The transportation industry, employing connectors in transit systems, aerospace, and shipboard applications, may subject connectors to corrosive environments, exposure to humidity, immersion in a wide range of conductive, corrosive fluids such as salt water, and petroleum products such as gasoline, engine oils, and jet fuel (JP-4).

Connectors specified for use in damp or wet environments can be exposed to dripping, splashing, spraying, or even submersion in liquids. Connectors used in wet environments must be adequately sealed by mechanical or other means to prevent moisture and other liquids from penetrating the housing and coming into contact with the energized pins. The seal must prevent moisture from seeping into the active area even under prolonged exposure.

Connectors used in the above applications are quite often sealed using a mechanism that will provide adequate environmental protection at the contact interface. The sealing mechanism can employ a variety of techniques including potting compounds, heat shrinkable boots, rubber grommets, thermoplastic elastomer seals, O-rings, and gaskets. Table 17.7 is a list of elastomers used by one manufacturer to seal connectors. Figure 17.11 is a cross-sectional view of a moisture-sealed cylindrical connector using silicone sealing grommets and a silicone gasket.

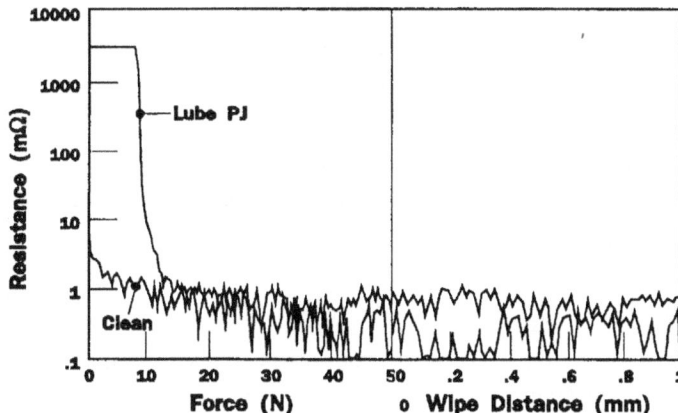

Figure 17.10   Load-wipe curves for gold-over-nickel surfaces *(reprinted with permission of AMP).*

Table 17.7 List of Elastomers *(courtesy of W.W. Fischer Electronic Connectors, Inc.)*

Compound and trade name	Chemical name	Temperature range (ASTM D-1414)	Excellent resistance to
NBR BUNA N	Acrylonitrile-butadiene rubber	−30 to +110° C	Weather, ozone, fuels, oils
NBR per MIL-P-25732	Acrylonitrile-butadiene rubber	−54 to +135° C	Weather, ozone, fuels, mineral oil based hydraulic fluids, military applications
FPM VITON	Fluoro elastomer	−20 to +200° C	Acids, weather, ozone, fuels, mineral and silicone oils, high vacuum, gamma rays
CR NEOPRENE	Polychloroprene elastomer	−40 to +100° C	Alcohol, ammonia, chlorine, ozone, Freon™, soda, vegetable and silicone oils
EPDM, EPM (EPR)	Ethylene-propylene diene elastomer	−50 to +160° C	Alcohol, weather, hot water, vapor, brake fluids, detergents, gamma rays
TPE-S Thermoplastic elastomer	Styrene-ethylene-butadiene-styrene	−70 to +130° C	Very resistant, except to aromated and chlorinated hydrocarbons

Mechanical Durability. In some applications, connectors may undergo a limited number of mechanical insertion/extraction cycles. However, in harsh mechanical applications, connectors may experience numerous insertion/extraction cycles and may be exposed to mechanical stress and vibration. Additionally, a multiple-circuit connector may be subjected to tangential (oblique) forces when connected to a multiconductor braid with a small bending radius.

Insertion/Extraction Force. During insertion, a connector is plugged into its mating surface; each male contact pin mating with its corresponding female pin. This requires an expenditure of force, the magnitude of which depends on several factors including the number of contact pins, individual pin size, pin construction and geometry, and the material used for plating the individual pins. During extraction, a negative force is expended. It is desirable to have minimal insertion force per contact pin to allow for easy insertion without mechanical damage to the contact pins and the connecting cables. Conversely, it is desirable to have a high extraction force to ensure that the connector pins have low contact resistance, even under mechanical stress, vibration, and extended mechanical insertion/extraction cycles. Connector insertion/extraction force ranges from a few ounces to several tens of pounds. Figure 17.12 exhibits an insertion/extraction force requirement versus displacement for a 340-A, high-current battery connector. The typical specified insertion/extraction force for this connector is 15 lb.

17.2.4 Terminal Blocks

Terminal blocks belong to a family of connectors with screw-type terminals mounted on rigid insulating material, permitting easy insertion and removal of the conductors. Terminal blocks may be of various types, such as printed circuit (PC) board type, rail-mounted, double- or triple-level type, and so forth. Smart terminal blocks may also have design options such as integral relays, replaceable fuses, surge protection, or thermocouple termi-

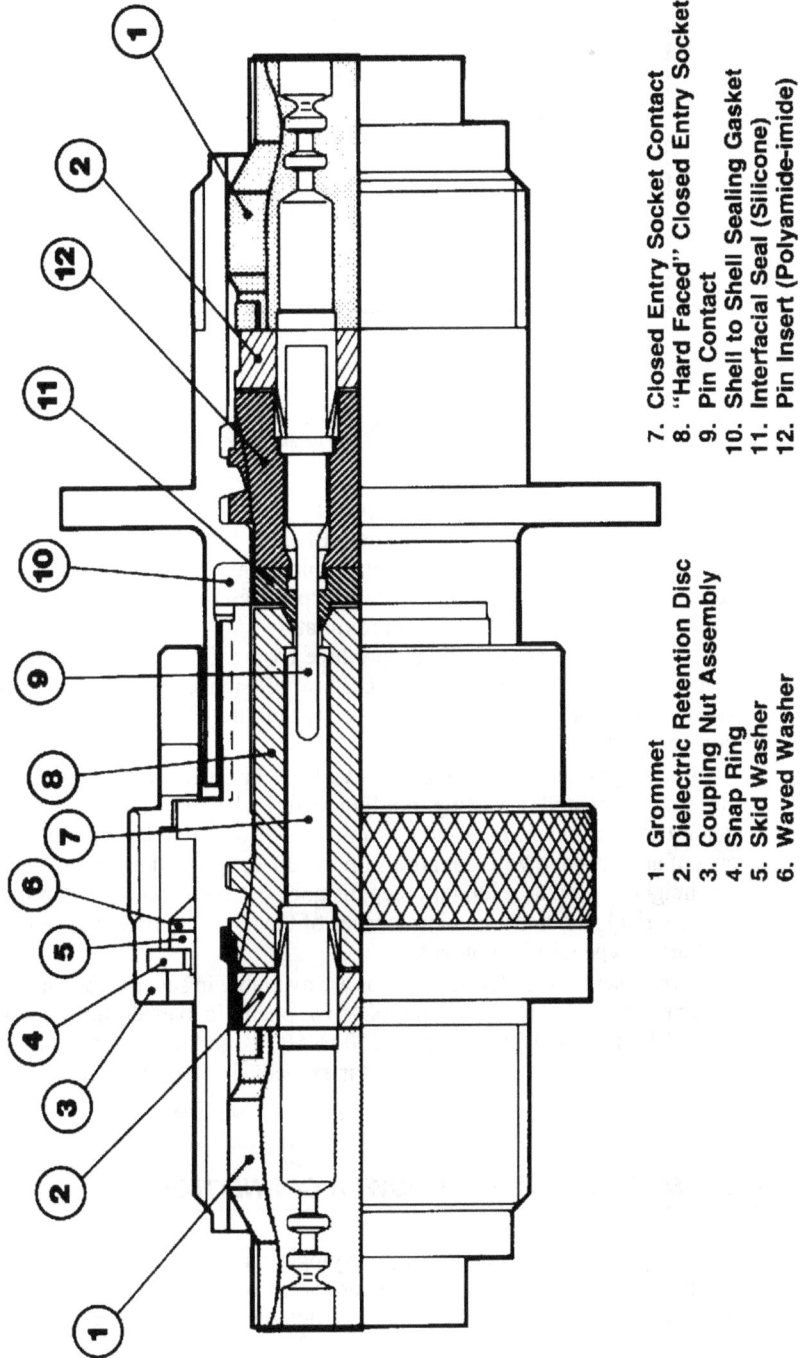

1. Grommet
2. Dielectric Retention Disc
3. Coupling Nut Assembly
4. Snap Ring
5. Skid Washer
6. Waved Washer

7. Closed Entry Socket Contact
8. "Hard Faced" Closed Entry Socket Insert
9. Pin Contact
10. Shell to Shell Sealing Gasket
11. Interfacial Seal (Silicone)
12. Pin Insert (Polyamide-imide)

Figure 17.11   Cross-sectional view of a moisture-sealed cylindrical connector *(reprinted with permission of Amphenol Corp.)*

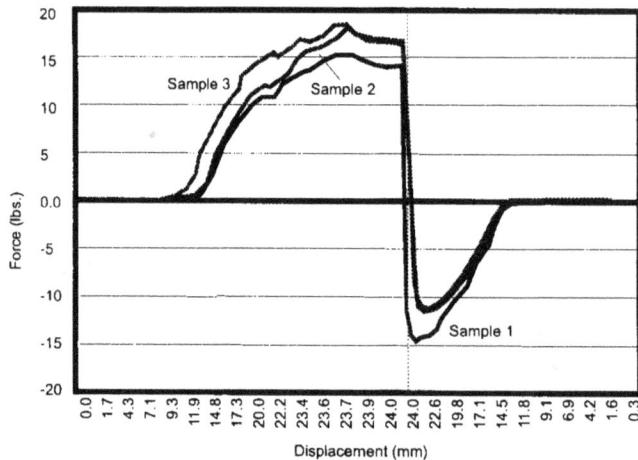

Figure 17.12   Insertion/extraction force requirement vs. displacement for a 340-A high-current battery connector *(reprinted with permission of KonneKtech).*

nals. DIN-rail terminal blocks, adopted by the IEC, have proliferated many markets such as the programmable logic controller (PLC) industry. Printed circuit terminal blocks for low- and medium-power applications typically use #8 to #28 AWG wire, and nylon insulating material (polyamide 6.6) with an upper temperature limit of 100° C. Typical terminal blocks exert a pressure of approximately 1000 lb/in^2 between the conductor and the current bar. Some manufacturers produce terminal blocks with a unique clamping action that provides a high extraction force, low resistance, and secure connection even in the presence of vibration and thermal cycling. Figure 17.13 shows a rail-mount interconnection system from one manufacturer.

Barrier-type terminal blocks are available in a variety of sizes, electrical voltages and current ratings, and insulating molding materials. Figure 17.14 is a cross section of a pluggable terminal block. High power units are also available with current ratings of several tens of amperes and voltage ratings up to 600 V. Selected units may have an increased barrier height to provide a longer creep distance between adjacent terminals. Table 17.8 gives the electrical and mechanical properties of the molding material used by one manufacturer for barrier type terminal blocks.

Barrier, screw-type terminal strips are available in several sizes and various voltage and current ratings. A range of double-row phenolic barrier terminal strips is shown in Table 17.9 for a voltage rating of 250 to 600 V, and a current rating of 10 to 50 A. Note that the specified withstand voltage far exceeds the normal voltage rating of the devices.

17.3   THERMAL MODEL OF TYPICAL POWER CONNECTOR

Hot pluggable power connectors carrying tens to hundreds of amperes are used extensively in high-power systems. The connector voltage drop and subsequent contact power dissipation are important variables that determine the temperature rise of the connection system.

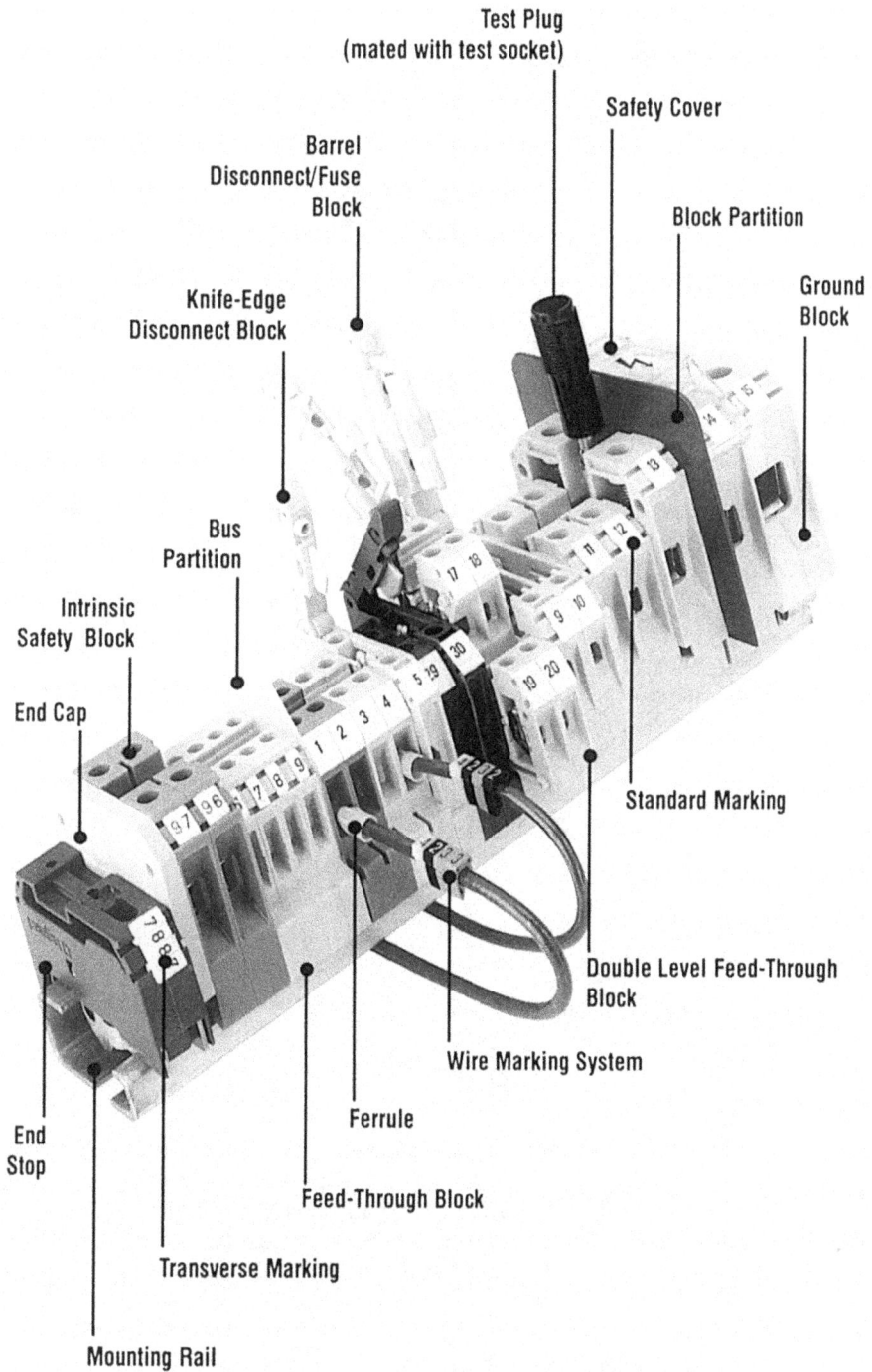

Figure 17.13 Rail-mount interconnect system overview *(reprinted with permission of PCD, Inc.).*

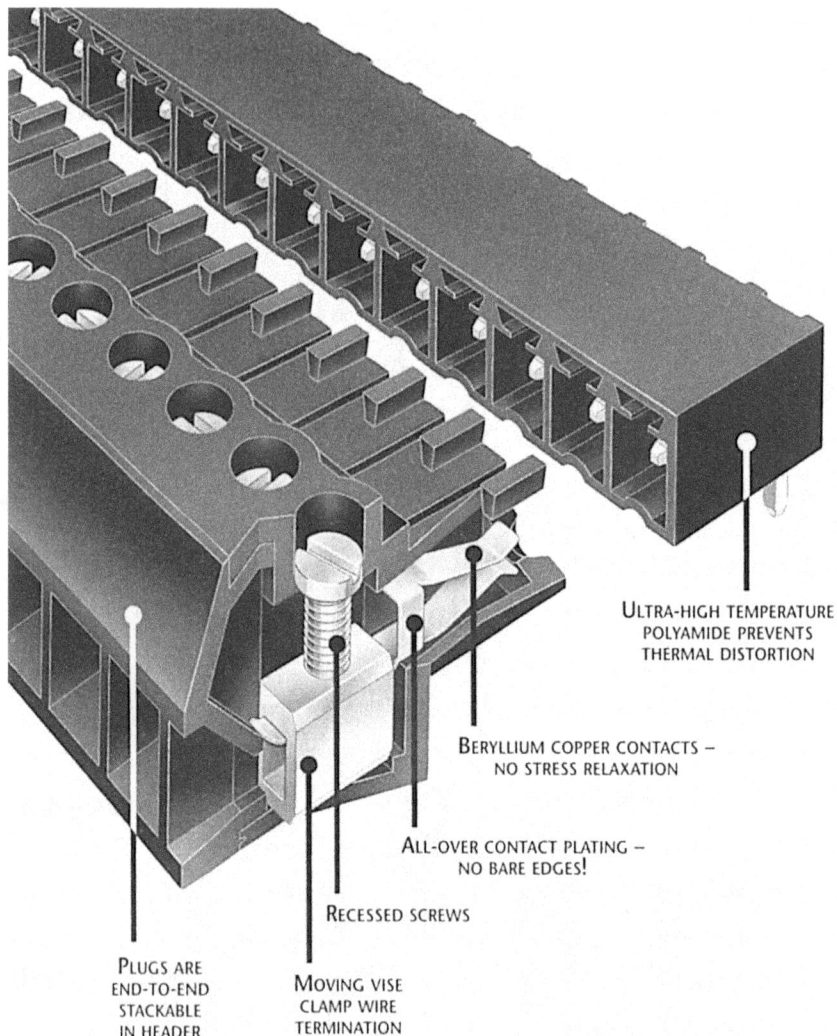

Figure 17.14 Cross section of a pluggable terminal block *(reprinted with permission of PCD, Inc.)*.

The methodology of thermal calculations for high-power connectors requires a good understanding and characterization of the connection system, with an exhaustive knowledge of the different modes of heat generation, distribution, and dissipation in the analyzed system. The complex nature of the connector elements and the prevailing degree of thermal coupling between adjacent circuits require an appreciation of heat transfer mechanics and a comprehensive knowledge of the mathematics of linear transformations. Determination of the elemental temperatures and the corresponding thermal governing relationships may require solutions to many interrelated heat transfer differential equations, taking into consideration the thermal resistances (conductive and convective) as necessary. Furthermore, obtaining a transient or steady-state solution may require numerous

Table 17.8 Kulka Molding Materials—Applications and Comparisons *(reprinted with permission of Marathon Special Products)*

Application of molding materials (terminal boards)

GP general-purpose phenolic
An economic material with good appearance, good electrical properties, and well balanced physical properties.

GDI-30F diallyl phthalate long glass fiber filled
Good strength characteristics combined with physical stability and good electrical properties. Retains high insulation resistance after prolonged exposure to humidity. Flame resistant and self-extinguishing.

MAI-60 alkyd glass fiber filled
For applications requiring high strength and good electrical properties. After prolonged exposure to high humidity, there is a reduction of insulation resistance.

Thermoplastic
A tough, flame-retardant plastic. Excellent electrical properties, flexibility, and toughness.

Comparisons of molding materials

The information below is compiled from various sources and is presented to facilitate comparison among the various materials. Values given are average and are not to be considered as absolute values.

Material Kulka code Military specification	General-purpose Phenolic GP	MAI-60 D MIL-M-14	GDI-30F JJ MIL-M-14	Thermoplastic RZ
Composition				
Filler	Cellulose	Glass	Glass	—
Resin	Phenolic	Alkyd	DAP	Thermoplastic
Color	Black	Grey	Gray	Translucent/black
Impact strength				
Ft lb/in	0.28	6.0	3.0	—
Tensile strength				
PSI $\times 10^3$	7.0	3.5	9.5	3.4
Flexural strength				
PSI $\times 10^3$	9.0	12.0	10.0	—
Volume resistivity				
Ω/cm	10^{11}	10^{12}	10^{13}	10^{12}
Dielectric strength				
Short time volts mil	350	150	325	Excellent
Arc resistance, seconds	—	130	125	—
Specific gravity	1.37	2.1	1.86	1.39
Water absorption–%24 hr	0.6	1.5	0.5	0.4
Burning rate	94 V-O	Self-ext.	Self-ext.	Slow
Heat resistance				
Continuous °F	290	300	400	250

iterations of the generated set of heat flow equations. Due to the high degree of complexity involved, manual computational procedures are cumbersome and time consuming. Modeling techniques are often used to assist in solving the heat transfer equations and to expedite a reasonable solution. The number of elements in the connection system model, and the degree of complexity of each element within the model, are determined by the computing time and effort available and by the accuracy required. A generalized thermal model is

Table 17.9   Double-Row Phenolic Barrier Terminal Strips (adapted from and reprinted with permission of Vernitron Corp., Beau Products Division)

Series	10	14	18	19	21	21CB	28
Terminal centers	0.250 in (6, 4)	0.375 in (9, 5)	0.438 (11, 1)	0.438 in (11, 1)	0.563 in (14, 3)	0.563 (14, 3)	0.688 in (17, 5)
UL current rating	10 A	15 A	20 A	20 A	30 A	30 A	50 A
Construction	Open-back and feedthrough	Open-back and feedthrough	Open-back and feedthrough	Closed-back and feedthrough	Open-back and feedthrough	Closed-back and feedthrough	Open-back and feedthrough
Voltage rating, rms							
1. UL recognized							
Class B: commercial equip.	250	250	250	250	600	600	250
Class C: General industrial	150	150	300	300	600	600	250
2. CSA certified	150 @ 5 A	300 @ 15 A	300 @ 20 A	300 @ 20 A	300 @ 30 A	300 @ 30 A	600 @ 50 A
3. Withstand voltage, Vdc	3,000	8,500	7,500	8,500	10,500	9,000	10,500
Insulator material	General-purpose phenolic UL temp. index 150° C Color: black	General-purpose phenolic UL temp. index 150° C Color: black	General-purpose phenolic UL temp. index 150° C Color: black	General-purpose phenolic UL temp. index 150° C Color: black	General-purpose phenolic UL temp. index 150° C Color: black	General-purpose phenolic UL temp. index 150° C Color: black	General-purpose phenolic UL temp. index 150° C Color: black
Width × height, in (mm)	0.63 × 0.31 (16 × 7.9)	0.88 × 0.41 (22.2 × 10.3)	1.12 × 0.50 (28.5 × 12.7)	1.12 × 0.55 (28.5 × 13.9)	1.31 × 0.63 (33.3 × 16)	1.31 × 0.66 (33.3 × 16.8)	1.81 × 0.75 (46.0 × 19.1)
Wire size recommended, max. AWG	22	16	14	14	12	12	No. 6 lugged
Terminal screws (standard)	No. 2-56 × 3/16 in binding head screws, nickel-plated brass	No 5-40 × 1/4 in binding head screws, zinc-plated steel	No. 6-32 × 1/4 in binding head screws, zinc-plated steel	No. 6-32 × 1/4 in binding head screws, zinc-plated steel	No. 8-32 × 5/16 in binding head screws, zinc-plated steel	No. 8-32 × 5/16 in binding head screws, zinc-plated steel	No. 10-32 × 3/8 in binding head screws, nickel-plated brass

shown in Fig. 17.15, and a fairly complex thermal model involving more than 125 elements is cited in Sec. 17.8.

17.3.1 Constructing the Electrical Equivalent Circuit of the Thermal Model

The following is a summary of the functional steps for generating a thermal model of a typical connection system.

1. Thermal resistance RTH, calculated from mechanical dimension or by other means, is modeled using an equivalent electrical resistance RTH.

2. Calculated thermal capacitance CTH is modeled using an equivalent electrical capacitor. For steady-state conditions, the thermal capacitance may be ignored.

3. The I^2R Joule thermal heat source is modeled as an equivalent current source.

4. Element temperatures in the thermal model have a one-to-one relationship to the respective voltages in the electrical model.

5. Convection losses may be modeled using equivalent electrical resistance elements.

6. Ambient temperatures, modeled using voltage sources at corresponding nodes, are used to establish the thermal boundary conditions for the connector thermal model.

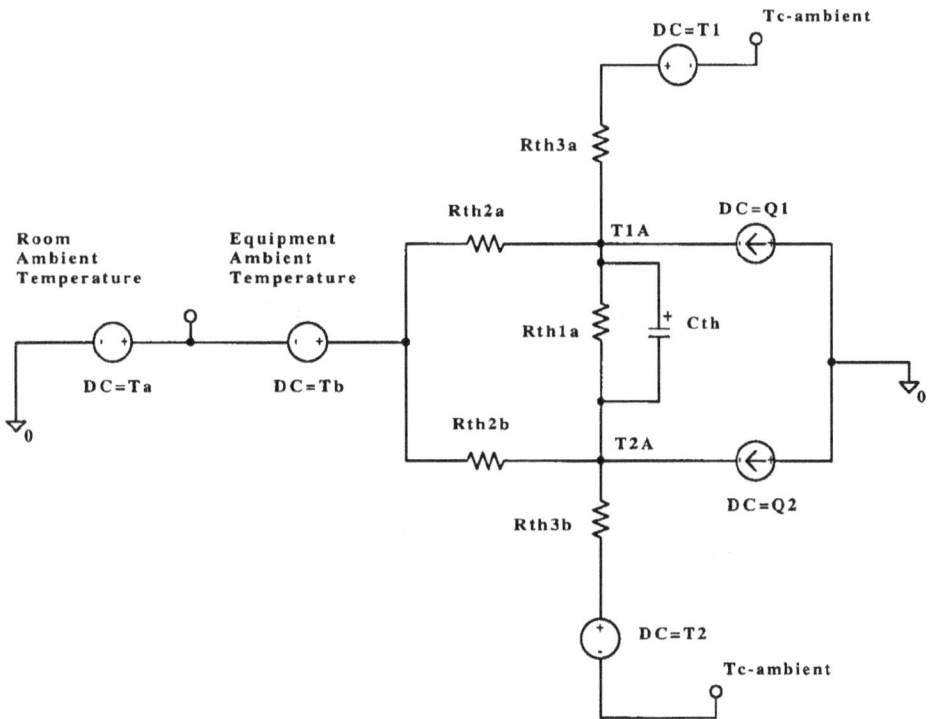

Figure 17.15   Generalized thermal model of a typical power connector *(courtesy of Failure Analysis Associates).*

7. Power contacts and conductors, circumferentially enclosed by other power contacts and conductors, may experience an increase in ambient temperature in the immediate vicinity of the enclosed contacts. This increase in localized ambient temperature can be modeled using mathematical equations to iteratively generate the forcing functions. Commercial SPICE versions that offer behavioral modeling options, permit the insertion of numerical equations into the electrical model and are appropriate for this modeling technique.

8. If necessary, thermal impedances of the insulation material may also be modeled as an equivalent electrical resistance.

9. In lieu of lumped circuit elements, distributed circuit elements may be used in the modeling with increased circuit complexity and computation time.

Depending on the degree of complexity and the required output format of the solution, sophisticated computer simulation programs such as MicroSim Corporation's PSpice® may be used to assist in providing the equivalent electrical solutions. The ultimate goal of this thermal audit is to generate a temperature profile of the connection system in an effort to determine the maximum temperatures of each element in the connection system.

17.4 CONNECTOR-TO-CONDUCTOR CONNECTIONS

In a typical connector, each contact pin comes into direct mechanical contact with its associated conductor or other connecting surface. The metal connection between the contact pin and the conductor may be mechanical such as with a crimped or wire-wrapped connection, or it may be metallurgical, such as with a soldered or welded connection. Crimped connections, briefly discussed below, are extensively used in low-, medium-, and high-power connectors and are made by employing a crimping tool that deforms the conductor and connector barrel in a predetermined manner.

17.4.1 Crimped Connections

One manufacturer, Molex,[5] defines crimping as follows:

> Crimping is the metallurgical compression of a terminal around the conductor of a wire, which creates a common electrical path of low resistance and high current carrying capabilities. A secondary crimp around the insulation of the wire provides wire support for insertion into a housing and allows the terminal to withstand shock and vibration.
> The technology of crimping was developed to provide a high quality connection between a terminal and a wire at a relatively low applied cost. This technology was developed as a substitute to soldering and the problems associated with that technology....

Figure 17.16 shows primary and secondary crimps on a conductor. Crimp tooling includes basic hand tools such as hand crimpers, a stripper, a crimper, a press with interchangeable dies, and a fully automated system. High-speed systems are available with speeds in excess of 3000 terminations per hour. Figure 17.17 shows selected hand crimping tools that can produce output forces up to 6.2 tons, for wire sizes up to 4/0.

Wire Brush
(Conductor Strands)

Conductor Form
(Conductor Grip)

Insulation Form
(Insulation Grip)

Contact Area
(Molex Terminal)

Strip Length

Transition Area
(Contact Area & Conductor Form)

Cut-off Tab

Wire Insulati

Transition Area
(Conductor & Insulation Forms)

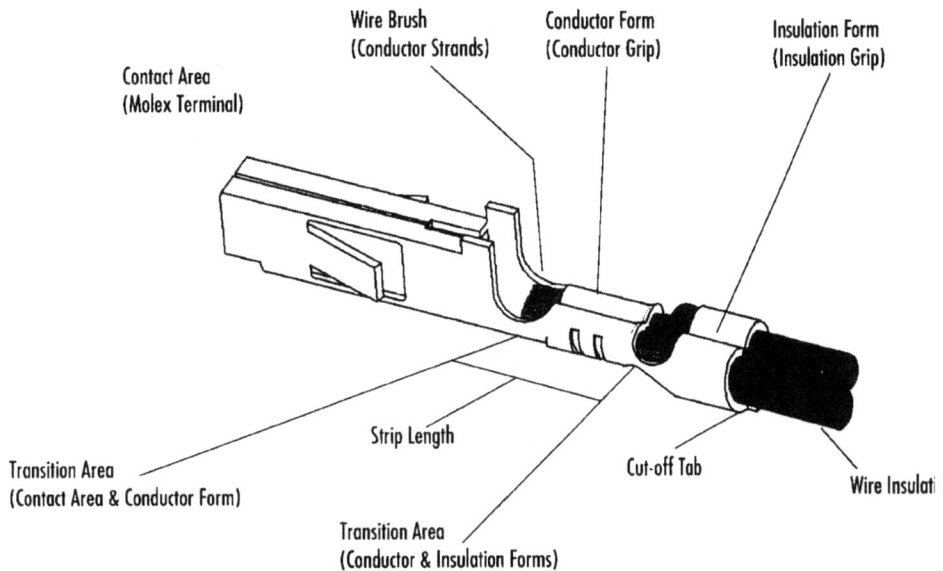

Figure 17.16 Primary and secondary crimp on a conductor *(reprinted with permission of Molex).*

17.5 CONNECTOR STANDARDS

Electrical connectors are used in many different applications involving high voltage, high current, high levels of power, and high mechanical cycling operations. Connector insulating material must be nonflammable, have high dielectric strength, and be suitable for operation at high temperatures. Connectors may be subjected to many insertion and extraction cycles and may be exposed to corrosive and/or electrically conductive contaminants. Standardized test methods have been established to determine the suitability of connectors under specified operating conditions. Selected connector standards, including U.S. and European standards, are covered in this section.

The extensive use of connectors in commercial, industrial, and military environments has led to a excess of information on the design, testing, and use of connectors in industry. The accumulated information is, in large part, provided as technical support to the design engineer as guidelines for connector design and specification.

Connector standards permit the designer to make an intelligent engineering selection of the particular electrical connector that will adequately meet the design requirements and conform to the specified operating and environmental conditions. Standards are often referenced in contractors' procurement specifications, and requests for proposals.

Standards present accepted test procedures, allow quality monitoring of components and equipment, and permit recognized testing agencies to ensure that established safety and reliability procedures are rigorously observed.

Due to the large volume of connector standards currently available to the engineer, it is not feasible to list all the individual standardizing agencies or the vast matrix of connector standards and specifications available. However, several connector standards that are currently in use include those developed by EIA, UL, and IEC.

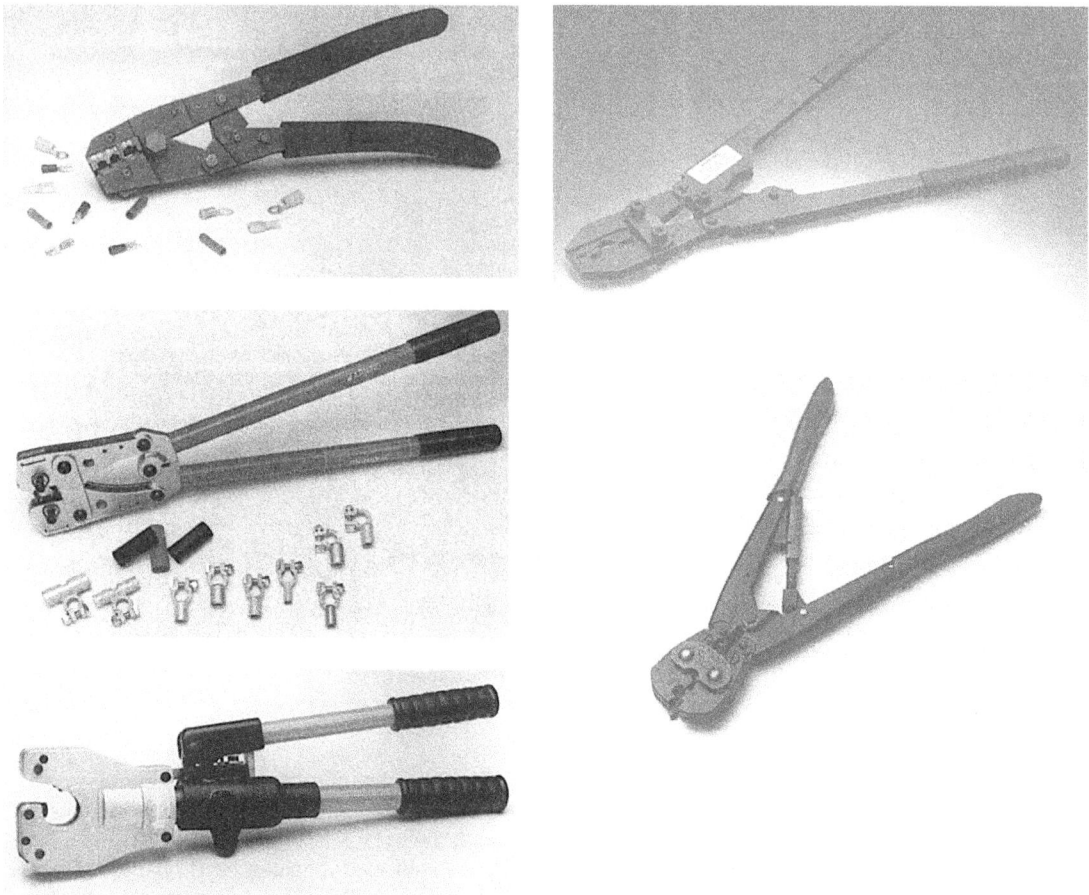

Figure 17.17 Crimping tools *(reprinted with permission of Molex)*.

17.5.1 Electronic Industries Alliance Standards

The Electronic Industries Alliance (EIA) (formerly Electronic Industries Association) generates many standards outlining testing procedures for connectors, including environmental classifications, manual force test procedures, contact insertion/removal force, contact resistance test procedures, crimp tensile, fluid immersion, resistance to solvents, withstand voltage, insulation resistance, aging, corrosion, capacitance, humidity, thermal shock, inductance measurements, cable pull out, corona, flame, sand and dust, current cycling, and temperature rise.

The following is a partial list of EIA standards covering established test procedures for connectors:

- EIA-364-B-1—Insulation Material Batch Acceptance Tests
- EIA-364-04—Normal Force Test Procedure
- EIA-364-05A—Contact Insertion, Release, and Removal Force Test Procedure

- EIA-364-06A—Contact Resistance Test Procedure for Electrical Connectors
- EIA-364-07A—Crimp Contact Deformation Test Procedure
- EIA-364-08A—Crimp Tensile Test Procedure for Electrical Connectors
- EIA-364-10—Fluid Immersion Test Procedure for Electrical Connectors
- EIA-364-11—Resistance to Solvents Test Procedure
- EIA-364-13A—Mating and Unmating Forces Test Procedure
- EIA-364-15—Contact Strength Test Procedure
- EIA-364-17A—Temperature Life with or without Electrical Load Test Procedure
- EIA-364-20A—Withstanding Voltage Test Procedure
- EIA-364-21B—Insulation Resistance Test Procedure
- EIA-364-23A—Low Level Contact Resistance Test Procedure
- EIA-364-24A—Maintenance Aging Test Procedure
- EIA-364-26A—Salt Spray (Corrosion) Test Procedure
- EIA-364-27B—Mechanical Shock (Specified Pulse) Test Procedure
- EIA-364-28B—Vibration Test Procedure
- EIA-364-30—Capacitance Test Procedure
- EIA-364-31A—Humidity Test Procedure
- EIA-364-32B—Thermal Shock Test Procedure
- EIA-364-33—Inductance of Electrical Connectors
- EIA-364-38A—Cable Pull-Out Test Procedure
- EIA-364-41B—Cable Flexing Test Procedure
- EIA-364-43A—Cable Clamping (Bending Moment) Test Procedure
- EIA-364-44—Corona Testing, Test Procedure
- EIA-364-45—Flame Test Procedure for Firewall Electrical Connectors
- EIA-364-50—Sand and Dust Test Procedure
- EIA-364-55—Current Cycling Test Procedure
- EIA-364-69—Low Level Inductance Measurement for Electrical Contacts of Electrical Connectors
- EIA-364-70—Test Procedure for Current vs. Temperature Rise

17.5.2 Underwriters Laboratories Standards

Underwriters Laboratories® Inc. (UL) evaluates materials, components, and systems and is one of the most prominent protection organizations in the industry. UL standards for safety provide regulations pertaining to fire and/or electrocution. The intent of the UL standard is listed on the title page of their catalog[6] as follows:

> UL Standards for Safety represent the judgment of Underwriters Laboratories Inc. as to the basic engineering requirements for the products covered by UL under each product category. These requirements are based upon sound engineering principles, research, records of tests and field experience....

A partial list of UL standards is listed below:

- UL 486-A—Connectors and Soldering Lugs for use with Copper Conductors, Wire
- UL 486-C—Splicing Wire Connectors
- UL 486-D—Connectors for use with Underground Conductors, Insulated Wire
- UL 218-A—Connectors for use in Diesel Engine Driving Centrifugal Fire Pumps, Battery
- UL 1059—Terminal Blocks
- UL 310—Terminals, Electrical Quick Connect

17.5.3 Navy Power Supply Standard

The Navy Power Supply Reliability Design and Manufacturing Guidelines, January 1989, designated as NAVMAT P-4855-1A, provides design and manufacturing guidelines for low- and high-voltage switch-mode power supplies. The standard presents component derating criteria for contact current (50 percent derating), and dielectric withstand voltage (25 percent derating) for connectors used in switch-mode power supplies.

17.5.4 International Electrotechnical Commission Standards

The International Electrotechnical Commission (IEC) has developed a set of global standards for connectors used in electronic equipment. The IEC standards are accepted worldwide and are often used for determining accepted test procedures for component specifications, acceptable environmental conditions, and safety. IEC 130 applies to connectors used in telecommunication equipment and in electronic devices. The catalog description[7] states

> Its object is to establish uniform requirements for the electrical, climatic and mechanical properties of connectors as well as safety aspects, to lay down test methods, to ensure interchangeability and compatibility and to classify connectors into groups according to their ability to withstand extremes of temperature and humidity....

IEC 603 provides standards for connectors used with PC boards. Selected IEC 130 and 603 series standards are listed below.

- 130-1 (1988)—Part 1: General Requirements and Measuring Methods
- 130-2 (1965)—Part 2: Connectors for Radio Receivers and Associated Sound Equipment; Amendment No. 1 (1969)
- 130-3 (1965)—Part 3: Battery Connectors
- 130-10 (1971)—Connectors for Coupling an External Low-Voltage Power Supply to Portable Entertainment Equipment
- 603-1 (1991)—Generic Specification—General Requirements and Guide for the Preparation of Detail Specifications, with Assessed Quality
- 603-2 (1995)—Detail Specification for Two-Part Connectors with Assessed Quality, for Printed Boards, for Basic Grid of 2.54 mm (0.1 in) with Common Mounting Features

- 603-7 (1990)—Detail Specification for connectors, 8-Way, Including Fixed and Free connectors with Common Mating Features
- 603-8 (1990)—Two-Part Connectors for Printed Boards for Basic Grid of 2.4 mm (0.1 in), with Square Male Contacts of 0.63 mm x 0.63 mm
- 603-9 (1990)—Two-Part Connectors for Printed Board, Back-Panels and Cable Connectors, Basic Grid of 2.54 mm (0.1 in)
- 603-10 (1991)—Two-Part Connectors for Printed Boards for Basic Grid of 2.54 mm (0.1 in), Inverted Type
- 603-11 (1992)—Detail Specification for Concentric Connectors (Dimensions for Free Connectors and Fixed Connectors)
- 603-12 (1992)—Detail Specification for Dimensions, General Requirements and Tests for a range of Sockets Designed for use with Integrated Circuits

17.5.5 American Society for Testing and Materials Standards

The American Society for Testing and Materials (ASTM) presents standards for measuring contact resistance, including ASTM Method B539, "Measuring Contact Resistance of Electrical Connections."

17.5.6 Department of Defense Standards

A series of military standards have been released and are currently available. These standards set very stringent guidelines for the connector industry and require that connector manufacturers thoroughly understand and adhere to the various testing protocols and parameters involved.

A partial list of military standards governing cylindrical connectors is given below.

- Cylindrical Connectors
  - MIL-C-5015MIL-C-10544MIL-C-12520
  - MIL-C-22539MIL-C-22992MIL-C-25955
  - MIL-C-26482MIL-C-26500MIL-C-27599
  - MIL-C-28840MIL-C-29600MIL-C-38999
  - MIL-C-55116MIL-C-55181MIL-C-55243
  - MIL-C-81511MIL-C-83723
- Coaxial Connectors
  - MIL-C-23329MIL-C-26637MIL-C-27434
  - MIL-C-39012MIL-C-55339
- Tools
  - MIL-C-22520MIL-I-81969
- Rectangular Rack and Panel Connectors
  - ARINC 404ARINC 600MIL-C-8384
  - MIL-C-26518MIL-C-28731MIL-C-28748
  - MIL-C-28804MIL-C-83527MIL-C-83733
  - MIL-C-85028

17.6   *CONNECTOR FAILURES*

Several conditions are known to be contributing factors in connector failures, including short-term and long-term overload conditions, presence of conductive contaminants, prolonged effect of moisture/humidity, and crimping problems. These critical factors may cause permanent degradation of the connector that could eventually culminate in an ignition/fire, electrocution, or both. In an effort to determine the root cause of connector failures, many such incidents have been investigated, with an additional agenda of analyzing regulation, risk, and economic impact.

This section details the customary steps involved in an investigation of connector failures, providing a generic, systematic approach in analyzing the incidents with emphasis on logic and totality. It also establishes a basis for understanding why components such as connectors fail, how a failure investigation is conducted to preserve the integrity of the evidence, and how the probability of future failures may be reduced.

17.6.1   Purpose of Failure Analysis

A failure analysis will serve to assess damages in the event that occurred and attempt to develop protection methods for the enhancement of the system, preventing future failures. It will also determine the culpability of the party and/or organization that contributed to the failure.

17.6.2   Steps in a Failure Investigation

Figure 17.18 presents, in summary form, a typical flow chart of the various steps in a failure investigation. The first step is to define the overall problem and to draft an operational plan. It is important to ensure that evidence preservation, disassembly, and examination protocols are vigorously adhered to at all times. Background information, including historical and maintenance data, must be gathered and carefully reviewed for use in determining a possible scenario. The field investigation requires meticulous scrutiny, including a nondestructive visual examination of the items involved in the failure/accident. It is extremely important to exercise caution when handling accident evidence, always maintaining an accurate and accountable chain of custody. Specimens and sample material of

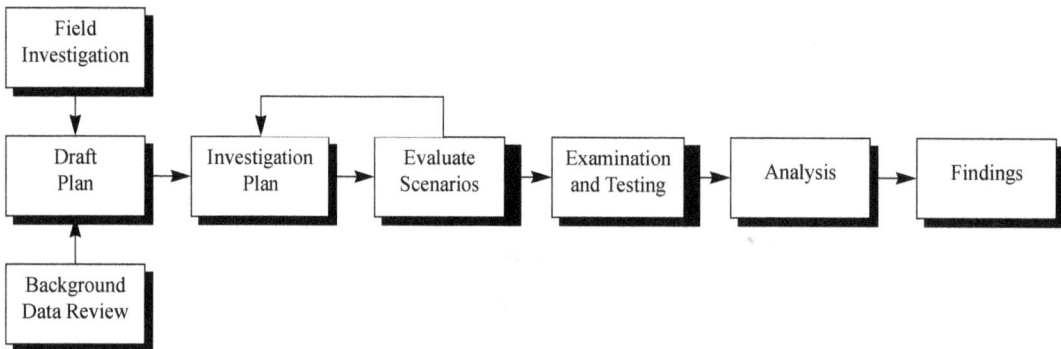

Figure 17.18   Typical flowchart of the various steps in a failure investigation *(courtesy of Failure Analysis Associates).*

failed components should be identified, categorized, and stored. Precise documentation with photographic/videographic corroboration of the evidence is essential.

The next step is the development of a formal investigation plan. Detailed laboratory examinations should be scheduled, which may include optical microscopic (OM) examination; scanning electron microscope (SEM) and energy dispersive spectroscopy (EDS); Fourier transform infrared spectroscopy (FTIR); chemical analysis; metallography; and physical, mechanical, and electrical testing. The investigation should include a careful evaluation of all possible failure scenarios and causes and, if necessary, electrical and thermal testing of one or more exemplar units under various environmental and electrical loading conditions. This task also calls for a complete design analysis and calculations, including computer simulations of the failed system if necessary.

Next, a compilation and preliminary analysis of the calculations and computer simulations should be performed. At this time, a review of the historical data and test results should be made in an attempt to determine the root cause of the failure. An additional task should include a risk assessment and the generation of a list of conclusions and recommendations with possible corrective actions to reduce or eliminate the cause of the failure.

17.6.3 Root Cause Investigation

A prime reason for performing a failure analysis is to determine the root cause of the failure. This determination may involve extensive field inspections, laboratory tests and experiments, hand calculations, detailed design analyses, and computer simulations. The evaluation may vary, depending on the past experience of the investigators.

Failures often occur due to design deficiencies, use of improper or low-grade materials, manufacturing defects, poor construction and workmanship, or a combination of factors. The cause may be due to the misapplication or inadequate maintenance of the equipment/system. Oftentimes, human error, such as misoperation, poor maintenance procedures, or management, is involved. Excessive duty, especially under severe overload conditions, or wear-out can cause devices to fail. Prolonged exposure to extreme conditions may result in loose or corroded connectors. Failures can also occur as a result of connector operation under high transient fault conditions for which the connector was not designed or specified. Natural events such as lightning or earthquakes, and known or unknown environmental effects, may also generate failures.

17.6.4 Basic Cause of Failure

The following are three leading causes of failure in connectors:

Excessive Current.   Failures may occur due to excessive current in the connector, resulting in a high operating temperature. High current may be present as a transient fault condition, or as a long-term or continuous overload condition, resulting in reduced operating life, overheating, and melting of the connector. In such cases, connector contact surfaces may be severely damaged resulting in immediate or eventual failure of the component.

Insulation Breakdown.   Insulation breakdown results in the failure of any component, including connectors. Insulation breakdown may be caused by increased leakage current due to overvoltage alone or may occur at normal voltage in association with electrically conductive contaminants (such as salt, moisture, and humidity) or with conditions that permit filament whisker formation.

Excessive Power Dissipation.   A third common cause of failure is excessive power dissipation at the contact interface. Excessive power can be due to reduced mechanical force at the contact mating surfaces, thus increasing contact resistance. Increased contact resistance, by itself, can cause higher operating temperatures. In the presence of environmental stress conditions such as high ambient temperatures, contamination, and excessive mechanical operating cycles, it may result in degradation of the plastic insulation. This can eventually result in reduced operating life, overheating and melting of the connector, and an ignition and fire. Additional contributors may be excessive mechanical stress from high tangential forces due to a small bending radius on the wires to the connector, excessive thermal cycling, and so on.

The consequence of the failure is either an open circuit on the connector or a resistive short. Typically, an open circuit results in a failed component without the risk of an ignition or electrocution. However, a resistive short condition may cause further overheating, propagating the failure with eventual system shutdown, fire, and possibly an electrocution. Figure 17.19 presents a terminal block which overheated, resulting in severe melting of the polyamide housing.

Figure 17.19   Terminal block that overheated, resulting in sever melting of the polyimide housing *(courtesy of Failure Analysis Associates)*.

17.6.5 Conditions Leading to Electrothermal Events

Electrothermal events contribute to the failure of many components, including connectors, and often occur as a result of a well defined set of operating conditions, including the following:

1. A loose connection between a contact pin and associated conductor. Loose connections cause excessive power dissipation in the component, oxide formation, and result in increased temperatures. The increased temperatures further degrade the insulating characteristics of the plastic supporting material, contributing to multiple short circuit conditions.

2. The presence of electrically conductive contaminants bridging two or more adjacent terminals, resulting in a high leakage current across the plastic insulating housing.

3. Transient short circuit conditions resulting in localized hot spots and the formation of an oxide layer over a portion of the contact surface. The electrically insulating oxide layer results in reduced contact area, higher contact current density, exceeding normal operating temperature, and degrading the plastic insulating material.

4. Sustained overload conditions resulting in high contact operating temperatures for long periods of time.

5. Arcing between the contact surfaces due to loose connections or mechanical vibration.

6. Dielectric breakdown between adjacent contact pins due to transient overvoltage conditions, resulting in arcing along the plastic insulation surface.

If a connector mounted on a printed circuit (PC) board is thermally stressed, the PC board laminate material immediately under the thermally stressed connector may make a transition from being an insulator to being a partial conductor. The partially electrically conductive PC board may cause a high leakage current to flow between adjacent energized nodes, further overheating the PC board. This may cause multiple short circuits on the PC board with possible layer-to-layer insulation failure. Such a cascading failure is very likely to result in an ignition and a fire.

17.6.6 Events Leading to Connector Failures

Investigation of connector failures often necessitates a thorough research of historical information, usage, manufacturing materials and techniques, management practices and procedures, installation, and workmanship. One or more of the following events may contribute to the eventual failure of the connector.

Degree of Compression. A critical aspect of connector installation is the contact conductor interface. Typically, the conductor is crimped, screwed, soldered, or otherwise mechanically secured to the contact pin of the connector. The degree of compression is a significant factor in connector reliability. It is imperative to use only the recommended degree of compression, as too low a compression may result in high contact resistance, and too high a compression results in reduced pull-out force and possible deformation and fracture of the conductor strands. A SEM photomicrograph of a crimp connection cross section is shown in Fig. 17.20.

A too-tight connection is almost as inauspicious as a connector that is too loose, and it will tend to deform the conductor strands. Overtightening the screws applies excessive

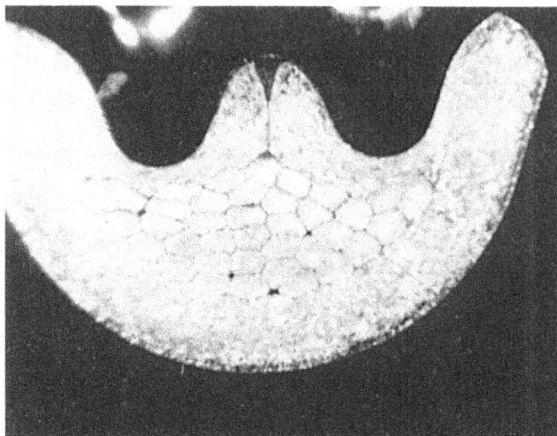

Figure 17.20 SEM photomicrograph of a cross section of a crimp (reprinted with permission of AMP).

torque and causes stress concentrations at the screw thread, with a tendency to strip the threads. If the threads are stripped, the conductor will have substantially reduced compression, with increased contact resistance. Overtightening also tends to pivot the screw, causing misalignment between the conductor and the terminal end of the screw.

Overtightening the conductor tends to deform the conductor strands, causing a partial fracture of some strands and completely fracturing others. Completely fractured strands may work loose, making contact with energized sections of the equipment, and could result in misoperation, failure, and a fire. The loose strands could make contact with a live part and the metal housing, creating a shock hazard. Partially fractured strands would be forced to carry the full load current of the system. The decreased conductor cross-sectional area increases contact resistance and contact power dissipation. To further complicate matters, the reduced cross section would result in poor heat conduction from the contact area.

Figure 17.21 shows the variation of crimp conductivity, tensile strength, and current-carrying capacity as a function of crimp deformation for a typical connector. It is observed that the recommended value of the sinusoidal pull-out force is reached just before the peak value. Any further increase in the deformation force results in a reduced cross-sectional area and failure, evidenced as conductor shearing.

Crimping Problems. Even a high-quality connector, if used improperly or installed incorrectly, will eventually result in a failure. For high reliability and increased mechanical operating cycles, the connector should be installed according to the manufacturers' specifications and established guidelines. For a good crimp connection, the wire should be correctly stripped to the required length with no damage to, or break of, the copper strands. Care should be taken not to damage the adjacent insulation while stripping the wire or to leave particles of insulation on the conductor. It is essential to use a smooth, gentle motion to crimp the wire and to carefully check the integrity of the crimp prior to insertion of the pin into the connector housing. Failure to follow these suggestions in the crimping operation may result in a poor crimp connection and lack of mechanical integrity, with the future possibility of a failed connector.

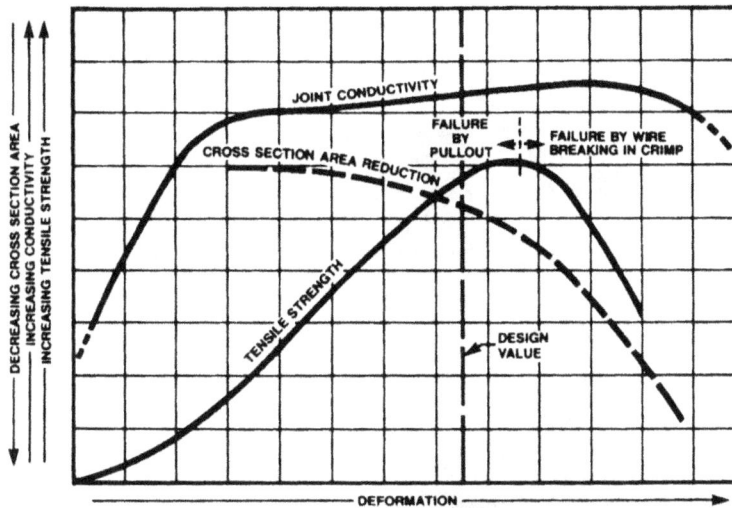

Figure 17.21 Variation of crimp conductivity, tensile strength, and current-carrying capacity with crimp deformation *(reprinted with permission of AMP)*.

Whisker Formation.   Whiskering is another phenomenon that can lead to contact failure. Mroczkowski[3] defines whiskering as a filamentary single-crystal growth morphology that can lead to shorting of contacts or contact pads. The article states that whiskering is caused by compressive stresses in the electroplate. Tin-lead alloy plating such as 93 tin-7 lead may be used to reduce whiskering. Tin whiskers may grow to be in excess of 0.1 in (2.5 mm) long and have current densities as high as 106 A/in^2. Figure 17.22 is an SEM photomicrograph of a tin contact surface showing tin whiskers.

Fretting Corrosion.   Fretting may be defined as submicron repetitive motions of the contact interface, and may result in breaks in the tin contact interface. If the break at the interface re-oxidize, the effect, termed fretting corrosion, may result in very high contact resistance. Figure 17.23 presents the mechanism of fretting corrosion. Fretting corrosion can be avoided by use of contact lubricants and also by high normal contact force. Mroczkowski[3] provides information on fretting corrosion.

Pore Corrosion.   Pore corrosion, caused by finish defects, is another phenomenon that can result in contact failure. Metal defects and defects in manufacturing create small holes in the product, exposing the base metal and resulting in corrosion. A nickel underplate forms an oxide film, which prevents pore corrosion. Information on pore corrosion can be found in Mroczkowski.[3] Figure 17.24 illustrates the formation of pore corrosion and the beneficial effects of nickel underplating.

Migration of Corrosion Products.   Corrosion products forming at some unused area of the contact may tend to migrate to the contact interface area. A nickel underplate can provide a measure of protection against migration of corrosion by reducing the migration distance. Mroczkowski[3] discusses migration of corrosion.

Long-Term Overload Conditions.   Electrical equipment, including connectors, may be called on to carry higher than normal currents under overload conditions until the overload

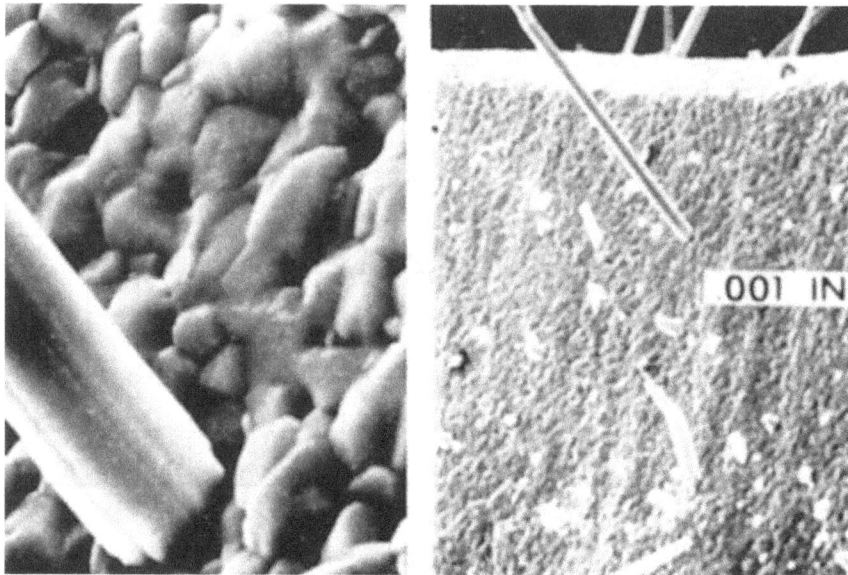

Figure 17.22   SEM photomicrograph of tin contact surface showing tin whiskers *(reprinted with permission of AMP).*

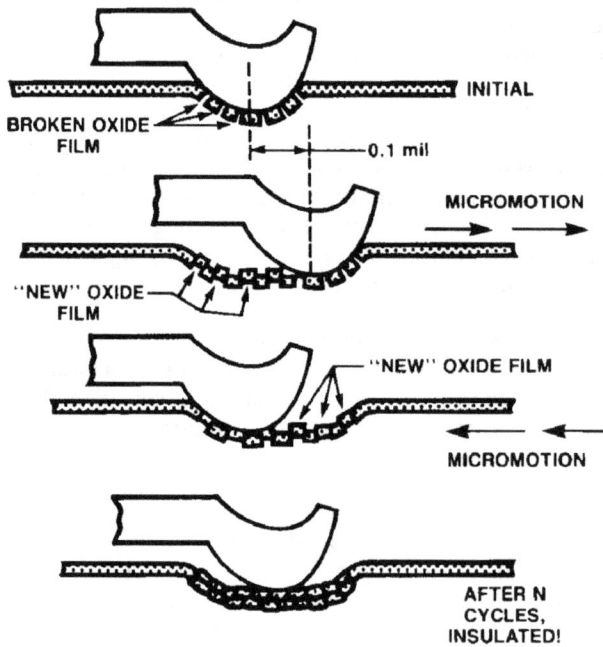

Figure 17.23   Schematic representation of the kinetics of fretting corrosion due to translational micromotions *(reprinted with permission of AMP).*

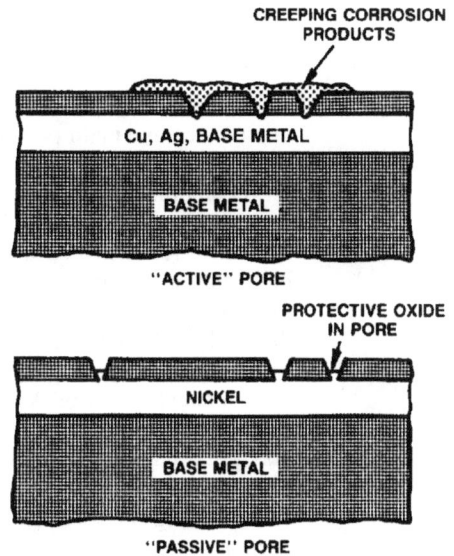

Figure 17.24 Schematic representation of the effects of underplates on pore corrosion *(reprinted with permission of AMP).*

is interrupted or cleared. Overload currents are typically 5 to 20 times normal current and include starting currents for large motors, inrush currents for incandescent lamps, and so forth. These currents may exceed the continuous current rating of the connectors. In most cases, if they are not rapidly reduced, they may cause permanent degradation of the connector. Consider, for example, a 12-Vdc application, where the steady-state current was specified as 10 A, and the short-term overload current was specified as 100 A. Under abnormal operating conditions, the measured current had a peak value of almost 600 A, decaying to 440 A after approximately 20 ms. If a connector designed to carry a short-term overload current of 100 A and a continuous current of 10 A was forced to carry a current of 440 A for even a short period of time, it would likely overheat and be permanently degraded. Consequently, long-term, high-current overload or fault conditions may not be safely withstood by the connector, and a means for overcurrent protection must be designed to interrupt the high current before the connector is permanently degraded or even destroyed.

Transient Fault Conditions. Connectors used in primary and secondary circuits may fail due to degradation of the contact area. Degradation may be due to a variety of causes such as overcurrent, overtemperature, operation in a corrosive environment, or the presence of contaminants. A short-term, high-current condition (e.g. during a fault) is a stress parameter that is often overlooked.

Under short-term transient fault conditions, peak currents are very high—on the order of 20 to 100 times or more than the normal steady-state current. The total time that the transient fault current occurs is relatively short. However, if the connector is not designed for this high short-circuit current, it may be damaged by this event.

Even though a high fault current is present for a very short duration, the abnormally high current density in the contact(s) results in the formation of hot spots on the contact surfaces, which causes overheating, oxidation, and possible degradation of the contact area. If the degraded contacts are then exposed to the rated normal contact current, the reduced active area results in the contact operating at higher than normal temperature, causing further degradation of the contact area. Eventually, the contact overheats, causing the plastic insulating material to melt, resulting in an ignition and fire.

High transient fault currents are often overlooked in board-to-board connectors used to interconnect the main power supply to the output load. This high fault-current capacity is readily available in capacitor input power supplies frequently used in audio and linear power amplifiers. These amplifiers often have dual power supplies, with the bipolar dc bus operating at ±30 V to ±100 V. Main bulk filter capacitors are chosen that have low equivalent series resistance (ESR) and low equivalent series inductance (ESL), with associated high short-circuit capacity. The total filter capacitance may range from 33,000 µF for the low-end voltage to 10,000 µF at the high-end voltage. The power supply PC board, with the bridge rectifier and filter capacitor, may be connected to the main amplifier circuit using heavy-gauge insulated conductors and 10- to 15-A plug-in connectors. A typical system may have a ±60-Vdc supply with 10-A load current, and 22,000 µF of capacitance on each dc bus. In such a system, the peak fault current may easily exceed 1500 A and may last for several milliseconds.

A PSpice® computer simulation of a fault current, for a fault resistance of 10 mΩ, is shown in Fig. 17.25. It is observed that the peak fault current is 1850 A, exponentially decaying to zero in approximately 3.5 ms. A fault current of such high magnitude could degrade the contacts of a typical 10- to 15-A connector.

Figure 17.26 presents the time-current characteristic for three values of fault resistance for a typical audio amplifier power supply. Observe that for a fault resistance as high as 100 mΩ (0.1 Ω), the fault current still exceeds 450 A after 1 ms, and exceeds 250 A even after 5 ms. Furthermore, as a comparison, the plot shows the fusing time-current curve for a typical 10A, fast-acting fuse. Note that the 10-A fuse time-current characteristic is below the time-current characteristic of even the 100-mΩ fault, indicating that the 10-A fuse would probably melt if subjected to the excessively high fault currents in the power supply connector circuit. If the fault is expected to be repeated a number of times over the life of the connector, the connector contacts would need to be derated accordingly. For high transient fault current applications, connector selection should include a thorough analysis of the magnitude of the fault current, and the corresponding effect on the contact surfaces.

Exposure to External Tangential Stress. The increased demand for high density connection systems requires that connectors with multiple pins be attached to a braid of conductors. If not properly aligned and secured, the multi-conductor braid can exert a substantial angular force. Studies show that forcing a plugged-in connector cable into a small bending radius, results in a substantial angular force on the contact pins of the connector. Electrical/thermal tests indicate a significant increase in temperature when connectors are subjected to such lateral forces. Results of lateral force tests are discussed in Sec. 17.8 (p. 17.47), which describes case studies.

Effect of Contamination. Connectors used in hostile environments may often be exposed to contaminants such as moisture, oil, dirt, dust, salt spray, carbon, and other electrically conductive deposits. Such contaminants impose a severe environmental stress condition on the connector. Corrosion products from these contaminants may be deposited at the contact interface and may contribute to its rapid failure. Failure may occur as a permanent increase in contact resistance, with resultant higher power dissipation, rise in temper-

* C:\MSIM63A\CIRCKT\CAP1.SCH

Date/Time run: 02/02/97 08:52:17   Temperature: 50.0

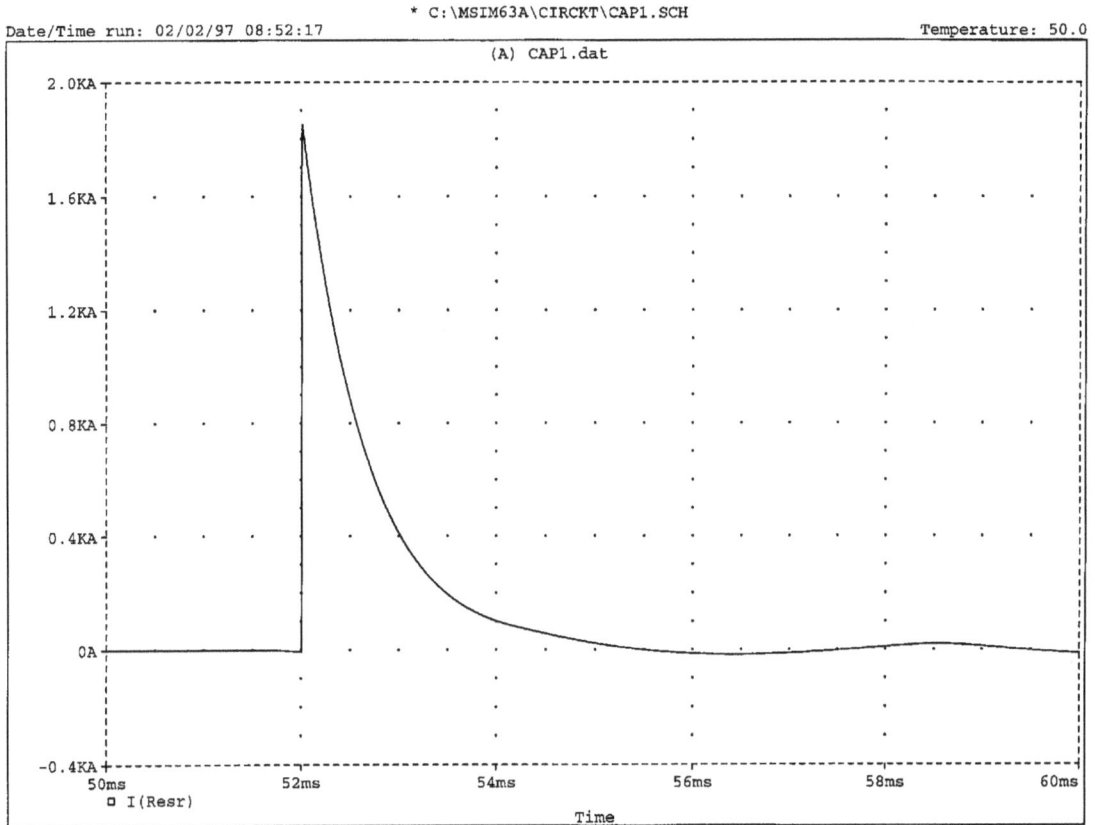

Figure 17.25 Computer simulation of a short-circuit current in a connector connected to a filter capacitor for fault resistance = 10 μΩ *(courtesy of Failure Analysis Associates).*

ature, and subsequent degradation of the insulating material. The increased temperature may further oxidize the contacts, thus raising the contact resistance and eventually melting the insulating material. The failure may also exhibit itself as an intermittent contact, especially in a high-vibration environment.

Failure may also occur due to the substantial increase in leakage current between adjacent pins caused by deposits of electrically conductive contaminants. Increased leakage current results in power dissipation along the surface of the plastic insulation, which may compromise its insulating characteristics. Eventually, the leakage current causes carbonization of the insulating material, with a corresponding increase in temperature, increased power dissipation in the plastic, plus smoke, ignition, and fire.

Excessive Mechanical Operating Cycles. Generally, contact resistance of connectors increases with increased mechanical cycles. For typical connectors, the contact resistance is minimal at start and gradually increases, reaching a steady-state value at approximately 500 to 1000 mating cycles. The contact resistance remains somewhat constant to 3000 mating cycles. At 3000 mating cycles, the contact resistance once again starts to increase.

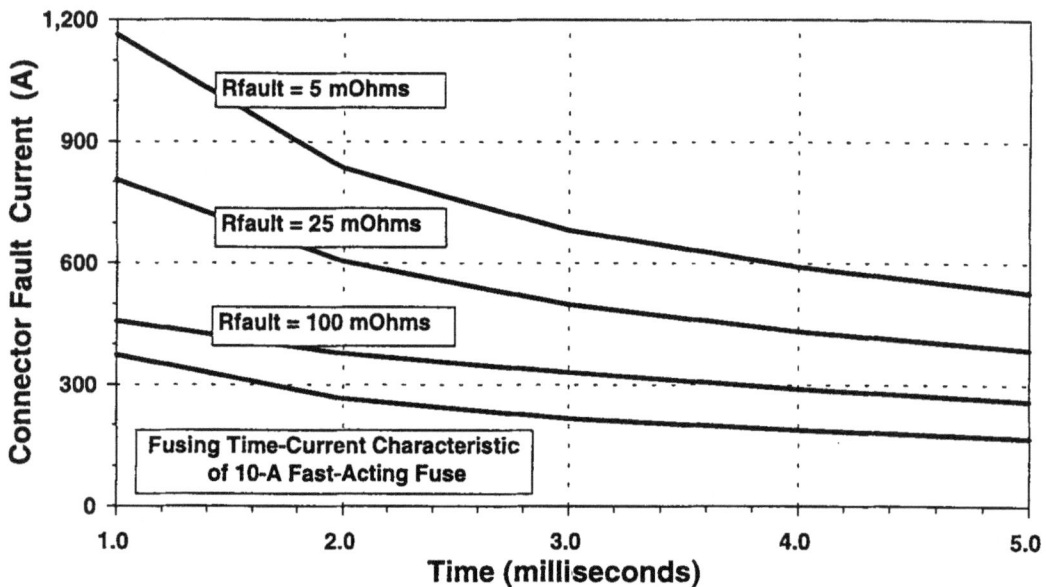

Figure 17.26   Time-current characteristics of the fault current for various fault resistances *(courtesy of Failure Analysis Associates).*

17.7   ELECTRIC SHOCK

Connectors are used extensively in all types of electrical equipment. Occasionally, design defects, manufacturing defects, improper installation, or environmental conditions may result in the leakage current across the plastic insulation presenting an electric shock hazard to the user. Even relatively small currents can cause dangerous electric shocks, particularly if the victim is wet and/or well grounded. Electrical equipment must be solidly grounded or double insulated to provide a high degree of protection to the user. An approved ground fault circuit interrupter (GFCI) may provide an additional level of protection, disconnecting power to the equipment if a ground fault condition is detected. Three-wire grounding detectors that dictate a specific plug/outlet orientation may be used to provide additional electrical protection to the user. A superior grounding detector, which utilizes a unique detection circuit, is listed in the reference.[8]

17.7.1   Solidly Grounded Equipment

For safety, electrical appliances and equipment such as clothes washers and dryers, bench tools, welders, and so on are designed to be solidly grounded by a grounding wire to the power supply system. A three-wire (black hot wire, white neutral wire, green ground wire) ac power cord is used for grounded equipment. Connectors or terminal strips are sometimes used to connect the three-wire ac power cord to the internal wiring of the equipment. The hot and neutral wires are used to supply electrical power to the equipment. The green safety ground wire is provided to trip the circuit breaker by overcurrent in the event that the hot wire comes into contact with the grounded metal enclosure. An improperly installed or contaminated connector can present a shock hazard to the user and is discussed further in Sec. 17.8 (p. 17.47).

17.7.2 Double-Insulated Equipment

Electrical equipment such as electric lawn mowers, vacuum cleaners, portable electric drills, and other power tools usually has a two-wire line cord, with hot and neutral conductors. The two-wire line cord may be connected to the internal wiring of the equipment by a terminal block. No ground wire is provided, and the outer metal housing is left electrically unconnected (floating) and galvanically isolated from the ac power system. Two separate layers of insulation (double insulation) are provided to ensure that the outer metal housing and all exposed metal parts will not be energized in the event of an insulation failure of the energized portion of the equipment, or by a break in the hot wire. An improperly installed or contaminated connector can compromise the integrity of the double insulation, presenting a shock hazard to the user. One such installation is presented in Sec. 17.8 (p. 17.47).

17.7.3 Conditions for Electric Shock

For electrical equipment to present a shock hazard to a person coming in contact with exposed metal parts, such as the housing, two conditions would have to exist:

1. The grounding connection would have to be deteriorated or compromised by (a) a break in the grounding wire, (b) a loose or corroded connection in the grounding circuit, either in the cable or within the equipment, or (c) the use of a three-to-two–wire ac adapter.

2. The hot wire of the 115-Vac line would have to make contact with, or leak current to, the metal housing. This hazardous condition may be brought about by (a) the failure (short circuit) of an electrical component such as a filter capacitor, connected between the 115-V hot wire and the metal housing, (b) the fracture of a solder connection of a component connected to the hot wire, with the resultant loose wire touching the metal housing, or (c) a leakage current path created by conductive contaminants across an insulating surface such as a connector, between the hot wire and the metal housing.

Under the above conditions, even equipment supplied from isolated power supplies may fail to provide electrical isolation when subjected to harsh environments.

17.7.4 Electric Shock Currents

Electric shock currents are classified according to their ability to cause bodily harm. The minimum value of 60-Hz detectable current is 0.5 to 1.0 mA. The accepted maximum harmless current to average humans through the body trunk for a one-second contact is 5 mA. Currents greater than 5 mA are considered onset shock currents. Figure 17.27 presents a curve of electrocution equations (current versus time) from ANSI/ISA Standard ANSI/ISA-S82.01-1988. Threshold levels of currents are listed in Table 17.10.

Due to the relatively low level of shock currents, the hot wire within electrical equipment does not necessarily have to be in solid contact with the metal housing to cause a painful or lethal electric shock. Leakage current across the surface of contaminated insula-

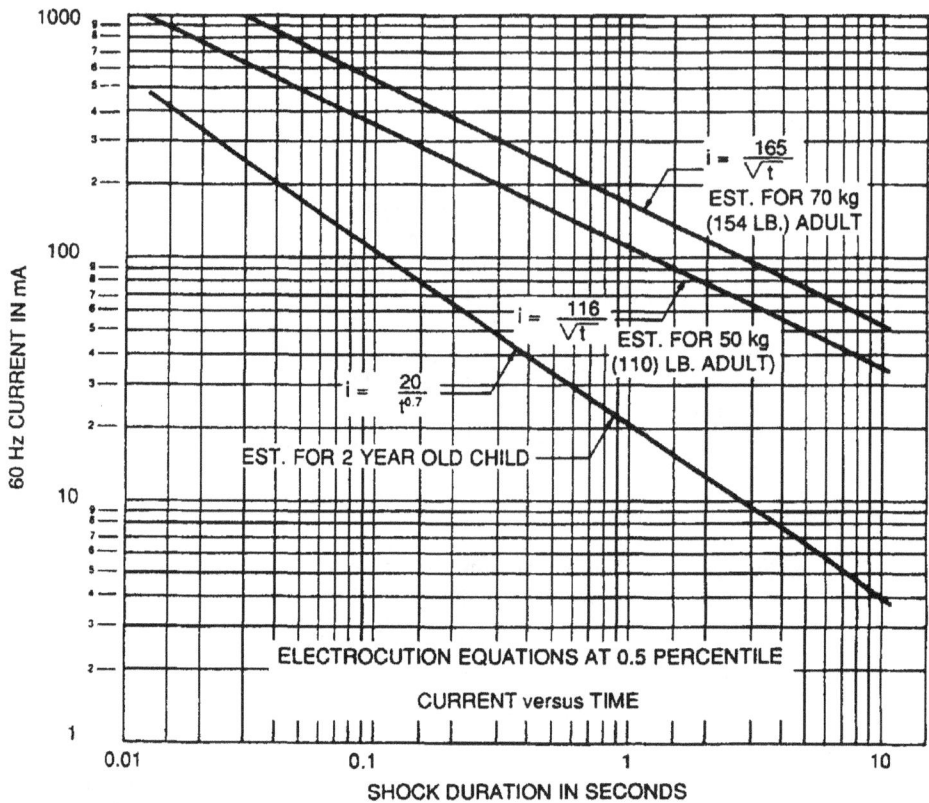

Figure 17.27   Curve of electrocution equations (current vs. time) obtained from ANSI/ISA Standard S82.01-1988 *(reprinted with permission of the Instrument Society of America).*

Table 17.10   Effects of 60-Hz Electric Current on an Average Human through the Body Trunk (Macroshock)*

Current intensity, mA (1-s contact)	Physiological effect
1	Threshold of perception
5	Accepted as maximum harmless current intensity
10–20	"Let-go" current before sustained muscular contraction
50	Pain, possible fainting, exhaustion, mechanical injury; heat and respiratory functions continue
100–300	Ventricular fibrillation will start, but respiratory center remains intact
6,000	Sustained myocardial contraction, followed by normal hear rhythm; temporary respiratory paralysis; burns if current density is high

* From ANSI/ISA-S82.01-1988, Safety Standard for Electrical and Electronic Test, Measuring, Controlling and Related Equipment, p. 58, reprinted with permission of the Instrument Society of America.

tion may be sufficient to cause a shock or even ventricular fibrillation. Thus, the small spacing between energized terminals of a connector and exposed metal may acquire a deposit of electrically conductive contaminants, compromising the insulation resistance and permitting current to leak between the hot 115-Vac line and the exposed metal chassis, resulting in a shock hazard.

17.8 CASE STUDIES

This section examines several case studies pertaining to spade connectors, terminal blocks, Euro-series plug connectors, and power connectors that have been implicated in occurrences of fire and electrocution.

17.8.1 Terminal Block Connector Used in an Oil-Filled Heater

A series of fires were attributed to oil-filled electric heaters manufactured abroad. The UL-listed oil-filled heaters, rated at 120 Vac, 60 Hz, 1500 W, had a manual thermostat temperature control and dual heating elements, and they were typically used in residential areas.

One incident involved a house fire in which three people were killed. Several exemplar heaters and the heater involved in the incident were visually inspected to determine a possible ignition source. Further investigation and preliminary tests indicated that a nylon terminal block in the heater could overheat under certain operating conditions. The terminal block was located in the control panel of the subject heater and was used to connect the two-wire ac power cord to the internal wiring of the heater. The terminal block was made of nylon and had four screws: two to secure the line cord, and two to secure the internal wiring. All electric current for the 1500-W heater passed through the terminal block. If the screws were not tight, the connections would have increased electrical resistance, resulting in hot spots, overheating, and the possibility of a fire.

Temperature measurements were performed on exemplar two-pole terminal blocks, similar to the one used in the subject oil-filled heater, to determine the magnitude of the temperatures involved under four operating conditions:

1. All four screws on the terminal block tightened

2. One screw loosened on one pole of the terminal block

3. Two screws loosened on one pole of the terminal block

4. All four screws on the terminal block loosened

The tests were conducted to identify an ignition source in the heater and to determine if an ignition would occur if any of the above conditions were present.

The terminal block under test had two poles for connecting the two-wire power cord to two internal heater wires with two screws on each pole, pole spacing 0.40 in. The terminal block was rated at 30 A, 300 V. The insulating housing material was polyamide 6.6 (nylon) with a melting temperature of 240 to 260° C (464 to 500° F). During the test, the terminal block was enclosed in a glass housing, mounted on a metal base plate, and placed on a 0.125-inch thick plastic sheet on top of the heater fins. This placement approximated the relatively high ambient temperature in the vicinity of the terminal block when located in the control panel of the accident heater unit, and also allowed for photodocumentation.

The terminal block was instrumented to display the pole temperatures and the voltage drop across the contact resistance caused by the loosened screws. The heater thermostat control was set to maximum, and the heater switches were set to operate the heater at its maximum rating of 1500 W. Heater operation was intermittent and was shut off at the end of each day of testing.

Figures 17.28 and 17.29 present data sheets that graphically display the measured temperature of pole 1 and pole 2 of the terminal block, with two screws loosened on one pole. The data sheets also show the melting temperature of nylon (shown as a dotted line on the graphs), which was the housing material used in the construction of the terminal block. During tests 2 through 4, listed above, the pole temperatures gradually increased, eventually exceeding the melting point of nylon. Occasionally, a red glow was observed in the vicinity of the pole with the loose screw. Finally, it was observed that there was an eruption of smoke, arcing, and a visible flame. Table 17.11 shows the maximum temperatures measured and compares these temperatures to the melting point of nylon. Figure 17.30 shows an ignition and flame on the terminal block under the one screw loose test. Figure 17.31 shows the remains of the terminal block at the conclusion of one of the tests.

17.8.2   Contact Power Dissipation in Spade Connector

A four-slice toaster was alleged to have caused a house fire due to an improperly seated connector. The UL-listed toaster, rated at 120 Vac, 12.5 A, 1500 W, had a manual thermostat temperature control and multiple heating elements.

The control panel of the subject toaster was equipped with lug-type disconnect terminal strips for making connections with push-on spade connectors. The connectors were used to connect the two-wire power cord to the internal wiring of the toaster. The male push-on

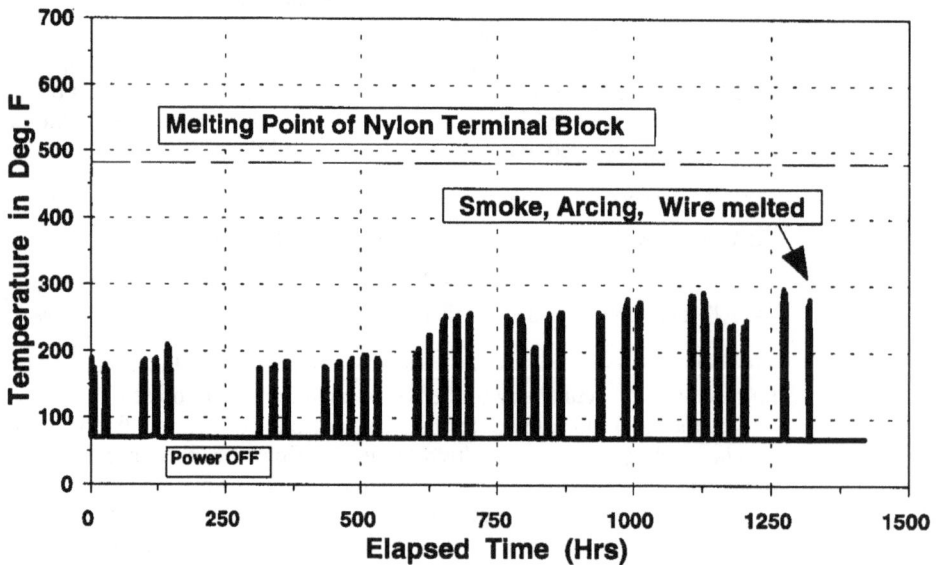

Figure 17.28   Terminal block temperature test, temperature of pole 1, two screws loose *(courtesy of Failure Analysis Associates).*

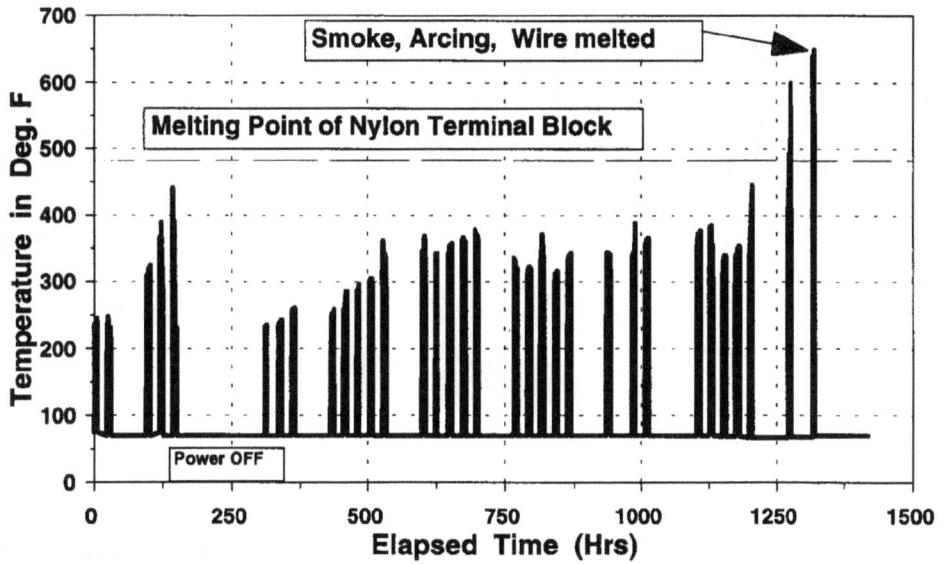

Figure 17.29   Terminal block temperature test, temperature of pole 2, two screws loose *(courtesy of Failure Analysis Associates)*.

Figure 17.30   Ignition of terminal block during one-screw-loose test *(courtesy of Failure Analysis Associates)*.

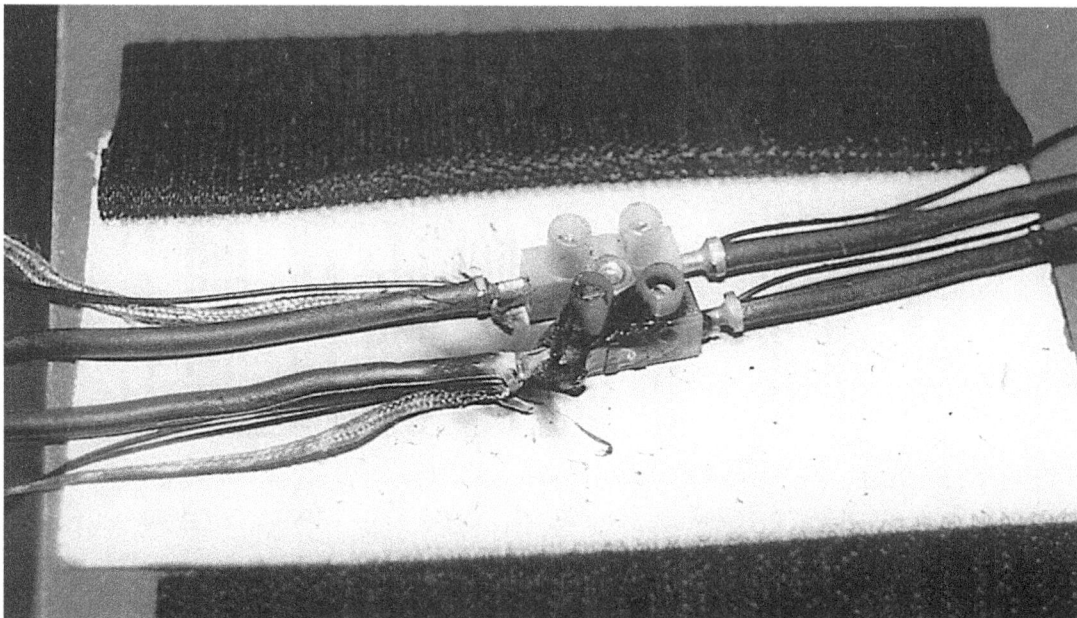

Figure 17.31   Remains of terminal block at conclusion of test *(courtesy of Failure Analysis Associates).*

Table 17.11   Maximum Temperatures on a Terminal Block for Four Test Conditions *(courtesy of Failure Analysis Associates)*

No.	Test conducted on terminal block	Melting point of nylon, °F	Maximum pole 1 temp., °F	Maximum pole 2 temp., °F
1.	All four screws tight on terminal block	482	138	132
2.	One screw loose on one pole of terminal block	482	297	965
3.	Two screws loose on one pole of terminal block	482	296	650
4.	All four screws loose on terminal block	482	930	1060

spade connectors, connected to the line cord, were plugged into the female terminal strips. It was alleged that the connectors were not plugged in completely. This resulted in the terminal strip/spade lug mating length to be reduced to 60 percent of nominal length.

A series of resistance measurements was made on an exemplar terminal strip, similar to the one used in the subject toaster, to determine the connector contact resistance and connector power dissipation as a function of connector length. Tests were conducted in an attempt to identify a possible ignition source in the 1500-W toaster, and furthermore to determine if an ignition would occur if the push-on spade connector was installed to only 50 percent of its normal length.

During testing, all electric current to the 1500-W toaster passed through the terminal strips. If the contact area was insufficient, the connector resistance could substantially increase, resulting in excessive power dissipation in the lugs, with possible overheating and a potential for fire. During the test, contact voltages were recorded as a function of connector distance.

Figures 17.32 and 17.33 present data sheets that graphically display the measured contact resistance and contact power dissipated on the terminal strip as a function of connector distance. The data sheets also show the threshold beyond which the contact resistance rapidly increases, resulting in excessive power dissipation. Test data indicated that when the connector was pushed all the way into the lug, the contact resistance was less than 1.5 mΩ, and the connector power dissipation was less than 0.25 W. When the connector was moved out from the lug such that the terminal strip/spade lug mating length was reduced to 60 percent of the nominal length, the connector resistance increased to over twice its original value, about 3.75 mΩ. The connector power dissipation was then about 0.6 W—still within acceptable limits. Furthermore, it was observed that the contact resistance rapidly increased after the connector mating length was reduced to less than 35 percent of its nominal value.

17.8.3 Contact Power Dissipation in a Connector Used in a Security System

A high-sensitivity security monitoring system was examined in an effort to determine the source of an ignition problem. The overall system incorporated a power distribution system supplying power to up to three distributed detection monitors. An investigation was conducted to determine the cause of melting and burning that had been experienced on the connectors and to recommend a solution to the problem. It was suspected that the connec-

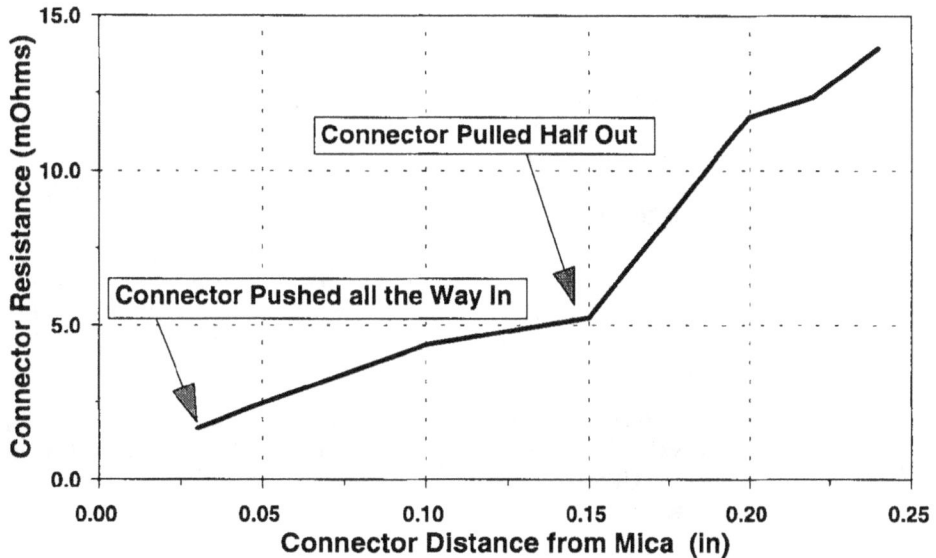

Figure 17.32 Spade connector contact resistance as a function of active length *(courtesy of Failure Analysis Associates).*

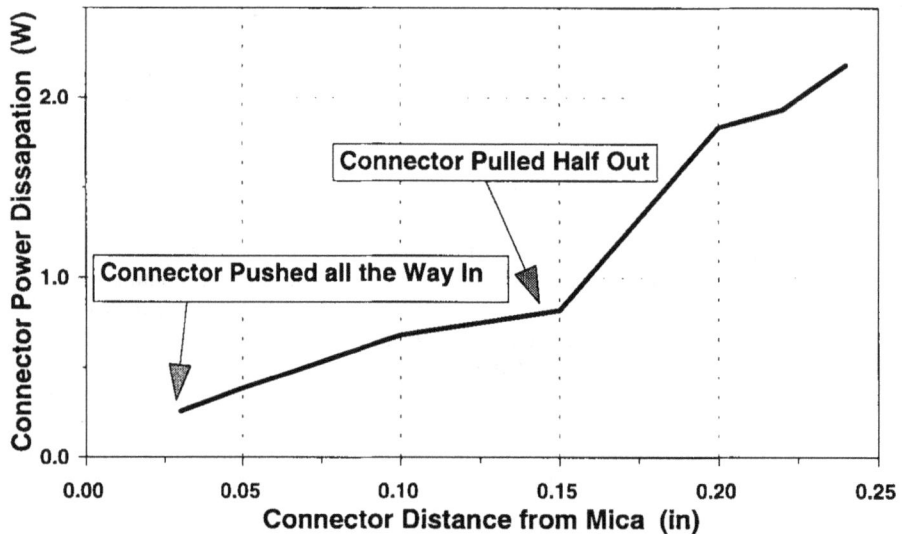

Figure 17.33 Spade connector contact resistance as a function of active length *(courtesy of Failure Analysis Associates).*

tor failures occurred due to (1) loose screws on the connector terminals, (2) excessive horizontal force applied by the cables connected to the connectors, caused by the small bending radius of the cables due to inadequate spacing, and/or (3) accidental short circuits occurring downstream of the connector.

Electrical/thermal tests were conducted to demonstrate the possible failure modes of the connectors. The three-pin connectors used on the power distribution boards were manufactured by two different companies and are described and compared as follows.

Comparison of Connectors. Connector #1, labeled ABC, was a three-pin Euro-series, PC-mount connector (terminal block) with a rated voltage of 250 Vac, 300 Vdc, and a rated current of 20 A. UL ratings were 30-12 AWG wire, rated voltage 300 V, rated current 20 A. The connector temperature range is –40° C to +100° C, with a short time peak temperature of 200° C.

Connector #2, labeled XYZ, was a three-pin vertical Euro-series plug-type terminal block, mated with a series vertical socket. The plug connector was rated at 250 Vac, 15 A, 12 AWG wire max. The vertical socket material was polybutylene terephthalate (PBTB), rated at 250 Vac, 15 A, 12 AWG wire max. Connector #2 had male and female sliding contacts, with PC board-type solder pins on the male, and screw-type terminals on the female plugs.

Housing material for connector #1 and the plug of connector #2 was polyamide 6.6 (nylon), with a melting temperature of 240 to 260° C (464 to 500° F) and a UL flammability rating of UL 94V0. Both polyamide 6.6, and polybutylene terephthalate (PBTB) are rated for continuous use to 105° C.

System Description. The power supply system consisted of a 30-Vac, constant-voltage power unit protected by a 12-A circuit breaker on the secondary side and a 4-A slow-blow fuse on the line side. The 30-Vac power was brought by a three-wire shielded cord to the three-pin connector on a power distribution board. The three-pin, 15-A connector also supplied power to a second three-pin connector for daisy chaining 30-Vac power to a second power distribution board (up to three could be used). The only electrical protection for the three-pin connectors and the copper trace was the 12-A circuit breaker on the output side of the 30-Vac power unit.

The power unit was required to supply 3.3 A per power distribution board, or approximately 10 A for a maximum of three boards. The power unit was capable of supplying up to 20 A continuously at 26.5 V, up to 100 A under short circuit conditions, and 42 A under overload conditions, when protected by the 12-A circuit breaker on the low-voltage side. The high-output capability of the power unit to supply currents far in excess of the 10-A load requirements, protected by a 12-A circuit breaker, meant that low-resistance faults between adjacent terminals of the connectors could become potential failures.

Analysis.   Detailed visual, microscopic, electrical, and thermal analyses of the system, boards, and connectors (including the failed connectors) were conducted to determine the cause of the failures. A visual inspection of the as-built security system, including the installation and mounting arrangement of the power distribution board, indicated that there was a relatively small space (~0.7 in) between the three-pin connectors and the metal sides of the enclosure. This poor mounting arrangement resulted in very sharp bends in the three-wire shielded cord from the 30-Vac power unit and produced excessive horizontal mechanical force on the three-pin connectors. The excessive horizontal force on the connectors resulted in misalignment, with correspondingly reduced contact pressures, resulting in increased contact resistance and higher connector dissipation.

Test Results.   Five thermal tests were conducted on the ABC and XYZ connectors to determine their failure modes. Table 17.12 shows the maximum temperatures measured and compares these temperatures to the maximum specified operating temperature of 221° F, and to the melting point of nylon, 482° F. Figure 17.34 presents the waveforms of the voltage drop on one pole during one- and two-screws-loose test.

Figures 17.35 and 17.36 present the measured temperature of poles 1 and 2 of the ABC three-pin connector with one/two screws loose on one pole. During the test, smoke, arcing, and melting of the connector were observed. The maximum recorded temperature of 847° F far exceeded 482° F, the melting point of nylon.

Figure 17.37 presents a data sheet that graphically displays the measured temperature of poles 1 and 2 of the three-pin XYZ connector/plug assembly for normal operating conditions, and for short circuit conditions. It was observed that, after two deliberate 30-Vac short circuits, the connector temperature was 358° F, exceeding the specified normal operating temperature of 105° C (221° F) and coming close to the 482° F melting point of nylon.

Figure 17.38 shows a data sheet that graphically displays the measured temperature of pole 1 of the slightly damaged three-pin XYZ connector with all screws fully tightened. A tangential horizontal force was applied to the connector. During the test, melting of the connector was observed, at which point the test was discontinued. Maximum recorded temperature was 525° F, exceeding the 482° F melting point of nylon.

17.8.4   Short Circuit Tests on Barrier Terminal Block

A series of short circuit tests were conducted on a three-pole barrier terminal block for use in an electronic system. In addition, temperature measurements were made on the three poles of the barrier terminal block.

Table 17.12 Maximum Temperatures on Two Different Connectors for Five Test Conditions (*courtesy of Failure Analysis Associates*)

No.	Test conducted on terminal block	Connector	Connector Current, A	Maximum pole 1 temp., °F	Maximum pole 2 temp., °F	Melting point of nylon, °F	Specified connector continuous max. operating temp., °F
1.	Ambient temp. measurement in the vicinity of the connectors	—	12.2	95*	—	482	221
2.	Temp. measurements on a new connector #1, all screws fully tightened	New connector #1, ABC	12.2	127	—	482	221
3.	Temp. measurements on a new connector #1, one and two screws loose	New connector #1, ABC	13.4	591	847	482	221
4.	Short-circuit test	New connector #2, XYZ	13.4†	218	363	482	221
5.	Temp. measurements on slightly damaged connector #2, normal mounting and horizontal force applied	New connector #2, XYZ	13.4	525	—	482	221

* Inside ambient temperature
† Short-circuit current, 100-A peak for 1 cycle and 42 A rms for 0.519 s exceeds 15 A connector rating

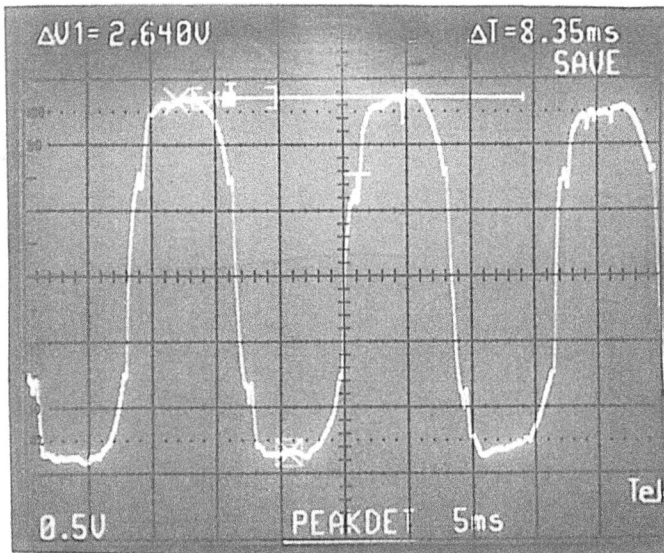

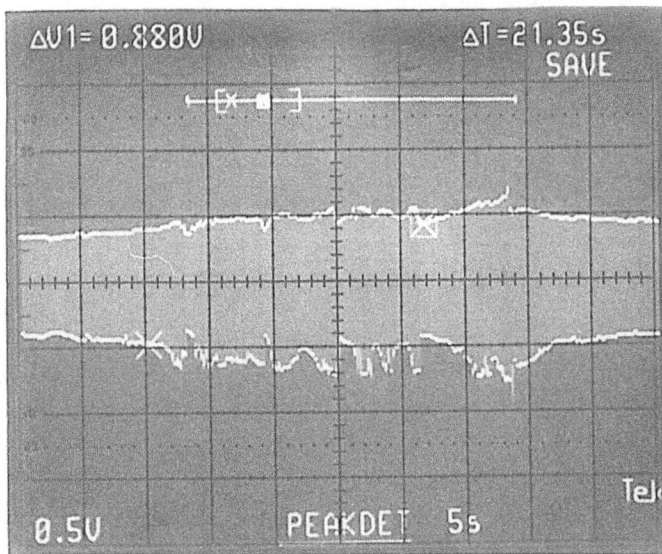

Figure 17.34 Voltage drop on one pole of ABC connector during one- and two-screws-loose tests *(courtesy of Failure Analysis Associates).*

Table 17.13 presents the temperature of poles 1A, 1B, and 3 of the barrier terminal block. Poles 1A and 1B had loosened screws. Pole 3 had tightened screws. The "stop testing" temperature was 400° F. Maximum measured temperature was 541° F, at which point the test was concluded. Figures 17.39 and 17.40 present pole 1A and pole 3 temperatures during the short-circuit test.

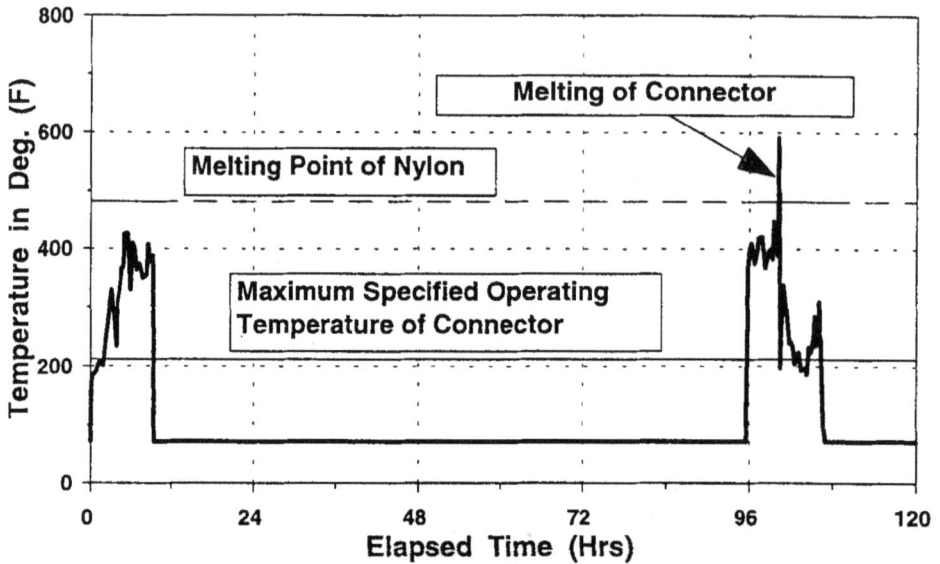

Figure 17.35 New ABC three-pin connector, temperature of pole 1 *(courtesy of Failure Analysis Associates).*

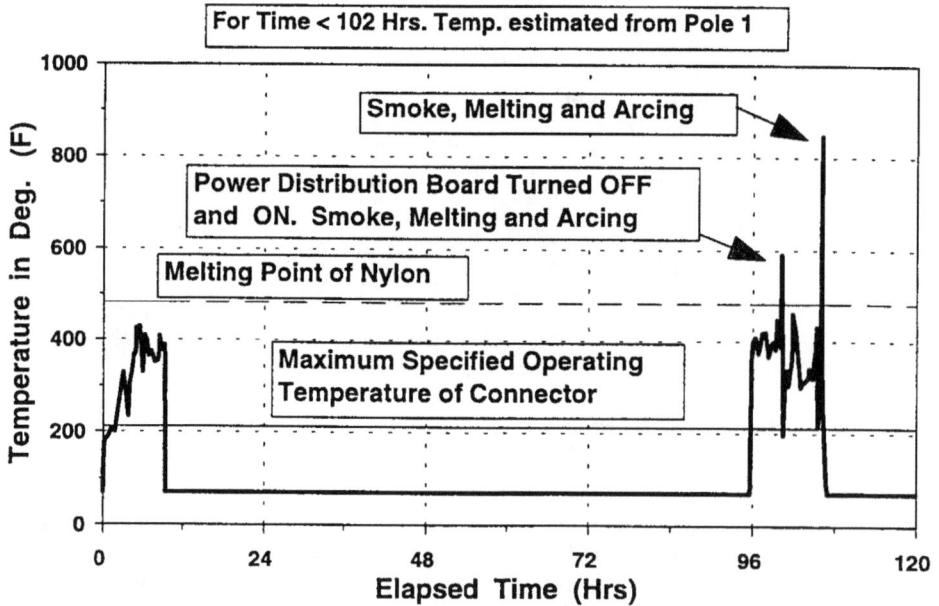

Figure 17.36 New ABC three-pin connector, temperature of pole 2 *(courtesy of Failure Analysis Associates).*

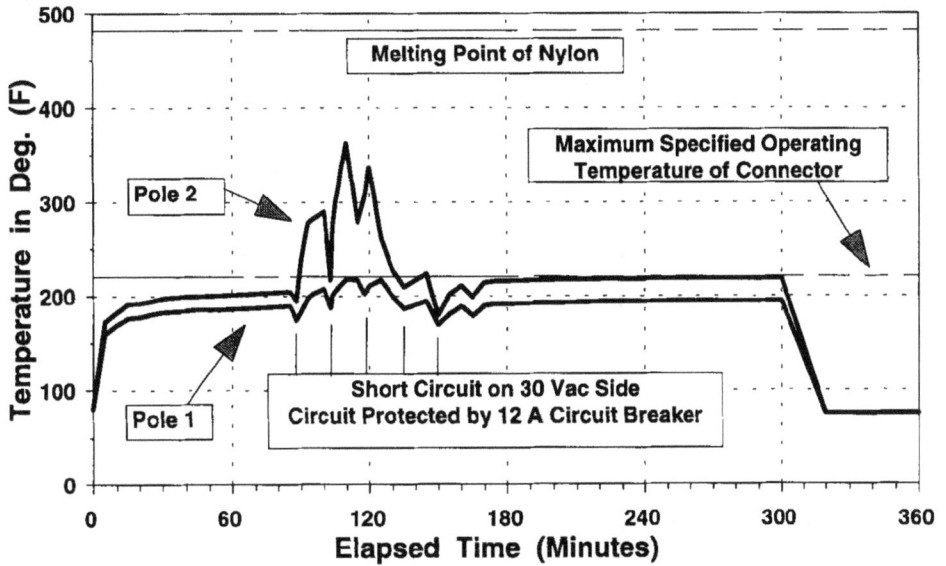

Figure 17.37   XYZ three-pin connector, normal operation/short-circuit test *(courtesy of Failure Analysis Associates).*

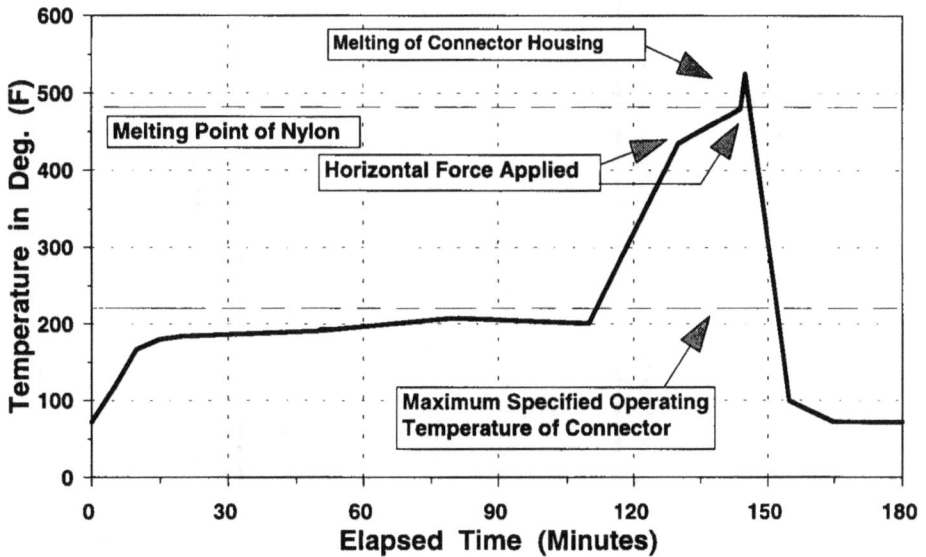

Figure 17.38   Used XYZ three-pin connector, normal operation/horizontal force test *(courtesy of Failure Analysis Associates).*

Table 17.13 Maximum Temperatures Recorded on Barrier Three-Pole Terminal Block *(courtesy of Failure Analysis Associates)**

No.	Test conducted on terminal block	Pole number	Pole current, A	Status of pole screw	Max. pole temperature, °F	Manufacturer-specified heat resistance (continuous) temp., °F	Measured pole temperature exceeded manufacturer-specified heat resistance (continuous) temp?
1.	Temp. measurements on new barrier terminal block Pole 1A temp.	1A	14.5A[†]	Screw loose	541	290	Yes
2.	Temp. measurements on new barrier terminal block Pole 2A temp.	1B	14.5A[†]	Screw loose	470	290	Yes
3.	Temp. measurements on new barrier terminal block Pole 3 temp.	3	14.5A[†]	Screw tight	205.5	290	No

* Inside ambient 95° F, minim. stop test temperature 400° F)
† Short circuit current, 60 A for 1 s exceeds 15-A connector rating.

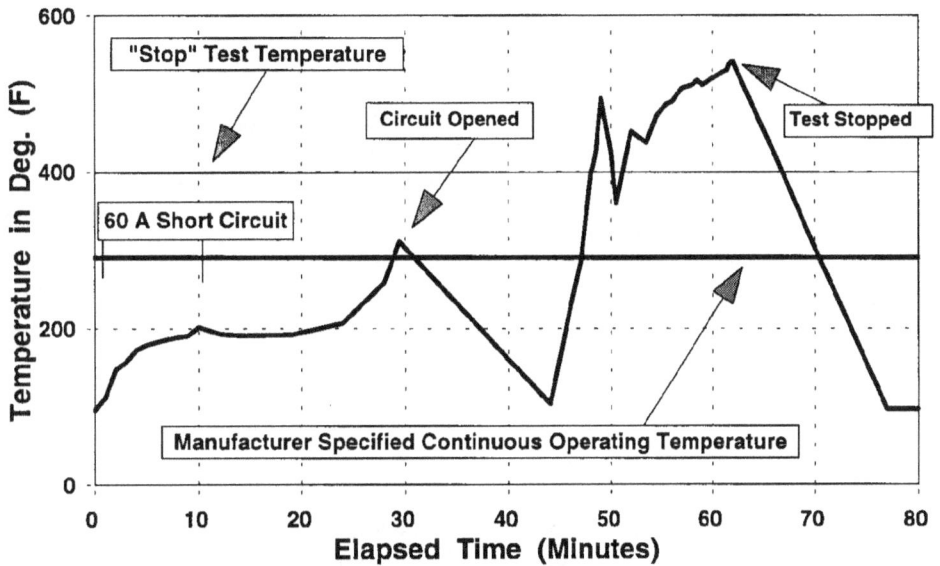

Figure 17.39   Temperature of pole 1A of barrier terminal block during short-circuit test *(courtesy of Failure Analysis Associates).*

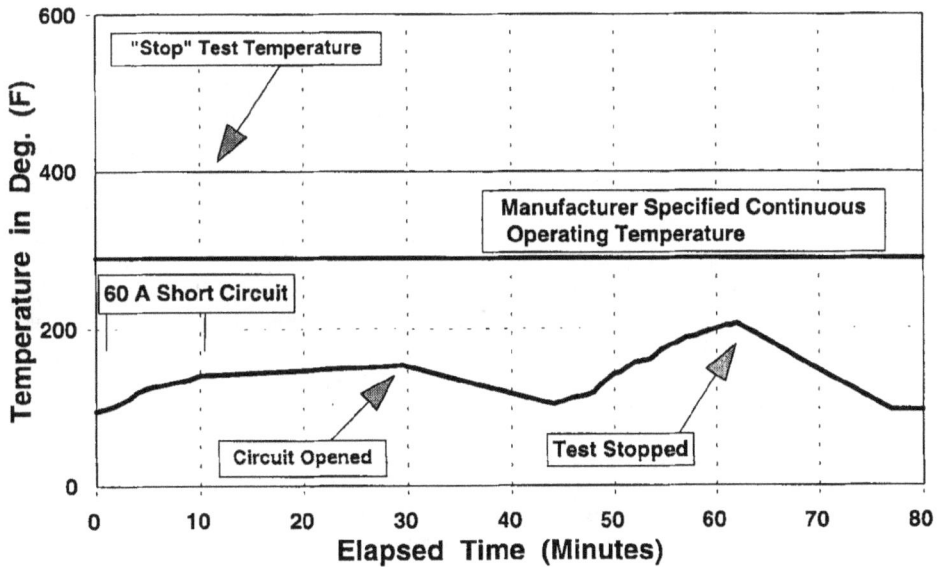

Figure 17.40   Temperature of pole 3 of barrier terminal block during short-circuit test *(courtesy of Failure Analysis Associates).*

17.8.5   Electric Shock Hazard in Energized Terminal Strip

Commercial welding equipment used for arc welding in the open hull of a ship is expected to be exposed to severe weather conditions and to be subjected to considerable electrical, mechanical, and environmental abuse. Equipment for use in the harsh environment of a shipyard should be built to operate in such environments without creating an electrical hazard to users. Both in storage and in use, these units are exposed to extremes in temperatures. Since the cabinet of the subject unit was not sealed against the elements, the atmosphere would have been expected to deposit salt, metal particles, and chemical contaminants on the parts of the welding unit that served as electrical insulation. Furthermore, these commercial units, used around the clock for three shifts, were subjected to a great deal of mechanical abuse (i.e., lifting, dropping, shock, and vibration). Finally, the welding equipment may have been subjected to electrical transients entering the 120-Vac socket from the electrical system.

For safety, the subject welding unit was designed to be solidly grounded by a grounding wire to the power supply system. However, it was determined that the grounding connection was broken, and there was a leakage current path created by conductive contaminants across an insulating surface between the hot wire and the cabinet. As discussed in a previous section, the human body is very sensitive to electric shock, and even relatively low levels of leakage current from conductive contaminants can cause painful or lethal shock.

In the case of the subject welding equipment, the lug-type terminal strip to which the line-filter capacitors were soldered had three 115-Vac line lugs. The two outer lugs were also connected to the hot and neutral wires. The center terminal was the ground terminal and was bolted to the case. Figure 17.41 is a photograph of the filter network and the three-terminal soldering strip. Due to the small spacing between the lugs, the strip presents an electric shock safety hazard.

In a shipyard environment, with the presence of humidity and salt, the buildup of conductive contaminants between the terminals decreased the surface resistance between the metal lugs of the terminal strip and permitted current to leak between the hot 115-Vac line and the cabinet. Due to vibration, mechanical movement, and lack of a strain relief *pigtail* on the metal clamp to the line cord, the safety ground wire of the ac cord became frayed, resulting in an intermittent grounding connection to the cabinet. With a defective grounding wire, the cabinet became a shock hazard. The conditions for an electric shock to occur were present due to (a) an intermittent ground connection, (b) leakage current due to inadequate clearance, and (c) contamination across the insulating surface. The arrow in Fig. 17.41 indicates the buildup of contaminants between the metal lugs of the three-position terminal strip.

When a worker made contact with the equipment, the frayed grounding conductor was displaced, breaking the intermittent ground connection. The leakage current passed from the 115-V hot lug through the contaminated surface of the lug-type terminal strip to the cabinet of the equipment and through the worker's body to ground, resulting in an electric shock.

17.8.6   Connector Failure at Military Installation

A case study involved performing cable fault tests using a low-voltage dc power supply on the 240 insulated wires of aircraft fueling station legs served by two pumphouses at a military installation. The installation was in a wet environment near the ocean and was subject to heavy mechanical vibration due to the close proximity of departing high-speed military aircraft.

Figure 17.41   Photo of filter network and three-terminal soldering strip *(courtesy of Failure Analysis Associates)*.

The analysis called for measuring the dc resistance between 240 insulated control wires. The control wires, installed in a four-inch underground conduit, were used to transmit the refueling, defueling, and emergency signals from individual hydrants located on the taxiway. The signals, transmitted by the control wires, activated 48-Vac control relays located in the pumphouse, which were used to energize the main contactors of the pump motors. Each hydrant had three single-pole switches: a normally open (NO) refueling switch, a NO defueling switch, and a normally closed (NC) emergency switch. Each switch had two control wires connecting the switch to the pumphouse.

During the inspection at the military installation, it was observed that several of the conduits at each pumphouse were filled with water. Consequently, the 300- to 1000-ft long control wires could be immersed in water during normal operation. In addition, installation problems were discovered, including poor wiring techniques and several others discussed below.

1. It was observed that several jumpers were screwed between the terminal lugs to connect adjacent terminals. These jumpers were made using a U-shaped solid wire placed under the screw terminals. The jumpers did not have a loop around the screw to hold them in place mechanically. This was not in accordance with good design practice, which requires that all wire connections under a screw terminal have a loop around the screw terminal. In a high-vibration environment, the absence of the wire loop could result in the jumper becoming dislodged, opening the connection, and resulting in pos-

sible misoperation of the system. Figure 17.42 presents some of the terminal blocks of the control panel at one pumphouse. Arrows indicate the U-shaped jumpers with no loop around the screw.

2. The control wires emanating from the conduits were crimped to spade lugs. It was observed that on several of these wires, the spade lugs were not adequately crimped and were loose, or had even come apart and were separate from the control wires.

3. During an inspection of the hydrants, it was observed that splices were used to connect the wires from the switches to the wires in the conduits. A polypropylene splice cartridge tube was used, which was not considered to be a suitable material in the presence of hydrocarbon contaminants such as jet fuel.

4. During an inspection of the hydrants at one pumphouse, it was observed that some of the splicing connections between the control wires and the switch wires were made using only a wire nut filled with an insulating compound, possibly in an attempt to make it waterproof. This may have resulted in the wires in the splice making contact with the heavily contaminated water in the conduit, resulting in misoperation of the system. Figure 17.43 is one of the hydrants with the cover removed, showing the high level of contamination on the switches, splices, and insulated wires. Figure 17.44 is a

Figure 17.42 Terminal blocks of control panel at a pumphouse. Arrows show U-shaped jumpers with no loop around the screw *(courtesy of Failure Analysis Associates)*.

Figure 17.43   Hydrant with cover removed, showing high level of contamination on switches, splices, and insulated wires *(courtesy of Failure Analysis Associates).*

close-up of one of the hydrant switches showing the splicing covered by wire nuts filled with an insulating compound. Figure 17.45 is a close-up of the high level of contamination on the splice cartridge, and on the insulated wires.

17.8.7   Contact Resistance of Power Connector

A 50-A, 125/250-V plug-in connector was used in a 240-Vac, 60-Hz application with a typical connected load of approximately 20 A. It was alleged that the plug-in connector failed to provide low ground terminal contact resistance. It was purported that the ground terminal contact resistance increased far beyond its nominal maximum value over a period of time and eventually, when a fault developed in an external circuit, resulted in a reduction in the fault current and prevented the circuit breaker from tripping. The instantaneous trip fault current measured for the circuit breaker was approximately 416 A.

Tests were conducted on the subject connector to determine the contact resistance as a function of current for the three phase terminals (W, X, and Y) and for the ground terminal. The terminal current was varied using a variac and an external wire-wound load resistor. Terminal current and contact voltage were measured at the terminal pocket screws.

As observed in Fig. 17.46, phase terminals W, X, and Y of connector ABC had a measured contact resistance of <0.8 mΩ at the maximum rated current of 50 A, resulting in a contact voltage drop at full load of <40 mV. The measured contact resistance of the ground terminal was <1.8 mΩ at 50 A, resulting in a contact voltage drop at full load of <90 mV.

Figure 17.44   Hydrant switch showing splicing covered by wire nuts filled with an insulating compound (bottom arrow). Yellow cartridges have silicone insulating gel and are used to moisture seal the connections *(courtesy of Failure Analysis Associates).*

It was determined that the instantaneous trip current measured for the circuit breaker was 416 A. Analysis showed that a fault current of 416 A would result in a maximum contact voltage drop of 333 mV on the contacts of phase terminals W, X, Y, and 749 mV on the ground terminal contact. This voltage drop was minimal compared to the source voltage. Consequently, during a fault condition, this minimal contact resistance and correspondingly minimal voltage drop would not have any measurable effect on the magnitude of the fault current.

17.8.8   Bus Bar Temperatures in a Tap Panel

An arcing fault followed by a fire occurred in a low-voltage tap panel located in the electrical room of an office building. The tap panel was mounted on a 480-V, low-impedance bus duct rated at 480 V, 2000 A, three phase, four wire. Eight 2.5- × 0.25-in aluminum bus bars, wrapped with varnish cambric tape, were located in close mutual proximity. The tap panel provided access to the bus bars via a series of switches and fuses that fed lighting and motor loads. The switches were connected to copper bus bars, which in turn were connected to the aluminum bus bars using bolts and Belleville washers. Calculations were performed to determine the temperature rise of the phase B and phase C bus bars prior to the fault using a thermal model of the system.

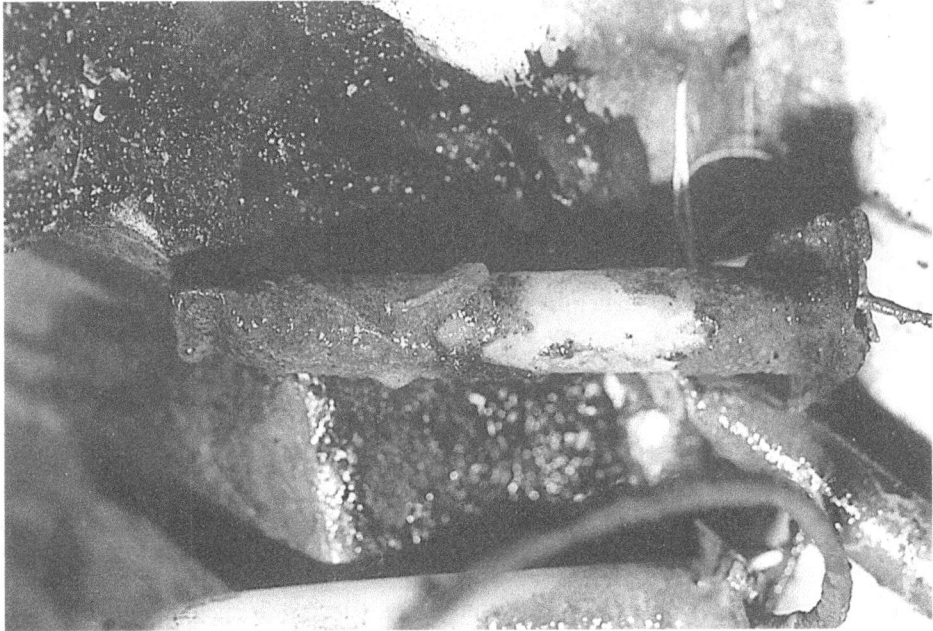

Figure 17.45   Close-up of contamination deposits on splice cartridge and insulated wires *(courtesy of Failure Analysis Associates)*.

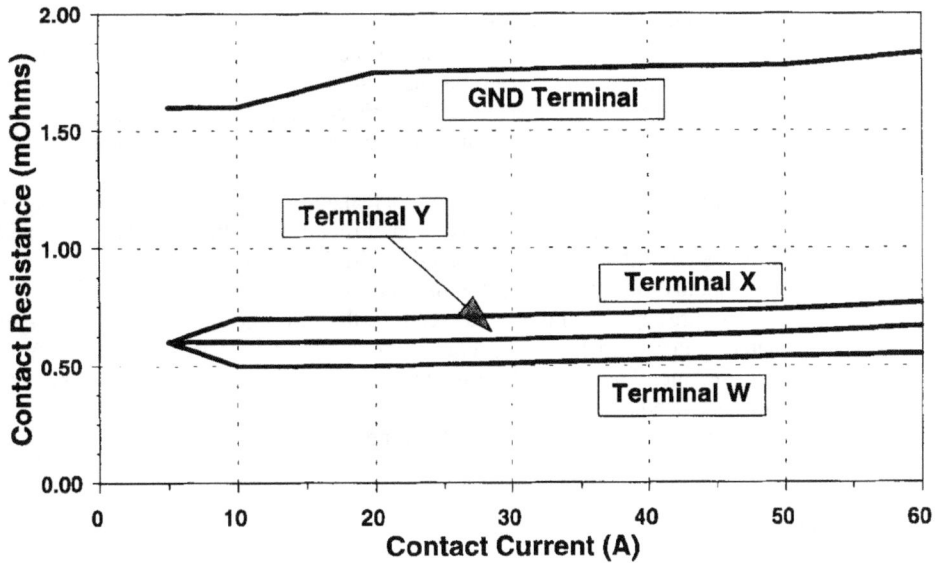

Figure 17.46   Contact resistance vs. contact current, ABC 50-A, 125/250-V plug/connector *(courtesy of Failure Analysis Associates)*.

Figure 17.47 presents the electrical circuit of the thermal model for the Phase A, B, and C bus bars in the tap panel. Voltage sources were used to set boundary conditions for the ambient temperatures in the tap panel. Thermal impedances of the bus bars at 11 predetermined sections were computed and included in the model for each phase. Additionally, each of the 11 bus sections in each phase dissipated power and was modeled by a calculated forcing current element. Furthermore, the thermal resistance of the varnish cambric tape was included to determine the rise in temperature at each of the temperature nodes T1 through T10. The high degree of complexity in the thermal model involving more than 125 individual elements required the use of a high-level computer simulation program. Micro-Sim Corporation's PSpice® computer simulation program was used to perform the calculations. The Analog Behavioral Modeling option of PSpice® was used to insert equation blocks into the model and generate forcing temperature functions determined by averaging two model temperatures. The result of the analysis was a profile of the temperatures along the lengths of the aluminum bus bars.

17.8.9 Electrocution by a Wet Vac

A 120-Vac commercial wet vacuum used at a sports facility was alleged to have caused an electrical shock to the operator. The wet vac was being used in the vicinity of an indoor pool, a wet environment.

The wet vac had double insulation between the hot energized parts and the exposed metal parts, and it had a two-wire cord. Visual inspection showed that a two-pole nylon terminal block was used to connect the 120-Vac power cord to the control panel of the wet vac. In addition, an 0.125-in (3.175 mm) dia hole was visible immediately adjacent to the terminal block. This hole provided a means for mounting a three-pole terminal block in a previous model of the wet vac and was no longer being used for the double-insulated version.

Leakage current measurements were conducted on the terminal block to determine the leakage resistance between the poles of the terminal block to the outer housing in the presence of contaminants. During testing, a spray of contaminated water was observed issuing from the hole. It was determined that this contaminated water spray could cause a leakage current to flow between the energized poles of the terminal block and the exposed metal housing.

Figure 17.48 shows the measured leakage current at 170 Vdc using a saturated salt solution (NaCl) as the conductive contaminant. It is observed that the maximum measured leakage current was 35 mA, exceeding the 5 mA threshold of safety as specified in ANSI/ISA S82.01, outlined in a previous section.

The presence of the hole adjacent to the terminal block, the visible spray of contaminated water from the hole, and the geometrical proximity of the terminal block to the exposed metal housing all show that the double insulation requirement of the wet vac was compromised. Figure 17.49 shows the terminal block and the buildup of contamination. An arrow shows the adjacent hole in the panel.

17.9 SUMMARY COMMENTS

Failures in connectors appear to occur for a large variety of "apparent" reasons, many of which have been listed in this chapter. However, the fundamental causes of failures are

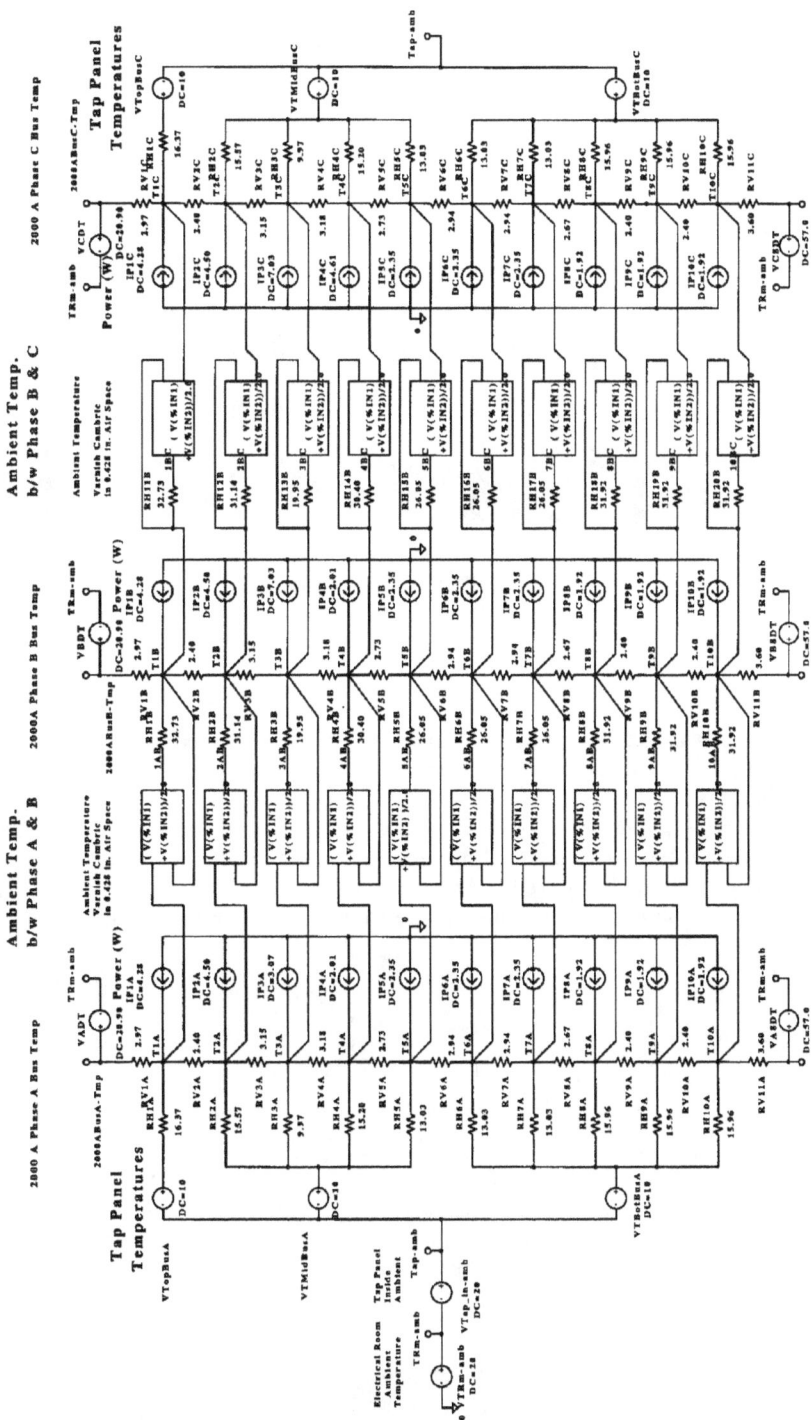

Figure 17.47 Electric circuit of thermal model for phase A, B, and C busbars in tap panel *(courtesy of Failure Analysis Associates)*.

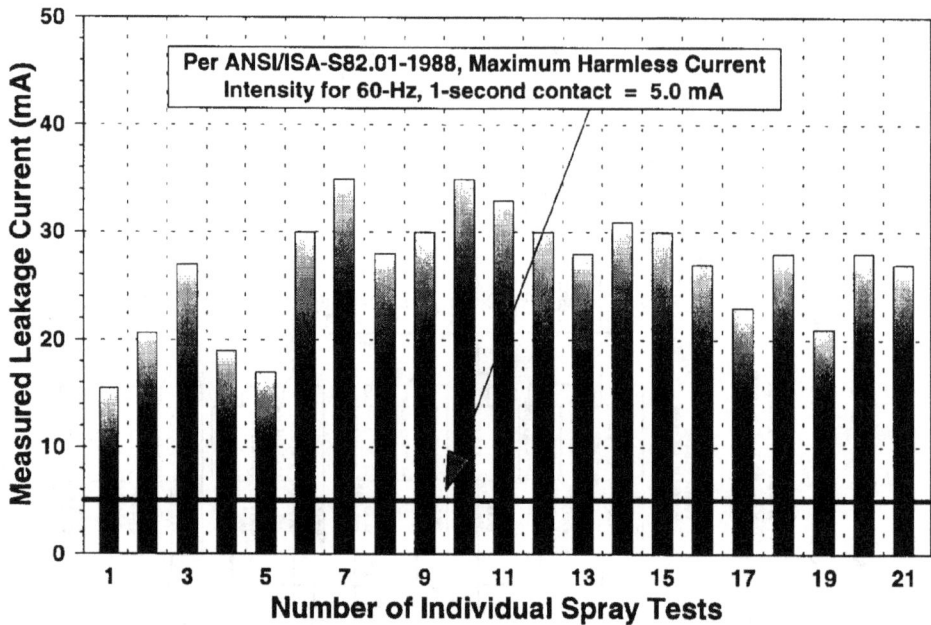

Figure 17.48 Leakage current measured at 170 Vdc using a saturated salt solution (NaCl) as the conductive contaminant *(courtesy of Failure Analysis Associates).*

few. The probability of failure can be significantly reduced by following a rigorous design philosophy substantiated by extensive testing. In all cases, the design should be in accordance with all established codes and standards.

Failure analysis and *root cause analysis* are complicated endeavors that require good structure, extensive care, and strong commitment to determine the cause of the failure. The analysis should be viewed as a valuable and powerful learning experience whose real benefits will be realized in improved performance indicators resulting from reduced failure rates.

17.10 REFERENCES

1. *The New IEEE Standard Dictionary of Electrical and Electronics Terms,* 6th ed. (1996). New York: Institute of Electrical and Electronics Engineers, Inc. The IEEE disclaims any responsibility or liability resulting from the placement and use in the described manner.

2. Horn, Jochen, Egenolf, Berhard, AMP Deutschland GmbH. Shape Optimization of Connector Contacts for Reduced Wear and Reduced Insertion Force, *AMP Journal of Technology* 2 (November 1992), 42–45.

3. Mroczkowski, R.S. Materials Considerations in Connector Design, AMP Technical Paper P310-88, 6–17, presented at ASM World Material Congress, Sept. 1988, Chicago, IL.

4. Van Dijk, Piet, AMP-Holland, "Some Effects of Lubricants and Corrosion Inhibitors on Electrical Contacts," AMP Journal of Technology, Vol. 2, November 1992, pp. 56-61. Reprinted with the permission of AMP.

5. Crimp Application Tooling, *950 Full Line Catalog.* Lisle, IL: Molex, W-2.

6. *Standards for Safety Catalog.* Underwriters Laboratories, Inc., July 1990.

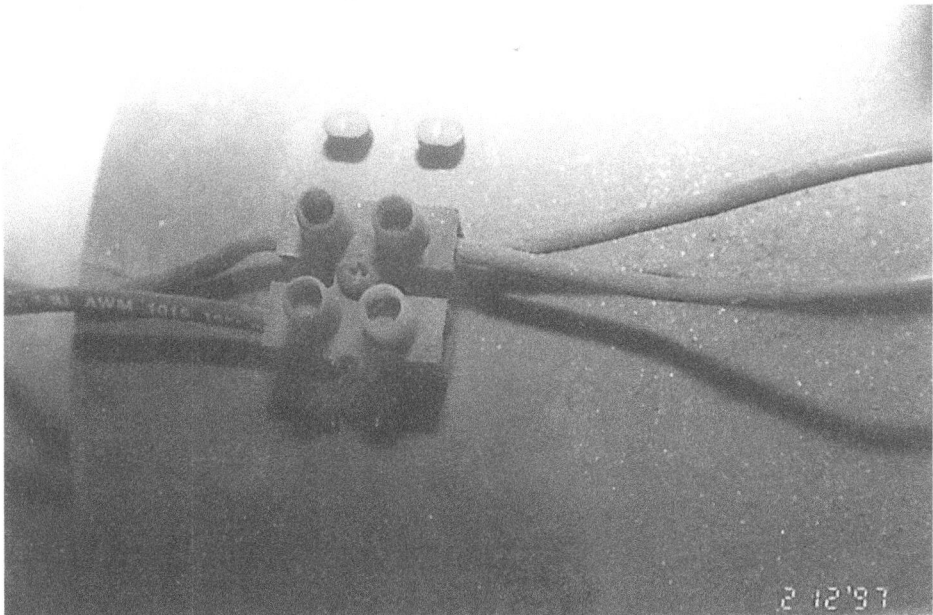

Figure 17.49   Terminal block and buildup of contamination *(courtesy of Failure Analysis Associates).*

7. *1996 Catalog of International Electrotechnical Commission (IEC) Publications, World Standards for Electrical and Electronic Engineering,* 34 (catalog description of IEC 130).
8. U.S.Patent No. 5,065,104, Fault Sensing with an Artificial Reference Potential Provided by an Isolated Capacitance Effect, November 12, 1991, A. Kusko, N. K. Medora.

CHAPTER 18
FAILURE ANALYSIS OF COMPONENTS

Donald Galler
Department of Materials Science and
* Engineering*
Massachusetts Institute of Technology
Richard A. Blanchard
Duncan Glover

Alexander Kusko
John D. Loud
Noshirwan K. Medora
Gregory J. Mimmack
Günter Müller
E^xponent

18.1   INTRODUCTION

This chapter discusses the methodology for determining the root cause of component failures that occur in circuits and systems. This subject is approached by first discussing the physical features and electrical characteristics that can be observed when examining failed components. The physical features, in conjunction with the electrical characteristics, can provide significant information about the specific root cause of the failure. A general approach for determining the physical features and electrical characteristics of components is discussed in the first section, along with the technique that can be used to obtain this information. Subsequent sections discuss the failures that occur for in the various components encountered in contemporary circuits and systems.

18.1.1   Failure Characteristics of Components

The physical features and electrical characteristics of a failed component may provide the only information about the specific root cause of the failure. This section discusses these two critical sets of information about failed devices.

Overvoltage.   Overvoltage refers to the presence of a voltage sufficiently high to damage or destroy a device by causing it to operate outside its allowed voltage range. Short-duration overvoltage events can cause device failures and usually result in some observable physical damage. Overvoltage conditions can be caused by a number of events that are external to the device in question. Some of the most common conditions are:

- Electrostatic discharge (ESD)
- Supply voltages transients
- Electromagnetic coupling between signal cables
- Connector and printed wiring board failures

There are three basic mechanisms for overvoltage failures:

1. The voltage can cause breakdown of the insulating material layers
   used in device construction.

2. The voltage permanently alters the characteristics of *p-n* junctions. This failure is usually caused by exceeding the reverse breakdown voltage of a component or IC.

3. The voltage causes a four-layer semiconductor device, called a silicon-controlled rectifier (SCR), to conduct excessive current, providing a low-impudence connection between the power supply and ground. This failure occurs almost exclusively in CMOS integrated circuits.

Overvoltages that cause device failures also cause some physical damage to the device. The amount of damage depends on the amount of energy available at the fault site. It also depends on how the device is used, and the voltage and current capability of the power supply.

One common error that occurs in analyzing component failures is to assume that the maximum available voltage that can occur is the rated voltage of the power line or power supply. Two specific sources of voltage transients are discussed below.

1. Conventional 115-Vac power lines carry transients that far exceed the nominal voltage, and the transients can be as large as ≈6000 V.[1] These transients may be caused by lightning, by the switching other loads on the power distribution system *on* or *off*, or by the power source itself. The value of ≈6000 V is the voltage required to arc across the spacing between the conductors in a 115-V service. These transients are usually dealt with by placing a transient suppression component (or network) across the power input leads of a product.

2. Power supply transients occur because of either transients on the input power line or transients from the operation of the circuitry of the power supply itself. In either instance, transients well in excess of the rated output voltage can occur, and the overall performance of power supplies must be considered when trying to determine the root cause of a failure. In power supplies, it is important to evaluate performance during turn-on and turn-off, as well as during normal operation.

Overcurrent. A common failure mode of bond wires and device metallization used to bond discrete semiconductor devices for integrated circuits is overcurrent failure, which tend to melt the conductor material.

Overcurrent failures are usually caused by electrical short circuits in load devices or cables. Overcurrent failures in microelectronics devices are usually secondary failures that can be used to determine the primary failure mechanism and location.

Overcurrent conditions can produce sufficient heat to melt or vaporize bond wires and metallization. Excessive heating occurs when the current density of the wire or trace is exceeded. The thermal power is expressed in watts as $P = I^2R$.

Both gold and aluminum bond wires exhibit overcurrent failures in similar ways. Both materials are excellent thermal conductors, and the attachments at the end will help cool the ends of the wires. As a result, the highest temperature and most common failure location will be near the middle of a span of the wire.

Gold wires will usually form a ball at the remaining ends following an overcurrent failure. The pressure of a ball is the key feature to seek. Aluminum wires usually melt and separate while taking on a wrinkled appearance. Aluminum does not usually form a ball at the failure site.

The magnitude and duration of the overcurrent event will determine whether the bond wire or the device metallization will melt first. Very short-duration current pulses will tend to melt the metallization first, because it has lower cross-sectional area and higher resistance than the bond wire. As the pulses become longer in duration and lower in magnitude, failure of the wire becomes more prevalent. This behavior occurs because the thermal coupling to the bulk of the die becomes effective in cooling the aluminum metallization[2] for longer-duration pulses.

The melting point and thermal conductivity of some materials used for bond wire and metallization interconnects are listed in Table 18.1 for reference.

Table 18.1   Melting Point for Bond Wire and Metallization Materials

Material	Melting point, °C	Thermal conductivity, W/cm°C
Aluminum	660	2.18
Gold	1063	2.96
Silicon	1410	0.84
Tungsten	3410	1.99

An approximate formula for the fusing current of an infinitely long 99 percent aluminum, 1 percent silicon wire is $I = 0.4D$, where D is the wire diameter in mils. This expression assumes that the length of the wire $L > 100D$, and this is a good approximation for many devices.[3]

Melting of conductors or device metallization will normally be visible at magnifications below 1000× so that optical microscopy and SEM may both be used effectively to investigate suspected overcurrent failures. The SEM is preferred for its depth of field.

Corrosion.   Corrosion causes the physical degradation of metals. It often leads to electrical and mechanical failures in microelectronic devices.[4] Corrosion requires a galvanic couple, a corrosive contaminant, and moisture for activation. Corrosion is the destructive attack of metals by a chemical reaction between the metals and a corrosive agent in the presence of moisture. Contaminants inside devices can come from several sources:

- Cleaning fluids used prior to packaging
- Package leaks
- Human contaminants, such as saliva and perspiration

The result of corrosion, regardless of its specific cause, is usually the same: destruction of the metal. The reduced metal cross section typically results in high-resistance electrical connections and mechanical failures. Operating voltages present in the device usually accelerate the corrosion process. However, these voltages are not required to cause catastrophic failures.

Moisture can enter the device during the packaging process. It can also enter the package after assembly, from failures caused by mishandling, vibration and thermal stress. Failure of lid seals and cracking of glass-to-metal connection seals are two packaging defects in hermetic packages that can result in moisture intrusion. Moisture can be sealed inside the device package during assembly or generated from material outgassing after sealing. Outgassing is a problem usually associated with molded plastic packages. No molded package is hermetic, so moisture from the environment may also reach levels that are high enough to cause corrosion in devices with molded packages.

Corrosion inside the device package usually occurs where exposed metal is present. Corrosion is most common in semiconductor devices that use aluminum for the interconnect metallization and either aluminum or gold for the bond wires. Corrosion may occur on the bond wires, the bonding pads, and any exposed metallization. In addition, some contaminants will diffuse through the thin glass layer often used for passivation and attack the underlying metallization. Corrosion outside the device package usually occurs on or near metal surfaces, especially on leads, metal packages, and glass-to-metal seals.

Elemental analysis typically reveals the presence of ionic contaminants in the products of aluminum metallization corrosion. Attack by saliva results in a corrosion product containing aluminum, potassium, chlorine, sodium, some calcium, and a trace of magnesium. Attack by perspiration sometimes contains zinc, which is a common ingredient in most antiperspirants.

Another source of corrosion is phosphorus in the glass passivation that combines with moisture in the package. If the phosphorus and moisture levels are high enough, then phosphoric acid can form and attack the device. Moisture levels above 5000 ppm are usually needed for this type of corrosion.

Cosmetics are also a possible source of contamination during manufacturing. Iron and aluminum have been found in mascara. Facial powder may contain titanium, iron, magnesium, aluminum, and potassium. Sodium and chlorine are usually not found in cosmetics.[5] Analytical techniques for identifying the presence and the causes of corrosion include optical microscopy, SEM, and EDS.

Lead or Wire Bond Failure. Mechanical and electrical failure of bond wires can be observed in device failures. Detached leads or bonds, shorts, corrosion, and signs of overcurrent may all be indicators of lead or bond failures. Lead or bond wire failures can be caused by contamination, mechanical stress, electrical overstress, or improper attachment parameters (pressure, temperature, time).

Contamination of the surface can prevent proper bonding of the lead or bond wire to the terminal bonding pad. Electrical overload can cause local melting of the lead attachment solder or bond wire.

Improper lead attachment or bonding parameters can cause incomplete bonding and mechanical fractures and promote conditions for favorable metallurgical failures. The resulting bond can fail mechanically from vibration, shock, and temperature cycling in normal operation.

Gold bond wires attached to aluminum metallization in discrete devices or integrated circuits can develop a condition called *purple plague*. This condition is caused by the formation of $AuAl_2$, which is a gold-aluminum intermetallic compound. The formation of gold-intermetallic compounds is essential to making a good bond. As the $AuAl_2$ intermetallic forms, it expands in the region between the gold wire and the aluminum, degrading the mechanical strength and electrical contact of the bond. The five gold-aluminum intermetallics are listed in Table 18.2.

The optical microscope or the SEM, often in conjunction with EDS, is the best tool for examining leads or wire bonds. Backscattered electron (BSE) imaging can be helpful in identifying intermetallics that may provide information about the presence of contaminants that contributed to the failure.

Mechanical Damage. Mechanical damage is caused by a variety of forces that occur during a mishap. The nature of devices makes them susceptible to damage from shock and

Table 18.2   Gold-Aluminum Intermetallic Compounds

Compound	Formation temperature, °C	Color
Au_5Al_2	100	Tan
Au_2Al	50	Metallic gray
$AuAl_2$	150	Deep purple
Au_4Al	150	Tan
$AuAl$	250	White

vibration. Pure shock loads are just one type of loading that can occur. Crushing, bending, shearing forces, or vibration may also be present.

Unlike some of the other failure characteristics discussed in this chapter, mechanical damage to devices can manifest itself in a variety of ways. Some of the more common indications are described below:

- Lead or bond wire deformations occur when the device is subject to mechanical shock. The mass of the wire or lead causes them to deform in the direction of the impact.

- Die fracture can be the result of shock loads or bending stresses applied from outside the device. The silicon die of a microelectronic device is brittle; it will crack easily if bending forces are applied. Under normal circumstances, the package and printed wiring board isolate the die from these forces. The forces encountered during high levels of mechanical shock, however, may cause die fractures.

- Package damage can be caused by crushing, scraping, shock vibration, and bending forces that occur during breakup and impact. Ceramic packaged devices are extremely brittle and are subject to fractures from impact. In contrast, metal packages can be dented and deformed while leaving the device operational.

The specific type of damage caused by shock or vibration depends on several factors. These factors include:

- *Device size.* Larger, heavier devices will generally be more susceptible to damage under the same conditions than lighter devices.

- *Package style.* Ceramic packages are more brittle than plastic and will chip and crack from shock loads.

- *Mounting.* Through-hole mounting is more rugged than surface mount technology. Some surface mount devices have been sheared off the printed wiring boards from impact shock loads.

Since components are usually mounted on printed circuit boards, a careful visual examination is always the first step. At this stage, damage to device packages and external leads should be apparent. Low-magnification (10 to 100×) optical microscopy will be useful when there is significant damage to the package.

Decapsulation or delidding of the package will be necessary to verify internal damage. Optical microscopy at higher magnifications may be needed. It may also be necessary to section the printed wiring board so that it can be handled conveniently under the microscope.

Thermal Damage. Thermal damage manifests itself in devices in a number of related ways. Sources of thermal damage include fire, cooling system failures, overheating of external electrical equipment, and excessive internal power dissipation.

The junctions of silicon microelectronic devices are formed by doping regions of the bulk silicon. Junction characteristics are altered at elevated temperatures, causing the device to fail. This type of failure occurs in the range of 150 to 250° C (302 to 482° F) for silicon. If the device is not energized, the condition is reversible and does not physically damage the device. Physical damage will occur if the device is energized. Junction failures will generally lead to catastrophic currents. The resulting damage usually causes a series of secondary failures that destroy the device.

Other events may occur in semiconductor devices as the temperature is elevated above 250° C (482° F). These include:

- Various gold/aluminum eutectic compounds begin to form at temperatures of 150 to 250° C (302 to 482° F). Thus, the interface of the gold bond wires and the aluminum metallization may be altered at these temperatures.

- A gold/silicon eutectic compound is used to attach the die to hermetically sealed ceramic packages. The eutectic forms at about 370° C (698° F). Temperatures of 225 to 275° C (437 to 527° F) should not lead to significant degradation of the bond.

- An aluminum/silicon eutectic compound forms at about 570° C (1058° F) at the interface between the metallization and the bulk silicon of the die. However, temperatures of 450 to 500° C (842 to 932° F) for short periods generally will not result in the formation of the eutectic.

This discussion excludes thermal damage in specific areas of a transistor or IC caused by electrical failures. Temperatures for damage of materials in microelectronic devices are listed in Table 18.3.

Table 18.3 Temperatures for Damage of Materials for Microelectronic Devices and Components

Temperature, °C	Phenomenon
150 to 250	Semiconductor actions fails
150 to250	Gold/aluminum eutectic forms
370	Gold/silicon eutectic forms
570	Aluminum/silicon eutectic forms
660	Aluminum melts
1063	Gold melts
1083	Copper Melts
1410	Silicon Melts

Thermal damage to other components (e.g., capacitors, inductors, transformers) is determined by the materials used in their construction. In general, external heat damages the exterior of the component and increases the current flow for a given applied voltage, which can ultimately result in device failure.

Analysis is best conducted by optical microscopy and SEM. Eutectics can be identified by SEM in BSE mode or by EDS since they have a composition that is distinctly different from the standard materials. It is often useful to compare an exemplar to the incident device during the analysis.

The interpretation of thermal damage is not well established. This state of affairs is partly due to the fact that most of the mechanisms are time and temperature dependent. Many combinations of these variables may result in similar physical damage to a device.

Altered Electrical Characteristics.   Altered electrical characteristics are encountered frequently during mishap investigation. They may be observed during testing of equipment and on isolated devices. Changes in device characteristics can be caused by failures that occurred prior to the failure or by thermal damage during the failure.

A variety of manufacturing, handling and circuit conditions can cause degradation of device characteristics or total failure during operation. These include:

- Contamination
- Package integrity failures
- Wire bond defects
- Die attach problems
- Device fabrication problems
- Overvoltage
- Overcurrent

Localized heating resulting from a failure can also cause many types of device failures, with distinctly different causes and failure conditions.

Electrical testing should be conducted with caution. Curve tracers and other test equipment can damage devices and destroy important information. Furthermore, partially damaged devices can be destroyed at much lower levels of electrical stress than good devices. Extreme caution should be exercised during the first stages of electrical testing.

Electrical testing should be conducted prior to destructive tests to document the condition of the device. Only limited electrical testing can be accomplished when a device has suffered great physical damage. Test methods that may be used depend on the type of device and the extent of damage. There are basically three types of testing that can be performed:

1. Basic electrical testing may be accomplished using an ohmmeter to identify device failures, shorts, and open circuits. This type of testing is useful for all types of devices. (Before using an ohmmeter, determine the maximum current and voltage that the ohmmeter will supply.)

2. Curve tracer testing may be used to determine characteristics of diodes, transistors, and their components. This type of testing is most useful on discrete devices but can be used on ICs where junctions or transistors can be isolated for testing. (If *p-n* junctions are present, keep the applied voltage below 0.5 V to avoid damaging junctions.)

3. Complete electrical performance testing of ICs may be accomplished using automatic test equipment. The test equipment is preprogrammed to provide a series of electrical inputs, called *test vectors,* and compares the output to the recorded characteristics of ideal devices.

The types of altered characteristics that a component can exhibit may be divided into the following two categories.

Catastrophic Failures

- Short circuit: The component exhibits an extremely low resistance.
- Open circuit: The component exhibits an extremely high resistance.

Parametric Failures

- The component demonstrates a significant change in an electrical parameter such as leakage, capacitance, and so on, which causes failure of the circuit or system in which it resides.

18.2   EQUIPMENT AND TECHNIQUES FOR COMPONENT FAILURE ANALYSIS

18.2.1   Failure Analysis Process Flow

The equipment and techniques used to perform a failure analysis vary depending on the type of component being investigated. However, the failure analysis process should include the following steps:

- Confirm that a failure has occurred. Studies show that 50 percent or more of components with reported failures commonly meet their performance specifications.
- Determine the characteristics of the failure.
- Determine the event (or possible events) that could have caused the observed component characteristics.
- If necessary, perform tests to determine which of the events is the root cause of the failure.

18.2.2   Laboratory Analysis

The types of components being analyzed fall into the following two broad categories.

- Discrete semiconductor devices or integrated circuits
- Other discrete components

The specific flow for analyzing semiconductor devices is longer, and the equipment required for this work is typically more complex, but the general flow for all component failure analysis is given below. Steps 1 through 8 apply to all components, while additional steps 9 through 11 apply to discrete semiconductor devices and integrated circuits.

1. Verify that all device data has been recorded. Include the manufacturer, model, date code, rating, and any other available data.

2. Perform a careful visual examination. Look for signs of corrosion, overheating, arcing and mechanical damage. Photograph or otherwise document these observations.

3. Perform in-circuit testing if applicable.

4. Remove the device or circuit from the assembly or printed wiring board for further examination or testing.

5. Perform more detailed visual examination and functional electrical testing.

6. Examine the device by optical microscopy or SEM. Look for cracks in package, lead corrosion, discoloration, or gross contamination.

7. Perform layer removal, decapsulation, or delidding as needed to allow further examination.

8. Perform optical microscopy or SEM. Look for any visible mechanical damage, conductor or wire bonds problems, die flaws or cracks, melted areas, discoloration, or gross contamination.

9. Remove passivation layer from die to allow further analysis.

10. Perform SEM/EDS to verify contaminants and other identified problems.

11. Perform voltage contrast, EBIC or other analysis to identify the source of electrical malfunctions.

More detailed information regarding laboratory analysis procedures and techniques is provided in sections that discuss each type of component.

18.3 GOALS OF COMPONENT FAILURE ANALYSIS

The goal of performing component failure analysis is to determine the root cause of the failure. The motivation to find the root cause of a given component failure may include one or more of the following reasons:

- High-volume production creates a large exposure if the failure turns out to be a systematic defect that is still being repeated in current production. This exposure can be significant in electronics manufacturing where the economies of scale have resulted in the production of vast unit quantities.

- It may be important to predict whether similar failures should be expected in the installed base of units containing the device in question.

- Low-volume production of a critical device demands an understanding of every failure so that it is known whether other critical devices presently in service, or devices yet to be assembled, will experience the same failure.

- A failure may be the result of the environment or the circuit in which the device is used, which is not a device issue.

Regardless of the specific motivation for accurately determine the root of a component failure, the alternative—not determining the root cause—can lead to recurring or ongoing problems, since the appropriate corrective action cannot be taken. The following sections of this chapter address the issue of determining the cause of the failure of specific types of component.

18.4 CAPACITOR FAILURES

A capacitor is an electrical charge storage device. Capacitors are used extensively in all electronic circuits. Capacitors perform many functions and are commonly used to:

- create a localized ideal voltage source for adjacent components

- provide charge storage for timing functions

- tune critical circuits

- provide an impedance that varies with frequency for signal selection

- store charge for lost power conditions

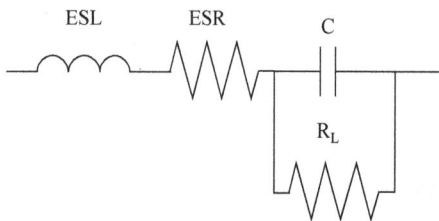

Figure 18.1 Equivalent circuit of a capacitor.

Commonly used capacitors have a vast range in size and packaging. From the pole-mounted, oil-filled power factor correction capacitors used in high-voltage distribution systems to the IC-resident capacitance on a semiconductor substrate, all capacitors have the same basic construction: two conductive surfaces separated by a dielectric material. The energy stored in a capacitor is stored in the dielectric medium. Transfer of charge into and out of a capacitor is achieved through the conductive plates and lead wires. An equivalent circuit for a capacitor is shown in Fig. 18.1.

There are numerous failure modes for capacitors including:

- Catastrophic dielectric breakdown resulting in an unintended release of energy

- Reduction (sometimes intermittent) in capacitance

- An increase in the equivalent series resistance (ESR)

- An increase in the leakage current that may lead to a thermal runaway condition

- Loss of an electrical connection resulting in a loss of capacitance

The results of each one of these failures varies from no apparent effect (as in the case of an *open* failure of a redundant bypass capacitor) to a catastrophic and spectacular explosion and arcing event when a high-voltage electrolytic bus capacitor fails.

The causes for capacitor failures include:

- Excessive applied voltage

- Voltage transients

- Surge currents

- Excessive power dissipation

- Thermal stress

- Mechanical stress and flexing

- Manufacturing defects

- Contamination

18.4.1 Ceramic Capacitors

Construction. Ceramic capacitors come in many different packages, including disk, tubular, and assorted surface mount designs. The dielectric material in a ceramic capacitor is made from earth fired under extreme heat.

Failure Modes. Ceramic capacitor failure mechanisms include:

- Mechanical cracking
- Surge currents
- Dielectric breakdown

Mechanical Cracking. The dielectric material in a ceramic capacitor is extremely inflexible. Combined with the minimal required spacing between capacitor plates to achieve the required capacitance, this inflexibility results in a somewhat fragile structure. Mechanical stresses on ceramic capacitors may result in mechanical cracking. The sources of mechanical stresses include different coefficients of expansion between a ceramic capacitor and PCB materials, mechanical flexing of a PCB, assembly-induced stress, and mechanical shock or vibration.

The effect of a mechanical crack may take time to show up in a ceramic capacitor. For example, if the stress caused by a flexed PCB causes a ceramic capacitor to crack, it may return to its normal position when the flexing force is relaxed. This return to normal can result in no noticeable loss in capacitance or degradation in performance as the severed capacitor plates may physically touch again. However, only a slight misalignment of the parallel interleaved plates would result in a short circuit.

Surge Currents. Excessive current (called *surge* current) exceeds the instantaneous power dissipation capability of localized regions of the dielectric, resulting in a thermal runaway condition.

Dielectric Breakdown. Dielectric breakdown can be caused by overvoltage conditions or manufacturing weaknesses that compromise the dielectric. Manufacturing weaknesses can rarely if ever be documented, as the evidence is inevitably destroyed in the failure event. Dielectric breakdown results in an unintended flow of current between the capacitor terminals, which leads to excessive power dissipation and may result in an explosive failure. Diagnosing a dielectric breakdown capacitor generally is readily apparent, as the failures are typically explosive and obvious to the eye. Should the energy available to a failed capacitor not result in a catastrophic failure, testing of a suspect capacitor should be performed using a light bulb in series with the capacitor under test to prevent supply energy to a failed capacitor. Leakage current can more safely be measured using this technique.

Failure Analysis

- A stereo microscope should be used to make observations of the failed component and the printed circuit board where it was mounted. The capacitor body should be visually inspected. A stereo optical microscope is recommended for inspecting for cracks, burns, or obvious contamination on the device surface.
- The suspect capacitor should be isolated from its circuit using mechanical means. The use of a soldering iron or heat gun could reflow failed solder joints or change the chemical composition of contamination in the region.

- Electrical measurement of the dc leakage current can be made using a curve tracer or four-wire ohmmeter.

- Measurement of the capacitance and dissipation factor should be made at the typical operating frequency of the circuit that the device was in. A capacitance meter or an impedance meter can be used to make this measurement.

Case Study: SMT Ceramic Capacitor.   A manufacturer of peripheral cards for use in personal computers found that many boards were failing shortly after being placed into service. Some of the failures resulted in substantial damage to the peripheral cards. A close examination of the assembly process and end-of-line testing resulted in no explanation for the field failures. Examination of the failed boards revealed large carbonized regions.

Using the techniques described in Chap. 14 to evaluate a propagating PCB failure, the fault sites on each failed board were mapped, and a pattern was established. The damage to each board was too great to conclusively determine the exact point of initiation. Next, a list of potential components was evaluated. Included in this list were ceramic decoupling capacitors. An inspection of failed boards that did not exhibit a charred region was performed. One of these boards was found to have a ceramic capacitor that appeared to have been overheated. Visual microscopy revealed that a small crack was visible on the side of this capacitor. Electrical measurements revealed that the overheated capacitor had a mechanically determined intermittent electrical *soft short*. Flexing of the PCB caused the short to appear and disappear. Further inspection of other noncharred PCBs revealed similar faults that would lead to the charring damage if power was applied. A revisit to the mapped fault regions revealed that ceramic capacitors were present in all known fault locations.

Having identified a cracked capacitor as the cause of the charring, it was now necessary to determine the root cause of the cracked capacitor. Since all of the boards were assembled together, failures should have occurred randomly in all computers in which they were installed—if it was a board fault. A detailed investigation of the returned boards revealed that only one computer manufacturer was experiencing this type of failure. Further investigation revealed that all of the failures were coming from one of many production lines. Only one significant difference was found between production lines at the computer manufacturer. One operator had decided that it was easier to install the peripheral card into the computer first, then to attach the ribbon cable, instead of following the installation instructions, which required the installation of the ribbon cable first, followed by installation into the computer. This alternative method of installing the cable resulted in significant flexing of the peripheral PCB, which occasionally cracked the ceramic capacitors. Very few of the cracked capacitors failed during testing of the PCs, which resulted in failures in the field over time. It now became apparent that the fault location pattern mapped the stress lines that intersected with unfortunately placed ceramic capacitors.

A relocation or reorientation of a number of the ceramic capacitors resulted in a more flex-tolerant PC board. In addition, ensuring compliance with the PCB installation instructions resulted in a virtual elimination of this problem on all future production.

18.4.2   Electrolytic Capacitors

Construction.   Capacitance (C) is calculated from the formula $C = K(a/d)$, where K is the dielectric constant, a is the conductive plate surface area, and d is the distance between the conductive plates. In a never-ending quest for higher capacitance per unit of volume, designers have always tried to reduce the distance between the conductive plates while maximizing the area. The aluminum electrolytic capacitor (most common) is made from a ribbon of aluminum with a thin layer of oxide on its surface. A water-based or other electrolyte, electrically isolated from the insulated aluminum plate, forms the other electrode as it is in contact with a nonoxidized aluminum plate. The oxide layer on the positive aluminum plate is typically 0.01 μm. A diagram showing typical construction of an electrolytic capacitor is shown in Fig. 18.2.

The oxide insulator behaves as a rectifier that conducts current in one direction while blocking current in the other. Electrolytic capacitors must therefore be used in applications where the voltage across the capacitor will maintain the correct polarity. The application of the incorrect polarity across an aluminum electrolytic capacitor will result in destruction of the capacitor.

Electrolytic capacitors are specified for a maximum dc working voltage. The peak of a ripple voltage riding on a dc level must not be allowed to exceed the maximum dc working voltage during normal continuous operation. However, electrolytic capacitors are specified to withstand an occasional voltage surge beyond their maximum dc working voltage. Exceeding the maximum surge rating will likely result in a catastrophic capacitor failure. Leakage current is relatively high on electrolytic capacitors due to the thin dielectric barrier and the large conductor surface area.

The capacitance of electrolytic capacitors typically is not specified very precisely, as production methods result in large variations in capacitance. Common capacitance values specified would include: −10/+50 percent, −10/+75 percent, and ±20 percent.

The equivalent series resistance (ESR) (Fig. 18.1) is a simplified model to account for losses including those caused by lead resistance, foil resistance, and dielectric losses.

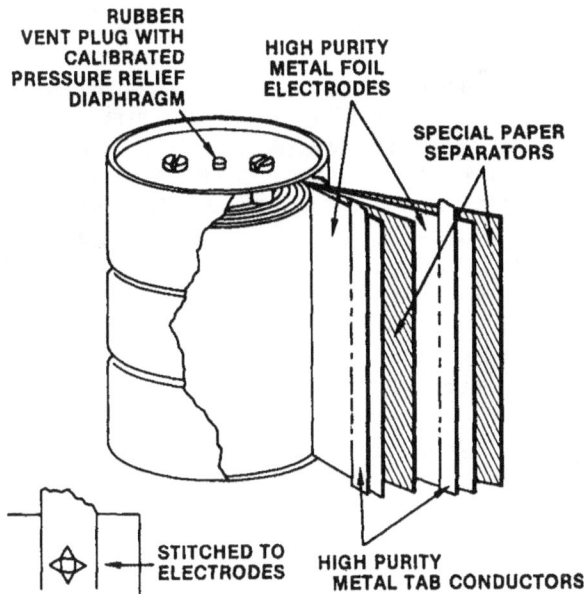

Figure 18.2   Typical electrolytic capacitor construction.[F1]

Depending on the application, this simplification may not be appropriate. The ESR model will allow calculation of the expected capacitor power dissipation for a specified ripple condition. Since heating is a major factor in reducing the expected life of an electrolytic capacitor, ripple current limitations must be maintained to avoid overheating of the capacitor. Ripple currents are especially significant (and harder to evaluate) in high-frequency switching applications. Not only do the fundamental frequency ripple currents contribute to losses (heating), but higher-frequency harmonic currents will also contribute.

Failure Modes.   The key factors in determining premature failure of electrolytic capacitors in service are:

- Elevated temperature (caused by ambient temperature or ripple current)
- Excessive voltage (continuous or momentary)

For example, for a given operating temperature, the life of a typical electrolytic capacitor will be a factor of 5 times longer if it is run at 50 percent of the rated voltage. Another example is that the life of a typical electrolytic capacitor will be a factor of 17 times longer if the capacitor core temperature is reduced from 65° to 25° C.

Electrolytic capacitors fail in four general ways:

- Dielectric breakdown resulting in a short between capacitor terminals
- Loss of capacitance
- Open circuit
- Reverse polarity

Dielectric Breakdown.   Dielectric breakdown can be caused by overvoltage conditions or manufacturing weaknesses that compromise the dielectric. Manufacturing weaknesses can rarely if ever be documented, as the evidence is inevitably destroyed in the failure event. Dielectric breakdown results in an unintended flow of current between the capacitor terminals, which leads to excessive power dissipation and may result in an explosive failure. Diagnosing a dielectric breakdown capacitor is generally readily apparent, as the failures are typically explosive and readily visible. Should the energy available to a failed capacitor not result in a catastrophic failure, testing of a suspect capacitor should be performed using a light bulb in series with the capacitor under test to prevent supply energy to a failed capacitor. Leakage current can more safely be measured using this technique.

Loss of Capacitance.   A loss of capacitance occurs when there is a loss of electrolyte. A loss of electrolyte typically occurs when there is a leak in the sealed case of the electrolytic capacitor. A leak in a seal can develop over time due to normal environmental processes. Seal degradation is generally accelerated by some cleaning solvents, increased temperature, vibration, or manufacturing weakness. A typical result of a loss of electrolyte is a loss of capacitance, an increase in ESR, and a corresponding increase in power dissipation.

Open Circuit.   An open circuit typically occurs when terminals of connections inside the capacitor degrade and fail. Loss of an electrical connection can occur as a result of corrosion, vibration, or mechanical stress. The result of an open-circuit failure is a loss of capacitance.

Failure Analysis

- A stereo microscope should be used to make observations of the failed component and the printed circuit board where it was mounted. The capacitor body should be visually inspected. A stereo optical microscope is recommended for inspecting for cracks, burns, or obvious contamination on the device surface.

- The suspect capacitor should be isolated from its circuit using mechanical means. The use of a soldering iron or heat gun could reflow failed solder joints or change the chemical composition of contamination in the region.

- Electrical measurement of the dc leakage current can be made using a curve tracer or four-wire ohmmeter.

- Measurement of the capacitance and dissipation factor should be made at the typical operating frequency of the circuit in which the device was mounted. A capacitance meter or an impedance meter can be used to make this measurement.

Case Study: Electrolytic Capacitor.   Failure of an electrolytic capacitor in a power supply filter application resulted in increased ripple to the power supply lines. The capacitors were found to have reduced capacitance, and the power supply lines started to show voltage ripple as the unregulated voltage dipped below the voltage regulator set point. The result of ripple on the regulated power supply lines resulted in an unanticipated output from the analog signal generating circuit. The unfortunate result of this failure was a false speed signal to a speedometer in a transit application.

Analysis of the failure data showed that only older products were failing. Dissection of the failed electrolytic capacitors revealed that the electrolyte was drying out. Analysis of the seals around the component leads revealed that this was the location of the electrolyte loss. The root cause of the electrolyte loss was determined to be a weak seal from the manufacturer around the capacitor leads that allowed a loss of electrolyte over time. Accelerating factors included a high-vibration environment with inadequate component support, elevated ambient temperatures, and significant voltage ripple.

18.4.3 Film Capacitors

Construction.   A metallized film capacitor is typically constructed of alternating electrode and dielectric layers that are tightly wound and pressed flat. The electrodes are made of a nonconductive film with a coating of zinc or aluminum that may be only 80 Å thick. Typically films include polyester, polystyrene, polycarbonate, polypropylene, and paper. The metallization is segmented with the intention that, if a dielectric breakdown occurs within a segment, a narrow region of the metallization will vaporize, interrupting the fault current. Each electrode is offset from the center so that electrical contacts can be formed on each end of the finished winding. By connecting the electrode to each layer of the winding, the inductance of the device is minimized. Terminals are welded onto the endcontacts, and the winding is encapsulated in epoxy. The metallized film capacitor is typically used in applications to "steer" electromagnetic interference. These capacitors are often installed across the ac power line in video monitors and power supplies to suppress voltage transients and high-frequency pulses.

Failure Modes.   Failure mechanisms include:

- Short-circuit failure
  - Overheating as a result of high-resistance contacts between the terminals and the electrodes

- Contact between the electrodes due to application of a high frequency energy pulse that causes the electrodes to move sufficiently to crack the epoxy package and contact in places

- Dielectric breakdown due to excessive applied voltage, or ionization of air pockets internal to the capacitor (Dielectric breakdown results in a short-circuit failure.)

- Open-circuit failure

  - No connection between a lead and an electrode due to physical stress on the lead

Failure Analysis

- A stereo microscope should be used to make observations of the failed component and the printed circuit board where it was mounted. The capacitor body should be visually inspected. A stereo optical microscope is recommended for inspecting for cracks, burns, or obvious contamination on the device surface.

- The suspect capacitor should be isolated from its circuit using mechanical means. The use of a soldering iron or heat gun could reflow failed solder joints or change the chemical composition of contamination in the region.

- Electrical measurement of the dc leakage current can be made using a curve tracer or four-wire ohmmeter.

- Measurement of the capacitance and dissipation factor should be made at the typical operating frequency of the circuit in which the device was mounted. A capacitance meter or an impedance meter can be used to make this measurement.

Case Study: Film Capacitor.   A computer monitor was in service in a children's hospital in the intensive care section. It was being used to display patient information. A failure occurred that resulted in a propagating failure that consumed most of the high-voltage scan control board and released smoke into the room. Using the techniques described in Chap. 14 to analyze a propagating failure, it was determined that the failure initiated within a small region of the scan control board. One of the components within that region was a large film capacitor. No other component failure in that region was a likely point of initiation.

Testing on exemplar film capacitors revealed that the typical breakdown voltage for these film capacitors was an order of magnitude higher than the operating voltage. UL 94V-0 testing revealed that the polymer capacitor outer covering would not sustain combustion once a flame was removed. Characterization of the normal operating environment revealed nothing unusual.

Dissection of a number of capacitors revealed a suspect connection between the component leads and the metal film on the outside layer of the film capacitor. A survey of field returned boards resulted in the discovery of one film capacitor with an intermittent connection between the component lead and the metal film. Testing with this capacitor demonstrated that arcing and overheating would result when this intermittent capacitor was placed into normal service. Further testing revealed that the film capacitor could be made to sustain combustion (although it was not easy) if electrical energy was continuously supplied to a failed capacitor. Using this failure mechanism, testing was conducted to demonstrate that a fault, initiated in the location of this film capacitor, could reproduce the incident damage.

18.4.4 Tantalum Capacitors

Construction.   The tantalum capacitor has high volumetric efficiency, so capacitors of several hundred microfarads are available in surface mount packages. The capacitor is formed from tantalum powder pressed in to a pellet. A tantalum wire embedded in the pellet and becomes the anode, or positive lead. A solid electrolyte layer is formed of manganese dioxide. A conductive layer, which becomes the cathode, or negative lead, is formed over the manganese dioxide. The pellet is then anchored inside a hermetically sealed metal can.

Tantalum capacitors are known for having a *self-healing* property in which the energy from a dielectric breakdown converts localized defects in the manganese dioxide layer to a more resistive oxide. Repeated self-healing events may lead to a slight increase in the measured capacitance of a device. The ability of a capacitor to self-heal depends on the available energy to the fault region. The lower the impedance of the voltage source to the capacitor, and the larger the device capacitance, the more energy is available to the fault region. These observations explain why large-value capacitors, or capacitors connected to voltage buses, can fail catastrophically instead of self-healing.

Failure modes

- Short-circuit failure
  - Dielectric breakdown (called "scintillation") due to degraded dielectric or excessive applied voltage
  - Excessive current (called "surge" current) exceeds the instantaneous power dissipation capability of localized regions of the dielectric, resulting in a thermal runaway condition
  - Applied voltage in the reverse polarity
  - Excessive ripple current resulting in thermal runaway
  - Overheating as a result of high-resistance contacts between the terminals and the electrodes
- Dielectric breakdown due to excessive applied voltage, or ionization of air pockets internal to the capacitor (Dielectric breakdown results in a short-circuit failure.)
- Open-circuit failure
  - Separation of an internal electrical connection due to temperature cycling or mechanical vibration
  - Melting or vaporizing an internal electrical connection due to high internal heat, or high current
  - No connection between a lead and an electrode due to physical stress on the lead

Failure Analysis

- A stereo microscope should be used to make observations of the failed component and the printed circuit board where it was mounted. The capacitor body should be visually inspected. A stereo optical microscope is recommended for inspecting for cracks, burns, or obvious contamination on the device surface.
- If the failed part is still attached to the board, fein-focus X-ray can be used to determine whether the part was mounted in the correct polarity due to its internal construction.

- The suspect capacitor should be isolated from its circuit using mechanical means. The use of a soldering iron or heat gun could reflow failed solder joints or change the chemical composition of contamination in the region.

- Electrical measurement of the dc leakage current can be made using a curve tracer or four-wire ohmmeter.

- Measurement of the capacitance and dissipation factor should be made at the typical operating frequency of the circuit that the device was in. A capacitance meter or an impedance meter can be used to make this measurement.

Case Study: Tantalum Capacitor

Component:	A 22 mF tantalum capacitor
Failure:	The capacitor and the printed circuit board in the region burned.
Observations:	Two tantalum capacitors were mounted in the same orientation in close proximity. One of these capacitors and the board underneath were burned during a 19-hr burn-in.
Analysis:	Due to the large amount of damage to the tantalum capacitor, its polarity could not be visually determined. A fein-focus X-ray showed that one capacitor was in the correct orientation, while the other had been installed backward. Figure 18.3 shows the fein-focus X-ray image of the two tantalum capacitors.[2]

18.4.5   Trim Capacitors

Construction.   Trim capacitors, which are typically in the picofarad range, allow the capacitance in a circuit to be adjusted to final value after a circuit has been assembled. Trimming capacitors are used to compensate for component variations and variable parasitic capacitance that may affect a sensitive circuit. A typical trimming capacitor is adjusted using a screwdriver as shown in Fig. 18.4. The rotation of one of the plates (or sets of plates) with respect to the other plate(s) changes the overlap of the plates, thus changing the capacitance.

Failure Modes.   The need to be able to physically adjust the position of one of the plates (or sets of plates) and the presence of a cavity inside the case can lead to two common failures in a trim capacitor:

- Parasitic resistance, and therefore current, between the two plates caused by contamination.

- Variability in the capacitance value caused by mechanical instability in the physical distance between the two plates.

These two failure mechanisms can be investigated as described below. The value of the capacitor should not be adjusted, as this procedure may interfere with the determination of the failure mechanism.

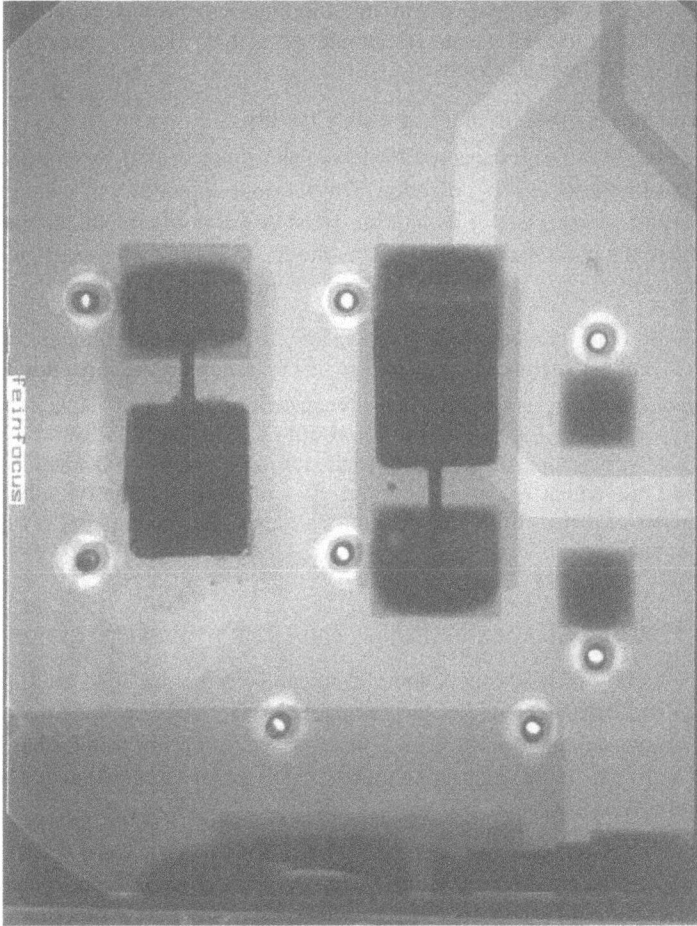

Figure 18.3 X-ray showing a tantalum capacitor installed in the wrong polarity (left) next to one installed in the correct polarity (right).[F2]

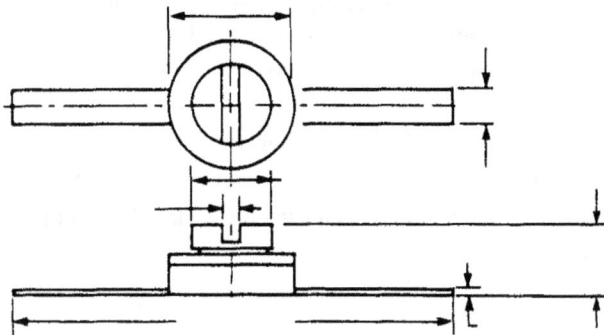

Figure 18.4 Typical trim capacitor.

Contamination. Contamination can enter the interior of a trim capacitor at any point in its manufacture or use. However, contamination is more likely to enter during the assembly operation, for several reasons:

- Contaminants are present in the assembly facility.

- The heating of the trim capacitor, followed shortly thereafter by cooling, is expected to occur during the assembly operation. This thermal sequence will increase then decrease the pressure of any gases in its interior. If the unit is not hermetic, contaminants may be drawn into the interior of the package by the relatively low pressure that results from the cooling step.

Contamination in a trim capacitor is most likely to affect the component by increasing the leakage current between the two plates (or sets of plates) in the capacitor.

Mechanical Instability. The physical construction of a trim capacitor may result in mechanical instability, particularly as the result of residual stress from the assembly process. It is also very important for a trim capacitor to be mounted on a stable platform where flexing and movement are at a minimum. Any presence of relative motion between the plates (or sets of plates) of this component will result in a change in the capacitance value of the trim capacitor.

Failure Analysis

- In-circuit analysis of a suspect component can be difficult due to multiple parallel components and therefore parallel current paths, as well as variations that may occur as a result of measurement probes. Care must be exercised not to eliminate the root cause information as the analysis is conducted. Individual circumstances will guide in the appropriate procedures to follow.

- Leakage measurements can be made using a DVM to measure the lead-to-lead resistance of an incident trim capacitor, but a curve tracer or and electrometer and a power supply will provide data on the actual leakage current as a function of voltage. It is important to apply a voltage close to the value across the component in normal circuit operation. The value of the leakage current should be compared with typical leakage current values of exemplar units.

- Mechanical instability may be detected by monitoring the incident capacitor's capacitance value while its mounting platform is subjected to mechanical stress. There are three methods of applying mechanical stress to evaluate a component's susceptibility to mechanical instability.

  - Mechanical stress may be applied by flexing or twisting the PC board on which the component is mounted.

  - Mechanical stress may be applied by pushing on the incident component with a know force at a number of locations.

  - Mechanical stress may be applied by heating or cooling the incident component.

The capacitance of the incident trim capacitor and an exemplar trim capacitors are compared to determine if the failure mechanism is mechanical instability.

18.5   *DISCRETE SEMICONDUCTOR DEVICES AND INTEGRATED CIRCUITS*

Microelectronic devices, which include both discrete semiconductors and integrated circuits, are found in almost all modern electrical and electronic equipment. In the more than 50 years since the invention of the bipolar transistor, microelectronic devices have become extremely varied in technology as well as pervasive. This chapter provides a basic understanding of the most accessible and frequently used failure analysis procedures.

18.5.1   Delidding

Overview.   Delidding is the procedure used to remove the lid or cover from a microelectronic device. Several b asic procedures are used for delidding:

- Grinding
- Cutting
- Desoldering

In most cases, a combination of these methods are used for optimum results.
The preferred method for metal-can ICs and discrete devices is outlined below:

- Use a small grinding tool to reduce the thickness of the cover in an area around the perimeter of the package. Packages with round covers may be mounted in a lathe and turned down using a small file, coping saw, or conventional metal-turning tools.
- Blow off the devices to remove any metal particles from the grinding operation. This should be done, of course, before the case is perforated.
- Peal off the lid with a needle-nose pliers or tweezers.

Alternately, a mechanical device with a revolving blade—often referred to as a *can opener* may be used.

Delidding Ceramic ICs.   In this method, a machine vice is modified by adding knife blades to the top surface of each jaw. The cutting edges of the knife blades face each other. The procedure below opens the ceramic package, either by penetrating the glass seal at the lead frame interface or by removing soldered lids on the top surface of the package.

1. Place the device in the modified vice with the knife edges contacting the seal. Use just enough pressure to hold the device in place.
2. Heat the lid for about five seconds with a small butane-oxygen blow torch.
3. Remove the heat and slowly close the vice to increase the pressure.
4. Repeat steps 2 and 3 until the lid is sheared off at the seal.

Precaution.   Wire bonds inside microelectronic devices are extremely fragile. Practice on sample devices before attempting to open devices obtained from a mishap.

18.5.2   Decapsulation

Overview.   Decapsulation is the procedure used to remove the molding material from plastic packaged ICs and transistors.

There are three basic methods for decapsulation:

- Mechanical abrasion and cutting
- Chemical etching
- Plasma etching

Each method has its advantages for specific types of devices.

Mechanical Abrasion. The plastic is removed by grinding or sandblasting. Mechanical abrasion is simple but appropriate only for larger devices. In addition, it generates dirt and dust that can contaminate the device.

Chemical Etching. In this process, the plastic molding material is dissolved with specialized solvents and acids. Chemical etching is difficult to control but works on a broad range of encapsulated devices. It can leave chemical residues that interfere with the analysis. Information on etchants and the specific chemical reactions can be found in Refs. 1 and 2, respectively.

Plasma Etching. A radio frequency (RF) energy source is used to ionize gas in a reaction chamber. The IC is placed in a chamber, and the ionized gas attacks the plastic and the IC layer materials. Plasma etchers are extremely controllable and do not leave chemical by-products behind as in chemical etching. Plasma etching is covered in Refs. 3 and 4.

Preparation. Although it is possible to etch devices with litter preparation, better results are obtained when the package is mechanically abraded. Usually, the package is ground down or milled to thin the molding material in the area above the die. This leaves less material to be removed by chemical action and allows the reactions to be controller better. Since the underground part of the package will remain partially intact, it helps hold the device together after etching.

Etchants. The two most common reagents are concentrated nitric and concentrated sulfuric acid. Many of the etching procedures in the literature are fuming nitric and fuming sulfuric acid. These acids are not merely concentrated version of the same chemical. They actually have a different chemical makeup, which makes them extremely dangerous but particularly well suited for dissolving plastics. The fuming versions of these acids are powerful oxidizers and have many unstable and explosive reactions with other common laboratory agents.

Several commercially available organic solvents are formulated specifically for dissolving encapsulants. Two sources for these solvents are Dynaloy, Inc., of Hanover, NJ, and Emercon and Cuming, Inc., of Woburn, MA.

Precaution. Decapsulation involves the use of some of the most dangerous and poisonous chemical agents known. Anyone attempting decapsulation activities should be trained in the use of these chemicals and should take proper safety precautions. Proper protective clothing, handling, and disposal procedures must be used. Review material safety data sheets (MSDSs) and consult with safety personnel before attempting to use the materials.

The waste products from decapsulation are considered hazardous materials. They must be disposed of in accordance with the legal disposal requirements for hazardous materials at your facility.

Procedure.   The procedure depends on the material being dissolved and the solvent. Several detailed procedures are presented in Refs. 1 and 2.

The procedure for concentrated sulfuric acid is outlined below.[1]

1. Mill a small well in the plastic above the die.

2. Boil the acid in a small beaker. The concentrated acid should be boiled down to one-half its original volume to remove as much water as possible.

3. Bake the device at 200° C (392° F) for 30 min to remove moisture.

4. Immerse the device in the acid until the die becomes visible. It will be necessary to remove the device from the acid to inspect it, as the acid becomes dark once some of the plastic has been dissolved.

5. Rinse in deionized water to remove the acid.

6. Rinse in methanol and dry gently in air.

Passivation Removal.   Most SEM examination techniques require that the passivation layer be removed from the device surface. Glass passivation can be removed by wet chemical etching.[5]

Etch Evaluation.   The glass removal rate should be evaluated using a trial exposure of 30 s. After each exposure, wash in deionized water for 30 to 60 s, rinse in isopropyl alcohol, and dry gently with dry nitrogen.

Observe the progress of the etching with an optical microscope. The die is examined for the appearance of a vivid color that is characteristic of the thermal oxide. If the colors are pastel shades or nonexistent, additional etching is needed. Exposed aluminum at the bond pads and conductors should be checked for damage or undercutting.

The device should undergo a functional test following each exposure. Power supply currents are a good indicator that device characteristics have not been altered. No further etching should be attempted if supply currents increase. Proper etching may be verified by observing the behavior of the IC with the SEM voltage contrast method. Observe dc voltage contrast imaging for metal conductors and ac voltage contrast imaging for diffused areas. SEM operating parameters can also be evaluated at this time.

The etch evaluation process should determine the following parameters:

- Appropriate etchant formulation
- Etch time
- Optical appearance when properly etched
- SEM operating parameters

Glass Etchants.   Two etchants are reported in Ref. 5. These are as follows:

Etchant Number 1	Etchant Number 2
125 mL 49% HF	280 mL 40% NH_4F
25 mil 70% HNO_3	35 mL 48% HF
250 mL glycerine	

Etchant number 1 works well for bipolar devices, and etchant number 2 works well for all devices.

Precautions. Etchants should be evaluated on sample ICs before use, due to the wide variation of etch rates on different types of ICs.

Glass characteristics can vary from lot to lot from the same manufacturer. This makes it difficult to develop consistent etch exposure times.

Etchants that are too aggressive will produce grooves adjacent to the metal conductors. They will also etch the aluminum and cause deterioration of device performance.

Plasma Etch. The increased use of silicon nitride as a passivating layer and the advent of multiple layers of metallization have increased the use of plasma etching techniques to selectively remove dielectric layers (as well as layers of polycrystalline silicon and conductors) from the surface of devices and ICs. Typically, a specific plasma will be used to remove a given layer. It is beyond the scope of this book to discuss these etch procedures in detail.

18.5.3 Optical Microscopy

Introduction. The primary advantages of an optical microscope are the minimum sample preparation required and the ability to inspect without damage or modification to the device. The observation and photographic recording of color is another distinct advantage. This technique may be used to identify manufacturing defects, mechanical damage, thermal damage, and corrosion.

Metallurgical microscopes are usually used for failure analysis examination of microelectronic devices because of their high-quality optics and the number of illumination options available. An excellent reference on optical microscopy is Ref. 6. Additional examples of optical microscopy applied to microelectronic devices can be found in Ref. 7.

Bright and Dark Field. In bright-field illumination, the image is formed by reflected light. The light source is internal to the microscope optics and is directed through the objective stage perpendicular to the surface of the sample. The light reflects off the surface of the sample and reenters the objective lens. Since most of the light that reflects off the sample is directed back through the optics, the illumination is very strong and accentuates reflective properties of the sample.

In dark-field illumination, the image is formed by light that is reflected off the sample at an oblique angle. Image contrast is generally greater than with bright field. For this reason, it is often possible to see features not visible with bright-field illumination.

18.5.4 SEM Voltage Contrast

Mechanism. In voltage contrast imaging, an IC is examined and electrically operated while in the SEM vacuum chamber. Special connectors are used to connect external electrical circuits to the IC through the wall of the vacuum chamber.

Secondary electron emission is affected by the voltage applied to the surface that the beam is striking. Positive potentials appear darker, while negative potentials appear brighter than those at ground potential.

The effect works well over the range of voltages used for IC operation. The voltage changes that occur in 5-V logic circuits can be easily distinguished.

Preparation.   The IC must be delidded or decapsulated so that the die can be imaged with the SEM. The electrical connections must be intact to operate the device.

The passivation layer must be removed to allow the electron beam to contact the metallization layer.

The device must be fitted with interconnections for the leads that allow the device to be modulated by external circuits. The device should not be connected to the SEM ground, as with most other samples. The appropriate leads should be brought out through the interconnection system so that grounding can be done externally. This provides additional flexibility in imaging.

It is possible to damage the IC if the beam sweep area is too small, or if the beam current becomes too high. Sensitivity to beam current depends on the particular IC technology, feature size, and other factors. Preliminary testing on samples should be conducted to verify the acceptable levels.

Application.   The technique works well for the identification of cracks, metallization problems, and other failures where circuit connections are affected. When the IC is statically activated, the metallization traces will appear darker or lighter, depending on their voltage levels.

In dynamic applications, such as counters and memory circuits, the circuit is clocked. One or more frequencies may be present on the traces. If the voltages changes during the beam scanning period, the traces will take on a striped appearance. This is sometimes called stroboscopic voltage contrast.

The striping effect obtained with this technique depends on the orientation of the trace with respect to the beam scanning direction and to the frequency of the modulation voltage. Circuit frequencies usually can be chosen to provide a useful spacing of the stripes. For example, if the frame rate is 60 Hz and 30 stripes on a particular trace are desired, then the frequency on that trace should be 30×60, or 1800 Hz.

If stroboscopic voltage contrast photographs are taken on an SEM with along photo sweep rate, the circuit frequency must be reduced accordingly.

Examples.   Stroboscopic voltage contrast is illustrated in Fig. 18.5, which shows an operating memory IC. The different stripe widths on the right side of the photo illustrate the various frequencies of the address lines used to activate the circuit.

Figure 18.6 shows a broken interconnect line. The break becomes very obvious, because the two sides of the connection are at different voltages.

18.5.5   Electron Beam Induced Current

Mechanism.   Electron beam induced current (EBIC) is an SEM imaging technique in which the electron beam alters the conducting characteristics of P-N junctions in the device being examined. The measurement is usually made on a reverse-biased junction. Although this is not always necessary to obtain a good image. When the electron beam strikes the junction, the junction develops a voltage. The voltage is amplified by an external amplifier, and the resulting signal is used to modulate the SEM image intensity. Beam voltage determines the penetration into the material and is used to gauge junction depth.

One side of the junction is usually connected to ground, either directly or through a power supply, which biases the conducting state of the junction. This connection allows beam current to return to the SEM electrical ground.

Figure 18.5   Stroboscopic voltage contrast of a memory IC showing different frequencies of operation, 500× *(courtesy of Martin Marietta Corp.)*.[F3]

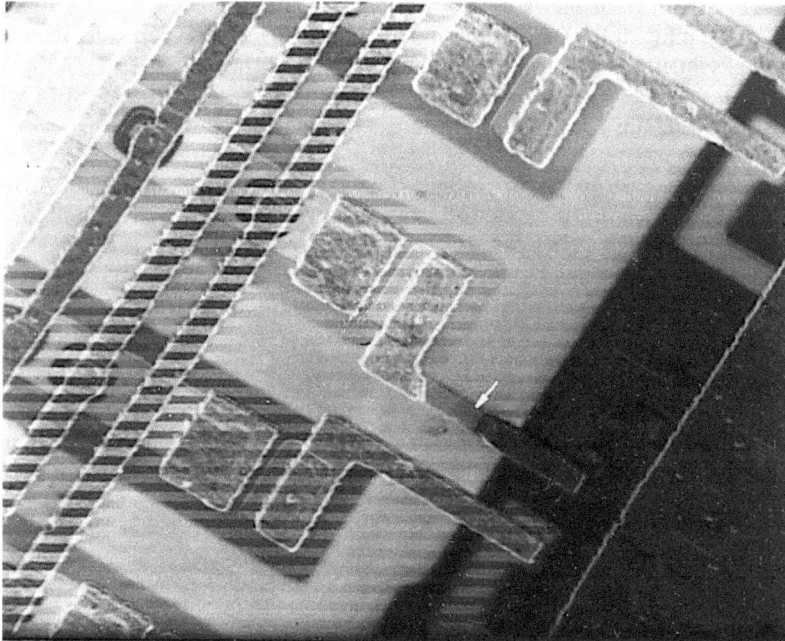

Figure 18.6   Broken interconnect made visible by the voltage difference across the break, ≈1000× (see arrow in center of photo) *(courtesy of Martin Marietta Corp.)*.[F3]

In P-N junctions, the anode is positively doped silicon and will assume a negative voltage (relative to the cathode) when the electron beam strikes the junction; that is, as if the beam causes a reverse bias of the junction. If external bias is applied, it is also in the reverse direction.

A detailed discussion and several good examples can be found in Ref. 7.

Preparation.   The device must be delidded or decapsulated. In some cases, the device junction is reverse-biased with an external power supply.

Application.   The junction usually appears as a bright area in the EBIC image. It is possible to damage the IC if the beam sweep area is too small or if the beam current becomes too high. Sensitivity to beam current depends on the particular IC technology, feature size, and other factors. Preliminary testing on samples should be conducted to verify the acceptable levels. Penetration depths for various beam voltages appear in Table 18.4.

Table 18.4   Penetration Depth for Various Beam Voltages

Material	Depth in microns		
	5 kV	10 kV	15 kV
Silicon	0.4	1.2	2.3
Aluminum	0.3	1.0	2.0
Gold	0.1	0.3	0.5

18.5.6 Other Techniques

Hermeticity Techniques.   In general, failure analysis techniques for package leaks will include a visible examination to localize the leakage site, early nondestructive examination by light microscope or SEM, device dismantling or cross-sectioning, and then metallurgical analysis.

Test methods and acceptance limits for microelectronic packages are included in Ref. 8. Method 1014 includes test methods using the following:

- Helium and radioisotope trace gases
- Fluorocarbon bubbles
- Dye penetrants
- Package weight gain tests

Package Ambient Gas Analysis.   This test analyzes the composition of the gas inside a hermetically sealed container. Gas volumes inside integrated circuit devices are in the order of 0.01 to 0.85 cc, and highly specialized equipment is required to perform the analysis. Microelectronic packages are usually back-filled with dry nitrogen at or near atmospheric pressure before being hermetically sealed. Contaminants in the nitrogen are detected using mass spectrometry.

Thermal Mapping Techniques.   These techniques analyze temperature patterns in operating semiconductor devices and integrated circuits. Their applications are limited to operating equipment and may not be suitable for accident investigation because of damage to the circuitry. The basic method involves the use of infrared (IR) scanning equipment.

Liquid Crystal Analysis Techniques.   Liquid crystal films create polarization and intensity changes on a microscopic level when applied directly to the surface of a microelectronic device.

Liquid crystals can be used in the following ways:

1. For locating areas of high power dissipation in microelectronic devices by causing color changes which indicate temperature gradients
2. By magnifying the effects of submicron oxide defects so that they can be observed optically
3. For displaying active areas of an operating device and detecting logic circuit failures

Scanning Acoustic Microscopy.   In the scanning acoustic microscope (SAM), an acoustic wave is transmitted through the device and the transmission characteristics are used to generate a map of the device features. SAM is especially useful in detecting cracks, flaws, and mounting anomalies inside packaged microelectronic devices.

Scanning Laser Acoustic Microscope.   In the scanning laser acoustic microscope (SLAM), the sample is subjected to a plane acoustic wave and illuminated with laser light. The sound is scattered and absorbed within the sample, according to the internal elastic microstructure of the material. The principle of imaging is based on the minute displacements that occur as the sound wave propagates in the sample. Useful information is morphological in nature. The technique can be used to sort and classify materials, detect and localize flaws, identify defects in optically opaque samples, and map compressibility and density variations on a microscopic scale.

Particle Impact Noise Detection.   Particle impact noise detection (PIND) tests are used to identify loose particles inside a device cavity. One procedure is described in MIL Standard 883B and is identified as Method 2020.1. The apparatus consists of a transducer and vibrator assembly, which is used to mechanically excite the device under test and detect the presence of vibrations caused by particles inside the device package.

Mechanical Testing.   Wire bond tests are used to determine the mechanical strength of wire bonds. The wires may be bonded to the die, to the lead frame, or to the ceramic substrate. Several tests have been devised. In each, a force is applied to the bond until failure occurs. Two common tests are described below. Failure criteria are given in Ref.s 1 and 8.

1. In the bond pull test, the wire is pulled in a direction perpendicular to the die. This test can be done on one bond by cutting the wire, or on both bonds simultaneously when the wire is not cut.
2. The bond peel test is usually applied to external leads, such as circuit board or hybrid IC leads. A peeling stress is applied to induce the failure.

Die attach tests verify the integrity of the die attachment to the package or hybrid substrate. This is done by applying a shear force and measuring the force at which failure occurs.

X-Ray.   X-ray techniques, including equipment that provides magnification of the part, can be useful when mechanical objects such as wire bonds. X-rays do not usually affect other aspects of the device under investigation, so they are generally considered non-destructive.

18.5.7   ESD/EOS Failure of Devices

The effects of the relatively small amounts of charge present on humans and other objects have been well documented. Electrostatic discharge (ESD) and electrical overstress (EOS) are the terms used to describe these phenomena. The goal of much design effort has been to develop techniques to reduce the susceptibility of semiconductor devices to charge that has accumulated on an object and is suddenly allowed to discharge through a device. The most frequent result of ESD/EOS is an increase in leakage current, which is the result of damage to a dielectric layer or to a P-N junction. The relatively small volume of material affected by ESD/EOS often makes it difficult or impossible to locate the actual failure location using optical microscopy or SEM techniques. If sufficient current is forced through a damaged area, it is sometimes possible to locate the area using thermal or image intensifying techniques. Additional information on ESD/EOS failures can be found in Ref. 9.

18.6   *INDUCTORS AND TRANSFORMERS*

18.6.1   Construction

Windings.   Inductors and transformers are composed of a series of parallel conductor windings surrounding a central core. The conductors used in winding is called "magnet wire," and may be round, square, or rectangular in shape. The magnet wire is usually composed of either copper or aluminum. The magnet wire has an insulating coating to prevent conduction between overlapping windings, or between windings and the core material. National Electrical Manufacturer's Association ("NEMA") standards publication number 1000 contains test procedures for assessing the compliance of magnet wire to requirements for flexibility, dielectric strength, continuity of insulation thickness, solderability, thermoplastic flow, thermal endurance and other characteristics. Other applicable standards include federal specification JW1177A, and military specification MIL W 583C. Some common types of magnet wire insulation are summarized in Table 18.5.

Table 18.5   Magnet Wire Insulation Summary[11]

Thermal Class	Insulation Type
105° C	Oleoresinous enamel (plain enamel), polyurethane
130° C	Polyurethane H.T., polyurethane-nylon
155° C	Polyester
180° C	Polyester-imide, polyester-imide-nylon
200° C	Polyester-imide-amide, polytetrafluoroethylene (Teflon[*])
220° C	Polyamide, Kapton[*] tape
500° C	Aluminum oxide, ceramic coated

* Registered trademarks of DuPont Corp.

Core Material.   The core material is chosen to have a high magnetic permeability ("μ") within the operating frequency range. Typical core materials include iron, powdered iron, and ferrites.

Iron core transformers are typically assembled of many parallel laminations that are electrically isolated to reduce eddy currents. Iron alloy cores are used in applications that require operation at 2 kHz or less because of eddy currents.[10]

Powdered iron cores consist of small particles of iron that are electrically isolated and held in place by a binder which may be organic. Iron powder experiences a phenomenon known as *magnetostriction,* which is a change in the material's physical dimensions that occurs when it becomes magnetized. In circuits that operate at audio frequencies (<20 kHz), a buzzing or whining noise is noticeable emanating from these components.[11]

Ferrite cores are typically brittle, consisting of a mixture of iron oxide and other magnetic elements. Ferrites have no significant eddy current losses, due to their high resistivity. However, ferrites also have relatively low saturation flux densities.[10]

Failure Modes

- Overheating caused by:
    - Winding cross section too small
    - Excessive core losses: excessive voltage, excessive current, excessive frequency
    - Defect (crack) in magnetic core
    - Core insulation breakdown
    - Winding insulation breakdown
- Permanent change in inductance caused by:
    - Overheating
    - Mechanical damage to core or winding
    - Degradation of core or winding insulation
    - Corroded winding
- Primary to secondary conductive path (transformers only) caused by:
    - Moisture in the presence of ionic contamination
    - Degradation of primary to secondary insulation or winding-to-core insulation

18.6.2   Failure Analysis

The physical and electrical characteristics of an incident device should always be compared to those of one or more known-good, unused devices that were manufactured at the same time.

- The suspect transformer or inductor should be isolated from its circuit using mechanical means. The use of a soldering iron or heat gun could reflow failed solder joints or change the chemical composition of contamination in the region.
- The winding resistance should be measured for continuity using a current which is well below the normal operating current of the device.
- Primary to secondary leakage on a transformer can be measured using a "megger," "hi-pot" or equivalent high voltage, high impedance test. The voltage at which this test is conducted should exceed the normal operating voltage of the device. Care should be taken not to damage an incident device while performing this test.

- The deviation in the inductance of the incident device can be measured using a sine wave generator with an appropriate power rating. Use the time varying voltage and current waveforms to calculate the device impedance at various frequencies and voltage levels. This testing should be performed with voltage and current excitations that approach the normal operating condition of the device.

- Following electrical testing, the winding may be inspected using an optical stereo microscope. The winding should be inspected for insulation defects, cuts, breaks, nicks, local or general overheating, and signs of arcing. It may be necessary to unwind the winding by hand to inspect it.

- With the winding partially removed from the core, it may be possible to measure the conductivity of the core using a four wire method.

- The core can be inspected using an optical microscope. The core should be inspected for signs of overheating, cracking and damage to the lamination.

- X-ray analysis can be performed on the core to identify cracks or defects internal to the core.

- Electrical testing can be performed on the core alone by rewinding the core with new wire and testing using a sinewave generator.

- The internal structure of the core can be examined by performing a metallurgical cross section and using an optical microscope. The core should be inspected for cracks and signs of overheating. The location and extent the overheating should be noted as internal to the core or on the surface of the core.

- Scanning electron microscopy can be performed on sections of the core to characterize the size and location of ferromagnetic domains.

Case Study 1: Inductor for Power Factor Correction.   The following section summarizes an investigation to determine the cause of the failure of custom-wound inductors in a power supply.

Component.   A power factor correction coil wound on a powdered iron core.

Failure.   The power factor correction coil and the printed circuit board in the region overheated.

Observations.   The coil had desoldered itself from the PCB. No permanent damage was observed on the PCB, other than discoloration. The core was a low-loss type.

Analysis.   Characterization of the failed coil showed that its electrical characteristics matched a known-good coil at the normal operating frequency and current. This test also showed that the onset of saturation occurred at approximately 40 percent of the rated power supply input current. A dc current source was used to determine the winding loss as a function of RMS current. The results of these two tests showed that the I^2R losses (also referred to as *copper losses*) in the windings accounted for over 65 percent of the total heating generated by the coil. Testing of a known-good power supply at rated load and reduced line voltage confirmed that this condition resulted overheating of the power factor correction coil. When the incident core was rewound with the same number of turns of larger diameter magnet wire (16 AWG replaced 18 AWG wire), the coil could not be made to overheat under the same conditions.

Case Study 2: Filter Inductor.   The following section summarizes an investigation to determine the cause of a failure that occurred on a regulator PCB in a power supply.

Component.   A filter inductor on the 48-V, 40-A output of the regulator.

Failure.   The inductor burned the printed circuit board.

Observations.   The burn damage included over 50 percent of the area of an 8 × 10 inch printed circuit board. More than 20 different components were mounted in the failure region.

Analysis.   Following photographic documentation, components were removed from the incident printed circuit board, starting at the input and progressing toward the output. Each component was characterized to determine the physical damage, direction of heating, and any change in electrical characteristics. The last components analyzed were a pulse transformer (T3), and the filter inductor (L2) on the output of the transformer.

Transformer T3 was removed from the incident board. The PCB was observed to be clean and undamaged in the area where T3 was mounted. The outer paper wrapping of T3 was removed using a knife. The interior of this component showed no evidence of heat damage. These observations indicate that the damage to component T3 was most likely due to smoke and heating which occurred as a result of the incident failure.

Component L2 was an iron E-core filter inductor that functioned to remove current ripple from the power supply output. This component was uniformly burned on all sides. The plastic bobbin on the top of the component was observed to be melted and partially burned. The component was observed to be physically *twisted* on its plastic base. This effect most likely occurred due to heating of the plastic and gravity, which caused the component to move relative to its base. The outermost ribbon winding of the inductor had partially broken away, and this layer was removed in order to examine the interior of the component. The inside of L2 exhibited uniform heat damage similar to the exterior. A pattern in the shape of a V is typically left in burned areas as a result of a fire. The point of the V indicates the approximate area of origin. When L2 was removed from the incident PCB, a V pattern was observed, which indicated that L2 was the origin of the heat. Characterization of L2 using an ac power supply showed that this component had a finite resistance and no measurable inductance. Testing confirmed a fault that allowed conduction between the windings of L2. Analysis of the circuit showed that all other observed component failure resulted in removing electrical power from L2. Therefore, the L2 failure preceded the other component failures. A summary of this analysis is shown in Fig. 18.7.

18.7   RESISTOR FAILURES

18.7.1   Construction

Resistors are often considered to be the simplest component used in circuits and systems. For this reason, the possibility that the failure of a resistor resulted in the subsequent failure of a circuit or system is often overlooked. In addition, the physical condition and electrical characteristics of a resistor may provide significant information about the root cause of a failure. Resistors that have failed may be open circuits or resistive shorts, or they may have experienced a change in resistance. Any of these three changes in device characteristics may lead to the failure of the circuit they are part of. Most resistors used today fall into one of the following three categories when grouped by shape:

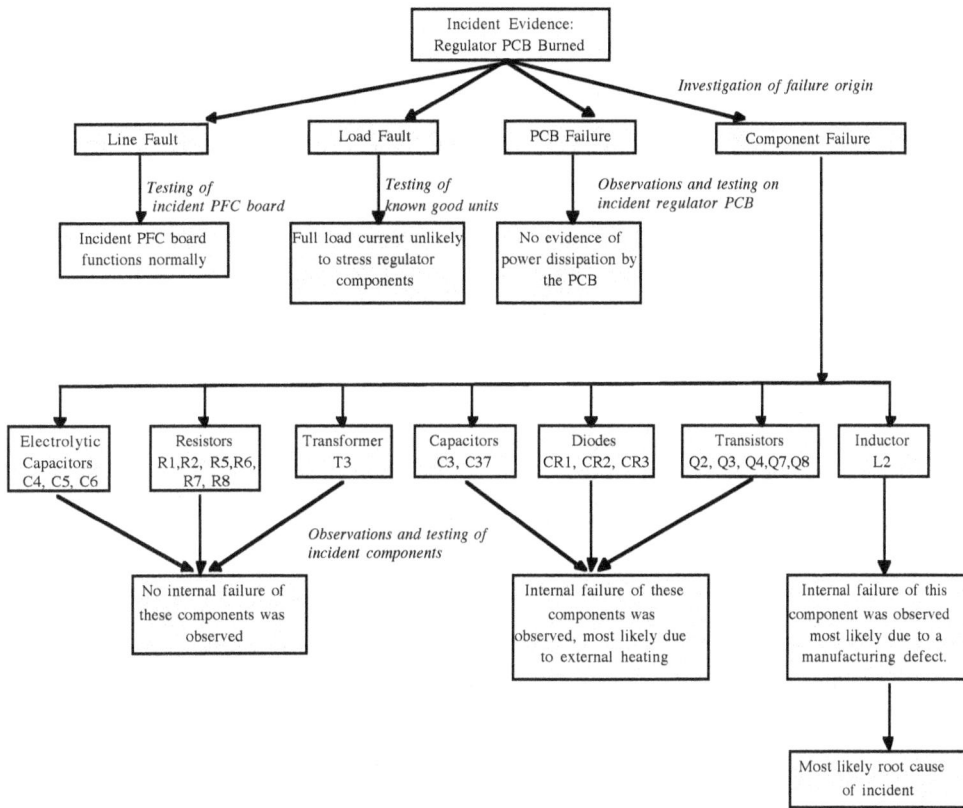

Figure 18.7   Summary of analysis performed on the failure of component L2.

1. Two leaded axial devices with the leads attached to the center of each end of the resistor. This group contains three types of resistors.
   - *Composition resistors.* These resistors are most often made from a carbon composition material. They are the most common resistor and may be obtained with a wide range of resistance values and proper ratings.
   - *Wirewound resistors.* A wire of an alloy selected for its behavior is would around an insulating core. Both high-precision and high-power resistors may be obtained.
   - *Film resistors.* A film of material is vacuum deposited (thin film resistor) or screened (thick film resistor) on an insulating substrate such as ceramic. A variety of materials, most, which usually contain a metal, are used.
2. Surface mount resistors having rectangular or cylindrical shape with metallized end regions.
3. Thin or thick film resistors that have the resistive material in a pattern on a substrate (typically ceramic), with leads attached to the ceramic that provide electrical contact.

The types of failures that occur in each type of resistor are similar, although the specifics of each failure may be unique.

18.7.2   Resistor Failure Modes

Resistors are affected in a variety of ways by extreme by physical or electrical environments. These environments and the specific effects are listed below.

- *Mechanical stress.* Exposing a resistor to conditions beyond its physical or thermal limits is the most likely cause of opens for resistors. The lead or metallization on one end of the resistor can become mechanically separated from the resistor body, or the resistor body may become cracked. In either instance, the mechanical stress results in a resistor that is either an intermittent "open" circuit or a continuous "open" circuit.

- *Overcurrent stress.* Resistors are designed to dissipate a specified amount of power. If excessive current flows through a resistor (or, alternatively, if the voltage across the resistor results in excessive power dissipation) for a period of time, the resistor overheats, which usually results in one of the following conditions:
  - *A change in resistance* (usually an increase)
  - *An open circuit*

  In either case, both the physical and the electrical characteristics of the resistor are altered.

- *Overvoltage stress.* Overvoltage stress is used in this chapter to refer to a voltage that is many times the normal operating voltage of a component, but is present for a period of time too short to alter device characteristics from power dissipation. Overvoltage conditions can cause dielectric films to *break down,* resulting in current flow through regions that were previously electrically insulating. Overvoltage stress can results in localized regions that show relatively large amounts of damage, while other nearby regions show little or no damage.

- *Corrosion.* Corrosion normally increases the resistance of a resistor by removing or reacting with the conductive material of the resistor or its leads. A change in the appearance of a portion of the resistor is often observable. However, if silver (or a similar material) is present, corrosion can lead to the formation of a conductive filament that can short-circuit some or all of the resistance in a resistor.[13]

- *Contamination.* Resistance decreases have been observed on large-value resistors (1 MΩ or greater) due to surface contamination.

18.7.3   Failure Analysis

A typical flow for determining the cause of a resistor failure is provided below:

1. The suspect resistor should be isolated from its circuit using mechanical means. The use of a soldering iron or heat gun could reflow failed solder joints or change the chemical composition of contamination in the region.

2. The resistor body should be visually inspected. A stereo optical microscope is recommended for inspecting for cracks, burns, and obvious contamination on the resistor surface.

3. Electrical measurements can be made using a precise ohmmeter, a curve tracer, or a four-wire method. Measurements should be made over a range of currents, up to the

current that is applied to the resistor during normal use. Care should be taken not to exceed the device ratings during measurement.

4. If contamination is suspected, appropriate analytical tests such as FTIR (Fourier transform infrared spectroscopy) can be used on the resistor to identify the contaminant.

5. The environmental coating on through-hole resistors can typically be removed chemically without altering the condition of the resistor. Once this coating is removed, the conductive body or film of the resistor can be visually and electrically inspected.

6. An *open* (i.e., a discontinuity in the conductive film) can be located by measuring the resistance between various points on the conductive body and one lead of the resistor.

7. Optical microscopy can be used to inspect the construction of the resistor and the body.

8. Scanning electron microscopy (SEM) can be used to detect discontinuities and variations in the conductive region of the resistor.

Duplication of the Resistor Failure. In some investigations, it may not be possible to determine the root cause of a failure without performing tests on known-good resistors to actually duplicate the failure. Failure duplication may require mechanical stress, high currents, high voltages, high humidity and temperature, the presence of contaminants, or other mechanical, electrical, or physical environments. The observations and results obtained as a result of an investigation as described are critical to the successful duplication of a resistor failure.

18.7.4 Case Study 1: Thick Film Resistor Failure

It is often possible to determine the root cause of a resistor failure through a combination of observations and electrical tests. An example of an investigation to determine the root cause of a thick film power resistor that failed in a communication system is covered in this section.

Component. A pair of matched 40-Ω thick film power resistors on a single ceramic substrate.

Failure. The resistor on the right side of the ceramic failed, cracking or fracturing the underlying ceramic.

Observations

- Some of the failed resistors on the ceramic had signs of burning or arcing present.
- The ceramic that the resistor was printed on cracked or fractured in a similar fashion on most of the failed resistors.

Testing. Two possible causes of the resistor failures were identified.

1. Failure of other electronic components, such as relays, which resulted in application of the full system voltage (typically 40 to 60 V) to the power resistor, causing excessive power dissipation in a short period of time

2. Unexpected high-voltage transients of unknown magnitude originating inside or outside of the system

Tests were designed and performed to determine if one or both of the above possibilities were the cause of the resistor failures.

Test Methodology to Duplicate a "System-Voltage" Failure. A step-stress test method was used to test the power resistors. Current pulses were applied to a resistor continuously until it failed. The initial current pulses were 0.5 A. The number of pulses for each current level was 100. After 100 current pulses were applied to a resistor, the current was increased by 0.25 A. The pulses had a duty cycle of 10 percent. Results for the step stress test are shown below in Table 18.6.

Table 18.6 Step-Stress to Failure Test Results on Power Resistors

Resistor number	Current stress level passed	Current stress level failed
1	100 cycles @ 1.25 A	Not tested
2	100 cycles @ 1.25 A	Not tested
3	Not tested	First pulse @ 1.50 A
4	Not tested	Fourth pulse @ 1.50 A
5	Not tested	Second pulse @ 1.50 A

The physical appearance of the resistors that failed the step-stress test closely matched the appearance of the failed resistors from the field.

Test Methodology to Duplicate a "High-Voltage" Failure. High-voltage discharge tests were conducted on the power resistor. Voltages of 1.0 kV, 2.0 kV, 3.0 kV, 3.5 kV, and 4.0 kV were applied to the power resistors. Table 18.7 shows the initial voltage and calculated initial current, $V_c(0)$, $I_{ic}(0)$ applied to each power resistor.

Table 18.7 Initial Voltage and Calculated Initial Current Applied to the Resistor

Initial voltage (kV)	Initial current (A)
1.0	25.0
2.0	50.0
3.0	75.0
3.5	87.5
4.0	100.0

Both the left and the right resistors on each substrate were subjected to the high-voltage pulses. Table 18.8 shows the high-voltage test results. The results shown in the table indicated that the power resistors will fail if a voltage in excess of 3.5 kV is applied.

Table 18.8 High-Voltage Test Results

	Left resistor	Right resistor
Applied voltage (kV)	Number of tests	Resistor failure
1.0	2	no
2.0	2	no
3.0	2	no
3.5	2	no
4.0	2	yes

The failure mechanism observed as a result of these tests was different from the fracture that resulted from thermal stress. Examination of the failed resistors revealed that these power resistors failed because of the *flashover* between the two connection pins.

The localized damage of flashover caused the insulation to break down, and the current flowed through a small segment (between points "a" and "b") of the resistor. However, no fractures were observed on any ceramic.

Conclusion as to the Cause of the Failed Resistors.   Based on the results of these tests, it was concluded that the most likely cause of the field failures was a single current pulse (or a small number of pulses) of 1.25 to 1.50 A. This corresponds to voltage condition of approximately 50 to 60 V.

18.7.5 Case Study 2: An Axial Lead Resistor on a Printed Circuit Board

Component.   A carbon film type resistor with metal end caps on a ceramic body.

Failure.   The resistor value was high.

Observations

- Resistor R125 was measured to have a resistance greater than 40 MΩ.
- The resistor was mechanically removed from the incident PCB and examined using a stereo microscope. No evidence of overheating was observed. Small cracks in the resistor's coating were observed on one end around the lead.
- The leads of R125, which has been left in place on the PCB, were measured to make continuous electrical contact to the PCB.

Determination of the Failure Location.   A chemical process was used to remove the protective coating from the incident resistor and several known-good resistors. The resistor was found to be a carbon film type with metal end caps on a ceramic body. A diagram of the resistor construction is shown in Fig. 18.8.

With the protective coating removed, the failure was located by probing along the resistive film from one end of the resistor to the other end, measuring the resistance using an ohmmeter. This measurement showed that a gap of less than 1 mm in length was present in the resistive film on the body of the resistor. A low-resistance electrical connection was measured from the film to both end caps on the resistor. Both failed resistors provided for

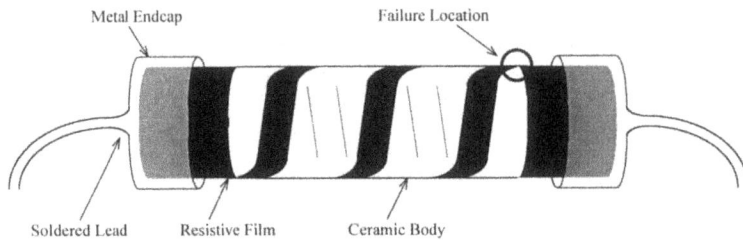

Figure 18.8 Resistor construction and failure location.

analysis were found to have failed in the same location. The failures both occurred at the beginning of the spiral cut in the resistive film. No evidence of overheating was observed in the location of the failure.

Scanning Electron Microscopy. A scanning electron microscope (SEM) was used to determine the distribution of conductive particles on the failed resistor in the region of the failure. A fast scan rate was used in the failure region to locate physical defects, such as cracks, nicks or voids in the conductive film. When no physical defects were observed, a backscatter image was obtained at a slow scan rate. This method characterized the population of conductive particles in the resistive film and on the insulator surface of the resistor. The failed region was found to be a region of film that contained no conductive particles. Excessive conductive particles were observed in the insulating region of the resistor. These observations demonstrated that the resistor failure was most likely a result of the migration of conductive particles away from a region of the resistor. The failure was observed to occur at the end of the spiral cut in the resistive film. This area is likely to have the highest electric field, which provides a force to cause the migration of conductive particles.

Testing to Characterize Stress in the Operating Environment. Excessive voltage applied to a resistor can cause failure due to overheating. The physical area of the resistive film determines the maximum power dissipation that can occur before the film melts or burns open. Measurements of the incident power supply and on known-good power supplies shown that R125 normally has 430 to 460 Vdc applied. This voltage corresponds to a power dissipation of 130 mW by the resistor. This power dissipation represents 13 percent of the rated power dissipation of 1 W. This calculation shows that the resistor is properly rated for the application.

Temperature Testing. Film resistors, and especially carbon film resistors, are known to experience a drift in their resistance value over temperature and time. Permanent changes in resistance are caused by the migration of conductive particles in the resistive film. Known-good resistors were characterized to determine their susceptibility to permanent changes in resistance due to temperature. A set of five resistors was placed in an environmental chamber that was cycled between 28° C and 140° C once per hour for a total of 120 hours. A potential of 650 Vdc was applied to the resistor terminals during this test. The resistance values were measured at intervals during this test. The results are summarized in Table 18.8. This test demonstrated that temperature stress in the presence of an applied voltage could cause the resistance increases in known-good components.

This testing showed that temperature stress in conjunction with applied voltage could cause an increase in the resistance of known good devices over time. However, no devices

Table 18.9   Summary of the Results of Temperature Cycling Resistors

	Measured Resistor Value (MΩ)				
	Resistor #1	Resistor #2	Resistor #3	Resistor #4	Resistor #5
Initial Resistance	1.62	1.59	1.61	1.59	1.59
Final Resistance	2.05	2.18	1.78	1.81	1.85
Percent Change	+26%	+37%	+11%	+14%	+16%

were observed to fail in the open circuit mode. This result indicates that an additional factor is required to cause the failures observed on the incident units. This factor is most likely a variation in manufacturing such as variations in the distribution of conductive particles in the resistive film.

Pulsed High-Voltage Testing.   A short-duration application of excessive voltage could cause dielectric breakdown between the bands of the resistive film. This effect could lead to excessive currents in a localized region on the resistor body, causing an open-circuit failure of the resistor. Several known-good resistors were tested to determine the effects of excessive voltage. Each resistor was measured using an ohmmeter, then a voltage was applied to the resistor for one second. The resistor was measured again to determine the change in resistance. Applied voltages of up to 10,000 V for one second were observed to cause no change in the resistor characteristics. This test showed that excessive voltage across resistor R125 was not likely to cause an open-circuit failure in a region of the conductive film.

Continuous Overvoltage Testing.   An excessive voltage applied for a long duration is known to result in resistor failures. In the power supply circuit, R125 controls the current through itself (but not the voltage across its terminals). To simulate an overvoltage situation in the circuit, a test was performed in which a constant current of 2.5 mA was applied to known-good resistors. The voltage across the terminals of R125 was measured to decrease continuously. The test was terminated when the voltage dropped to 1,200 V. The resistor under test was measured to have experienced a permanent change in resistance to 474 kW due to excessive temperature. The resistor was noticeably discolored from overheating following this test. These tests demonstrated that an excessive applied voltage was not a likely cause of the incident resistor failures.

Testing of Exemplar Resistors.   A total of 17 exemplar resistors were provided for analysis. These resistors were removed from the R125 location in units that had been returned from the field. The physical and electrical characteristics of the resistors were recorded to assess the likely variations in the field population. The environmental coating on the resistors was chemically removed, and the resistors were examined using a stereo microscope. The average width of the spiral bands of conductive film was measured using a stage micrometer. The average width of the spiral cut in the film was also measured.

This testing showed that resistor failures correlated with physical defects in the resistors from the manufacturing process. The resistors were grouped according to their physical characteristics. This work indicated there are likely four or more manufacturers (or manufacturing lines) of the component used in location R125.

The resistors were grouped as shown below:

1. Manufacturer A: resistor number 1

2. Manufacturer B: resistors number 2, 3, 4, 5, 6, 7, 8, and 14

3. Manufacturer C: resistors number 9, 10, and 15

4. Manufacturer D: resistors number 11, 12, 13, 16, 17, and the incident resistors

This grouping shows that only resistors from manufacturer D tended to fail "open."

Summary and Conclusions

Resistor (R125) failures.   Testing and analysis, including scanning electron microscopy, showed that these devices fail where a narrow region of the resistive film is found to contain no conductive particles. This failure was most likely due to manufacturing variation and subsequent migration of conductive particles due to temperature and applied electric field. Examination of 17 resistors removed from field returned units has shown that resistors from at least four manufacturers or manufacturing processes are used to produce this part. Examination of the incident resistors has shown that the failed resistors were most likely all produced by the same manufacturer or process.

18.8   OTHER PASSIVE COMPONENTS

18.8.1   Metal Oxide Varistors (MOVs)

A metal oxide varistor (MOV) is a highly nonlinear device designed to limit the voltage across the terminals in case of surges and high-voltage transients. Symmetrical, sharp breakdown characteristics as shown in allow operation in ac and dc circuits. Typical applications include line power surge reduction in power supply circuits or other electronic equipment. Typical characteristics are shown in Fig. 18.9.

Impedance changes of many orders of magnitude from near-open to near-short allow clamping of high-voltage transients by dissipating the power in the MOV.

In general, MOVs can be understood as series-parallel arrangements of semiconducting diodes. The basic conduction mechanism results from multiple semiconductor junctions at the boundaries of sintered zinc oxide grains.

Typical MOV ratings extend from 2.5 to 3000 V and currents up to 70,000 A. Energy capacity extends beyond 10,000 Joules for larger units. It is possible to connect multiple MOVs in parallel to increase current ratings or to connect them in series to provide higher or special voltage ratings. Typical response times are in the order of 500 ps.

Construction.   An MOV is a bulk semiconductor device. The device is constructed of a large number of zinc oxide *grains.* Each grain boundary acts as a P-N junction with an avalanche breakdown at approximately 2 to 3 V.

The material of a metal oxide varistor is primarily zinc oxide, with small additions of other metal oxides. Zinc oxide powder is sintered into ceramic parts, which are then electroded with either thick film silver or arc/flame sprayed metal. Figure 18.10 shows a schematic of the microstructure of a MOV. The ZnO grains are separated by intergranular

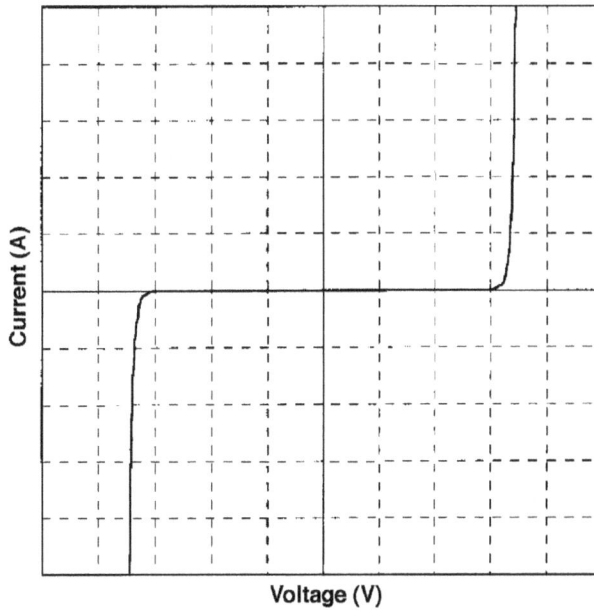

Figure 18.9   Typical varistor V-I characteristics.

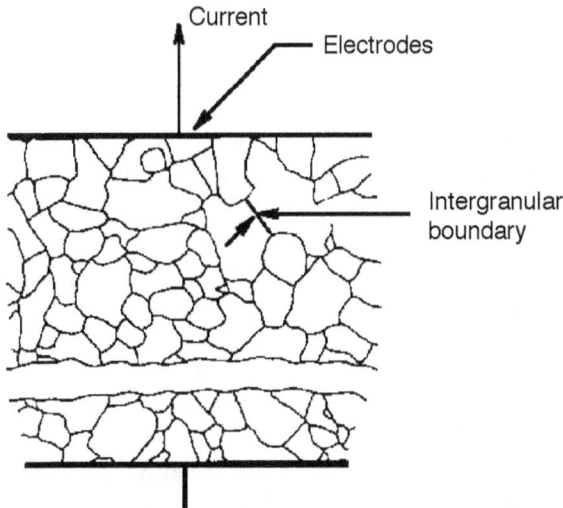

Figure 18.10   Grain structure of sintered zinc oxide.[F4]

boundaries. In brief, the thickness of the device determines the voltage rating and the cross sectional area determines the current rating. A typical through-hole MOV is shown in Fig. 18.11. An equivalent circuit model of a MOV is shown in Fig. 18.12. The lead inductance plays an important rule in terms of fast transient suppression.

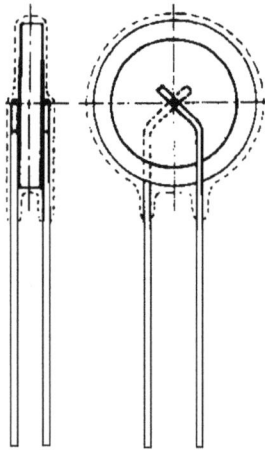

Figure 18.11 Typical through-hole MOV for medium ratings.[F4]

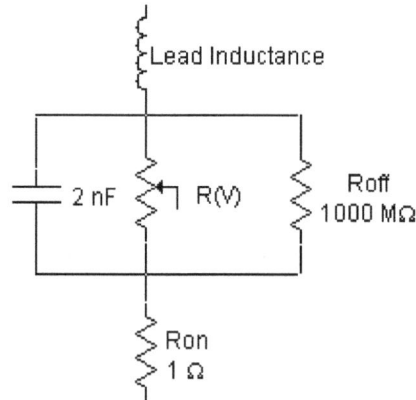

Figure 18.12 Equivalent circuit model of metal oxide varistor.

Failure Modes

- *Short Circuit.* Similar to diodes, MOVs usually fail short in a thermal runaway when exposed to excessive voltages and/or currents. Additional devices, such as fuses, are required to clear the fault current.

- *Open Circuit.* MOVs can explode and fail open when the fault energy is of excessive value. This often results in an open circuit. A number of cases are known when MOVs exploded and led to injuries or substantial damage of equipment. Burn testing of an MOV also showed that excessive temperatures can cause the device to fail open. Since typical applications of MOVs shunt fault currents to ground or neutral and do not affect normal operation, an open circuit fault may not be detected by the user of the MOV protected instrument.

- *Degradation.* Degradation of MOVs during service due to frequent surges results in limited service life and hazards from leakage current. Surge currents generate heat in the MOV that can exceed values that fuses intergranular boundaries. This results in gradual lowering of the clamping voltage. In this manner, that MOV sacrifices itself in service, increasing the leakage current and lower the clamping voltage subsequently. This can lead to explosions of the MOV.

- *Resistive Failure.* Following an explosion, a MOV can also fail resistive. This typically leads to overheating of the MOV and may result in either open or short circuit, depending on the power provided to the failed MOV.

Failure Analysis

- A suspect MOV should be isolated from its circuit using mechanical means. The use of a soldering iron or heat gun could reflow failed solder joints or change the chemical composition of contamination in the region.

- Electrical measurements can be made using a curve tracer. Care should be taken not to exceed the device ratings during measurements.
- Leakage currents can also be evaluated with an appropriate power supply and a precise current meter. Four-wire measurements are preferred.
- Performance tests with a high-voltage-transient or burst generator and a monitoring device such as an oscilloscope and a current shunt can provide useful information about the failure mode.
- The MOV should be visually inspected. A stereo optical microscope is recommended for inspecting for cracks.
- Leakage of an MOV can be measured using a *megger, hi-pot,* or equivalent high-voltage, high-impedance test. The voltage at which this test is conducted should exceed the normal operating voltage of the device. Care should be taken not to damage an incident device while performing this test.

18.8.2 Crystals

Construction.   A piezoelectric crystal is typically constructed from a quartz crystal. Electrodes are applied to the crystal, mounted in a holder, and placed in a hermetically sealed assembly.

Observed Failure Modes.   The most likely failure mode for a crystal is a permanent change in its resonant frequency. This change may be due to one or more of the following problems.

1. Absorption or deabsorption of contamination which changes the crystal mass
2. Stress relief between the interface of the electrodes and the crystal face
3. Leakage of air into the sealed assembly which dampens the crystal oscillation
4. Excessive voltage applied to the crystal
5. Exposure to radiation which causes ionization of the crystal
6. Mechanical vibration, shock, or acceleration (Excessive force can break or deform the electrode bonds or the holding structure.)

Failure Analysis

- Isolate the suspect crystal from its circuit using mechanical means. The use of a soldering iron or heat gun could reflow failed solder joints or change the chemical composition of contamination in the region.
- Due to the high Q of these devices, electrical testing should be performed using an accurate impedance meter designed for testing crystals.
- A leak in the hermetic seal of the crystal package can be identified by performing electrical testing under decreased and increased atmospheric pressure.
- Mechanical deformation of the holder structure can be identified using fein-focus X-ray.
- Examination of the electrode contacts and the crystal surface may be possible using Scanning Electron Microscopy.

18.8.3 Relays

Construction.   The two types of relays are electromechanical relays and solid state relays. Solid state relays consist of back-to-back thyristors or other semiconductor devices. A solid state relay should be analyzed according to the discrete components from which it is constructed. Electromechanical relays use an electrical signal to "open" and "close" mechanical contacts to complete a second electrical circuit. A magnetic field created by the control circuit exerts a force that causes the contacts to mechanically bend, rotate on a hinge, or travel linearly. These devices provide significant electrical isolation between the control circuit and the contact circuit. Common types of electromechanical relays are listed below.

1. *Latching.* Contacts lock in position until reset.
2. *Polar.* Operation depends on the polarity of the control coil current.
3. *Stepping.* Contacts are stepped to successive positions when the control coil is energized by current pulses.
4. *Interlock.* The relay state can only be changed depending on the state of a different relay.
5. *Time delay.* The time between the application of the control signal and the change of state of the relay can be programmed.

Failure Modes

- Contact circuit fails "open"
- Contact armature mechanical damage such as breaking or bending
- Armature fixed in place due to adhesion from contamination or overheating
- Film formation on the contact surface due to organic or inorganic contamination
- Contact surface erosion
- Contact circuit fails "closed"
- Cold welding of contacts due to excessive contact current
- Armature fixed in place due to adhesion from contamination or overheating
- Control coil fails "open"
- Excessive coil current melted the coil wire
- A mechanical break occurred in the coil wire
- Control coil "shorts" to contacts

Failure Analysis

- The suspect relay should be isolated from its circuit using mechanical means. The use of a soldering iron or heat gun could reflow failed solder joints or change the chemical composition of contamination in the region.
- The control coil resistance should be measured for resistance and inductance using a current that is well below the normal operating current of the device.
- The contact resistance should be measured so that the state of the contacts is known.
- Fein-focus X-ray can be used to inspect the armature and contacts in a reed-type or other relay where they can not be easily inspected by visual means.

- Fein-focus X-ray can be used to inspect the control coil.

- Coil-to-contact leakage on a transformer can be measured using a *megger, hi-pot,* or equivalent high-voltage, high-impedance test. The voltage at which this test is conducted should exceed the normal operating voltage of the device. Care should be taken not to damage an incident device while performing this test.

- Following electrical testing, the winding may be inspected using an optical stereo microscope. The winding should be inspected for insulation defects, cuts, breaks, nicks, local or general overheating, and signs of arcing. It will likely be necessary to chemically remove any potting material encasing the winding to inspect it.

- The contact condition can be inspected for pitting, welding, or the presence of contamination once the relay is disassembled. An optical microscope is recommended for inspection of the contact surface. SEM may provide additional information about the condition of the surface.

Case Study: Relay in Communication Equipment.   The following section summarizes an investigation to determine the cause of failures that occurred on telephone switching equipment.

Component.   A miniature 1.25 A reed-type relay.

Failure.   The contacts were intermittently "sticking" on units in the field.

Observations.   The relay contacts were confirmed to be "closed" with no voltage applied to the control coil. An analysis of the circuit showed that an R-C snubber was installed in parallel with the relay contacts.

Analysis.   Fein-focus X-ray was performed and verified that the contacts of failed units were touching. The relays were depotted using nitric acid. Microscope inspection of the contacts confirmed that they were welded. Analysis of the R-C snubber showed that this circuit could provide a current in excess of the relay rating if the relay "opened" when its corresponding telephone circuit was ringing.

18.9   REFERENCES

1. Doyle, Jr., E., B. Morris, eds.*Microelectronics Failure Analysis Techniques.* Rome Air Development Center. No publication date.

2. T.W. Lee. Mechanical and Chemical Decapsulation. *ISTFA/90 Microelectronic Failure Analysis Desk Reference,* 2nd ed. 1990 ISTFA Conference.

3. Beall, J. Plasma Etching. *ISTFA/90 Microelectronic Failure Analysis Desk Reference,* 2nd ed., p. 47, 1991. Materials Park, OH: ASM International.

4. Kiefer, D.S. Reactive Ion Etch Recipes For Failure Analysis. *ISTFA/90 Microelectronic Failure Analysis Desk Reference,* 2nd ed., p. 55, 1991. Materials Park, OH: ASM International.

5. Beall, J.R. *SEM Analysis Techniques for LSI Microcircuits.* Martin Marietta Corporation. J.R. Beall, W.R. Echols, D.D. Wilson. RADC report no RADC-TR-80-250.

6. *ASM Handbook, Volume 9, Metallography and Microstructures.* Materials Park, OH: ASM International, 1985.

7. Devaney, J., G. Hill, R. Seippel. *Failure Analysis Mechanisms, Techniques and Photo Atlas.* A guide to the performance and understanding of failure analysis. Failure Recognition & Training Service, Inc. 1988.

8. MIL-STD-883B, Test Methods and Procedures for Microelectronics.

9. Song, M., D.C. Eng, K.P. MacWilliams. Quantifying ESD/EOS Latent Damage and Integrated Circuit Leakage Currents. *Electrical Overstress/Electrostatic Discharge Symposium Proceedings,* 1995.

10. Mohan, Underland, Robbins. *Power Electronics Converters, Applications, and Design.* New York: John Wiley and Sons, 1995.

11. Power conversion and line filter applications. *Micrometals,* issue F

12. Mertol, A. Estimation of Aluminum and Gold Bond Wire Fusing current and Time. *IEEE Transactions on components, Packaging, and Manufacturing Technology* – Part B, Vol. 18, No. 1, Feb. 1995, p. 210.

13. Rida, B., D. Jennings. Tutorial Failure Mechanism for Conductive-Filament Formation. *IEEE Transactions on Reliability,* Vol. 43, No. 3, Sept. 1994

14. IEEE Recommended Practice on Surge Voltages in Low-Voltage AC Power Circuits, IEEE C62.41-1991, Nov. 1991, page 31.

18.10 FIGURE SOURCES

F1. *1985/86 Resistor Capacitor Data Book,* Mepco/Electra, p. 62.

F2. *Exponent Failure Analysis,* 1998.

F3. *Aircraft Mishap Investigation Handbook for Electronic Hardware.* D. Galler, D. Glover, A. Kusko, Failure Analysis Associates, Inc. 1995. WL-TR-95-4004.

F4. Harris Semiconductor Application Note: http://www.semi.harris.com/data/an/an9/an9767.

F5. MWS Wire Industries, 31200 Cedar Valley Drive, Westlake Village, CA 91362.

18.11 RECOMMENDED READINGS

Gallium Arsenide. E. Doyle, Jr., B. Morris, eds. Rome Air Development Center. No Publication Date.

Physics of Semiconductor Failures, 3d ed. H. Dicken, ed. Scottsdale, Arizona: DM Data, Inc., 1989.

ISTFA Microelectronic Failure Analysis Desk Reference, Suppl. 2, Nov. 1991. T. Lee, ed. Metals Park, OH: ASM International.

Microchip Fabrication. Peter Van Zant. 1986, Semiconductor Series, San Jose, CA.

Electronic Materials Handbook, Volume 1, Packaging. Metals Park, OH: ASM International, 1989.

Limiting Phenomena in Transistor and Interpretation of EOS damage. J/ T. May. *Microelectronic Failure Analysis Desk Reference,* 2d ed. 1990 ISTFA Conference.

Electronic Packaging and Interconnection Handbook. Charles A. Harper, ed. New York: McGraw-Hill, Inc., 1991.

CHAPTER 19
SEMICONDUCTORS

Perry L. Martin
National Technical Systems (NTS)

19.1 INTRODUCTION

Failure analysis has played a preeminent role in the success of the semiconductor industry and has a primary responsibility for the monumental increases in semiconductor reliability. The importance of semiconductor, and especially microelectronic, failure analysis has spawned multiple conferences and symposia which have generated volumes of work pertaining to microelectronic failure analysis guidelines and case studies. In addition are many independent works on microelectronic failure analysis, some of which are noted in the references. This chapter will present an outline of the many procedures and techniques performed during semiconductor failure analysis and a few of the more prevalent failure modes that they uncover. Figure 19.1 identifies some of the primary device reliability concerns.

19.2 SEMICONDUCTOR FAILURE ANALYSIS PRELIMINARIES

Semiconductor devices are almost always part of a larger, more complex piece of electronic equipment. The semiconductor device will work in concert with other circuit elements and be subject to system, subsystem, and environmental influences. It often happens that, after equipment failure, a technician will troubleshoot the unit and make a determination that a particular device is at fault. The device will be removed from the assembly, using less-than-controlled thermal, mechanical, and electrical [e.g., electrostatic discharge (ESD)] stresses and submitted to a lab for analysis. This is not the optimal failure analysis path.

For the analysis of field failures, the failure investigation initially should be directed toward gaining an acquaintance with all pertinent details relating to the failure; collecting the available information regarding the manufacturing, processing, and service histories of the failed equipment; and reconstructing, to the extent possible, the sequence of events leading to the failure. The collection of background data on the manufacturing and fabrication history of a failed item will begin with obtaining specifications and drawings and will encompass all of the design aspects of the failed item. Service histories should be collected and special attention given to environmental details. Environmental concerns include normal and abnormal loading, accidental overloads, cyclic loads, variations in pressure and temperature, pressure and temperature gradients, and operation in a corrosive environment, including the concentration and/or flow of a liquid environment.

Once the system-level failure has been verified, investigation will then be directed toward the isolation of the subsystem responsible for its failure. This investigation will be a thorough step-by-step study of the failure with full documentation. A complete photographic record of the investigation can be important, because a failure that appears almost inconsequential in a preliminary investigation later may be found to have serious conse-

PACKAGE

Integrity
Plating Quality
Dimensions
Electrical Leakage
Thermal Resistance
Resistance to Heat and Humidity
Resistance to Corrosion
Resistance to Mechanical Stress
Resistance to Temperature Change
Electrical Orientation
Changes in Design, Materials or Construction
General Quality and Workmanship
Protection During Shipment
External Damage

SEAL

Integrity
Hermeticity
Width
Solder Splash
Susceptibility to Thermal Stress
Susceptibility to Mechanical Stress

DIE CONCERNS

Cracks, Chips, Splits
Die Undercutting
Die Thickness
Surface Absorption
Gold Backing Thickness
Rework Induced Faults
Changes in Design, Material or Process
General Quality and Workmanship
Cleanliness
Surface Contamination
Thermal Stability
Foreign Material
Scribing Defects
Diffusion Faults
Oxide Faults

INTERNAL CAVITY

Moisture Content
Loose Particles
Plating

DIE ATTACH MEDIUM

Strength (Adhesion)
Consistency and Uniformity
Coverage
Wetting
Die Orientation
Excessive Buildup
Potentially Loose Particles
Dissipation Characteristics
Resistance to Temperature Stress
Rework
Type of Medium
General Quality and Workmanship
Resistance to Mechanical Stress

BONDING PADS

Clearance
Bridging
Probe Damage
Residual Glassivation
Scratches and Voids

MARKING

Permanency
Accuracy
Control

BONDING WIRES

Strength
Placement
Height and Loop
Clearance
Adhesion
Uniformity
Size
Material
Current Density
Rework and Overbonding
Bimetallic Contamination (Kirkendall Voids)
Resistance to Stress
Dissipation Characteristics
Nicking and Other Damage
General Quality and Workmanship
Resistance to Mechanical Stress

DIE METALLIZATION

Thickness
Overetching (Underetching)
Bridging
Scratches and Voids
Alignment
Step Coverage
Lifting or Peeling
Current Density
Hillock
Materials Composition

GLASSIVATION

Integrity
Thickness
Mechanical Defects

LEADS

Strength
Solderability
Hermeticity
Materials and Finish
Resistance to Heat and Humidity
Resistance to Corrosion
Spacing and Length
Damage
General Quality

ELECTRICAL PERFORMANCE

Continuity and Shorts
Parametric Stability
Parametric Performance
Temperature Stability
Long Term Reliability
Storage Degradation
Susceptibility to Electrostatic Discharge
Susceptibility to Radiation Damage

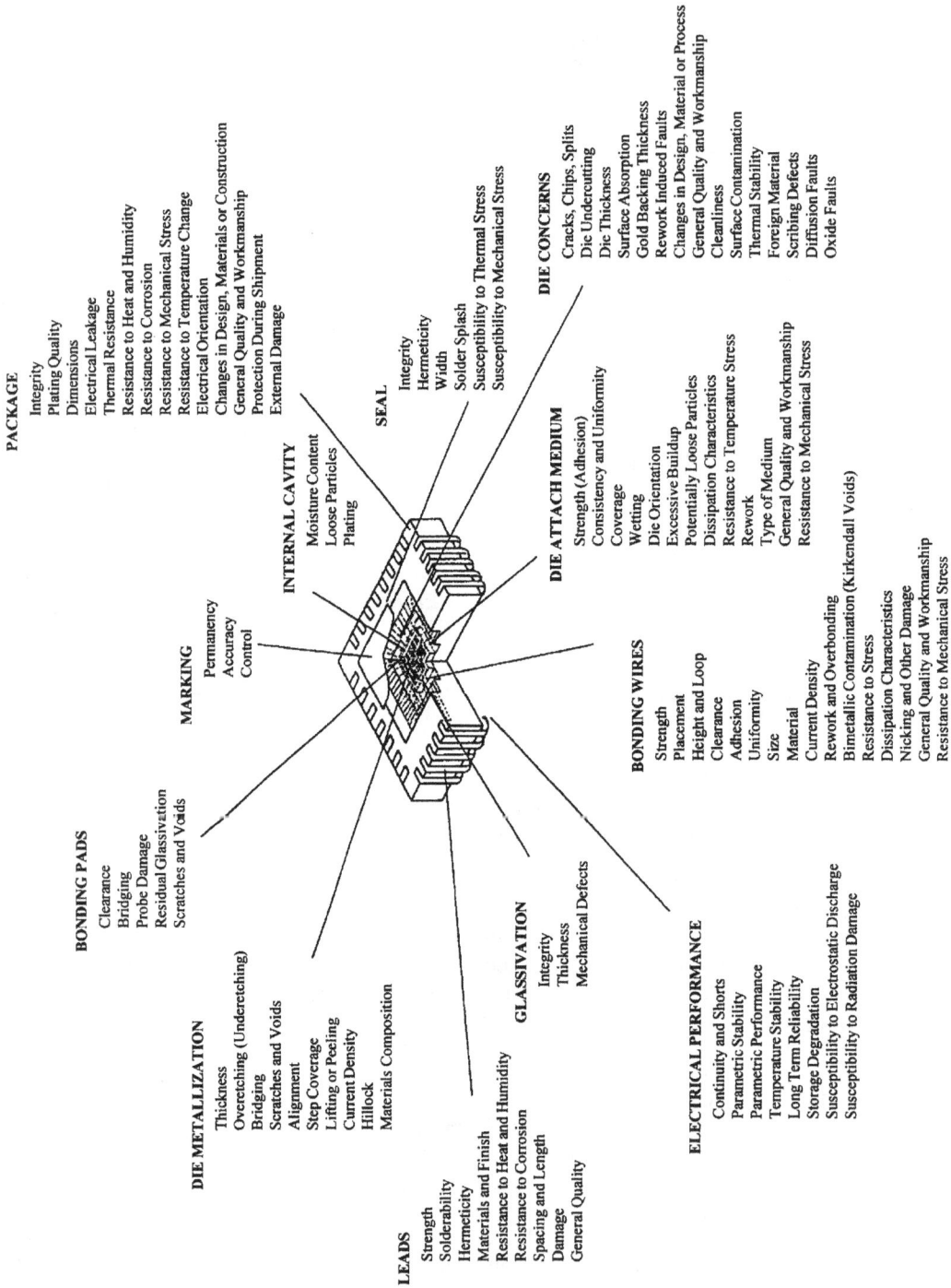

Figure 19.1   Primary device reliability concerns.

quences. During the course of the investigation, the investigator(s) will obtain drawings, plans, blueprints, diagrams of operation procedures, repair and maintenance manuals, and any pertinent information about the subsystems involved. Depending on the complexity of the support equipment system, isolation of its failed subsystem may involve either the turning *off*, turning *on*, replacement, or bypass of other subsystems within the system itself. When the responsible subsystem has been identified, the failed section will then be further isolated to the board level, card level, or smallest mechanical structural level containing all related parts and/or components. Previous chapters of this handbook have detailed the factors involved in the failure analysis of these assemblies. The discussion from this point on will assume that the device responsible for the failure has been isolated be it by optimal or suboptimal failure analysis path.

19.3   STAGES OF A FAILURE ANALYSIS[1]

The ultimate goal in failure analysis is to arrive at an accurate determination of the cause of failure. In electrical component failure analysis, destructive testing is necessary in a large portion of the analysis efforts. Such tasks as decapsulation, scribing metal, and cross-sectioning are all accepted techniques. Performed prematurely, however, these procedures can result in irreversible damage and a ruined analysis. Thus, the analyst must consider the potential damage and purpose of each task and remember the rule followed by carpenters: *measure twice and cut once.*

19.3.1   Electrical Component Failure Verification

Electrical test verification is an essential step necessary to characterize the suspect device and to establish its role in the circuit malfunction. Electrical testing is also performed to compare its present condition with specified parameters and operation at different temperature extremes. Often, electrical testing or verification may not be possible due to the device being badly damaged. Other times, validation of the failure cannot be made because the parts all perform within the electrical specifications or required limitations of their manufacturing data sheets. A judgment is then made as to the appropriate direction of testing and/or examining the failed device. Care must be used in the removal of electronic components so as not to introduce secondary damage, such as the case when leads are clipped and packages are pried off circuit boards by repair technicians who are interested only in trying to get the equipment to work again in a hurry. If the device cannot be patched up to perform an electrical check, then optical examination of the remains may be the only alternative. Noted anomalies must be analyzed with cause and effect in mind. An electrical overstress site on a cracked die in a broken integrated circuit package most probably occurred prior to lid removal and could be the cause of failure. It is essential that repair persons be educated or at least informed of the importance of component care. Also, ESD precautions should be exercised in handling all electrical components so as to not introduce electrostatic damage to the components during handling and analysis (see Figs. 19.2 through 19.4).

Electrical/electronic component types are unique in their construction and purposes, and so is the equipment used to test them. Test equipment has evolved around component types, and both test software and hardware is modified as the components change. As new

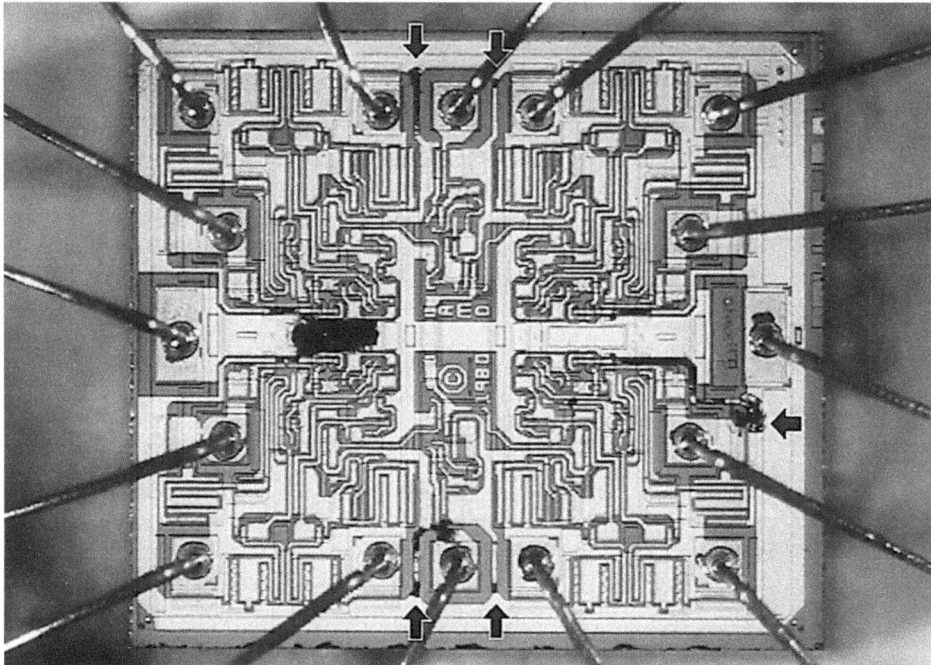

Figure 19.2 50× optical photo illustrating the typical conditions of catastrophic EOS between V_{CC} and ground pins. Notice that the input pins are also affected, due to the large overstress voltage applied *(courtesy of DPA Labs, Inc.)*.

technology becomes available, test equipment follows in its footsteps. A well equipped failure analysis laboratory should have the following on hand for analysis purposes:

- Probing equipment (mechanical or e-beam)
- Digital integrated circuit test systems
- Linear integrated circuit test systems
- Discrete semiconductor test systems
- Microwave device test systems
- Passive device testers such as
  - Digital capacitance bridges
  - Voltmeters, ohmmeters, ammeters
  - Inductance bridges, Q-meters
  - Resistance bridges
- General bench test equipment
  - Power supplies, dc/ac
  - Oscilloscopes, pulse generators, waveform generators
  - Thermal-stream units, temperature baths
  - ESD testers, hi-pot testers

Figure 19.3   100× optical photo of the catastrophic EOS of the output driver section of an IC *(courtesy of DPA Labs, Inc.).*

Figure 19.4   80× SEM photograph of 1.0-mil aluminum bond wires "blown" from excessive current *(courtesy of DPA Labs, Inc.).*

19.3.2 Physical Analysis Tests

A properly conducted electrical part analysis considers all factors that could result in deviation from normal observed electrical behavior. A rigorous analysis of the device will be performed mechanically or chemically prior to decapsulation. This analysis will include any test or series of tests which may provide significant empirical data concerning the internal and/or external workings of the part. The tests or examinations that may be performed on a device are listed in Table 19.1. The number and types of these tests to be performed are determined by preliminary electrical indicators and their evaluation. All tests and examinations are performed with the goal of resolving some particular electrical failure mode. The chosen tests are those that relate to mechanisms in the device that could cause the observed electrical malfunction.

Table 19.1   Tests Performed before Package Opening or Decapsulation

External visual inspection, microscopic, low/high power
Particle impact noise detection (PIND) for devices with cavities
Radiography, X-ray analysis
Hermeticity testing, fine and gross leak test, dye penetrant test
Residual gas analysis (RGA)
Vacuum bake, electrically retest
High-temperature reverse-bias burn-in (HTRB), electrical retest
External cleaning, preservation, photodocumentation
Analysis of external surface anomalies

19.3.3 Low-Power Stereo Zoom Optical Microscopy

Failure modes readily identified by the low-power stereo microscope are as follows:

- Contaminants on the package surface located between leads (can cause electrical leakage or shorts)
- Fractured or broken dielectrics or glass seals
- Fractures in weld seams, blow holes, voids, etc.
- Broken leads or loose feed-through pins
- Arc-over or burns across the dielectrics

19.3.4 Medium- to High-Power Optical Microscopy

The medium- to high-power optical microscope is generally used to accentuate a failure mode already detected by other means. Because of limitations due to field of view, lighting difficulties, and depth of focus, this type of microscopy is generally used only for inspection specific to a previously detected fault. It is especially effective for examining the following:

- Fractures in leads, plating, and glass-to-metal seals
- Chemical damage to nonconductive glasses
- Small defects in weld zones

19.3.5   Particle Impact Noise Detection (PIND)

PIND test systems are used to detect loose particles within a device that has an unfilled internal cavity. The analyst would typically perform a PIND test on a device whenever the suspected failure mode is high leakage, intermittent, or a short. An open usually would not be caused by loose particles but may be detected if the open is due to a loose bond wire. However, if the part fails PIND test, it must be decapped to verify what particle caused the PIND test failure, since the PIND test cannot differentiate between conductive and non-conductive particles (see Figs. 19.5 and 19.6).

19.3.6   X-Ray Radiography

X-Ray radiography is a good method for nondestructive analysis in failure analysis of most device types. The failure analyst should inspect for encapsulated foreign material, internal opens and shorts, and changes in alignment due to the encapsulation process. When properly performed, X-ray analysis will not alter or affect either the device or its failure mode. For this reason, the failure analyst may routinely direct all failed devices to

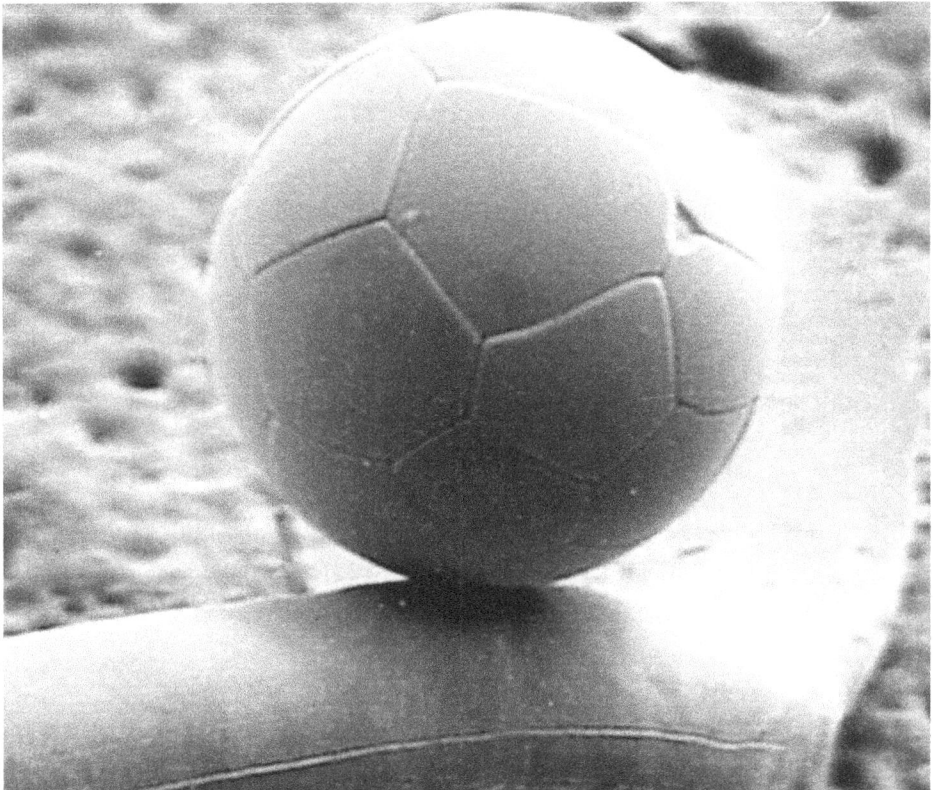

Figure 19.5   100× SEM photograph of a gold flame-off droplet that formed during the ball-bonding operation. Grain boundaries formed during the recrystallization give it a "soccer ball" appearance. The diameter of the wire beneath the ball is 1.0 mil *(courtesy of DPA Labs, Inc.)*.

Figure 19.6 200× SEM photograph of a chipped-out die corner that occurred during the eutectic die-attach process. If the chipped piece breaks off later, it can cause intermittent electrical shorts *(courtesy of DPA Labs, Inc.)*.

X-ray as part of preliminary photographic documentation prior to any destructive tests. X-ray documentation can include standard wet film (type R or type M, double or single emulsion) for high definition, or Polaroid® instant film where definition or high resolution is not critical. Real-time X-ray with digital image capture is available to many labs to speed analysis. X-Ray radiography can provide two major benefits for the failure analyst:

1. It provides a graphic representation of internal or covered characteristics of the device or sample under inspection prior to any cutting, depotting, or handling. As an example, where a device exhibits a defect, misalignment, or foreign material after disassembly, examination of an X-ray made previously, when the sample was whole, will usually allow the analyst to determine whether the fault existed prior to disassembly.

2. The second major benefit of X-ray radiography is that it provides a graphic representation of how the device is constructed. This enables the analyst to formulate effective disassembly procedures.

19.3.7 Scanning Acoustic Microscopy

Scanning acoustic microscopy (SAM) uses the absorption and reflection of ultrasonic waves in a sample. This technique is especially sensitive to any change in acoustic impedance, such as a disbond or delamination as can occur in a plastic encapsulated device. These types of faults are very difficult to resolve using X-ray techniques, and SAM has a definite advantage in these areas. Typical uses for SAM include the evaluation of die attach integrity, detection of voids in the molding compound, characterization of wire bonds, and the identification of cracks in the die or molding compound (see Fig 19.7).

19.3.8 Hermeticity Testing

Hermeticity testing is used to determine the integrity of the device encapsulation. The purpose of the encapsulation is to seal gases or fluids inside the device package and also to prevent gases or fluids from leaking into the device. A capacitor, such as a wet slug tantalum, needs to keep its electrolyte inside the package. A solid-state device would need to be protected from the outside environment. Moisture may enter into the cavity interior of electrical devices due to various conditions. Once in the cavity, it is condensed out or absorbed onto internal surfaces, resulting in many possible reactions and, ultimately, device failure. A good example of this phenomena is the thin-film nichrome resistor. These devices are attacked by high moisture and gases on internal surfaces that have no protective plating, resulting in corrosion damage and failure. Moisture also attacks semiconductor devices by physical corrosion, electrical leakage, and shorts. To determine the package integrity of these devices and other components, failure analysts may use a gross-leak tester or a fine-leak (helium tracer gas) tester. Gross-leak, or bubble-leak, testers use fluorocarbon liquids for indications. Fine-leak testing is performed using digital helium mass spectrometers. The device under examination is pressurized in helium gas then tested for outgassing helium in the detector. Note that a leak test may not always be required. A careful analysis of the electrical test data and a knowledge of part construction may preclude the need for leak testing. Hermeticity testing is a secondary failure analysis tool and is used to aid the analyst in postulating the probable cause of failure.

19.3.9 Residual Gas Analysis

Knowledge of the gas type inside a package can be extremely important in the accomplishment of an effective failure analysis. Many electrical components are hermetically sealed in dry nitrogen. If a salt moisture atmosphere is found inside the package, then corrosion and electrical leakage failures are very possible. Water vapor content and residual gas analysis are performed using the same basic technique. Outgassing of lubricants that condense on relay contacts can increase contact resistance on the relays at low currents and voltages. Outgassing of epoxies in hybrid circuits, which subsequently condense on die surfaces, lead to electrical leakage failures. Depending on materials and systems involved, residual gas analysis can often pinpoint poor processing or manufacturing techniques. RGA should be performed only by those certified in the technique, and even then there are concerns about inaccuracies and inconsistencies.[2]

19.3.10 Vacuum Bake

Low-temperature vacuum drying of a package can be performed to dry out suspected external moisture or evaporate volatile contaminants without exposing internal components to elevated temperatures. A recovery of the device after a low-temperature vacuum dry would be strongly suggestive of absorbed or trapped moisture on the external surfaces. A low-temperature bake with a hole punctured in the package allow removal of moisture and volatile gases from the package interior. Recovery after this procedure would suggest trapped internal moisture or volatile contaminants.

(b)

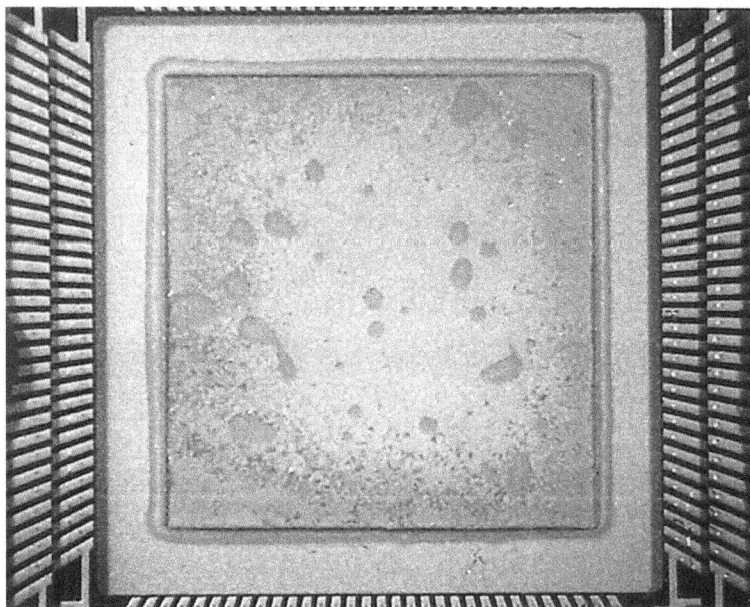

(b)

Figure 19.7   (a) Photo of the underside of a failed die attach. The die separated during centrifuge testing due to the lack of wetting during the eutectic die-attach process. This type of defect can result in thermal overstress from insufficient heat transfer from die to package. (b) Photo of the corresponding package floor where the die lifted off. Lack of wetting caused insufficient die-attach area *(courtesy of DPA Labs, Inc.)*.

A bake at high temperature on semiconductor devices can "heal" or reverse degraded electrical characteristics when the failure is caused by ionic contamination or slow state/bound charge leakage paths such as inversion layers, by dispersing the charges on the die. This would indicate that the device failed as a result of manufacturing fault rather than electrical damage from external means. Parts that are irreversibly electrically damaged by external overstress conditions do not heal by a simple bake.

19.3.11 External Cleaning

There are several primary reasons for performing external washing of a component during device failure analysis:

1. When examination of the package has found evidence of chemical deposits
2. When package materials appear to have been chemically altered

   or degraded
3. When external vacuum bake seems to have a beneficial affect on device performance

Wash solutions are selected from results of previous evaluations and the type of contaminant to be removed. Several methods and types of wash are used, including deionized or distilled water wash, acetic acid wash, solvent washes, and plasma cleaning. Solvent washes are recommended when minimal affects on metals and glasses are desired. Simple immersion with a light agitation is the recommended methodology. Be advised that the use of ultrasonics can impart damage. Fluorine-based plasma cleaning is recommended for glasses. Oxygen-based plasma cleaning is used for organics.

19.3.12 Package Opening, Decapsulation

After all nondestructive analysis external to the device has been explored and documented, the device package is ready for opening and internal examination. Opening of identical packages requires a variety of procedures predicated by the suspected failure mode. It is often necessary to sacrifice a portion of the device or its internal structure to facilitate examination of the area of interest. The appropriate technique must be selected with regard to the suspected failure mode. Chemical depotting of the encapsulant may or may not damage the internal interconnect wire of an integrated circuit, but it can easily damage the die metallization, for example. The two general categories or electrical/electronic device opening techniques are *chemical* and *mechanical*.

Mechanical techniques usually apply to metal, glass, and ceramic packages. Special tools and equipment for mechanical decapsulation include round-style can openers, low-speed diamond saws, grinding wheels, motorized grinding hand tools, jeweler's files, jeweler's hack saws, vices, dip package openers, hot plates, dental picks, pliers, metal cutters, tweezers, diagonal cutters, heat guns, and—most importantly— the Exacto® knife.

Chemical techniques generally apply to plastic or epoxy encapsulated devices. The goal in decapsulation is to expose the failure and internal construction without altering the failure mode. Depending on the specific materials involved, chemicals can selectively remove glass, metal, or plastic encapsulants. Chemicals are thus the logical choice for removing die coatings and for decapsulating potted modules and encapsulated devices of all kinds. Acid etching may involve the use of hot sulfuric acid, fuming nitric acid, hydrofluoric acid, phosphoric acid, or hydrochloric acid. Most acids generally are not very selective in that they will attack materials indiscriminately. Commercial depotting chemicals such as Ure-solve®, Dynasolve®, and Decap® are available for more selective decapsulation.

19.3.13 Internal Visual Examination

After decapsulation, internal examination can be achieved by the use of optical micro-scopes or a scanning electron microscope (SEM) to evaluate physical anomalies, damaged areas, or electrically overstressed areas. SEMs have the additional ability to perform volt-age contrast and electron beam induced current (EBIC) analysis of devices under biased conditions. Biased devices can also be examined with high magnification infrared (IR) thermography. This technique can identify hot spots that command interest as possible failure sites. An ultraviolet (UV) microscope can be used for the evaluation of organic contamination. More detailed descriptions of these techniques can be found in other chap-ters of this handbook (see Figs. 19.8 through 19.12).

19.3.14 Surface Analysis

An energy dispersive X-ray analysis (EDXA) attachment to the scanning electron micro-scope is a valuable tool in failure analysis in that it can use the generated X-rays of the SEM to analyze the material composition. When the sample under observation is bom-barded by a high-energy beam of electrons, X-rays are given off. The generated X-rays impinge on the silicon surface of the EDXA detector. The penetration depth of the X-ray into the silicon is a direct function of the energy of the X-ray. Along the penetration track, interaction occurs between the X-ray and silicon atoms and creates hole-electron pairs. The currents generated are sampled, and the magnitude of pulses related to signal output is

Figure 19.8 250× photo of a chip-out at the corner of the silicon substrate that resulted in an electri-cal short between the bond pad and the die *(courtesy of DPA Labs, Inc.).*

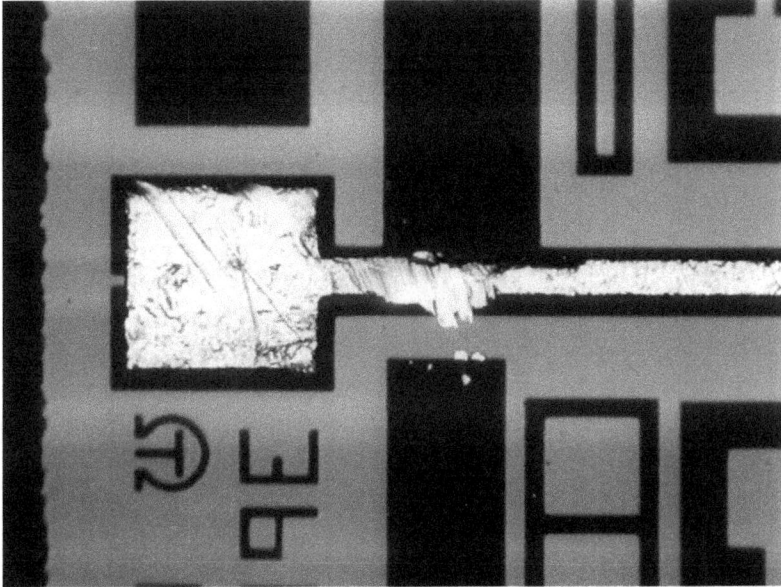

Figure 19.9   250× photo of mechanical damage (smear/scratch) on metallization interconnect resulting in reduced cross-sectional area. This will reduce the current-carrying capability of this path and is a reliability risk for hi-rel applications *(courtesy of DPA Labs, Inc.)*.

Figure 19.10   420× photo of an open metallization trace. This was a processing defect *(courtesy of DPA Labs, Inc.)*.

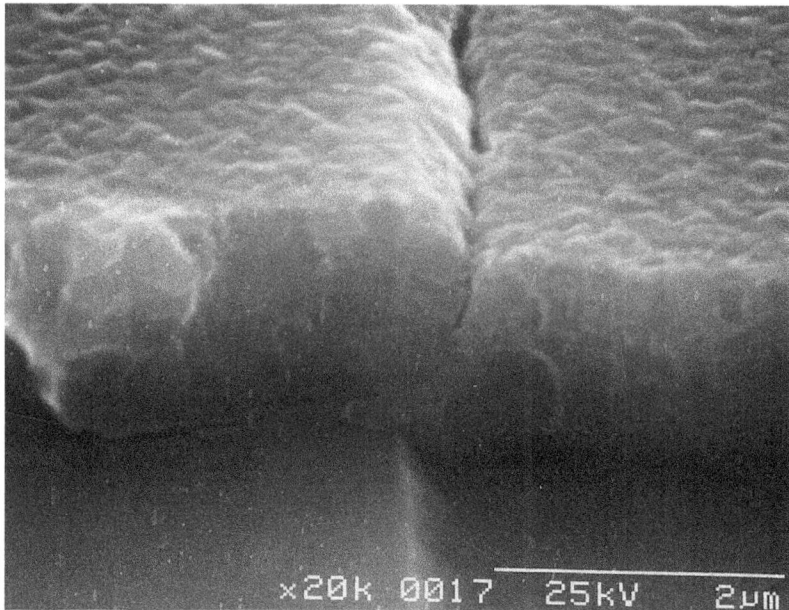

Figure 19.11 20,000× SEM photograph showing inadequate metallization step coverage at an oxide step. Thickness of the metallization shown is ≈14,000 Å *(courtesy of DPA Labs, Inc.).*

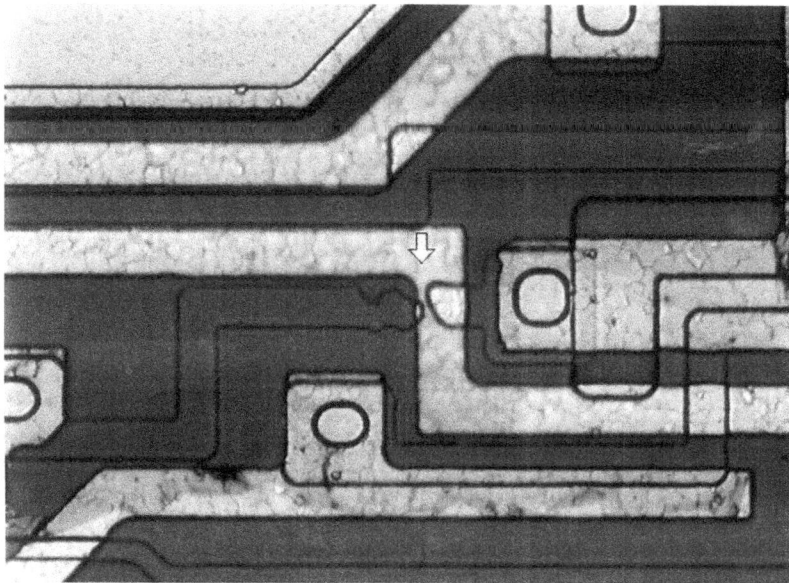

Figure 19.12 500× photo. The arrow points to a discontinuity in the diffusion interconnect that resulted in a high-resistance or open path *(courtesy of DPA Labs, Inc.).*

usually a multichannel spectral output with peaks at specific energies representing X-rays for the various elements present in the specimen being observed in the SEM. The greatest advantage of the EDXA system is the detection of the entire energy/elemental spectrum simultaneously.

Other more specialized contaminate surface analysis techniques include auger electron spectroscopy (AES), energy spectroscopy chemical analysis (ESCA), secondary ion mass spectroscopy (SIMS), wavelength dispersive X-ray (WDX) analysis, and electron micro-probe (EMP) analysis. Results from EMP are often quantitative, as compared to the EXDA. Determination of oxides and nitrides are a significant advantage of the EMP over the EDXA. AES may involve ion etching of the surface, resulting in a depth profile of the contaminant being analyzed. ESCA uses valence state information of the material present on the surface of the device and provides excellent resolution of the various carbon compounds. ESCA is an excellent analytical tool for determining the molecular structure of polymer coatings and the identification of chemical states. SIMS is the most sensitive of all techniques and is the only instrument capable of directly measuring dopant profiles in a semiconductor.

19.3.15 Microsectioning

During the course of failure analysis of microelectronic and other electrical/electronic components, there are occasions when it is appropriate or mandatory to cross-section the sample. The concept of the cross section is obtained from the metallurgist who has, of necessity, been forced to cross-section weld joints, rivet interfaces, solder joints, and so on. The goal of cross-sectioning or microsectioning in any electrical failure analysis is to expose internal features of electrical components and their packaging. Sometimes the sectioned sample will be stained or etched to enhance certain features to make them more visible. Almost all common silicon etches are composed of hydrofluoric and nitric acids, along with various buffering agents such as acetic acid and stain such as copper sulfate.

Several factors tend to differentiate the cross-sectioning of electrical and semiconductor devices from the more typical metallurgical samples: size, complexity, and target. Many of the physical features and subcomponents to be examined are quite small. The complexity lies in the wide variety of materials to be encountered in a single specimen. Very hard or brittle materials include glass or silicon bonded to ceramics or metals. The metals encountered may be quite soft and ductile, such as gold or solder. Last, a specified target could be a shorted junction area with a feature size of 1 μm. It is recommended that the experience required to hit such targets be acquired on practice samples. A cross-sectional photo of a diode is presented in Fig. 19.13. Typical microsectioning uses in electrical/electronic failure analysis are given in Table 19.2.

19.3.16 Analyzing the Evidence & Determination of Failure Mode

After a certain amount of evidence has been revealed during the analysis, and preliminary conclusions are formulated, the pattern and extent of subsequent investigation will be directed toward confirmation of the probable cause and the elimination of other possibilities. The findings at each stage may determine the manner in which the investigation proceeds. As new facts modify first impressions, different hypotheses of failure will develop and will be retained or abandoned as dictated by the findings. During an examination, it is important to distinguish between work that is unnecessary and that does not produce use-

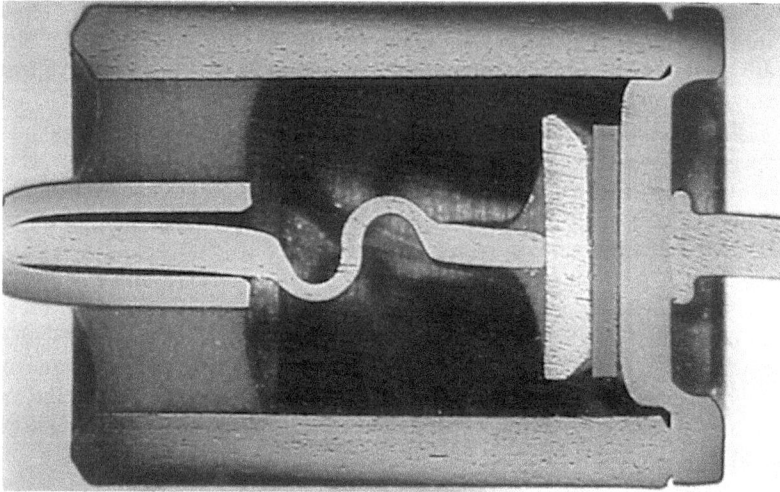

Figure 19.13   5× cross-sectional optical photo of a diode *(courtesy of DPA Labs, Inc.)*.

Table 19.2   Typical Microsectioning Uses in Electrical Failure Analysis

Plating thickness, sequences, quality
All types of bonds/wire bonds
Seals—lid, glass-to-metal
Die attach conditions
Step coverage
Alloying—contact windows
Welds/microwelds
Printed circuit boards, printed wiring boards, multilayer boards
Ceramic substrates
Diodes, transistors, capacitors, inductors, ICs
Junction depths, oxide thickness, diffusions

ful results. Some negative evidence may be helpful in dismissing some causes of failure from consideration. Some of the more prevalent failure modes and mechanisms are presented in the case studies of Sec. 19.5.

19.4   FAILURE ANALYSIS REPORTS[3]

The microelectronic failure analysis report may include the following sections:

- Description of the failed component(s)
- Service conditions at the time of failure
- Prior service history
- Manufacturing and processing/handling history of component(s) (if applicable)

- Electrical characterization and physical study of failure
- Internal visual inspection results of the component(s)
- Summary of the mechanism(s) that caused failure
- Recommendations for prevention of similar failure or for correction of similar components in service

JEDEC Committee 14.6 is working the issue of standardized failure reporting in microelectronics and may publish guidelines soon.

19.5   CASE STUDIES[4]

19.5.1   Die Fabrication Failure No. 1

Background

Device type:	telecommunications IC
Date code:	unknown
Package type:	plastic quad flat pack
Chip technology:	CMOS
Failure location:	IC performance testing

Failure Mode.   *Single-bit failures in a static RAM array within the IC.* The failing cells were programmed as logic 0s but were found to be logic 1s when read. The cells failed at V_{CC} equal to 8 V. Operation of the devices at 5 V was normal. SPICE modeling indicated that a high resistance at the V_{SS} contact of the failing cell would create this failure mode.

Cause of Failure.   Silicon mound growth in the V_{SS} contact (metal 1-to-silicon) of the failing cells. The presence of the silicon mounds resulted in a high resistance between the metal 1 and the transistor source/drain.

Analysis Techniques Employed

Electrical Test.   The locations of the failing cells were determined by testing at the manufacturer's site.

Optical Inspection.   Optical inspection of the intact dice did not reveal any evidence of the cause of failure.

Selective Delayering.   The dice were selectively delayered and inspected optically and by SEM. Inspection levels were metal 2 intact, metal 2 removed, metal 1 intact, metal 1 removed and delayered to the poly gate level. SEM inspection after removal of metal 1 revealed evidence of silicon mound growth in the V_{SS} contact of the cell.

Die Cross Section.   SEM examination of a 90° cross section through the suspect V_{SS} contact also showed the presence of silicon mound growth (see Fig. 19.14).

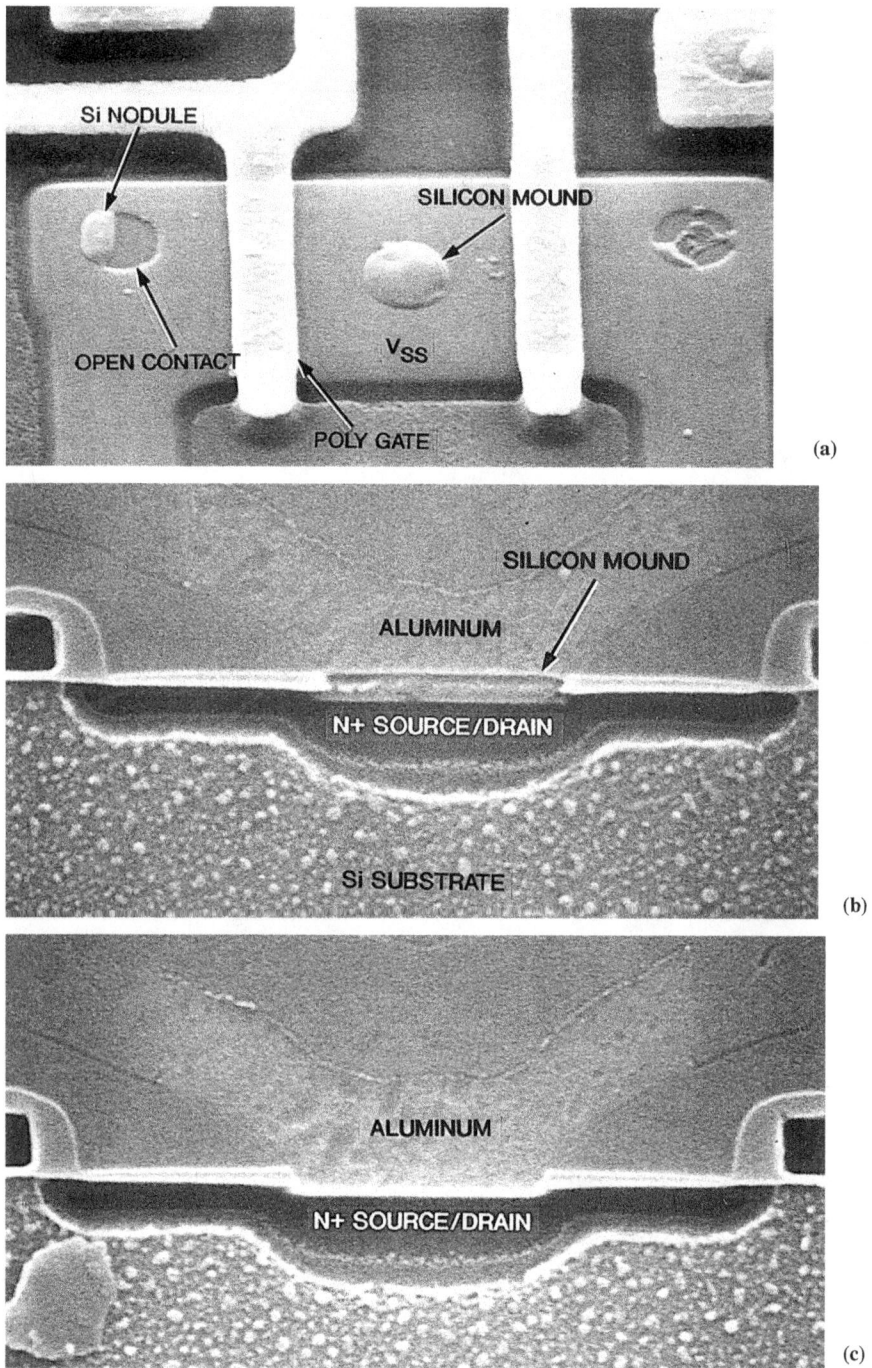

Figure 19.14 Perspective and cross-sectional SEM views illustrating the mound in the suspect contact: (a) 8,500×, (b) suspect V_{SS} contact, 20,000×, and (c) normal V_{SS} contact, 20,000× *(courtesy of ICE, Inc.)*.

19.5.2 Die Fabrication Failure No. 2

Background

Device type:	series 2000-16 RISC CPU
Date code:	8927, 8932, 9002
Package type:	144-pin PGA
Chip technology:	CMOS
Failure location:	field

Failure Mode. *Functional failures in the field.* In-depth testing by the user and the manufacturer indicated that one or two memory cells in an internal memory array on the RISC CPU were stuck *high.*

Cause of Failure.   SEM inspection of the failed cells showed evidence of defects in the gate oxide of the P-channel transistors on all of the samples inspected.

The damage to the gate oxide suggests that a contaminant was present either during growth of the gate oxide or possibly early in the poly deposition. The contaminant either did not allow gate oxide growth at the sites or its presence resulted in a poor quality oxide.

These failures occurred over several package date codes. However, consultation with the manufacturer revealed that the devices were assembled from a die bank, and tracking of wafer lots was not employed.

Analysis Techniques Employed

External Inspection.   No evidence of mechanical or electrical damage was seen.

Internal Inspection.   The physical location of the failing cells was located using information and plots provided by the manufacturer. Optical inspection of the intact cells showed no evidence of damage or defects.

Selective Delayering.   The oxide and metal layers on the die were selectively removed, and the suspect cells were inspected optically and by SEM after removal of each layer. No anomalies were noted until removal of the poly gate level.

SEM Inspection.   Inspection of the failed cells after removal of the polysilicon gate showed evidence of small pinhole-type defects (less than 1 µm) in the gate oxide. The pinholes were found only in the failed cells, indicating that the delayering process did not create the pinholes (see Fig. 19.15).

19.5.3 Die Fabrication Failure No. 3

Background

Device type:	telecommunications IC
Date code:	unknown
Package type:	28-pin PLCC, received decapsulated
Chip technology:	CMOS
Failure location:	device qualification

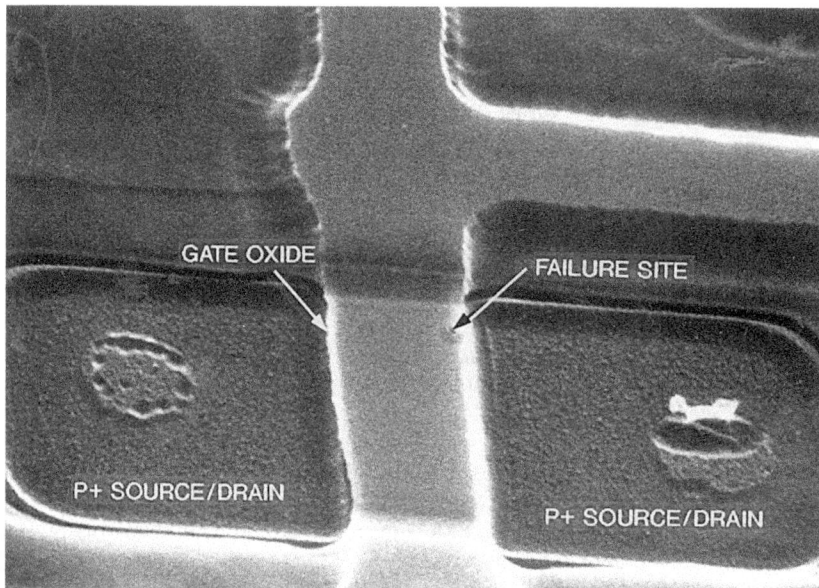

(a)

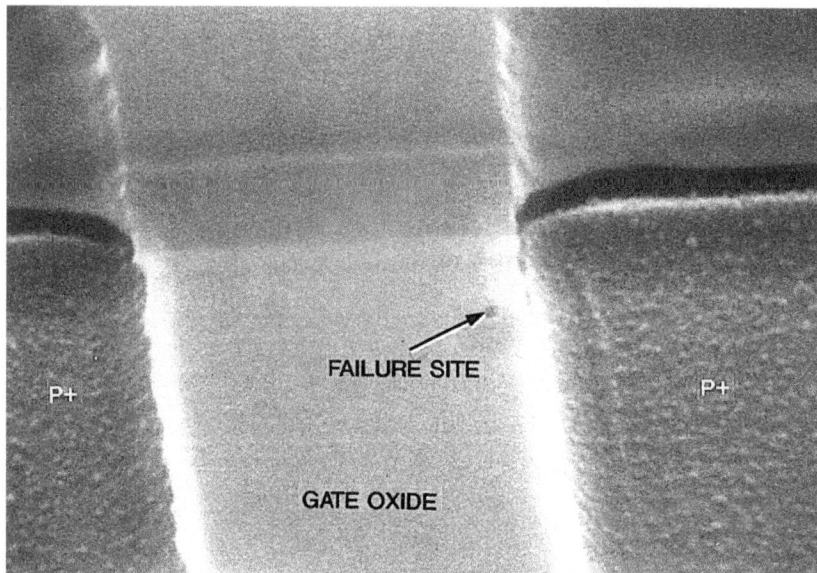

(b)

Figure 19.15 SEM views of the failure in the gate oxide of the memory cell: (a) 10,000× and (b) 30,000× *(courtesy of ICE, Inc.)*.

Failure Mode.   Testing by the client indicated that a P-channel transistor in a buffer circuit was shorted from the gate to the source and drain. Two devices were evaluated.

Cause of Failure.   A nearly identical patterning defect was found in the failing transistor of both samples. This suggests that the mask for the poly level had a defect or foreign particle, resulting in a repeating patterning defect.

Analysis Techniques Employed

Optical Inspection.   High-power optical inspection of the suspect transistor (P1) with all layers intact indicated that the poly gate fingers in the suspect transistor had variations in gate length.

Microprobing.   Probing metal 2, after removal of the final passivation, between the gate to source and gate to drain revealed resistive shorts between all terminals

Selective delayering.   Optical inspection of the poly gate level after removal of the various metal and oxide layers revealed the presence of nearly identical patterning defects on both samples. Due to the defect, the poly was not completely cleared from the source/drain areas. This resulted in metal 1 source/drain interconnect being connected to the gate (see Fig. 19.16).

19.5.4   Die Fabrication Failure No. 4

Background

Device type:	ASIC
Date code:	n/a
Package type:	none
Chip technology:	CMOS
Failure Location:	wafer-level test

Failure Mode.   Many of the dice on the wafer had high I_{DD} current.

Cause of Failure.   Particle defects resulted in shorts between adjacent metal lines.

Analysis Techniques Employed

Optical Inspection.   The initial optical inspection of the wafer did not reveal any clear evidence of damage or defects.

Microprobing.   Probing the power supply pins of the failing dice showed high leakage current in the peripheral power supply pins, as compared to good dice.

Liquid Crystal.   Biasing of the peripheral power supply circuits with liquid crystal on the die surface revealed distinct hot spots. Optical inspection of the hot-spot areas showed the presence of a foreign particle that had resulted in adjacent metal lines being shorted together. The defects were seen at both metal 1 and metal 2 levels (see Fig. 19.17).

(a)

(b)

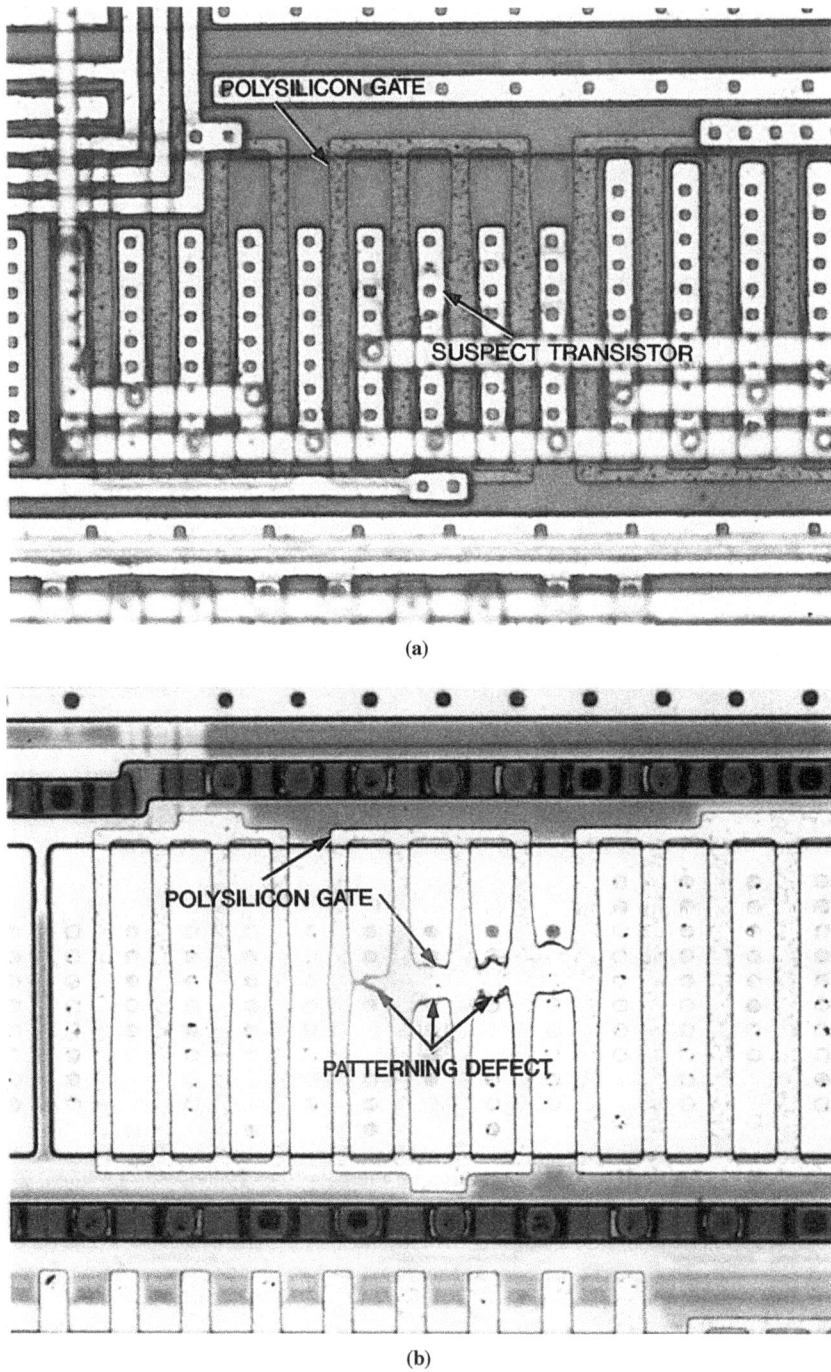

Figure 19.16 Optical views of the defective transistor, 800×: (a) metallization intact, and (b) delayered to poly *(courtesy of ICE, Inc.)*.

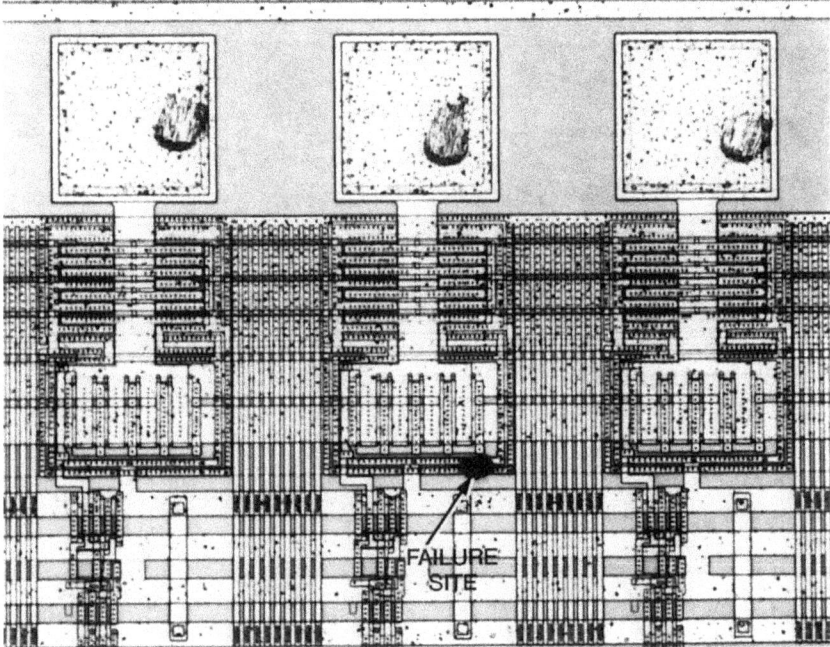

(a)

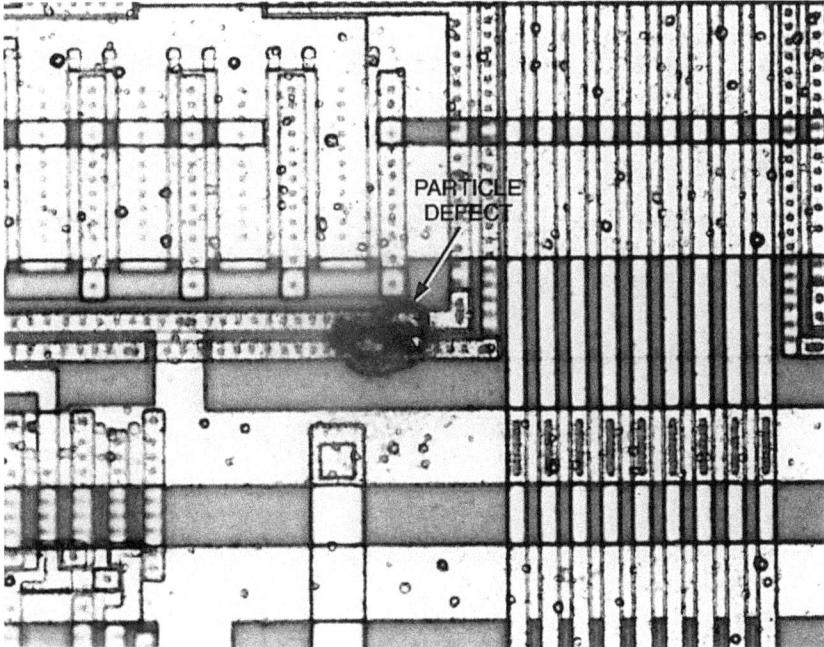

(b)

Figure 19.17   Optical views of a particle defect responsible for high leakage current between the peripheral power supply pins: (a) 160×, and (b) 800× *(courtesy of ICE, Inc.).*

19.5.5 Die Fabrication Failure No. 5

Background

Device type:	unidirectional transient surge suppressor
Date code:	9210
Package type:	modified TO-202
Chip technology:	SCR-type thyristor
Failure location:	lot qualification

Failure Mode. The device initially functioned but failed over time due to high leakage current. Bench testing showed that the device could be made to fail by high-temperature reverse bias (HTRB) and then recovered by an unbiased bake.

Cause of Failure. Mobile ionic contamination was found on the surface of the die. EDX analysis showed the presence of sodium, chlorine, potassium, and calcium.

Analysis Techniques Employed

External Inspection. No evidence of mechanical or electrical damage was noted.

Electrical Test. The sample had originally failed during HTRB testing (at 48 hr) by the client. The part was then subjected to a high-temperature unbiased bake. Testing after the bake showed that the unit recovered. Duplication of these tests by ICE revealed the same results.

Package Decapsulation. The ability to induce and remove the leakage current indicated the presence of mobile ionic contaminants, which requires special consideration when choosing a decapsulation method. The sample was mechanically decapsulated by carefully grinding away the plastic on the front surface until the cathode wire was reached. The backside of the plastic package was then ground away to expose the lead frame. The sample was then placed on a hotplate, and the remaining plastic was carefully pried from the die surface.

Internal Inspection. Optical inspection of the die surface and the imprint of the die surface that was present in the plastic revealed evidence of a contaminant site. The replication of the contaminant site in the plastic imprint was evidence that the suspect material was present prior to decapsulation. The suspect material was brown in color and appeared semitransparent. The contaminant was located on oxide between the cathode metallization and a metal guard ring at the die perimeter.

Microprobing. Probing of the die after decapsulation showed that the leakage current was now within specification. The die was placed on a hot chuck (125° C) and reverse biased for 2 hr. The leakage current increased by several orders of magnitude during this time.

Energy Dispersive X-Ray Analysis (EDX). EDX evaluation of the contaminant showed high levels of sodium, chlorine, potassium, and calcium. These are common constituents of human contamination (spittle, etc.).

Surface Cleaning. The surface of the die was cleaned with distilled water and acetone. The majority of the contaminant was removed from the die surface. Probing showed that the leakage current was within specifications. The die was subjected to HTRB for five hours. No increase in leakage current was noted (see Fig. 19.18).

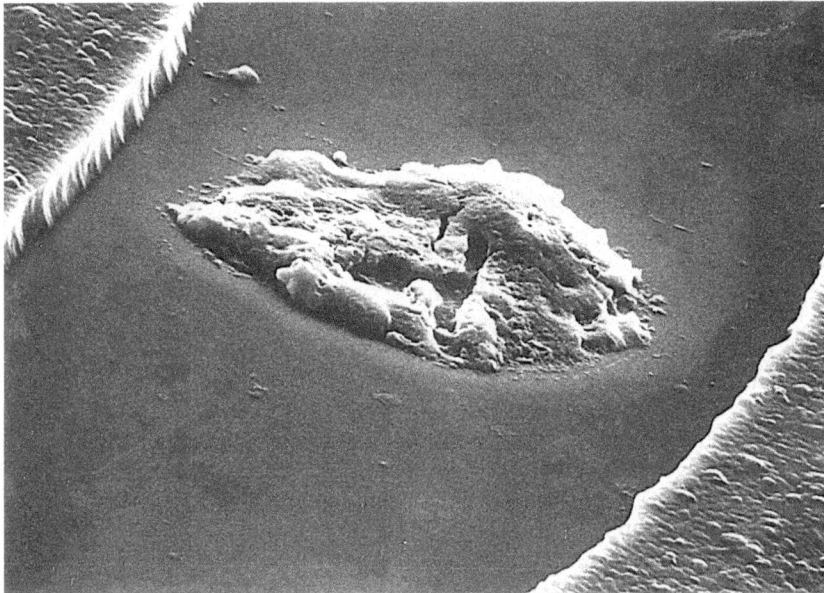
(a)

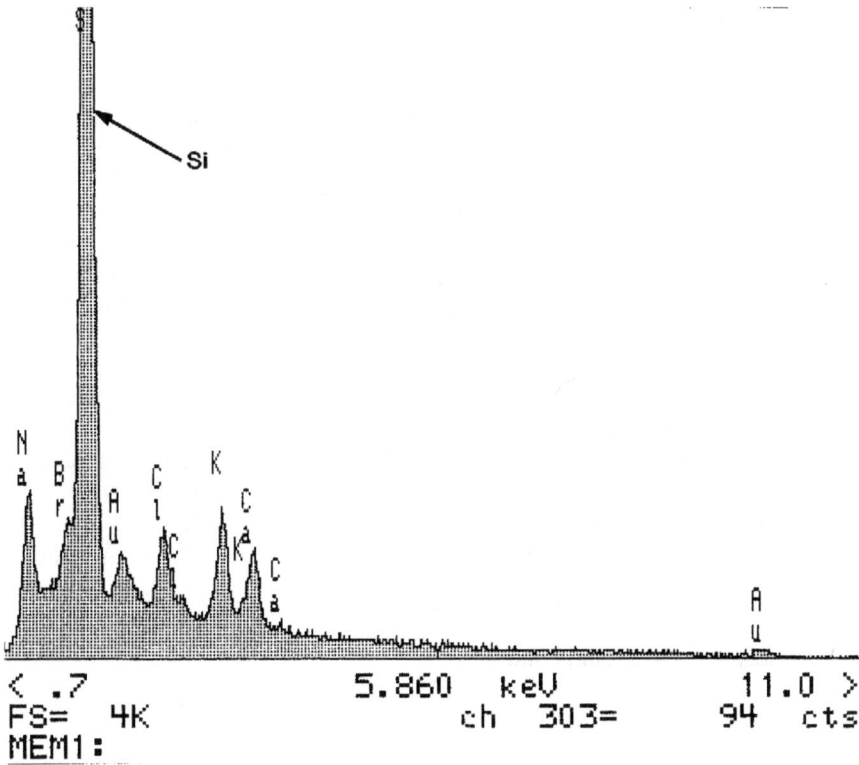
(b)

Figure 19.18   (a) 1500× SEM view and (b) EDX spectra of the contaminant *(courtesy of ICE, Inc.).*

19.5.6   Die Fabrication Failure No. 6

Background

Device type:	unidirectional transient surge suppressor
Date code:	9206
Package type:	modified TO-202
Chip technology:	SCR-type thyristor
Failure location:	board test during system assembly

Failure Mode.   Eight devices failed due to high leakage current after board-level testing, during system assembly.

Cause of Failure.   All of the failures were due to manufacturing defects (pinholes) in the oxide beneath the die metallization, which resulted in leakage current paths between the anode and cathode. Pinholes in oxide are usually created by particulate contamination on the surface of the wafers. The particles either do not allow oxide to be grown or deposited in these areas, or the particles do not allow proper deposition of photoresist and the oxide in these areas is etched during subsequent manufacturing steps.

Analysis Techniques Employed

External Inspection.   No evidence of mechanical or electrical damage was noted.

Electrical Test.   The suppressor consists of a thyristor with a zener diode integrated into the gate region. The zener diode initially clamps the voltage transients until sufficient current flows, activating the thyristor. Curve tracing of the devices showed high leakage current as compared to the good units. Baking of one sample, unbiased and at high temperature, did not result in recovery of the failure.

Internal Inspection.   The devices were decapsulated by immersion in hot acid. Optical inspection of the dice did not reveal any evidence of electrical overstress.Optical inspection of the dice using a mercury filter accentuated anomalies in the oxides on the dice. The filter highlights interference fringes in the oxide, indicating that the oxide is not of uniform thickness.

Liquid Crystal.   Evaluation of the samples with liquid crystal showed definite hot spots at specific pinhole locations on each sample.

SEM Inspection.   Inspection by SEM before and after removal of the die metal confirmed the presence of pinholes in the oxide beneath the die metallization.

Die Cross-Sectioning.   A pinhole was cross-sectioned on one sample. The oxide deposited beneath the metal thinned from 1 μm to approximately 700 Å in the pinhole sectioned (See Figs. 19.19 and 19.20).

19.5.7   Assembly and Packaging Failure No. 1

Background

Device type:	8k × 8 EEPROMs
Date code:	9032

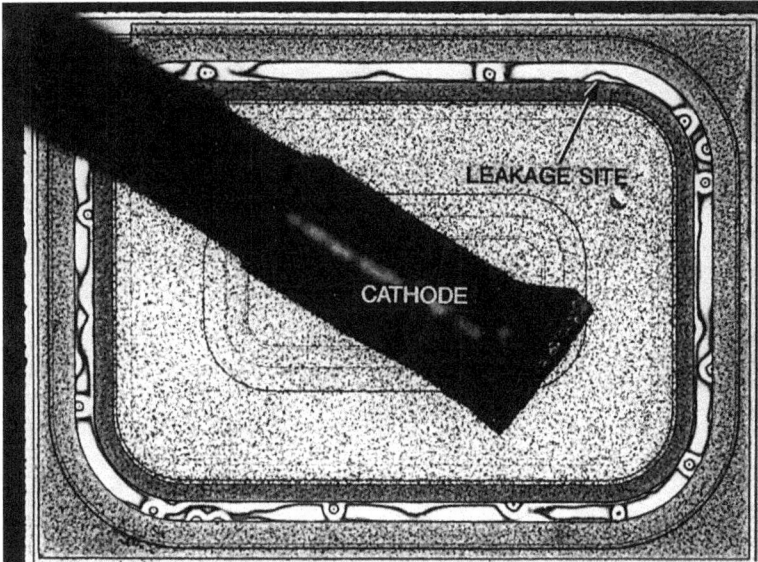

(a)

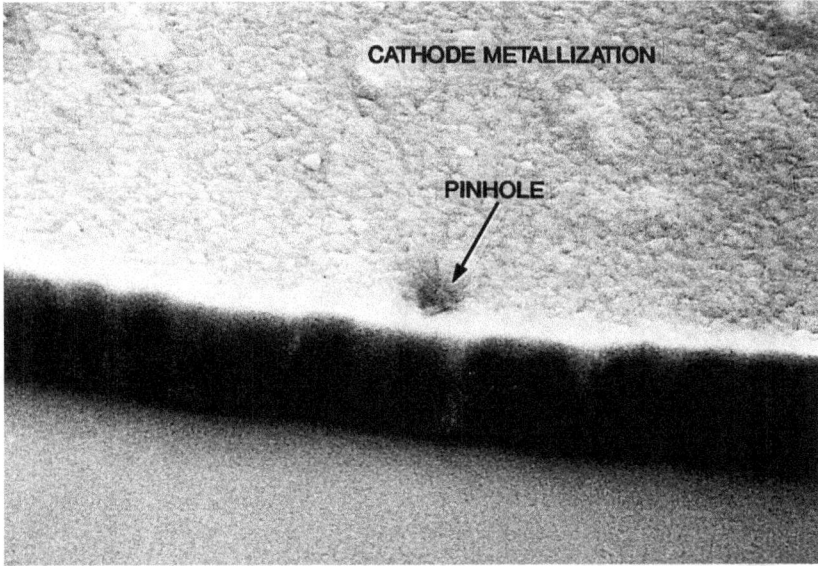

(b)

Figure 19.19 Optical and SEM views of the dice showing evidence of pinholes in the oxide: (a) intact die, optical view at 50× and (b) metal intact, SEM view at 2,500× *(continues) (courtesy of ICE, Inc.)*.

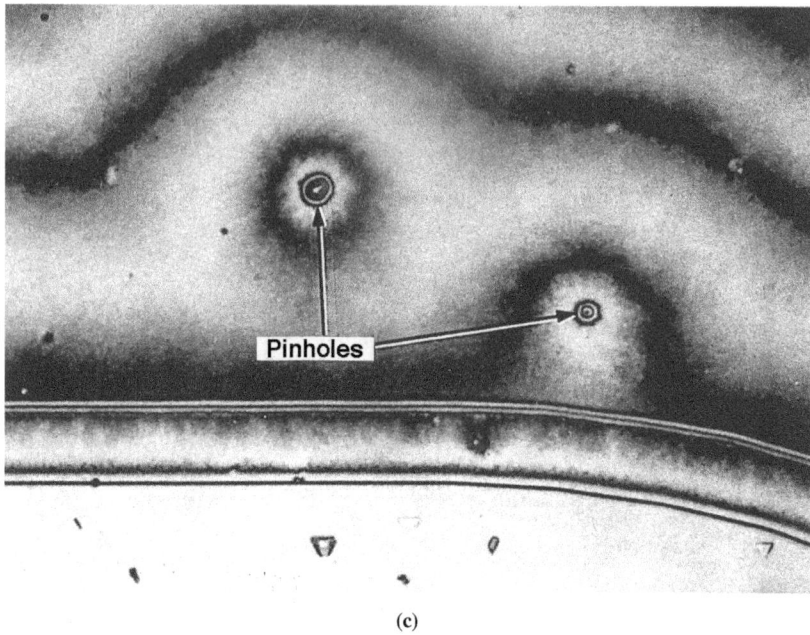

(c)

Figure 19.19 Optical and SEM views of the dice showing evidence of pinholes in the oxide: (c) metal removed, optical view at 800× *(courtesy of ICE, Inc.)*.

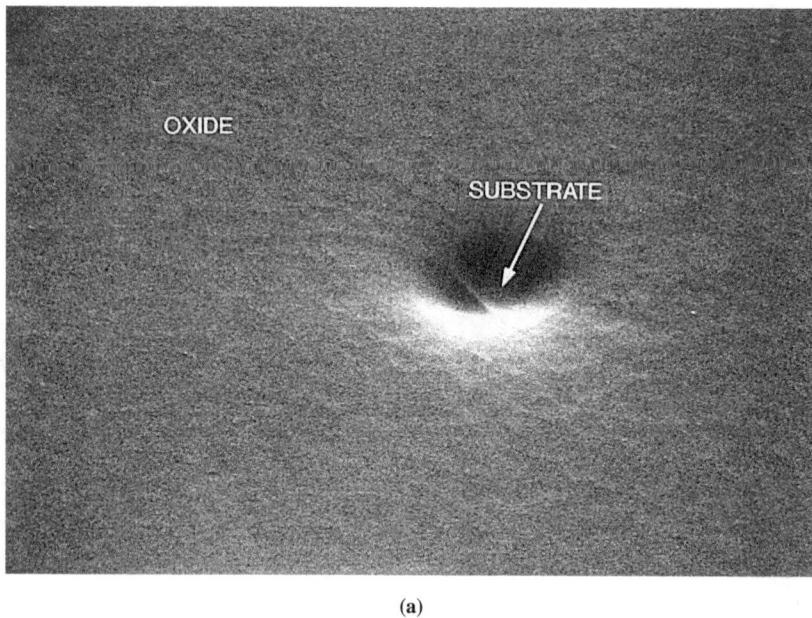

(a)

Figure 19.20 Perspective and cross-sectional views of the pinholes: (a) metal removed, SEM view at 5000× *(continues) (courtesy of ICE, Inc.)*.

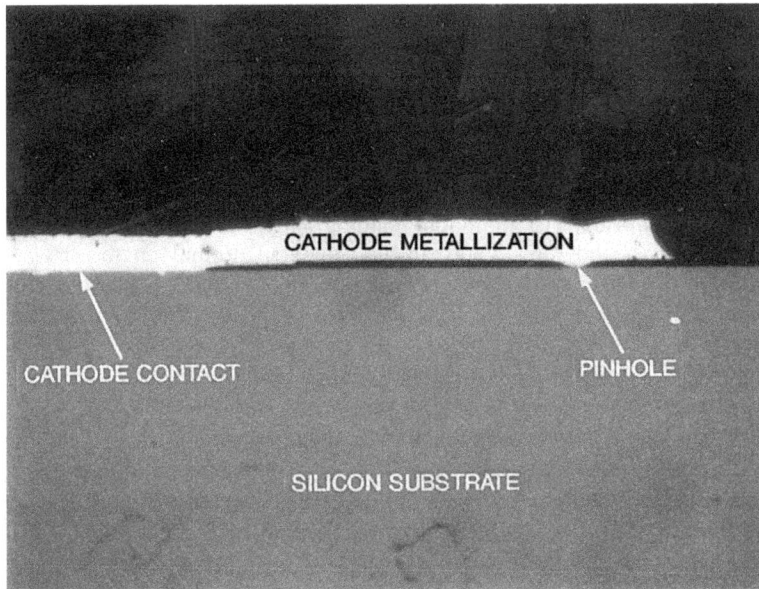

(b)

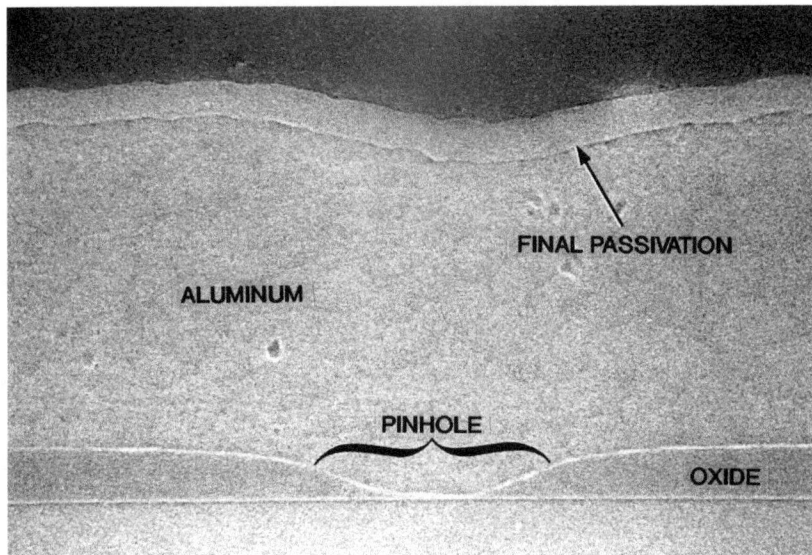

(c)

Figure 19.20   Perspective and cross-sectional views of the pinholes: (b) metal removed, SEM view at 8000×, and (c) SEM view at 8,000× *(courtesy of ICE, Inc.).*

Package type: ceramic leadless chip-carrier

Chip technology: CMOS

Failure location: testing during board assembly

Failure Mode.   Open input pins.

Cause of Failure.   The aluminum bonding wires were broken at the heel of the package land bond. The bonding workmanship appeared normal on the devices. The damage appeared similar to that which is created by high-frequency vibration, such as ultrasonic cleaning.

Analysis Techniques Employed

External Inspection.   No anomalies were noted with the units intact.

Electrical Test.   Curve tracing confirmed the open input pins.

Internal Inspection.   Internal inspection revealed that the aluminum bonding wire of each suspect pin was broken at the heel of the bond to the package land. The breaks were clean and sharp. The breaks were not due to electrical overstress.

SEM Inspection.   SEM inspection revealed that additional wires had cracks in heel of the bonding wire at the package land. The bonding workmanship appeared normal on the samples. The arc and length of the bonding wires was normal and the bonds did not appear overcompressed or malformed. Furthermore, forward bonding was employed, and cracking due to poor machine setup would most likely occur at the bond to the die surface (see Fig. 19.21).

19.5.8   Assembly And Packaging Failure No. 2

Background

Device type: diode array

Date code: 9225

Package type: modified 8-pin plastic SOIC

Chip technology: bipolar

Failure location: high-temperature reverse bias (HTRB) testing

Failure Mode.   All of the samples showed increased leakage current on one or more of the diode pairs.

Cause of Failure.   All of the samples failed due to cracked dice. The cracks were created by excessive stress, induced by poor workmanship during rework of the thermocompression wirebonding.

Analysis Techniques Employed

External Inspection.   No anomalies were noted with the units intact.

Electrical Test.   The devices consisted of eight discrete unipolar diodes that were configured into four back-to-back diode pairs. The diodes were fabricated in a planar bipolar

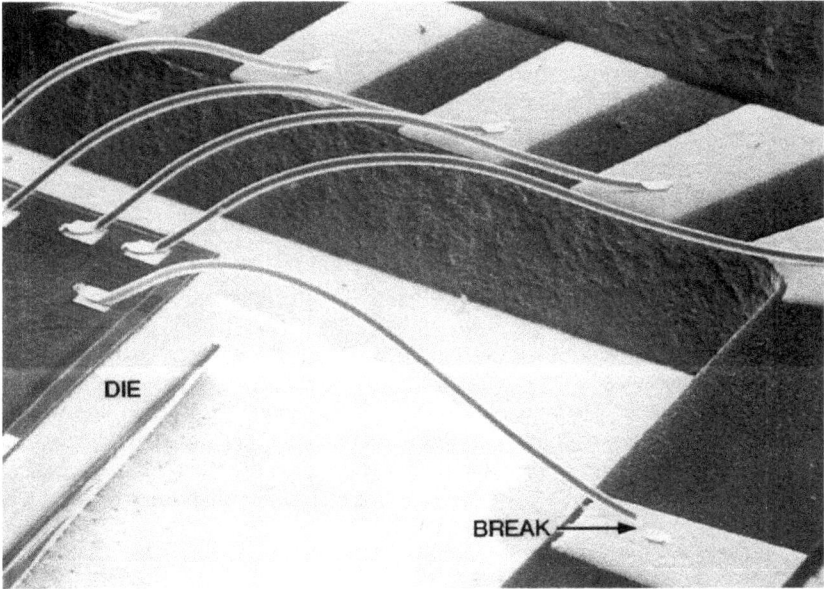

(a)

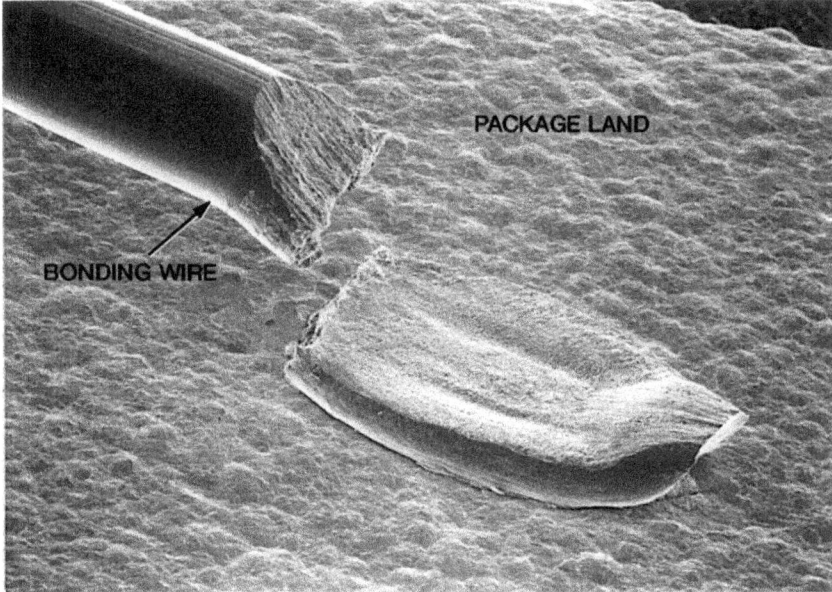

(b)

Figure 19.21   SEM views of broken bonding wire: (a) 40× and (b) 600× *(courtesy of ICE, Inc.)*.

process. Curve tracing showed that one or more of the diode pairs had excessive leakage current in one or both directions.

Internal Inspection. Inspection of the diodes after decapsulation showed significant tool damage to the die metallization and evidence that rework of the wirebonding had been conducted on a number of the diode pairs. Curve tracing showed that decapsulation did not affect the leakage current.

Microprobing. Probing of the individual diodes identified which of the diodes had the increased leakage current.

Selective Delayering. The die metallization was etched away, except for the contact area occupied by the wire bonds. Probing afterward showed no change in the leakage currents. No evidence of pinholes in the oxide, which could short the die metal to the substrate, were seen. The remaining die metallization and the bonding wires were removed by chemical etching. Inspection afterward showed evidence of cracks and chipouts in the silicon on the failing dice. The damage occurred in the areas occupied by the wirebonds. All of the devices showed evidence that the wirebonds had been reworked on the failing diodes. It was also noticed that the second bond was typically placed in the area where the first bond had been scraped away. In some cases, a significant amount of the die metal had been scraped away (see Figs. 19.22 and 19.23).

19.5.9 Assembly And Packaging Failure No. 3

Background

Device type: diode

Date code: unknown

Package type: modified J-bend (DO-214 AB)

Chip technology: bipolar

Failure location: device qualification

Failure Mode. All samples showed very high leakage current.

Cause of Failure. Excessive mechanical stress resulting in cracks in the die and the plastic package. It was later determined that the user was employing improper assembly techniques that were inducing high mechanical stress on the devices.

Analysis Techniques Employed

External Inspection. Examination of the outside of the packages revealed cracks in the plastic. The cracks always occurred in the bottom of the package, at the end of one of the leads.

X-ray. X-ray confirmed the presence of the cracks and showed that the cracks extended from the outside of the package to the internal leadframe. No other assembly anomalies were seen.

Electrical Test. Curve tracing of the diodes confirmed the high leakage current.

Internal Inspection. Optical inspection of the samples revealed cracks in die corresponding with the cracks in the plastic package (see Fig. 19.24).

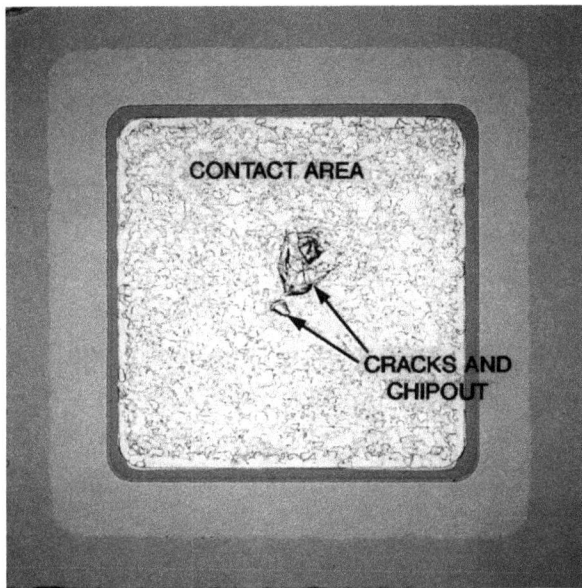

(a)

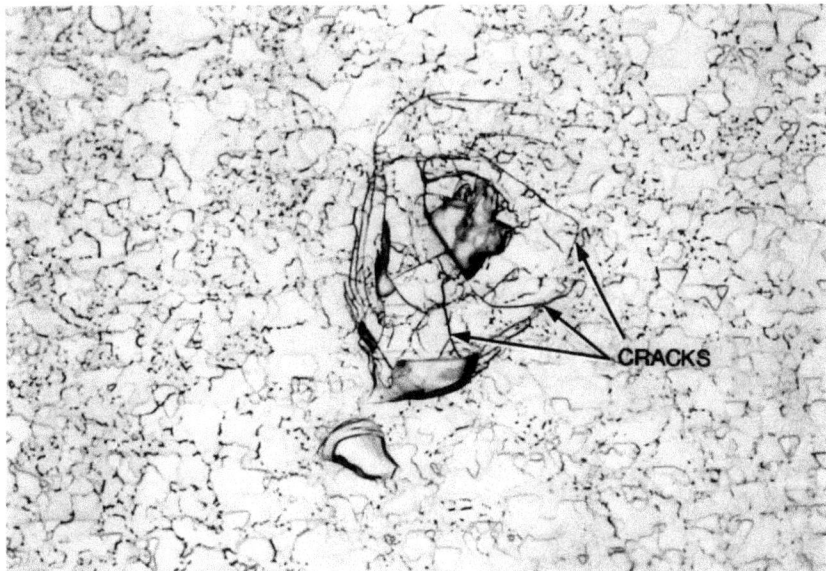

(b)

Figure 19.22   Optical views of the cracks found in the silicon after removal of the die metallization: (a) 130× and (b) 500× *(courtesy of ICE, Inc.).*

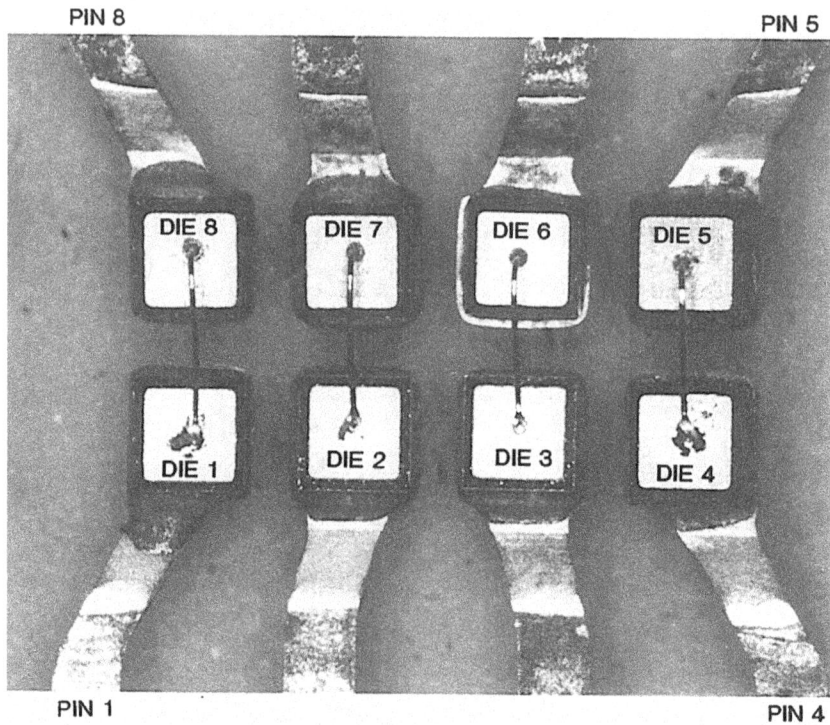

(a)

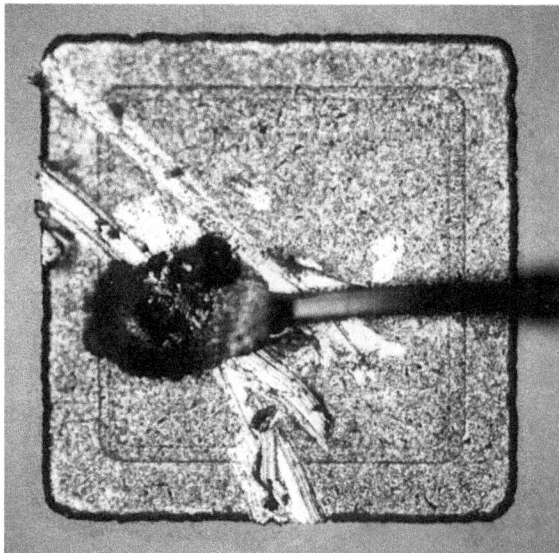

(b)

Figure 19.23 (a) Optical view of the die package layout, 25×, and (b) detail optical view showing smeared die metal and evidence of rebonding—close-up of die 3 in (a), 130× *(courtesy of ICE, Inc.).*

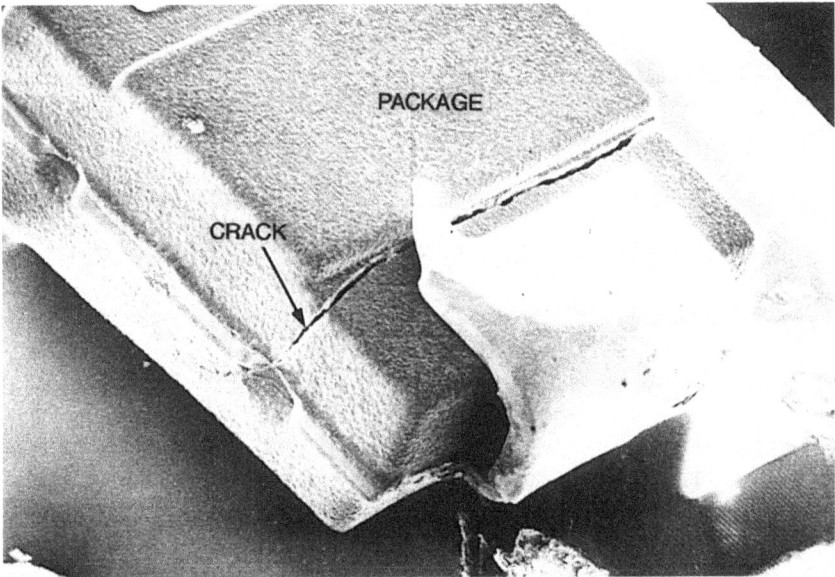

(a)

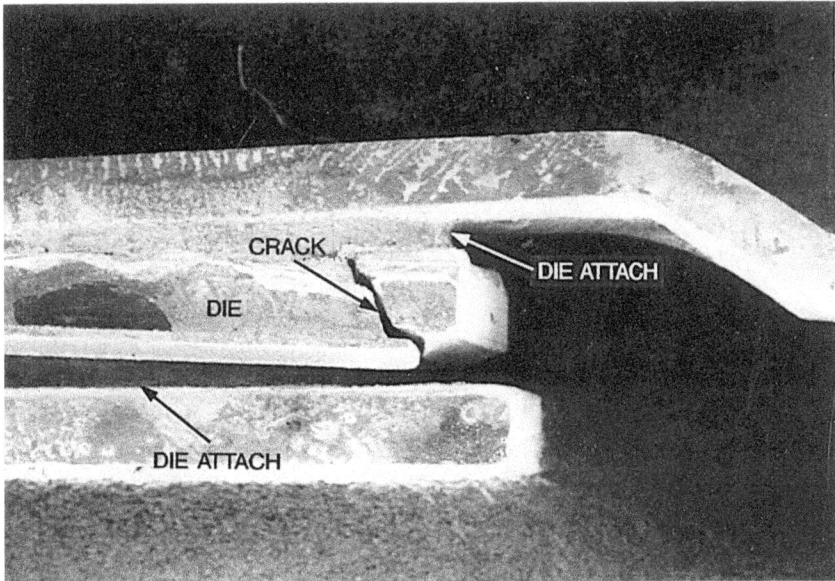

(b)

Figure 19.24 (a) SEM view illustrating a typical crack in the plastic package, 20×, and (b) SEM view of a decapsulated device illustrating crack in die, 50× *(courtesy of ICE, Inc.)*.

19.5.10 ESD/EOS Failure No. 1

Background

Device type:	dual switch-mode solenoid driver
Date code:	9133
Package type:	15-pin plastic, single-in-line power tab
Failure location:	final system level test

Failure Mode.   No output to either channel. Bench testing showed that power supply current was only 2 mA as compared to 20 mA on a good unit.

Cause of Failure.   The device failed due to electrical overstress that damaged an internal segment of the $+V_{SS}$ power supply line. The location and appearance of the damage indicates that a latch-up condition may have occurred, resulting in excessive current flow through the $+V_{SS}$ bus line.

Analysis Techniques Employed

External Inspection.   No evidence of mechanical or electrical damage was seen on the intact unit.

X-ray.   X-ray of the good and failed samples did not reveal any anomalies.

Electrical Test.   Pin-to-pin curve tracing showed increased resistance between the power supply pins.

Internal Inspection.   The device was decapsulated by immersion in hot acid. Optical inspection of the die showed evidence of carbonized plastic and reflowed aluminum in a segment of the $+V_{SS}$ bus line. The damaged metal line was of minimum width and was the only connection between $+V_{SS}$ and most of the die circuitry.

Microprobing.   The final passivation was removed and the $+V_{SS}$ line was probed beyond the damage site. The I-V characteristic was identical to the I-V characteristic on the good sample. Power was applied between $+V_{SS}$ (beyond the damage) and ground, and the I_{SS} current was monitored. The I_{SS} current was the same as the good sample, indicating that there were no internal shorts or damage on the die. The location of the damage in the minimum width interconnect line between the $+V_{SS}$ pad and the internal circuitry suggests that the device may have experienced a latch-up condition, which resulted in high current flow through the $+V_{SS}$ pin (see Fig. 19.25).

19.5.11 ESD/EOS Failure No. 2

Background

Device type:	high-voltage, high-current Darlington driver array
Date code:	8929
Package type:	16-pin CERDIP
Failure location:	field

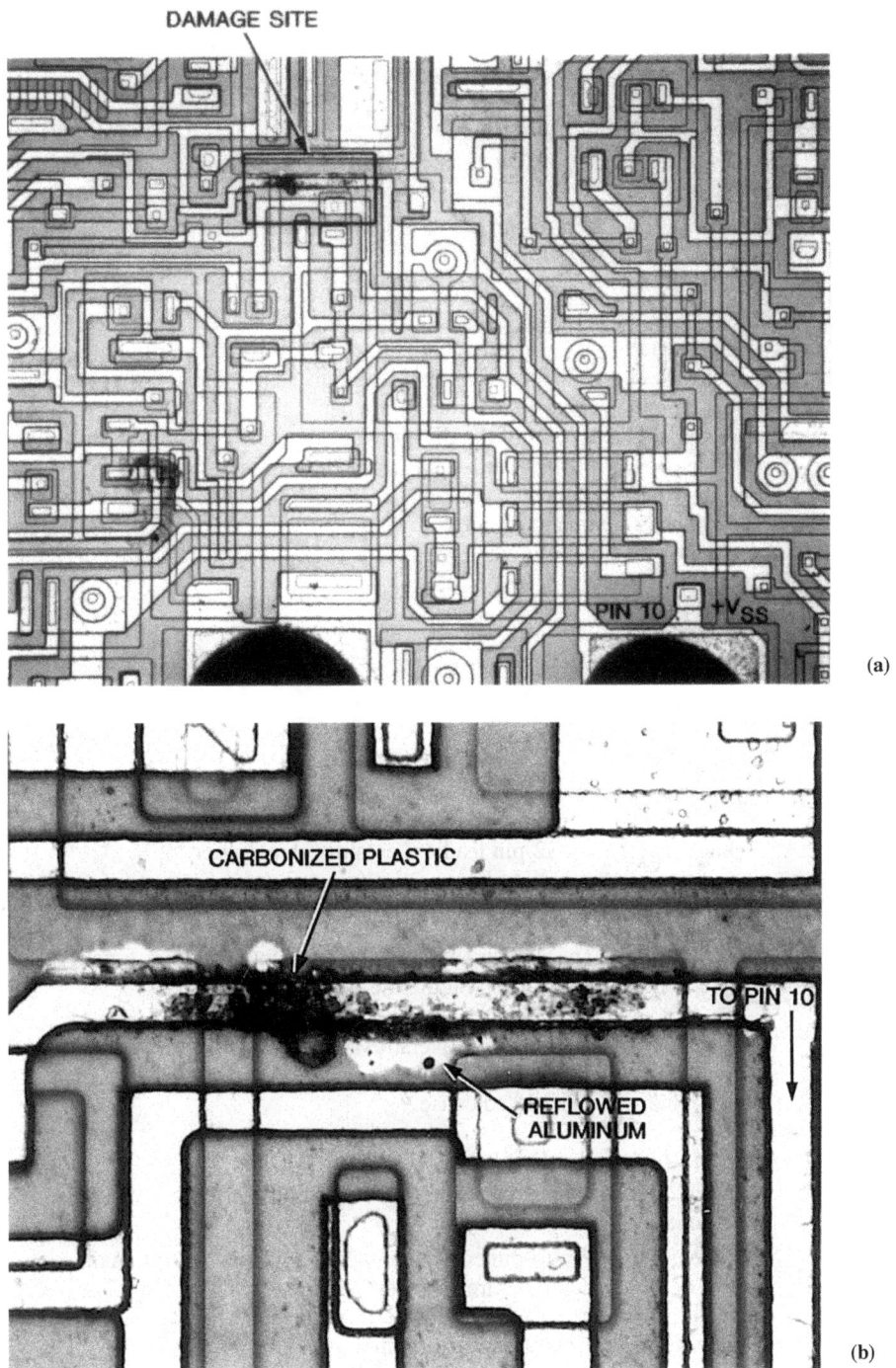

Figure 19.25 Optical views of the damage site: (a) 130× and (b) 640× *(courtesy of ICE, Inc.)*.

Failure Mode.   Pin 16 (output 1) was open to all pins.

Cause of Failure.   Electrical overstress.

Analysis Techniques Employed

External Inspection.   No evidence of mechanical or electrical damage was seen on the intact unit.

Electrical Test.   Pin-to-pin curve tracing showed that pin 16 was open to all pins. No other anomalies were noted.

Internal Inspection.   The device was delidded by mechanically separating the package halves. Low-power optical inspection of the package cavity revealed that the aluminum bonding wire of pin 16 was fused open. A ball of melted aluminum had accumulated at the end of the wire attached to the package pin. This indicated that the direction of current flow during the overstress was out of the package and that the potential on pin 16 was positive. Optical inspection of the die showed significant burning of the aluminum in the output transistor at pin 16. A large flashover of aluminum, alloyed into the silicon, was also present from the collector contact to the emitter. This indicated that the transistor conducted a large amount of current. It is suspected that pin 16 was connected to a low-impedance load that allowed excessive current to flow when this driver was enabled (see Fig. 19.26).

19.5.12   ESD/EOS Failure No. 3

Background

Device type:	32k × 8 EEPROM
Date code:	9114
Package type:	32-pin leadless ceramic chip carrier
Failure location:	circuit card testing

Failure Mode.   The device could not perform any program functions, either page write or byte write, to any location on the die.

Cause of Failure.   Electrical overstress damage was found in several of the capacitors within the internal charge pump circuit used for programming. An electrical transient on the V_{CC} line is the suspected cause.

Analysis Techniques Employed

External Inspection.   No evidence of mechanical or electrical damage was seen on the intact unit.

Electrical Test.   Pin-to-pin curve tracing did not show any anomalies. The parts were then sent back to the manufacturer for complete electrical testing. The results of these tests showed that the devices could not perform any program function. This indicted that the failure was in the programming circuitry.

Internal Inspection.   The metal lids of the samples were mechanically removed. Optical inspection of the intact dice did not show any evidence of damage.

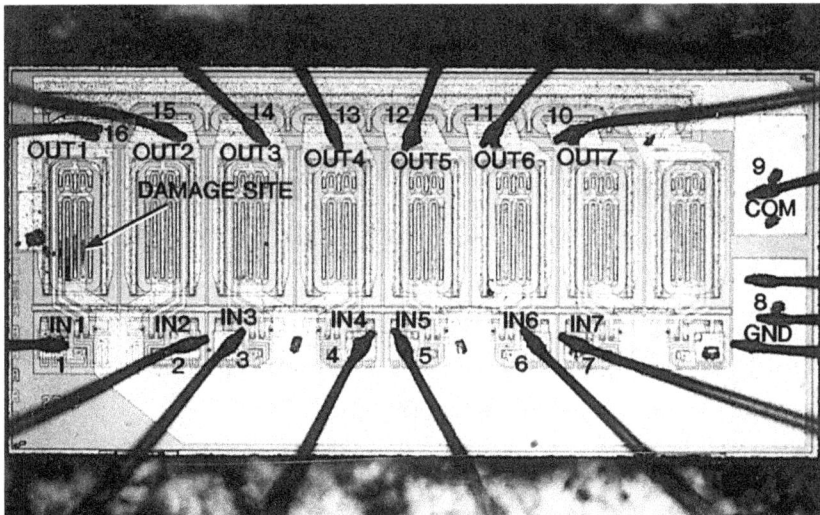

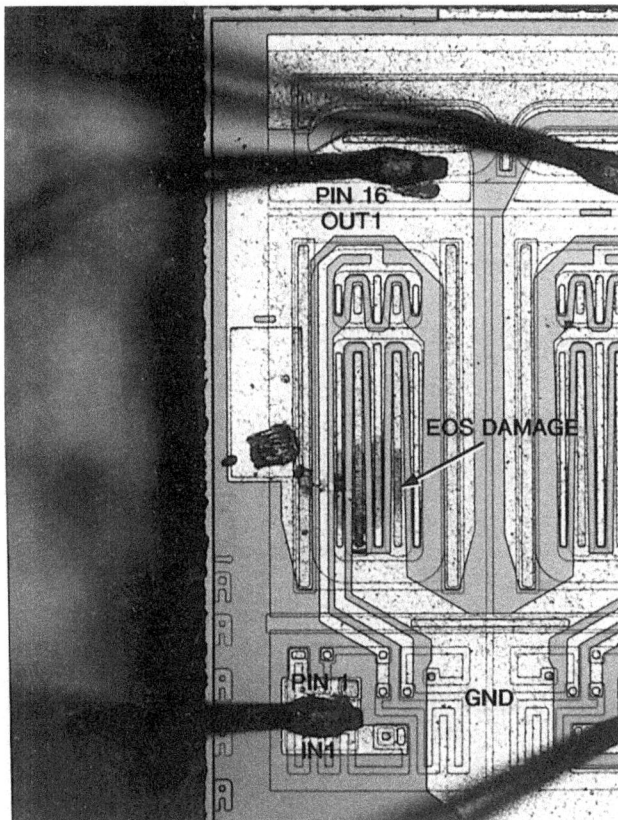

Figure 19.26 Optical views of EOS damage site: (a) whole die photo illustrating location of site, 30× and (b) 80× *(continues) (courtesy of ICE, Inc.)*.

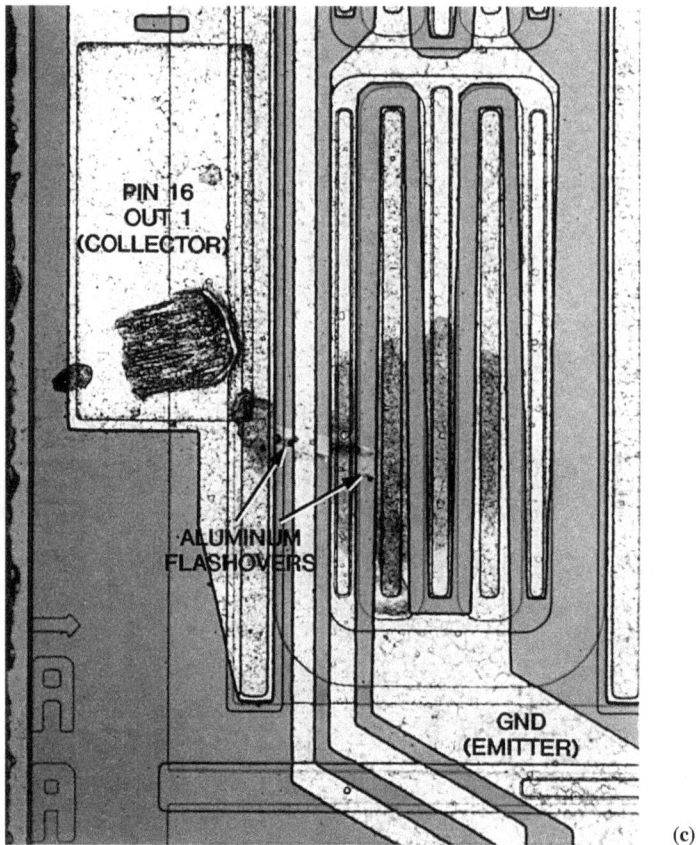

Figure 19.26 Optical views of EOS damage site: (c) 200× *(courtesy of ICE, Inc,),*

Microprobing.   The final passivation was thinned and the high-voltage programming circuits were probed. Probing of the charge pump circuit showed that the output voltage was approximately 10 V lower that the output on a good device. The units were delayered to the first-level metal and the capacitors within the charge pump were isolated and probed. It was determined that four to six of the capacitors in the charge pump circuits were resistively shorted.

SEM Inspection.   Inspection of the suspect capacitors after delayering to the polysilicon level revealed evidence of electrical overstress damage at the edges of the capacitors. This type of damage is typically caused by a high voltage transient (see Fig. 19.27).

19.5.13   ESD/EOS Failure No. 4

Background

Device type: N-channel power MOSFET

Date code: 9123

(a)

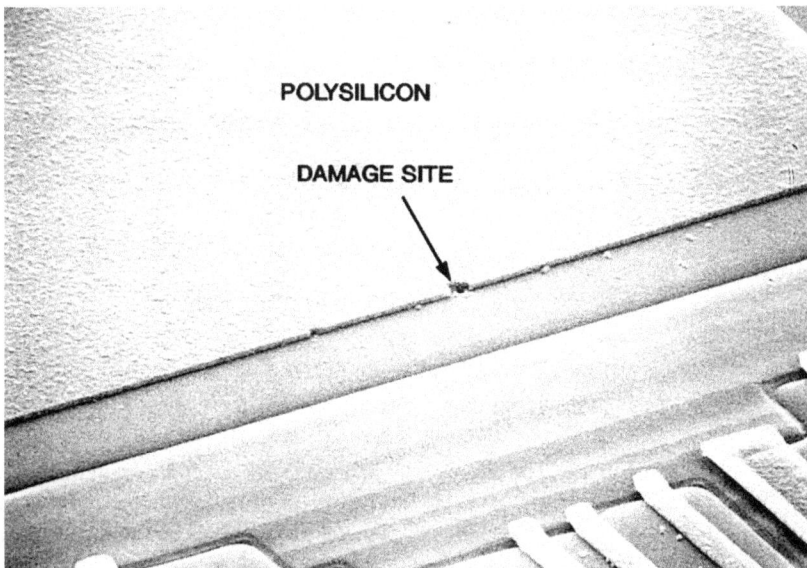

(b)

Figure 19.27 (a) Optical view of the charge pump, 65×, (b) SEM view illustrating EOS damage at capacitor edge, delayered to poly, 2,300× *(continues) (courtesy of ICE, Inc.)*.

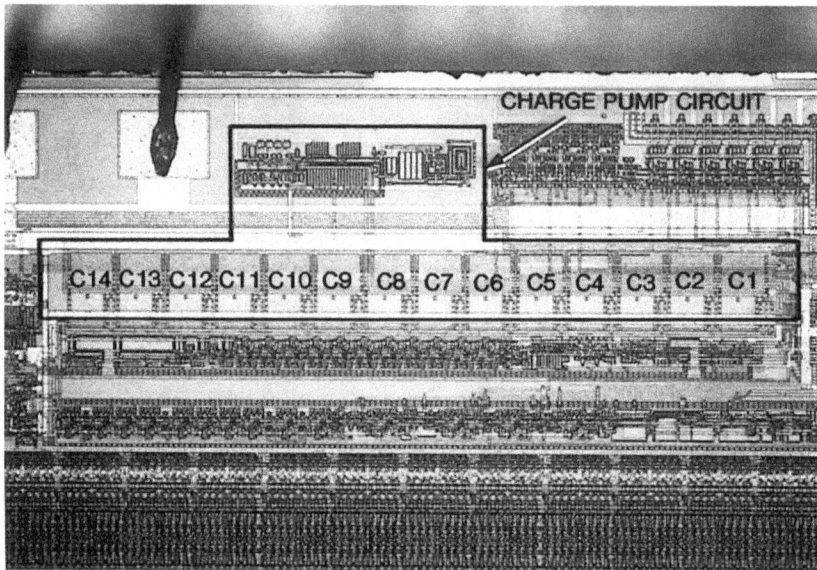

(c)

Figure 19.27   (c) Same site as in (b) at 2,800× *(courtesy of ICE, Inc.).*

Package type: TO-205AF metal can

Failure location: device qualification

Failure Mode.   A resistive short between the gate and the drain.

Cause of Failure.   The device failed due to electrical overstress between the gate and drain.

Analysis Techniques Employed

External Inspection.   No evidence of mechanical or electrical damage was seen on the intact unit.

X-ray.   X-ray of the good and failed samples did not reveal any anomalies.

Electrical Test.   Pin-to-pin curve tracing showed a resistive short between the gate and the drain.

Internal Inspection.   The device was delidded with a standard can opener. Optical inspection of the intact die did not reveal any evidence of physical or electrical damage.

Liquid Crystal.   Liquid crystal solution was applied to the surface of the die, and bias was applied between the gate and the drain. A single, well defined hot spot, indicating high, localized current flow, was noted.

Delayering.   The sample was selectively delayered and inspected. No evidence of damage was noted until removal of the polysilicon gate. SEM inspection of the failure site showed a circular thermal melt through the gate oxide, accompanied by a ragged trace of what appeared to be reflowed polysilicon and oxide. Traces such as this have been seen before and are believed to be due to partial melting of the poly gate due to the high current flow during the EOS event (see Fig. 19.28).

19.5.14 Contamination Failure

Background

Device type:	bipolar gate array
Date code:	9320
Package type:	132-pin ceramic, metal-lid flat-packs
Failure location:	pre–board-assembly inspection

Failure Mode.   Cracked/corroded leads.

Cause of Failure.   Chlorine contamination between the Kovar leads and the gold-over-nickel plating. Improper cleaning before plating is indicated.

Analysis Techniques Employed

External Inspection.   Showed damage to consist of cracks and blisters.

SEM Inspection.   Showed that cracks appeared to coincide with blistered areas and Kovar material had been attacked.

Cross Sectioning.   Confirmed that foreign material including chlorine was present between Kovar and plating.

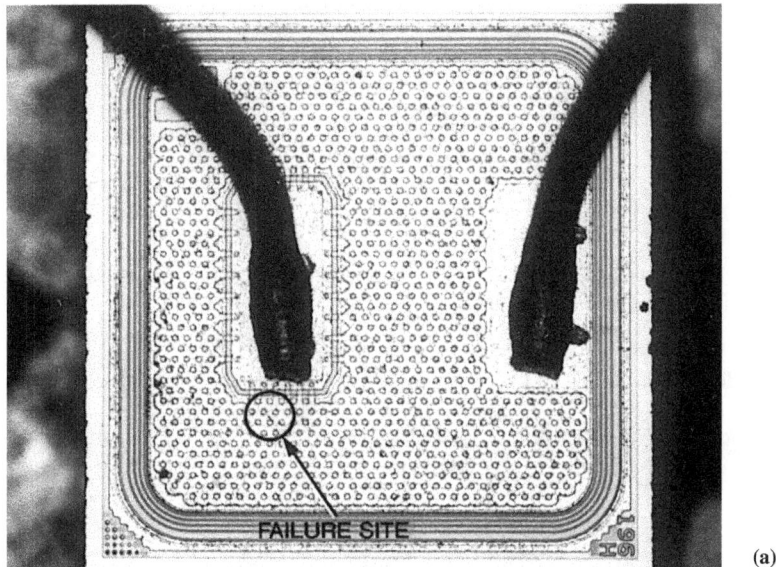

(a)

Figure 19.28   (a) Optical view of die illustrating location of failure, 50× *(continues) (courtesy of ICE, Inc.).*

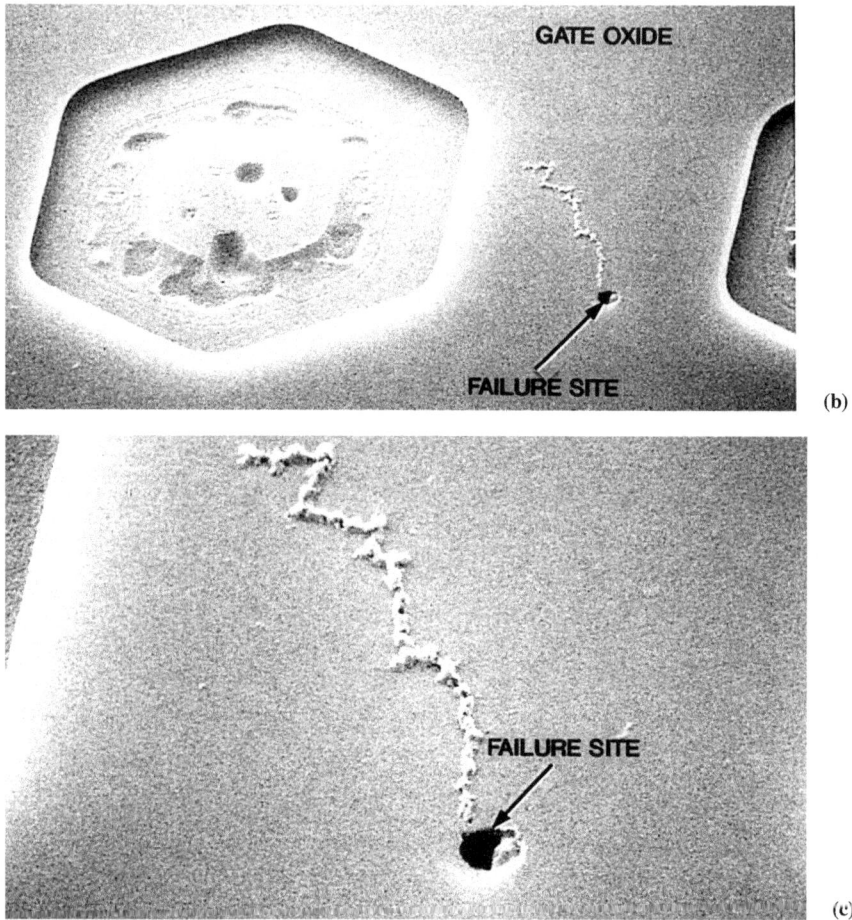

Figure 19.28 (b) SEM view of the EOS damage, 2,800×, and (c) 8,800× *(continues) (courtesy of ICE, Inc.).*

Material Analysis. EDX revealed large concentration of chlorine at blisters and cracks (see Figs. 19.29 and 19.30).

19.5.15 Wirebonding Failure

Background

Device type: HE 1612

Date code: 9317

Package type: 16-pin PDIP

Failure location: burn-in

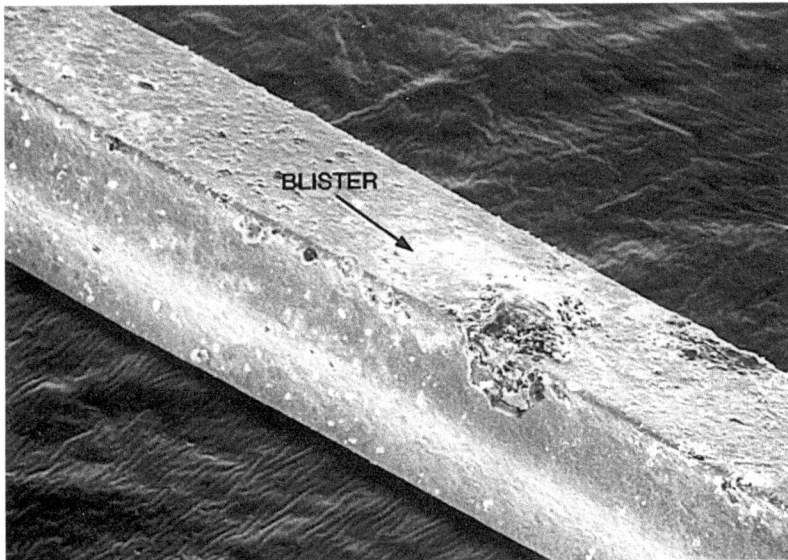

(a)

(b)

Figure 19.29 SEM views of a blister and crack defects: (a) Intact, 120×, (b) 75× *(continues) (courtesy of ICE, Inc.).*

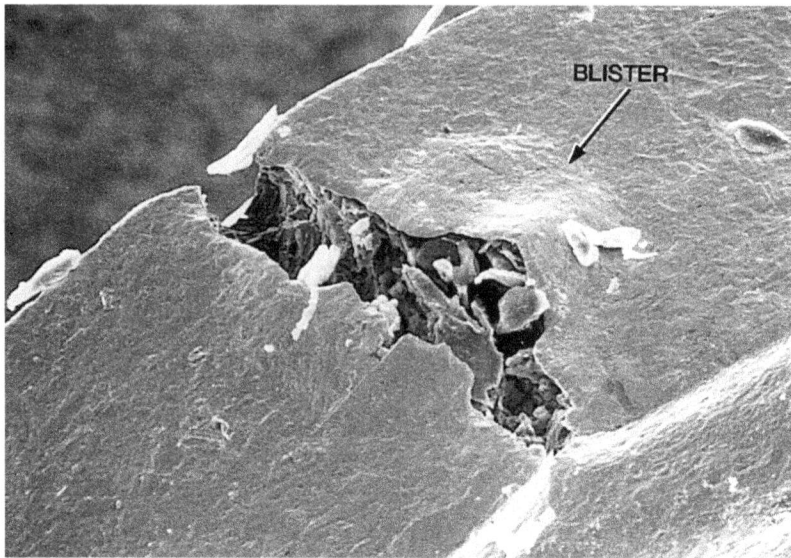

(c)

Figure 19.29 SEM views of a blister and crack defect: (c) 300× *(courtesy of ICE, Inc.)*.

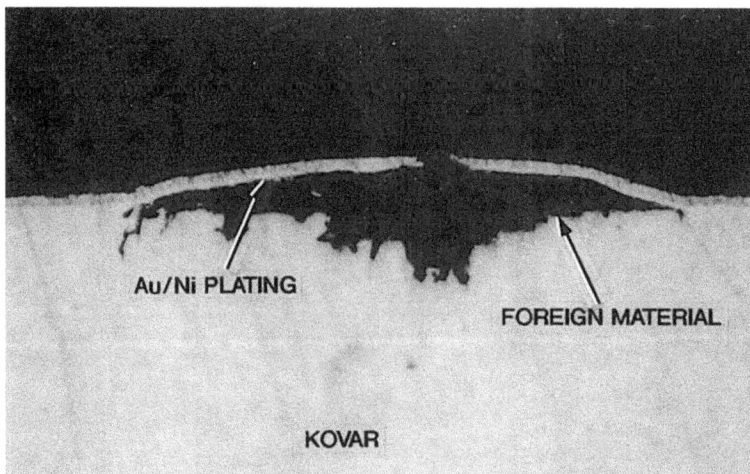

(a)

Figure 19.30 Cross-sectional views through a blister defect: (a) Optical view, 400× *(continues)* *(courtesy of ICE, Inc.)*.

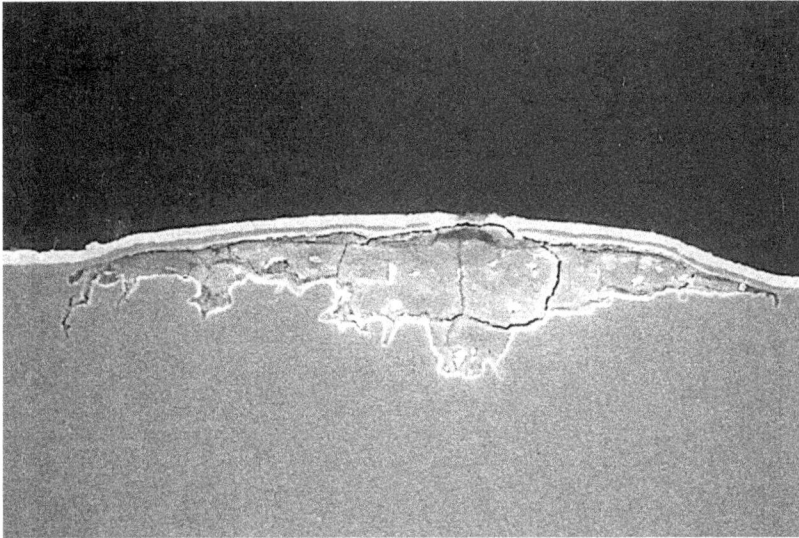

(b)

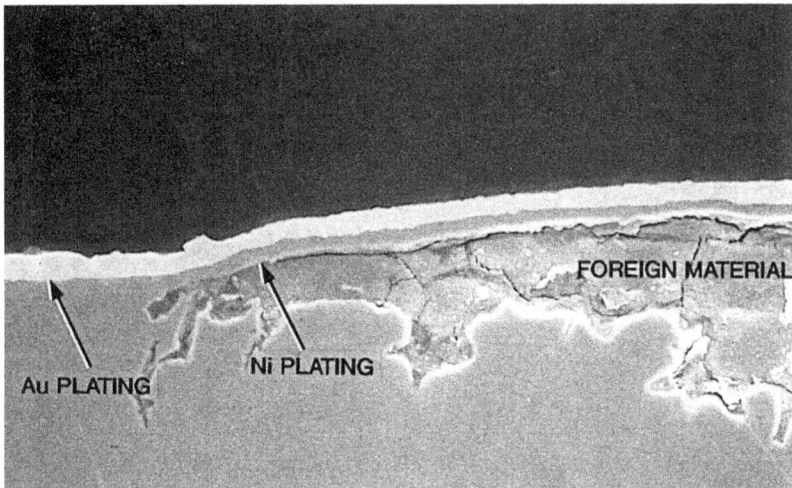

(c)

Figure 19.30   Cross-sectional views through a blister defect: (b) SEM view, 500×, (c) SEM view, 1000× *(courtesy of ICE, Inc.).*

Failure Mode.   Open pin 4.

Cause of Failure.   Improper wirebonding cratered the oxide and silicon, causing an intermittent connection.

Analysis Techniques Employed

External Inspection.   Normal.

X-ray.   Normal.

Electrical Test.   Showed intermittent continuity at pin 4 depending on temperature (25° C to 60° C).

SEM Inspection.   Revealed bond at pin 4 lifted from pad and cratered oxide and silicon underneath (see Fig. 19.31).

19.6 ACKNOWLEDGMENTS

The author would like to thank Mr. Philip Young and Mr. Chris MacMartin, both of DPA Labs Inc., for providing general guidance in the development of the text for this chapter and the photographs that illustrate many of the key points. I would also like to thank Mr. Greg Spawn, of Integrated Circuit Engineering (ICE) Corp., for providing the case studies. This chapter could not have been developed without the excellent contributions of these failure analysis professionals.

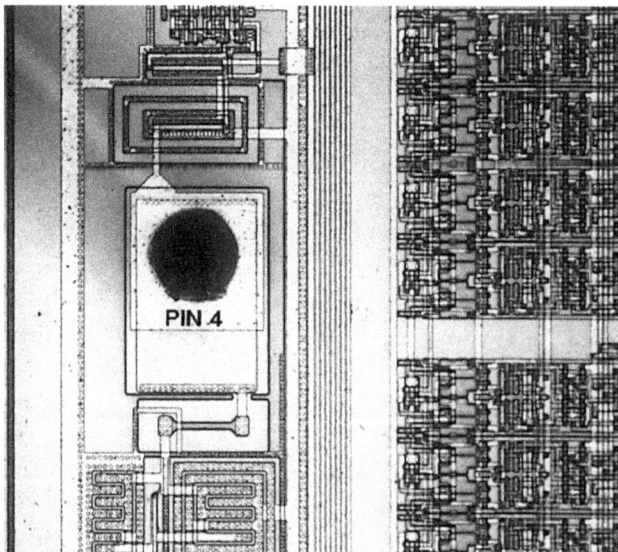

(a)

Figure 19.31   (a) Optical view of the lifted ball bond at the failed pad, 130× *(continues) (courtesy of ICE, Inc.)*.

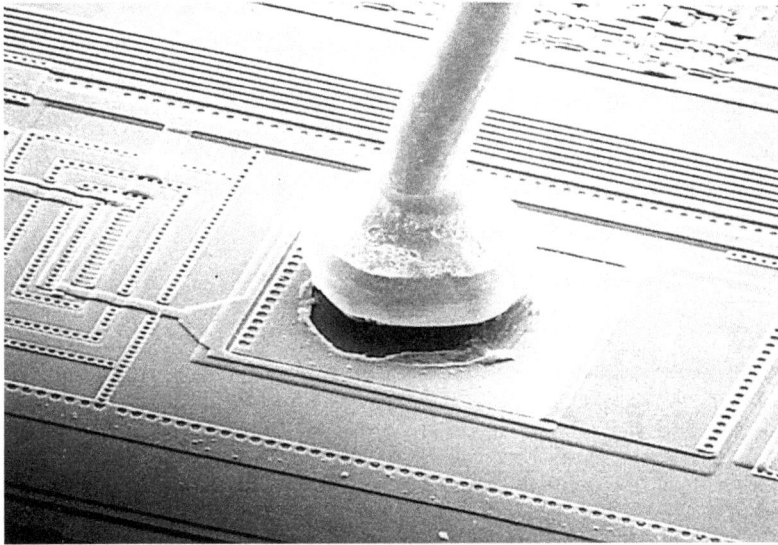

(b)

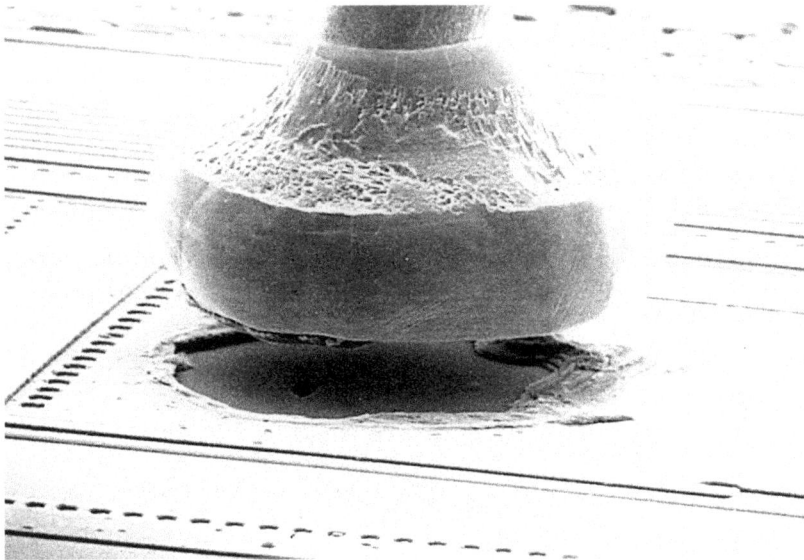

(c)

Figure 19.31   (b) SEM view of the same area shown in (a), 340×, and (c) SEM view with increased magnification, 775× *(courtesy of ICE, Inc.).*

19.7   REFERENCES

1. Private communication with DPA Labs

2. David, Shores, and Friedlander. A Critical Evaluation of RGA Testing, *IEEE Trans. on Components, Hybrids, and Manufacturing Technology* 12(4), December 1989.

3. Private communication with DPA Labs.

4. Case studies provided by ICE Corp.

19.8   RECOMMENDED READINGS

Doyle, E., Jr., and B. Morris, eds. *Microelectronics Failure Analysis Techniques.* Rome Air Development Center. No publication date. Contract Nos.: F30602-78-C0339, F30602-78-C-0281.

Doyle, E., Jr., and B. Morris, eds. *Gallium Arsenide.* Rome Air Development Center. No publication date.

Devaney, J., G. Hill, and R. Seippel. *Failure Analysis Mechanisms, Techniques, & Photo Atlas.: A guide to the performance and understanding of failure analysis.* Failure Recognition & Training Services, Inc., 1988.

Dicken, H., ed. *Physics of Semiconductor Failures*, 3d ed. Scottsdale, AZ: DM Data, Inc., 1989.

Lee, T., ed. *ISTFA Microelectronic Failure Analysis Desk Reference, Supp. 2.* ASM International, 1991.

Van Zant, P. *Microchip Fabrication.* San Jose, CA: Semiconductor Services, 1986.

Electronic Materials Handbook. Volume 1, Packaging. Metals Park, OH: ASM International, 1989.

MIL-STD-883B, Test Methods and Procedures for Microelectronics. Washington, DC: U.S. Government Printing Office.

May, J.T. Limiting Phenomena in Power Transistors and the Interpretation of EOS Damage, in *Microelectronic Failure Analysis Desk Reference*, 2d ed. 1990 ISTFA Conference.

May, J.T. Electrical Overstress in ICs. *Microelectronics Desk Reference Supplement 2.* ASM/ISTFA, 1991.

Survey of Microbeam Analysis Applications for IC Reliability, in *Microelectronic Failure Analysis Desk Reference*, 2d ed. ISTFA Conference, 1990.

Lee, T.W. Mechanical and Chemical Decapsulation, in *ISTFA/90 Microelectronic Failure Analysis Desk Reference*, 2d ed., p. 35. Materials Park, OH: ASM International, 1991.

Beall, J. Plasma Etching, in *ISTFA/90 Microelectronic Failure Analysis Desk Reference*, 2d ed., p. 47. Materials Park, OH: ASM International, 1991.

Keifer, D.S. Reactive Ion Etch Recipes For Failure Analysis, in ISTFA/90 Microelectronic Failure Analysis Desk Reference, 2d ed., p. 55. Materials Park, OH: ASM International, 1991.

Beall, J.R., W. R. Echols, and D.D. Wilson. *SEM Analysis Techniques for LSI Microcircuits.* RADC report no. RADC-TR-80-250. Martin Marietta Corporation.

ASM Handbook, Volume 9, Metallography and Microstructures. Materials Park, OH: ASM International, 1985.

Harper, C.A., ed. *Electronic Packaging and Interconnection Handbook.* New York: McGraw-Hill, 1991.

CHAPTER 20
POWER AND HIGH-VOLTAGE CONSIDERATIONS

Perry L. Martin
National Technical Systems (NTS)
William G. Dunbar
Consultant

20.1 INTRODUCTION

In earlier chapters of this handbook, the reader was introduced to some of the various stresses that drive electronic and electrical equipment to failure, and by now the reader is familiar with the many various techniques for ferreting out information about *low-voltage* failures that often have insignificant or no visual trace. Failure of power and high-voltage equipment almost often leaves evidence, consisting of a large blackened area, that can appear intimidating to the inexperienced investigator charged with finding the reason for failure. The authors have read many reports that simply announce that the failure occurred either by an arc between adjacent parts or between a part and ground. Can the objective of the failure analysis, which is to prevent failures, be achieved with this type of analysis?

Failure analysis requires a multidisciplinary approach, and the failure analysis techniques that should be employed will depend on the skill and training of the assigned personnel, the equipment available to them for the analysis, and when in the product life cycle the analysis is done. It must be realized that each piece of power or high-voltage equipment is a composite of many electrical and structural parts. Each part has an aging characteristic that is unique to the design and utility of the part and will fail based on a schedule determined by the design, materials, parts, workmanship, environment, and maintenance program associated with the equipment.

This chapter presents some of the information necessary for a successful failure analysis of power and high-voltage equipment, which must perform reliably under conditions of high current density, large voltage gradients, large thermal excursions, and other stress factors. Common deficiencies and failure indicators are presented. Obviously not all types of electrical power equipment can be referenced in this handbook and so these factors will be illustrated through the equipment types most familiar to the authors.

20.2 BACKGROUND

20.2.1 Failure History

Electronic failures are not new or allocated to one specific type of equipment. When failures occur during the design and development cycle, failure analyses are performed to identify the problems. The application of the solutions developed will result in design and manufacturing improvements and longer equipment life. It is hoped that the majority of these failures will surface during design verification and be solved prior to production; however, there are a multitude of failure mechanisms and failure modes inadvertently designed and/or built into the typical power or high-voltage package. If testing or screening finds these defects, the only effect is on the manufacturing yield. If the defects are not found, they will be fielded and will show up as field failures if the unit remains in service long enough.

It can be concluded that electronic system and component failures usually fall into one or more of the categories: of (a) design, (b) packaging, (c) materials, and (d) processes. In the experience of the authors, engineering design and packaging together account for approximately half of the failures. Manufacturing materials and processes account for the remainder of the failures. These categories may be subdivided as shown in Tables 20.1 and 20.2.

Table 20.1 Failure Analysis Distribution Summary

Item	Percentage	Item	Percentage
Design	25	Materials	15
Packaging	24	Processes	36

Table 20.2 Summary of Majority Failures for Each Category

Category	Failure mechanisms	Category	Failure mechanisms
Design	High electrical stresses Insufficient funding and design time Parts selection and screening Selection of void-prone encapsulants	Materials	Parts and potting materials incompatibility Improper materials storage Reversion Poor heat transfer materials
Packaging	Poor heat rejection Electromagnetic coupling Location of structural members Screw and bolt threads in high field areas Spacing and flashover	Processes	Cleanliness Voids and materials cracking Improper curing techniques Wire and terminal breakage Workmanship and human error

20.2.2 Commercial Electronics

A commercial electronic package must be developed as a cost-effective product. The design and development, parts selection, and manufacturing philosophy must be adjusted to keep the package visually acceptable to the user, with weight and volume within the handling acceptability of the potential product purchaser. However, the packaging design must include excellent quality parts and workmanship to result in long life and high-quality electrical performance. This philosophy may result in larger printed circuit boards, selection of parts with known high quality, least failure history, and a manufacturing facility known for high-quality workmanship. As a result, changes in product performance, size, and visual appearance are limited to yearly or multiyear changes to keep the manufacturing costs acceptable to customer satisfaction.

Commercial electronic hardware manufacturers have found that no one design or material is adequate for every application but that a large spectrum of applications can use a few standardized electrical parts, modules, and potting materials. Thus, a spectrum of applications covered by a few standardized parts and potting materials limits the number of processes required to successfully produce acceptable products for several product lines or applications. Such a standardization program will decrease time to market and prove very cost-effective, in addition to providing high-reliability benefits.

20.2.3 Aerospace Electronics

Aerospace electronic packages must be lightweight, have minimal volume by the use of high-density packaging design practices, and be cost effective to satisfy the system design requirements. Unfortunately, high-density packaging requires very strict coordination between the many engineering skills required for the design, development, and manufacture of high-voltage and high-power equipment. Minute flaws during the design or manufacture of an electronic module may result in low yield, which increases the manufacturing costs. Lack of attention to electrical and mechanical field stresses and the chemical properties of encapsulating materials may result in higher production yield, but the end product may have aging limitations or higher failure rates. In either case, the system costs are high. To eliminate problems associated with poor yield and high failure rate concerns, it is necessary to select designs, materials, parts, and manufacturing procedures that result in an end product that will operate within the environmental, electrical, and mechanical constraints of the system design specification. This implies that electrical and mechanical materials and processes must be developed, along with high-quality workmanship, to ensure high yield and long life.

A high-voltage power supply or power converter is a critical part of every high-voltage electronic subsystem. The power converter may have an input voltage ranging from 6 Vdc to 270 Vac, with outputs of a few volts to tens of thousands of volts, and power outputs ranging from a few watts to several megawatts. Several applications using high-voltage power supplies are shown in Fig. 20.1, with typical power output ranges shown in Fig. 20.2. The need for upgraded performance and, more importantly, enhanced reliability has been continually identified as an area of concern as the demand continues for higher-power electronic equipment in smaller packages. Thus, highly skilled technical personnel must be motivated to develop the innovative high-density packaging required for most aircraft electronics and space vehicle applications.[1]

A review of many aerospace field failure reports indicate that high-voltage electronic systems are frequent replacement items. This is especially true where high-density packaging is essential in moderately controlled thermal and mechanical environments. A list of the most common causes of electronic failures is shown in Table 20.2. In addition, some electronic units are susceptible to damage by improper integration with the electrical power system or ancillary electronic equipment. Even when each electronic unit is protected from short circuits, overvoltage or undervoltage transients, environmental conditions, and improperly identified input and output cabling, they can be severely damaged if improperly connected to interfacing components or subsystems.[2]

20.2.4 Failure Investigation Preliminaries

A failure analysis should start with an examination of the historical record of the failed unit. If available, review both the acceptance and qualification data on the prototype and

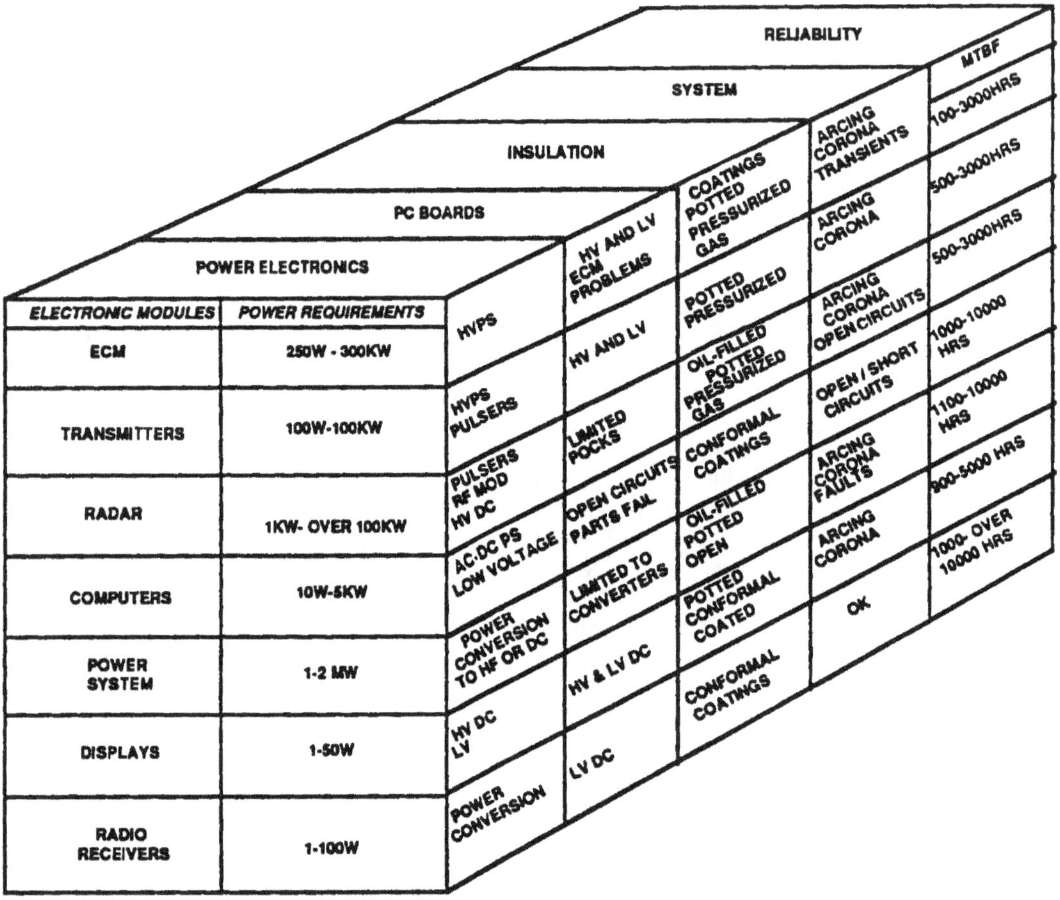

Figure 20.1 System Power Requirements

first production item and the acceptance test data on production items. Many times during production, circuits or parts are modified to improve electrical performance. Although these changes will not affect the original qualification performance, they may affect the high-voltage circuit compatibility. Thus, accurate production modification records are a major source for historical data.

Production assembly drawings and repair manuals are an important source of information for the failure analyst. This information will detail the specified routing of wiring and should include termination information such as the proper torque for bolt terminals. Common failure problems for power equipment are the improper routing of wires and loose terminals, as can be seen in Fig. 20.3.

Data for circuit evaluation should include the results from both electrical and environmental property tests. Some important electrical property tests for electrical insulation are listed in Table 20.3, the environmental tests are listed in Table 20.4. Additional data that should be evaluated includes dielectric withstand voltage (DWV), resistivity, corona, and the effects of altitude and temperature.[3] Note whether just the parts or the entire circuit is

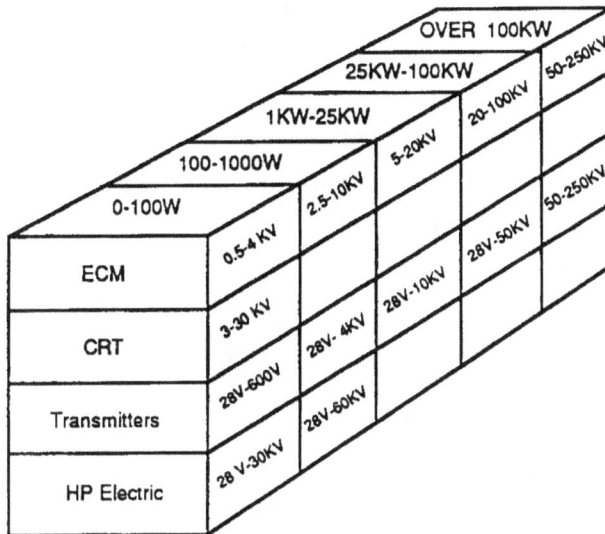

Figure 20.2   Applications for HV power supplies.

to be tested. If the circuit is tested, note whether the procedure requires full voltage testing or a voltage or frequency other than that specified for the circuit. Furthermore, most test operators will only record *pass* or *fail* results. *Pass* often may not indicate the various test parameters used—only that the test article operated before and after the test. Quite often, the failure analysis will require a refinement of the original test or the use of a specific new test procedure.

Table 20.3   Tests of Electrical Properties of Insulation

Tested property	Test condition	Evaluated	Test Method
Dielectric strength	dc/ac, 1/4-in electrodes	When received and following environmental stress	ASTM D-149-61 (modified)
Tracking	dc/ac	Following environmental stress	ASTM D-495 or D-2302
Dielectric constant	1 kHz	When received	ASTM D-150-59T
Dissipation factor	1 kHz	When received	ASTM D-150-59T
Volume resistivity	125 V	When received and following environmental stress	ASTM D-257-61 (modified)
Surface resistivity	dc	When received and following environmental stress	ASTM D-257-61 (modified)
Insulation resistance	dc	Following environmental stress	Based on 0.05 mfd wound parallel-plate capacitor
Life	dc/ac	Vacuum (plasma)	ASTM D 2304-64T (modified)

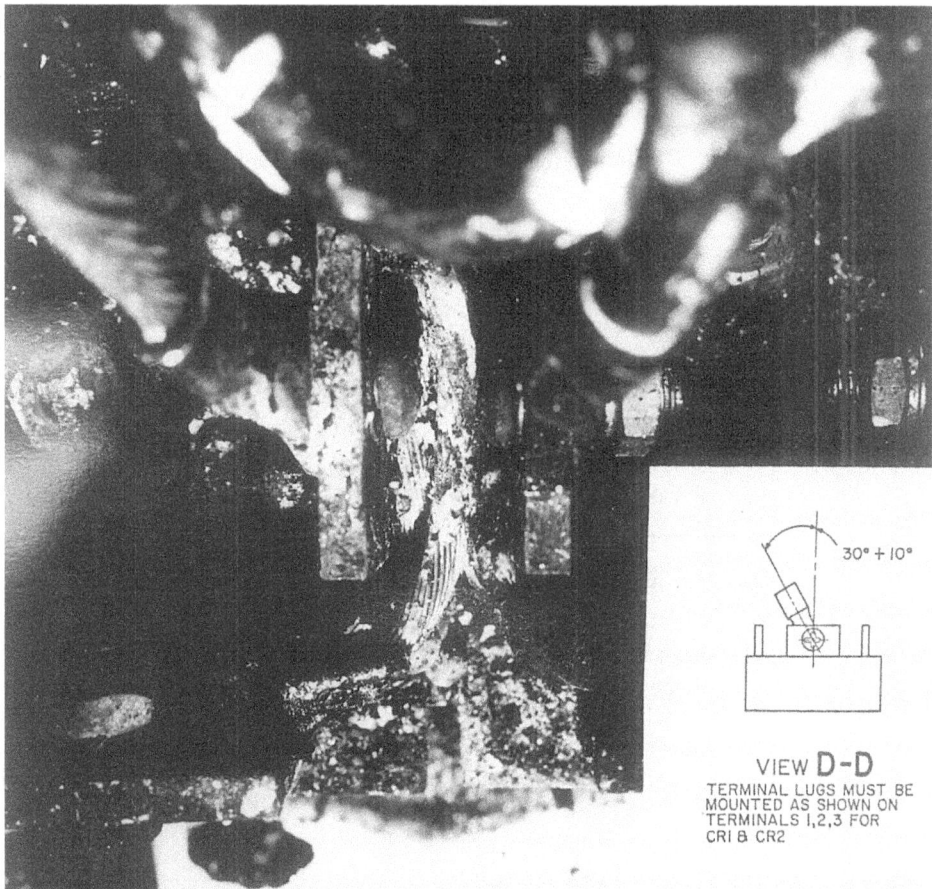

Figure 20.3 Power supply failure due to incorrect routing. Inset shows correct orientation of cables from the terminals.

Finally, the preliminary investigation should include the average operating time for identical devices, the operating conditions or environment, the qualifications of the operating personnel, and what they observed at the time of failure. With this information, plus accurate historical and test data, the investigator can do a reasonably thorough failure analysis. The authors wish to note that, in their collective experience, they have found that even for well documented, critical military systems, there are often few operational performance records and little test data available for fielded units. In these cases, the experience of the investigator, repair personnel, and others must be substituted for the material mentioned above.

20.3   FAILURE FUNDAMENTALS

This section will serve to introduce the reader to some of the fundamental concepts relating to the failure of power and high-voltage equipment and the associated terminology.

Table 20.4 Materials Properties and Test Methods for Potting Materials

Properties	Test method
Dielectric strength	ASTM D-149-87
Dielectric constant	ASTM D-150-87
Dissipation factor	ASTM D-150-87
Volume/surface resistivity	ASTM D-257-78
Dry arc resistance	ASTM D-495-84
Viscosity	ASTM D-1824-83
Pot life	Correlate measured viscosity or flow rate
Coefficient of thermal expansion	ASTM E-831-86
Coefficient of thermal conductivity	ASTM D-1674-67
Brittleness temperature	ASTM D-746-79
Tensile strength	ASTM D-638-84
Lap shear strength	ASTM D-429-81
Peel strength	ASTM D-429-81
Tear strength	ASTM D-624-86
Thermal shock resistance (cracking)	ASTM D-1674-67
Thermal aging	ASTM D-2304-85
Cure shrinkage	ASTM D-1917-84
Moisture absorption	MIL-I-16923G
Specific gravity	MIL-I-16923G
Flammability	MIL-I-16923G

20.3.1 Corona

Corona is defined as a luminous discharge due to ionization of the gas surrounding a conductor around which exists a voltage gradient exceeding a certain critical value. Corona may also be visible in the dielectric of an insulation system. This is caused by an electric field and is characterized by the rapid development of an ionized channel that does not completely bridge the insulation between electrodes. Corona can be continuous or intermittent. It is not a material property but is related to the system, including electrodes. Corona should be differentiated from partial or micro discharges.[4]

A *partial discharge* is an electric discharge that only partially bridges the insulation between conductors when the voltage stress exceeds a critical value. These partial discharges may or may not occur adjacent to a conductor. Partial discharge is often referred to as corona, but the term is preferably reserved for localized discharges in cases around a conductor, bare or insulated, remote from any other solid insulation. A partial discharge pulse is a voltage or current pulse that occurs in some designated location in the test circuit as a result of a partial or micro discharge.[5–7]

Corona and partial discharges usually occur due to insulation imperfection in insulation systems, such as entrapped gas voids and cracks within the insulation, and insufficient space between an insulated conductor and ground or other insulated parts. As examples of defects, a transformer may have small voids within the insulation between turns, between the turns and bobbin, between the active turns and the magnetic core, or between the turns of the two coils.

20.3.2 Paschen's Law

The breakdown voltage of a uniform-field gap in a gas can be plotted to relate the voltage to the product of the gas pressure times the gap length. This is known as the *Paschen's law curve*.[8,9] The law may be written in the general form:

$$V = f(pd) \tag{14.1}$$

where p = gas density
     d = distance between parallel plates.

In words, Paschen's law states:

> As gas pressure is increased from standard temperature and pressure, the voltage break-down is increased, because at higher densities the molecules are packed closer, and a higher electric field is required to accelerate the electrons to ionizing energy within the mean free path. The voltage breakdown decreases as gas density is decreased from standard pressure and temperature because the longer mean free path permits the electrons to gain more energy prior to collision. As density is further decreased, the voltage decreases until a minimum is reached.

As density is further reduced to values less than the Paschen law minimum, the voltage breakdown rises steeply because the spacing between gas molecules becomes so large that although every electron collision produces ionization, it is difficult to achieve enough ionizations to sustain the chain reaction. Finally, the pressure becomes so low that the average electron travels from one electrode to the other without colliding with a molecule. This is why the minimum breakdown voltage varies with gas density and spacing. Examples of Paschen-law curves for several gases are shown in Fig. 20.4.

The pressure corresponding to minimum breakdown depends on the spacing of the electrodes; for a 1-cm spacing at room temperature, this pressure occurs at approximately

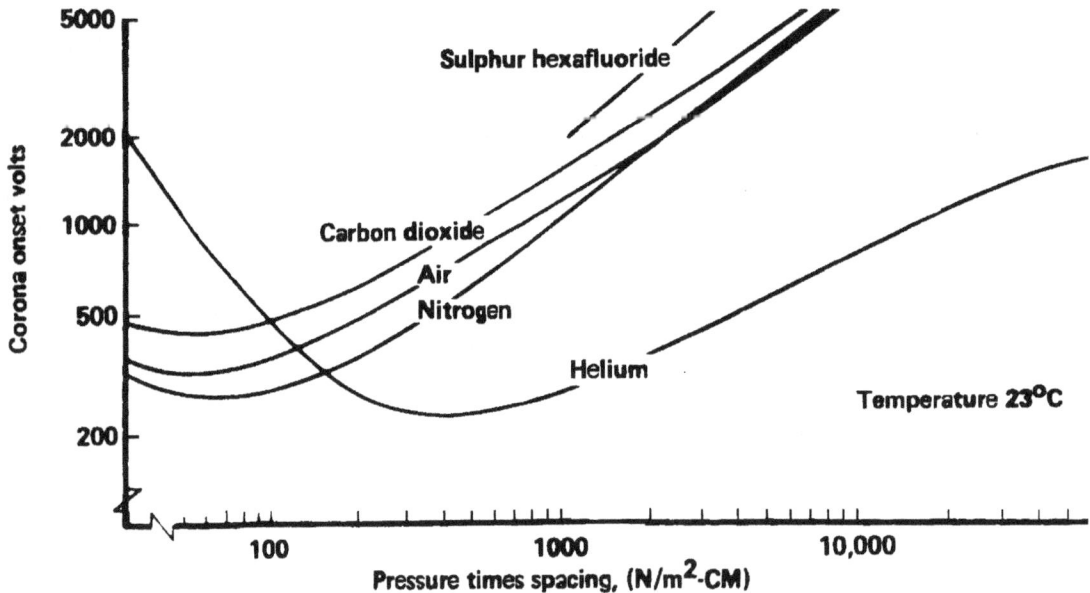

Figure 20.4   Direct current breakdown voltage between parallel plates.

100 pascals (Pa). One pascal is equal to one newton per square meter (N/m^2) or 7.5×10^{-3} torr. A representative minimum for air is 326 Vdc. For a contact spacing of 1 cm at standard atmospheric conditions, the breakdown voltage of air is 31 kV.

Voltage breakdown, under normal conditions, has no sharply defined starting voltage, because initiation depends on an external source of ionization. There is generally a finite time delay between the application of voltage and breakdown. This time delay varies statistically and is a function of the difference between the applied voltage and the "critical" voltage. Ultraviolet light and higher-energy radiation will reduce the time delay considerably. Paschen-law curves for nonuniform fields become difficult to predict, because the effective gap length is not easily defined.

20.3.3 Treeing

The best example of a laboratory treeing test was first demonstrated by D. W. Kitchin in 1954.[10] In the experiment, a sharpened sewing needle was partially embedded in a block of polyethylene. The base of the 6.5 cm thick by 2.5 cm square block of polyethylene was placed on a metal plate. The needle was then connected to an electrical high-voltage source. The voltage was set at a value to initiate a "tree" at the point of the needle in one hour. This was named the *characteristic voltage.*

Over the years, investigators have argued about the mechanism of starting an electrical tree. A probable cause is the bombardment of the surface of the insulation adjacent to a high-electrical-stress field by high-speed electrons. The source of the electrons may be high-intensity ionization in a void, the field emission from the surface of a point, or the emission from a dirt particle embedded in a nonhomogeneous insulation system. An example of treeing is shown in Fig. 20.5.

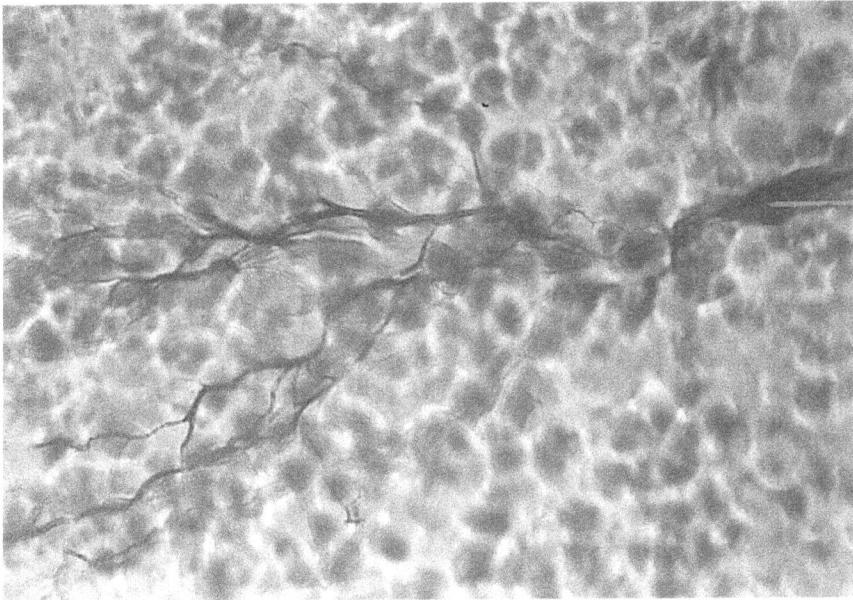

Figure 20.5 Example of treeing *(courtesy of Mississippi State University High Voltage Laboratory).*

Treeing is usually associated with a solid organic insulation. Therefore, the tree initiates and spreads along the molecular interfaces rather than atomically.[11] Thus, the tree branches may disperse throughout the insulation, eventually resulting in electrical failure. Experimentally, it has been determined that the tree is often initiated by a large electrical pulse exceeding 200 picocoulombs (pC). Furthermore, partial discharge pulses are recorded as tree branches extending from the initiating point toward the base plate. The branches may defy gravity and move upward as well as downward in their continual path of least resistance toward the ground plane.[12,13,14]

There are several types of treeing. Much activity has been given to water tree growth in power cables in recent years. In addition, the chemical inclusions of sulfur and other debris have been assessed as initiators for treeing.[15, 16,17]

20.3.4   Surface Flashover

The term *surface flashover* means that the surface of a solid insulation, between two adjacent electrodes, has become so conductive that it can sustain a flow of substantial current from one high-voltage electrode to the other. Current flowing across a surface of an insulator, especially when slightly wetted and containing a conductive contaminant, may produce enough heat to generate a track of carbon, which becomes a conductive path tending to reduce the capability of the insulator to resist the voltage.[18] With some materials, the surface erodes, but no *track* is produced.[19] Some fillers effectively reduce the tracking tendency of organic materials. Eroding materials such as acrylics do not require filler protection. Obviously, no tracking is the ideal requirement for an organic insulator. Tracking can also be controlled by reducing the volts per millimeter stress on the surface. Petticoat insulation configurations lengthen the surface creepage path to reduce stresses tending to cause tracking.[20]

When new, cycloaliphatic epoxy inorganic filler is applied to the surface of a laminate, the finished product will withstand higher voltage stress than porcelain. Surface erosion and exposure to UV radiation will degrade the epoxy to where it is inferior to the porcelain. In one application having a glass-cloth epoxy-based laminate coated with cycloaliphatic epoxy, the surface was stressed at a voltage exceeding 45 kV/cm impulse and 35 kV/cm dc. However, the atmosphere was sulfur hexafluoride, and such a high-voltage stress is not recommended for long-life equipment.

As an example, the flashover voltage was measured between 1.9-cm dia. washers on an uncoated glass-epoxy laminate as illustrated in Fig. 20.6. The washers were spaced 1 to

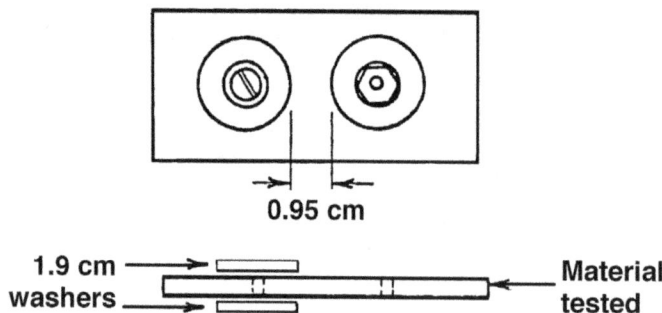

Figure 20.6   Flashover fixture.

4 cm apart. Shown in Fig. 20.7 is the flashover initiation voltage as a function of spacing at three frequencies. The impulse and steady-state flashover voltage stress is shown for the same configuration in Table 20.5.[21]

It is both interesting and useful to determine the relationship between flashover strength at 25° C and that which would prevail at some other temperature (T). For gaseous breakdown in a uniform field, this relationship involves the ratio of the gas densities at the two temperatures. To test this relationship, it is necessary only to multiply the 25° C value by the factor $(25 + 273)/(T + 273)$, which is the inverse ratio of the absolute temperatures involved. This ratio is part of the well known air density correction factor, which is commonly used in spark-gap measurements over a considerable range of density and gap lengths. The broken lines in Fig. 20.8 show the values obtained when this factor is applied to the 25° C flashover values.

Most materials have lower flashover at higher frequencies. The example in Fig. 20.9 illustrates the magnitude of change. Likewise, high dielectric constant materials have much lower resistance to surface voltage creep than the low dielectric constant materials. Figure 20.10 illustrates the advantage in selecting the correct dielectric constant insulation to prevent failures. The "breakdown factor" in the illustration represents the results of many measurements, showing how a decreasing flashover voltage can be expected across a dielectric when insulations with progressively higher dielectric constants are tested.[22]

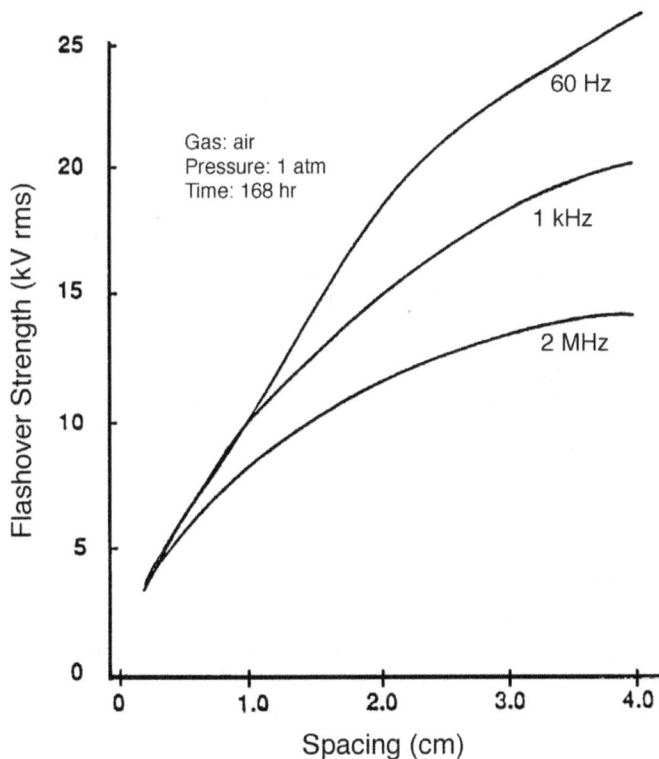

Figure 20.7   Effect of spacing on the initial values of strength for the fixture shown in Fig. 20.6.

Table 20.5   Variations in Flashover Strength with Environment and Electrical Stress

Flashover strength comparison		
Glass cloth		
Test	Maximum (kV)	Minimum (kV)
60 Hz	15.3	13.4
DC positive	15.5	13.9
DC negative	15.9	14.9
Impulse positive	17.9	16.2
Impulse negative	22.0	16.0

Environment effect on flashover, 60 Hz

Glass cloth (kV)

	1 hour	1 day	1 week	1 month	6 months
85° C (dry)	7.8				
125° C (dry)	7.0				
25° C (100 RH)	9.5	3.6^*	2.5^*	2.5^*	
50° C (100 RH)	8.8	2.4^*	2.4^*	2.4^*	

Polystyrene (kV)

25° C (100 RH)	10.7				5.5
50° C (100 RH)	10.3			6.2	5.1

Flashover safety factors

Humidity	A safety factor of 2 is recommended where humidity can go to 100 percent with condensation.
Contamination	Where dirt, dust, and residual ions from plating baths and so on may contaminate the surface between electrodes, the flashover (creepage) path should be two or three times the minimum air spacing.

* Internal flashover

20.4 POWER AND HIGH-VOLTAGE PACKAGING

Each designer has developed unique methods and techniques for designing compact, light-weight equipment. Some critical design and packaging problems include heat transfer, voltage stress, electromagnetic coupling between circuits, and mechanical integrity. The failure analyst must be aware that, as system power demand increases, the electrical and electronic devices and modules of less than one kilowatt input power may be packaged using either modular construction, conformally coated parts and surfaces on the boards, or encapsulated modules and submodules. For space applications, the heat easily can be

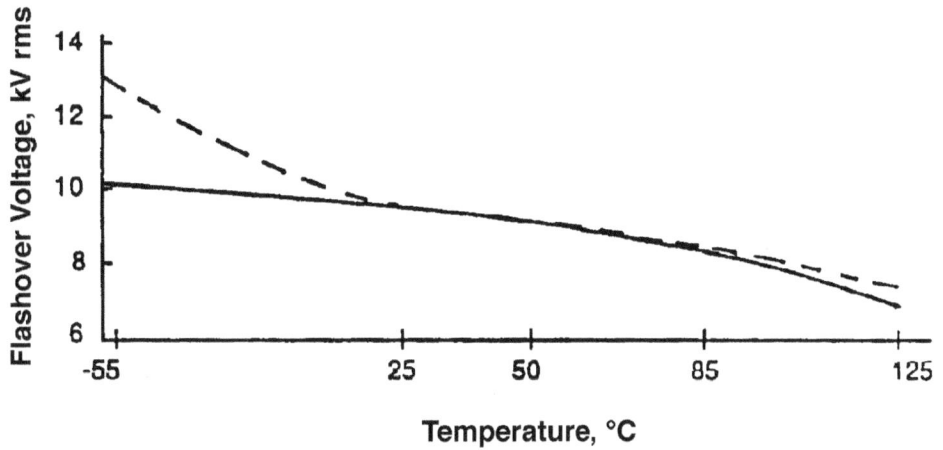

Figure 20.8 Effect of temperature on 60-Hz flashover stress.

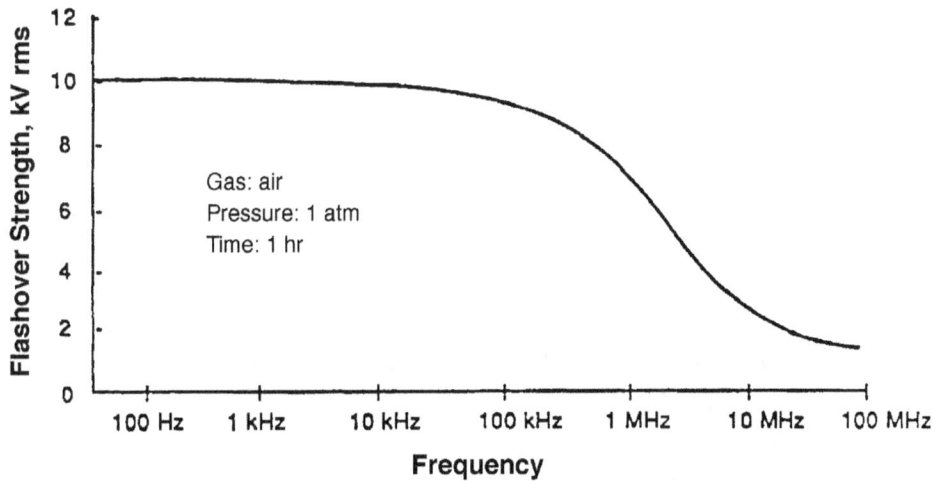

Figure 20.9 Effect of frequency on flashover strength for configurations shown in Fig. 22.6.

directed toward a cold plate and radiator. For aircraft and commercial applications, the heat may be dissipated by either a refrigerated coolant or by passing air through cooling fins. As the power input to electronic devices (e.g., power transistors, voltage dividers, vacuum tubes, and other high-power critical parts and components) increases, it may become necessary to actively cool the package using either a closed-loop liquid or gas coolant system. The collection or deposition of dirt, debris, and oils must be minimized to prevent early-life failures. This can be accomplished by using lightweight filters.

The packaging design may use one of the three major insulation techniques: (1) open construction, where all parts and components are surrounded with air, an insulating gas, or a liquid, (2) solid construction, where all electronic circuitry is encapsulated into a block of solid insulation, or (3) a modular construction, where the low-voltage circuitry, circuit

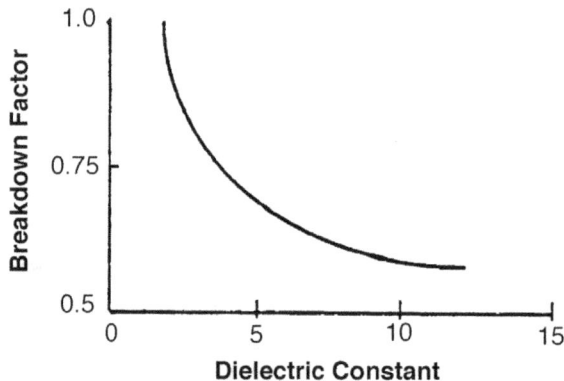

Figure 20.10 Variation of flashover voltage with changing insulation dielectric constant.

boards, and parts are conformally coated, and the high-voltage circuitry is encapsulated in solid insulation or contained in a liquid-filled container. The type of construction chosen will depend on the application, surrounding environment, and heat transfer technique available at the application site.

20.4.1 Gas Insulation Systems

Electrical and electronic equipment must be designed to operate at either normal sea level pressure and temperature or at higher altitudes and temperatures. Most terrestrial equipment with voltages to 30 kV can be designed to meet normal operational requirements by using conformal coatings on all parts and boards, and properly spacing parts and circuits. Aerospace equipment designs must consider the rarefied air as an essential packaging design criterion. That is, lightweight, small-volume packaging dictates that circuitry operating above 300 V peak require special consideration—either pressurization or encapsulation. Gas-filled packages require conformal coatings to prevent corrosion, voltage breakdown, and flashover between closely spaced electrodes with more than 300 V peak applied. Many failures in aerospace equipment are caused by corroding circuitry due to exposure to wide temperature and humidity ranges and the collection and deposition of debris, oils, and dirt.

The potentials required for voltage breakdown in gases at the minimum pressure/spacing dimension (Paschen law minimum) and between parallel plates spaced 1 cm apart at the critical pressure are listed in Table 20.6. Of these gases, conditioned air is used whenever possible. When air cannot be used, sulfur hexafluoride is the preferred pressurization gas because it is stable, is electronegative, has excellent heat transfer properties, and is easily obtained. Sulfur hexafluoride is used in compact switching equipment, substations, cables, and many other commercial high-voltage equipment applications. It should be the first gas considered for high-voltage aircraft and spacecraft equipment when component density and other high-voltage criteria suggest that a pressurized gas environment is best. Mixing sulfur hexafluoride with other gases can improve some characteristics with little change to the direct voltage uniform field strength.[23,24] Air and nitrogen are also used for pressurized circuits. These gases have very similar breakdown characteristics, are readily available, and require little special handling. Nitrogen, when mixed with as little as 1 per-

Table 20.6 Breakdown Voltage between Bare Electrodes Spaced at 1 cm

Gas	Minimum @ critical pressure spacing		Breakdown voltage @ 1 atm	
	Vac rms	Vdc	kVac	kVdc
Air	223–230	315	23	33
Ammonia	—	—	18.5	26
Argon	196	280	3.4	4.8
Carbon dioxide	305	430	24	28
Freon 14	340	480	22.87	32
Freon 114	295	420	64	90
Freon 115	305	430	64	90
Freon 116	355	500	—	—
Freon C 138	320	450	—	—
Helium	132	189	1.3	1.63
Hydrogen	205	292	12	17
Nitrogen	187	265	22.8	32
Oxygen	310	440	—	—
Sulfur hexafluoride	365	520	63	89

cent air or similar gas, has nearly the same breakdown characteristic as air at the Paschen law minimum.

Gases with oxygen and halogen atoms are electronegative and hence good insulators, in contrast to hydrocarbon and noble gases. Electronegative gases have outer rings deficient of one or two electrons. These molecules and compounds are able to capture free electrons, forming heavy and relatively immobile negative ions. The negative charge of such an ion equals the number of free electrons captured. Electronegative gases have high dielectric strength, because the heavy ions arrest the formation of electrical discharges normally initiated by mobile electrons. Some electronegative gases are sulfur hexafluoride, dichlorodifluoromethane, perfluoropropane, perfluorobutane, hexafluoroethane, chloropentafluorethane, dichlorotetrafluoroethane, tetrafluoromethane, and sulfur hexafluoride-nitrogen mixtures or fluorocarbon mixtures. These gases are chemically inert and have good thermal stability, but they can decompose chemically when exposed to partial discharges or arcs.[25] The products of decomposition are often toxic and corrosive. In addition, a small quantity of water decomposes the sulfur hexafluoride to form hydrofluoric acid when in the presence of a partial discharge or arc. Once formed, the hydrofluoric acid etches into crevices and requires special cleaning of all parts within the pressurized module.

The dewpoint of a gas mixture is important for pressurizing gases. Most military applications require a dew point less than −55° C, otherwise heaters must be installed in the equipment to prevent condensation of the gas.

20.4.2 Liquid Insulation Systems

Liquids have the advantage of self-circulation within the module. They are also self-healing, in contrast to solid dielectrics, since the effective area of a failure caused by a temporary overvoltage is immediately reinsulated by the fluid flow back into the former cavity. The disadvantages include weight and leakage. Liquid dielectric can leak and collect on critical surfaces, thereby rendering nearby low-voltage equipment ineffective. Liquids must also be kept pressurized so that bubbles will not be formed. A gas bubble in a high-voltage electric field will generate partial discharges and cause loss of the device. Liquid dielectrics may be used as insulators and as a heat transfer media, and can be used in conjunction with solid insulations such as papers, films, and composite materials.

The general classes of liquid dielectrics used in commercial and aerospace applications are mineral oils, silicone oils, fluorocarbons (fluorinated liquids), vegetable oils, organic esters including castor oil, and polybutanes (polyhydrocarbon oils). When selecting a liquid dielectric, its properties must be evaluated in relation to the application. Most important are dielectric strength, dielectric constant, conductivity, flammability, viscosity, thermal stability, and—very important—compatibility with other materials of construction and the local environment. Parameters affecting dielectric breakdown in insulating liquids include electrode materials, electrode surface area and configuration, manufacturing treatments, contamination, and deterioration. Only the latter two will be discussed in this section.

Disadvantages that always accompany the use of liquid dielectrics are cost, weight, and temperature limit. Other disadvantages with many liquids are combustibility, oxidation, and contamination, and deterioration of materials in contact with the liquid. Deterioration of materials generates moisture, evolves gas, forms corrosive acids, produces sludge, increases dielectric loss, and decreases dielectric strength. The selected liquid should provide the best required properties, consistent with keeping disadvantages within the acceptable limits. Typical properties are shown in Table 20.7. The temperature of a liquid dielectric affects its life and stability, since chemical deterioration reactions usually proceed faster at higher temperatures. Temperature also effects the conductivity of a liquid. As temperature increases, fluid viscosity decreases and the higher mobility of the ions permits increased conduction.[26,27]

Water is soluble to some extent in all insulating liquids. Water usually decreases dielectric strength and increases dielectric loss. Moisture dissolved in pure mineral oil does not effect dielectric strength until it separates from the oil solution and deposits on conductors, solid insulating surfaces, or solids floating in the oil. Polar contaminates dissolved in the oil give moisture its greatest degradation effect on dielectric strength. The effect of moisture varies among other liquid dielectrics. For example, in using liquid dielectrics to impregnate cellulosic insulations in transformers, capacitors, wire, and cables, the rate of increase of water solubility in the liquid with increasing temperature is important. When the rate increase in water solubility in the liquid is different from that in the cellulosic insulation, changes in temperature can make the dissolved water separate from the liquid. Such a separation leads to the formation of liquid-water emulsions and severe dielectric degradation.

The effect of gas absorption and liberation in a liquid dielectric must be considered for long-term successful operation. This is especially true when the liquid is used to impregnate solid dielectrics, as in capacitors and cables. Changes in pressure can make dissolved gases evolve from a liquid. Also, temperature affects the solubility of a gas, so heating can cause dissolved gas to evolve from the liquid. Corona will start in the evolved gas bubbles, leading to eventual dielectric breakdown. Thus, liquids used as impregnates must have low, stable gas content.[28]

20.4.3 Solid Insulation Systems

An ideal solid insulation has no conductive elements, voids, or cracks, and has uniform dielectric properties. Practical insulations have thickness variations, and many shrink with

Table 20.7 Properties of Dielectric Fluids

Property	Insulating liquid							
	Mineral oil	Dimethyl silicone, 50 csec	Hydrocarbon distillates MW 500–700	Isopropyl biphenyl	Isobutyl* monochloro biphenyl oxide	Di-2-ethylhexyl phthalate-25 w/o trichloro-benzene	Phenyl xylyl ethane	Di-1-ethyl-hexyl phthalate
Relative dielectric constant @ 60 Hz, 100° C	2.1	2.6	2.1	2.6	4.0	4.5	2.5	4.3
Dissipation factor @ 60 Hz, 100° C, %	0.1	0.01	0.4	0.2	1.0	1.0	0.2	0.5
Dielectric strength ASTM-D877, kV	35	35	38	60	35	35	55	42
Gas absorption coeff. ASTM D-2300 μL/ min	21	–13	–5	180	80	38	120	25
Viscosity @ 100° F (37.8° C), csec	10	41	310	5.8	10.5	13	6.1	29
Pour point, °C	–55	–55	–20 to –30	–55	–45	–52	–48	–45
Fire point, °C	150	345	300	155	200	255	148	235
Use	Transformers, cables, circuit breakers	Fire-resistant transformers	Fire-resistant transformers	High-voltage capacitors	High-voltage capacitors	High-voltage capacitors	High-voltage capacitors	Low-voltage capacitors

* No longer available

curing temperature and age. Furthermore, they may have some deposited conductive elements, and their dielectric properties may change with temperature, frequency, and mechanical stress. Most commercial equipment operates in a reasonably quiescent, thermally controlled environment. However, in aerospace applications and some mobile commercial applications, the environmental and electrical stresses vary as a function of time. Some parameters vary independently while others are synergistic. These variations make it difficult to select an ideal insulation for a specific application. It is important to note that it is impossible to extrapolate the operation of a second- or third-generation device based on the performance of a first-generation device. For instance, the composition of materials varies from batch to batch. The cleanliness, manner of handling, and manufacturing in a production facility are not the same as in a prototype shop. All these factors must be addressed by a failure analyst who is investigating a high-voltage product failure.

Table 20.8 summarizes the chemical, electrical, mechanical, and thermal properties that need to be considered for solid insulation. Sometimes, solid potted insulation is specified to be transparent so that the packaging engineer can assess parts stability, movement, stressing, and bonding. Weight, water absorption, and outgassing are often specified for aerospace applications. Most important for all categories of high-voltage insulation is aging, which depends on the electrical stress and environment.[29–33]

Table 20.8   Properties of Interest for Insulating Materials

Mechanical	Electrical	Thermal	Chemical	Miscellaneous
Tensile, compressive, shearing, and bending strengths	Electric strength	Thermal conductivity	Resistance to reagents	Specific gravity
	Surface breakdown strength	Thermal expansion	Effect on adjacent materials	Refractive index
Elastic moduli	Liability to track	Primary creep	Electrochemical stability	Transparency
Hardness	Volume and surface resistivities	Plastic flow	Stability against aging and oxidation	Color
Impact and tearing strengths	Permittivity	Thermal decomposition,	Solubility	Porosity
Viscosity	Loss tangent	Spark, arc, and flame resistances	Solvent crazing	Permeability to gases and vapors
Extensibility	Insulation resistance	Temperature coefficients of other properties		Moisture adsorption
Flexibility	Frequency coefficients of other properties	Melting point		Surface adsorption of water
Machinability		Pour point		Resistance to fungus
Fatigue		Vapor pressure		Resistance to aging by light
Resistance to abrasion				
Stress crazing				

An important area of solid insulations for high voltage applications is encapsulating materials. For convenience, some of the typical problems are summarized in Table 20.9. An encapsulating material is usually required for high-density packaging of high-voltage electronic units. A dielectric has two conflicting characteristics: excellent electrical insulating properties when properly processed, but poor heat transfer characteristics that can lead to heat buildup and premature failure if not addressed in the design. We emphasize the insulation processing because it is very important. Inadvertent voids, cracks, or improper bonding can lead to partial discharges and eventually arcing and failure. The potential for arcing of a high-voltage unit is high when associated with mediocre design practices and poor workmanship.

Table 20.9   Typical Encapsulating Material Problems

Below 10,000 V	Incomplete bonding of materials
	Insufficient outgassing of potting compounds during vacuum impregnation
	Insulation cracking during thermal/mechanical stress
Above 10,000 V	Gaseous ionization and partial discharges within small voids and cracks
	Creepage and tracking
	Arcing and flashover
	Insulation treeing during long periods of operation

20.5   EQUIPMENT CATEGORIES AND ASSOCIATED FAILURE MECHANISMS

20.5.1   Aerospace Power Systems

The power quality for an aerospace vehicle power system is controlled by rigorous specifications and standards. Deviations from the specified standards are rare but can occur during inadvertent bus faults, short circuits, and switching surges because of the loading and unloading of high-impedance circuits. Once an investigator has eliminated these possibilities, other areas can be explored, and some areas for consideration are presented in the following paragraphs.

A spacecraft power system solar array or thermal-electric power source may be built using several (<1 V) series and parallel connected solar cells or thermal-electric devices integrated into a panel with a voltage output ranging from 3 to 200 Vdc to meet the system power requirement. Booster, missile, and experimental rocket power systems are designed using either batteries or auxiliary power units having outputs from 12 to 6,000 Vdc. The very high-voltage (6,000 Vdc) experiment power systems are time and energy limited.

Aircraft electric power system voltages are usually 28 Vdc, 270 Vdc, or 115/200 Vac @ 400 Hz. Studies for higher voltages for very large aircraft are in progress and will be developed for special applications. For systems with nominal operating voltages equal to or less than 270 V peak and frequencies less than 40 kHz, corona, partial discharges, and glow discharges do not play a significant role when operated in either a pressurized or depressurized air atmosphere, as is the case for most commercial airplanes. However, transients and system overvoltage excursions that exceed 327 V peak may initiate discharges between closely spaced bare electrodes at low pressure for supersonic airplanes, missiles, rockets, and spacecraft that must operate from Earth's surface to space. That is, voltages greater than 327 V peak may be plagued with partial discharges or corona when the Paschen law minimum pressure spacing/voltage parameter is exceeded. Therefore, the power quality imposed on power system designers must be adhered to, and the insulation for wire, parts, and cables must be designed such that the initiation voltages are not exceeded. It is true that the discharges will extinguish as the voltage returns to normal operating voltage, but the time between the discharge initiation and extinction is cumulative and, when enough time is accumulated, a system fault may occur at the weakest insulation point. An example of insulation effectiveness as a function of altitude or decreasing pressure is shown for electrically insulated wiring in Figs. 20.11 through 20.13. First, it should be noted that the initiation voltage at high pressure is much greater than the transient voltage peaks. Second, as the pressure decreases for very small spacing, the initiation voltage decreases to the maximum transient initiation value. Finally, as the pressure is further

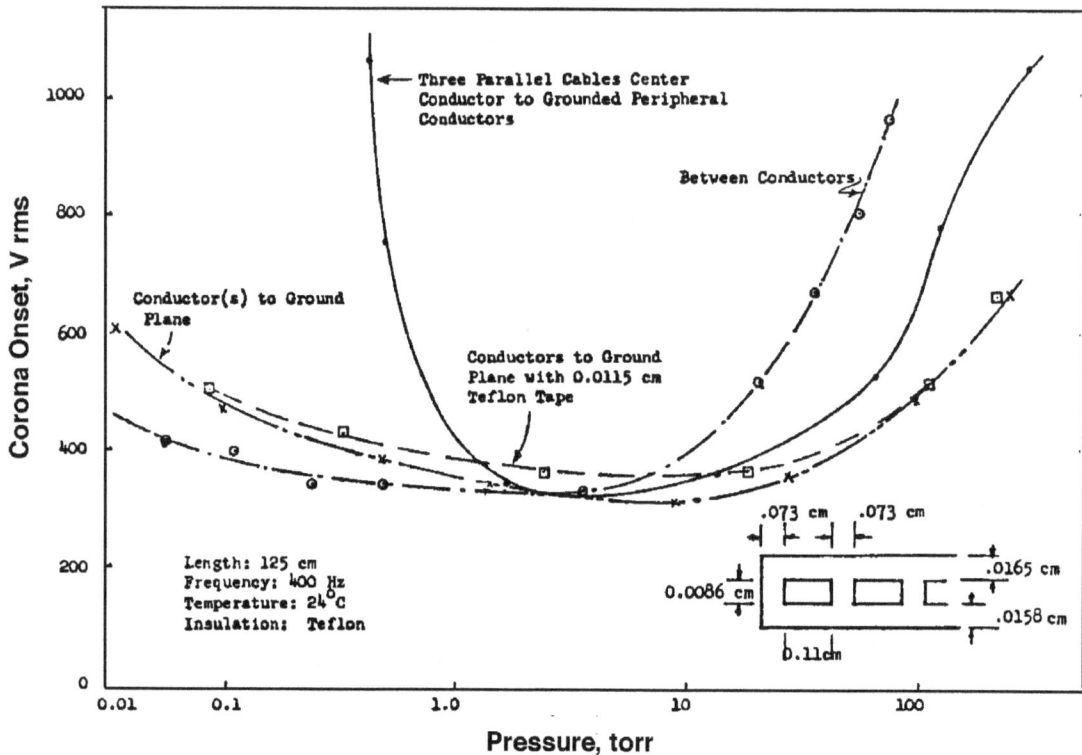

Figure 20.11   Corona onset voltage (COV) of Teflon®-insulated flat conductor cable.

decreased, the wider spacings, in accordance to the Paschen law criteria, allow the electrical insulation to become insignificant with relation to the air gap. This does not imply that the unit will fail immediately. It shows that very low-current discharges will occur during the time the voltage transient exceeds the Paschen law minimum. As the discharge time accumulates, the electrical insulation will deteriorate due to the ozone produced, the heating during the discharge time, and the electric field overstress imposed on the insulation. The deterioration time may take from a few accumulated hours for very large discharges to several hundred accumulated hours for very small discharges. The quantity and quality of the electrical insulation will determine the aging process of the insulation. Electrical insulation that is being slowly deteriorated usually becomes brittle (dried), rough to the touch, and discolored. The color change may be white rather than black, depending on the chemical change within the outer surface of the insulating material.

Other negative design impacts to aerospace and other high-voltage electronic packages are irregular shapes, unreasonably high packaging densities, and insufficient design development time. For instance, it has been demonstrated that a second model of a design with the same fit, form, and function may have better than ten times the first model's reliability. This will be due to minor reductions in electrical stresses, use of lower failure rate parts, and utilization of better materials and improved processing. Usually, the cost associated with improved subsystem performance is insignificant when compared to the gain in the aging characteristic.

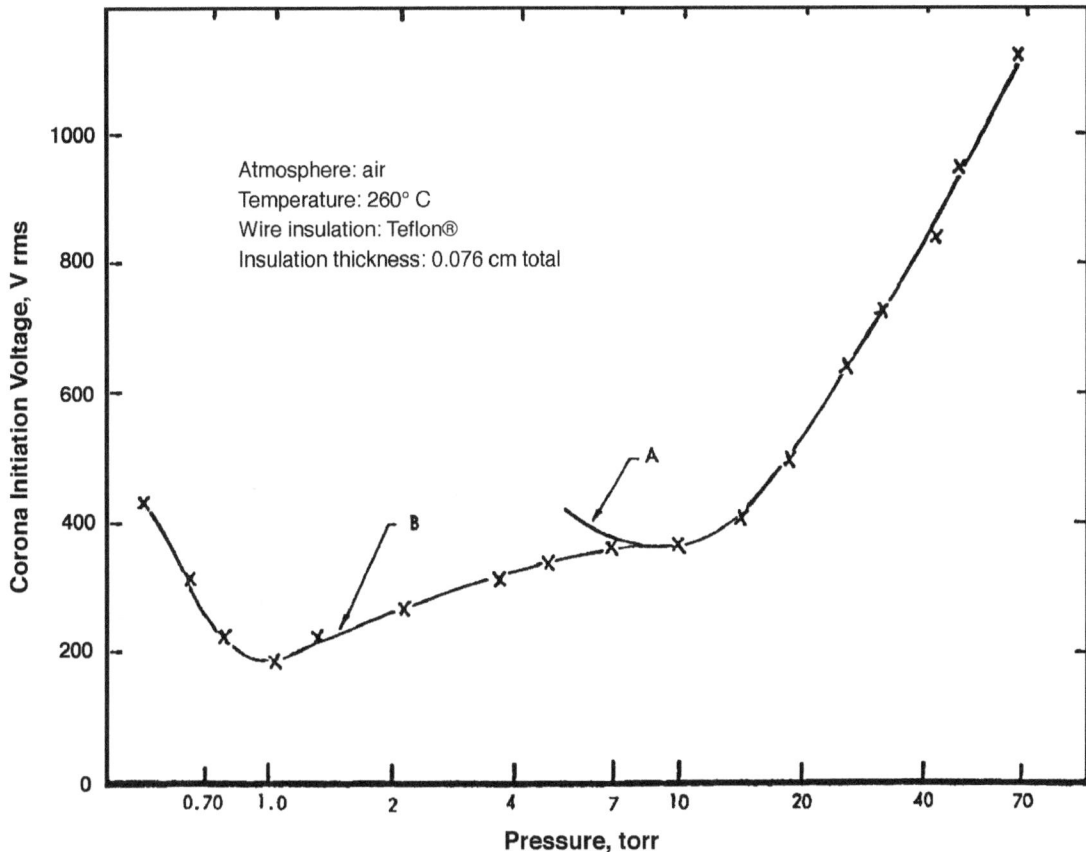

Figure 20.12   Corona initiation voltage of Teflon®-insulated twisted wire pairs.

20.5.2   Switching Circuits and Pulse Power

There are very low-power switching circuits and very high-power switching circuits. For megawatt and greater switching and pulse power circuits, the reader is referred to the *Pulse Power Conference Transactions* for the many specialties required to generate and insulate very high-power mechanisms. This section is devoted to low- and medium-power switchers as found in small power supplies and radar or communication systems for aero-space applications.

Modern high-voltage power supplies use high-frequency switching circuits for size, weight, and cost reduction. High-frequency transformers and ancillary part volumes can be reduced significantly by using switching frequencies of 40 kHz or higher. Likewise, the voltage multiplier diodes and capacitors can be reduced in size. However, there are areas of concern associated with steady-state and pulsed high-frequency switching. High-frequency transformers may use the flat-pack design, consisting of the windings etched onto the surfaces of a thin printed circuit board. The printed circuit windings are centrally

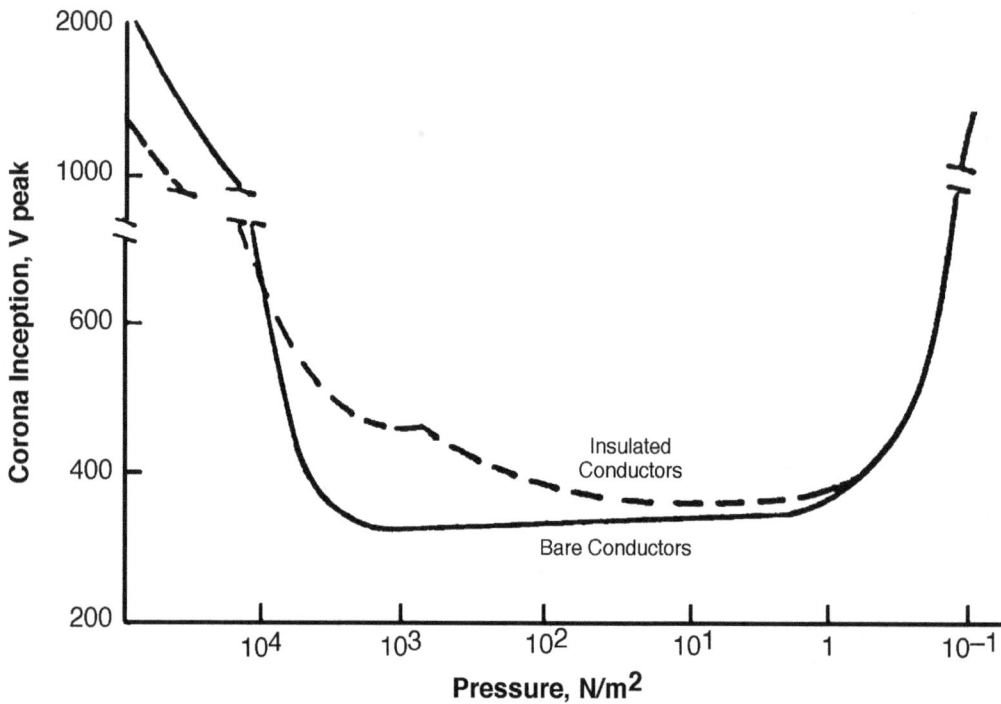

Figure 20.13   Critical pressure region for high-voltage equipment failure.

located in a flat-pack ferrite core with little spacing between the windings and the ferrite core surfaces. To protect the windings, a conformal coating of parylene or equivalent material is deposited over the windings. When this design is completed, it has excellent electrical characteristics with little hysteresis, eddy current, or resistance losses, making it an excellent design. The concern is the application of these circuits in aerospace equipment where the pressure may vary from that at the surface of the Earth to the virtual vacuum of space. The corona initiation voltage (CIV) between points, rods, and plates is shown in Fig. 20.14.[34]

High-frequency square-wave pulse switchers have three primary frequencies of concern: the repetitive pulse (squarewave) frequency, the frequency associated with the pulse rise time, and the frequency associated with the pulse fall time. Although the repetitive frequency may be acceptable as a design frequency, the other two frequencies may be the initiators of partial discharges or a glow discharge between the transformer winding printed circuit board and ferrite core inside surface. Using normal field theory design practices, the engineers and technologists only consider the repetitive squarewave frequencies. This assumes that a squarewave and a sinewave voltage have identical voltage characteristics. The assumption is true for most parallel plate configurations spaced less than 0.5 mm apart (separated by a gaseous insulation) when the frequency spectrum contains frequencies below 1 MHz. The assumption is not true, and the corona initiation voltage will be significantly modified, for parallel plates with spacing greater than 0.4 mm, points, wires and edges when the frequency is greater than 40 kHz. Indeed, the wider spaced parallel plate configurations may decrease the CIV by as much as 20 percent, and the pointed con-

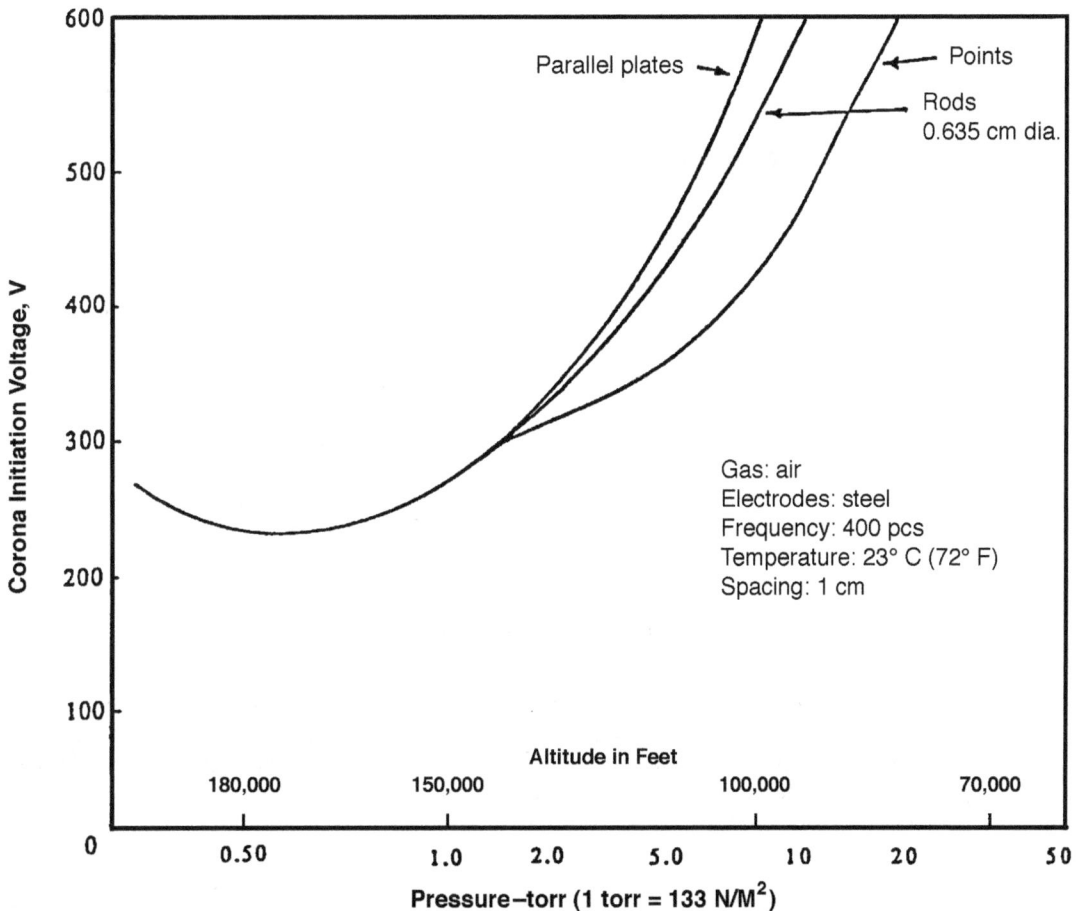

Figure 20.14 Corona initiation voltage between points, rods, and plates.

figurations by as much as 50 percent. It has been shown experimentally that the partial discharge voltage is approximately 327 V peak for frequencies from dc to 40 kHz.[35] As the frequency is increased above 40 kHz, the initiation voltage may drop to a value as low as 39 V at 1 GHz for nonconformal (pointed, rough, or sharp-edged) electrode configurations. For very smooth, closely spaced parallel plates (Rogowski electrode configurations), the initiation voltage will remain at 327 V peak from dc to ≈1 MHz, then fall abruptly to a minimum value of 39 V at 1 GHz as shown in Fig. 20.15. This implies that the squarewave pulse train frequency may have a partial discharge initiation voltage of 327 V peak, but the frequencies associated with the squarewave rise time and fall time may have significantly lower partial discharge initiation voltage due to the part or component electrode surface condition or configuration. A prefailure analysis associated with continual high-frequency glow discharge, partial discharge, and corona discharges usually are observed as very dry, brittle, or a slight hazy discoloration of the surface material. At the time of failure, there will be either an arc path or a significant burn between the two electrodes with high stress,

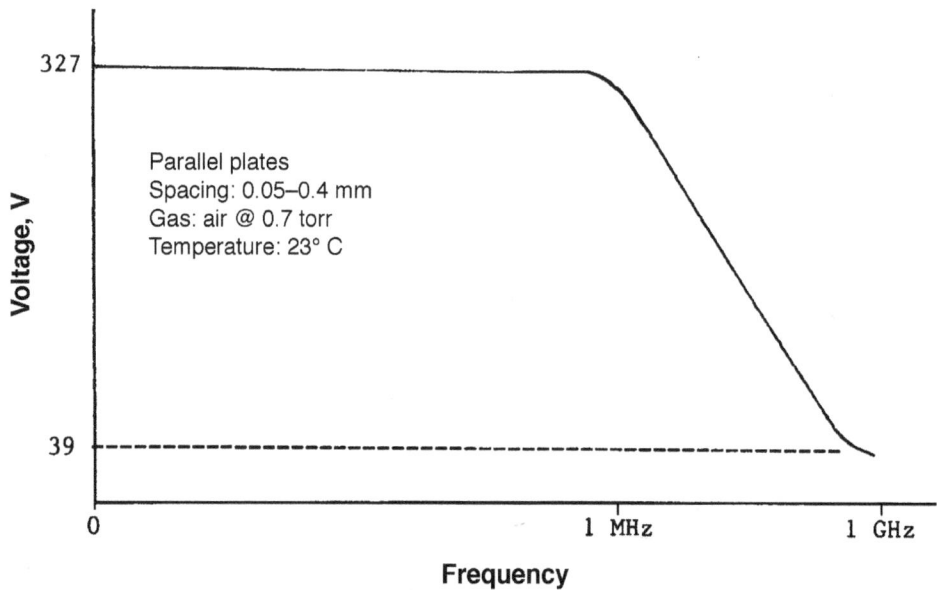

Figure 20.15   Breakdown initiation voltage changes with frequency for Rogowski electrode configurations.

but not necessarily a clear initiation point. The initiation may be associated with two adjacent electrode configurations that are separated with sufficient gas to prevent the completion of a sustained arc during the time allocated by the pulse width high-voltage time. This implies that the complete circuit must be examined for voltage overstress, frequencies, and electrode configurations and conditions.[36]

20.5.3   Direct Voltage Circuits

Direct voltage circuits should be easier to analyze than alternating voltage circuits, because voltage and resistance are the main parameters of concern for the analysis. Unfortunately, all parameters of voltage, frequency, resistance, inductance, capacitance, and time are involved. That is, a direct voltage may be designed to have a maximum value just below the initiation voltage for partial discharges. Transients generated by the circuit or coupled into the circuit from other sources can initiate partial discharges due to the rapid rise time associated with the pulse. Another cause for concern is the presence of voids within the encapsulating materials. A void must be treated as a group of parallel series resistors and capacitors with a switch across the center of the capacitor representing the void. The switching capacitor is triggered by the voltage rise of the resistor capacitance or inductance time constant of the encapsulating material surrounding the void. That is why dc partial discharge detection is very time consuming. Furthermore, it is more difficult to determine the source of the discharges within the encapsulating material for direct voltage circuits, especially voltage multiplier circuits.

Just when the problems appear to be solved for normal direct voltage circuits for earthbound and commercial airplane applications, space and supersonic aircraft applications appear. Very high-altitude aircraft (60,000 feet altitude and above) have problems associated with space, such as solar light intensity, higher temperature, and (when at maximum altitude) many more space-associated particles such as X-rays, alpha and beta particles, and space plasmas.[37] All are initiators for partial discharges in high-voltage circuits. The continuous cycling from negative to positive voltage for ac circuits is not as severely affected by space particles as are direct voltage circuits. Very small voids in dc circuits can initiate corona and partial discharges at a more rapid rate when space plasma surrounds the circuit. With a plasma environment, the initiation voltage may be as low as 39 V peak or lower. A plasma environment usually surrounds many low- to medium-altitude orbiting vehicles and aerospace craft such as the shuttle.[38]

20.5.4 Commercial Power Considerations

Commercial power considerations will be limited to industrial applications, household electronic systems and appliances, and aerospace ground support equipment. The input voltages for such appliances and equipment are well understood and, with reasonable maintenance and handling, have several years of continuous operating life. However, normal aging is a factor. For instance, the wiring in many commercial residential buildings has never been inspected. Yet, it is well known that many home and commercial fires are the result of faulty wiring. As a rule, it is not the conductor; instead, the electrical insulation has aged and become brittle, cracked, and in many cases fallen off the conductors. As long as the area of concern is free of carbonaceous materials and the bare wires are separated and mechanically undisturbed, the air between the conductors will be sufficient insulation to prevent arcing.

Electronic appliances and equipment failures are usually caused by insufficient maintenance. That is, high voltage parts and circuits must be free of dirt and debris. Accumulated dirt, greases, and debris between conductors and across printed circuit boards decrease the surface resistance between conductors, resulting in surface flashovers. High-voltage transformers and voltage multipliers can be cleaned of foreign matter by simply removing the material with the correct application of forced air or cleaning fluids. However, when using cleaning fluids, time must be allowed for the residual fluid to be removed by evaporation. A good indicator of debris buildup on the HV circuits of a display (television screen) is the noise generated and displayed on the screen as hash marks, dots, or dashes.

In the commercial world, power supplies are sometimes expected to operate for 15 to 20 years or longer. If the quality of the supply output is not critical, there are usually no problems in meeting the expected life. If a clean output is critical for the particular application, then increased failure rates are observed. This is usually caused by excessive *ripple* (1 percent is a typical limit placed on the outputs of switching supply ripple), which is a term used to describe periodic variations in the output voltage that are usually fed through an output filter. The ripple frequency is the operating frequency of the primary transformer, which is 60 Hz for a conventional linear supply, or at the switching frequency of a switching power supply. Most supplies have a 60 Hz ripple and a higher-frequency ripple, both of which are seen to increase with the age of the power supply. Commercial power supplies typically use electrolytic capacitors for filtering applications to save space and reduce cost. Electrolytic capacitors have a finite life relating to the very close internal spacing and the aging quality of the dielectric. Filter capacitor replacement as part of the power supply cleaning and maintenance program will eliminate this problem.

20.5.5 HV/LV Solid-State Device Coupling Failures

Many complex circuits contain both high-voltage and low-voltage parts. Some low-voltage parts may be designed to withstand high-voltage transients, whereas other parts may have high-voltage breakdown voltage limits. Of special interest are the solid-state devices that can be destroyed by voltage transients with magnitudes less than 25 V peak. Occasionally, a circuit containing devices of this type may be packaged alongside a high-voltage circuit, with no direct connection.

The frequency bandwidth for partial discharges is several megahertz wide, as shown in Fig. 20.16. Some typical low-voltage failures initiated by coupling from high-voltage circuits are presented in Table 20.10. Should the electronic circuitry containing the solid-

Table 20.10   High-Voltage Coupling Effects

Electromagnetic interference	Logic circuit upset
	Alteration of electrical component parameters
	Damage or burnout of electrical components
Thermal control degradation	Burnout of ground straps
	Deposition of burnout materials on surfaces
	Alteration of surface absorption and emittance properties
Optical sensor degradation	False optical signals due to partial discharges
	Deposition of arc materials on optical surfaces
Mechanical effects	Coulomb forces due to electrostatic charging
	Couloumb forces due to electrostatic arc discharges
	Mechanical distortions due to long-term changes, arc, and transients

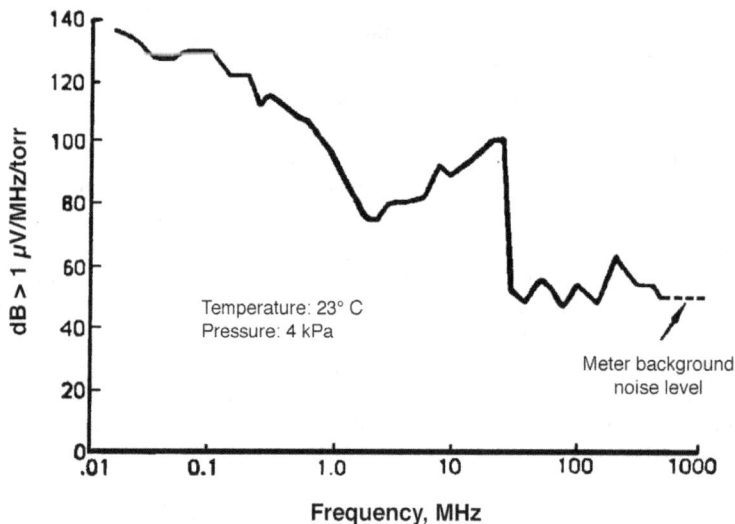

Figure 20.16   Frequency spectrum at corona discharge.

state devices also contain resonant circuits, and should the partial discharge bandwidth include the resonant frequency(s), there is an excellent probability that some of the solid-state devices will fail due to RF coupling through air or insulation with the discharge circuit. The highest probability of failure in high-voltage direct current (HVDC) circuits will occur when the voltage is turned on and when it is turned off. During charging and discharging of a HVDC circuit, there are many more partial discharges with very high pulse heights than there are during steady-state operation. Some low-voltage circuit designs are deenergized during the HV start period. To solve half the problem, they should be energized only during steady-state operation.

20.6 CRITICAL FACTORS FOR EVALUATION

High-voltage and high-power electronics manufacturing and design are greatly influenced by cost, time, and workmanship. Some hurriedly designed and manufactured equipment, developed at low cost, may result in overstressed insulation and low MTBF, which in turn results in higher costs at a later time. This section discusses major causes of overstress and failure that occur due to the influences of the design, the manufacturing process, and environmental effects. Included in this discussion are remedial guidelines to minimize the probability of overstress and catastrophic failures.

20.6.1 Stress Interactions

Design stress interactions lead to low MTBF or catastrophic failures and are associated to the circuit design, selection of materials and components, the packaging design, the design of the manufacturing fixtures, and the testing parameters.

Many stress interactions are caused by the preparation and application of the encapsulant. Thus, the material characteristics normally associated with the encapsulant are degraded. These degrading stresses are occasionally contributed to poor manufacturing procedures caused by improper mixing, potting, and curing of the materials; component and module assembly; workmanship and manufacturing facilities; and environment. Some areas of concern in the packaging of high-voltage systems are detailed in Table 20.11.

20.6.2 Material Characteristics

Materials characterization should be conducted by the design or materials and process engineers. Solid insulations have electrical, mechanical, thermal, and chemical properties, all of which can affect reliability. Sometimes transparent materials are specified so the packaging engineer can visually assess parts stressing and bonding. Weight, water absorption, and outgassing are often specified. Most important for all categories of electronic insulation is life, which depends on the level of electrical stress and the environment.

Materials characteristics that have contributed to stress include:

- High viscosity, which prevents complete filling of magnetic devices and densely packaged electronics.

- Service temperature, which must be compatible with the operating and storage temperature for the application.

Table 20.11   Areas of Concern in Packaging High-Voltage Solid Systems

Potential problem	Effect*	Prevention method
Voltage stress across surfaces	Corona creepage	Layout to minimize surface voltage stress Creepage barriers Select arc-resistant material
Voltage stress exceeding material breakdown	Corona arcing	Maintain spacing between routed wires Increase package size Select higher-strength dielectric Solder balls Hardware selection to avoid exposed sharp edges Use of voltage shields Large-diameter conductors Insulation dielectric strength compatibility (ac) Insulation resistivity compatibility (dc)
Excessive temperature	Component failure and insulation degradation	Layout to minimize thermal paths within the module Use "loaded" insulating materials (higher conductivity) Use of thermal spreaders Minimize thermal resistance from module to ambient Increase package size Select thermally stable material
Voids in insulation	Corona arcing	Packaging geometry to allow "easy fill" and gas escape Packaging design to accommodate cure shrinkage Process control to avoid bubbles
Cracks and delamination during thermal cycling	Corona arcing	Package configuration to accommodate insulation shrinkage and expansion (physical constraints) Coefficient of expansion compatibility Material compatibility to promote adhesion Low bulk modules if material is severely constrained
Part or solder joint failure during thermal cycling	Shorts or opens, corona arcing	Component mounting to provide stress relief Soft conforming coat of sensitive components
Particulate contamination	Corona arcing	Packaging design to allow easy cleaning Process control

* Corona and creepage lead to material degradation, formation of carbon deposits, etc., thus producing arcing and circuit failure. Corona also causes EMI, which affects the operation of sensitive circuits.

- Brittle and very hard materials that add stress to component structures, solder joints, platted-through holes, and printed circuit boards during thermal cycling.
- Very low thermal conductivity, which prevents proper cooling of power circuitry and may cause hot spots within the insulation system.
- Excessive differential thermal expansion between the potting material and electronic components, which mechanically stresses and cracks the insulation and components.
- Low dielectric strength, which leads to catastrophic failure for materials between high-voltage, small-radii conductors and grounded surfaces or small, low-voltage conductors.

- Dielectric constants greater than 6 and dissipation factors greater than 0.1 may cause very high stress across microvoids and macrovoids, causing partial discharges and eventual voltage breakdown. High dissipation factors at higher frequencies cause dielectric heating and a rapid decrease in dielectric strength, resulting in short life.
- Low volume and surface resistivities tend to increase dielectric heating and probability of surface arc-over.
- A high glass transition point, e.g., the temperature where a material changes from a plastic to solid structure, adds stress to all electrical components within the potted material when subjected to low temperatures. In addition, it enhances the probability of cracking of large volumes at temperatures at or below the glass transition point.

Undesirable properties for a potting material to be used in power supplies or electronic components include:

- High shrinkage during cure
- Gas release during cure, causing bubbles (voids) within the potted volume
- Molecular sieving of catalysts and fillers, i.e., removal of the catalyst from the encapsulant
- Reversion back to the original chemical compounds
- Sieving of fillers by dense sleevings, tapes, and wrappings

20.6.3 Dielectric Parameters

Critical parameters that affect the life of electrical insulation are:

- Temperature
- Frequency
- Voltage
- Thickness
- Mechanical stress
- Incompatibility with electrical parts and materials
- Dielectric constant
- Dielectric strength
- Insulation thickness
- Polarization
- Humidity
- Dissipation factor
- Gradation of dielectrics between surfaces
- Configurations
- Improper selection of materials with respect to resistivity for direct voltage applications and dielectric constants for alternating voltage applications
- Voids
- Cracks

Some manufacturing problems associated with dielectric parameters include:

- Incomplete outgassing of air from the wiring and parts before application of the potting material
- Displacing components, modules, or wiring in a way that decreases the insulation thickness between high and low voltage conductors
- Wrong formulation of encapsulation or filler
- Debris or foreign matter on the components or in the material
- Contamination when silicones and epoxies are mixed in the same potting facility
- Improper use of primers when they are required
- Variations in formulations

20.6.4 Parts and Component Configurations

Parts and small assemblies have their operating voltages and power dissipation levels derated for particular applications to ensure that the parts will operate at required reliability levels under specified environmental conditions. Voltage and power derating are separate, independent procedures. Voltage derating reduces the possibility of electrical breakdown, whereas power derating is done to maintain the component material below a specified maximum temperature. The minimum derating criteria recommended for designs of power supplies and electronic equipment are shown in Table 20.12. Some electronic equipment requires more derating because of the ambient temperature environment. Those deratings should be controlled by the detailed specification.[39]

Problems associated with parts and components are as follows:

- Coatings: For example, wax-filled ceramic surfaces are difficult to clean and will not bond.
- Surface coating materials that unbond during cleaning.
- Surface coating materials that are incompatible with the potting compound unbond or form gas voids.
- Pot cores: Sealing and filling may crack the seals, making a gap between transformer halves.
- Small, high-voltage leads arc to the ferrite core.
- An air void between the outer winding and the ferrite core can result in corona and arcing.
- Ferrite cores are easily cracked during environmental testing.
- Tapes within the transformers and other parts may be a source for voids.
- Connectors: Deep-seated pins with gas-filled wells around the pins are very difficult to pot and are a source of partial discharge and corona.
- Voids or gaps along the insulation interfaces are a source of partial discharges and eventual breakdown.
- Wiring: Use of stranded wire for very high voltage applications have high electrical stress at the conductor.

Table 20.12 Component Derating Guidelines

Component type	Derating parameter	Derating to percentage rating (or absolute value indicated)[*]
Resistors		
Carbon composition	Power/voltage	50/80
Film high-stability	Power/voltage	50/80
Wirewound accurate	Power/voltage	50/80
Wirewound power	Power/voltage	50/80
Wirewound chassis mounting	Power/voltage	50/80
Variable wirewound	Power/voltage	50/80
Variable non-wirewound	Power/voltage	50/80
Thermistor	Power/voltage	50/80
Tantalum nitride chip	Power/voltage	50/80
Capacitors		
All types	Ripple voltage	50
All types	Ripple current	70
All types	Core temperature	20° C less than rated
Ceramic	Voltage	50
Glass	Voltage	50
Mica	Voltage	60 dipped, 40 molded
Film dielectric	Voltage	50
Tantalum solid	Voltage	50[†]
Tantalum wet	Voltage	60
Tantalum foil	Voltage	50
Aluminum electrolytic	Voltage	70[‡]
Relays		
All	Use arc suppression	
	Contact current (cont.)	See note[§]
	Contact current (surge)	80
	Coil energize voltage	Manf. nom. rating
	Coil dropout voltage	75 (incl. Q of mounting)
	Vibration	0.005 min.
	Contact gap	Opening
Switches	Contact current	30
	Voltage	50
Connectors	Contact current	50
	Voltage (dielectric withstand)	25
Magnetic devices	Current (cont.)	60
	Current (surge)	90
	Voltage (cont.)	60
	Voltage (surge)	90
	Hot-spot temperature	30° C below insulation rating
	Insulation breakdown voltage	25
	Temperature rise	40° C maximum
RF coils	Current	50

Table 20.12 Component Derating Guidelines *(continued)*

Component type	Derating parameter	Derating to percentage rating (or absolute value indicated)[*]
Transistors	Power[**]	50
	Forward current (cont.)	60
	Voltage[††]	75
	Transient peak voltage	75
Diodes		
Switching, general purpose, rectifier	Current (surge)	70
	Current (cont.)	70
	Peak inverse voltage	65
Zener	Current (surge)	70
	Current (cont.)	60
	Power	50
SCR	Current (surge)	70
	Current (cont.)	70
	Peak inverse voltage	65
All	Junction temperature	110° C maximum
Microcircuits		
Linear	Current (cont.)	70
	Current (surge)	60
	Voltage (signal)	75
	Voltage (surge)	80
	Voltage reverse junction (signal)	65
	Voltage reverse junction (surge)	85
	Junction temperature	110° C maximum
Digital[‡‡]	Supply voltage	Manf. nom. rating
	Junction temperature	100° C maximum
	Fanout	80

[*] Passive high-voltage components have both voltage and electric field limitations.
[†] Special 100% current surge test
[‡] Resin end seal protected and 99.96% aluminum foil purity
[§] 50 resistive, 25 inductive
[**] The safe operating area (SOA) curves, adjusted for junction temperature, should not be exceeded under any transient condition
[††] Power devices exhibiting "punch-through" characteristics should be derated to 50% on voltage parameters.
[‡‡] Many families of digital microcircuits exhibit additional characteristics that may require derating (e.g., toggle frequency, hold times)

- Ties: Loose fabric or ties are a source of corona because of dielectric charging.
- Mounting brackets and screws: Any sharp-edged or pointed bracket or screw associated with a void or gap will fail on a high-voltage system. Plastic and metal screws contribute to failure whenever sharp edges are present.
- Sleeving over components traps air next to the component.

20.6.5 Miscellaneous Design Criteria

Large power supplies, 300 W or higher, sometimes limit surge currents to prevent false tripping of circuit breakers or interference with other supplies on the same line. Supplies with large input filter capacitors can draw peak currents of over 10 times the steady-state current at startup. A typical specification will limit surge current somewhere in the range of 2 to 4 times the steady-state current. This can be accomplished by using a thermistor in series with the input, a current limiting resistor with a bypass switch, or a large input filter inductance.

Hold-up time is an input parameter specified for power supplies that run off line voltage and where the load is a digital system that can be interrupted by a momentary power glitch. The majority of commercial power supplies have hold-up times specified at 20 ms. This means that, for critical applications, an uninterruptible power supply (UPS) is required for the system. Since the vast majority of line interruptions are less than five ac cycles, a manufacturer may attempt to eliminate the need for a UPS by specifying a long hold-up time of 100 ms for the power supply. The failure analyst should note that power supplies with longer hold-up times cost more and should be taken as an indicator to check for the presence of a UPS on critical systems.

Some design criteria that cause problems associated with components are as follows:

- Matching material stresses within large components such as transformer coils, high-voltage solid-state power devices, transformer cores, and base plates. These can cause cracks and disrupt or fail the circuit.
- Placing low-voltage, sensitive circuits in a high-voltage compartment.
- High-voltage and low-voltage wire separation.
- Lack of corona shields around nuts, screws, sharp-edged components, and terminations located in the high-voltage power supply area.
- Bonding of multidielectric systems with epoxies and silicones.
- Contamination with waxes, greases, silicones, and oils after cleaning.
- Mounting parts on circuit boards with exceptionally small gaps, thus preventing filling with encapsulating material.
- Circuit board configuration: potted, densely populated, long, wide circuit boards that crack at low temperature.

The packaging design should contribute to system reliability by:

- Minimizing thermal stress
- Minimizing dielectric stress
- Facilitating initial achievement of high dielectric strength
- Maintaining high dielectric strength for the required life

A packaging design that fails to provide for these items is inadequate.

20.6.6 Assembly and Test Methods

Some problems associated with assembly and test caused by manufacturing procedures developed by engineering and in-process assembly and acceptance tests are:

- *Cleanliness.* A number of items contribute to the bond worthiness of a potting material, and many involve cleaning and personnel handling. Foremost are greases, oils, and residues that contaminate the hardware during assembly when personnel handle them. Secondary items include air pollution such as particulates, flaking of particles from parts, or smoke from hot soldering devices. Last is the handling or lack of cleaning after environmental testing, before high-voltage testing.

- *Solvents.* Dirty solvents leave residues. If not used properly, electronic cleaning products will simply redistribute dirt. In addition, some solvents fill small pores in the surfaces of the cleaned component and outgas for a long time after cleaning. This can result in high voltage surface arcing during testing.

- *Moisture.* Condensates collect on parts and materials during environmental testing and result in lowered surface resistivity.

- *Soldering.* Incomplete solder joints, cold solder joints, or solder flow between conductors.

- *Parts.* Misplaced or misaligned parts. Lack of stress relief on part terminals.

- *Spacers.* Use of multiple spacers rather than a single spacer, allowing voids to exist in the spacer string. Loose fitting spacers on plastic or metal bolts.

- *Bolts.* Threaded section of bolts extending into the potting material or board. Overstressing bolted components on boards.

- *Tabs.* Use of glued tabs on layered insulation causes voids. The glue may not be compatible with the potting material.

- *Sleeving.* Solid sleeving traps air between the conductor and sleeve, causing partial discharges and breakdown. Some porous sleeves do not fill properly with the encapsulant. Sleeves on small wires should be kept short to avoid voids.

- *Mold release agents.* These can result in silicone contamination of epoxies and urethanes.

- *Materials aging.* Stored improperly or used after recommended storage life.

- *Batch variations.* Variations in manufacturer's quality control. Variations during mixing, outgassing, and application of catalyst.

- *Coatings.* Part coating materials incompatibility with the potting materials.

- *Testing.* Insufficient quality control test of incoming materials to control viscosity and aging. Inadequate part burn-in, subassembly continuity test, and module and system acceptance test. Lack of accelerated or aging tests. Overstressing materials by acceleration and temperature tests.

20.6.7 Environmental Constraints

The environmental constraints that contribute to low MTBF are:

- Contaminated cooling gases and liquids
- Thermal shock
- Excessive vibration and mechanical shock at high or low temperature extremes

- High humidity in areas near unsealed high-voltage connectors and terminations causing frost and water droplets to form after low-temperature soaks
- Misapplication of solvents or chemicals during cleaning and general maintenance
- Non-filtering of incoming cooling or circulating gases
- Admission of engine exhaust fumes or fuel tank fumes to the high-voltage equipment compartments
- Admission of loose fibers insulation or metal flakes to the surfaces of high-voltage insulated and noninsulated circuitry
- Use of nutrient materials that may promote fungus growth near or on high-voltage parts or equipment

Designers, operators, and maintenance personnel must be made to understand these environmental problems to obtain the best performance and long life from equipment. Feedback from the failure analyst will help to accomplish this.

20.7 REFERENCES

1. Dunbar, W.G. *High Voltage Design Guide: Aircraft,* AFWAL-TR- 88-4143, Vol. II, Materials Laboratory, Wright-Patterson AFB, Ohio, August, 1988.
2. Dunbar, W.G. *Design Guidelines: Designing and Building High Voltage Power Supplies.* Auburn University, AL: Space Power Institute, August, 1988.
3. Ibid.
4. Loeb, L.B. *Fundamental Processes of Electrical Discharges in Gases.* New York: John Wiley & Sons, 1939, p. 550.
5. Loeb, L.B. *Electrical Coronas.* Berkeley, CA: University of California Press, 1965.
6. Penning, F.M. *Electrical Discharges in Gases.* New York: Macmillan, 1957
7. Cobine, J.D. *Gaseous Conductors.* New York: Dover Press, 1958
8. Paschen, F. Ueber die Zum Funkenuebergang in Luft, Wasserstoff und Kohlensaeure bei Verschiedenen Drucken Erfoderliche Potentialdifferenz, *Ann. Phys. Chem.* 37, 69–96, 1889.
9. Meek, J.M., and J.D. Craggs. *Electric Breakdown of Gases.* New York: John Wiley and Sons, 1978.
10. Kitchin, D.W., and O.S. Pratt. An Accelerated Screening Test for Polyethylene High-Voltage Insulation, *AIEE Transactions,* Paper 62-54.
11. Billings, M.J., A. Smith and R. Wilkins. Tracking in Polymeric Insulation, *IEEE Trans. Electrical Insulation* EI-2(3), 131–136.
12. McMahon, E.J. A Tutorial on Treeing, *IEEE Trans. Electrical Insulation* EI-13(4), 277.
13. Mason, J.H. Dielectric Breakdown in Solids, in *Progress in Dielectrics,* Vol.1, New York: John Wiley and Sons, 1959, 1–59.
14. Olyphant, M. *Corona and Treeing Breakdown of Insulation.* Lake Publishing Co., Vol. 9, Nos. 2, 3, and 4, February, March, and April 1963.
15. Tobata, T., et al., Sulfide Attack and Treeing of Polyethylene Insulated Cables-Cause and Prevention, *IEEE Trans.* Paper 71 TP 551-PWR.
16. Dissado, I.A., et al. An Analysis of Field-dependent Water Tree Growth Models, *IEEE Trans. Electrical Insulation* EI-23(3), 345–356.
17. Steennis, E.F., and F.H. Kreuger, Water Treeing in Polyethylene Cables, *IEEE Trans. Electrical Insulation* EI-25(5), 989–1028.
18. Konig, D., et al. Surface Discharges on Contaminated Epoxy Insulators, *IEEE Trans. Electrical Insulation* 24(2), 229–237.

19. Kreuger, F.H., et. al. Optical Detection of Surface Discharges, *IEEE Trans. Electrical Insulation* 23(3), 447–449.

20. Shunyuan, L., et al. Measurement of Dynamic Potential Distribution During the Propagation of a Local Arc Along a Polluted Surface, *IEEE Trans. Electrical Insulation* 25(4), 757–761.

21. Chalmers, I.D., et al., Surface Charging and Flashover on Insulators in Vacuum, *IEEE Trans. Electrical Insulation* 2(2), 225–230.

22. Starr, W.T. *High Altitude Flashover and Corona Problems, Part 1, Electro Technology,* 124–127, 1962.

23. Naidu, S.R., et. al. Volt-time Curves for a Coaxial Cylindrical Gap in SF6-N2 Mixtures, *IEEE Trans. Electrical Insulation* EI-22(6), 755–762.

24. Qiu, Y., et al. Improved Dielectric Strength of SF6 gas With a Trichlorotrifluoroethane Vapor Additive, *IEEE Trans. Electrical Insulation* EI-22(6), 63–768.

25. Van Brunt, R.J., and J. T. Herron. Fundamental Processes of SF6 Decomposition and Oxidation in Glow and Corona Discharges, *IEEE Trans. Electrical Insulation* 25(5), 75–94.

26. Sharbaugh, A.H., and J. C. Devins. Progress in the Field of Electrical Breakdown in Dielectric Liquids, *IEEE Trans. Electrical Insulation* EI-13(4), 249.

27. Rzad, S.J., J. C. Devins, and R.J. Schwabe. Transient Behavior in Transformer Oils: Prebreakdown and Breakdown Phenomena, *IEEE Trans. Electrical Insulation* EI-14(6), 289–296.

28. Sierota, A. and Rungis. Electrical Insulating Oils, Part I: Characterization and Pre-treatment of New Transformer Oils, *IEEE Electrical Insulation* 11(1), 8–20.

29. Von Hippel, A. *Dielectric Materials and Applications.* New York: John Wiley & Sons, 1954, 63–122.

30. Bruins, P.F. *Plastics for Electrical Engineering.* New York: Interscience Publishers 1968, 24–58.

31. Bartnikas, R., and E. J. McMahon. *Engineering Dielectrics, Vol. 1, Corona Measurements and Interpretation,* ASTM Special Publication 669. Philadelphia, PA: ASTM, 1979.

32. Whitehead. *Dielectric Breakdown of Solids,* 2nd. ed. London: Oxford University Press, 1953, 157–213.

33. Kreuger, F.H. *Discharge Detection in High Voltage Equipment.* New York: American Elsevier, 1965.

34. Lewis, T.G. *Analysis of the Frequency Characteristics of Corona Discharge at Low Pressure.* Arizona State Univ., 1991.

35. Dunbar, W. G., "Corona Onset Voltage of Insulated and Bare Electrodes in Rarefied Air and Other Gasses," AFAPL-TR- 65-122, Air Force Aero-Propulsion Laboratory, Wright-Patterson Air Force Base, Ohio, June 1966.

36. Bilodeau, T., W. Dunbar, and W. J. Sarjeant. High-Voltage and Partial Discharge Testing Techniques for Space Power Systems, *IEEE Electrical Insulation* 5(5), 14.

37. Cohen, H.A., et. al. *Spear II High Power Space Insulation.* Lubbock, TX: Texas Tech University Press, 1995, 10-1–10-33.

38. Purvis, C.K., et al. Active Control of Spacecraft Charging on ATS-5 and ATS-6, *Proc. Spacecraft Charging Technology* (NASA-TMS-73537), 1977.

39. *Navy Power Supply Reliability—Design and Manufacturing Guidelines,* NAVMAT P1955-2, 1988.

GLOSSARY

Accelerated Stress Test (AST)   A screen conducted at a higher level of stress and for a shorter time than a conventional screen. Usually not called a screen, because the acronym would then be unacceptable.

Ambient   The surrounding environment that contacts the system, assembly, or component of interest.

Anode   An electrode or terminal through which current enters a metallic conductor or electrical device. Usually the positive terminal.

AOI   Automated optical inspection.

Arc   An electrical discharge through a gas. Normally characterized by a voltage drop in the immediate vicinity of the cathode approximately equal to the ionization potential of the gas.

Arc Tracking   A mechanism that occurs on insulating materials in which electrical discharges break down the chemical structure of the material and leave behind more conductive materials, which tend to degrade the material's insulating properties.

ATE   Automated test equipment.

Azeotrope   A mixture of two or more liquids that have a constant boiling point different from that of any of the component liquids.

Backscattered Electrons (BSEs)   Electrons arising from the elastic collision between the electrons produced by the electron gun of the scanning electron microscope and the nuclei of the specimen. BSE yield is strongly dependent on atomic number.

Bed-of-Nails   A fixture featuring spring-loaded contact pins that engage points on the board to perform in-circuit tests. These fixtures are usually dedicated to one PCB design.

Bimetallic Element   An actuating element consisting of two strips of metal with different coefficients of thermal expansion bound together in such a way that the internal strains caused by temperature changes bend the compound strip.

BSE Image   The image coming from the backscattered electrons that reveals both relative atomic number and topographical information about the specimen.

Burn-In   Parts or assemblies are subjected to operation at elevated temperature for long periods in order to weed out *marginals* as failures.

Capacitor   An electric circuit component or device in which the voltage developed across the device results in the storage of energy in an electric field.

Cathode   An electrode or terminal through which current leaves a metallic conductor or electrical device. Usually a negative terminal.

Chafed   Possessing frictional wear damage, usually caused by two parts rubbing together with limited motion.

Chip Carrier   A low-profile surface mount component package with a large chip mounting cavity. The external connections may be leads or pads located on all four sides.

COB   Chip-on-board. A technology for mounting dice directly on substrates with subsequent attachment usually by wire bonding. It is usually followed by plastic globule encapsulation.

Coefficient of Thermal Expansion   The ratio of the increase in length, area, or volume of a body per degree rise in temperature compared to its original length, area, or volume respectively, at some specified temperature.

Cold Solder Joint   A solder connection that exhibits poor wetting and a dull greyish (porous) appearance. Can be due to insufficient heat, poor cleaning, substandard solder, etc.

Component Lead   The solid or stranded, formed conductor that extends from a component and serves as an easily formed mechanical and/or electrical connection.

Contact   A conducting part that acts with another conducting part to make or break a circuit.

Conventional Current   An arbitrary convention that represents current flow from positive (anode) to negative (cathode) potentials.

Coplanarity   The relationship between the leads of a component and the lands on a printed wiring board. The vertical spread between the lowest and highest contact of a device's package.

Crazing   Fine cracks near the surface of a material. Can be due to solvents or mechanical stress in some polymeric materials. In a circuit board, the exhibiting of white crosses or spots where separation from the resin occurs between the glass fibers at the weave intersections *(measling)*.

Defect   The nonconformity that should not be present in the product if it had been designed, produced, and handled in a correct manner.

Defendant   The supplier, designer, manufacturer, distributor, or seller of the product that is alleged to be defective and against whom action is being taken.

Dendrite   A crystal that has a tree-like branching pattern.

Dielectric Breakdown   The voltage required to cause electrical failure of an insulator.

Dielectric Withstand   The ability of insulating materials and spacings to withstand specified overvoltages for a specified time without flashover or puncture.

DIP   Dual in-line package; an integrated circuit package with two rows of pins.

Dot Map   The imaging of sites of X-ray emission from a specimen produced with an electron microscope. Sometimes called an *X-ray map.*

Double-Sided Assembly   A packaging and interconnecting structure with components mounted on both sides.

Energy-Dispersive Spectroscopy (EDS)   A method of X-ray analysis, usually used in conjunction with a scanning electron microscope (SEM), which discriminates among the energy levels of characteristic X-rays produced during electron-beam irradiation. Also called *energy dispersive X-ray analysis.*

Environmental Testing   Testing that utilizes external environmental factors such as temperature or vibration to determine the effects on the structural and electrical functions of a component, PCB, or system.

Express Warranty and Misrepresentation   Where a seller expressly warrants or claims that a product will perform in a specific manner or has certain characteristics and the product fails to fulfill these claims, the buyer can take action to recover, as the seller is then liable, based on his misrepresentation of the product.

Fillet   A smooth, concave junction where two surfaces meet. For a solder joint, the quality of the fillet can determine the strength of the joint.

Fine Pitch ≥ 0.025-inch spacing of leads.

Flashover   The sudden catastrophic failure of a insulation system separating two energized metal parts or conductors. Flashover is usually accompanied by a visible electrical discharge.

Flatpack   An integrated circuit package, commonly surface mounted, with gull-wing shaped or flat leads extending from the sides.

Footprint   The land or pad pattern on the printed circuit board to which the leads on a surface mounted device are mated.

Grain Size   For metals, a measure of the areas or volumes of grains in a polycrystalline material, usually expressed as an average when the individual sizes are fairly uniform.

Ground Plane   A conductor layer used as a common reference point for circuit returns, shielding, or heat sinking.

Gull-Wing Lead   A package lead configuration, typically used on small outline or flatpack packages, with formed leads that resemble a gull in flight.

Hertz (Hz)   A unit of frequency. One hertz equals one cycle per second.

Hipot   An electrical test that utilizes high voltage potential to measure the voltage breakdown of a substrate or dielectric.

Hygroscopic   The capacity of a material or compound to absorb and retain moisture from the ambient surroundings.

Implied Warranty   The theory of implied warranty of merchantability implies a contractual right. However, this has been broadened so that it is not simply the immediate purchaser who can take action; action for recovery in the event of a failure leading to loss can be taken by the ultimate user against the original supplier or manufacturer.

Inductor   An electric circuit component or device in which current flows through the device, resulting in the storage of energy in a magnetic field.

Infrared   The region of the electromagnetic spectrum between the long-wavelength extreme of the visible spectrum (about 0.7 μm) and the shortest microwaves (about 1 mm).

Intermetallics (i.e., Intermetallic Compounds)   Metallic compounds that form at the surfaces between different metals. Intermetallics usually have significantly different properties from those of the original metals.

Interpackage Spacing   The distance between two or more components on a printed circuit board.

J-Lead   A lead configuration typically used on plastic chip carrier packages in which leads are rolled underneath the package body. A side view of the formed lead resembles the shape of the letter *J*.

Land   The metal conductive pattern on a substrate or PWB usually used for electrical connection or component attachment. Sometimes called a *footprint* or a *pad*.

Leakage Current   The current which flows between two energized metal parts that are separated by an insulating air space or solid insulating material.

Mean Time Between Failure (MTBF)   A stated reliability level of a component or assembly. Usually measured in the average hours between failures of the unit when operating continuously.

Measling   A noncatastrophic defect of PCBs characterized by white spots on the board that represent a separation of glass fibers from the resin at the weave intersections. Sometimes called *crazing*.

Microsectioning   A destructive testing procedure in which a sample part is cross-sectioned. The procedure may involve encapsulation, polishing, and etching and is usually used in conjunction with microscopic examination.

Multichip Modules (MCMs)   A modular package designed to provide multiple functions by housing both active and passive devices in a single unit.

Negligence   This has been defined as *conduct that involves an unreasonably great risk of causing damage*.

Nonwetting   The characteristic of a metal surface where the solder comes in contact but does not form an intermetallic or metallurgical bond.

Open   The result of two electrically connected points becoming separated.

Pad   The metal portion of a printed circuit board where the leads of a surface mount component are fixed. Sometimes called a *footprint* or a *land*.

Passive Components   Devices that do not provide the active device functions of rectification, amplification, or switching. Passive components, such as resistors and capacitors, simply react to voltage and current.

Pitch   The centerline spacing of the leads on an electronic interconnect.

Plaintiff   The individual or party who is making the complaint that a deficiency in a product has caused injury, damage, or both.

Plastic Leaded Chip Carrier (PLCC)   A surface mount package that has anywhere from 18 to 84 J-leads on four sides, with uniform spacing between leads.

Polymer   A natural or synthetic chemical compound or mixture of compounds formed by polymerization and consisting essentially of repeating chemical units.

Potting Compound   An insulating material used to seal components and associated conductors in high-voltage units. Also provides protection from contaminants.

Printed Circuit Board (PCB)   An epoxy glass and metal composite on which circuits are etched and to which components are attached to form a functioning electronic circuit. Sometimes called a *printed wiring board (PWB)* or *printed wiring assembly (PWA)*.

Privity   *Privity of contract* relates to the existence of a direct relationship between buyer and seller. In the past, this provided an almost absolute immunity to manufacturers, because users normally dealt with or purchased products through an intermediary.

Product Liability   This refers to the liability of a manufacturer, supplier, or designer for the consequences of malfunction of the product.

Radiograph   A photograph shadow image resulting from uneven absorption of radiation in the object being examined.

Recrystallization Temperature   The approximate minimum temperature at which complete recrystallization of a cold-worked metal occurs within a specified time.

Res Ipsa Loquitur   This Latin phrase, meaning "the matter speaks for itself," refers to the doctrine that, if a failure has occurred, then this is a sufficient demonstration of the product's having a fault and of the liability of the manufacturer or supplier.

Resistance   The opposition of current flow in a conductor. In dc circuits, it is the ratio of voltage to current.

Rheology   The science of viscous materials and their flow properties.

RMS   Root mean square. The square root of the mean value.

SIR   Surface insulation resistance.

Small Outline Integrated Circuit (SOIC)   A surface mount integrated circuit package with two parallel rows of gull-wing leads.

Small Outline J-Leaded Package (SOJ)   An integrated circuit surface mount package with two parallel rows of J-leads.

Solder   Usually a eutectic or near-eutectic alloy of tin and lead used for the electrical and mechanical connection of components.

Solder Balls   Small spheres of solder that remain on a PCB after soldering. These spheres affect reliability by moving into positions that cause a short circuit.

Solder Bridging   An unwanted path of solder that forms an electrical short circuit.

Solder Fillet   A normally concave solder surface formed at the intersection of mating metal surfaces.

Soldermask   A PCB manufacturing technique in which everything is coated with a plastic except the contacts to be soldered.

SOT   Small outline transistor.

Strict Liability   This doctrine has been developed and applied on the basis that a manufacturer, lessor, or other supplier of a defective product (including a service) is held to be responsible for any damage resulting, independent of the care taken and the inspection procedures used during its production.

Substrate   The supporting insulating material upon which parts or substances are deposited or attached.

Surface Mount Technology (SMT)   A packaging method for assembling PCBs or hybrid circuits in which components can be connected electrically and mechanically to the surface of a conductive pattern.

Surface Tension   The molecular force existing in the surface film of all liquids, which tends to contract the volume into a form using the least surface area.

TAB   Tape-automated bonding. A bonding technology in which precisely etched leads, supported by a flexible tape, are automatically positioned over bonding pads on a chip then thermocompression bonded.

Thixotropic A property of certain materials which are paste or gel-like at rest but fluid when stressed.

Tombstoning Refers to a chip capacitor or chip resistor that is standing on one solder termination, at an angle of 90° from the plane of the PCB. It can be called *drawbridging* when at an approximate 45° angle.

Tort This can be defined as a civil wrong. An action under the law of tort involves an action for recovery for damages inflicted on a plaintiff arising from a failure of a defective product.

Ultra Fine Pitch ≥0.015-inch spacing of leads.

Via A plated-through hole used to connect different conductive layers. It is not intended for a component lead.

Viscosity A measure of a material's resistance to flow or change in shape. Measured in centipoises or millipascals.

Wetting A physical phenomenon of liquids (such as molten solder), usually in contact with solids, where the surface tension of the liquid has been reduced so that the liquid flows and makes intimate contact in a thin layer over the substrate surface. In the case of soldering, it allows formation of a metallurgical bond.

Wicking The flow of a liquid from one area to another by capillary action or wetting of the surface.

Index

INDEX